Stable Isotopes as Indicators of Ecological Change

A volume in the Academic Press | TERRESTRIAL ECOLOGY SERIES

Published Books in the Series

T. E. Dawson and R. T. W. Siegwolf, *Stable Isotopes as Indicators of Ecological change, 2007*

Stable Isotopes as Indicators of Ecological Change

Edited by

Todd E. Dawson

Department of Integrative Biology
University of California, Berkeley
Berkeley, California, USA

Rolf T. W. Siegwolf

Lab of Atmospheric Chemistry
Paul Scherrer Institute
Villigen, Switzerland

ELSEVIER

AMSTERDAM • BOSTON • HEIDELBERG • LONDON
NEW YORK • OXFORD • PARIS • SAN DIEGO
SAN FRANCISCO • SINGAPORE • SYDNEY • TOKYO
Academic Press is an imprint of Elsevier

ERRATUM

Back Cover Credit: These images show a portion of the state of Rondônia, Brazil, in which tropical deforestation has occurred. The 1975 image is MSS data. The 1992 image is TM data. The data is made available from U.S. Geological Survey/EROS, Sioux Falls, SD.

Front Cover Credit: A clear-cut in Salmon National Forest, Idaho.
Photographer Joel Sartore/National Geographic Image Collection (C)

Spine Image Credit: 200-year old Scots Pine
(*Pinus sylvestris* L.) with fire scar from northern Russia (C) N. J. Loader & A. Ratcliffe

Academic Press is an imprint of Elsevier
84 Theobald's Road, London WC1X 8RR, UK
Radarweg 29, PO Box 211, 1000 AE Amsterdam, The Netherlands
Linacre House, Jordan Hill, Oxford OX2 8DP, UK
30 Corporate Drive, Suite 400, Burlington, MA 01803, USA
525 B Street, Suite 1900, San Diego, CA 92101-4495, USA

First edition 2007

ISBN: 978-0-12-373627-7
ISSN: 1936-7961

For information on all Academic Press publications
visit our website at books.elsevier.com

Printed and bound by CPI Group (UK) Ltd, Croydon, CR0 4YY

Transferred to Digital Printing, 2011

Contents

Contributors

Numbers in parentheses indicate the pages on which the authors' contributions begin

NATÀLIA ALONSO (319) Department of History, University of Lleida E-25003 Lleida, Spain

JOSÉ LUIS ARAUS (319) Department of Plant Biology, University of Barcelona E-08028 Barcelona, Spain

E. BARKAN (111) Institute of Earth Sciences, Hebrew University, Givat Ram, 91904 Jerusalem, Israel, e-mail: eugenib@cc.huji.ac.il

GABRIEL J. BOWEN (163, 285, 399) Department of Biology, University of Utah, Salt Lake City, Utah 84112, USA; Department of Earth and Planetary Sciences, Purdue University, Layafyette, Indiana 47907, USA; Earth and Atmospheric Sciences and Purdue Climate Change Research Center, Purdue University, West Lafayette, Indiana 47907, USA, e-mail: gabe@purdue.edu; gjbowen@purdue.edu

NINA BUCHMANN (347) Institute of Plant Sciences, ETH Zurich, Universitaetsstr. 2, LFW C56, Zurich C56, 8092, Switzerland, e-mail: nina.buchmann@ipw.agrl.ethz.ch

THURE E. CERLING (163, 285) Department of Biology, Geology and Geophysics, University of Utah, Salt Lake City, Utah 84112, USA, e-mail: tcerling@mines.utah.edu

TODD E. DAWSON (3, 19) Department of Integrative Biology, University of California, Berkley, California 94720, USA, e-mail: tdawson@berkeley.edu

DANIEL DEMAIFFE (333) Laboratoire de Géochimie Isotopique et Géodynamique Chimique, Université Libre de Bruxelles (ULB), Brussels, Belgium

TOMAS FERREIRA DOMINGUES (301) CENA, University of São Paulo, Piracicaba-SP 13416-000, Brazil

MARK F. DREIER (235) Department of Geological Sciences and Environmental Studies Program, University of Colorado, Boulder, Colorado 80309, USA

THOMAS DROUET (333) Laboratoire de Génétique et d'Ecologie Végétales, Université Libre de Bruxelles (ULB), Brussels, Belgium, e-mail: tdrouetd@ulb.ac.be

JAMES R. EHLERINGER (19, 163, 285, 301, 383) SIRFER, Department of Biology, University of Utah, Salt Lake City, Utah 84112, USA, e-mail: Ehleringer@biology.utah.edu

DAVID M. ETHERIDGE (235) CSIRO Atmospheric Research, Private Bag 1, Aspendale, Victoria 3195, Australia

R. DAVID EVANS (383) School of Biological Sciences, Washington State University, Pullman, Washington 99164-4236, USA

DOMINIC F. FERRETTI (235) Department of Geological Sciences and Environmental Studies Program, University of Colorado, Boulder, Colorado 80309, USA; National Institute of Water and Atmospheric Research Ltd., Wellington, New Zealand

JUAN PEDRO FERRIO (319) Department of Crop and Forest Sciences, University of Lleida E-25198 Lleida, Spain

BRIAN FRY (173) Department of Oceanography & Coastal Sciences and Coastal Ecology Institute, School of the Coast and Environment, Louisiana State University, Baton Rouge, Louisiana 70803, USA

MARY GAGEN (27) Department of Geography, University of Wales Swansea, Singleton Park, Swansea SA2 8PP, United Kingdom, e-mail: m.h.gagen@swansea.ac.uk

FELIPE GALVÁN-MAGAÑA (173) Centro Interdisciplinario de Ciencias Marinas-Instituto Politécnico Nacional La Paz, Baja California Sur, C.P. 23000, México, USA

JALEH GHASHGHAIE (193) Laboratoire d'Écologie, Systématique et Evolution, CNRS-UMR 8079, IFR 87, Bâtiment 362, Université de Paris XI, 91405 Orsay, France

GERD GLEIXNER (249) Department of Biogeochemical processes, Max-Planck Institute for Biogeochemistry, D07701 Jena, Germany, e-mail: ggleix@bgc-jena.mpg.de

BRITTANY S. GRAHAM (173) Department of Oceanography, University of Hawaii, Honolulu, Hawaii 96822, e-mail: grahamb@hawaii.edu

H. GRIFFITHS (361) Department of Plant Sciences, University of Cambridge, Cambridge CB2 3EA, United Kingdom

HENRI D. GRISSINO-MAYER (63) Department of Geography, University of Tennessee, Knoxville, Tennessee 37996, USA

CECELIA C. S. HANNIDES (173) Department of Oceanography, University of Hawaii, Honolulu, Hawaii 96822

BRITTA HARTARD (77) Department of Ecology, University of Kaiserslautern, Kaiserslautern 67653, Germany; Stable Isotopes Laboratory, ICAT/CEBV, FCUL University of Lisbon, Campo Grande, 1749-016 Lisbon, Portugal, e-mail: hartard@rhrk.uni-kl.de

DEBBIE HEMMING (361, 399) Hadley Centre, Met Office, Exeter. Devon. EX1 3PB, United Kingdom, e-mail: debbie.hemming@metoffice.gov.uk

JACQUES HERBAUTS (333) Laboratoire de Génétique et d'Ecologie Végétales, Université Libre de Bruxelles (ULB), Brussels, Belgium

KEITH A. HOBSON (129) Environment Canada, Saskatoon, Saskatchewan, Canada S7N 3H5, e-mail: Keith.Hobson@EC.GC.CA

FRANÇOISE YOKO ISHIDA (301) CENA, University of São Paulo, Piracicaba-SP 13416-000, Brazil

RISTO JALKANEN (27) Finnish Forest Research Institute, Rovaniemi Research Station, Rovaniemi FIN-96301, Finland, e-mail: risto.jalkanen@metla.fi

ANSGAR KAHMEN (347, 399) Department of Integrative Biology, University of California–Berkeley, Berkeley, California 94720, USA, e-mail: akahmen@berkeley.edu

ALEXANDER KNOHL (399) ETH Zürich, Institute of Plant Sciences, Universitaetsstr. 2, LFW C30, 8092 Zurich, Switzerland, e-mail: alexander.knohl@ipw.agrl.ethz.ch

CHUN-TA LAI (399) Department of Biology, San Diego State University, San Diego, California 92182, e-mail: lai@biology.sdsu.edu

MICHAEL LAKATOS (77) Department of Ecology, University of Kaiserslautern, Kaiserslautern 67653, Germany; Stable Isotopes Laboratory, ICAT/CEBV, FCUL University of Lisbon, Campo Grande, 1749-016 Lisbon, Portugal, e-mail: lakatos@rhrk.uni-kl.de

A. LANDAIS (111) Institute of Earth Sciences, Hebrew University, Givat Ram, 91904 Jerusalem, Israel, e-mail: amaelle.landais@cea.fr

KEITH R. LASSEY (235) National Institute of Water and Atmospheric Research Ltd., Wellington, New Zealand

MARKUS LEUENBERGER (211) Climate and Environmental Physics, Physics Institute, University of Bern, 3012 Bern, Switzerland, e-mail: leuenberger@climate.unibe.ch

NEIL J. LOADER (27, 361) Department of Geography, University of Wales Swansea, Singleton Park, Swansea SA2 8PP, United Kingdom, e-mail: N.J.Loader@swansea.ac.uk

GLADIS A. LÓPEZ-IBARRA (173) Centro Interdisciplinario de Ciencias Marinas-Instituto Politécnico Nacional La Paz, Baja California Sur, C.P. 23000, México, USA

MICHAEL J. LOTT (173) Department of Biology, University of Utah, Salt Lake City, Utah 84112, USA

DAVID C. LOWE (235) National Institute of Water and Atmospheric Research Ltd., Wellington, New Zealand

B. LUZ (111) Institute of Earth Sciences, Hebrew University, Givat Ram, 91904 Jerusalem, Israel, e-mail: Boaz.Luz@huji.ac.il

CECELIA M. MACFARLING (235) CSIRO Atmospheric Research, Private Bag 1, Aspendale, Victoria 3195, Australia; School of Earth Sciences, University of Melbourne, Victoria 3010, Australia

CRISTINA MÁGUAS (77, 193) Stable Isotopes Laboratory, ICAT/CEBV, FCUL University of Lisbon, Campo Grande, 1749-016 Lisbon, Portugal; Centro de Ecologia e Biologia Vegetal, Faculdade de Ciências, Universidade Lisboa, Campo Grande, P-1749-016 Lisbon, Portugal, e-mail: cristina.maguas@icat.fc.ul.pt

A. MARCA (361) Stable Isotope Laboratory, School of Environmental Sciences, University of East Anglia, Norwich NR4 7TJ, United Kingdom

LUIZ ANTONIO MARTINELLI (301) CENA, University of São Paulo, Piracicaba-SP 13416-000, Brazil, e-mail: martinelli@cena.usp.br

DANNY McCARROLL (27) Department of Geography, University of Wales Swansea, Singleton Park, Swansea SA2 8PP, United Kingdom, e-mail: D.McCarroll@swansea.ac.uk

DANA L. MILLER (63) Department of Earth and Planetary Sciences, University of Tennessee, Knoxville, Tennessee 37996, USA

JOHN B. MILLER (235) NOAA/ESRL, Global Monitoring Division, Institute of Arctic and Alpine Research, Colorado 80305, USA

ANDY MOORE (145) Lowestoft Laboratory, Centre for Environment, Fisheries and Aquaculture Science, Lowestoft, Suffolk NR33 0HT, United Kingdom

CLAUDIA I. MORA (63) Department of Earth and Planetary Sciences, University of Tennessee, Knoxville, Tennessee 37996, USA, e-mail: cmora@utk.edu

INES MÜGLER (249) Department of Biogeochemical processes, Max-Planck Institute for Biogeochemistry, D07701 Jena, Germany, e-mail: imuegler@bgc-jena.mpg.de

GABRIELA BIELEFELD NARDOTO (301) CENA, University of São Paulo, Piracicaba-SP 13416-000, Brazil

RAFAEL SILVA OLIVEIRA (301) CENA, University of São Paulo, Piracicaba-SP 13416-000, Brazil

ROBERT J. OLSON (173) Inter-American Tropical Tuna Commission, La Jolla, California 92037, USA, e-mail: rolson@iattc.org

JEAN PIERRE HENRY BALBAULD OMETTO (301) CENA, University of São Paulo, Piracicaba-SP 13416-000, Brazil

TAS VAN OMMEN (235) Department of the Environment and Heritage, Australian Antarctic Division, and Antarctic Climate & Ecosystems CRC, Hobart, Tasmania 7001, Australia

ANNA PAPROCKA (267) Institute of Geological Sciences, Polish Academy of Sciences, Twarda 51/55, 00-818 Warszawa, Poland, e-mail: paprocka@pf.pl

JOÃO S. PEREIRA (193) Instituto Superior de Agronomia, Universidade Técnica de Lisboa, 1349-017 Lisbon, Portugal, e-mail: jspereira@isa.utl.pt

BRIAN N. POPP (173) Department of Geology and Geophysics, University of Hawaii, Honolulu, Hawaii 96822, e-mail: popp@hawaii.edu

IAIN ROBERTSON (27, 361) Department of Geography, University of Wales Swansea, Singleton Park, Swansea SA2 8PP, United Kingdom, e-mail: i.robertson@swan.ac.uk

MATTHIAS SAURER (49) Paul Scherrer Institute, 5232 Villigen, Switzerland, e-mail: matthias.saurer@psi.ch

ROLF T. W. SIEGWOLF (3, 49) Lab for Atmospheric Chemistry, Paul Scherrer Institute, CH-5232 Villigen PSI, Switzerland, e-mail: rolf.siegwolf@psi.ch

JED P. SPARKS (93) Department of Ecology and Evolutionary Biology, Cornell University, 149E Corson Hall, Ithaca, New York 14853, USA, e-mail: jps66@cornell.edu

MATT SPONHEIMER (163) Department of Biology, University of Utah, Salt Lake City, Utah 84112, USA; Department of Anthropology, University of Colorado, Boulder, Colorado 80309, USA, e-mail: matt.sponheimer@colorado.edu

CATHY M. TRUDINGER (235) CSIRO Atmospheric Research, Private Bag 1, Aspendale, Victoria 3195, Australia

CLIVE N. TRUEMAN (145) School of Ocean and Earth Science, National Oceanography Centre, Southampton, University of Southampton Waterfront Campus, Southampton SO13 4ZH, United Kingdom, e-mail: trueman@noc.soton.ac.uk

KEVIN P. TU (399) Center for Stable Isotope Biogeochemistry, Department of Integrative Biology, University of California-Berkeley, CA 94720, USA, e-mail: kevintu@berkeley.edu

STEPHAN UNGER (193) Experimental and Systems Ecology, University of Bielefeld, Universitätsstr. 25, D-33615 Bielefeld, Germany, e-mail: stephan.unger@biologie.uni-bielefeld.de

DENA M. VALLANO (93) Department of Ecology and Evolutionary Biology, Cornell University, 149E Corson Hall, Ithaca, New York 14853, USA e-mail: dmv24@cornell.edu

JORDI VOLTAS (319) Department of Crop and Forest Sciences, University of Lleida E-25198 Lleida, Spain

CHRISTIANE WERNER (193, 399) Experimental and Systems Ecology, University of Bielefeld, Universitätsstr. 25, D-33615 Bielefeld, Germany, e-mail: c.werner@uni-bielefeld.de

JASON B. WEST (383) Department of Biology, University of Utah, Salt Lake City, Utah 84112, USA

JAMES W. C. WHITE (235) Department of Geological Sciences and Environmental Studies Program, University of Colorado, Boulder, Colorado 80309, USA e-mail: james.white@colorado.edu

DAVID G. WILLIAMS (361, 383) Departments of Renewable Resources and Botany, University of Wyoming, Laramie, Wyoming 82071, USA e-mail: dgw@uwyo.edu

L. WINGATE (361) Institute of Atmospheric and Environmental Science, School of GeoSciences, University of Edinburgh, Edinburgh EH9 3JN, United Kingdom, e-mail: lwingate@ed.ac.uk

D. YAKIR (111, 361) Department of Environmental Sciences and Energy Research, The Weizmann Institute of Science, Rehovot 76100, Israel, e-mail: dan.yakir@weizmann.ac.il

Acknowledgments

Foremost, we would like to express our thanks to the many reviewers who provided excellent input for our authors and whose comments no doubt improved the volume. The 2006 UC Berkeley "Isotopics" graduate discussion group also provided important critiques, discussions, and many helpful suggestions for how to improve the chapters of the volume; we thank everyone. We also want to express our gratitude and thanks to the 125 participants from 26 different countries who attended the Tomar workshop and to the authors who have contributed to this volume. A very special thanks is owed to Cristina Maguas, João Pereira, and "team Tomar" (Rodrigo Maia, Carla Rodrigues, Hugo Evangelista, Duarte Ferreira, Sofia Cerasoli, Cristina Silva, and Ângela Pereira) who helped with all of the onsite meeting arrangements and logistics. Our thanks also to Sónia Tenreiro for her artwork that became the meeting logo featured in this volume preface and on our conference t-shirts and to EuroVector for sponsoring the t-shirt printing. Many thanks also go to the staff of the Hotel dos Templários for all of their hard work in making the meeting run so smoothly. Our thanks also go to the organizing committee, Nina Buchmann, Diane Pataki, Jim Ehleringer, Cristina Maguas, and João Pereira for helping to craft an exciting program and to decide on all of the meeting details. Special thanks to April Siegwolf for drafting together the book of abstracts and to Renee Brooks, Howard Griffiths, and Diane Pataki for judging the student posters at the meting and making the awards. Nina Buchmann and Claudine Hofstettlers provided their expertise with the web-based meeting registration details and Jeff Sherlock, our webmaster, designed and setup the website for the Tomar meeting; thanks to them all. Financial support was provided by the US National Science Foundation who funded BASIN (DEB-0090135 and DEB-0541849) and the European Science Foundation who funded SIBAE, the two programs that have coordinated and supported stable isotope workshops like the Tomar meeting. Many thanks go to Kevin Tu for helping to organize and move along the chapter review process. Many thanks go to Andy Richford and Jim Ehleringer for encouraging us to draw together the chapters that have become this special volume and for helping us with the volume process, and to our production editor Kirsten Funk of Elsevier Academic Press for her persistence, assistance, and unwavering support in seeing that the volume really happened; it was a pleasure to work with her. And finally, we thank Stefania and April for their patience and support during the entire volume process.

Preface

The twentieth century experienced environmental changes that appear to be unprecedented in their rate and magnitude during the Earth's history. As human influences on the environment continue through the twenty-first century, alterations to the functions and services that ecosystems provide to society are expected to accelerate. Land use and land cover change, atmospheric change, and losses of or transformations to ecosystem biodiversity have and will continue to have both predictable and unpredictable impacts on the environments microbes, plants and animals depend on. These effects will be further exacerbated by the deposition of pollutants caused by increased urbanization and industrialization. During the past century or more, the types of environmental changes cited above have been paralleled by changes in the isotopic composition of a wide variety of substrates in the environment. Stable isotope data obtained from the analysis of such substrates therefore serves as important recorder, indicator, or tracer of change. The stable isotope composition of animal and plant tissues, soil organic matter, the carbonates of teeth, corals, soils, and sediments, as well as in water sources moving through the hydrosphere and the atmosphere have all recorded diverse changes to modern environments. Many of these isotope records demonstrate nonlinear, threshold responses to change that would be difficult or impossible to detect using other methods. Stable isotope data can therefore serve as a tool for long-term monitoring of environmental change as well as an indicator of the types of ecological changes that could be used to predict future transformations to a wide range of Earth systems (physical, chemical, and/or biological).

Considerable progress has been made in the past two decades toward understanding the processes that lead to isotopic variation in a diversity of different materials, many with biological origins that allow a more reliable interpretation of the environmental record. Based on short-term records (10–200+ years), we now have the foundations to test, validate, and refine our current understanding of this isotopic information. In addition, we possess abundant archives and inventories of various sample types in museum specimens, bands in sediment cores, and from rings sampled from wood or carbonates that could serve as isotope records of change over timescales from centuries to millennia. These archives also hold the potential to reveal how ecosystems have responded to environmental and biotic fluctuations in the past. Such records could serve modeling efforts aimed at predicting how the terrestrial ecosystems might respond to the ongoing rapid alterations and enhance the prognostic power of new models to evaluate future scenarios involving the ongoing anthropogenic impacts.

This special volume contains chapters that are the result of an international conference-workshop entitled "*Isotopes as Tracers of Ecological Change*" held in Tomar, Portugal, March 13–25, 2006 (see the meeting logo below; artwork by Sónia Tenreiro).

The goal of this special volume is to draw together a wide range of perspectives and data that speak directly to the issues of ecological change using stable isotope samples. The information presented in each of the chapters originates from a range of biological and geochemical sources and from research fields within biological, climatological, and physical disciplines covering timescales from days to centuries. This special volume therefore highlights where isotope data can detect, record, trace and help to interpret environmental change.

The volume highlights examples from talks, poster presentations, and discussions that occurred at the conference-workshop and emphasize how isotope data are demonstrating changes in processes we can see in ecological time at different spatial scales. A diverse suite of authors with equally broad expertise provides conceptual frameworks for using stable isotopes to reconstruct ecological change and then present supporting data derived from soils, sediments, plants, animals, water, specific compounds, carbonates, and gases that each clearly demonstrate modifications and transformation to a wide range of terrestrial and aquatic systems at scales relevant to ecologists. While each chapter examines and discusses case studies and ongoing investigations, they also "draw the map" and define where new research directions exist or are needed that will facilitate a deeper understanding about how alterations to ecosystems have occurred. The chapters and their authors provide an international perspective as well and so many of the chapter discussions highlight possible research initiatives both within and across national boundaries. This special volume also marks the final workshop of the first 5-year phase of the BASIN and SIBAE coordinated networks. No other volume has ever been produced on this particular topic that we know of.

As we organized the meeting that lead to the chapters contained in this volume we considered several elements we wanted to include. The volume's audience could be researchers, scientists, and educated laypeople interested in environmental change. The topics discussed touch on the disciplines of biology, ecology, climatology, geology, paleontology, oceanography, and atmospheric science and cover timescales from hours to days, to centuries. One goal for the volume is to try and achieve broad appeal for students, researchers, and agency staff who desire to read the latest findings. As mentioned, our authors are the leading researchers in the field largely from universities, research institutes, and government agencies.

Todd E. Dawson
Rolf T. W. Siegwolf

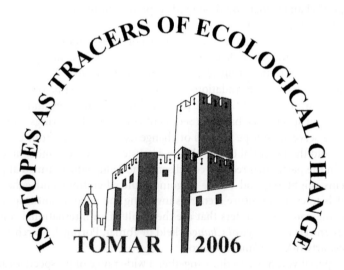

ISOTOPES AS TRACERS OF ECOLOGICAL CHANGE

TOMAR 2006

Section 1

Ecological Isotope Archives

CHAPTER 1

Using Stable Isotopes as Indicators, Tracers, and Recorders of Ecological Change: Some Context and Background

Todd E. Dawson* and Rolf T. W. Siegwolf†

*Department of Integrative Biology, University of California
†Lab for Atmospheric Chemistry, Paul Scherrer Institute

Contents

I. CONTEXT FOR BOOK

Even set against the backdrop of natural disturbances, anthropogenically induced impacts have left their fingerprints on the ecosystems of the world. By good fortune, many different substrates subjected to or resulting from these impacts "record" change in the ratio of the stable isotopes contained in the material. We now know that impacts and alternations that result in changes to modern environments are also seen as changes to the H, C, N, O, S, and Sr isotope ratios of atmospheric gases, animal and plant tissues, soil organic matter and its diverse chemical substrates, the carbonates of teeth, corals and soils, the organic matter deposited in sediments, as well as water sources in the hydrosphere and the atmosphere. Any type of sample that is analyzed for its isotope composition, particularly over some period, can therefore serve as an archive of change. This is the basis of this volume and the conference-workshop that lead to it.

A valuable body of past research that has employed stable isotope data has been *process oriented*, satisfying the need to know how living organisms interact with one another and with their environment and how these interactions can be better understood using stable isotope data (Rundel *et al.* 1989, Ehleringer *et al.* 1993, Griffiths 1998, Flanagan *et al.* 2005). Today we face a new challenge, we must extend what we know about processes to track, trace, record, and predict what may result as environments have and will continue to change. The environmental changes that ecosystems are experiencing today appear to be unprecedented in their rate and magnitude during Earth history (Millennium Ecosystem Assessment 2001, Ruddiman 2004, Flannery 2005, Gore 2006, IPCC 2007). Human influences on the environment will continue through the twenty-first century, and there is no doubt that they will increase due to the growth of industrialization in rapidly developing countries like China and India. Also, as human populations grow throughout every corner of the planet and resources are consumed, ecological changes are inevitable. Thus, alterations to the functions and services that ecosystems provide to society are expected to accelerate, some unabated because of public ignorance, flawed regulatory policies, and/or blatant disregard for the very planet that sustains human existence. A logical next step therefore in applying stable isotope analyses to address the critical issue of ecological change is to make use of the knowledge and methodologies that have been developed in the last two decades. We believe that these advances can produce the data that will not only document the nature and magnitude of change but also help highlight the solutions for how we can slow or mitigate further ecological impacts that clearly threaten the future of all organisms. By drawing together examples from contemporary research, we hope that the information presented in the chapters to follow can and will be used to better understand how stable isotope data and research can inform present and future changes to a wide variety of ecological systems.

This volume grew out of the presentations and discussions that took place at a scientific conference by the same name held in Tomar, Portugal from March 13 to 15, 2006. The conference-workshop drew together a diverse community of atmospheric, life and earth scientists that are using stable isotope data to archive and also elucidate the ways in which modern environments have changed and are changing. During the course of the conference and as the authors wrote the chapters that follow, we examined, critically, and then discussed the nature of existing data. We identified knowledge gaps and defined where new research directions that could provide needed insights and therefore where essential research must still be done. We hope that the chapters contained in this focused volume will serve as an introduction for the newcomer to become acquainted with some of the current information about using stable isotopes to trace ecological change and for the broader community of scientists and citizens who are both concerned about and researching ecological change and its many dimensions.

II. ISOTOPES AS TRACERS, RECORDS, AND INTEGRATORS OF CHANGE

The applications of stable isotope analyses, first in the fields of chemistry and geochemistry (Faure 1986, Hoefs 1997, Criss 1999, Sharp 2007) and later in biogeochemistry (Boutton and Yamasaki 1996, Clark and Fritz 1997, Griffiths 1998, Kendall and McDonnell 1998, Flanagan *et al.* 2005) and ecology (Ehleringer *et al.* 1993, Lajtha and Michener 1994, Rundel *et al.* 1989, Unkovich *et al.* 2001, Dawson *et al.* 2002, Fry 2006, and this volume), have proven to be an extremely valuable and powerful tool for indicating (sourcing), tracing, and recording various changes to the Earth's diverse terrestrial, aquatic, marine, and atmospheric systems. In the past 20 years, a large number of new ideas have been developed and a wide range of questions have been addressed using stable isotope information. This information in turn has helped spawn the development and expansion of powerful methods, including many that now make use of continuous-flow isotope analyses pioneered by John Hayes. Isotope data generated with these new and diverse methods has been used to unravel complex processes taking place in a diversity of ecosystems, from food chain relationships to carbon fluxes, from tracking metabolic change in leaves to large-scale net ecosystem flux partitioning. It is now clear that stable isotope information has had and still has a very significant impact on enhancing research being done in various disciplines at a range of temporal and spatial scales and therefore for addressing ecological change issues.

Understanding the utility and power of using stable isotope data requires a grasp of not only some fundamental principles of isotope behavior but also the underlying basis of observed variation in the isotope composition of diverse materials. The fundamentals of how stable isotopes behave in chemical, physical, and biological reactions are generally well understood owing to the pioneering work beginning in the 1930s by scientists like Alfred Nier, Harold Urey, Malcolm Dole, Samuel Epstein, Harmon Craig, and others [Sharp (2007) provides a brief history of how stable isotope science developed]. Of relevance to the information presented in this volume, we will focus on how the ratios of the lighter isotopes of H, C, N, O, and S as well as Sr in the biogeosphere vary and are measured using mass spectrometers and modern laser-based spectroscopic techniques. These stable isotopes are in fact among the elements that vary the most on Earth, constitute the bulk of all living matter, and are arguably used most effectively to track changes in the Earth's biogeochemical cycles. Therefore, we have placed this information in the context of the water, carbon, nitrogen, and the mineral element cycles impacting the isotope composition of S and Sr since it is these cycles that have been and will continue to be perturbed by human beings. This chapter therefore serves as a reference for the volume's content and is intended to act as an introductory primer to the language, notation, and general background on the levels and types of isotope variation in these major cycles. For the newcomer or seasoned user it also serves as the framework on which all of the more detailed chapters are based. We have not been comprehensive but instead have provided only the necessary background information for understanding the basics underlying the information presented in the chapters that follow.

III. STABLE ISOTOPE NOTATION

Because all of the elements discussed in this volume's chapters are composed of at least two different stable isotopes (Table 1.1) (one at very high relative abundance, most often the lighter isotope, and the other(s) at much lower relative abundances), the ratio of the rare-to-common (or heavy-to-light) stable isotope in any particular material can very often contain valuable information about both processes and sources. Also, because of the very small absolute abundances of each isotope in any particular material, by convention we express the stable isotope composition as the difference in isotope abundances relative to an international standard. Relative abundances can then be discussed more precisely than absolute isotope abundance ratios, which often only vary in the third decimal place. These facts have led to the now widely accepted use of the so-called "delta" (δ) notation (after McKinney *et al.* 1950), where the isotope ratio of the analyzed sample (SA) is expressed relative to an internationally accepted standard (STD; Table 1.1) as:

$$\delta^{XX}E = (R_{SA}/R_{STD} - 1) \times 1000$$

where *E* is the element of interest (H, C, N, O, S), "XX" is the atomic mass of the heaviest isotope in the ratio, *R* is the absolute ratio of the element of interest (*e.g.*, $^{13}C/^{12}C$), and the subscripts SA and STD are as noted above. As the isotope abundances are often very small, the δ value is multiplied by 1000 to allow the expression of small differences in units that are convenient to use. Thus, δ values are in units of parts per thousand (ppt) or the more commonly used "per mil" and given by the symbol ‰.

TABLE 1.1 Abundances of the stable isotopes from terrestrial sources that have been used to trace ecological change

Element	Isotope	Abundance	Ratio measured	Reference standard
Hydrogen	^{1}H	99.984	$^{2}H/^{1}H$ (D/H)	VSMOW[a]
	^{2}H (D)[b]	0.0156		
Carbon	^{12}C	98.982	$^{13}C/^{12}C$	PDB[c]
	^{13}C	1.108		
Nitrogen	^{14}N	99.630	$^{15}N/^{14}N$	N_2-atm[d]
	^{15}N	0.366		
Oxygen	^{16}O	99.763	$^{18}O/^{16}O$	VSMOW, PDB[e]
	^{17}O	0.0375	$^{18}O/^{17}O$[f]	VSMOW
	^{18}O	0.1995		
Sulfur	^{32}S	95.02	$^{34}S/^{32}S$	CDT[g]
	^{33}S	0.756		
	^{34}S	4.210		
	^{36}S	0.014		
Strontium	^{84}Sr	0.560	$^{87}Sr/^{86}Sr$	NBS-987[h]
	^{86}Sr	9.860		
	^{87}Sr	7.020		
	^{88}Sr	82.56		

[a] The original standard SMOW or standard mean ocean water is no longer available so the International Atomic Energy Agency (IAEA) (http://www.iaea.org/) "builds" an equivalent water sample in Vienna of a similar isotope value now known as VSMOW.
[b] The correct notation for the heavy isotope of hydrogen is ^{2}H though a commonly convention used is "D" standing for the hydrogen isotope with mass 2 called "deuterium."
[c] The original carbon isotope standard, the fossil belemnite from the PeeDee geological formation is no longer available and instead the IAEA "builds" an equivalent carbon standard in Vienna of a similar isotope value (VPDB) though for carbon isotope analyses it is still referred to as PDB.
[d] The IAEA standard is N_2 gas in the atmosphere because N_2 comprises ~78% of the Earth's atmosphere and there is no known additional source of N_2 of significance to dilute this atmospheric source, it is assumed that N_2-atm is not changing enough to warrant developing a different standard.
[e] In the case where investigators desire to know the $\delta^{13}C$ of a carbonate, the standard VPDB is used instead of VSMOW.
[f] The ^{17}O composition of air or water is also referenced to VSMOW; see Chapter 8 for further discussion.
[g] Sulfur isotope values are expressed relative to the FeS in a meteoritic troilite from Meteor Crater in Arizona (US) known as the Cañon Diablo Troilite or CDT.
[h] A widely used standard for strontium isotope analyses is the National Bureau of Standards No. 987 (now called the National Institute of Standards and Technology, NIST; http://www.nist.gov/), a carbonate power. $^{87}Sr/^{86}Sr$ and $^{86}Sr/^{88}Sr$ ratios are determined with a TIMS and unlike the other light isotopes the ratio measured is not expressed as the rare-to-abundant ratio (or heavy-to-light ratio) but as the ratio of the two isotopes that are most easily measured, $^{87}Sr/^{86}Sr$. A common practice is to normalize $^{87}Sr/^{86}Sr$ values with the $^{86}Sr/^{88}Sr$ (a light-to-heavy ratio of 0.1194) present in seawater because of fractionations that occur during thermal ionization; Chapters 21 provides further discussion of this.

By definition, the accepted standard (STD) has a δ value of 0‰. Therefore, any substance with a positive δ value has a ratio of the heavy to light isotope, R_{SA}, that is higher than the standard, R_{STD}. By analogy, a negative δ value has the opposite meaning. Substances with positive δ values are often said to be "heavier" or "enriched" relative to the standard, although this convention can often lead to confusion or errors if not expressed carefully [*e.g.*, heavier or lighter relative to what?—Sharp (2007) provides a more detailed discussion of this issue and the various errors commonly made when using δ notation or expressing isotope language]. As such, a sample of water with a $\delta^{2}H$ of −55‰ means that the ratio of $^{2}H/H$ is 55 per mil or 5.5% lower than that in the standard, V-SMOW. In the case of Sr, expressions of isotope abundance and change are shown as the absolute ratio of $^{87}Sr/^{86}Sr$ (not δ notation); these numbers range in value from a low of near 0.700 to highs near 1.200 (Table 1.1). Differences on the order of a few thousandths can be significant and this will be discussed further in Section IV.D (Chapter 23).

There are many useful and detailed discussions about the practical ways in which δ values are obtained using both modern gas-phase isotope ratio mass spectrometers (IRMS) or laser-based spectroscopic absorption methods. We direct you toward these references and their use in geochemistry (Faure 1986, Hoefs 1997, Sharp 2007), hydrology (Clark and Fritz 1997, Kendall and McDonnell 1998), and ecology (Fry 2006) for full details. For the purposes of this volume the data presented were obtained using either dual inlet or continuous-flow IRMS methods (Dawson and Brooks 2001) unless otherwise discussed in a particular chapter.

The reason underlying different observed abundances between the heavier and lighter isotopes is that heavier atoms have a lower vibration frequency than lighter ones. Therefore, heavier atoms/molecules react more slowly than their lighter counterparts because the bond strength of the heavier substances is greater. Thus more energy is needed to break the bonds that contain heavier isotopes, since the potential energy for the heavier isotope is lower than that for the lighter element. This results in lower turnover rates as soon as a heavier substance is involved. This ultimately leads to uneven isotope distributions that we term fractionation. Isotope fractionation occurs during a chemical reaction or during diffusion of a substance along a concentration gradient. As a result the abundance of the heavy isotopes in the *substrate* is greater than the abundance of the heavy isotopes in the *product.* And for most isotopic fractionation processes, we can distinguish between those that occur in one direction (no back reactions), or "kinetic" fractionation, and those that are reversible, or "equilibrium" fractionation (*e.g.*, diffusion or dissociation). For biological systems, kinetic fractionation is generally associated with enzyme-mediated processes (*e.g.*, photosynthesis) and is commonly referred to as *isotope discrimination.* The kinetic or equilibrium characteristics of different isotopes involved in physicochemical or biologically mediated (enzymatic discrimination) reactions is what leads to the changes in the isotope abundances we measure. And while in absolute terms such changes are only on the order of a few percent, in relative, or "delta" terms, some of these changes are very specific and quite large and can therefore be used to track process, determine source, and therefore record change. It is these facts that the authors in this volume have put to use and by example will show that they can be extremely informative for determining the magnitude and trajectories of ecological change.

IV. THE STABLE ISOTOPE COMPOSITION OF MATERIALS IN BIOGEOCHEMICAL CYCLES

What follows is a broad outline of what is known about the magnitude and manner by which carbon, hydrogen, oxygen, nitrogen, and sulfur isotope ratios vary in the major biogeochemical cycles on Earth. We also highlight some general details about how the strontium isotope ratio, $^{87}Sr/^{86}Sr$, can vary and its relevance to research focused on ecological change. This is not a comprehensive overview but instead we highlight important information that is relevant to understanding the detailed chapters that follow. The elements presented here are intended to provide a general background and introductory framework that is relevant to each chapter that follows. The chapters provide examples and case studies that build from these fundamental elements. What we present is placed in the context of the known processes leading to variation in the stable isotope composition of the molecules and compounds composed of C, H, O, N, and S. For the information included in this volume, mean processes are associated with the major biogeochemical cycles. It is clear that understanding the baseline isotope values of materials moving through these biogeochemical cycles must first be known before we can then explore how they can and do change when a particular cycle is perturbed. Therefore, the research examples shown link isotopic changes with ecological changes that are in some way ultimately associated with modification having occurred to these major biogeochemical cycles.

A. Variation in the $\delta^{13}C$ in Carbon Cycle Processes

Since the early-to-mid 1950s we began to grasp the underlying reasons for how and why $\delta^{13}C$ of organic and gas samples varies during the processing of carbon in the biogeosphere. The basis for much of the observed variation derives from the two major metabolic processes, photosynthesis and respiration (Farquhar *et al.* 1989, Brugnoli and Farquhar 2000) with additional variation being expressed during biosynthetic, anabolic, or catabolic reactions that utilize carbon-based substrates (Duranceau *et al.* 1999, Ghashghaie *et al.* 2001, Tcherkez *et al.* 2003 and reviewed by Tu and Dawson 2005). Additional geochemical reactions that take place in the Earth's crust or mantel can also impact the $\delta^{13}C$ of different types of rock; these details are not presented here but we direct you toward two references that can provide this important information (Hoefs 1997, Sharp 2007). For the most part, the range of $\delta^{13}C$ values is generally greater in biological materials than in geological samples.

Major differences in the $\delta^{13}C$ of plant carbon is observed when comparing marine and terrestrial plants as well as among terrestrial plants that possess different photosynthetic pathways [*e.g.*, C3, C4, and crassulacean acid metabolism (CAM); Ehleringer and Rundel 1989, Ehleringer *et al.* 1993 (Figure 1.1)]. The basis for the characteristic carbon isotope ratios observed among these groups is largely explained by the different ways they assimilate carbon and their associated fractionations (Farquhar *et al.* 1989, Keeley and Sandquist 1992, O'Leary *et al.* 1992, Raven 1992, Brugnoli and Farquhar 2000) as well as the responses different plant taxa have in relation to ecological changes in resource availability. For example, C3 plants can vary in their $\delta^{13}C$ from −20‰ to nearly −35‰ in response to water, light, and nutrient availability (Dawson *et al.* 2002). Water availability in particular can be especially important in inducing changes in stomatal physiology and/or biochemical discrimination that in turn are expressed in the $\delta^{13}C$ of the photosynthetic products as well as the tissues and compounds made from these products (Sternberg 1989, Gleixner *et al.* 1993, 1998); several examples of this appear in several of the chapters in this volume. Plants, such as tropical grasses including agricultural species like corn and sugar cane, that possess what is known as the C4 photosynthetic pathway assimilate carbon dioxide in a two-stage process that involves two different cell types within the leaf and two carboxylating enzymes. The C4 carbon assimilation process leads to carbon products that are "heavier" in $\delta^{13}C$ compared to C3 plants with values that range from highs near −10‰ to lows near −19‰. This difference allows us to distinguish these photosynthetic groups from one another using the $\delta^{13}C$ values from leaf tissues. If these leaf tissues become soil organic matter, one can use the

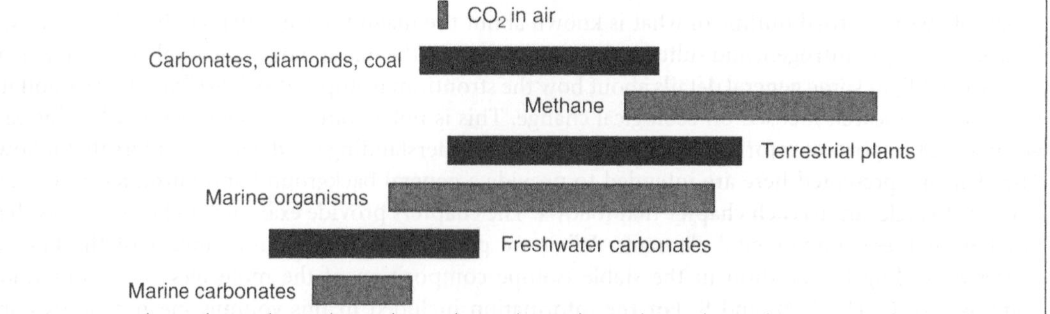

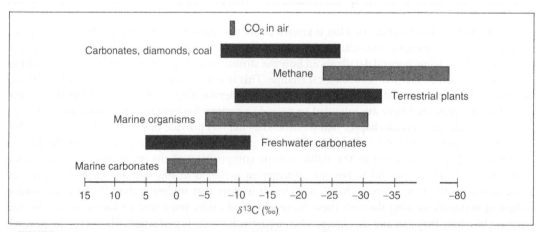

FIGURE 1.1 The observed ranges in the carbon isotope ratios of various materials on Earth expressed in delta units ($\delta^{13}C$) relative to PDB (modified and/or adapted from Ehleringer and Rundel 1989, Hoefs 1997, Criss 1999, Coplen *et al.* 2002, Fry 2006, and literature cited in the text).

$\delta^{13}C$ values of soil cores to document land-use or vegetation changes over time as shown in Chapter 19. Finally, we know that there is a third group of plants (*e.g.*, pineapple and many tropical epiphytes) that possess what is called the CAM photosynthetic pathway. CAM photosynthesis functions much like it does in C4 species but in relation to the time of the day as well as in response to water stress. The degree of nocturnal carbon fixation in a particular CAM plant and in facultative species, the balance between nighttime CAM activity and daytime C3 metabolism, is the basis for the observed variation in the tissue $\delta^{13}C$ values; they range from the highest $\delta^{13}C$ values near −11‰ to the lowest near −30‰ or close the full range of values for C3 and C4 plants combined (Griffiths 1992) (Figure 1.1).

If photosynthetic pathway and different forms of resource stress like low soil water or nitrogen availability, low light or salinity stress can be accounted for it has been shown that the $\delta^{13}C$ of tree rings (either bulk wood or purified cellulose) can serve as a useful archive of information. Analysis of the $\delta^{13}C$ of tree rings is therefore one of the more powerful ways to reconstruct ecological and physiological changes in plants and is the subject of chapters 3–5 (see also Section IV.B on $\delta^{18}O$ in tree rings). Beyond the bulk plant tissue values cited above one observes additional variation in the $\delta^{13}C$ of various C-based materials or specific compounds that are caused by additional plant and microbial metabolic fractionations and/or geochemically mediated C processing. For example, the wide variation observed in the $\delta^{13}C$ of methane owes to the very diverse ways in which plants, microbes, temperature, and pressure impact C substrates (Figure 1.1). Human-induced impacts can also modify the $\delta^{13}C$ of methane, carbon dioxide, and other trace gases that contain carbon (isoprene, methylbromide, and so on). For example, it is now well known that the combustion of C-based substrates gives rise to products like CO and CO_2 with $\delta^{13}C$ values that carry their characteristic identity (*e.g.*, from the C3 or C4 vegetation they were originally associated with). These products modify the "bulk" carbon isotope composition of CO_2 in the atmosphere and as a consequence the $\delta^{13}CO_2$ has declined in the past several decades (because the $\delta^{13}C$ value of the vegetation is "lighter" than the current atmospheric CO_2 of ca. −8‰). Land-use practices and deforestation that liberate organic matter and induce shifts in ecosystem metabolism (decomposition) also lead to changes in the $\delta^{13}CO_2$ of bulk air. Finally, geological processes alone that involve the utilization of carbon as well as the biogeochemically mediated formation of carbonates lead to some of the observed variations seen in diverse materials analyzed for their $\delta^{13}C$ (Figure 1.1). The wide range of $\delta^{13}C$ values within biological and geological materials suggests that multiple and very different types of processes can lead to this observed variation. As environments change, it is therefore no surprise that we would expect the $\delta^{13}C$ of many different types of materials to also occur. As such, $\delta^{13}C$ can serve as an important indicator of change.

B. Variation in the $\delta^{18}O$, $\Delta^{17}O$, and $\delta^{2}H$ in Hydrologic Processes

Perhaps the best-understood cycle from an isotope perspective is the cycling (and recycling) of water on planet Earth. For decades we have known that the oxygen and hydrogen isotope composition of so-called meteoric waters is extremely variable (Figures 1.2 and 1.3) and the processes that lead to this variation have been well studied. However, it is only recently that information on the third and rarest stable isotope of oxygen, ^{17}O, has also become of use in understanding and linking plants in particular (Angert *et al.* 2004) to hydrologic processes (Chapter 8).

A vast literature exits on the diversity of ways in which meteoric waters can vary in their $\delta^{2}H$ and $\delta^{18}O$ as they move through the hydrologic cycle (Dansgaard 1953, 1954, 1964, Epstein and Mayeda 1953, Craig 1961, Friedman *et al.* 1964, Gonfiantini 1982, Rozanski *et al.* 1993, Jouzel *et al.* 2000, chapters in Gat 1996, Kendall and McDonnell 1998, Bowen and Revenaugh 2003, Bowen and Wilkinson 2003, Bowen *et al.* 2005). Among the most important effects that have been shown to influence the $\delta^{2}H$ and $\delta^{18}O$ of meteoric waters are seasonality and therefore the changes in condensation temperatures of precipitation (Craig 1961, Dansgaard 1964), latitude, and altitude which like seasonality impact condensation temperature as well as the orographic/topographic lapse rates at

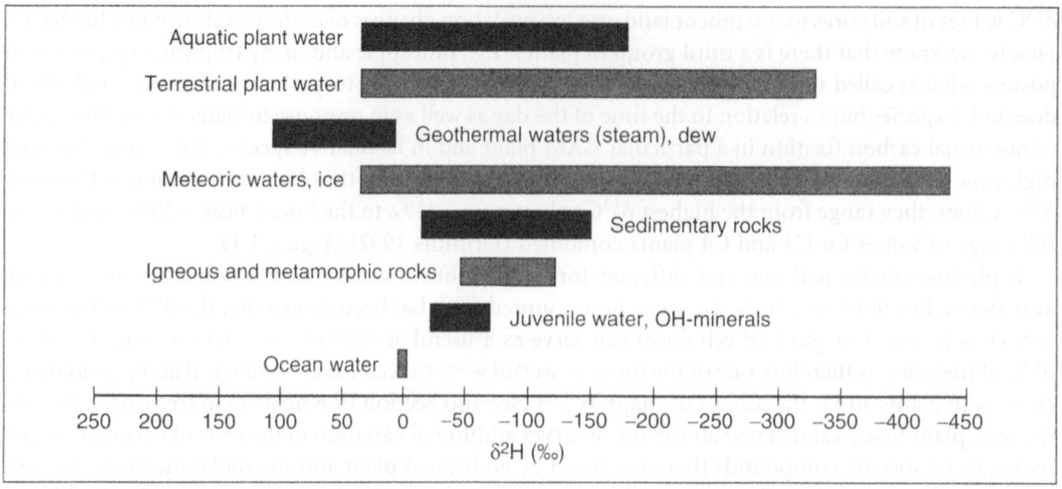

FIGURE 1.2 The observed ranges in the hydrogen isotope ratios of various materials on Earth expressed in delta units (δ^2H) relative to VSMOW (modified and/or adapted from Ehleringer and Rundel 1989, Hoefs 1997, Criss 1999, Coplen *et al.* 2002, Fry 2006, and literature cited in the text).

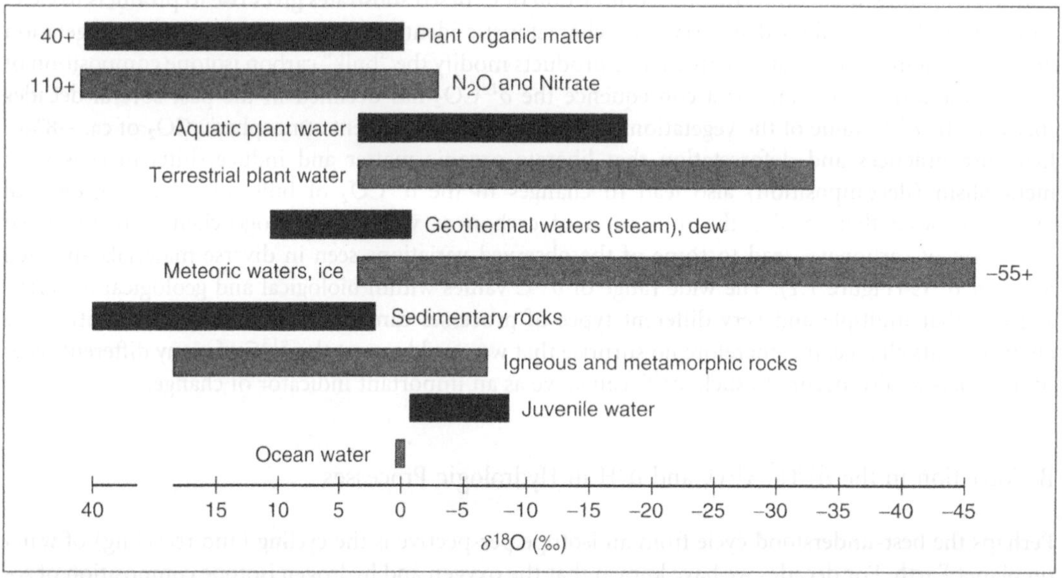

FIGURE 1.3 The observed ranges in the oxygen isotope ratios of various materials on Earth expressed in delta units (δ^{18}O) relative to VSMOW (modified and/or adapted from Ehleringer and Rundel 1989, Hoefs 1997, Criss 1999, Coplen *et al.* 2002, Fry 2006, and literature cited in the text).

which air masses that contain water move across the Earth (White 1989, Ehleringer and Dawson 1992, Dawson 1993). The magnitude of water loss or "rainout" as well as the phase of the condensate (liquid or solid) have a marked and important influence on the δ^2H and δ^{18}O of meteoric waters (Rozanski *et al.* 1993, Ingraham 1998; Figures 1.2 and 1.3). Another known effect that influences the isotope composition of meteoric waters is "continentality," or the isotope effect caused by the distance an air mass containing water has traveled over land from coastal zones where most rain storms have their origin (Ingraham and Taylor 1991). As moisture-laden air masses move and interact with the topographic features below them along their pathways, water condenses from them and the isotope composition of the remaining vapor within the air mass becomes more depleted in ^{2}H and ^{18}O. This leads to the general finding that water collected in continental interiors is more negative in $\delta^{2}H$ and $\delta^{18}O$ (see maps showing this pattern in Chapter 18). In addition, during heavy rainfall events there is a so-called "amount effect" that is known to influence the $\delta^{2}H$ and $\delta^{18}O$ of precipitation in storm systems; as greater quantities of precipitation are lost from an air mass during one of these heavy rain storms, the precipitation becomes progressively more negative in both $\delta^{2}H$ and $\delta^{18}O$ (Ingraham and Taylor 1991, Dawson and Ehleringer 1998). There are also several secondary processes like the evaporation of water from soil surfaces (Dawson *et al.* 1998) or standing water bodies like lakes, rivers, or streams (Craig and Gordon 1965), transpirational water loss from plants (Wershaw *et al.* 1966, Dongmann *et al.* 1974, Flanagan *et al.* 1991), as well as local and thermally driven processes like condensation, advection, and convection that lead to dew or fog formation, respectively. All of these processes have associated isotope effects on the water involved in the process (Gat 1996, Dawson 1998, Ingraham 1998, Ingraham and Matthews 1990, Dawson *et al.* 2002).

Within the Earth, geothermal activities that force subsurface waters back into the atmosphere or interact with carbon or mineral substrates in soils or rocks can lead to large variation in the $\delta^{2}H$ and $\delta^{18}O$ of igneous, metamorphic, and sedimentary rock (Figures 1.2 and 1.3). Water contained in aquifers or that interacts with zones in the Earth or water bodies containing salts (Mazor 1991) can also influence the $\delta^{2}H$ and $\delta^{18}O$ of both minerals and mineral waters that contact or contain these salts or rock (Hoefs 1997).

Once water is taken up by plants, animals, or microbes, the $\delta^{2}H$ and $\delta^{18}O$ of "body/source" water and biosynthetic compounds that incorporate H or O may or may not also have associated fractionations that are "recorded" in the organic molecules that contain these elements. Materials such as cellulose, hair and fingernails, feathers as well as carbonates composing bone and teeth can all show different degrees of fractionation (Chapter 11) as can the many specific compounds like *n*-alkanes and lipids (Krull *et al.* 2006, Sachse *et al.* 2006) that compose this tissues. A majority of these fractionation effects are the largest known in biological systems, leading to highly enriched organic matter. In recent years the processes that lead to enrichment in O and H of organic matter are becoming better understood and as such these materials are becoming increasingly more valuable as "biomarkers" of ecological change largely because they are known to record temperatures, water sources, and even levels of relative humidity that were present at the time these tissues were synthesized (Chapter 16). Therefore, the H and O stable isotope analysis of this broad suite of organic molecules is now providing one of the most useful archives of ecological change (Chapters 3, 4, and 9). Many examples can be cited, such as when $\delta^{18}O$ of tree-ring cellulose (Sternberg *et al.* 1986, Yakir 1992, Borella and Saurer 1999) changes with climate or land-use change, and these types of isotope changes provide a powerful way to not only document ecological transformations but do so over long periods and vast areas.

C. Variation in the $\delta^{15}N$ in the Nitrogen Cycle Processes

One of the more important, intensively studied, yet poorly understood biogeochemical cycles from a stable isotope perspective is the nitrogen cycle (Natelhoffer and Fry 1988, Evans 2001, Robinson 2001, Dawson *et al.* 2002). Understanding how nitrogen (N) cycles is critical because it has long been known that N is a key limiting nutrient for both plants and animals and therefore changes in its concentration and availability can have marked influences on organismal and ecosystem functions (Persson *et al.* 2000). It is also well documented that the global N cycle has been severely altered by human activities (Vitousek *et al.* 1997) and this is reflected in the nitrogen isotope composition of a wide range of substances (Durka *et al.* 1994, Kendall 1998). Despite knowing that the $\delta^{15}N$ composition of diverse materials is changed by human-induced ecological change, our understanding of N isotope behavior in the N cycle has only recently received the same sort of attention as the other major biogeochemical cycles. This is mostly because N in rocks is in very low concentration and was therefore of little interest to isotope geochemists and isotope fractionation is in many cases large and the processes that lead to this variation poorly understood. When, however, biological reactions that are known to move or transform N are closely studied via integrated sets of observations and experiments, we begin to understanding what leads to variation in the $\delta^{15}N$ of various substances (Robinson 2001).

Some of the earliest investigations that used stable isotopes to investigate the N cycle did so with ^{15}N-enriched substances, such as ammonium or nitrate, that were more easily used as tracers. Their large signal-to-noise ratio allows us to investigate particular processes like nitrification or denitrification with their often large but well-characterized fractionations (Table 1.2). By adding a nonradioactive label in the form of a commonly used nutrient, many useful investigations were accomplished without the need to fully characterize each fractionation factor (Handley and Scrimgeour 1997). As natural abundance studies that have used $\delta^{15}N$ data increased in the areas of physiology, ecology, and biogeochemistry, it became clear that new challenges had to be faced in applying N isotopes to trace, integrate, or record a particular process (Shaerer and Kohl 1988, Evans 2001, Robinson 2001, Dawson *et al.* 2002; Chapters 7, 12, 19). Of particular importance is the challenge of characterizing the many and varied fractionation factors associated with the transformation, utilization, and immobilization of N substances as they move through the N cycle (Table 1.2). Today, many of these fractionation factors are known, some are even well characterized, yet many more remain unknown. Those that have been studied have revealed that it is the transformation of N-rich substances that results in the range of $\delta^{15}N$ we measure (Figure 1.4 and Table 1.2). Because many challenges for using natural abundance $\delta^{15}N$ remain, easy or straightforward interpretations of $\delta^{15}N$ data in the soil, in gasses, and within a wide range of organisms are still poorly understood.

Ecological research that has made more and more use of nitrogen stable isotopes at natural abundance levels began in the 1950s with the work of Hoering (1955) and later in the 1980s with Shaerer and Kohl (1988) and many others. Using $\delta^{15}N$ has grown in popularity as knowledge of how the isotopes of N fractionate during catabolic reactions (Macko *et al.* 1986), in soils and plants in relation to N utilization, transformation, and N fixation (Högberg 1997, Evans 2001), and in food web and trophic interaction studies aimed at elucidating the pathways and interactions among producers and consumers as well as predators and their prey in a diversity of marine, aquatic (Chapter 12), and terrestrial ecosystems (Chapter 19). In general, animals are more enriched in 15-N than plants and at each trophic level the consumers are commonly 3–4‰ more enriched than the foods they consume. As such, in a food chain with three trophic levels one can expect the herbivores to be 3–4‰ more enriched than the plants on which they feed but 3–4‰ more depleted in 15-N than the primary carnivores that

TABLE 1.2 The processes and associated fractionation factors (ε) within the nitrogen cycle

Processes	ε (‰)
Nitrogen fixation	−3.7 to 3.9
Mineralization (organic N to NH_4)	−0.8 to 5.0
Nitrification	5.4 to 34.7
Denitrification ($NO_3 \rightarrow N_2O \rightarrow N_2$)	17.3 to 40.0
Ammonia volatilization (equilibrium)	25 to 35
Ammonium assimilation	
Aquatic	−9.7 to 22.0
Terrestrial	1.1 to 14.8
Nitrate assimilation	
Aquatic	0.7 to 23.0
Terrestrial	1.1 to 4.9
N_2O production: Nitrification	34.9 to 68.4
N_2O reduction	27 to 39
Ammonium ion exchange	−1.3 to 11
Nitrate ion exchange	1.6 to 5.5

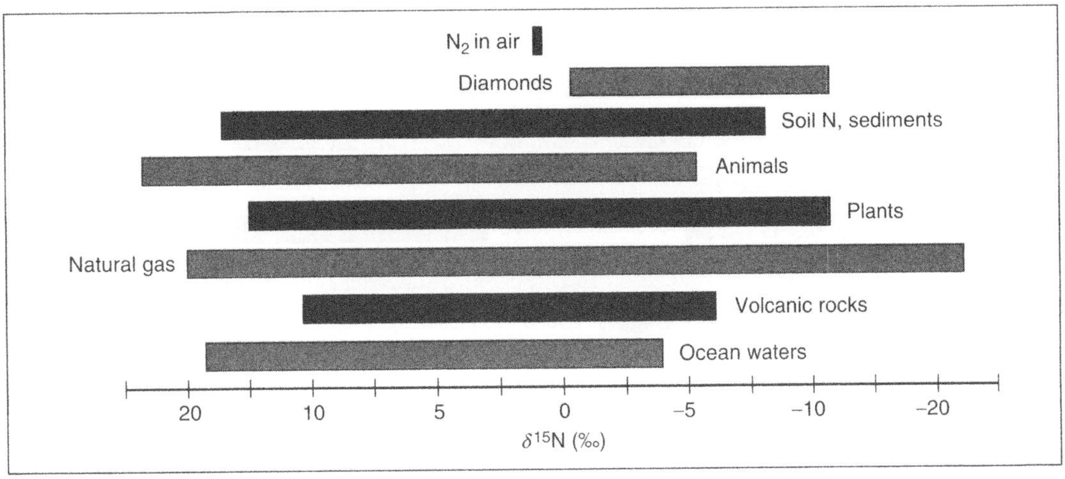

FIGURE 1.4 The observed ranges in the nitrogen isotope ratios of various materials on Earth expressed in delta units ($\delta^{15}N$) relative to N_2 in the atmosphere (modified and/or adapted from Ehleringer and Rundel 1989, Hoefs 1997, Criss 1999, Coplen *et al.* 2002, Fry 2006, and literature cited in the text).

prey on them. This so-called trophic enrichment in $\delta^{15}N$ is also reflected in the specific compounds extracted from tissues like fatty acids or other organic molecules that can serve as powerful biomarkers to trace and unravel ecological complexities as discussed in Chapters 9–12.

D. Variation in the $\delta^{34}S$ and $^{87}Sr/^{86}Sr$ in Mineral Cycle Processes

Recently, it has become apparent that the use of both sulfur (S) and strontium (Sr) isotopes hold a great deal of promise for detecting and therefore understanding the nature and magnitude of ecological change. For some years now $\delta^{34}S$ information has been used very effectively in both food web and some pollution research (Krouse *et al.* 1984, Michel and Turk 1996, Mitchell *et al.* 1998). Data from $^{87}Sr/^{86}Sr$ has been used to inform more ecologically oriented questions, particularly as geochemists, paleontologists, and biologists have begun to work more closely (Capo *et al.* 1998, English *et al.* 2001, Drouet *et al.* 2005, Chapter 21).

Sulfur stable isotopes have been useful in both pollution studies and trophic chain investigations (Fry 2006). The $\delta^{34}S$ of soils and plants are relative large (Figure 1.5), can be readily linked to marine or terrestrial sources (Fry 1991), and can therefore help constrain the feeding preferences or biases among the members of complex food webs. Organisms exposed to sulfur-based pollutants are commonly more enriched (and more variable) in their $\delta^{34}S$ and this allowed early investigators to use S isotopes to trace pollutants from their anthropogenic sources into and through various ecosystems as well as to characterize the degree of impacts experienced by the biota (Trust and Fry 1992). Recently, sulfur isotopes have also been useful in enhancing our understating of the roles that sulfur bacteria and other microorganism like archea play in cycling these substances as well as heavy metals like iron (Fe) through anoxic or high-temperature environments. And while the study of $\delta^{34}S$ in soils, microbes, and plants has not yet seen much research focused on ecological change, these recent advances suggest that there may be some informative work along these lines waiting to be done.

With regards to strontium (Sr) isotope ratios, they are not commonly expressed in standard "delta notation" but instead as the ratio of ^{87}Sr to ^{86}Sr in a sample; this ratio has a wide range. For years geochemists have researched Sr isotope systematics, but until very recently there was very little on

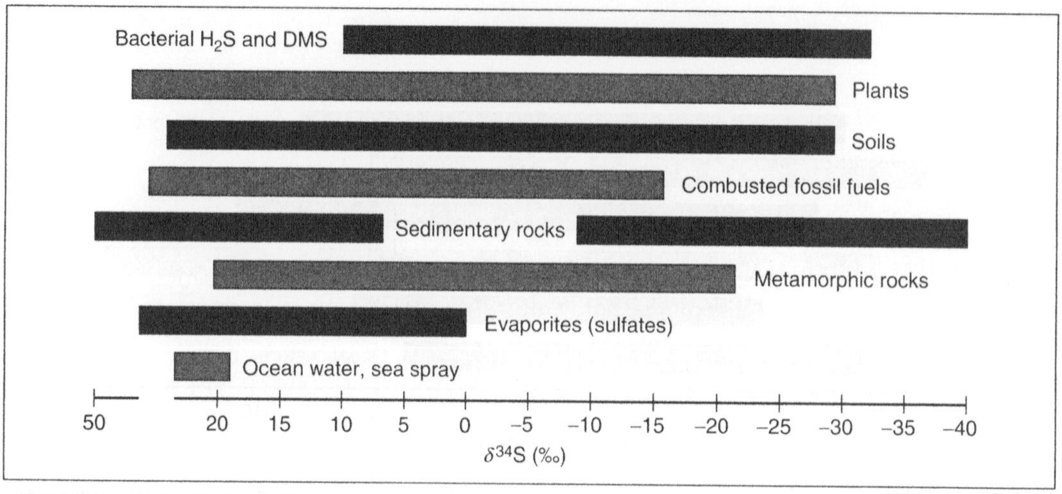

FIGURE 1.5 The observed ranges in the sulfur isotope ratios of various materials on Earth expressed in delta units ($\delta^{34}S$) relative to CDT. DMS is dimethyl sulfide (modified and/or adapted from Ehleringer and Rundel 1989, Hoefs 1997, Criss 1999, Coplen *et al.* 2002, Fry 2006, and literature cited in the text).

ecological applications. For all Sr isotope research, it is important to understand that the $^{87}Sr/^{86}Sr$ ratio depends on what the parent–daughter rubidium (Rb) to strontium ratio (specifically, $^{87}Rb/^{86}Sr$) in the source is, and how long ago in time it fractionated. For example, rocks derived from the Earth's mantle have low $^{87}Rb/^{86}Sr$, and so over time the mantle evolves very slowly, from values on the order of 0.699 at 4.5 billion years ago to about 0.7025 today. Oceanic islands like Hawaii have slightly higher $^{87}Rb/^{86}Sr$ ($\approx$0.703; Capo *et al.* 1998). On the basis of meteorite studies Earth as a "bulk material" is estimated to have a $^{87}Rb/^{86}Sr$ of about 0.7045. The process of melting in the mantle preferentially releases Rb isotopes into the melt phase, so that continental rocks that are derived from mantle melts even if reprocessed have high $^{87}Rb/^{86}Sr$ and usually much higher $^{87}Sr/^{86}Sr$. Granites that exceed 1 billion years in age can have $^{87}Sr/^{86}Sr > 1.0$ because of the commonly high $^{87}Rb/^{86}Sr$ ratio seen early during a fractionation event or during events such as melting and subsequent metamorphism that have then has long periods for radioactive decay to proceed. Over timescales of a few million years, changes in $^{87}Sr/^{86}Sr$ are minor, but beginning at 10 million years ago time begin to matter more to the ratios measured. For example, Himalayan rocks have very high $^{87}Sr/^{86}Sr$ largely because they are old and were formed from crustal material with high $^{87}Rb/^{86}Sr$ ratios. On average, continental values lie near 0.72 but are quite heterogeneous. The Amazon River, which drains a large area containing old limestone rock, has a $^{87}Sr/^{86}Sr$ ratio of $\approx$0.711, while some Himalayan rivers are >0.74. The oceans have a well-defined value presently at 0.70917 (DePaolo 2004).

For ecological investigation it is important to recognize that surface weathering preferentially releases Sr from certain minerals. Carbonates, for example, have high levels of Sr, near seawater values, and dissolve easily. So in any system with carbonates, they tend to dominate the Sr systematics and it is usually estimated that roughly 75–80% of "typical" river Sr is carbonate derived. For studies using Sr in plant materials like tree-ring cellulose (English *et al.* 2001, Chapter 21), there seems to be no significant fractionation during uptake or ion exchange processes.

To summarize, one can expect the Earth's mantle to possess a $^{87}Sr/^{86}Sr$ of $\approx$0.7025, volcanic arcs $\approx$0.705, the oceans to be $\approx$0.7092, the average river to be $\approx$0.712, and the continental crust to have a highly variable and skewed distribution with the mode at about $\approx$0.72 with a long upper tail. As noted in the footnotes of Table 1.1, the analytical standard, NBS 987, is often used when obtaining $^{87}Sr/^{86}Sr$ data but it has no special significance, it is simply a well-characterized source of Sr. Also, during analysis there is likely some small fractionation of Sr isotopes. To account for this, not unlike other stale isotopes, when the $^{87}Sr/^{86}Sr$ is measured we also simultaneously measure $^{88}Sr/^{86}Sr$, and normalize to a "known" absolute 88/86 value. Such procedures are applied routinely during Sr isotope analyses so that possible fractionation such as those that might be inadvertently produced on the ion exchange column used to isolate Sr or in the source of the mass spectrometer are accounted for. If this is not done then the $^{87}Sr/^{86}Sr$ value would vary continuously through a thermal ionization mass spectrometer (TIMS) run and preferentially "evaporate" the light isotope off the hot filament.

Advances that have emerged where Sr isotopes have been informative in an ecological context are highlighted in the tree-ring research of English *et al.* (2001) and in the information presented by Drouet *et al.* (Chapter 21) in this volume. Both are good examples of how Sr can be used to document changes over time from various types of ecological archives like tree rings.

V. SUMMARY

Stable isotope investigations have grown steadily in the past 50 years or so and have become a cornerstone for many ecological investigations since the 1980s (Rundel *et al.* 1989). Our understanding of the many processes that lead to isotopic variation in natural environments is now often well understood and is improving each year. But as stated above, the challenge we are now confronted with is to extend what we know about these processes and the products that result from them to see how changes have been "recorded" and how we can interpret these changes (what do such changes really mean?). This has required a different approach that many past ecological investigations have not taken and one that has forced ecologists to find materials such as carbonates, feathers, hard tissues, hair, tree-ring cellulose, and the many chemical compounds that compose them, so that they can be used to trace and therefore record change. By linking known processes to these changes, the growing field of "ecological forensics" that has emerged is now documenting changes in many novel and powerful ways. Many of these stories are presented in the chapters contained in this volume.

As ecologists, earth and atmospheric scientists, and citizens who are concerned about the pace and magnitude of the ecological and environmental changes that are happening around us everyday, we have a challenge and an opportunity to inform the debate and the knowledge base about how these changes have come about. Through innovative stable isotope methods and their novel application we are now documenting the manner and magnitude of ecological transformations across a wide range of natural systems. Environmental and ecological changes will continue and in many instances will accelerate. Applying stable isotope analyses in these ways, we believe can lead to solutions that will help mitigate future ecosystem damage and environmental degradation. In the chapters that follow the authors present some of the best examples showing how stable isotope research done in the context of ecological forensics can inform us about ecological change.

VI. ACKNOWLEDGMENTS

We want to thank the 2006 U.C. Berkeley "Isotopics" graduate discussion group for the critiques and discussions that lead to this chapter and to Lou Derry and Oliver Chadwick for providing feedback and a great deal of the material used in the section on strontium isotopes. Financial support from the Biogeosphere-Atmosphere Stable Isotope Network (BASIN) and Stable Isotopes in Biospheric and Atmosphere Exchange (SIBAE) programs in the United States and Europe, respectively, is greatfully acknowledged.

VII. REFERENCES

Angert, A., C. D. Cappa, and D. J. DePaolo. 2004. Kinetic O-17 effects in the hydrologic cycle: Indirect evidence and implications:. *Geochimica et Cosmochimica Acta* **68:**3487–3495.

Borella, S. M. L., and M. Saurer. 1999. Analysis of δ^{18}O in tree rings: Wood-cellulose comparison and method dependent sensitivity. *Journal of Geophysical Research* **104:**19267–19273.

Boutton, T. W., and S.-I. Yamasaki (Eds.). 1996. *Mass Spectrometry of Soils.* Marcel Dekker, Inc., Publishers, New York.

Bowen, G. J., and J. Revenaugh. 2003. Interpolating the isotopic composition of modern meteoric precipitation. *Water Resources Research* **39:**1299, doi:10.129/2003WR002086.

Bowen, G. J., L. I. Wassenaar, and K. A. Hobson. 2005. Global application of stable hydrogen and oxygen isotopes to wildlife forensics. *Oecologia* **143:**337–348.

Bowen, G. J., and B. Wilkinson. 2003. Spatial distribution of δ^{18}O in meteoric precipitation. *Geology* **30:**315–318.

Brugnoli, E., and G. D. Farquhar. 2000. Photosynthetic fractionation of carbon isotopes. Pages 399–434 *in* R. C. Leegood, T. D. Sharkey, and S. von Caemmerer (Eds.) *Advances in photosynthesis and Respiration: Photosynthesis, Physiology and Metabolism.* Springer, The Netherlends.

Capo, R. C., B. W. Stewart, and O. A. Chadwick. 1998. Strontium isotopes as tracers of ecosystem processes: Theory and methods. *Geoderma* **82:**197–225.

Clark, I. D., and P. Fritz. 1997. *Environmental Isotopes in Hydrogeology.* CRC Press, Boca Raton.

Coplen, T. B., J. A. Hopple, J. K. Böhlke, H. S. Peiser, S. E. Rieder, H. R. Krouse, K. J. R. Rosman, T. Ding, R. D. Vocke, Jr., K. M. Révész, A. Lamberty, P. Taylor, *et al.* 2002. Compilation of minimum and maximum isotope ratios of selected elements in naturally occurring terrestrial materials and reagents. USGS Water-Resources Investigations Report 01–4222, Reston Virgina. 98 p.

Craig, H 1961. Isotopic variation in meteoric waters. *Science* **133:**1702–1703.

Craig, H., and L. I. Gordon. 1965. Deuterium and oxygen-18 variations in the ocean and the marine atmosphere. Pages 9–130 *in* E. Tongiorgi (Ed.) *Proceedings of the Conference on Stable Isotopes in Oceanographic Studies and Paleotemperatures.* Laboratorie Geologia Nuclear, Pisa, Italy.

Criss, R. E. 1999. *Principles of Stable Isotope Distribution.* Oxford University Press, New York. 254 p.

Dansgaard, W. 1953. The abundance of ^{18}O in atmospheric water and water vapor. *Tellus* **5:**461–469.

Dansgaard, W. 1954. The ^{18}O abundance of fresh water. *Geochimica et Cosmochimica Acta* **6:**241–260.

Dansgaard, W. 1964. Stable isotopes in precipitation. *Tellus* **16:**436–468.

Dawson, T. E. 1998. Fog in the California redwood forest: Ecosystem inputs and use by plants. *Oecologia* **117:**476–485.

Dawson, T. E. 1993. Water sources of plants as determined from xlyem-water isotopic composition: Perspectives on plant competition, distribution, and water relations. Pages 465–496 *in* J. R. Ehleringer, A. E. Hall, and G. D. Farquhar (Eds.) *Stable Isotopes and Plant Carbon-Water Relations.* Academic Press, Inc., San Diego.

Dawson, T. E., and J. R. Ehleringer. 1998. Plants, isotopes, and water use: A catchment-level perspective. Pages 165–202 *in* C. Kendall and J. J. McDonnell (Eds.) *Isotope Tracers in Catchment Hydrology.* Elsevier Sceince Publications, The Netherlands.

Dawson, T. E., and P. D. Brooks. 2001. Fundamentals of stable isotope chemistry and measurement. Pages 1–18 *in* M. Unkovich, A. McNeill, J. Pate, and J. Gibbs (Eds.) *The Application of Stable Isotope Techniques to Study Biological Processes and the Functioning of Ecosystems.* Kluwer Academic Publishers, Dordrecht, Boston, London.

Dawson, T. E., R. C. Pausch, and H. M. Parker. 1998. The role of H and O stable isotopes in understanding water movement along the soil-plant-atmospheric continuum. Pages 169–183 *in* H. Griffiths (Ed.) *Stable Isotopes: Integration of Biological, Ecological and Geochemical Processes.* BIOS Scientific Publishers, Oxford, UK.

Dawson, T. E., S. Mambelli, A. H. Plamboeck, P. H. Templer, and K. P. Tu. 2002. Stable isotopes in plant ecology. *Annual Review of Ecology and Systematics* **33:**507–559.

DePaolo, D. J. 2004. Calcium isotopic variations produced by biological, kinetic, radiogenic and nucleosynthetic processes. *Reviews in Mineralogy and Geochemistry* **55:**255–288.

Dongmann, G., H. W. Nurnberg, H. Förstel, and K. Wagner. 1974. On the enrichment of $^{2}H^{18}O$ in the leaves of transpiring plants. *Radiation and Environmental Biophysics* **11:**41–52.

Drouet, T., J. Herbauts, and D. Demaiffe. 2005. Long-term records of strontium isotopic composition in tree-rings suggest changes in forest calcium sources in the early 20th century. *Global Change Biology* **11:**1926–1940.

Duranceau, M., J. Ghashghaie, F. Badeck, E. Deleens, and G. Cornic. 1999. δ^{13}C of CO_2 respired in the dark in relation to δ^{13}C of leaf carbohydrates in *Phaseolus vulgaris* L. under progressive drought. *Plant, Cell and Environment* **22:**515–523.

Durka, W. E., D. S. Schultze, G. Gebauer, and S. Voerkelius. 1994. Effects of forest decline on uptake and leaching of deposited nitrate determined from ^{15}N and ^{18}O measurements. *Nature* **372:**765–769.

Ehleringer, J. R., and P. W. Rundel. 1989. Stable isotopes: History, units and instrumentation. Pages 1–15 *in* P. W. Rundel, J. R. Ehleringer, and K. A. Nagy (Eds.) *Stable Isotopes in Ecological Research. Ecological Studies*, Vol. 68, Springer-Verlag, Heidelberg.

Ehleringer, J. R., and T. E. Dawson. 1992. Water uptake by plants perspectives from stable isotope composition. *Plant, Cell and Environment* **15:**1073–1082.

Ehleringer, J. R., A. E. Hall, and G. D. Farquhar (Eds.). 1993. *Stable Isotopes and Plant Carbon-Water Relations*. Academic Press, Inc., San Diego. 555 p.

English, N. B., J. L. Betancourt, J. S. Dean, and J. Quade. 2001. Strontium isotopes reveal distant sources of architectural timber in Chaco Canyon, New Mexico. *Proceedings of the National Academy of Sciences* **98**(21):11891–11896.

Epstein, S., and T. Mayeda. 1953. Variation in the ^{18}O content of water from natural sources. *Geochimica et Cosmochimica Acta* **42:**213–224.

Evans, R. D. 2001. Physiological mechanisms influencing plant nitrogen isotope composition. *Trends in Plant Science* **6:**121–126.

Farquhar, G. D., J. R. Ehleringer, and K. T. Hubick. 1989. Carbon isotope discrimination and photosynthesis. *Annual Review of Plant Physiology and Plant Molecular Biology* **40:**503–537.

Faure, G. 1986. *Principles of Isotope Geology*. 2nd ed. John Wiley and Sons, Publishers, New York.

Flanagan, L. B., J. P. Comstock, and J. R. Ehleringer. 1991. Comparison of modeled and observed environmental influences in stable oxygen and hydrogen isotope composition of leaf water in *Phaseolus vulgaris*. *Plant Physiology* **96:**588–596.

Flanagan L. B., J. R. Ehleringer, and D. E. Pataki (Eds.). 2005. *Stable Isotopes and Biosphere-Atmosphere Interactions*. Elsevier Academic Press, San Diego.

Flannery, T. 2005. *The Weather Makers*. Grove Press, New York. 359 p.

Friedman, I., A. C. Redfield, A. Schoen, and J. Harris. 1964. The variation of deuterium content of natural waters in the hydrologic cycle. *Reviews in Geophysics* **2:**177–224.

Fry, B. 1991. Stable isotope diagrams of fresh-water food webs. *Ecology* **72:**2293–2297.

Fry, B. 2006. *Stable Isotope Ecology*. Springer-Verlag, New York. 370 p.

Gat, J. R. 1996. Oxygen and hydrogen stable isotopes in the hydrologic cycle. *Annual Review of Earth and Planetary Science* **24:**225–262.

Ghashghaie, J., M. Duranceau, F. W. Badeck, G. Cornic, M. T. Adeline, and E. Deleens. 2001. $\delta^{13}C$ of CO_2 respired in the dark in relation to $\delta^{13}C$ of leaf metabolites: Comparison between *Nicotiana sylvestris* and *Helianthus annuus* under drought. *Plant, Cell and Environment* **24:**505–515.

Gleixner, G., H.-J. Danier, R. A. Werner, and H.-L. Schmidt. 1993. Correlations between the ^{13}C content of primary and secondary plant products in different cell compartments and that in decomposing basidiomycetes. *Plant Physiology* **102:**1287–1290.

Gleixner, G., C. Scrimgeour, H. L. Schmidt, and R. Viola. 1998. Stable isotope distribution in the major metabolites of source and sink organs of *Solanum tuberosum* L.: A powerful tool in the study of metabolic partitioning in intact plants. *Planta* **207:**241–245.

Gonfiantini, R. 1982. On the isotopic composition of precipitation. *Rendiconti Society of Italian Mineralogy and Petrology* **38:**1175–1187.

Gore, A. 2006. *An Inconvenient Truth*. Rodale Books, Emmaus, PA. 325 p.

Griffiths, H. 1992. Carbon isotope discrimination and the integration of carbon assimilation pathways in terrestrial CAM plants. *Plant, Cell and Environment* **15:**1051–1062.

Griffiths, H. (Ed.). 1998. *Stable Isotopes: Integration of Biological, Ecological and Geochemical Processes*. BIOS Scientific Publishers, Oxford. 438 p.

Handley, L. L., and C. M. Scrimgeour. 1997. Terrestrial plant ecology and ^{15}N natural abundance: The present limits to interpretation for uncultivated systems with original data from a Scottish old field. *Advances in Ecological Research* **27:**133–212.

Hoefs, J. 1997. *Stable Isotope Geochemistry*. Springer, Berlin. 201 p.

Hoering, T. 1955. Variation of nitrogen-15 abundance in naturally occurring substances. *Science* **122:**1233–1234.

Högberg, P. 1997. ^{15}N natural abundance in soil-plant systems. *New Phytologist* **137:**179–203.

Ingraham, N. L. 1998. Isotopic variations in precipitation. Pages 87–118 *in* C. Kendall and J. J. McDonnell (Eds.) *Isotope Tracers in Catchment Hydrology*. Elsevier Science Publications, The Netherlands.

Ingraham, N. L., and B. E. Taylor. 1991. Light stable isotope systematics of large-scale hydrologic regimes in California and Nevada. *Water Resources Research* **27:**77–90.

Ingraham, N. L., and R. A. Matthews. 1990. A stable isotopic study of fog: The Point Reyes Peninsula, California, U.S.A. *Chemical Geology (Isotope Geoscience Section)* **80:**281–290.

Intergovernmental Panel on Climate Change (**IPCC**) fourth assessment report. 2007. http://www.ipcc.ch/.

Jouzel, J., G. Hoffmann, R. D. Koster, and V. Masson. 2000. Water isotopes in precipitation: Data/model comparison for present-day and past climates. *Quaternary Science Reviews* **19:**363–379.

Keeley, J. E., and D. R. Sandquist. 1992. Carbon: Freshwater plants. *Plant, Cell and Environment* **15:**1021–1036.

Kendall, C. 1998. Tracing nitrogen sources and cycles in catchments. Pages 519–576 *in* C. Kendall and J. J. McDonnell (Eds.) *Isotope Tracers in Catchment Hydrology*. Elsevier Science B.V., Amsterdam.

Kendall, C., and J. J. McDonnell (Eds.). 1998. *Isotope Tracers in Catchment Hydrology*. Elsevier Science B.V., Amsterdam. 839 p.

Krouse, H. R., A. Legge, and H. M. Brown. 1984. Sulfur gas emissions in the boreal forest: The West Whitecourt Case Study V: Stable sulfur isotopes. *Water, Air, and Soil Pollution* **22:**321–347.

Krull, E., D. Sachse, I. Mügler, A. Thiele, and G. Gleixner. 2006. Compound-specific $\delta^{13}C$ and $\delta^{2}H$ analyses of plant and soil organic matter: A preliminary assessment of the effects of vegetation change on ecosystem hydrology. *Soil Biology and Biochemistry* **38:**3211–3221.

Lajtha, K., and R. H. Michener (Eds.). 1994. *Stable Isotopes in Ecology and Environmental Science.* Blackwell Scientific Publications, Oxford. 316 p.

Macko, S. A., M. L. F. Estep, and M. H. Engle. 1986. Kinetic fractionation of stable nitrogen isotopes during amino acid transamination. *Geochimica et Cosmochimica Acta* **50:**2143–2146.

Mazor, E. 1991. Stable hydrogen and oxygen isotopes. Pages 122–146 *in Applied Chemical and Isotopic Groundwater Hydrology.* Halsted Press—Division of John Wiley and Sons, Inc., London.

McKinney, C. R., J. M. McCrea, S. Epstein, H. A. Allen, and H. C. Urey. 1950. Improvements in mass spectrometers for the measurement of small differences in isotope abundance ratios. *The Review of Scientific Instrument* **21:**724–730.

Michel, R. L., and J. T. Turk. 1996. Use of sulphur-35 to study sulfur migration in the Flat Tops Wilderness Area, IAEA Symposium on Isotopes in Water Resources Management, Vienna, 20–24 March, 1995, 10 p.

Millennium Ecosystem Assessment. 2001. http://www.maweb.org/en/index.aspx.

Mitchell, M. J., H. R. Krouse, B. Mayer, A. C. Stam, and Y. Zhang. 1998. Use of stable isotopes in evaluating sulfur biogeochemistry of forest ecosystems. Pages 489–518 *in* C. Kendall and J. J. McDonnell (Eds.) *Isotope Tracers in Catchment Hydrology.* Elsevier, Amsterdam.

Natelhoffer, K. J., and B. Fry. 1988. Controls on natural nitrogen-15 and carbon-13 abundances in forest soil organic matter. *Soil Science Society of America* **52:**1633–1640.

O'Leary, M. H., S. Madhavan, and P. Paneth. 1992. Physical and chemical basis for carbon isotope fractionation in plants. *Plant, Cell and Environment* **15:**1099–1110.

Persson, T., A. Rudebeck, J. H. Jussy, M. Colin-Belgrand, A. Priemé, E. Dambrine, P. S. Karlson, and R. M. Sjöberg. 2000. Soil nitrogen turnover—mineralization, nitrification and denitrification in European forest soils. Pages 297–331 *in* E.-D. Schulze (Ed.) *Ecological Studies*, Vol. 142, Carbon and Nitrogen Cycling in European Forest Ecosystems. Springer-Verlag, Berlin.

Raven, J. A. 1992. Present and potential uses of natural abundance of stable isotopes in plant science, with illustrations form the marine environment. *Plant, Cell and Environment* **15:**1083–1090.

Robinson, D. 2001. $\delta^{15}N$ as an integrator of the nitrogen cycle. *Trends in Ecology and Evolution* **16:**153–162.

Rozanski, K., L. Araguás-Araguás, and R. Gonfiantini. 1993. Isotopic patterns in modern global precipitation. *Geophysical Monograph* **78:**1–36.

Ruddiman, W. F. 2004. *Earth's Climate: Past and Future.* W.H. Freeman Publishers, New York.

Rundel, P. W., J. R. Ehleringer, and K. A. Nagy (Eds.). 1989. Stable Isotopes in Ecological Research. Ecological Studies, Vol. 68, Springer-Verlag, Heidelberg. 525 p.

Sachse, D., J. Radke, and G. Gleixner. 2006. δD values of individual *n*-alkanes from terrestrial plants along a climatic gradient—Implications for the sedimentary biomarker record. *Organic Geochemistry* **37:**469–473.

Shaerer, G., and D. H. Kohl. 1988. Estimates of N_2 fixation in ecosystems: The need for and basis of the ^{15}N natural abundance method. Pages 342–374 *in* P. W. Rundel, J. R. Ehleringer, and K. A. Nagy (Eds.) *Ecological Studies*, Vol. 68, Stable Isotopes in Ecological Research. Springer-Verlag, New York.

Sharp, Z. 2007. *Principles of Stable Isotope Geochemistry.* Person Prentice Hall, New Jersey. 344 p.

Sternberg, L. S. L., M. J. DeNiro, and R. A. Savidge. 1986. Oxygen isotope exchange between metabolites and water during biochemical reactions leading to cellulose synthesis. *Plant Physiology* **82:**423–427.

Sternberg, L. D. S. L. 1989. Oxygen and hydrogen isotope ratios in plant cellulose: Mechanisms and applications. Pages 124–141 *in* P. W. Rundel, J. R. Ehleringer, and K. A. Nagy (Eds.) *Stable Isotopes in Ecological Research. Ecological Studies*, Vol. 68, Springer-Verlag, Heidelberg.

Tcherkez, G., S. Nogue's, J. Bleton, G. Cornic, F. Badeck, and J. Ghashghaie. 2003. Metabolic origin of carbon isotope composition of leaf dark-respired CO_2 in french bean. *Plant Physiology* **131:**237–244.

Trust, B. A., and B. Fry. 1992. Stable sulfur isotopes in plants: A review. *Plant, Cell and Environment* **15:**1105–1110.

Tu, K. P., and T. E. Dawson. 2005. Partitioning ecosystem respiration using stable carbon isotope analyses of CO_2. Pages 125–153 *in* L. B. Flanagan, J. R. Ehleringer, and D.E Pataki (Eds.) *Stable Isotopes and Biosphere–Atmosphere Interactions: Processes and Biological Controls.* Elsevier Academic Press, San Diego, CA.

Unkovich M., A. McNeill, J. Pate, and J. Gibbs (Eds.). 2001. *The Application of Stable Isotope Techniques to Study Biological Processes and the Functioning of Ecosystems.* Kluwer Academic Press. Boston.

Vitousek, P. M., J. Aber, R. W. Howarth, G. E. Likens, P. A. Matson, D. W. Schindler, W. H. Schlesinger, and G. D. Tilman. 1997. Human alteration of the global nitrogen cycle causes and consequences. *Issues in Ecology* **1:**1–15.

Wershaw, R. L., I. Friedman, and S. J. Heller. 1966. Hydrogen isotope fractionation of water passing through trees. Pages 55–67 *in* F. Hobson and M. Speers (Eds.) *Advances in Organic Geochemistry.* Pergamon Press, New York.

White, J. W. C. 1989. *Stable hydrogen isotope ratios in plants: A review of current theory and some potential applications*. See Rundel *et al.* 1989. **68:**142–162.

Yakir, D. 1992. Variations in the natural abundance of oxygen-18 and deuterium in plant carbohydrates. *Plant, Cell and Environment* **15:**1005–1020.

CHAPTER 2

Stable Isotopes Record Ecological Change, but a Sampling Network Will Be Critical

James R. Ehleringer* and Todd E. Dawson†

*SIRFER, Department of Biology, University of Utah
†Department of Integrative Biology, University of California

It is often said that the only constant is change. From seasonal dynamics to decadal oscillations, natural variations in abiotic and biotic forcings known to impact both ecosystems and organisms have resulted in changes in both patterns and processes. Historically, traditional measures of ecological changes focused on describing these patterns using the presence or absence of a marker or of the abundance of that marker in sequentially deposited materials, such as sediments in a pond (Davis *et al.* 1991). Stable isotope ratio analyses have complemented these traditional ecological approaches and have played a major role in improving our interpretations of these dynamics and in providing a mechanistic basis for understanding the drivers of change (Rundel *et al.* 1988, Ehleringer *et al.* 1993, Griffiths 1998). This is particularly true when the deposited material is homogenous in its composition, such as thick carbonate deposits. In most cases, fundamental progress and ecological advances were made only after theoretical advances provided a basis for isotopic interpretation, such as the models to understand the basis of carbon isotope changes in plants (Farquhar *et al.* 1989).

With a focus in this volume on stable isotopes in modern ecological systems, it is important to recognize that human impacts on ecosystems are of sufficient magnitude to both directly and indirectly impact the trajectories of biogeochemical dynamics, resulting in transitions and changes that are also recorded in the stable isotopic composition of both biotic and abiotic materials. Classic examples at the local scale would include the role of human behavior in changing the abundances of C_3 and C_4 photosynthetic systems across the landscape through forest harvesting (Eshetu and Hogberg 2000) or through agricultural practices (Balesdent *et al.* 1990, Rochette *et al.* 1999). At the continental and global scales, changes in the carbon isotope ratios of atmospheric carbon dioxide and methane associated with fossil-fuel burning represent another example of the human influence on a natural cycle (Keeling *et al.* 1979).

Stable isotope analyses have proved to be a powerful tool in ecological studies to trace, record, source, and integrate different ecological parameters of interest (West *et al.* 2006). Perhaps this is most easily illustrated considering animal hair. Scalp hair in humans (O'Connell and Hedges 1999, Ehleringer *et al.* 2007), body hair in cattle (Schwertl *et al.* 2003), and tail hair in an elephant (Cerling *et al.* 2004, 2006) grow continuously, each recording the most recent dietary inputs. Under natural conditions, the environment can have a major role influencing the abundances of C_3 and C_4 food items or how these food items might vary spatially. Since hair is a linear recorder of dietary inputs and there is no subsequent exchange of proteins in hair with other animal tissues, the hair becomes a linear, quantitative recorder of dietary inputs. Analogous examples are found in feathers (Chapter 9), bone, or fish scale collagen (Chapter 10), amino acids (Chapter 12), in tree-ring cellulose (Chapters 3–5) and lichen biomass (Chapter 6).

This volume focuses on how stable isotope analyses have contributed to understanding ecological changes within atmospheric gasses and a vast array of materials collected from aquatic, terrestrial, and marine ecosystems. It is noteworthy that biochemical, physiological, and ecological processes, as well as climatological cycles, operate on different temporal and spatial scales. This requires that an isotopic proxy correctly match and record the changes on interest. How often is this the case? One of the key challenges in using isotopes as recorders of ecological change is to ensure that activities of the isotope recorder properly match both the process as well as the temporal and spatial scales of the process of interest. Consider again the metabolism of an animal that gets recorded sequentially in hair. On shorter timescales, changes in dietary inputs are reflected in the carbon dioxide of breath and the isotopic of blood (Ayliffe *et al.* 2004, Passey *et al.* 2005). Yet, because here the time constant of change is so short, the isotopic signal may be quite dynamic, responding to individual meals, and making it challenging to interpret an ecological pattern. In this volume, Cerling *et al.* address ways in which shorter-term dietary dynamics can be extracted from longer-term records (hair).

Consider three kinds of ecological processes that might be measurable using stable isotopes: spatial patterns, sourcing the origin of a material (Chapter 21), and a nutrient cycle (Table 2.1). Choosing the appropriate component to measure will be greatly influenced by the time period integration of interest (Chapter 1). Spatial patterns of water can be measured on a scale of hours to days, such as is the case for a sample of river water (Kendall and Coplen 2001). On the other hand, a broader annual measure of water is reflected in the keratin of bird feathers (Hobson 1999). An even longer-term measure of terrestrial water is reflected in the *n*-alkanes in lake sediments (Sachse *et al.* 2004). Similar temporal sequences can be suggested for both tracing the origins of animals and ecological nutrient cycles (Chapters 19 and 7) (Table 2.1).

One important attribute that is commonly shared when making temporal comparisons, such as in Table 2.1, is that the short-term isotopic signals are labile and are generally not preserved over time. These parameters are often of significant interest and require establishment of a sampling network in order to preserve this labile signal. Rather than consider the range of examples, let us focus on just one example—carbon dioxide.

Short-term dynamics of the modern carbon cycle are measured by frequent collections of atmospheric CO_2 at remote or isolated locations where a near-homogeneous signal can be obtained (*e.g.*, NOAA global CO_2 monitoring, http://www.esrl.noaa.gov/gmd/). These large-grid, largely marine-based measures of global atmospheric CO_2 become the basis of inverse modeling to estimate carbon sources and sinks. Here, the spatial variations in atmospheric CO_2 are relatively small because of the

TABLE 2.1 Ecological and environmental parameters often measured that reflect processes integrated over different temporal periods

Ecological interest	Short term (days)	Medium term (months)	Long term (years)
Spatial patterns of water	River water	Feathers	*n*-Alkanes
Recording the dietary inputs of animals	Organic matter in blood or CO_2 in breath	Hair	Bones
Carbon cycles	Atmospheric CO_2	Leaves	Tree rings or soil organic matter

strong mixing among atmospheric layers (Trolier *et al.* 1996). However, at the ecosystem scale, short-term vertical and temporal variations of atmospheric CO_2 that are of greater magnitude are of keen interest because they reflect the contributions of ecosystem components to respiratory fluxes (Ehleringer and Cook 1998, Bowling *et al.* 2003, Pataki *et al.* 2003b).

The ecosystem-scale measures of CO_2 and isotopes, known as Keeling plots, can be integrated across continental scales to provide a broad picture of how ecosystem carbon and oxygen isotope fractionation varies across ecosystems and in response to climate drivers. Consider the small terrestrial network underway and supported by the US Department of Energy TCP and contributing to the North American Carbon Cycle Program (Figure 2.1). Here, we see that the carbon isotope ratio of ecosystem respiration can exhibit strong cyclic patterns and that the values among different ecosystems are not constant (Figure 2.2). From the limited numbers of multisite studies to date, we have learned that there can be some common-response patterns, such as ecosystem photosynthesis and respiration responses to water stress (Lai *et al.* 2005, Alstad *et al.* 2007). The large, positive excursions in the carbon isotope ratio of ecosystem respiration are tightly correlated with stomatal closure and reductions in net primary productivity.

Extending short-term atmospheric CO_2 measurements to human-dominated landscapes, such as the urban ecosystems, reveals even larger CO_2 fluctuations in response to diurnal changes in both natural and anthropogenic activities (Pataki *et al.* 2005, 2006). Using the Salt Lake Valley in Utah as an example (Figure 2.2), we see that the short-term fluctuations provide a strong temporal picture of carbon-source dynamics, and these isotopic signals can be used to partition anthropogenic from natural CO_2 sources (Pataki *et al.* 2003a). Current online measurement of both the concentration and isotope of atmospheric CO_2 are possible but will require networks in order to understand the ecological patterns. At the moment, only a small, regional network is operational (http://co2.utah.edu), but through the activities of both the National Ecological Observatory Network (http://neoninc.org) and the Biogeosphere-Atmosphere Stable Isotope Network (http://basinisotopes.org) larger networks should emerge in the near future that will be of benefit to the entire environmental and ecological communities (Chapter 23).

In order to consider how stable isotopes can be used to understand major ecological processes when there are no "permanent recorders," a network of collections is the only viable solution at the moment (*e.g.*, Figure 2.1). While the previous discussion focused on CO_2 networks, there are also integrated

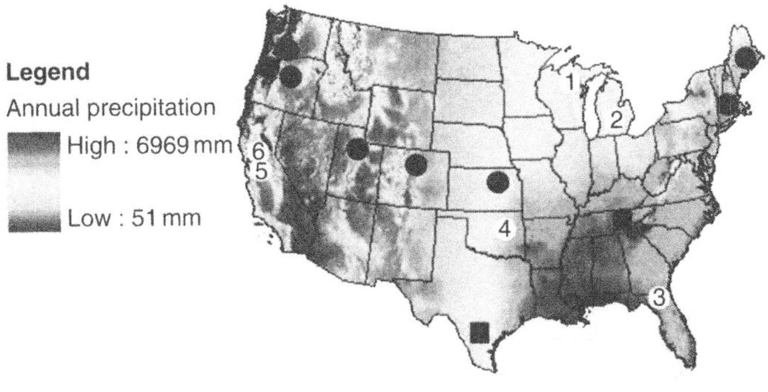

FIGURE 2.1 AmeriFlux sites with ongoing flask-based isotope measurements for Keeling-plot analysis of ecosystem respiration. Color shading shows spatial gradients of total annual precipitation across the United States. Solid symbols indicate measurements sites supported by the Terrestrial Carbon Processes Program at the US Department of Energy; these sites include deciduous and coniferous forests, grasslands, and an urban ecosystem. Open circles indicate studies maintained by other BASIN partners in: (1) deciduous forest, (2) deciduous forest, (3) slash pine, (4) grassland, (5) grassland, and (6) oak savanna ecosystems. (**See Color Plate.**)

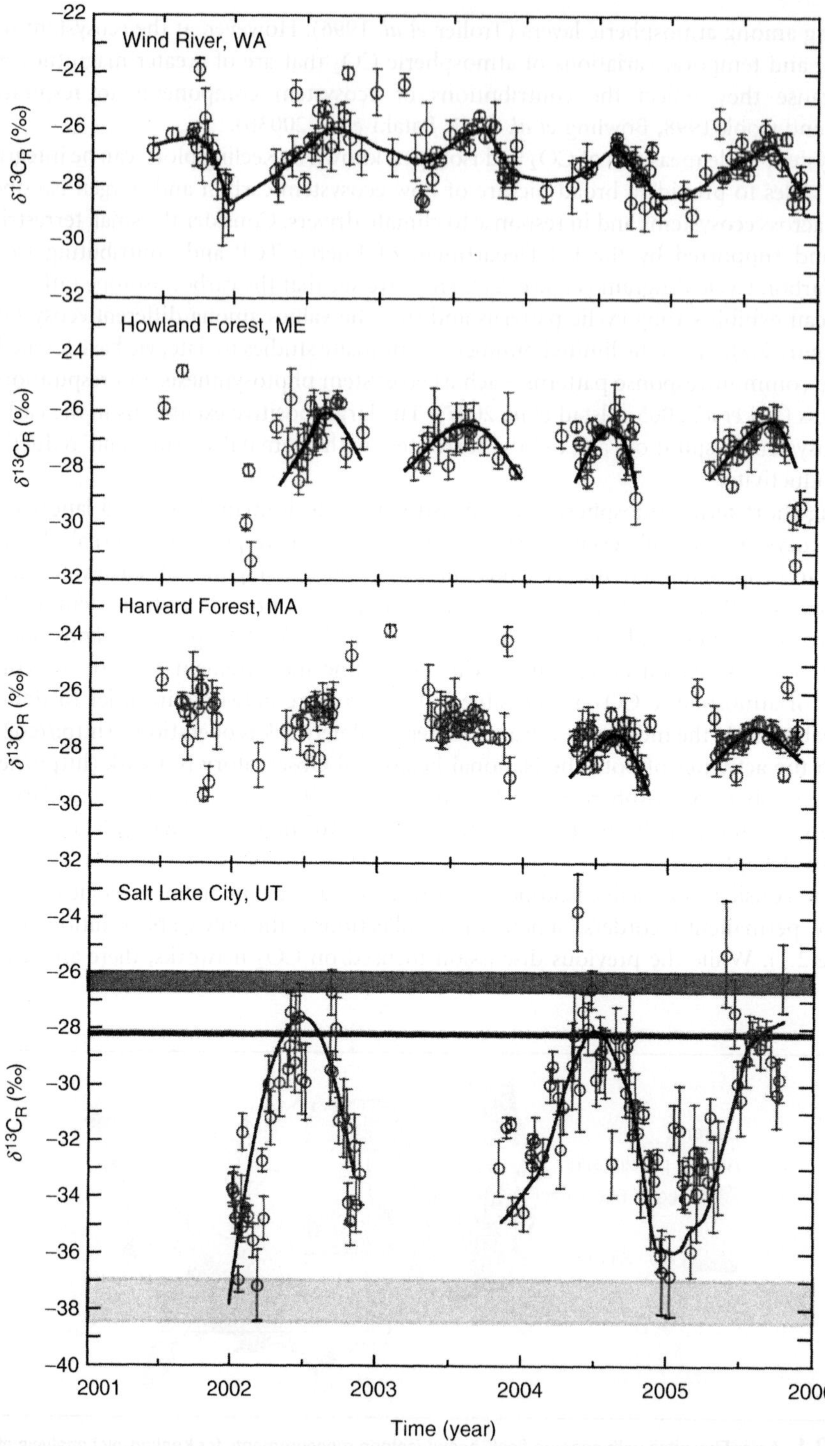

FIGURE 2.2 A time series of weekly measurements of carbon isotope ratio of ecosystem respiration ($\delta^{13}C_R \pm$ S.E.) at three US temperate forests and one urban site. These data could not have been collected without a network of sampling sites because there is no permanent, high resolution preserved in organic matter to go back and sample. Smooth curves are superimposed on weekly values to highlight seasonal patterns. Intraseasonal $\delta^{13}C_R$ variations are greatest in urban ecosystems (Salt Lake City), followed by coniferous forests (Wind River and Howland) and deciduous forests (Harvard). This multiseasonal $\delta^{13}C_R$ observation also reveals a long-term decreasing trend in $\delta^{13}C_R$ values that would not be evident with a single-year observation. $\delta^{13}C_R$ measurements in the forest sites are from C.T. Lai *et al.* (unpublished data). For the urban site, the shaded bars show the 95% confidence interval for the $\delta^{13}C_R$ values due to emissions from respiration (dark gray), gasoline combustion (black) and natural gas combustion (gray). 2002 data are redrawn from Pataki *et al.* (2003a).

networks sampling isotopes in precipitation (Global Network of Isotopes in Precipitation, International Atomic Energy Agency, http://iaea.org), in rivers (Global Network of Isotopes in Rivers, International Atomic Energy Agency, http://iaea.org), and in leaf water and atmospheric water vapor (Measurements of Isotopes in the Biosphere-Atmosphere, International Atomic Energy Agency, http://iaea.org). Chapter 24 in this volume discusses INEWS as a new network that could be part of the NEON effort.

Many ecological patterns are only revealed by long-term sampling efforts on individual organisms. While the isotopic signal preserved in the individual may be of immediate interest, such as muscle tissue in a fish, it is the long-term or population-level changes in the signal that are of community-scale ecological interest (Chapter 22). For instance, how do trophic-level patterns in either marine or terrestrial ecosystems shift in response to direct and indirect human pressures? Here, there is a need for the ecological community to recognize the benefit of associations and networks for collections and preservations of biological materials (Chapter 25). Museums have been an extremely valuable source of materials to reconstruct temporal patterns (Chapter 20), but a taxonomically diverse, geographically distributed network of collections will better serve the ecological community if isotopes are to prove useful in looking at process-level changes, such as changes in trophic-level complexity in response to invasive species or human activities.

REFERENCES

Alstad, K. P., C.-T. Lai, L. B. Flanagan, and J. R. Ehleringer. 2007. Environmental controls on the carbon isotope composition of ecosystem respired CO_2 in contrasting forest ecosystems in Canada and USA. *Tree Physiology* (in press).

Ayliffe, L. K., T. E. Cerling, T. Robinson, A. G. West, M. Sponheimer, B. H. Passey, J. Hammer, B. Roeder, M. D. Dearing, and J. R. Ehleringer. 2004. Turnover of carbon isotopes in tail hair and breath CO_2 of horses fed an isotopically varied diet. *Oecologia* **139:**11–22.

Balesdent, J., A. Mariotti, and D. Boisgontier. 1990. Effect of tillage on soil organic carbon mineralization estimated from 13C abundance in maize fields. *Journal of Soil Science* **41:**587–596.

Bowling, D. R., N. G. McDowell, J. M. Welker, B. J. Bond, B. E. Law, and J. R. Ehleringer. 2003. Oxygen isotope content of CO_2 in nocturnal ecosystem respiration: 1. Observations in forests along a precipitation transect in Oregon, USA. *Global Biogeochemical Cycles* **17**(4):1120.

Cerling, T. E., B. H. Passey, L. K. Ayliffe, C. S. Cook, J. R. Ehleringer, J. M. Harris, M. B. Dhidha, and S. M. Kasiki. 2004. Orphans' tales: Seasonal dietary changes in elephants from Tsavo National Park, Kenya. *Palaeogeography, Palaeoclimatology, Palaeoecology* **206:**367–376.

Cerling, T. E., G. Wittemyer, H. B. Rasmussen, F. Vollrath, C. E. Cerling, T. J. Robinson, and I. Douglas-Hamilton. 2006. Stable isotopes in elephant hair document migration patterns and diet changes. *Proceedings of the National Academy of Sciences of the United States of America* **103:**371–373.

Davis, M. B., M. W. Schawartz, and K. D. Woods. 1991. Detecting a species limit from fossil pollen in sediment. *Journal of Biogeography* **128:**653–668.

Ehleringer, J. R., and C. S. Cook. 1998. Carbon and oxygen isotope ratios of ecosystem respiration along an Oregon conifer transect: Preliminary observations based on small-flask sampling. *Tree Physiology* **18:**513–519.

Ehleringer, J. R., A. E. Hall, and G. D. Farquhar (Eds.). 1993. *Stable Isotopes and Plant Carbon/Water Relations.* Academic Press, San Diego.

Ehleringer, J. R., G. J. Bowen, L. A. Chesson, A. G. West, D. W. Podlesak, and T. E. Cerling. 2007. Hydrogen and oxygen isotopes in human hair are related to geography. *Proceedings of the USA National Academy of Sciences*: In review.

Eshetu, Z., and P. Hogberg. 2000. Reconstruction of forest site history in Ethiopian Highlands based on ^{13}C natural abundance in soils. *Ambio* **29:**83–89.

Farquhar, G. D., J. R. Ehleringer, and K. T. Hubick. 1989. Carbon isotope discrimination and photosynthesis. *Annual Review of Plant Physiology and Plant Molecular Biology* **40:**503–537.

Griffiths, H. (Ed.). 1998. *Stable Isotopes Integration of Biological, Ecological, and Geochemical Processes.* BIOS Scientific Publishers, Oxford.

Hobson, K. A. 1999. Tracing origins and migration of wildlife using stable isotopes: A review. *Oecologia* **120:**314–326.

Keeling, C. D., W. G. Mook, and P. P. Tans. 1979. Recent trends in the $^{13}C/^{12}C$ ratio of atmospheric carbon dioxide. *Nature* **277:**121–123.

Kendall, C., and T. B. Coplen. 2001. Distribution of oxygen-18 and deuterium in river waters across the United States. *Hydrological processes* **15**:1363–1393.

Lai, C.-T., J. R. Ehleringer, A. J. Schauer, P. P. Tans, D. Y. Hollinger, K. T. Paw, U. J. W. Munger, and S. C. Wofsy. 2005. Canopy-scale $d^{13}C$ of photosynthetic and respiratory CO_2 fluxes: Observations in forest biomes across the United States. *Global Change Biology* **11**:633–643.

O'Connell, T. C., and R. E. M. Hedges. 1999. Investigations into the effect of diet on modern human hair isotopic values. *American Journal of Physical Anthropology* **108**:409–425.

Passey, B. H., T. F. Robinson, L. K. Ayliffe, T. E. Cerling, M. Sponheimer, M. D. Dearing, B. L. Roeder, and J. R. Ehleringer. 2005. Carbon isotope fractionation between diet, breath CO_2, and bioapatite in different mammals. *Journal of Archaeological Science* **32**:1459–1470.

Pataki, D. E., D. R. Bowling, and J. R. Ehleringer. 2003a. Seasonal cycle of carbon dioxide and its isotopic composition in an urban atmosphere: Anthropogenic and biogenic effects. *Journal of Geophysical Research* **108**:NO. D23, 4735, doi:4710.1029/2003JD003865.

Pataki, D. E., J. R. Ehleringer, L. B. Flanagan, D. Yakir, D. R. Bowling, C. J. Still, N. Buchmann, J. O. Kaplan, and J. A. Berry. 2003b. The application and interpretation of Keeling plots in terrestrial carbon cycle research. *Global Biochemical Cycles* **17**:22-21–22-24.

Pataki, D. E., B. J. Tyler, R. E. Peterson, A. P. Nair, W. J. Steenburgh, and E. R. Pardyjak. 2005. Can carbon dioxide be used as a tracer of urban atmospheric transport? *Journal Of Geophysical Research D: Atmospheres* **110**: D15102.

Pataki, D. E., D. R. Bowling, J. R. Ehleringer, and J. M. Zobitz. 2006. High resolution atmospheric monitoring of urban carbon dioxide sources. *Geophysical Research Letters* **33**: L03813.

Rochette, P., L. B. Flanagan, and E. G. Gregorich. 1999. Separating soil respiration into plant and soil components using analyses of the natural abundance of carbon-13. *Soil Science Society of America Journal* **63**:1207–1213.

Rundel, P. W., J. R. Ehleringer, and K. A. Nagy (Eds.). 1988. *Stable Isotopes in Ecological Research* Springer-Verlag, New York.

Sachse, D., J. Radke, and G. Gleixner. 2004. Hydrogen isotope ratios of recent lacustrine sedimentary n-alkanes record modern climate variability. *Geochimica et Cosmochimica Acta* **68**:4877–4889.

Schwertl, M., K. Auerswald, and H. Schnyder. 2003. Reconstruction of the isotopic history of animal diets by hair segmental analysis. *Rapid Communications in Mass Spectrometry* **17**:1312–1318.

Trolier, M., J. W. C. White, P. P. Tans, K. A. Masarie, and P. A. Gemery. 1996. Monitoring the isotopic composition of atmospheric CO_2: Meaurements from the NOAA Global Air Sampling Network. *Journal of Geophysical Research* **101**:25897–25916.

West, J. B., G. J. Bowen, T. E. Cerling, and J. R. Ehleringer. 2006. Stable isotopes as one of nature's ecological recorders. *Trends in Ecology and Evolution* **21**:408–414.

Section 2

Plant-based Isotope Data as Indicators of Ecological Change

CHAPTER 3

Extracting Climatic Information from Stable Isotopes in Tree Rings

Neil J. Loader,* Danny McCarroll,* Mary Gagen,* Iain Robertson* and Risto Jalkanen†

*Department of Geography, University of Wales Swansea
†Finnish Forest Research Institute, Rovaniemi Research Station

Contents

I. INTRODUCTION

A. Fundamentals of the Tree Ring Proxies

Trees provide within the physical characteristics of their rings (width, relative density, reflectance) a record of past environmental changes which, when expressed strongly, may be used successfully to extract palaeoclimatic information. Such approaches are now well established and have been applied globally (Fritts 1976, Schweingruber 1988, Briffa 2000). Trees can live for many hundreds, or even thousands of years, it is therefore possible using these physical parameters to reconstruct climatic change throughout the life of the tree. Where trees coexist they experience similar external (climatic) forcing, and as a consequence of this "shared experience," it is possible to cross-match or synchronize the tree ring series of living trees with those collected from historical or ancient archives to extend the tree ring record beyond the life span of modern trees and to construct well-replicated tree ring chronologies covering many thousands of years, with absolute dating precision and annual resolution (Pilcher *et al.* 1984, Briffa 2000, Eronen *et al.* 2002, Hantemirov and Shiyatov 2002, Friedrich *et al.* 2004). The long tree ring chronologies provide the framework for much of our understanding of high-resolution climate variability during the Holocene (Jones and Mann 2004). Such archives are essential if we are to explore the nature and significance of recent climatic changes and to provide information as to the behaviour of the natural climate system and associated plant response during the past. A better understanding of baseline variability, thresholds and extremes will prove valuable in reducing the uncertainties of climate change models and assessing the impacts of predicted climate change scenarios.

Further to these established physical proxies, the stable carbon, hydrogen, and oxygen isotopic analyses of tree ring series provide a powerful suite of additional climate proxies. Early studies based on a range of species, materials, and methods and constrained by the technical limitations of the time produced unreplicated records (Craig 1954, Farmer and Baxter 1974, Epstein and Yapp 1976, Libby *et al.* 1976). Interpretation of these was often optimistically oversimplistic, treating trees as simple palaeothermometers recording the meteoric (source) water signal with little additional modification. At that time, resources and methods were not conducive to the higher levels of sample throughput and replication that is currently possible. As the field developed however, it became clear that the tree ring isotope archive is complex, containing physiological controls (*e.g.*, a variable component reflecting stored or remobilized photosynthates) and ecological factors (*e.g.*, disturbance, nutrient availability, disease, insect damage, and flowering). As a consequence, the isolation and extraction of a reliable and robust palaeoclimate record from tree ring isotopes remains a challenge.

In comparison with the measurement of the physical proxies, the analysis of stable isotope ratios in tree rings is demanding in terms of both personnel and resources. Consequently, stable isotope analysis of tree rings can only be justified if the resulting data can provide additional, reliable climate information that cannot be obtained through alternative methods such as ring width or relative density.

B. Scope

This chapter provides an introduction to the application of modern stable isotope techniques for the reconstruction and study of past climate from tree rings. Current challenges and limitations will be discussed with specific emphasis placed on the development of robust isotope-based palaeoclimate time series and their potential for both isotopic and multiproxy analysis using examples from a well-replicated site located close to the Boreal tree line. For a more complete review of the history of the scientific and methodologic developments in this field, we direct the reader to a number of comprehensive reviews (Switsur and Waterhouse 1998, Hughes 2002, McCarroll and Loader 2004, 2005).

Further to this recommendation, stable isotope dendroclimatology remains closely allied with isotopic work in the field of plant physiology. This chapter can only provide an outline of the general plant physiological processes controlling isotopic fractionation in tree ring cellulose and their relation to climatic forcing. For more detailed discussion of isotopic plant physiological studies which operate across a range of spatial and temporal scales and which fall beyond the scope of this chapter, we refer the reader to comprehensive reviews in this field (Rundell *et al.* 1988, Griffiths 1998, Barbour *et al.* 2001, 2002, and references therein).

II. SIGNAL PRESERVATION

An advantage of stable isotope analysis of tree rings over the physical tree ring proxies is the clearer mechanistic understanding of the influence of climate on the processes and pathways through which the carbon, hydrogen and oxygen pass during assimilation and wood formation. These processes have been characterized (Vogel 1980, Farquhar *et al.* 1982, Edwards and Fritz 1986, Barbour and Farquhar 2000, Roden *et al.* 2000) and the resulting models (described below) provide a robust framework for interpreting the isotopic composition of tree rings.

A. Carbon Isotopes

Trees assimilate carbon dioxide (CO_2) from the atmosphere, which enters and diffuses out of the leaf through stomata. During diffusion into the intercellular space within the leaf, the CO_2 is fractionated against ^{13}C in favor of the light isotope ^{12}C by ~4.4‰. This effect relates to the differential mobility of the isotopically heavy and light CO_2 and is largely independent of factors such as temperature or vapor pressure and even to all but the most extreme changes in stomatal aperture ($\leq$0.1 μm) below which interactions between the gas molecules and guard cells become more significant (Farquhar and Lloyd 1993). During photosynthesis, the internal CO_2 (c_i) is combined enzymatically with leaf water to produce sugars via carboxylation. This results in a biochemical fractionation against the heavy isotope, the net effect of which is a further isotopic depletion by ~27‰. The additive effect of fractionation is such that if photosynthesis consumes CO_2 at a rate much greater than it can be replenished, then the relative pool of $^{12}CO_2$ would decrease, resulting in an apparent decrease in fractionation against ^{13}C and a more enriched $\delta^{13}C$ value. Changes in the assimilation rate inside the leaf will influence c_i through changes to the rate at which the CO_2 is utilized to form sugars. Similarly, an increase or decrease in stomatal conductance will affect the rate at which this internal CO_2 can be replenished. In this manner, where external factors (in this case climatic forcing) exert a direct influence on either of these controls then it will be integrated by the plant during photosynthesis and expressed as a change in the intrinsic water use efficiency (WUE_i) of the plant and the $\delta^{13}C$ of the resulting product. When these sugars are converted to form the different components of the tree, there are additional fractionations such that cellulose and lignin exhibit lower $\delta^{13}C$ than the sugars formed in the leaf (Barbour *et al.* 2001). The nature of these fractionation processes, which have been modeled at leaf level, are not precisely characterized for wood formation in the trunk. In particular, the role of mesophyll conductance and the influence of respiration [presented in more complex versions of the Farquhar *et al.* (1982) model] remain areas of ongoing study.

The impact of storage and remobilisation of photosynthates on the isotopic composition of the resulting tree ring during or following periods of resource demand (*e.g.*, fruiting/flowering), extreme stress, or dormancy is also not completely understood and is a subject of ongoing research. For example, serial autocorrelation has been identified in tree ring isotope time series, and this has been supported through labeling experiments which clearly demonstrated the storage and remobilization of isotopically labeled carbon between and within individual growth rings (Kagawa *et al.* 2005, 2006). The influence of storage on our ability to reconstruct climate from tree ring isotopes at annual or coarser resolution is likely to be small, since this relationship can be quantified and the necessary adjustments made.

In climatic terms the dominant signal in the carbon isotope ratios of tree rings reflects changes in the balance between the supply and demand of CO_2 at the site of photosynthesis. This relates directly to the balance between stomatal conductance and the rate of photosynthesis. This relationship, described by Farquhar *et al.* (1982), served to unify the interpretation of tree ring carbon isotopes and explain, mechanistically, the wide ranging and often conflicting information from tree ring carbon isotopes at that time (Figure 3.1). The "Farquhar equation" for discrimination against ^{13}C (Δ) during carbon fixation is expressed in its simple form as:

$$\Delta‰ = a + (b - a)\left(\frac{c_i}{c_a}\right) \qquad (3.1)$$

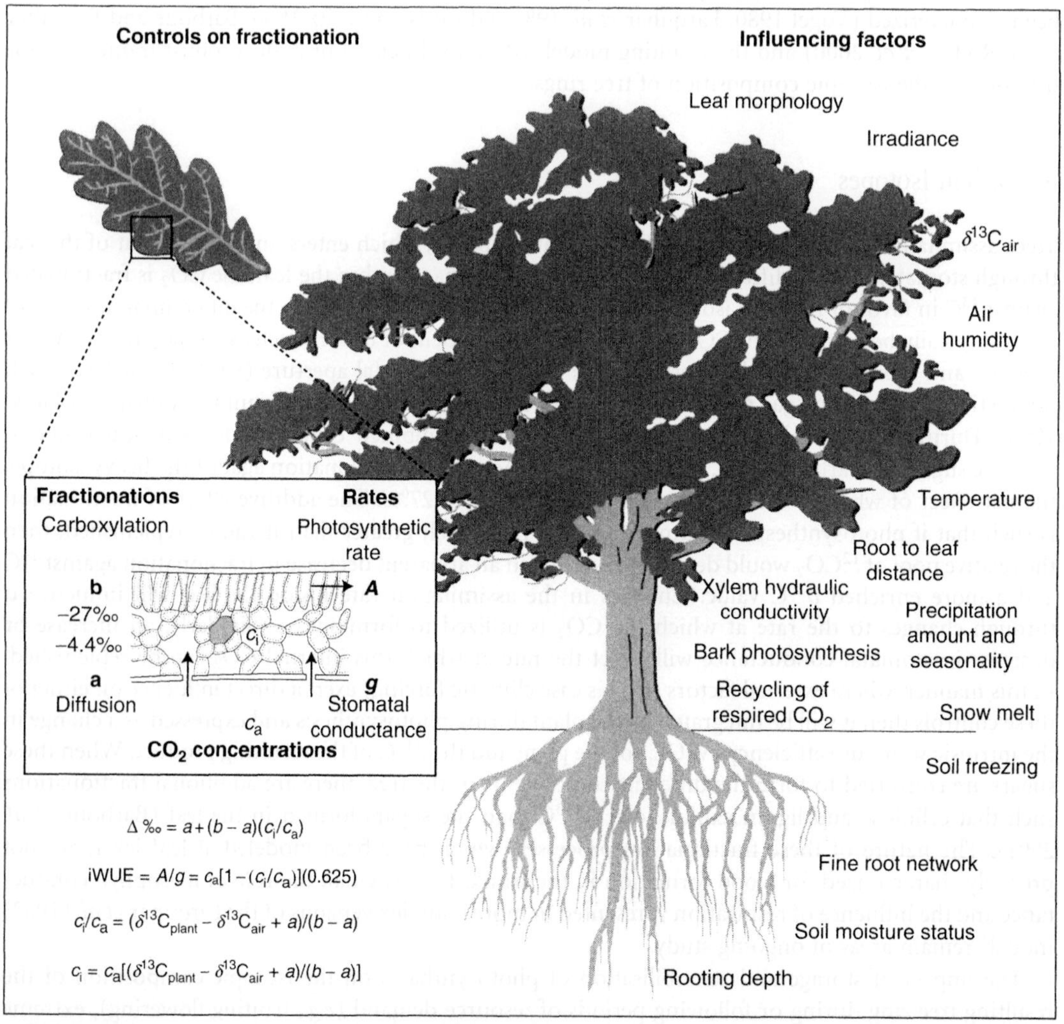

FIGURE 3.1 Diagram of a broad-leaved tree showing the main controls on the fractionation of carbon isotopes and the environmental factors that influence them. The equations are described in the text. [After McCarroll and Loader (2005: p.74); reproduced with kind permission of Springer Science and Business Media.]

where a is discrimination against $^{13}CO_2$ during diffusion and b is the net discrimination due to carboxylation, and c_i and c_a are intercellular and ambient CO_2 concentrations.

The dominant *climatic* controls on carbon isotopic fractionation are thus, those factors that control stomatal conductance which are dominated by air relative humidity and soil moisture status, and those that control photosynthetic rate which are dominated by light levels and leaf temperature (McCarroll and Loader 2004).

Importantly, there are additional *indirect* climatic associations between these dominant forcing mechanisms and a range of quantifiable meteorologic variables (including solar irradiance, cloud cover, temperature, rainfall, and relative humidity). Although not directly controlling isotopic variability per se, these related variables may correlate sufficiently strongly with both the forcing mechanisms and the tree ring isotope time series so as to provide useful palaeoclimate information.

B. Oxygen and Hydrogen Isotopes

The water used during photosynthesis enters the tree via the root system. Where trees sample soil water dominated by precipitation, the water has an isotopic signature relating to local air mass characteristics (Darling *et al.* 2005). Trees growing adjacent to water bodies may not necessarily draw their water from the closest or most likely source (White *et al.* 1985, Dawson and Ehleringer 1991). Similarly, trees with very shallow root systems may sample water dominated by local precipitation and evaporation regimes, whereas trees with deeper roots may be able to access winter recharge or deeper groundwater, to the extent that in both cases the source water may potentially bear little resemblance to meteoric water characteristics from the growing season (Robertson *et al.* 2001). At present, it would be difficult to determine the exact nature of water sourcing for archaeological samples or felled timbers; so when constructing series, it is generally considered that best results are likely to emerge from studies where material is well replicated (to identify or minimize the influence of outliers) and, where possible, of known (hydrologic) provenance.

Water taken up by the roots is transported through the xylem without isotopic fractionation, but in the leaf it is evaporatively enriched. Photosynthesis incorporating the enriched leaf water further fractionates the oxygen and hydrogen isotopes by ~27‰ (Sternberg *et al.* 1986, Barbour *et al.* 2001) and between −100% and −171‰ (Yakir 1992, Roden *et al.* 2000), respectively, such that the resulting sugars are enriched in ^{18}O and depleted in $^{2}H(D)$. At the point of cellulose synthesis additional biochemical fractionation (between +144‰ and +166‰) in favour of ^{2}H returns the isotopic signal of the cellulose to close to that of the source water.

Barbour *et al.* (2001) and Barbour and Farquhar (2000) have presented evidence for a Péclet effect (the process of diffusion of enrichment from the sites of evaporation opposed by convection of replacement of unenriched water via the transpiration stream) operating within the leaf which could potentially modify the effective enrichment of sugars fixed prior to cellulose formation. Similarly, the potential for some of the hydrogen and oxygen to exchange with xylem water post-photosynthesis has also been identified (Sternberg *et al.* 1986, Hill *et al.* 1995, Roden *et al.* 2000). This has been estimated to be as much as 40% of the oxygen in the cellulose molecule. In spite of these rather complex and, in part, uncertain processes, a mechanistic model has been proposed which explains the principal processes occurring during wood formation and which serves as an important starting point in extracting climatic information. The degree of ^{18}O enrichment of leaf water at the sites of evaporation ($\Delta^{18}O_e$) is given by:

$$\Delta^{18}O_e = \varepsilon^* + \varepsilon_k + (\Delta^{18}O_v - \varepsilon_k)^* e_a/e_i \qquad (3.2)$$

where ε^* is the proportional depression of water vapor pressure by the heavier water molecules, ε_k is the fractionation as water diffuses through the stomata and leaf boundary layer, $\Delta^{18}O_v$ is the oxygen isotope composition of water vapour in the atmosphere (relative to source water), and e_a and e_i are the ambient and intercellular vapour pressures (Craig and Gordon 1965, Barbour *et al.* 2001, 2002).

In a similar manner to carbon isotopes, the isotopes of oxygen and, although less studied, hydrogen in tree rings may preserve a record of past climate through the isotopic composition of the source (meteoric) water which is overprinted by an evaporative–transpirative signal in the leaf dominated by vapour pressure deficit (relative humidity) (Figure 3.2). The relative strength of these two signals, as

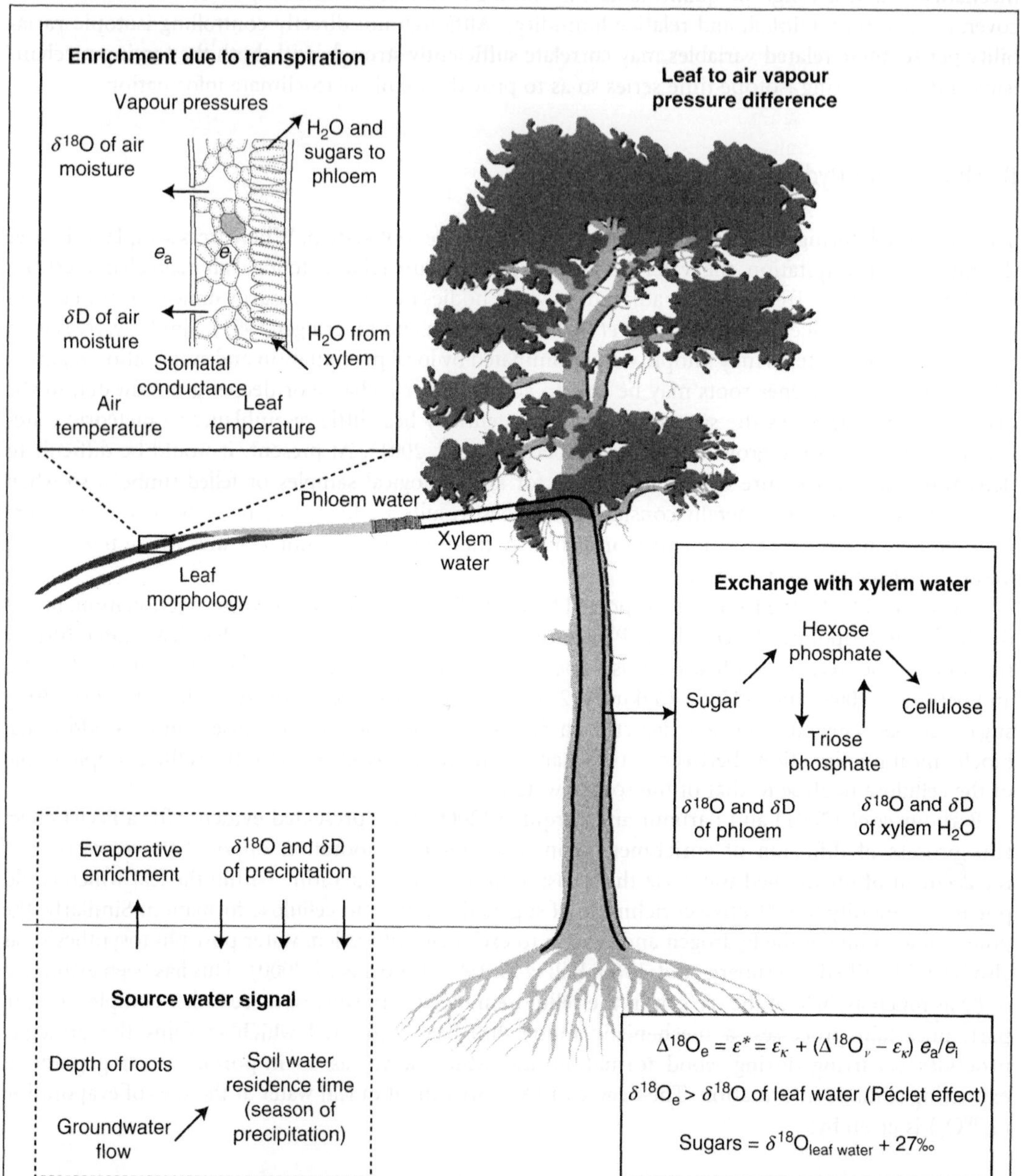

FIGURE 3.2 Diagram of a needle-leaf tree showing the main controls on the fractionation of the water isotopes and the environmental factors that influence them. The equations are described in the text. [After McCarroll and Loader (2005: p.74); reproduced with kind permission of Springer Science and Business Media.]

for oxygen, is governed by the extent to which phloem waters are capable of post-photosynthetic exchange with xylem waters at the site of cellulose synthesis (Barbour and Farquhar 2000).

Very high resolution sampling and analysis of isotopic ratios within individual tree rings suggest that there may exist a significant input from stored photosynthates or remobilization and isotopic exchange at some stages of the tree ring formation process, or during cambial dormancy (Hill *et al.* 1995, Robertson *et al.* 1997, Helle and Schleser 2004). This effect is likely to be species specific and could occur throughout the growing season; however, it appears to be more strongly expressed in the earlywood which may initiate prior to bud-break in some species (Pilcher 1995) rather than the late (summer) wood (which is formed largely from directly photosynthesized materials). Alternatively, the differences in earlywood and latewood observed may reflect initial enrichment of the water used to form these early growing season cells (Phillips and Ehleringer 1995). The impact of these processes and the mechanisms behind them are still the subject of significant debate and could have implications for the climate sensitivity of tree ring reconstructions and future sampling strategies.

III. SAMPLE PREPARATION AND ANALYSIS

A. Site Selection and Sampling

When sampling trees for stable isotope analysis, site selection is a critical consideration and should be informed by the aims of the study (ecological, climatological, physiological, and so on). Sampling should follow the general guidelines proposed in the scientific literature (Fritts 1976, Schweingruber 1988) and consider aspect, slope, altitude, species, hydrology, stand dynamics, and so on. However, if the aim of the scientist is the recovery of a robust and regionally representative record based on stable isotopes, then additional factors should be considered. Of these, site hydrology represents an important consideration. Where trees are analyzed for climatic purposes, samples should originate from well-constrained regions or watersheds, as significant isotopic variability in source waters (expected if timbers originate across a large geographical area) could adversely affect the reliability of the resulting climate reconstructions. Related to the isotopic composition of the water used by the trees are questions relating to the physical accessibility of that water by the organism and include rooting depth, species, soil type, drainage, and structure and which could potentially affect the trees' relative sensitivity to moisture variability (Dawson and Ehleringer 1991). Of course, there will always be variability between trees owing to genetic, local site, and microclimatic effects, but the aim of the isotope dendroclimatologist should be to reduce or quantify this variability.

In apparent conflict with the need for site selection is the consideration that when moving from modern living trees to the palaeoarchive, we may not be able to precisely locate the origin or source of the material sampled. It is therefore prudent to avoid material transported long distances in favour of trees preserved *in situ* (standing deadwood, submerged lake timbers) to minimise uncertainties associated with sourcing timbers from isotopically different climatic or hydrologic regimes. Even under such conditions we have little or no idea of the relative sensitivity of these palaeosamples to climate change in the past so, in order to address this uncertainty replication and dense sampling is perhaps the preferred solution.

B. Sample Preparation

Dendrochronology, the science of dating timber artefacts from tree rings relies on preparation of a clear surface for precise measurement prior to cross-dating (synchronization) with local or regional master chronologies. Absolute dating is essential for successful palaeoclimatology using tree ring stable isotopes however; since the chemical composition rather than the physical characteristics of the wood will ultimately be studied, it is important to reduce potential sources of contamination. Protocols vary between laboratories, and there is debate as to the relative risk or impact associated with different methods of preparation. However, we recommend that cores may be surfaced using a sharp razor blade or scalpel, or gently cleaned with abrasive paper, and excess dust removed manually or with compressed air prior to measurement. Cores may be mounted for measurement in precut blocks or holding clamps constructed to enable measurement without the risk of contamination associated with fixatives such as glue, or adhesive gums (*e.g.*, Blu-Tack^{TM}). Use of measurement markers (pens, pencils) and contrast materials (chalk, gypsum) should also be avoided. We have found that a satisfactory aid to dating may be provided by pin marks or scoring. While the relative influence of these factors on the final isotope measurement is not fully quantified, and may even be minimal, it makes good sense to reduce these potential uncertainties in the final results.

Tree rings are usually subdivided manually using a scalpel or razor blade. However, additional mechanical methods, such as microtome, micro-mills, and laser ablation methods, have been successfully employed enabling very high resolution sampling (ca. 10–100 μm) (Schleser *et al.* 1999, Schulze *et al.* 2004, Patterson *et al.* 2006). However, in all but the most difficult of samples an experienced scientist should be able to subdivide manually the tree ring at an annual (or finer) resolution.

Trees provide perfect annual resolution, so where possible this record should be exploited, although time, resources, and the nature of the archive itself can limit the application of this approach. Samples can be pooled between trees, or more commonly, blocked into absolutely dated 5- or 10-year ring groups, but the variable relative contribution of the individual rings within each block (ring width effect), the inability to separate out latewood, and loss of error estimation mean that, inevitably, some potentially useful climatic information is lost (McCarroll and Loader 2005). Clearly, while one can ultimately degrade (smooth) an annually resolved signal into 5- or 10-year "blocks," it is impossible to recreate annual resolution if the initial sampling strategy is based on blocks of years. Some estimation of error and signal strength can be retained if the pooled records are "broken" into individual measurements at regular intervals or if the blocks of rings are offset between trees (although the number of observations (degrees of freedom) is reduced relative to equivalent cores analyzed annually). A significant limitation of the pooling approach is the effect of differing ring contributions to the pool (ring-width effect), and in the "broken pool" approach the inability to detect errant trees, incorrect dating, or pooling errors. Where resources, time, and expertise are not limited we recommend series to be analyzed individually with annual resolution to preserve inter-tree variability and annual resolution.

Sometimes, under conditions of external stress or periodic changes in the distribution of rainfall during the growing season, trees can form partial (multiple or false) rings or fail to produce a ring at all (partially absent or missing rings). Such tendencies in the growth of an individual can usually be identified and characterized by the dendrochronologist, compared with cores from the same tree and with highly replicated regional chronologies, and accounted for in the dating of the wood. As such, "false" or "missing" rings should not represent a serious problem to the isotope dendroclimatologist when working at annual resolution.

When sampling trees and shrubs intra-annually or when working with samples that exhibit no apparent tree ring structure, such as in many species of the tropics, researchers have traditionally used a microtome to separate individual time slices with high spatial resolution (Poussart *et al.* 2004, Verheyden *et al.* 2004). The sample dimensions, ring structure, and analytical limits of the mass spectrometer or preparation method govern the ultimate spatial resolution attainable. Care should also be taken to avoid sections with excessive curvature that could effectively smooth the enviromental signal and "smear" the temporal resolution of each slice (the latter varying also with tree age). It should also be noted that the observed temporal resolution based on observed wood formation may not directly reflect the date that the sugars were synthesized.

As a consequence of its variable composition and chemistry, a single component of the wood is normally isolated prior to mass spectrometry. Cellulose has been the preferred sample material as this is unambiguously linked to the chronology of the sample and forms the framework around which lignin and resin are subsequently deposited. Alternative materials have also been studied, including acid-insoluble lignin and whole wood (Barbour *et al.* 2001, Loader *et al.* 2003, Robertson *et al.* 2004).

To extract α-cellulose is a time-consuming task compared with the analysis of whole wood, and there are advantages and disadvantages in both approaches. Analysis of wholewood has the advantage that many samples can be analyzed without major sample losses, compared with the number of analyses possible and reduction in sample mass (ca. 60%) if samples are purified to α-cellulose (Loader *et al.* 1997). For many species, however, the mobile resin component needs to be removed prior to analysis as it could introduce a signal of reduced temporal integrity compared to that obtained from lignin or cellulose. Any wholewood study should always be accompanied with a robust test, demonstrating the relationship between whole (resin extracted) wood and cellulose prior to proceeding, particularly where the species being studied is "new" to isotope dendroclimatology. Similarly, during sample preservation, changes in the lignin:cellulose ratio may alter due to diagenesis or decay. If this process cannot be identified and quantified then it is possible that the resulting isotope time series may contain lower frequency trends within them which could be misinterpreted as changes in external (climatic) forcing. One solution to this problem could be the periodic analysis of cellulose and (resin extracted) wholewood throughout the time series. Ultimately, the choice of component and indeed ring element (earlywood, latewood, whole ring) will likely depend on the species, the aims of the study, the ring-width characteristics of the wood, signal strength, the resources and time available, and the state of preservation or diagenesis of the sample (McCarroll and Loader 2005).

C. Isotopic Analysis

Prior to the advent of continuous flow methods, sample preparation and analysis were limited to as few as 10 samples per day. Now, using online methods interfacing an elemental analyser and isotope ratio mass spectrometer, analysis is more rapid and sample size requirements greatly reduced (Preston and Owens 1985).

For carbon isotopic analysis, samples are weighed into tin foil cups and then combusted with additional oxygen in a flow of helium. The resulting CO_2 is resolved gas chromatographically prior to mass spectrometry. Results are expressed as per mille deviations from the Vienna-Pee Dee Belemnite (VPDB) standard. Typical analytical precision for this method based on replicate analysis of a standard cellulose powder should be expected to be 0.1 per mille (Schleser *et al.* 1999).

For oxygen, samples are weighed into silver foil cups and pyrolised over glassy carbon at >1080 °C. The cellulose is then quantitatively converted to carbon monoxide (CO) which is analyzed in the mass spectrometer following gas chromatography. Online preparation has revolutionized the measurement of $\delta^{18}O$ in tree rings. Precision is typically between 0.25 and 0.3 per mille based on replicate analysis of fine grained cellulose (Werner *et al.* 1996, Farquhar *et al.* 1997, Koziet 1997, Saurer *et al.* 1998). For both oxygen and carbon, cellulose and wholewood have been analyzed. Results suggest that while for carbon isotopes cellulose and wholewood produce very similar signals, for oxygen the relationship is less clear, reflecting perhaps a more complex pathway for the oxygen atoms during lignification (Barbour *et al.* 2001).

Analysis of the hydrogen isotope ratios in cellulose is complicated by the propensity for a proportion of hydroxyl-bound hydrogen to exchange with external water. To overcome this effect and possible source of error, the exchangeable hydrogen (ca. 30%) is either exchanged with hydrogen of a known isotopic composition (and the ratio of the nonexchangeable hydrogen determined by a subsequent mass balance equation) or replaced with nitro groups prior to pyrolysis at >1400 °C (Prosser and Scrimgeour 1995, Kelly *et al.* 1998). The resulting hydrogen gas is separated from the CO using a gas chromatograph and the results typically expressed, as for $\delta^{18}O$, in parts per mille (‰) relative to the

Vienna standard mean ocean water (VSMOW). Typical reproducibility for this method is better than 2–3 per mille based on replicate analysis of cellulose nitrate.

IV. REPLICATION AND QUANTIFICATION OF SIGNAL STRENGTH

A. Replication

To extract an annually resolved series from the tree ring archive is far from a routine or simple task, and aside from the initial building of the chronology, a process which can take many years of expertise and dedication, the tasks of manual subdivision, cellulose extraction, homogenization, and analysis require significant effort. Nevertheless, following the initial proof of concept research, as studies became more ambitious it became clear that there was significant isotopic variability both within and between trees which could contribute to the overall amount of information attainable from the tree ring archive and potentially set it apart from many of the other terrestrial proxies.

One of the benefits of the tree ring archive over many other natural proxies is the widespread distribution of suitable trees and the possibility for series to be replicated. Replication is a key to robust palaeoclimate reconstruction and has been long recognized by dendroclimatologists working with the physical proxies, because through replication one may identify, with increased confidence, the characteristics of climate change and distinguish these from site or organism-dependent signals caused by non-climatic disturbance factors. Series replication also provides the only means for obtaining a measure of the natural variability within an archive. This variability can be significant between and within trees and emphasizes the requirement for representative sampling (Table 3.1). Where the aim is to reconstruct the climate of a region, a representative average value for the isotope ratio of the trees in that region for each year is required, and the difference between the trees is the largest source of variability (McCarroll and Pawellek 1998).

If a single time series or pooled curve has no expression of error or natural variability around the mean, then one must question the extent to which it can be used to quantify palaeoenvironmental

TABLE 3.1 Approximate values for inter- and intra-tree isotopic variability observed in various tree species

	Carbon (per mille)	Oxygen (per mille)	Hydrogen (per mille)
Intertree xylem cellulose (nitrate)	3.0 [a]*Pinus sylvestris* 3.2 [b]*Quercus* spp.	0.6 [b]*Quercus* spp.	5.7 [c]*Quercus* spp.
Inter-radii difference			
$n = 3^d$	1.0 [d]*Quercus robur*	1.6 [d]*Quercus robur*	26 [d]*Quercus robur*
$n = 2^e$ radii	1.3 [e]*Abies pindrow*	1.8 [e]*Abies Pindrow*	33 [e]*Abies pindrow*
Interring variability cellulose (nitrate) σ_{n-1} ($n = 104$)	0.48 [f]*Pinus sylvestris* 0.67 [f]*Quercus* spp.	0.65 [f]*Pinus sylvestris* 0.60 [f]*Quercus* spp.	14.48 [f]*Pinus sylvestris* 5.59 [f]*Quercus* spp.
Intraring xylem cellulose (nitrate)	1.0 [g]*Quercus petraea*	1.0 [g]*Quercus petraea*	50 [g]*Quercus petraea*

[a] This study.
[b] Robertson *et al.* (2001), Robertson Personal communication (maximum inter-mean difference 4 trees, 100 years).
[c] Waterhouse *et al.* (2002) (maximum inter-mean difference 4 trees, 100 years).
[d] Robertson *et al.* (1995).
[e] Ramesh *et al.* (1985) (2 radii, 30-year series).
[f] Loader Personal Communication (mean standard deviation 4 trees, 104 years).
[g] Robertson *et al.* (1996) (maximum inter-ring difference 1 tree, 4 rings).

change or identify thresholds, trends, and extremes. However, if the time series can be measured individually then it is possible to identify not only the natural variability about the mean, but also the signal strength and its stability through time, as well as missing rings and dating errors (Figure 3.3). The need for and benefits of replication, long recognized in isotope dendrochronology, have remained practically unavailable to the field until recently. Naturally, resources and practicalities continue to limit the ability to replicate proxy time series in many remote and extreme circumstances. This is not always the case, however, and perhaps more useful information might be retrieved, not just from the tree ring proxies, but from other (*e.g.*, sedimentary) archives, through multiple coring of site(s) with analysis at a lesser temporal resolution than from a more concentrated high resolution study of a single long core alone.

B. Signal Strength

No single tree will preserve a "perfect" isotope record, and there will always be some noise masking the desired signal. Changes in microclimate, local hydrology, tree physiology, and stand dynamics all influence the extent to which a climate signal can be extracted. The difference with tree ring isotope time series is that these can be readily replicated with the aim of enhancing the dominant environmental signals while reducing noise. Climate associations developed from measurement of tree ring widths alone can yield strong palaeoclimate signals at stressed sites; however, beyond such regions where growth is limited, the strength of the correlation with climate parameters even on very well-replicated time series is typically much reduced. From studies conducted so far, it would appear that stable isotopes are capable of explaining a similar proportion of climatic information based on a much reduced level of replication (Robertson *et al.* 1997, McCarroll and Pawellek 1998) and that in maritime regions, in particular, this signal is stronger than can be obtained from tree ring widths alone. This is not to say

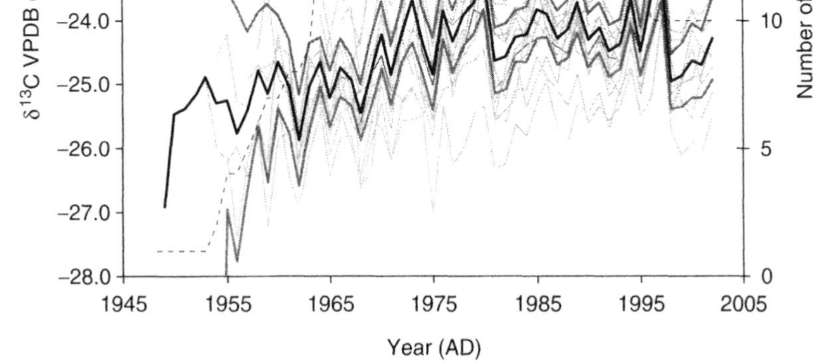

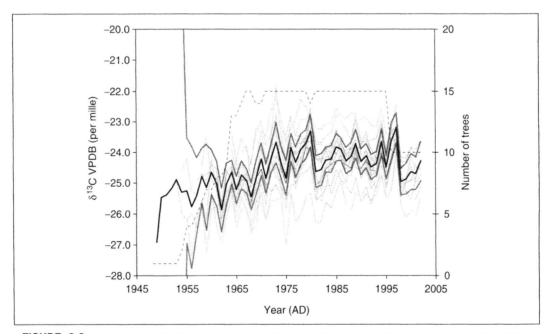

FIGURE 3.3 Composite plot of individual carbon isotope time series from northern Fennoscandia (fine grey lines). Black line represents the mean of the records with 95% confidence limits (thick grey lines). A measure of the number of trees in the record is plotted on the secondary ordinate. The EPS for this series is 0.97 (n = 15 trees).

that isotopic proxies are superior to the physical parameters, but instead that they provide additional information and simply perform better in some regions where physical proxies are less successful.

To characterize the common signal preserved within the dataset, the expressed population signal (EPS) (Wigley *et al.* 1984, Briffa and Jones 1989) can be employed to ascertain the reliability of the common signal through time and the likely sample depth required to explain a threshold level of common signal (normally an EPS > 0.85). Where one time series has been developed per tree the EPS is based on the mean correlation calculated between all possible pairs of trees ($\bar{r}_{bt}$) multiplied by the number of trees in the sample (t), where:

$$\mathrm{EPS}(t) = \frac{t \cdot \bar{r}_{bt}}{t \cdot \bar{r}_{bt} + (1 - \bar{r}_{bt})} \tag{3.3}$$

Such an approach is valuable when reconstructing palaeoclimate back through time, especially where available sample material might be limiting. Calculation of mean signal strength offers an objective method of determining the proportion of the common signal explained. The EPS is correlation based and consequently reflects in a large part the common forcing at an annual/interannual level. To what extent the EPS statistic can be used to characterise lower frequency records has yet to be fully explored. Although EPS is applicable to other palaeoarchives, it is rarely applied in the same way, owing to the absence of replication.

A measure of the stability of the common signal may also be obtained from composite chronologies of differing lengths by calculating a "running EPS" where a window of ca. 25 years passes through the time series. The resulting signal can then be used to identify those sections of the chronology where additional replication might be required to enhance signal strength to above the critical threshold to ensure a minimum standard of reliability throughout the record. It is important to note that the EPS is insensitive to absolute differences between the time series. It defines only the strength of the interseries (high-frequency) signal, and not the confidence that can be placed in estimates of the absolute mean isotopic value (McCarroll and Pawellek 1998).

V. NONCLIMATIC TRENDS

A. Age-Related Trends

One of the major limitations of climatic reconstructions based on the physical proxies is the loss of signal associated with the standardization and removal of nonclimatic biological or age-related growth trends. Curves can be statistically fitted to the time series to remove the underlying (growth) trend and, in so doing, preserve a series of growth indices more closely linked to climate variability. However, in removing this trend, not only is a proportion of the low-frequency climate signal lost, but also as a function of this detrending, the lowest frequency climatic signal retrievable from the archive is limited by the length of the detrended record; the so-called "segment-length curse" (Cook *et al.* 1995). Additional methods [*e.g.*, regional curve standardisation (RCS)] have been developed to detrend statistically, while reducing this sequence-specific effect (Esper *et al.* 2002), alternatively, analysis of ancient, very long-lived trees such as the millennial age bristlecone pines requires a less dramatic detrending throughout much of their series length. However, in all cases, when standardization is attained through curve fitting, this will result in some loss of environmental information. This effect is most pronounced in the low-frequency domain, and can compromise detection of climate variability at centennial and millennial scales.

Partly as a consequence of the cost and effort required to analyze stable isotopes in tree rings, it is not yet known exactly to what extent the stable isotopes of carbon, hydrogen, and oxygen preserve similar age-related trends in their records. A pilot investigation into the behaviour of the carbon isotopes was carried out on *Pinus sylvestris* trees from northern Fennoscandia. Individual annually resolved latewood α-cellulose samples were prepared from each tree to explore systematic trends related to tree age. The results demonstrate that as the tree matures the $\delta^{13}C$ values increase, tending toward convergence with the mean site chronology (in both absolute value and covariance) within a period of 40–50 years. These preliminary findings are supportive of previous observations (Freyer 1979a,b; Schleser and Jayasekera 1985) present new highly replicated data for a number of age cohorts, and suggest that beyond this initial "juvenile effect" (Figure 3.4), which may be easily identified and removed if individual series are analyzed separately, there is no apparent long-term trend or tendency in the data and thus no need for statistical detrending of age-related effects. Stable carbon isotopes could, therefore, preserve environmental signals across all temporal scales. The exact cause of this "juvenile effect" remains a subject of further study, but it likely reflects changes in plant CO_2 use, hydraulic conductivity, or lower light levels (Freyer and Belacy 1983, Schleser and Jayasekera 1985, Heaton 1999, McCarroll and Loader 2005). The behaviour of oxygen and hydrogen isotopes requires similar characterization, but together such results may help to establish the origins of this effect which is likely to differ between individuals, sites, and species.

B. Correction for Atmospheric $\delta^{13}C$ and CO_2

Following the onset of industrialisation (post-1800 AD) there has been a steady increase in the emission of ^{13}C-depleted CO_2 into the atmosphere. The impact of this over time has been a progressive lowering of $\delta^{13}C$ of atmospheric CO_2. Most trees reflect this change in atmospheric composition as a steady decrease in $\delta^{13}C_{tree\ ring}$ over time. If these series are to be compared against meteorologic data then this trend needs to be removed prior to calibration and verification studies. Some studies have

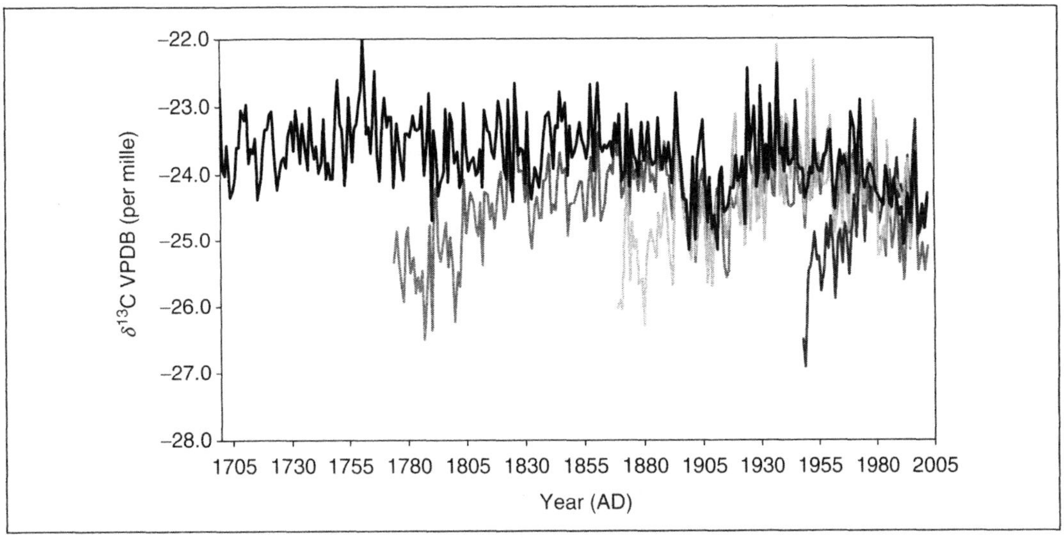

FIGURE 3.4 The "juvenile effect" represented in multiple age cohorts of *P. sylvestris* latewood $\delta^{13}C$ from Laanila, northern Finland. The site "master" curve is plotted in black. Three age cohorts are presented (light, medium, and dark grey) which exhibit an increase in $\delta^{13}C$ values until they approach the level of the site "master," after which point they exhibit a high level of covariance and common mean. It is interesting to note that this effect lasts for ~40 years and it cannot be accurately determined using inflection points, but may be identified in the manner presented using a representative mean time series.

attempted to "correct" the tree ring $\delta^{13}C$ curve through statistical fitting of a smooth curve in a similar manner to removal of growth trends (Freyer and Belacy 1983, Freyer 1986). The consequence of this is a loss in low-frequency signal and a correction specific to the site or series being analyzed. Evidence from ice cores and modern observations enable past variations in the $\delta^{13}C_{atmosphere}$ to be relatively well characterized, thus it is possible to mathematically correct the dataset without inadvertently removing lower frequency trends (McCarroll and Loader 2004). This approach should be favoured, as it is not dependent on series length or location. Datasets have been compiled that contain the correction values. These can be added to without having to recalculate the whole dataset, with the advantage that the correction is mathematical rather than statistical, the original dataset can easily be restored without loss of information (Figure 3.5).

Less well characterized is the response of trees to the increase in atmospheric CO_2 concentrations that has occurred during the same time. The increase in CO_2 may alter the way a tree assimilates carbon. If more carbon is available at a given stomatal conductance, the balance between water loss and carbon gain is altered. Several authors have identified changes in intrinsic water-use efficiency which may be related to changing atmospheric composition or attainment of a compensation point and serve as evidence for such an effect (Kürschner 1996, Waterhouse *et al.* 2004). To what extent this response is "plastic" remains to be seen; however, it is likely that over recent years rapid changes in CO_2 have influenced plant carbon assimilation and water use efficiency. Researchers have proposed a range of scaling factors with which to adjust for this increasing CO_2, although there is significant debate as to their magnitude and spatial relevance (Feng and Epstein 1995a, Treydte *et al.* 2001). These adjustments work by adding an incremental $\delta^{13}C$ correction per unit increase in atmospheric CO_2, and assume that the plant response is constant over time. Very different results can be obtained depending on which correction is applied (Figure 3.5), and it is likely that the response will differ between individuals, species, and sites. An improved understanding of changes in plant response to increasing CO_2 levels will be important in our quantification and modeling of carbon sequestration in a rapidly changing "greenhouse" world.

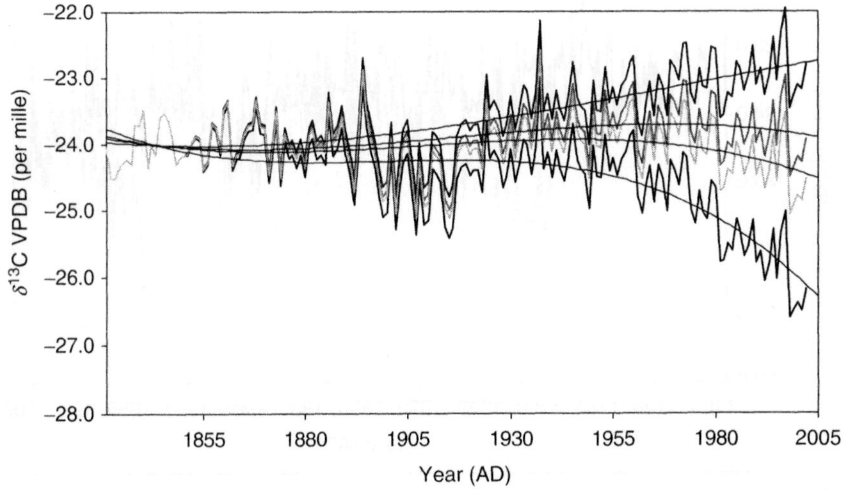

FIGURE 3.5 The "Industrial effect" as recorded in the mean latewood stable carbon isotope time series in *Pinus sylvestris* from Laanila, northern Finland. The raw uncorrected data (black line) is mathematically corrected for changing $\delta^{13}C$ of atmospheric CO_2 using the dataset of McCarroll and Loader (2005) (Light grey). Medium grey lines demonstrate the effects of mathematically correcting the stable isotope time series for changes in CO_2 (after Feng and Epstein 1995a, Kürschner 1996, Treydte *et al.* 2001).

VI. CALIBRATION AND MECHANISTIC MODELING

Traditionally, climate associations between tree ring parameters and meteorologic data have been established using a variety of correlation-based approaches (Fritts 1976, Schweingruber 1988, Cook and Kairiukstis 1990). Where factors controlling isotopic fractionation are well constrained or strongly expressed, it is possible to use these methods to reconstruct past climate, based on transfer function methodologies. Composite correlation diagrams can be used to demonstrate the relative dominance of monthly climate parameters on isotope fractionation that can serve to inform model development relating to the sensitivity of the trees to the climate of specific months or combinations of months (Fritts 1976, Loader *et al.* 2003). Where correlations are sufficiently strong, these records preserve sufficient climatic information to enable useful climate reconstruction, though the strength of the correlation required remains uncertain (McCarroll and Pawellek 2001, McCarroll *et al.* 2003, Lucy *et al.* in press). In all situations the climate associations should be independently verified or treated with a replicate resampling method (*e.g.*, bootstrap, jackknife) to establish the reliability of the model. As with any reconstruction based on calibration–verification methods, the reconstructions are confined by the limitations of uniformitarianism (stability of forcing and response), the strength of the correlation which influences the level of bias toward the mean (Birks 1995, McCarroll *et al.* 2003, von Storch *et al.* 2004) and by the range and characteristics of the calibration dataset against which the record is tested. Nevertheless, such statistical approaches can provide important and valid information on past climates.

A. Laanila, Northern Finland: A Carbon Isotope Case Study

A well replicated, annually resolved time series of stable carbon isotopes was developed from a northern Fennoscandian site near Laanila (68°30′N, 27°30′E; 220 masl). The individual time series were combined into a master curve which exhibits a high degree of common forcing and signal coherence (EPS = 0.96) calculated on the six trees covering the full 250-year record. The mean series and 95% confidence intervals were calculated prior to mathematical correction for changes in atmospheric $\delta^{13}CO_2$. The resulting time series was correlated against local meteorologic parameters from the nearest meteorologic station in Ivalo. Strong, highly significant statistical relationships with summer climate variables were identified (Table 3.2). Strongest relationships were observed with summer sunshine (the primary control on photosynthetic rate), relative humidity, precipitation amount, and temperature. Verification data for the correlations with sunshine and relative humidity were not available at this site; however, to test the structure of the correlation exercise and the stability of the relationship, the historical Tornedalen temperature record (Klingbjer and Moberg 2003) was used to test the related indirect relationship identified (summer temperature and summer sunshine being highly correlated at Ivalo).

Although the Tornedalen is a composite, distant record there is an obvious and positive relationship and coherence between the record of summer temperature and reconstructed summer temperature

TABLE 3.2 Pearson correlation coefficients for Laanila mean stable carbon isotope curve (corrected) and Ivalo meteorologic parameters (1958–2002 AD)

	Temperature	Relative humidity	Sunshine	Precipitation
June	0.30	−0.29	0.42	−0.35
July	0.42	−0.55	0.73	−0.52
August	0.62	−0.40	0.58	−0.43
June–August Mean	0.58	−0.62	0.81	−0.66

(through statistical transfer) developed from the Ivalo data (Figure 3.6). The relationship observed is not a perfect match, nor should we expect it to be. These differences are likely to reflect the limitations of the calibration exercise (simple least squares regression with temperature) and of the remoteness and composite nature of the verification dataset from Tornedalen. A low-frequency deviation from the 95% confidence limit is apparent in the most recent decade of the Tornedalen record. It is likely that this deviation is related more to an "over correction" for changing CO_2 or differences in local climate than a loss of climate sensitivity in tree response, as this deviation follows a trend and higher frequency trends are preserved. Improved calibration and verification of this record (with sunshine and more local data), as well as improvements in the nature of the mathematical corrections employed will likely enhance this already promising and sensitive record further.

Reports of a recent change in climate sensitivity of some tree ring parameters at a number of high-latitude sites (Briffa *et al.* 1998, Vaganov *et al.* 1999) have emphasised that care needs to be taken when applying statistical models calibrated over short recent periods back through time. Application of the mechanistic understanding of isotopic fractionation may provide a key to climate reconstruction where instrumental data may be limited. Models developed and refined by Roden *et al.* (2000) explain a simplified mechanism of isotopic imprinting and fractionation from source water to tree ring cellulose. Together the results from the carbon, hydrogen, and oxygen isotopes might theoretically be combined with great potential to describe past changes in photosynthetic rate, relative humidity, and source water to provide information on palaeotemperature, humidity, and air mass characteristics. The performance of such models can be tested against "real" tree ring oxygen and hydrogen isotope data (Roden *et al.* 2000, Robertson *et al.* 2001, Waterhouse *et al.* 2002). Results from such studies, although based on a limited dataset of source water measurements, generally demonstrate similarities in the behavior of modeled and observed time series, which offers encouragement for the future development and application of this approach. However, it should also be noted that although the trends are generally similar, in many cases the absolute values and sensitivity differ. These differences are most likely due, in part, to variations in local source water, poorly characterized storage or remobilisation

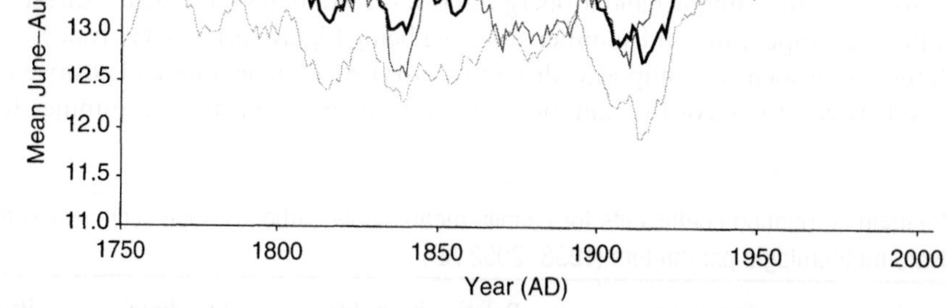

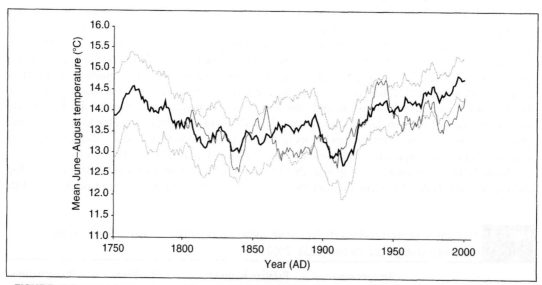

FIGURE 3.6 Reconstruction of mean June–August summer temperature (5-year moving average) based on the mean latewood stable carbon isotope time series in *Pinus sylvestris* at Laanila plotted with 95% confidence intervals. The distant composite record from Tornedalen (Klingbjer and Moberg 2003) is presented to demonstrate the close association between the two time series and the potential for signal preservation at all temporal scales. The EPS for this reconstruction is high (>0.91) based on six trees of >250 years old.

effects, and nonoptimal estimates of leaf characteristics. It is likely that such parameters will need to be constrained more accurately for each site or region and species before these models can be more widely applied in palaeoclimate studies.

B. Climate Reconstruction from Oxygen and Hydrogen Isotopes

In comparison to carbon isotopes there remain fewer sensitivity studies and climate reconstructions based on well-replicated oxygen and hydrogen isotope data, which reflects the difficulties of obtaining this information [Epstein and Yapp 1976, Ramesh *et al.* 1986, Feng and Epstein 1995b, Saurer *et al.* 1997, Feng *et al.* 1999, Robertson *et al.* 2001 (Oxygen), Waterhouse *et al.* 2002 (Hydrogen)]. Recent developments in oxygen isotope analysis (Farquhar *et al.* 1997, Koziet 1997, Saurer *et al.* 1998) and hydrogen isotope data (Filot *et al.* 2006) as well as international initiatives such as ISONET (http://www.isonet-online.de) are actively addressing this and it is anticipated that in the coming years our understanding of the climate sensitivity of these isotopes will be enhanced. Saurer *et al.* (1997) demonstrate a close association between tree ring cellulose isotopic composition and the isotopic composition of local meteoric waters. Similarly, Robertson *et al.* (2001) were able to assess signal coherence in oxygen isotopes from annually resolved oak latewood cellulose from the United Kingdom. A link was identified with temperature, meteoric water, and relative humidity, which supports the general structure of the mechanistic models discussed above. Saurer *et al.* (2002) progressed this work, exploring the spatial and temporal trends in tree ring oxygen isotopes at the northern Eurasian tree line using a network of 130 trees and successfully demonstrated the regional relevance and value of the network approach to stable isotope analysis. Treydte *et al.* (2006) presented a 1000-year record of precipitation from *Juniperus* growing in the Karakorum Mountains in Pakistan. This precipitation-sensitive record provides evidence for changing precipitation amount during the latter stages of the last millennium and represents one of the longest annually resolved stable isotope records currently in existence.

C. Multiparameter Dendroclimatology

So far this chapter has considered the potential and limitations preserved within the isotopic proxies alone; however, through combinations of additional independent proxies such as relative density (X-ray absorption), ring width, and annual height increment (Briffa *et al.* 2002, Pensa *et al.* 2005), it is possible to extract an even greater range of palaeoclimate variables and, in so doing deconvolve or simplify the complex signals of the tree ring (isotopic) record. This multiproxy approach to dendroclimatology (McCarroll *et al.* 2003) may take one of two paths; either the combination of independent proxies sensitive to a common parameter, or the combination of proxies sensitive to different forcing mechanisms to extract additional detail on the nature of climate forcing preserved in the tree ring record. Two examples of this approach are given below. In the first example a correlation diagram is presented for ring width, latewood density, carbon isotope, and height increment variability at a tree line site in northern Finland. Each parameter is correlated to a range of summer (June–August) climate parameters. Although each is relatively sensitive to climate variability in this region, by combining the records, the proportion of the variance explained is increased further (Table 3.3). In the second example of a multiproxy approach McCarroll *et al.* (2003) combined two tree ring proxies; relative density and stable carbon isotopes (Figure 3.7). On the basis of the calibration and study of tree ring formation, McCarroll *et al.* (2003) proposed that at high latitudes, the maximum latewood density relates primarily to late summer temperature, whereas stable carbon isotope variability was controlled at their study site primarily by photosynthetic rate (summer sunshine) and that it would only be during very dry years that controls from stomatal conductance would become sufficiently significant as to modify carbon isotopic ratios. Using a biplot analysis of these two parameters, the researchers were able

TABLE 3.3 The effect of arithmetic combination of multiple standardised tree ring proxies to enhance a common summer signal and to reduce nonclimatic "Noise"

Correlation coefficient (r) $n = 44$	Ring width	Relative density	Height increment	Carbon isotope ratios	Composite 4-proxy index
July–August mean temperature	0.54	0.68	0.76	0.64	0.84

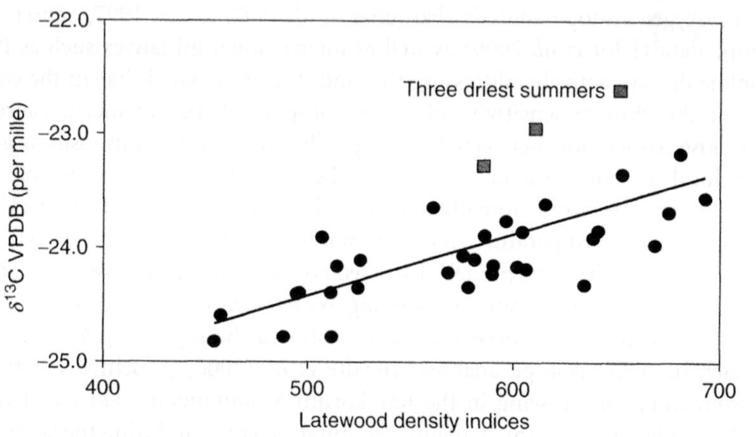

FIGURE 3.7 In *Pinus sylvestris* from northern Finland, both latewood density and stable carbon isotope ratios primarily respond to the amount of energy received during the summer. Soil moisture status can also influence the isotopic ratios (via stomatal conductance) but not wood density. When stable carbon isotope ratios are regressed on latewood density, the resulting residuals (squares) are found to represent the driest summers (based on June–August total precipitation). It is hypothesised that during these relatively dry years stomatal conductance becomes a more significant control on isotopic fractionation. After McCarroll *et al.* © 2003 Edward Arnold, Ltd. (Publishers) (www.hodderarnoldjournals\.com).

to demonstrate the relationship between the factors controlling the two proxies and were able to identify clearly three outliers which relate to the 3 driest years within the coverage of their record. This distinction between hot-dry and cool-moist years could not have been recovered using a single proxy alone and demonstrates the potential of a multiparameter approach (McCarroll *et al.* 2003). It also demonstrates that stable isotopes do not provide in isolation the key to reconstructing climate in all cases and that there may be more to be gained by a multiparameter approach.

VII. CONCLUSIONS

The stable isotope analysis of tree rings provides a virtually unrivaled terrestrial archive for reconstructing past climate changes at an annual resolution. The absolute timescale and potential for replication between trees and sites make tree ring proxies uniquely testable candidates for calibration and evaluation against the instrumental meteorologic record. Series replication has exposed the natural variability of the record between individual trees, cores, and sites, which has implications for their use and interpretation within the field of climate change study. However, this variability can be exploited to obtain a more realistic estimate for a mean regional value with statistically defined confidence intervals and measures of signal strength. Such variability is present in all climate proxies, yet replication of this order is rarely practiced to the same degree in other terrestrial or marine proxies.

Stable isotope analysis of tree rings also provides a means through which it is possible both statistically and mechanistically to determine the characteristics of the source CO_2 and water assimilated by the plant and the climatic factors influencing the tree that may be combined to enable study of changing air mass characteristics, water-use efficiency, or drought stress through both space and time.

Importantly, it would appear that tree ring stable carbon isotopes do not preserve any long-term age trend beyond an initial "juvenile effect," which may be objectively determined. Consequently, since no statistical detrending of the data is required, there is no loss of low-frequency signal or segment length effect. This would suggest that tree ring stable carbon isotopes could preserve a signal of climate variability across all temporal scales, which offers immense potential for palaeoclimatology. If this relationship holds for the other isotopes then this will set apart tree ring stable isotopes from the physical proxies and provide a reliable means for reconstructing climate trends and an improved technique for detection of past climate changes that is currently lacking from the existing powerful suite of tree ring proxies.

The development of replicated, annually resolved tree ring stable isotope time series is no trivial task and should only be carried out where a real benefit can be gained beyond existing tree ring methods. Multiparameter analyses to combine these records likely provide the key to future climate change studies based on tree ring stable isotopes, where a more complete picture of environmental change is to be gained through combinations of physical and isotopic proxies than might be provided by a single archive alone.

VIII. ACKNOWLEDGMENTS

This work was supported by an award to NJL from the UK NERC NE/B501504/1 and NE/C511805/1, and from the European Union EVK2-CT2002-00136 (PINE), EVK2-CT2002-00147 (ISONET), 017008-2 GOCE (MILLENNIUM). The authors thank the ESF (SIBAE) programme for their support and our colleagues at the Needle trace laboratory (Metla) and P. Santillo, J. Woodman-Ralph, N. Jones, and A. Ratcliffe (UWS) for their invaluable assistance.

IX. REFERENCES

Barbour, M. M., and G. D. Farquhar. 2000. Relative humidity- and ABA-induced variation in carbon and oxygen isotope ratios of cotton leaves. *Plant Cell & Environment* **23:**473–485.

Barbour, M. M., T. J. Andrews, and G. D. Farquhar. 2001. Correlations between oxygen isotope ratios of wood constituents of *Quercus* and *Pinus* samples from around the world. *Australian Journal of Plant Physiology* **28:**335–348.

Barbour, M. M., A. S. Walcroft, and G. D. Farquhar. 2002. Seasonal variation in $\delta^{13}C$ and $\delta^{18}O$ of cellulose from growth rings of *Pinus radiata*. *Plant Cell & Environment* **25:**1483–1499.

Birks, H. J. B. 1995. Quantitative paleoenvironmental reconstructions. Pages 161–254 *in* D. Maddy and J. S. Brew (Eds.) *Statistical Modelling of Quaternary Science Data. Technical Guide 5.* Quaternary Research Association, Cambridge.

Briffa, K. R. 2000. Annual climate variability in the Holocene: Interpreting the message of ancient trees. *Quaternary Science Review* **19:**87–105.

Briffa, K. R., and P. D. Jones. 1989. Basic chronology statistics and assessment. Pages 137–152 *in* E. Cook and L. Kairiukstis (Eds.) *Methods of Dendrochronology: Applications in the Environmental Sciences.* Kluwer Academic Publishers, Dordrecht.

Briffa, K. R., F. H. Schweingruber, T. J. Osborn, S. G. Shiyatov, and E. A. Vaganov. 1998. Reduced sensitivity of recent tree-growth to temperature at high northern latitudes. *Nature* **391:**678–682.

Briffa, K. R., T. J. Osborn, F. H. Schweingruber, P. D. Jones, S. D. Shiyatov, and E. A. Vaganov. 2002. Tree-ring width and density data around the Northern Hemisphere: Part 1, local and regional climate signals. *The Holocene* **12:**737–757.

Cook, E. R., and L. A. Kairiukstis. 1990. *Methods of Dendrochronology.* Kluwer Academic Publishers. Dordrecht, Boston, London.

Cook, E. R., K. R. Briffa, D. M. Meko, D. A. Graybill, and G. Funkhouser. 1995. The "segment length curse" in long tree ring chronology development for palaeoclimatic studies. *Holocene* **5**:229–237.

Craig, H. 1954. Carbon-13 variations in Sequoia rings and the atmosphere. *Science* **119**:141–144.

Craig, H., and L. I. Gordon. 1965. Deuterium and oxygen-18 variations in the ocean and marine atmospheres. Pages 9–130 *in* E. Tongiorgi (Ed.) *Proceedings of a Conference on Stable Isotopes in Oceanographic Studies and Palaeotemperatures.* Lischi and Figli, Pisa, Italy.

Darling, W. G., A. H. Bath, J. Gibson, and K. Rozanski. 2005. Isotopes in water. Pages 1–66 *in* M. Leng (Ed.) *Isotopes in Palaeoenvironmental Research.* DPER Springer, The Netherlands.

Dawson, T. E., and J. R. Ehleringer. 1991. Streamside trees that do not use stream water. *Nature* **350**:335–337.

Edwards, T. W. D., and P. Fritz. 1986. Assessing meteoric water composition and relative humidity from ^{18}O and ^{2}H in wood cellulose: Paleoclimatic implications for southern Ontario, Canada. *Applied Geochemistry* **1**:715–723.

Epstein, S., and C. J. Yapp. 1976. Climatic implications of the D/H ratio of hydrogen in C-H groups in tree cellulose. *Earth Planetary Science Letters* **30**:252–261.

Eronen, M., P. Zetterberg, K. R. Briffa, M. Lindholm, J. Meriläinen, and M. Timonen. 2002. The supra-long Scots pine tree ring record for Finnish Lapland: Part 1, chronology construction and initial inferences. *Holocene* **12**:673–680.

Esper, J., E. R. Cook, and F. H. Schweingruber. 2002. Low-frequency signals in long tree ring chronologies for reconstructing past temperature variability. *Science* **295**:2250–2253.

Farmer, J. G., and M. S. Baxter. 1974. Atmospheric carbon dioxide levels as indicated by the stable isotope record in wood. *Nature* **247**:273–275.

Farquhar, G. D., and J. Lloyd. 1993. Carbon and oxygen isotope effects in the exchange of carbon dioxide between terrestrial plants and the atmosphere. Pages 47–70 *in* J. R. Ehleringer, A. E. Hall, and G. D. Farquhar (Eds.) *Stable Isotopes and Plant Carbon-Water Relations.* Academic Press, New York.

Farquhar, G. D., M. H. O'Leary, and J. A. Berry. 1982. On the relationship between carbon isotope discrimination and intercellular carbon dioxide concentration in leaves. *Australian Journal of Plant Physiology* **9**:121–137.

Farquhar, G. D., B. K. Henry, and J. M. Styles. 1997. A rapid on-line technique for the determination of oxygen isotope composition of nitrogen containing compounds and water. *Rapid Communication in Mass Spectrometry* **11**:1554–1560.

Feng, X., and S. Epstein. 1995a. Carbon isotopes of trees from arid environments and implications for reconstructing atmospheric CO_2 concentration. *Geochimica et Cosmochimica Acta* **59**:2599–2608.

Feng, X., and S. Epstein. 1995b. Climatic temperature records in δD from tree rings. *Geochimica et Cosmochimica Acta* **59**:3029–3037.

Feng, X., H. Cui, K. Tang, and L. E. Conkey. 1999. Tree ring δD as an indicator of Asian monsoon intensity. *Quaternary Research* **51**:262–266.

Filot, M. S., M. Leuenberger, A. Pazdur, and T. Boettger. 2006. Rapid online equilibration method to determine the D/H ratios of non-exchangeable hydrogen in cellulose. *Rapid Communication in Mass Spectrometry* **20**:3337–3344.

Freyer, H. D. 1979a. On the ^{13}C record in tree rings. Part 1. ^{13}C variations in northern hemispheric trees during the last 150 years. *Tellus* **31**:124–137.

Freyer, H. D. 1979b. On the ^{13}C record in tree rings. Part 2. Registration of micro-environmental CO_2 and anomalous pollution effect. *Tellus* **31**:308–312.

Freyer, H. D. 1986. Interpretation of the northern hemispheric record of $^{13}C/^{12}C$ trends of atmospheric CO_2 in tree rings. *In* J. R. Trabalka and D. E. Reichle (Eds.) *The Global Carbon Cycle.* Springer-Verlag, London.

Freyer, H. D., and N. Belacy. 1983. $^{13}C/^{12}C$ records in northern hemisphere trees during the past 500 years—anthropogenic impact and climatic suppositions. *Journal of Geophysical Research* **88**:6844–6852.

Friedrich, M., S. Remmele, B. Kromer, J. Hofmann, M. Spurk, K. F. Kaiser, C. Orcel, and M. Küppers. 2004. The 12,460 year Hohenheim oak and pine tree ring chronology from central Europe. A unique annual record for radiocarbon calibration and paleoenvironment reconstructions. *Radiocarbon* **46**(3):1111–1122.

Fritts, H. C. 1976. *Tree Rings and Climate.* Academic Press, New York. 567 p.

Griffiths H. (Ed.). 1998. *Stable Isotopes.* BIOS Scientific Publishers Ltd, Oxford. 438 p.

Hantemirov, R. M., and S. G. Shiyatov. 2002. A continuous multimillenial ring-width chronology from Yamal, northwestern Siberia. *Holocene* **12**:717–726.

Heaton, T. H. E. 1999. Spatial, species, and temporal variations in the 13C/12C ratios of C3 plants: Implications for palaeodiet studies. *Journal of Archaeological Science* **26**:637–649.

Helle, G., and G. H. Schleser. 2004. Beyond CO_2-fixation by Rubisco—an interpretation of C-13/C-12 variations in tree rings from novel intra-seasonal studies on broad-leaf trees. *Plant, Cell & Environment* **27**(3):367–380.

Hill, S. A., J. S. Waterhouse, E. M. Field, V. R. Switsur, and T. ap Rees. 1995. Rapid recycling of triose phosphates in oak stem tissue. *Plant, Cell & Environment* **18**:931–936.

Hughes, M. K. 2002. Dendrochronology in climatology—the state of the art. *Dendrochronologia* **20**:95–116.

Jones, P. D., and M. E. Mann. 2004. Climate over past millennia. *Reviews of Geophysics* **42**:RG2002, doi:10.1029/2003RG000143.

Kagawa, A., A. Sugimoto, K. Yamashita, and H. Abe. 2005. Temporal photosynthetic carbon isotope signatures revealed in a tree ring through (CO_2)-C-13 pulse-labelling. *Plant, Cell & Environment* **28**:906–915.

Kagawa, A., A. Sugimoto, and T. C. Maximov. 2006. $^{13}CO_2$ pulse-labelling of photoassimilates reveals carbon allocation within and between tree rings. *Plant, Cell & Environment* **29:**1571–1584.

Kelly, S. D., L. G. Parker, M. Sharman, and M. J. Dennis. 1998. On-line quantitative determination of $^2H/^1H$ isotope ratios in organic and water samples using an elemental analyzer coupled to a mass spectrometer. *Journal of Mass Spectrometry* **33:**735–738.

Klingbjer, P., and A. Moberg. 2003. A composite monthly temperature record from Tornedalen in northern Sweden, 1802–2002. *International Journal of Climatology* **23:**1465–1494.

Koziet, J. 1997. Isotope ratio mass spectrometric method for the on-line determination of oxygen-18 in organic matter. *Journal of Mass Spectrometry* **32:**103–108.

Kürschner, W. M. 1996. Leaf stomata as biosensors of palaeoatmospheric CO_2 levels. LPP Contribution Series 5, 152 p.

Libby, L. M., L. J. Pandolfi, P. H. Payton, J. Marshall III, B. Becker, and V. Giertz-Siebenlist. 1976. Isotopic tree thermometers. *Nature* **261:**284–290.

Loader, N. J., I. Robertson, A. C. Barker, V. R. Switsur, and J. S. Waterhouse. 1997. An improved method for the batch processing of small wholewood samples to α-cellulose. *Chemical Geology* **136:**313–317.

Loader, N. J., I. Robertson, and D. McCarroll. 2003. Comparison of stable carbon isotope ratios in the whole wood, cellulose and lignin of oak tree rings. *Palaeogeography Palaeoclimatology Palaeoecology* **196:**395–407.

Lucy, D., I. Robertson, R. G. Aykroyd, and A. M. Pollard. Estimates of uncertainty in the prediction of past climatic variables. *Applied Geochemistry,* in press.

McCarroll, D., and F. Pawellek. 1998. Stable carbon isotope ratios of latewood cellulose in *Pinus sylvestris* from northern Finland: Variability and signal strength. *Holocene* **8:**693–702.

McCarroll, D., and F. Pawellek. 2001. Stable carbon isotope ratios of *Pinus sylvestris* from northern Finland and the potential for extracting a climate signal from long Fennoscandian chronologies. *Holocene* **11:**517–526.

McCarroll, D., and N. J. Loader. 2004. Stable isotopes in tree rings. *Quaternary Science Review* **23:**771–801.

McCarroll, D., and N. J. Loader. 2005. Isotopes in tree rings. Pages 67–116 *in* M. Leng (Ed.) *Isotopes in Palaeoenvironmental Research.* DPER Springer, The Netherlands.

McCarroll, D., R. Jalkanen, S. Hicks, M. Tuovinen, F. Pawellek, M. Gagen, D. Eckstein, U. Schmitt, J. Autio, and O. Heikkinen. 2003. Multi-proxy dendroclimatology: A pilot study in northern Finland. *Holocene* **13**(6):829–838.

Patterson, W. P., J. P. Dodd, J. M. Brasseur, and B. M. Eglington. 2006. The use of robotics in deriving high-resolution climate/environmental information from tree ring cellulose. *Geological Society of America Abstracts with Programs* **38**(7):377.

Pensa, M., H. Salminen, and R. Jalkanen. 2005. A 250-year-long height-increment chronology for *Pinus sylvestris* at the northern coniferous timberline: A novel tool for reconstructing past summer temperatures? *Dendrochronologia* **22:**75–81.

Phillips, S. E., and J. R. Ehleringer. 1995. Limited uptake of summer precipitation by bigtooth maple (*Acer grandidentatum* Nutt.) and Gambel's oak (*Quercus gambellii* Nutt). *Trees* **9:**214–219.

Pilcher, J. R. 1995. Biological considerations in the interpretation of stable isotope ratios in oak tree-rings. Pages 157–161 *in* B. Frinzel, B. Stauffer, and M. M. Weiss (Eds.) *Paläoklimaforschung/Paleoclimate Research 15.* European Science Foundation, Strasbourg, France.

Pilcher, J. R., M. G. L. Baillie, B. Schmidt, and B. Becker. 1984. A 7272 year tree ring chronology for western Europe. *Nature* **312:**150–152.

Poussart, P. F., M. N. Evans, and D. P. Schrag. 2004. Resolving seasonality in tropical trees: Multi-decade, high-resolution oxygen and carbon isotope records from Indonesia and Thailand. *Earth and Planetary Science Letters* **218:**301–316.

Preston, T., and N. J. P. Owens. 1985. Preliminary ^{13}C measurement using a gas chromatograph interfaced to an isotope ratio mass spectrometer. *Biomedical Mass Spectrometry* **12:**510–513.

Prosser, S. J., and C. M. Scrimgeour. 1995. High precision determination of $^2H/^1H$ in H_2 and H_2O by continuous flow isotope ratio mass spectrometry. *Analytical Chemistry* **34:**1992–1997.

Ramesh, R., S. K. Bhattacharya, and K. Gopalan. 1985. Dendroclimatological implications of isotope coherence in trees from Kashmir Valley, India. *Nature* **317:**802–804.

Ramesh, R., S. K. Bhattacharya, and K. Gopalan. 1986. Climatic correlations in the stable isotope records of silver fir (*Abies pindrow*) trees from Kashmir, India. *Earth and Planetary Science Letters* **79:**66–74.

Robertson, I., E. M. Field, T. H. E. Heaton, J. R. Pilcher, M. Pollard, V. R. Switsur, and J. S. Waterhouse. 1995. Isotope coherence in oak cellulose. *In* B. Frenzel (Ed.) *Problems of Stable Isotopes in Tree Rings, Lake Sediments and Peat-Bogs as Climatic Evidence for the Holocene. Palaeoclimate and Man* **10:**141–156.

Robertson, I., A. M. Pollard, T. H. E. Heaton, and J. R. Pilcher. 1996. Seasonal changes in the isotopic composition of oak cellulose. Pages 617–626 *in* J. S. Dean, D. M. Meko, and T. W. Swetnam (Eds.) *Tree Rings, Environmental and Humanity: Proceedings of the International Conference, Tucson Arizona, 17–21 May 1994.* Radiocarbon Arizona, USA.

Robertson, I., V. R. Switsur, A. H. C. Carter, A. C. Barker, J. S. Waterhouse, K. R. Briffa, and P. D. Jones. 1997. Signal strength and climate relationships in $^{13}C/^{12}C$ ratios of tree ring cellulose from oak in east England. *Journal of Geophysical Research* **102:**19507–19519.

Robertson, I., J. S. Waterhouse, A. C. Barker, A. H. C. Carter, and V. R. Switsur. 2001. Oxygen isotope ratios of oak in east England: Implications for reconstructing the isotopic composition of precipitation. *Earth and Planetary Science Letters* **191:**21–31.

Robertson, I., N. J. Loader, D. McCarroll, A. H. C. Carter, L. Cheng, and S. W. Leavitt. 2004. $\delta^{13}C$ of tree ring lignin as an indirect measure of climate change. *Water, Air, and Soil Pollution* **4:**531–544.

Roden, J. S., G. Lin, and J. R. Ehleringer. 2000. A mechanistic model for interpretation of hydrogen and oxygen isotope ratios in tree ring cellulose. *Geochimica et Cosmochimica Acta* **64:**21–35.

Rundell, P. W., J. R. Ehleringer, and K. A. Nagy (Eds.). 1988. *Stable Isotopes in Ecological Research.* Springer-Verlag, London. 530 p.

Saurer, M., S. Borella, and M. Leuenberger. 1997. $\delta^{18}O$ of tree rings of beech (*Fagus sylvatica*) as a record of $\delta^{18}O$ of the growing season precipitation. *Tellus* **49B:**80–92.

Saurer, M., F. H. Schweingruber, E. A. Vaganov, S. G. Shiyatov, and R. T. W. Siegwolf. 2002. Spatial and temporal oxygen isotope trends at northern tree-line in Eurasia. *Geophysical Research Letters* **29:**10.1–10.4, doi:10.1029/2001GL013739.

Saurer, M., I. Robertson, R. T. W. Siegwolf, and M. Leuenberger. 1998. Oxygen isotope analysis of cellulose: An interlaboratory comparison. *Analytical Chemistry* **70:**2074–2080.

Schleser, G. H., and R. Jayasekera. 1985. $\delta^{13}C$ variations in leaves of a forest as an indication of reassimilated CO_2 from the soil. *Oecologia* **65:**536–542.

Schleser, G. H., G. Helle, A. Lücke, and H. Vos. 1999. Isotope signals as climate proxies: The role of transfer functions in the study of terrestrial archives. *Quaternary Science Review* **18:**927–943.

Schweingruber, F. H. 1988. *Tree Rings: Basics and Applications of Dendrochronology.* D. Reidel, Boston.

Schulze, B., C. Wirth, P. Linke, W. A. Brand, I. Kuhlmann, V. Horna, and E.-D. Schulze. 2004. Laser ablation-combustion-GC-IRMS—a new method for online analysis of intra-annual variation of $\delta^{13}C$ in tree rings. *Tree Physiology* **24:**1193–1201.

Sternberg, L., M. De Niro, and R. Savidge. 1986. Oxygen isotope exchange between metabolites and water during biochemical reactions leading to cellulose synthesis. *Plant Physiology* **82:**423–427.

Switsur, V. R., and J. S. Waterhouse. 1998. Stable isotopes in tree ring cellulose. Pages 303–321 *in* H. Griffiths (Ed.) *Stable Isotopes.* BIOS Scientific Publishers Ltd, Oxford.

Treydte, K., G. H. Schleser, F. H. Schweingruber, and M. Winiger. 2001. The climatic significance of $\delta^{13}C$ in subalpine spruces (Lötschental, Swiss Alps). *Tellus* **53B:**593–611.

Treydte, K., G. H. Schleser, G. Helle, D. C. Frank, M. Winiger, G. H. Haug, and J. Esper. 2006. Millennium-long precipitation record from tree ring oxygen isotopes in northern Pakistan. *Nature* **440:**1179–1182.

Vaganov, E. A., M. K. Hughes, A. V. Kirdyanov, F. H. Schweingruber, and P. P. Silkin. 1999. Influence of snowfall and melt timing on tree growth in subarctic Eurasia. *Nature* **400:**149–151.

Verheyden, A., G. Helle, G. H. Schleser, F. Dehairs, H. Beeckman, and N. Koedam. 2004. Annual cyclicity in high-resolution stable carbon and oxygen isotope ratios in the wood of the mangrove tree *Rhizophora mucronata. Plant, Cell & Environment* **27:**1525–1536.

Vogel, J. 1980. Fractionation of carbon isotopes during photosynthesis. Sitzungsberichte der Heidelberger Akademie der Wissenschaften Mathematisch-naturwissenschaftliche Klasse Jahrgang, 3. Abhandlung. Springer-Verlag, Heidelberg.

von Storch, H., E. Zorita, J. Jones, Y. Dimitriev, F. González-Rouco, and S. Tett. 2004. Reconstructing past climate from noisy data. *Science* **306:**679–682.

Waterhouse, J. S., V. R. Switsur, A. C. Barker, A. H. C. Carter, and I. Robertson. 2002. Oxygen and hydrogen isotopes in tree rings: How well do models predict observed values? *Earth and Planetary Science Letters* **201:**421–430.

Waterhouse, J. S., V. R. Switsur, A. C. Barker, A. H. C. Carter, D. L. Hemming, N. J. Loader, and I. Robertson. 2004. Northern European trees show a diminishing response to enhanced atmospheric carbon dioxide concentrations. *Quaternary Science Review* **23:**803–810.

Werner, R. A., B. E. Kornexl, A. Rossmann, and H. L. Schmidt. 1996. On-line determination of $\delta^{18}O$ values of organic substances. *Analytical Chimica Acta* **319:**159–164.

White, J. W. C., E. R. Cook, J. R. Lawrence, and W. S. Broecker. 1985. The D/H ratios of sap in trees: Implications for water sources and tree ring D/H ratios. *Geochimica et Cosmochimica Acta* **49:**217–246.

Wigley, T. M. L., K. R. Briffa, and P. D. Jones. 1984. On the average value of correlated time series, with applications in dendroclimatology and hydrometeorology. *Journal of Applied Meteorology* **23:**201–213.

Yakir, D. 1992. Variations in the natural abundance of oxygen-18 and deuterium in plant carbohydrates. *Plant, Cell & Environment* **15:**1005–1020.

CHAPTER 4

Human Impacts on Tree-Ring Growth Reconstructed from Stable Isotopes

Matthias Saurer and Rolf T. W. Siegwolf

Paul Scherrer Institute

Contents

I. INTRODUCTION

Carbon and oxygen isotope ratios in tree rings have been widely applied for the reconstruction of climatic conditions (Schleser *et al.* 1999, McCarroll and Loader 2004). The carbon isotope ratio is sensitive to warm and dry conditions and enables the retrieval of information on past temperature and precipitation variations during the growing period. The oxygen isotope ratio in tree rings reflects the oxygen isotope ratio of precipitation and is sensitive to temperature, humidity, and the atmospheric circulation patterns. Beside these climatic factors, however, the response of the trees to other environmental conditions must also be considered. Over the past two centuries, numerous anthropogenic influences have been observed in many components of the biogeochemical cycle. Often referred to collectively as "global change," the most prominent is undoubtedly the increase in the atmospheric CO_2 concentration from about 280 ppm in preindustrial times to currently about 380 ppm (Francey *et al.* 1999, Allison *et al.* 2003). This increase in the main substrate of photosynthesis affects plant growth directly and indirectly in many ways. The direct effects include a stimulation of the assimilation rate by increasing the gradient of CO_2 between the canopy air and the intercellular leaf spaces and co-occurring reduction in stomatal conductance, which was often observed in controlled small-scale and large-scale artificial CO_2-fumigation experiments (Körner 2000). Indirect effects include changes in the hydrologic regime caused by reduced transpiration and climatic changes, most notably the temperature increase due to enhanced greenhouse forcing, on a global scale identified to about 0.5 °C during the twentieth century (IPCC 1996). Temperature increase, hydrologic changes, and CO_2 increase all influence tree growth simultaneously. In addition to this, anthropogenic emissions of many other compounds affect tree growth. In particular, in regions near industrialized areas, the effects of air pollutants may be important. For example, the emissions of nitrogen-bearing compounds, in reduced and oxidized forms, have led to substantial addition of nitrogen to ecosystems, with both potentially toxic and fertilizing effects (Norby 1998). While car emissions contain NO and NO_2 (the sum denoted as NO_x), emissions from intensive agriculture are observed mainly as NH_4. Other pollutants potentially affecting tree growth include sulphur, acid deposition, and ozone. Even if governmental regulations have partly curbed the dramatic increase and deposition of these pollutants to ecosystems in the recent 20 years, the problem is continuing and raises the question of threshold levels, where sensitive ecosystems may seriously change due to imbalances in the nutrient cycles (Schulze 1989).

One could, therefore, summarize that global change does influence the growth patterns of forest trees mainly by changing three parameters, namely climate, CO_2, and air pollution. While it may seem very challenging to determine the influence of each factor separately, this task is critical for estimating future forest response. Models need to be based on a sound understanding of processes, to go beyond simple linear projections of the current status to the future. Besides small-scale growth chamber experiments, which may be useful for studying the influence of different factors under controlled conditions, it is also important to be able to interpret the observed variations in real-world ecosystems and determine how and in which ways forests may have already been affected by global change. In this chapter, the potential of the combined analysis of different stable isotope ratios is explored to improve our knowledge of tree response to the changes of environmental conditions that occurred in the last two centuries. This chapter is divided into three main parts, which discuss the following topics: (A) carbon isotopes and water-use efficiency reconstructed from tree rings, referring to the effect of increasing atmospheric CO_2, (B) description of a conceptual model for the combined analysis of $\delta^{13}C$ and $\delta^{18}O$, and (C) a case study with the application of the findings in (A) and (B) to assess local environmental impacts on trees growing near a motorway, using also the analysis of $\delta^{15}N$.

II. SITES AND SAMPLE PREPARATION

The investigated sites in Switzerland are situated in the Swiss Central Plateau and on the southern borders of the Jura mountain chain with a temperate-moist climate. The annual precipitation sum is about 1000 mm and the average annual temperature ca. 9 °C. At the site "Twann" (7°10′E, 47°5′N; 600 m a.s.l.), four beech trees (*Fagus sylvatica*) were investigated (Saurer *et al.* 1995, 1997). This site is on a southeastern slope on shallow soil and is relatively dry. Samples from this site are labeled FS. The site "Bettlachstock" (7°25′E, 47°13′N; 1150 m a.s.l.) is a Swiss long-term forest ecosystem research plot (WSL Birmensdorf), also located on a well-drained southern slope (Saurer *et al.* 2000). Species investigated include *Fraxinus excelsior* (FE) and *Abies alba* (AA). The site, "Koppigen", is located on flat terrain (47°8′N, 7°35′E, 480 m a.s.l), soils are pseudo-gley, and *Picea abies* (PA) trees were investigated (Saurer *et al.* 2004a). Sites from the Northern Eurasian Tree-Ring Project (Schweingruber *et al.* 1993) were located along a transect from Norway to Siberia (in the latitude range from 60° to 70 °N) and investigated species are *Larix sibirica, Larix gmelinii, Larix dahurica, Pinus sylvestris, Picea abies,* and *Pinus obovata.*

Tree-ring cores were dated and individual rings separated with a razor blade under a microscope. The wood was milled and homogenized in a centrifugal mill. Samples from Twann were subjected to cellulose extraction. The isotope offset of 1.2‰ between cellulose and wood was corrected as all other samples were analyzed as wood. Samples for $^{15}N/^{14}N$ analysis were treated with alcohol/toluene and hot water to remove soluble constituents. Isotope analyses were done either by combustion (for $^{13}C/^{12}C$ and $^{15}N/^{14}N$) or by pyrolysis (for $^{18}O/^{16}O$; Saurer *et al.* 2000) in an elemental analyser which was connected to an isotope ratio mass spectrometer (delta-S, Finnigan). The precision of the analysis was 0.1‰ for $\delta^{13}C$ and 0.3‰ for $\delta^{18}O$ and $\delta^{15}N$.

III. ISOTOPE THEORY

The isotope discrimination Δ, observed in plants as a result of the preferential use of ^{12}C over ^{13}C during photosynthesis, is defined as

$$\Delta = \frac{\delta^{13}C_a - \delta^{13}C_{tree}}{1 + \delta^{13}C_{tree}/1000} \approx \delta^{13}C_a - \delta^{13}C_{tree} \quad (1)$$

where $\delta^{13}C_a$ is the isotopic value of atmospheric CO_2, the input signal for the plant, and $\delta^{13}C_{tree}$ the isotope value of the organic matter or tree ring. Hence, for calculating the discrimination, the influence of varying isotopic composition of CO_2 in the past has to be considered. Farquhar *et al.* (1982) described the dependence of Δ on plant physiological properties, in particular on c_i/c_a, the ratio of intercellular to ambient CO_2 concentrations:

$$\Delta \cong a + (b - a)\frac{c_i}{c_a} \quad (2)$$

where a (=4.4‰) is the fractionation associated with the gaseous diffusion of CO_2 through the stomata and b (=27‰) the fractionation resulting from enzymatic C-fixation. This equation is true for any time point of the past when considering the respective tree ring. From Eqs. (1) and (2), it follows that c_i can be calculated from the discrimination:

$$c_i = c_a \frac{\delta^{13}C_a - \delta^{13}C_{tree} - a}{b - a} \quad (3)$$

using literature data for a, b, $\delta^{13}C_a$, and measured data for $\delta^{13}C_{tree}$.

In the next step, we consider the instantaneous leaf-level water-use efficiency WUE_{inst}, which is defined as the ratio of net photosynthesis (A) to transpiration (E):

$$WUE_{inst} = \frac{A}{E} \quad (4)$$

with E defined as:

$$E = g_{H_2O}(e_i - e_a) \quad (5)$$

whereby g_{H_2O} is the conductance for water vapor, and e_i and e_a the vapor pressures in the leaf cellular air space and ambient air, respectively. WUE_{inst} is a component of the long-term water-use efficiency, the ratio of carbon uptake to water loss at the plant level, which is also influenced by respiratory losses.

In stable isotope work, however, often a simpler concept of water-use efficiency is used, which is the intrinsic water-use efficiency WUE_i, the ratio of net photosynthesis A to conductance for water vapor g_{H_2O}, given by Ehleringer and Cerling (1995) as:

$$WUE_i = \frac{A}{g_{H_2O}} \tag{6}$$

usually indicated in units of μmol mol^{-1}. The advantage of Eq. (6) over Eq. (4) is that it is independent of the vapor pressure deficit, and therefore can be calculated from the isotope ratios alone, as shown in the following. On the basis of the equation for net photosynthesis

$$A = g_{CO_2}(c_a - c_i) \tag{7}$$

with g_{CO_2} as the conductance for CO_2, and using the relationship between the conductances for CO_2 and H_2O

$$g_{H_2O} = 1.6 g_{CO_2} \tag{8}$$

we obtain the following relationship, by combining Eqs. (6–8):

$$WUE_i = \frac{c_a - c_i}{1.6} \tag{9}$$

Finally, by replacing c_i with the expression from Eq. (3), we find:

$$WUE_i = \frac{c_a}{1.6} \frac{b - \delta^{13}C_a + \delta^{13}C_{tree}}{b - a} \tag{10}$$

Information on reconstructed, past changes in WUE_i calculated by this equation can give valuable insights into the response of trees to increasing CO_2 and temperature and help to determine whether the amount of carbon gain per unit water loss has changed. An ambiguity in the interpretation of WUE_i and Eq. (4), however, is that changes in WUE_i can be the result of changes in A or g. In other words, the question remains unresolved whether CO_2 assimilation is more strongly affected by biochemical constraints (usually expressed by A_{max}, the average maximum net photosynthesis at ambient CO_2 concentration under optimal environmental conditions) or diffusional limitations (stomatal conductance $= g_s$). For example, an increase of $\delta^{13}C$ in organic matter, interpreted as a reduction of c_i and an improvement of WUE_i in the Farquhar-model, can be the result of either (1) reduced stomatal conductance (at a constant A_{max}) or (2) increased photosynthetic capacity (at a constant g_s).

Here, the analysis of the oxygen isotopes is helpful. The oxygen isotope ratio of organic matter ($\delta^{18}O$) is mainly determined by the isotopic composition of the soil water, the leaf water enrichment due to transpiration, and biochemical fractionations during incorporation. According to Dongmann *et al.* (1974), the degree of leaf water enrichment depends on the ratio of the vapor pressures in the atmosphere and the intercellular spaces of the leaves ($e_a/e_i \approx$ rH), whereby humid air (low relative humidity, rH) causes a reduction of $\delta^{18}O$ in the leaf water. Further, high transpiration rates also tend to reduce the enrichment due to the Péclet effect (Farquhar and Lloyd 1993). The leaf water isotope signal is also reflected in the organic matter and tree rings (Roden and Ehleringer 1999, Yakir and Sternberg 2000). This information on evaporative enrichment contained in the $\delta^{18}O$ of organic matter is useful for distinguishing between possible causes of a change in c_i as derived from $\delta^{13}C$. Scheidegger *et al.* (2000) evaluated in detail this hypothesis in a conceptual model that we apply here to tree-ring studies. The model enables the deduction of changes of g_s and A_{max} from different isotope reaction patterns. Instead of changes between two environments (as in Scheidegger *et al.* 2000), we consider here changes between two time periods (represented by tree-ring values). The possible combinations of changes in $\delta^{13}C$ and $\delta^{18}O$ values are shown in Figure 4.1. The different environmental conditions in the two time points can cause higher (↑), lower (↓), or similar (≈) values in $\delta^{13}C$ and $\delta^{18}O$ ("similar" indicates statistically not significant). The results are depicted as eight possible arrows in the central scheme shown in Figure 4.1.

How can we now derive the consequences for stomatal conductance (g_s) and photosynthetic capacity (A_{max})? In a first step, a change in rH is derived from the change in $\delta^{18}O$, using an inverse relationship, while the Péclet effect may further enhance this relationship (dry condition results in reduced stomatal conductance, which further increases the enrichment and vice versa). This means that the change in $\delta^{18}O$ of the tree ring is assumed to be primarily due to a change in leaf water enrichment, caused by different air humidity in the two investigated time periods. A prerequisite for the validity of this assumption is a similar isotope signature of source water and water vapor in the two time periods compared. Otherwise, the difference has to be quantified and corrected for. In the next step, the change in c_i is derived from the change in $\delta^{13}C$, assuming a negative relationship between the two parameters as explained in Eqs. (2) and (3). In the final step, we evaluate "the most likely case" using the information we found in the first step (rH derived from $\delta^{18}O$) as a decision tool. As an example, we investigate the scenario with increasing $\delta^{13}C$ and $\delta^{18}O$. We know from the increase in $\delta^{18}O$ that rH must have decreased, where a decreasing Péclet effect also may have contributed, while from increasing $\delta^{13}C$ we infer decreasing c_i. To explain the reduction in c_i there are, therefore, two possibilities: case 1, A_{max}↑ and g_s≈; case 2, A_{max}≈ and g_s↓. Because plants in dry air tend to close their stomata, we choose case 2 as it is physiologically more plausible than case 1. Accordingly, the appropriate g_s–A_{max} plot is shown. Using the same procedure, we select a case for all scenarios, resulting in the different g_s–A_{max} plots (see Scheidegger *et al.* 2000 for more details).

Since this is not a quantitative model, the arrows in the g_s–A_{max} plots should be understood as an indication only which factor responded more strongly and should be used as a guideline in distinguishing among the different types of gas exchange response to environmental effects. The following is also

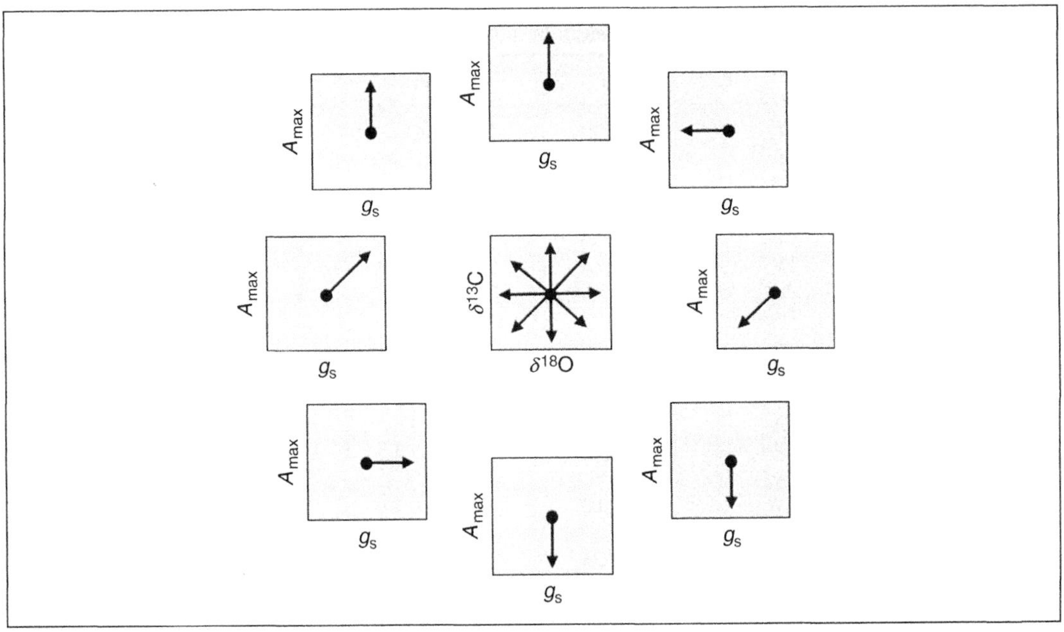

FIGURE 4.1 Schematic representation of the model explained in the Section III. The central scheme indicates the observed change in $\delta^{18}O$ and $\delta^{13}C$ due to an environmental impact as an arrow. In each of these eight scenarios, the arrow points toward the respective response in A_{max} and g_s as derived from the model.

noteworthy: When we assume that the arrow in the $\delta^{18}O$–$\delta^{13}C$ plot is rotating clockwise, the corresponding arrow in the g_s–A_{max} plot will be rotating counterclockwise. This emphasizes the existence of a predictable relationship between the isotope pattern and the physiological response.

IV. RESULTS AND DISCUSSION

A. $\delta^{13}C$ and Water-Use Efficiency

A summary of carbon isotope values of the Swiss sites and derived physiological parameters is shown in Table 4.1. Values are indicated as 50-year averages for the early periods (1825–1875, 1876–1925, 1926–1975), and 25-year averages for the most recent period (1976–2000). By considering averages of relatively long periods, the variations from year-to-year due to the short-term variable climatic conditions are omitted, while the long-term changes from the nineteenth century to the present can be more clearly seen. The $\delta^{13}C$ values as shown are corrected for the atmospheric decrease of $\delta^{13}C$ by adding the difference between actual and preindustrial value of $\delta^{13}C_{atm}$ (data used from Francey *et al.* 1999). After correction, the data are clearly more positive in the most recent period compared to earlier periods, indicating a systematic decrease in the isotope discrimination. Overall, the values for the deciduous trees, beech and ash (FS, FE), are in the range from −25.22‰ to −23.51‰, thus in general more depleted than the conifers, fir and spruce (AA, PA), with a range from −24.45‰ to −22.72‰. Accordingly, the inferred values of c_i are lower for the conifers than the deciduous species, whereas the WUE_i values are higher. This result is consistent with many studies from the literature where relatively positive $\delta^{13}C$ values and high WUE_i for conifers compared to deciduous species were reported, often related to lower stomatal density in conifers (Brooks *et al.* 1997).

TABLE 4.1 The carbon isotope values reported versus vpdb (corrected for the change in atmospheric $\delta^{13}C$) and derived physiological parameters for the swiss sites

		1850	1900	1950	2000
$\delta^{13}C$	FE	−25.22	−24.21	−23.92	−23.51
	FS			−25.07	−24.47
	AA	−24.45	−23.90	−23.31	−22.78
	PA		−24.06	−23.47	−22.72
c_i	FE	187.0	180.0	186.7	199.1
	FS			204.2	211.5
	AA	176.8	175.7	177.9	188.8
	PA		181.9	180.1	190.4
WUE_i	FE	62.0	72.8	79.6	91.6
	FS			69.7	80.9
	AA	68.2	75.4	85.1	99.4
	PA		75.6	83.7	101.1
WUE_i change	FE	100.0	117.4	128.3	147.8
	FS			100.0	116.1
	AA	100.0	110.5	124.7	145.7
	PA		100.0	110.8	133.7
c_i/c_a	FE	0.653	0.607	0.595	0.576
	FS			0.647	0.620
	AA	0.618	0.593	0.567	0.543
	PA		0.601	0.574	0.540

Species code as given in the text: FE = *Fraxinus excelsior*, FS = *Fagus sylvatica*, AA = *Abies alba*, PA = *Picea abies*. The time periods represented are "1850" = 1825–1875, "1900 = 1876–1925, "1950" = 1926–1975, "2000" = 1976–2000.

The calculated WUE_i increased strongly during the investigated period for all species. For sites going back to 1850, the increase was similar for ash and spruce (overall increase from 1850 to 2000 between 45.7% and 47.8%). About half of the increase appears to have occurred in the last 50 years, as the increase in WUE_i for the above mentioned species was only 24.7–28.3% until 1950. The results thus show that the trees nowadays are using clearly less water for the production of the same amount of biomass compared to preindustrial times, which was found for all investigated species and sites.

For comparison with results from other regions, an overview of published values is shown (Table 4.2), considering mainly the Northern Hemisphere, but also one study from the tropics (Hietz *et al.* 2005). In these studies, usually an age range of trees from 50 to 150 years was investigated, including quite a large number of different species. Climatic conditions vary strongly between the investigated regions from cold-dry in Siberia to warm-dry in Western United States, temperate-moist in Central Europe, and humid-warm in the tropics. Different altitudinal ranges were also covered, mountainous sites particularly prevalent in the Western United States. As indicated in Table 4.2, the results mostly show increases in WUE_i over the twentieth century similar in the range as observed in this study for the Swiss sites. As an exception, relatively low increases or even no increase was observed in the Western United States. Marshall and Monserud (1996) found that for Douglas-fir, Ponderosa pine, and western white pine the difference between c_a and c_i tended to remain constant, and therefore the authors suggested this difference as a homeostatic set point for photosynthesis under varying environmental conditions. This study was carried out in the foothills of the Rocky Mountains of northern Idaho, 800–950 m a.s.l., where climate conditions are quite moderate, and therefore the difference in response to the European sites may rather be related to the investigated species. A relatively small, but significant increase in WUE_i (around 20%) was observed in Northern Eurasia for a network of sites in the latitude range from 60 °N to 70 °N for larch, spruce, and pine trees (Saurer *et al.* 2004b). Here, the change in discrimination was small, resulting in stability of the ratio c_i/c_a, when comparing the periods 1861–1890 and 1961–1990 (Figure 4.2). This relationship was particularly strong for pine and larch, but only significant for spruce when the outlying value for one site was omitted. The trees in Siberia are exposed to very low temperatures and a short growing period, which may explain the relatively small increase in WUE_i. The parameter c_i/c_a was also suggested as a set point for photosynthesis, and was found to be constant under a range of environmental

TABLE 4.2 An overview of published values for the changes in WUE_i over the last two centuries, in comparison with the results of the swiss sites discussed in this chapter

Region	Species	WUE_i increase	Data source
England, Finland	*Quercus robur, Fagus sylvatica, Pinus sylvestris*	24–48%	Waterhouse *et al.* 2004
France	*Abies alba*	30%	Bert *et al.* 1997
France	*Fagus sylvatica*	23–44%	Duquesnay *et al.* 1998
Switzerland	*Abies alba, Picea abies, Fraxinus excelsior, Fagus sylvatica*	33–48%	This chapter
Western United States	*Pseudotsuga menziesii, Pinus ponderosa, Pinus monticola*	0%	Marshall and Monserud 1996
Western United States	*Pinus edulis, Picea sitchensis, Pinus ponderosa, Juniperus phoemicea, Pinus monophylla, Picea sitchensis, Pinus jeffreyi, Pinus longaeva, Pinus coulteri, Quercus lobata*	5–25%	Feng 1999
Northern Eurasia	*Larix sibirica, Larix gmelinii, Larix dahurica, Pinus sylvestris, Picea abies, Pinus obovata*	20%	Saurer *et al.* 2004b
Brazil	*Swietenia macrophylla, Cedrela odorata*	34–52%	Hietz *et al.* 2005

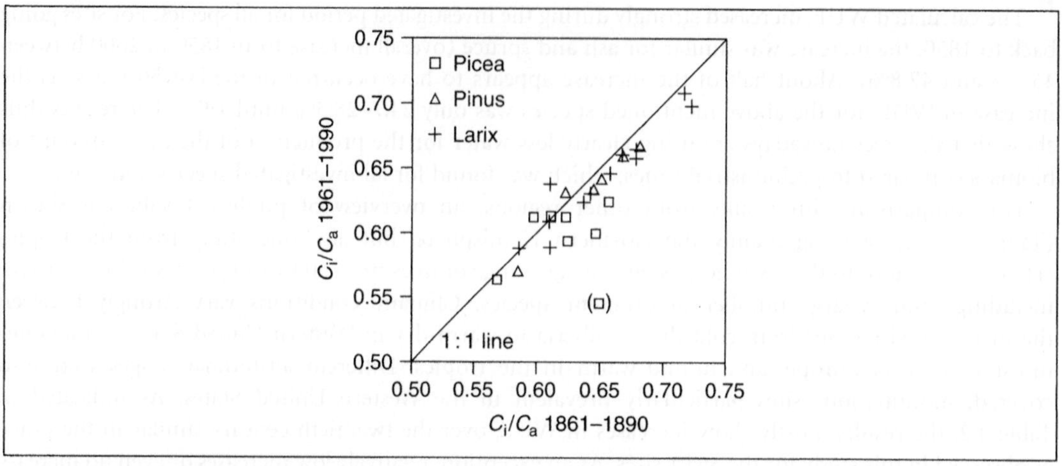

FIGURE 4.2 Relationship between c_i/c_a at the end of the nineteenth to the end of the twentieth century for *Picea*, *Pinus*, and *Larch* trees growing in Northern Eurasia. Values are close to the 1:1 line, with the exception of one value shown in brackets (site Murmashi, *Picea abies*, see Saurer *et al.* 2004b for details).

conditions, including enriched CO_2 (Ehleringer and Cerling 1995). This is not the case, however, for the Swiss sites, where the strong increase in WUE_i is accompanied by a reduction of c_i/c_a (Table 4.2).

We can therefore divide the findings into three situations:

1. Strong increase in WUE_i, such that c_i/c_a drops, for example Swiss sites, other European sites, and the tropics
2. Intermediate increase in WUE_i, such that c_i/c_a is kept constant, for example Northern Eurasia
3. No increase in WUE_i, such that c_a/c_i remained constant (Marshall and Monserud 1996)

Apparently, there is no uniform response to the increasing atmospheric CO_2 in different ecosystems. The above three response scenarios, however, may be used as a guideline to distinguish between weak, medium, and strong responses. While it is difficult to extrapolate regional studies to the global scale, the relatively large number of tree-ring carbon isotope investigations is now starting to provide patterns of change. Such results are important in global modeling efforts, where discrimination should not be considered as a constant, but differences dependent on the region, species, and site conditions should be taken into account. A better implementation of these variations would clearly improve carbon cycle models that use isotopes as a constraining variable (Ciais *et al.* 1995).

B. Combining $\delta^{13}C$ and $\delta^{18}O$

As mentioned above, useful physiological information can be derived from carbon isotope trends in tree rings, but there remains an uncertainty in the interpretation of calculated c_i and WUE_i values. It is not known from this analysis alone whether the trees achieved the improved WUE_i by reducing stomatal conductance or by increasing photosynthesis. Both processes, however, are likely to have occurred during the last 200 years as a response to increasing CO_2 concentrations in the atmosphere, as inferred by artificial CO_2 fumigation experiments (Drake *et al.* 1997). It would be crucial to determine which factors have led to the widespread improvement of WUE_i and to what percentage a response of stomata or photosynthesis is responsible. This would greatly improve our understanding of past forest response to the changing environmental conditions.

In the following, we explore the possibility using the combined carbon and oxygen isotope analysis to distinguish between stomatal or photosynthetic response. In Figure 4.3 we show oxygen isotope data of Swiss sites, analyzed for the same periods as for the carbon isotopes discussed above, and the data of $\delta^{18}O$ versus $\delta^{13}C$, plotted as scatter plots. Before use in this analysis, the oxygen isotope values were first correlated to summer temperatures, and then the respective regression lines were used to correct the data for the climatic signal. This removes from the data the effects of differences in source water $\delta^{18}O$ related to temperature, and retains only the signal related to evaporative enrichment. The data in Figure 4.3 are shown such that the temporal development can be observed. The data reveal interesting differences between the species. There is an increase in both $\delta^{13}C$ and $\delta^{18}O$ for spruce (PA) during the investigated period, resulting in a positive relationship. In contrast to spruce, for ash (FE) and fir (AA), there is little change in $\delta^{18}O$ resulting in an upward trend in the graph. The beech trees (FS) also show upward trends, if any, but the changes are small and not significant.

These patterns of change can now easily be interpreted with the aid of the scheme developed in Section III (Figures 4.1 and 4.4): The positive correlation corresponds to a situation where there is a reduction in g_s, but not much change in A_{max}, whereas the upward trend reflects an increase in A_{max}, with no or minor reactions in g_s. Accordingly, we can conclude that the improvement in WUE_i observed for all species in Switzerland seems to be mostly due to a reduction in stomatal conductance for spruce, but due to enhanced photosynthesis in ash and fir, possibly also in beech. These results are only qualitative. At present, we cannot quantify these changes in percentage. Nevertheless, this information helps greatly in improving our understanding of the processes involved in tree response

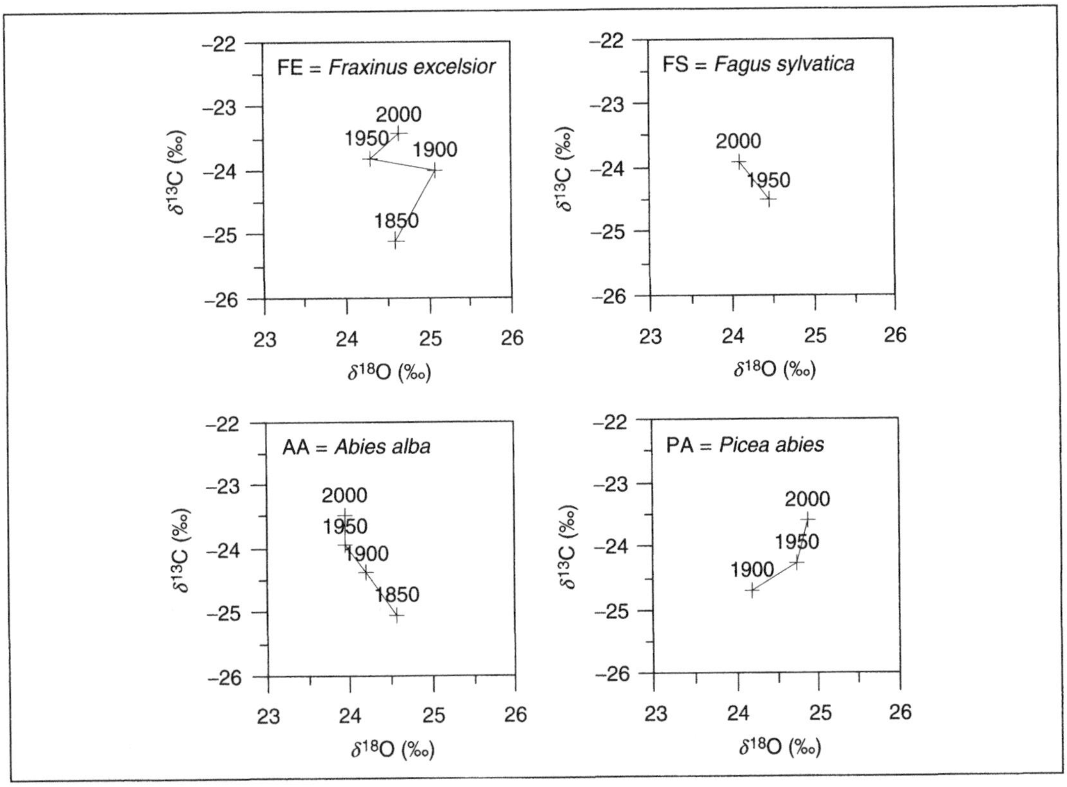

FIGURE 4.3 Relationship between $\delta^{18}O$ and $\delta^{13}C$ over the last two centuries for the Swiss tree-ring sites. Values are reported versus VPDB for $\delta^{13}C$ and versus VSMOW for $\delta^{18}O$. Carbon isotope values were corrected for the change in $\delta^{13}C$ of atmospheric CO_2.

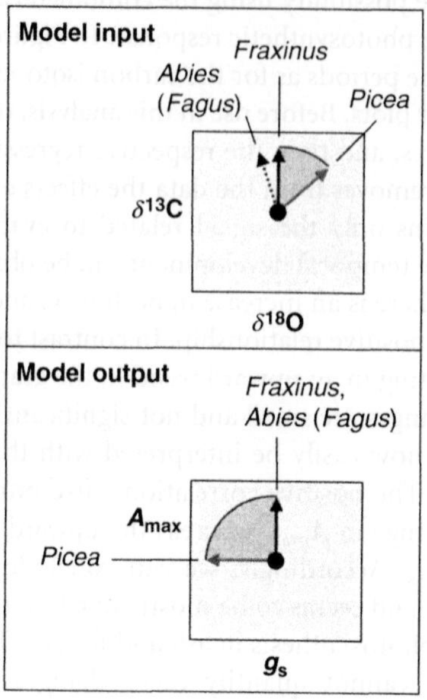

FIGURE 4.4 The A_{max}–g_s response of the different species from Swiss sites as derived with the conceptual model. The upper panel reflects the "model input," whereas the lower panel reflects the "model output."

to environmental change. The results also clearly show species-specific differences. The stronger response of A_{max} for ash and fir may well indicate that these species were able to enhance biomass accumulation due to increasing CO_2 during the twentieth century, whereas, spruce responded more strongly with reduced stomatal conductance and, accordingly, less transpiration and water loss.

C. Case Study in Air Pollution Research

Nitrogen deposition through the twentieth century is one of the main factors of global change that affects the ecosystem. Interpretations of tree-ring records that do not consider this influence may therefore be incomplete. While it is obvious that pollution effects (from traffic, industry, or agriculture) are most important near the sources, there is clear evidence that N-bearing compounds are distributed quite far throughout the atmosphere and are also significant in rural areas, in particular in nitrogen-limited ecosystems, for example for peat bogs (Aerts *et al.* 1992). For these case studies, though, it is more practical, at least initially, to sample close to pollution sources, where the strongest effects can be expected, while keeping control sites with low pollution loads under consideration for comparison. Nitrogen isotopes have been shown to be useful for detecting incorporation of N from pollution sources into plant organic material (Ammann *et al.* 1999, Siegwolf *et al.* 2001), although this is less evident for tree rings due to the potential of lateral nitrogen transport in the stem (Elhani *et al.* 2003). In our study near a freeway in Switzerland, we investigated several spruce trees growing at different distances from the freeway (10–1000 m). From the analysis of needles, it was known for this location that the $\delta^{15}N$ of NO_x from car emissions was about 5‰ higher than the background soil N (Ammann *et al.* 1999). In the tree-ring analysis, we could clearly detect the relatively enriched N from traffic emissions (Figure 4.5). In this figure, we show the range of values observed for four trees measured at the closest site (A1), as well as at a more distant site (A2; 150 m from the freeway) for the period from about 1920 to 2000. The values at site A1 increased clearly after the construction of the freeway in 1965, but not all of the four trees responded clearly (Saurer *et al.* 2004a). Accordingly, the range of minima to maxima values of $\delta^{15}N$ is wide at A1 and overlaps with values of the more remote site (Figure 4.5). These data show that N from NO_x was taken up by spruce trees, but the tree-ring width curves did not differ between sites, either indicating that the additional N-uptake through the needles had not any fertilizing effect on growth or reflecting a balance between fertilizing and toxic effects from other pollutants.

For a better understanding of the physiological response of the trees, the combined oxygen and carbon approach as introduced in the last section is useful. We considered differences only between the sites A1 and A2 to account for climatic influences which are not related to pollution sources. $\delta^{18}O_{diff}$ and $\delta^{13}C_{diff}$ values therefore denote the differences of $\delta^{18}O$ (and $\delta^{13}C$, respectively) between A1 and A2. For 10-year-filtered data, we found a strong negative correlation between $\delta^{18}O_{diff}$ and $\delta^{13}C_{diff}$ (Figure 4.6). This means that an increase in $\delta^{13}C$ (which was observed after the construction of the freeway)

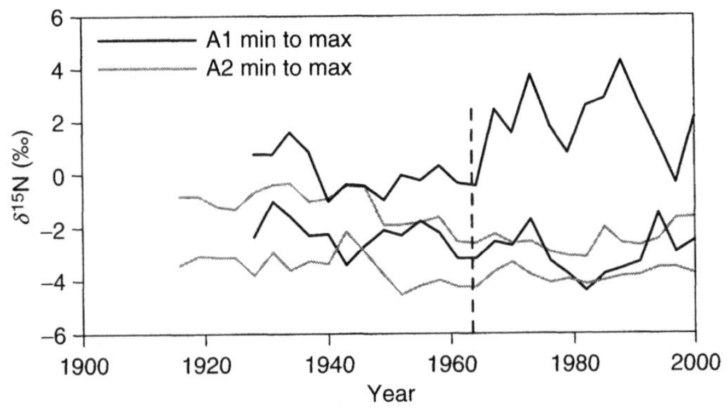

FIGURE 4.5 $\delta^{15}N$ values for two sites near a motorway in Switzerland shown as a time course over the twentieth century. The two bold lines indicate the minimum and maximum values observed at site A1 (the site more exposed to the traffic), the thin lines indicate the minimum and maximum values at site A2. The dashed line indicates the time when the motorway was constructed.

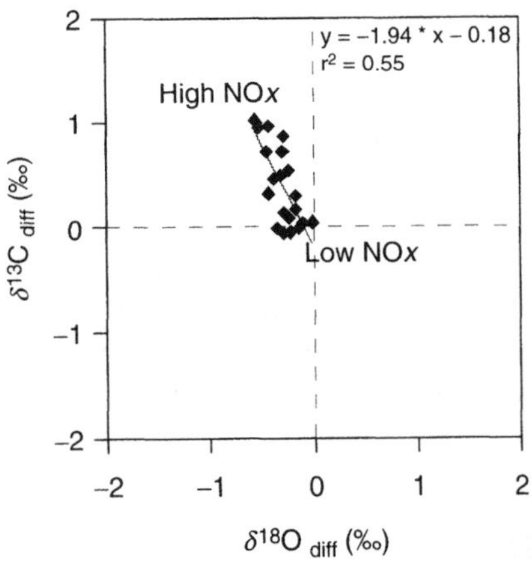

FIGURE 4.6 Relationship between the carbon and oxygen isotope signals of trees near the motorway. The data are shown as differences between the signals at sites A1 and A2, that is $\delta^{18}O_{diff} = \delta^{18}O$ (A1) − $\delta^{18}O$ (A2), and $\delta^{13}C_{diff} = \delta^{13}C$ (A1) − $\delta^{13}C$ (A2). Trees under high NO_x load near the motorway exhibit higher $\delta^{13}C$ and lower $\delta^{18}O$ compared to trees growing further away.

was accompanied by a decrease in $\delta^{18}O$. This trend can then be interpreted with the help of the scheme presented in Figure 4.1, which indicates that under polluted (*i.e.*, high NO_x) conditions, water-use efficiency and photosynthesis were stimulated, while stomatal conductance remained almost unchanged. This result is well in line with results from laboratory experiments that showed the same effects of the foliar uptake of NO_x as we found near the freeway, in contrast to N supplied through the roots which results in lower water-use efficiency (Siegwolf *et al.* 2001). The detailed analysis of the data from this laboratory study also revealed the toxic effect of NO_2. Although an increase in N supply should result in an increase of stomatal conductance, the N supply via NO_2 was not as effective as if the same amount of N had been added to the soil. The reason for this response is the toxicity of NO_2, which during its transformation to NH_4^+ releases an O radical with high chemical reactivity. These experiments show that detailed physiological information on the tree's responses to air pollutants can be obtained by the combined analysis of different isotopes. The isotope data are useful in combination with gas exchange measurements in controlled experiments and provide new information in retrospective tree-ring studies that may not be accessible by any other means.

V. CONCLUSIONS AND OUTLOOK

For assessing global change and ecosystem function in a changing environment, a historic perspective is most useful. Two different, but complementary research approaches may be compared: an approach assessing change by looking into the past, and an approach that monitors ongoing change. Monitoring activities and building networks of sites, where ecosystem and air parameters are continuously investigated, are not only extremely useful but also represent a huge task and years of sustained effort and investment before changes may be detected, and sometimes this may be too late to take adequate measures. If on the other hand, current investigations are viewed with a longer term perspective, the actual situation may be compared to the situation 50 or 100 years ago, or even before industrialization started. This could facilitate the earlier discovery of changes in the ecosystem. Although the drawback of historic methods is that they sometimes rely on indirect evidence, their potential to provide a reference point is invaluable. They may tell us how an ecosystem once used to be. Studying tree rings is one way to assess forest ecosystem function in the past. By looking at different ring parameters, including isotopes, it is possible to extract not only climatic but also physiological information. Accordingly, it is possible to gain insight into apparently transient and long-past properties, like the intercellular CO_2 concentration or the water-use efficiency of trees in preindustrial times. It may even be possible to gain insight into changes in stomatal conductance or photosynthesis due to past global change, using the combined carbon–oxygen approach as discussed in this chapter. This would greatly further our understanding of the response of adult trees to increasing CO_2; their ability, for instance, to assimilate more carbon or preserve water by reducing their stomatal conductance. Further, a deeper understanding of the isotope fractionations occurring during photosynthesis may also make carbon cycle models more effective and reliable. Accordingly, we may greatly enhance our understanding of forest response to global change by extracting the full information stored in tree-ring isotopes.

VI. ACKNOWLEDGMENTS

This work was supported by the EU projects EVK2-CT-2002-00147 (ISONET) and FP6-2004-GLOBAL-017008-2 (MILLENNIUM). We are grateful for numerous helpful discussions with P. Cherubini, M.-R. Guerrieri, and M. Jaeggi.

VII. REFERENCES

Aerts, R., B. Wallen, and N. Malmer. 1992. Growth-limiting nutrients in *Sphagnum*-dominated bogs subject to low and high atmospheric nitrogen supply. *Journal of Ecology* **80:**131–140.

Allison, C. E., R. J. Francey, and P. B. Krummel. 2003. $\delta^{13}C$ in CO_2 from sites in the CSIRO Atmospheric Research GASLAB air sampling network. "*Trends: A Compendium of Data on Global Change.*" Carbon Dioxide Information Analysis Center, Oak Ridge National Laboratory, US Department of Energy, Oak Ridge, TN, USA.

Ammann, M., R. T. W. Siegwolf, F. Pichlmayer, M. Suter, M. Saurer, and C. Brunold. 1999. Estimating the uptake of traffic-derived NO_2 from ^{15}N abundance in Norway spruce needles. *Oecologia* **118:**124–131.

Bert, D., S. W. Leavitt, and J. L. Dupouey. 1997. Variations of wood $\delta^{13}C$ and water-use efficiency of *Abies alba* during the last century. *Ecology* **78:**1588–1596.

Brooks, J. R., L. B. Flanagan, N. Buchmann, and J. R. Ehleringer. 1997. Carbon isotope composition of boreal plants: Functional grouping of life forms. *Oecologia* **110:**301–311.

Ciais, P., P. P. Tans, J. W. C. White, M. Trolier, R. J. Francey, J. A. Berry, D. R. Randall, P. J. Sellers, J. G. Collatz, and D. S. Schimel. 1995. Partitioning of ocean and land uptake of CO_2 as inferred by $\delta^{13}C$ measurements from the NOAA climate monitoring and diagnostics laboratory global air sampling network. *Journal of Geophysical Research* **100:**5051–5070.

Dongmann, G., H. W. Nürnberg, H. Förstel, and K. Wagener. 1974. On the enrichment of $H_2{}^{18}O$ in the leaves of transpiring plants. *Radiation and Environmental Biophysics* **11:**41–52.

Drake, B. G., M. A. Gonzalez-Meler, and S. P. Long. 1997. More efficient plants: A consequence of rising atmospheric CO_2? *Annual Review of Plant Physiology and Plant Molecular Biology* **48:**609–639.

Duquesnay, A., N. Breda, M. Stievenard, and J. L. Dupouey. 1998. Changes of tree-ring $\delta^{13}C$ and water-use efficiency of beech (*Fagus sylvatica* L.) in north-eastern France during the past century. *Plant, Cell and Environment* **21:**565–572.

Ehleringer, J. R., and T. E. Cerling. 1995. Atmospheric CO_2 and the ratio of intercellular to ambient CO_2 concentrations in plants. *Tree Physiology* **15:**105–111.

Elhani, S., B. F. Lema, B. Zeller, C. Brechet, J. M. Guehl, and J. L. Dupouey. 2003. Inter-annual mobility of nitrogen between beech rings: A labelling experiment. *Annals of Forest Science* **60:**503–508.

Farquhar, G. D., and J. Lloyd. 1993. Carbon and oxygen isotope effects in the exchange of carbon dioxide between terrestrial plants and the atmosphere. Pages 47–70 *in* J. R. Ehleringer, A. E. Hall, and G. D. Farquhar (Eds.) *Stable Isotopes and Plant Carbon-Water Relations.* Academic Press, San Diego.

Farquhar, G. D., M. H. O'Leary, and J. A. Berry. 1982. On the relationship between carbon isotope discrimination and the intercellular carbon dioxide concentration in leaves. *Australian Journal of Plant Physiology* **9:**121–137.

Feng, X. H. 1999. Trends in intrinsic water-use efficiency of natural trees for the past 100–200 years: A response to atmospheric CO_2 concentration. *Geochimica et Cosmochimica Acta* **63:**1891–1903.

Francey, R. J., C. E. Allison, D. M. Etheridge, C. M. Trudinger, I. G. Enting, M. Leuenberger, R. L. Langenfelds, E. Michel, and L. P. Steele. 1999. A 1000-year high precision record of $\delta^{13}C$ in atmospheric CO_2. *Tellus Series B—Chemical and Physical Meteorology* **51:**170–193.

Hietz, P., W. Wanek, and O. Dunisch. 2005. Long-term trends in cellulose $\delta^{13}C$ and water-use efficiency of tropical *Cedrela* and *Swietenia* from Brazil. *Tree Physiology* **25:**745–752.

IPCC. 1996. Climate Change 1995: The Science of Climate Change—Contribution of WGI to the Second Assessment Report of the Intergovermental Panel on Climate Change.

Körner, C. 2000. Biosphere responses to CO_2 enrichment. *Ecological Applications* **10:**1590–1619.

Marshall, J. D., and R. A. Monserud. 1996. Homeostatic gas-exchange parameters inferred from $^{13}C/^{12}C$ in tree rings of conifers. *Oecologia* **105:**13–21.

McCarroll, D., and N. J. Loader. 2004. Stable isotopes in tree rings. *Quaternary Science Reviews* **23:**771–801.

Norby, R. J. 1998. Nitrogen deposition: A component of global change analyses. *New Phytologist* **139:**189–200.

Roden, J. S., and J. R. Ehleringer. 1999. Observations of hydrogen and oxygen isotopes in leaf water confirm the Craig-Gordon model under wide-ranging environmental conditions. *Plant Physiology* **120:**1165–1173.

Saurer, M., U. Siegenthaler, and F. Schweingruber. 1995. The climate-carbon isotope relationship in tree-rings and the significance of site conditions. *Tellus* **47:**320–330.

Saurer, M., S. Borella, and M. Leuenberger. 1997. $\delta^{18}O$ of tree rings of beech (*Fagus silvatica*) as a record of $\delta^{18}O$ of the growing season precipitation. *Tellus Series B—Chemical and Physical Meteorology* **49:**80–92.

Saurer, M., P. Cherubini, and R. T. W. Siegwolf. 2000. Oxygen isotopes in tree rings of *Abies alba*: The climatic significance of interdecadal variations. *Journal of Geophysical Research* **105:**12461–12470.

Saurer, M., P. Cherubini, M. Ammann, B. De Cinti, and R. T. W. Siegwolf. 2004a. First detection of nitrogen from NOx in tree rings: A $^{15}N/^{14}N$ study near a motorway. *Atmospheric Environment* **38:**2779–2787.

Saurer, M., R. T. W. Siegwolf, and F. Schweingruber. 2004b. Carbon isotope discrimination indicates improving water-use efficiency of trees in northern Eurasia over the last 100 years. *Global Change Biology* **10:**2109–2120.

Scheidegger, Y., M. Saurer, M. Bahn, and R. T. W. Siegwolf. 2000. Linking stable oxygen and carbon isotopes with stomatal conductance and photosynthetic capacity: A conceptual model. *Oecologia* **125**:350–357.

Schleser, G. H., G. Helle, A. Lucke, and H. Vos. 1999. Isotope signals as climate proxies: The role of transfer functions in the study of terrestrial archives. *Quaternary Science Reviews* **18**:927–943.

Schulze, E. D. 1989. Air pollution and forest decline in a spruce (*Picea abies*) forest. *Science* **244**:776–783.

Schweingruber, F., K. R. Briffa, and P. Nogler. 1993. A tree-ring densitometric transect from Alaska to Labrador. *International Journal of Biometeorology* **37**:151–169.

Siegwolf, R. T. W., R. Matyssek, M. Saurer, S. Maurer, M. S. Günthardt-Goerg, P. Schmutz, and J. B. Bucher. 2001. Stable isotope analysis reveals differential effects of soil nitrogen and nitrogen dioxide on the water use efficiency in hybrid poplar leaves. *New Phytologist* **149**:233–246.

Waterhouse, J. S., V. R. Switsur, A. C. Barker, A. H. C. Carter, D. L. Hemming, N. J. Loader, and I. Robertson. 2004. Northern European trees show a progressively diminishing response to increasing atmospheric carbon dioxide concentrations. *Quaternary Science Reviews* **23**:803–810.

Yakir, D., and L. D. L. Sternberg. 2000. The use of stable isotopes to study ecosystem gas exchange. *Oecologia* **123**:297–311.

CHAPTER 5

Oxygen Isotope Proxies in Tree-Ring Cellulose: Tropical Cyclones, Drought, and Climate Oscillations

Claudia I. Mora,* Dana L. Miller,* and Henri D. Grissino-Mayer†

*Department of Earth and Planetary Sciences, University of Tennessee
†Department of Geography, University of Tennessee

Contents

I. INTRODUCTION

Tropical cyclones pose a grave threat to life and property along the heavily populated eastern and southern US coasts, where each year there is at least a one in six chance of $10 billion in tropical cyclone-related losses (Pielke and Landsea 1998). This number was exceeded many times over by the devastating human and economic impact of Hurricane Katrina in August, 2005. North Atlantic hurricane frequency and intensity have increased significantly since 1995 (Landsea *et al.* 1998, Elsner *et al.* 2000, Goldenberg *et al.* 2001). Controls on hurricane frequency and intensity are fiercely debated, in particular their relationship to global warming. Studies implicate global warming in the increased intensity of tropical cyclone events (Knutson and Tuleya 2004, Emanuel 2005, Webster *et al.* 2005), but others suggest no causative relationship (Pielke *et al.* 2005). An increase in hurricane frequency is thought to reflect, at least in part, natural and multidecadal scale variations in North Atlantic sea surface temperatures (SST), such as the Atlantic Multidecadal Oscillation (AMO) (Elsner and Kara 1999, Goldenberg *et al.* 2001). Although some climate modes affect the formation of North Atlantic hurricanes [*e.g.*, the tendency for an El Niño Southern Oscillation (ENSO) event to suppress their formation; Gray *et al.* (1997)], there is no simple theoretical basis for predicting changes in hurricane frequency (Pielke *et al.* 2005). The very short record of reliable, systematic, instrumental measurements, in many cases only 50–60 years long, makes it difficult to discern long-term trends and fluctuations in tropical cyclone activity, particularly on the multidecadal scale on which they are thought to vary.

In the absence of an instrumental record, a long-term temporal and spatial record of hurricane activity must be developed through geological, biological, and written documentary evidence. This is the challenge of the emerging field of *paleotempestology.* Historical archives, including newspapers, diaries, naval—and plantation logs, can be exploited to reconstruct sometimes detailed records of tropical cyclone activity several centuries past (Mock 2004). These records are limited in time and geography by human observation, and the observations are often subjective, so that qualitative descriptions must be scaled and interpreted. The records themselves require significant effort to locate. Biological and geological proxies for tropical cyclone activity provide an independent tool to assess variation in their occurrence and constrain the relationship between tropical cyclones and climate modes that may control their development. Millennial-scale records of tropical cyclone activity are archived in storm deposits in coastal ponds and lagoons (Liu and Fearn 1993, Donnelly *et al.* 2001, Liu 2004). These sedimentary records are useful to document the frequency of catastrophic strikes at given sites along the coasts and to define long-term climate phases resulting in the predominance of hurricane strikes along the Gulf coast versus the Eastern coast of the United States (Liu 2004). But, their temporal resolution is coarse and the studies rely on the unique identification of sand layers as the result of a hurricane which vary as a function of storm energy and trajectory.

Stable isotopic records captured in biological and geological materials which grow incrementally through time, such as tree rings, speleothems, and coral skeletons, potentially provide higher resolution proxies to examine multidecadal variation in tropical cyclone occurrence (Lawrence 1998). Studies have tested the viability of these records at very different scales of resolution from daily (*i.e.*, coral skeletons by ion microprobe; A. Cohen, unpublished data), approximately weekly to monthly (speleothems, Frappier *et al.* 2007), to seasonal (tree rings; Miller *et al.* 2006). Very high resolution coral analysis by ion microprobe is currently impractical to establish long time series. The speleothem record is better demonstrated, but, to capture an isotopic record of tropical cyclone-related precipitation, the epikarst hydrologic system must operate within conditions that favor supersaturation of storm precipitation with calcite, without homogenization with other soil water or groundwater. The reported speleothem record is still quite short (23 year), and precise dating of significantly older or inactive speleothems can be difficult. Notwithstanding these constraints, speleothem isotope records of hurricanes potentially extend well into the *Quaternary.*

This chapter describes the development of a proxy for tropical cyclone occurrence based on the oxygen isotope compositions of tree-ring cellulose and demonstrates the potential use of tree-ring isotope time series to elucidate the relationship between tropical cyclone occurrence and climate-forcing mechanisms operating at semi- to multidecadal scales, such as the AMO and ENSO. The approach can be applied to create a seasonally resolved record of hurricanes extending up to 500 years across a wide region of the southeastern US coast. Such proxy-derived data are necessary to clarify the relationship between tropical cyclone frequency and global climate change and to improve predictive modeling of tropical cyclone development.

II. A TREE-RING ISOTOPE RECORD OF TROPICAL CYCLONES AND CLIMATE

A. Climate Modes Influencing Tropical Cyclone Occurrence

The formation of Atlantic Basin tropical cyclones occurs predominantly between 10° and 20° N, and especially during the months of August to October, when the region has both the warm SST and low vertical wind shear that create favorable conditions for tropical cyclone formation (Goldenberg and Shapiro 1996). Tropical cyclone development varies dramatically on both annual and decadal scales, most likely in response to variability in SST, tropospheric wind patterns, sea level pressure, and disturbances in the development region (Goldenberg *et al.* 2001). These factors may themselves respond to remote climate modes, including the North Atlantic Oscillation (NAO), AMO, ENSO, Pacific Decadal Oscillation (PDO), among others. For example, ENSO events occur on ~3- to 5-year frequency and result in anomalously strong westerly upper tropospheric wind patterns over the Caribbean and equatorial Atlantic. The vertical wind shear thus produced suppresses the development of tropical cyclones in the North Atlantic basin.

One climate mode of particular interest is the AMO, defined as a low-frequency oscillation in North Atlantic SST (Delworth and Mann 2000). The AMO has warm and cool phases which last from ~15 to 40 years. Its most recent warm phases (1930–1960 and 1995–present) were associated with a greater number of major tropical cyclones, and the AMO is invoked as a natural, multidecadal climate control on hurricane occurrence and intensity (cf. Goldenberg *et al.* 2001). Further, the AMO may affect Atlantic hurricane formation indirectly by modulating the strength of El Niño events and the west African disturbances that may spawn hurricanes (Knight *et al.* 2006). Studies, however, challenge the influence of the AMO on SST and hurricane frequency or intensity (Webster *et al.* 2005, Mann and Emanuel 2006).

Climate modes, such as the AMO or ENSO, may also influence temperature, precipitation, seasonality, and other climate factors and are therefore expected to be expressed in (*i.e.*, can be correlated to) the oxygen isotope time series in tree-ring cellulose. The frequency of hurricane events, a record of which is superimposed on the isotope time series (see below), may correlate to particular phases of these climate modes. Thus, in addition to a proxy record of hurricane occurrence, long, seasonally resolved oxygen isotope time series in tree rings also hold potential for evaluating the relationship between hurricanes and climate factors that may influence their development or frequency. While instrumental or historical documentation may be useful to very precisely define a hurricane track and intensity, tree-ring isotope data may offer a glimpse of the overall climate picture governing hurricane development.

B. Isotopic Compositions of Tropical Cyclone Precipitation

Hurricanes (*i.e.*, tropical cyclones) are well-organized and long-lived mesoscale convective systems which produce precipitation that is markedly lower in ^{18}O than normal tropical rain systems (by as much as 10‰; Lawrence *et al.* 2002). The $\delta^{18}O$ values of hurricane precipitation cannot be readily modeled by simple Rayleigh fractionation, as almost perfect precipitation efficiency (>93%) would be required (Lawrence *et al.* 1998). The unusually low isotope ratios of hurricane precipitation results largely from the cyclone's precipitation efficiency (*i.e.*, amount effect), but is compounded by large T-dependent fractionations in the very high altitude cloud tops and the exchange of low altitude, inflowing vapor with falling rain ("diffusive exchange;" Figure 5.1). The isotopically depleted vapor is lofted in the eyewall and redistributed in the high-altitude outflow (Figure 5.1). As a result of these processes, isotopic compositions in hurricane rain can approach the composition of the source vapor,

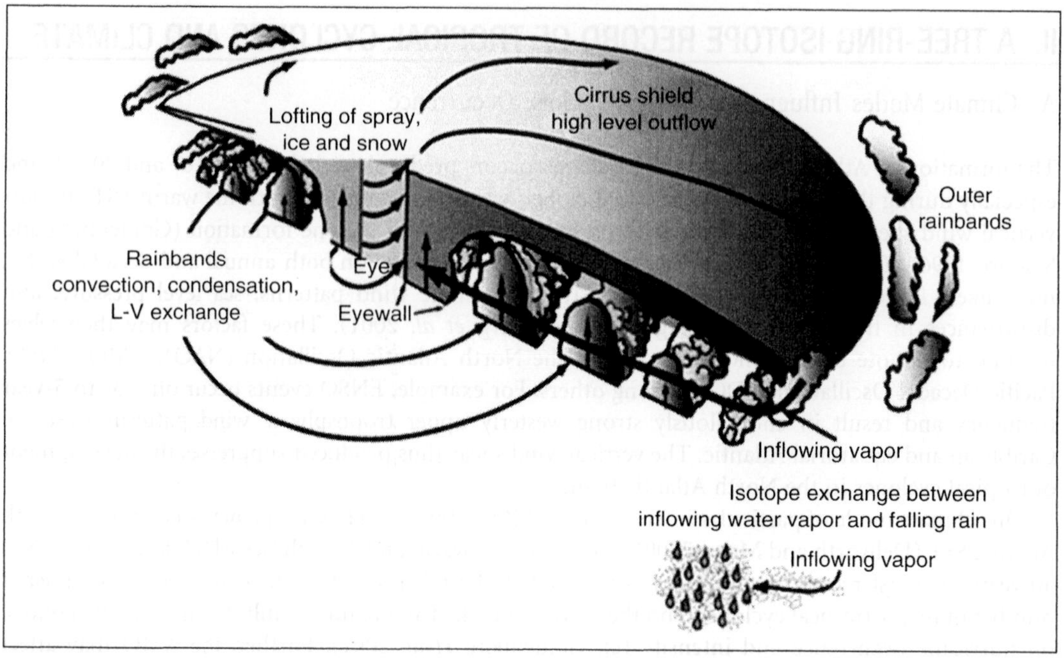

FIGURE 5.1 Structure of a tropical cyclone (after www.srh.weather.gov). The large size, tall, thick clouds, and longevity of hurricanes result in extreme precipitation efficiency and $\delta^{18}O$ values of precipitation that approach the values of the source water vapor. Isotopic exchange between inflowing vapor and falling rain leads to increased ^{18}O depletion toward the eye.

will generally decrease toward the eye, and can be significantly depleted even hundreds of kilometers from the eye (Lawrence and Gedzelman 1996, Lawrence *et al.* 1998).

Measurement and modeling of the isotopic compositions of tropical cyclone precipitation indicate complexities in the spatial and temporal variability of tropical cyclone precipitation due to rain and wind patterns within the storm, and lofting of sea spray, particularly in the strong updrafts at the eyewall (Gedzelman *et al.* 2003). These details will vary as a function of the size and evolution of the storm. Therefore, although better organized and longer-duration tropical cyclones, such as major hurricanes (class 3 or higher, Saffir-Simpson Scale), are likely to be associated with larger and geographically more extensive isotope depletions, the isotopic compositions of hurricane rain cannot be used to establish the storm's intensity or position relative to the proxy capturing the isotopic evidence.

C. Oxygen Isotope Compositions of Tree-Ring Cellulose

Oxygen isotopes in tree-ring α-cellulose mainly reflect the isotopic composition of the source water and a number of biophysiological factors, including biosynthesis, xylem water–sucrose exchange, and leaf water evaporative enrichment, as summarized in Figure 5.2 (Saurer *et al.* 1997, Anderson *et al.* 2002, Weiguo *et al.* 2004). Although the magnitudes of the biophysiological effects are large (~30‰), they tend to be similar for a given species grown in the same environment (Anderson *et al.* 2002). Thus, inter- and intraannual differences in the oxygen isotope composition of cellulose from an individual tree, or individuals of a given species grown under similar conditions, largely reflect changes in source water (precipitation) compositions. For conifers with shallow root systems, such as longleaf

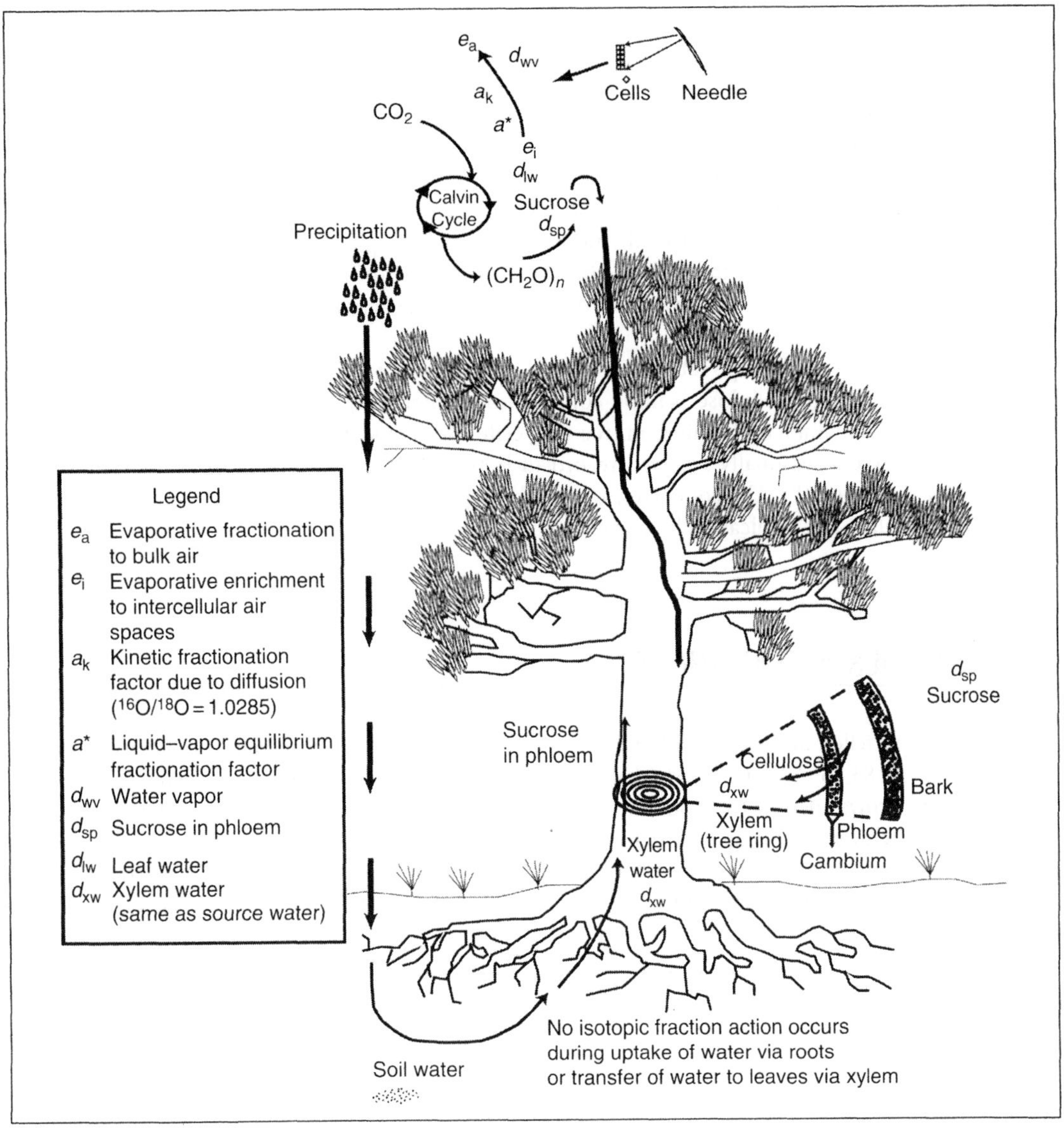

FIGURE 5.2 Major processes contributing to oxygen isotope compositions in tree-ring cellulose, redrawn after Roden *et al.* (2000) with permission from Elsevier. Biosynthesis and biophysiological processes impart the largest isotope fractionations, but isotopic variability in a single individual or individuals of a given species grown under same environmental conditions reflects changes in the isotope composition of source water (precipitation).

(*Pinus palustris* Mill.) pine, precipitation is captured mainly through soil water, which is drawn into the tree without isotopic fractionation (Anderson *et al.* 2002).

Isotopically depleted precipitation from tropical cyclones may persist in surface and soil waters for several weeks after a large event (Lawrence 1998, Tang and Feng 2001). Thus, hurricane precipitation, identified by its unusual ^{18}O-depleted composition, will be incorporated into cellulose during tree growth, capturing an isotopic record of tropical cyclone activity. The magnitude of hurricane-related depletions in cellulose will depend on many factors, including the size and proximity of the storm, soil type, and preexisting soil moisture conditions, and cannot be used as a measure of tropical cyclone intensity. Evaporative enrichment of soil water will eventually ameliorate the low ^{18}O signal

(Lawrence 1998, Tang and Feng 2001). The ephemeral nature of the light, hurricane-related soil water suggests that it is captured only in cellulose produced in the weeks following a storm event. This may be a relatively small proportion of annual cellulose production, and the isotopic anomaly may not be readily detected in an averaged sampling of an annual ring. Longleaf pine tree rings preserve distinct earlywood (EW; growth in the early portion of the growing season) and latewood (LW; growth in the later portion of the growing season) components that can be separately analyzed to obtain seasonally resolved isotope compositions. Tropical cyclones most typically (>90%) impact the southeastern United States during typical LW growth months of July through October (especially, August and September; Landsea 1993). For that reason, the EW and LW portions of annual tree rings are separately sampled and analyzed and isotopic evidence of tropical cyclone activity should be especially prevalent in LW cellulose.

Enrichment of oxygen isotope ratios in soil and leaf water due to high evaporative demand may also be recorded by tree-ring cellulose isotope compositions (Sternberg *et al.* 1989, Saurer *et al.* 1997), resulting in unusually high $\delta^{18}O$ values in tree-ring cellulose. Thus, the tree-ring cellulose isotope time series may preserve a seasonally resolved record of both hurricane occurrence and high evaporative demand (local "drought"). The concurrence of both, opposing, signals in a single season may impact our ability to detect the tropical cyclone event.

III. MATERIALS AND METHODS

Large slabs from felled and subfossil longleaf pine (*P. palustris* Mill.) near Valdosta, Georgia, United States (30.84° N; 83.25° W) and nearby Lake Louise (30.43° N; 83.15° W) were collected (Figure 5.3). Longleaf pine grows in well to moderately well-drained loamy sand soils and has been shown to produce consistent annual rings (Grissino-Mayer *et al.* 2001). The trees were chronologically dated using standard cross-dating techniques (Stokes and Smiley 1996), and the EW and LW portions of each annual ring were separately sampled using a scalpel. Because of the unusually resinous character of

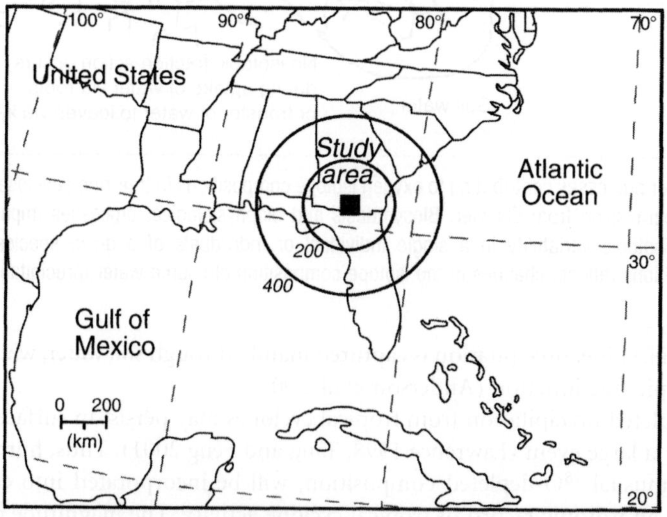

FIGURE 5.3 Location of the study area near Valdosta, Georgia (*shaded box*) showing target area of ~200 or ~400 km from study area. Isotopic evidence of tropical cyclones tracking within this target area was detected in tree-ring cellulose.

these samples, pine resins were first removed by accelerated solvent extraction using 3:1 toluene and reagent alcohol at 125 °C and 1500 psi, then α-cellulose was extracted from resin-free whole wood using established soxhlet extraction methods (Loader *et al.* 1997). Oxygen isotope compositions of α-cellulose (80–100 μg) in EW and LW samples were analyzed by pyrolysis using a TC/EA interfaced with a Finnigan MAT Delta Plus XL continuous flow mass spectrometer (Werner *et al.* 1996, Saurer *et al.* 1998) at the University of Tennessee Knoxville and are reported relative to V-SMOW. Both an internal standard and NBS-19 were routinely analyzed and α-cellulose samples were run in triplicate. Reproducibility of the standard runs is ± 0.15 2σ. The ± 0.33 2σ standard deviation for our samples most likely reflects some natural variation at the subseasonal scale, as discussed in a later section.

Variations in the cellulose isotopic compositions on an annual and decadal scale preclude direct observation of anomalous isotopic compositions in the time series. To better detect isotopic anomalies in the time series, we employed a 1 year autoregressive moving average model (AR-1). This approach essentially calculates a regression between isotopic values ± 1 year of a given sample, calculating the residuals for that sample (where residual = observed − model predicted value for $\delta^{18}O$). Negative residuals (residual values ≤ -0.5) result from anomalously low $\delta^{18}O$ values in cellulose. AR-1 modeling may result in smaller residual values for successive years recording storm-related isotope depletions (cf. 1880 decade). In a similar way, positive residuals (residual $> +1.0$) results from unusually high $\delta^{18}O$ values such as those resulting from periods of unusually high evaporative demand (hereafter, local "drought"). The calculated LW residuals for 1770–1990 are shown in Figure 5.4. The methodology used to test the efficacy of the hurricane proxy is described in greater detail in the following section. To evaluate drought conditions in the sample area, we compared the results to instrumental measurement of the Palmer Drought Severity Index (PDSI, a measure of meteorologic drought, determined from temperature and precipitation measurements) for 1895–1995. (http://www.ncdc.noaa.gov/paleo/usclient2.html; cell ID 131) and tree-ring reconstructed PDSI for 1770–1894 (same reference). The latter are derived from only a few sample locations in the southeastern United States, all well-removed from the study site. Spectral analysis of the isotope time series was performed using free software available from the National Center for Atmospheric Research that is available and fully documented at: http://www.ncl.ucar.edu/Applications/spec.shtml.

To evaluate the efficacy with which tree-ring $\delta^{18}O$ values capture a record of tropical cyclone activity, the isotopic results were considered in light of 50-year instrumental (or instrumentally derived) records of tropical cyclone intensity and storm track and local precipitation (precipitation records for Valdosta, GA, or other nearby recording stations at http://lwf.ncdc.noaa.gov/). Hurricane intensity and track data are compiled in the HURDAT database (http://www.nhc.noaa.gov/pastall.shtml; Jarvinen *et al.* 1984, Landsea *et al.* 2004). The instrumental data were used to identify dates between 1940 and 1990 in which: (1) a tropical cyclone tracked within ~200 or ~400 km of the study area (Figure 5.3) and (2) the local meteorologic station recorded a precipitation event on the days in which the storm track passed through the geographic target area. Where there was a concomitant record of both a hurricane and rain event, we hypothesized that LW tree-ring cellulose would be ^{18}O-depleted, and the LW residual would be negative. The target zone was arbitrarily set, and is meant to capture all tropical cyclones that had a reasonable likelihood of producing rain at the study site.

IV. RESULTS AND DISCUSSION

A. Testing the Tree-Ring Isotope Proxy Record of Tropical Cyclone Activity

Over the test period, 1940–1990, negative isotope anomalies (≤ -0.5) are identified in 21 years, 18 of which are associated with tropical cyclone events fitting our test criteria. Of the three remaining years, two (1961 and 1974) are associated with tropical cyclones passing outside the 400-km test zone (tropical cyclones Carla and Carmen, respectively), but which nevertheless dropped significant rain

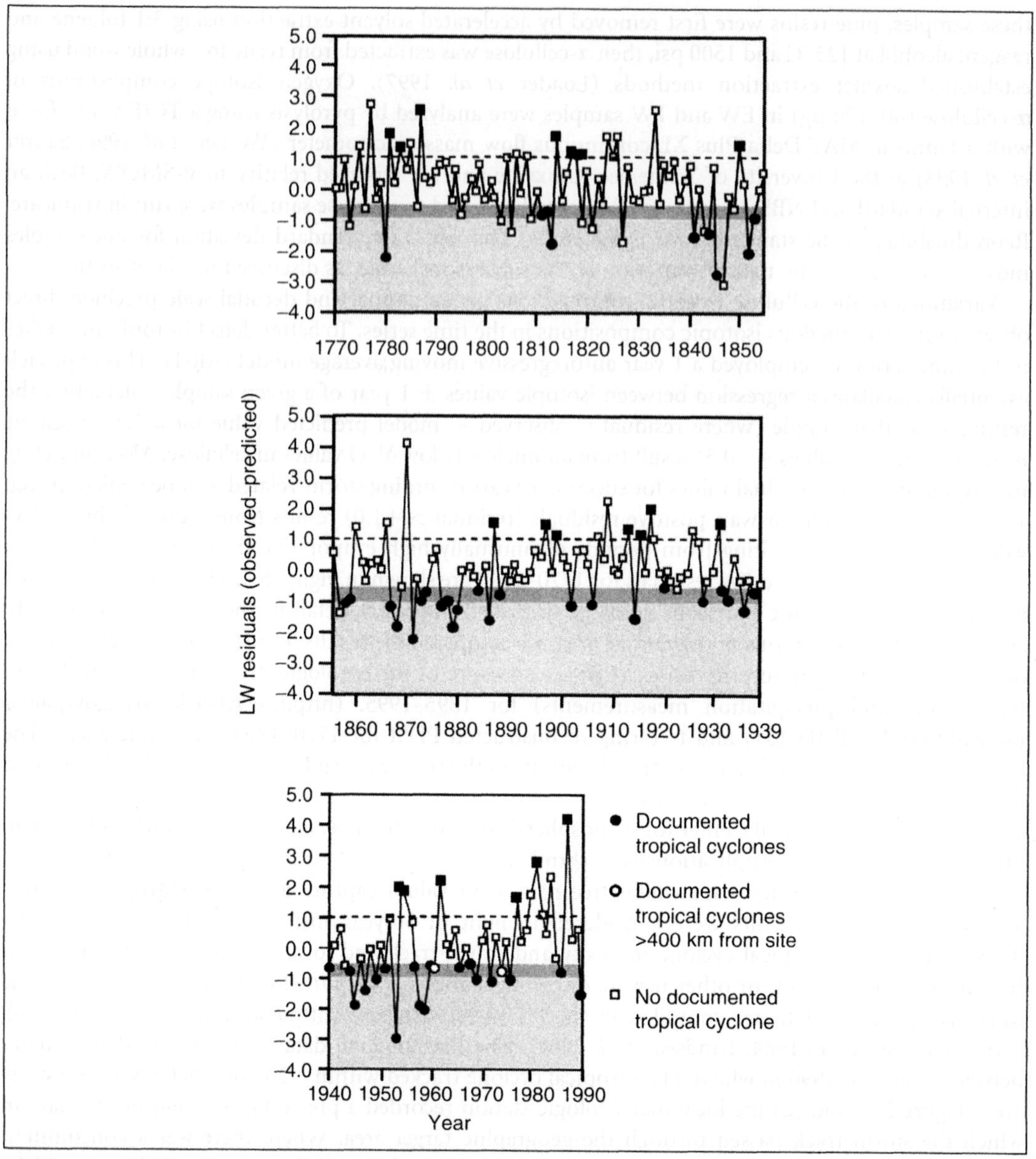

FIGURE 5.4 AR(1) modeling of the LW cellulose isotope compositions (see text). A tropical cyclone event affecting the area is detected by anomalously low oxygen isotope compositions, which appears as a negative LW residual values, ≤−0.5. The proxy was tested using 1940–1990 records of storm tracks and precipitation. Tropical cyclones for which there are other types of records are shown in solid circles. Droughts lead to enrichment of isotopic compositions and positive residual values. Droughts for which there are instrumental or other records are shown in solid squares.

(2.47, 3.37 cm) at the study site. Thus, for the test period, the tree-ring isotope proxy identified every storm known to have affected the study area, with only one "false positive" event, in 1943.

We cannot similarly test the findings for 1770–1940, as daily precipitation data are not available. Comparing the isotope proxy results to HURDAT and/or historical records (cited above), the isotope proxy identified 28 tropical cyclones from 1855 to 1940, only 3 of which cannot be matched to a documented storm. The tree-ring isotope proxy data confirm significant tropical cyclone activity for the southeastern United States between 1865 and 1890, in particular the 1870 decade (http://www.ncdc.noaa.gov; Mock 2004). No event had yet been historically documented for the 1847 and 1857 events indicated by the tree-ring proxy; however, examination of 1857 documents reveals evidence of a previously unreported event (Mock, personal communication), indicating the potential use of isotope proxy data to identify years especially deserving of historical reevaluation. A negative LW residual in 1879 may record Tropical cyclone no. 7, which was a tropical storm at the time of impart near lower Georgia, or it may reflect the effects of a major storm event of unspecified origin (Sandrik and Landsea 2003), which the tree-ring oxygen isotopes indicate was a tropical cyclone. The LW residuals for 1770–1855 suggest 21 events, 7 of which can be matched to the historical record (*e.g.*, events in 1811–13). The period from 1770 to 1800 was marked by relative quiescence, with greater storm activities in the 1810 and 1820 decades and the 1840 decade. The large number of storms suggested by the isotope proxy but not reported in the historical record is not surprising given the much more limited availability of historical documentation for this period. For example, although there is no historical record of a 1780 event in southern Georgia, the tree-ring data suggest that the "Great Tropical Cyclones of 1780," known to have affected Cuba, may also have impacted the US Gulf Coast near the study area (Sandrik and Landsea 2003).

We do not expect perfect correspondence with all known tropical cyclone events for several reasons. Above all, tropical cyclones are dynamic systems and factors affecting tropical cyclone precipitation isotope ratios (*e.g.*, the intensity, duration, and proximity of the storm to the study area; Lawrence *et al.* 1998) change throughout the life of the storm. Tree rings capture only a point in time and space of the event (not unlike a single historical document or instrumental measurement!). Hurricane rains are not concentrically distributed about the eye and may simply miss the sampling site. Because of the ephemeral nature of the tropical cyclone-related isotope depletion in soil water, only a portion of the LW cellulose may incorporate ^{18}O-depleted soil water. The present study utilized material representative of the entire seasonal portion of a ring (*i.e.*, average LW, average EW). Miller *et al.* (2006) showed that further subdividing the LW resulted in a much larger residual value for one single portion of the ring. Thus, averaged seasonal isotope ratios may mask less intense storm activity and, in trees with significant growth rates, higher-resolution sampling may improve the accuracy of the proxy. Finally, concurrent, and opposing, isotope signals associated with drought (see following section) may effectively cancel a storm-related signal. Thus, for the portion of the record without direct instrumental records of precipitation, this proxy best provides positive, rather than negative evidence of an event.

B. A Proxy for Seasonal Drought

Years for which meteorologic drought in the study area is established, based on precipitation records or the PDSI (Section III), are identified as solid squares in Figure 5.4. The high evaporative demand in these years is associated with a greater degree of soil water evaporation, and these events are marked by enriched ^{18}O cellulose compositions and positive residual values (residual value $\geq +1.0$). The data agree reasonably well with instrumental PDSI values, but much worse with tree-ring-reconstructed PDSI values. Some of the conflicting results may reflect the regional scale of the instrumental and reconstructed PDSI data and the more locally controlled evaporative demand represented by the tree-ring compositions.

Tropical cyclone-related isotope effects and moisture stress effects act in opposite directions on the oxygen isotope compositions. Their superposition may lead to "canceling" isotope effects as recorded in cellulose, and can complicate interpretation of the isotope proxy tropical cyclone record, principally by causing the proxy to miss an event. For example, several notable tropical cyclones in the 1890 decade, such as the Sea Islands Tropical cyclone of 1893 and tropical cyclones in 1896 and 1898 (Sandrik and Landsea 2003), are not detected in the proxy record. This decade coincides with PDSI tree-ring reconstruction of mild to severe drought in the study area (http://www.ncdc.noaa.gov/paleo/usclient2.html).

C. Decadal to Multidecadal Scale Variations in Tree-Ring Oxygen Isotopes

Spectral analysis of the EW and LW time series reveals no significant periodicities in EW isotope compositions and a ~5-year periodicity in LW compositions (Figure 5.5). Elsner *et al.* (1999) identify a periodicity of ~5.1 years for North Atlantic tropical hurricanes (*i.e.*, originating in the tropics and not derived from extratropical baroclinic dynamics). As expected, because of the predominant occurrence of tropical cyclones in the LW growing season, the 5-year periodicity is present only in the LW isotope time series.

A portion of the EW and LW time series (1895–2000) are compared to positive (warm North Atlantic SST) and negative (cool North Atlantic SST) phases of the AMO in Figure 5.6. The isotope time series were detrended, smoothed, and low pass filtered to capture only low-frequency variations in the time series (>20 year) and we show a smoothed AMO index (10-year running mean). For the period 1895–1950, a strong negative relationship exists between EW and LW compositions and the smoothed AMO index ($r = -0.66$, $p < 0.001$ for EW and $r = -0.35$, $p < 0.001$ for LW). The inverse relationship between EW oxygen isotope values and AMO appears to breakdown around 1950, after

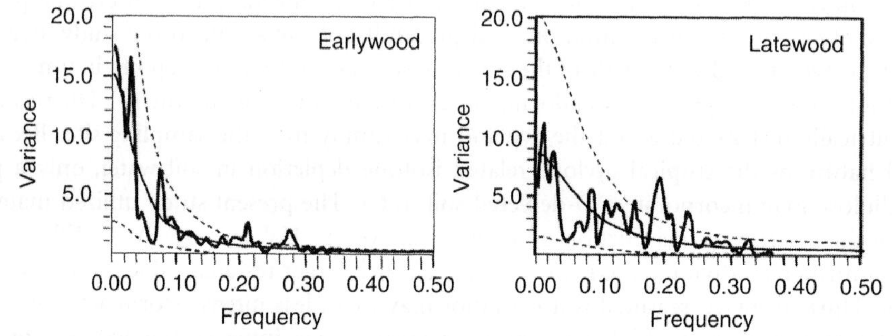

FIGURE 5.5 Spectral analysis of the LW and EW oxygen isotope time series identifies a 5-year periodicity in LW only, corresponding to North Atlantic tropical cyclone occurrence

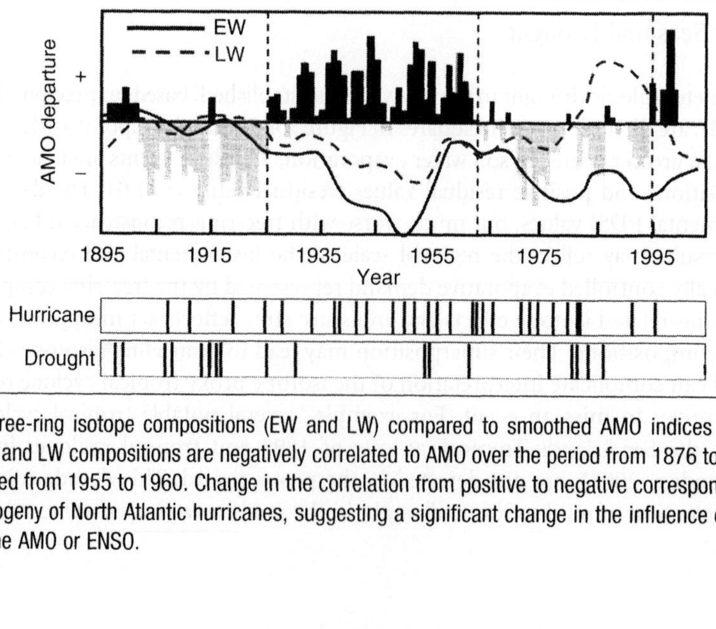

FIGURE 5.6 Tree-ring isotope compositions (EW and LW) compared to smoothed AMO indices (10-year running average). The EW and LW compositions are negatively correlated to AMO over the period from 1876 to ~1960 and then positively correlated from 1955 to 1960. Change in the correlation from positive to negative corresponds to a change in the dominant ontogeny of North Atlantic hurricanes, suggesting a significant change in the influence of remote climate factors such as the AMO or ENSO.

which EW compositions positively correlate to AMO ($r = +0.32$; $p < 0.05$) and diverge in behavior from LW compositions. These correlations suggest that the AMO exerted some influence on climate in the southeastern United States for the period 1895–1955, but was less influential on southeastern climate during 1955–1995, when other climate modes, such as the NAO, PDO or ENSO, may have had greater influence on regional climate and tropical cyclone formation.

It is interesting to note that, despite the strong correlation with the overall time series, there is no apparent relationship between the AMO and periods of hurricane activity. A relatively unusual period (1970–1995) where EW and LW behavior appear to be relatively "decoupled" corresponds to a period of relative quiescence for major North Atlantic hurricanes. The EW and LW time series become similar again in ~1995, when hurricane activity began to pick up again. The physical reasons for these correlations, and why they change, are still largely unknown but they indicate at least a circumstantial relationship between low-frequency climate modes and hurricane occurrence. In contrast, drought events appear most common when the AMO is in a negative (cool) phase.

V. CONCLUSIONS

Few climate proxies offer the exactly datable, intraannual resolution of stable isotope compositions in tree-ring cellulose. This application of stable isotopes suggests that seasonally resolved tree-ring oxygen isotope compositions yield a rich archive of information on long term; natural variability in tropical cyclone occurrence offers a glimpse of the overall climate picture governing storm occurrence. Such results will better inform our understanding of the effects of global climate change on tropical cyclone frequency, and may allow us to differentiate the natural versus anthropogenic components of these changes. The isotope record of tree-ring cellulose presented here supports its use as a new proxy for tropical cyclones and drought extending beyond historical records for the southeastern United States.

It is clearly evident that one site, such as presented here, is not sufficient for a realistic evaluation of tropical cyclones affecting the entire southeast region. Ongoing studies by our group have developed several tree-ring chronologies up to 500 years in length from longleaf pines in several coastal plain sites in the southeastern United States, from the Texas Gulf coast to South Carolina. These chronologies and ongoing isotopic analyses suggest that the isotope proxy works well at other sites, displaying similar low-frequency variation related to climate, but site-specific hurricane records. Thus, the tree-ring proxy method has great potential to develop spatially and temporally extensive records of climate and hurricane occurrence, allowing us to better differentiate natural variation in hurricane frequency from that caused by global warming.

VI. REFERENCES

Anderson, W. T., S. M. Bernasconi, J. A. McKenzie, M. Saurer, and F. Schweingruber. 2002. Model evaluation for reconstructing the oxygen isotopic composition in precipitation from tree ring cellulose over the last century. *Chemical Geology* **182:**121–137.

Delworth, T. L., and M. E. Mann. 2000. Observed and simulated multidecadal variability in the Northern Hemisphere. *Climate Dynamics* **16:**661–676.

Donnelly, J. P., S. Roll, M. Wengren, J. Butler, R. Lederer, and T. Webb. 2001. Sedimentary evidence of intense hurricane strikes from New Jersey. *Geology* **29:**615–618.

Elsner, J. B., and A. B. Kara. 1999. *Hurricanes of the North Atlantic: Climate and Society.* Oxford University Press, New York, 488 p.

Elsner, J. B., A. B. Kara, and M. A. Owens. 1999. Fluctuations in North Atlantic hurricane frequency. *Journal of Climate* **12:**427–437.

Elsner, J. B., T. Jagger, and X. Niu. 2000. Changes in the rates of North Atlantic major hurricane activity during the 20th century. *Geophysical Research Letters* **27:**1743–1746.

Emanuel, K. A. 2005. Increasing destructiveness of totropical cyclones over the past 30 years. *Nature* **436:**686–688.

Frappier, A. B., D. Sahagian, S. J. Carpenter, L. A. Gonzalez, and B. R. Frappier. 2007. Stalagmite stable isotope record of recent tropical cyclone events. *Geology* **35:**111–114.

Gedzelman, S., J. Lawrence, J. Gamache, M. Black, E. Hindman, R. Black, J. Dunion, H. Willoughby, and X. Zhang. 2003. Probing hurricanes with stable isotopes of rain and water vapor. *Monthly Weather Review* **131:**1112–1127.

Goldenberg, S. B., and L. J. Shapiro. 1996. Physical mechanisms for the association of El Nino and West African rainfall with the Atlantic major hurricane activity. *Journal of Climate* **9:**1169–1187.

Goldenberg, S. B., C. W. Landsea, A. M. Mestas-Nunez, and W. M. Gray. 2001. The recent increase in Atlantic hurricane activity: Causes and implications. *Science* **293:**474–479.

Gray, W. M., J. D. Sheaffer, and C. W. Landsea. 1997. Climate trends associated with multidecadal variability of Atlantic hurricane activity. Pages 15–53 *in* H. F. Diaz and R. S. Pulwarty (Eds.) *Hurricanes: Climate and Socioeconomic Impacts.* Springer-Verlag, New York.

Grissino-Mayer, H. D., H. C. Blount, and A. C. Miller. 2001. Tree-ring dating and the ethnohistory of the naval stores industry in southern Georgia. *Tree-Ring Research* **57:**3–13.

Jarvinen, B. R., C. J. Neumann, and M. A. S. Davis. 1984. *A Tropical Cyclone Data Tape for the North Atlantic Basin, 1886–1983: Contents, Limitations, and Uses.* NOAA Technical Memo. NWS NHC **22:**21 p.

Knight, J. R., C. K. Folland, and A. A. Scaife. 2006. Climate impacts of the Atlantic Multidecadal Oscillation. *Geophysical Research Letters* **33:**L17706, doi:10.1029/2006GL026242.

Knutson, T. R., and R. E. Tuleya. 2004. Impact of CO_2-induced warming on simulated hurricane intensity and precipitation: Sensitivity to the choice of climate model and convective parameterization. *Journal of Climate* **17:**3477–3495.

Landsea, C. W. 1993. A climatology of intense (or major) Atlantic hurricanes. *Monthly Weather Review* **121:**1703–1713.

Landsea, C. W., G. D. Bell, W. M. Gray, and S. B. Goldenberg. 1998. The extremely active 1995 Atlantic hurricane season: Environmental conditions and verification of seasonal forecasts. *Monthly Weather Review* **126:**1174–1193.

Landsea, C. W., C. Anderson, N. Charles, G. Clark, J. Dunion, J. Fernandez-Partagas, P. Hungerford, C. Neumann, and M. Zimmer. 2004. The Atlantic hurricane database re-analysis project: Documentation for the 1851–1910 alterations and additions to the HURDAT database. Pages 177–221 *in* R. J. Murname and K.-B. Liu (Eds.) *Hurricanes and Typhoons: Past, Present and Future.* Columbia University Press, New York.

Lawrence, J. R. 1998. Isotopic spikes from tropical cyclones in surface waters: Opportunities in hydrology and paleoclimatology. *Chemical Geology* **144:**153–160.

Lawrence, J. R., and S. D. Gedzelman. 1996. Low stable isotope ratios of tropical cyclone rains. *Geophysical Research Letters* **23:**527–530.

Lawrence, J. R., S. D. Gedzelman, X. Zhang, and R. Arnold. 1998. Stable isotope ratios of rain and vapor in 1995 hurricanes. *Journal of Geophysical Research* **103:**11381–11400.

Lawrence, J. R., S. D. Gedzelman, J. Gamache, and M. Black. 2002. Stable isotope ratios: Hurricane Olivia. *Journal of Atmospheric Chemistry* **41:**67–82.

Liu, K. B. 2004. Paleotempestology: Principles, methods, and examples from Gulf Coast lake sediments. Pages 13–57 *in* R. J. Murnane and K.-B. Liu (Eds.) *Hurricanes and Typhoons: Past, Present, and Future.* Columbia University Press, New York.

Liu, K.-B., and M. L. Fearn. 1993. Lake-sediment record of late Holocene hurricane activities from coastal Alabama. *Geology* **21:**793–796.

Loader, N. J., I. Robertson, A. C. Barker, V. R. Switsur, and J. S. Waterhouse. 1997. An improved technique for the batch processing of small wholewood samples to cellulose. *Chemical Geology* **136:**313–317.

Mann, M. E., and K. A. Emanuel. 2006. Global warming, the AMO, and North Atlantic tropical cyclones. *Eos* **87:**233–244.

Miller, D. L., C. I. Mora, H. D. Grissino-Mayer, C. J. Mock, M. E. Uhle, and Z. D. Sharp. 2006. Tree ring isotope records of tropical cyclone activity. *Proceedings of the National Academy of Sciences of the United States of America* **103:**14294–14297.

Mock, C. J. 2004. Tropical cyclone reconstructions from documentary records: Examples from South Carolina. Pages 121–148 *in* R. J. Murnane and K.-B. Liu (Eds.) *Hurricanes and Typhoons: Past, Present, and Future.* Columbia University Press, New York.

Pielke, R. A., and C. W. Landsea. 1998. Normalized hurricane damages in the United States: 1925–1995. *Weather and Forecasting* **13:**621–631.

Pielke, R. A., Jr., C. Landsea, M. Mayfield, J. Laver, and R. Pasch. 2005. Hurricanes and global warming. *Bulletin of the American Meteorological Society* **86:**1571–1575.

Roden, J. S., G. Lin, and J. R. Ehleringer. 2000. A mechanistic model for interpretation of hydrogen and oxygen isotope ratios in tree-ring cellulose. *Geochimica et Cosmochimica Acta* **64:**21–35.

Sandrik, A., and C. W. Landsea. 2003. Chronological listing of tropical cyclones affecting north Florida and coastal Georgia 1565–1899. *NOAA Technical Memorandum* NWS SR-224 Southern Region, National Weather Service, NOAA, 76 p.

Saurer, M., S. Borella, and M. Leuenberger. 1997. ^{18}O of tree rings of beech (*Fagus sylvatica*) as a record of ^{18}O of the growing season precipitation. *Tellus B* **49:**80–92.

Saurer, M., I. Robertson, R. Seigwolf, and M. Leuenberger. 1998. Oxygen isotope analysis of cellulose: An interlaboratory comparison. *Analytical Chemistry* **70:**2074–2080.

Sternberg, L., S. S. Mulkey, and S. J. Wright. 1989. Oxygen isotope ratio stratification in a tropical moist forest. *Oecologia* **81:**51–56.

Stokes, M. A., and T. L. Smiley. 1996. *An Introduction to Tree-ring Dating.* The University of Arizona Press, Tuscon. 73 p.

Tang, K., and X. Feng. 2001. The effect of soil hydrology on the oxygen and hydrogen isotopic compositions of plants' source water. *Earth and Planetary Science Letters* **185:**355–367.

Webster, P. J., G. J. Holland, J. A. Curry, and H.-R. Chang. 2005. Changes in tropical cyclone number, duration, and intensity in a warming environment. *Science* **309:**1844–1846.

Weiguo, L., F. Xiahong, L. Yu, Z. Qingle, and A. Zhisheng. 2004. $\delta^{18}O$ values of tree rings as a proxy of monsoon precipitation in arid Northwest China. *Chemical Geology* **206:**73–80.

Werner, R. A., B. E. Kornexl, A. Rossman, and H. L. Schmidt. 1996. On-line determination of ^{18}O values of organic substances. *Analytica Chimica Acta* **319:**159–164.

CHAPTER 6

The Stable Isotopes $\delta^{13}C$ and $\delta^{18}O$ of Lichens Can Be Used as Tracers of Microenvironmental Carbon and Water Sources

Michael Lakatos,[*,†] Britta Hartard,[*,†] and Cristina Máguas[†]

*Department of Ecology, University of Kaiserslautern
†Stable Isotopes Laboratory, ICAT/CEBV, FCUL University of Lisbon

Contents

I. INTRODUCTION

During the past 10 years, the application of stable isotope techniques has provided the scientific community with new insights into understanding physiological and ecological processes. Many of these studies have focused on higher plants (reviewed in Yakir and Sternberg 2000, Dawson *et al.* 2002) whose discrimination processes during photosynthetic and water exchange are now well understood (Brugnoli and Farquhar 2000; Chapters 1 and 2). In contrast, only a few number of publications focus on algae, cyanobacteria, and lichens, and several uncertainties still exist despite the global distribution of these photoautotrophic organisms in aquatic and terrestrial ecosystems. Within this large group of organisms, lichens are one of the most important. For example, ~8% of the earth's land surface is covered by vegetation types quantitatively dominated by lichens (Larson 1987). Lichens occupy various microhabitats such as leaves, bark, rocks, soil, or even anthropogenic substrates.

Due to their direct dependency on environmental conditions, lichens' stable isotope compositions reflect changes of carbon and water as an integral over a long period and on a microenvironmental scale. Their poikilohydric nature enables them to settle under environmental conditions where higher plants are unable to survive and also to assimilate extraordinary substrates such as vapor or carbon microresources which are not commonly utilized by higher plants. This chapter introduces how lichens' $\delta^{13}C$ can be used as tracers for carbon acquisition, environmental change of CO_2 sources, and global change, while $\delta^{18}O$ of thallus water and respired CO_2 operates as a tracer for varying water sources to serve as an environmental integrator and recorder for soil–atmosphere exchange processes.

Lichens are generally regarded as one of the most outstanding examples of symbiosis between a fungal "mycobiont" and a photosynthetic algal and/or cyanobacterial "photobiont" partner. Most of the lichen mycobionts, ~85%, are symbiotic with green algae, 10% with cyanobacteria, and 3–4% are simultaneously associated with both green algae and cyanobacteria (Honegger 1997). The photobionts are located extracellularly within the lichen thallus, arranged either as a distinct algal layer in internally stratified (heteromerous) thalli or randomly distributed in nonstratified (homeomerous) thalli. The morphology and anatomy of heteromerous thalli is highly diverse, but basically they are composed of three layers (Figure 6.1A): outer cortex (conglutinate zones), photobiont cells (algal layer), and aerial hyphae (medulla). They create foliose (leaf-like) and fruticose (shrub-like) growth forms, whereas homeomerous thalli are mainly found as crustose growth forms (reviewed in Honegger 1997).

Lichens are also an interesting example of a photoautotrophic system because even though some lichen physiological behavior is quite similar to that of higher plants, other processes (*e.g.*, water exchange mechanisms) are fundamentally different. As they are unable to control their water supply, their water status varies passively with surrounding environmental conditions. Hence, lichens repeat frequent cycles of drying and wetting (poikilohydry) depending on the availability of rain, dew, or in some cases even high humidity (Lange *et al.* 2001). In equilibrium with humid air, both green algal and cyanobacterial lichens reach approximately the same water content (WC) (Schlensog *et al.* 2000), but cyanobacterial lichens require much higher WC to gain positive photosynthesis, which cannot be attained by water vapor uptake (reviewed in Kappen and Valladares 1999). Thus, cyanobacteria require liquid water to become photosynthetically active (Bilger *et al.* 1989), whereas green algae can be activated by high humidity alone (Lange *et al.* 1986). In general, many lichens are able to maintain photosynthesis despite wide changes in thallus WC, but above certain WCs, supersaturation of the thallus causes a decrease of net photosynthesis (Lange *et al.* 2001) due to CO_2 diffusion limitation (Cowan *et al.* 1992). Carbon acquisition in lichens is directly related to the photobiont's photosynthesis (reviewed in Palmqvist 2000) which supplies the mycobiont with glucose or polyols (reviewed in Honegger 1997). The photosynthetic capacities of the photobionts are adapted to environmental conditions of light, temperature, and nutrition (Palmqvist 2000). To reduce CO_2 limitation, some photobionts have the ability to accumulate inorganic carbon to increase the CO_2 concentration near

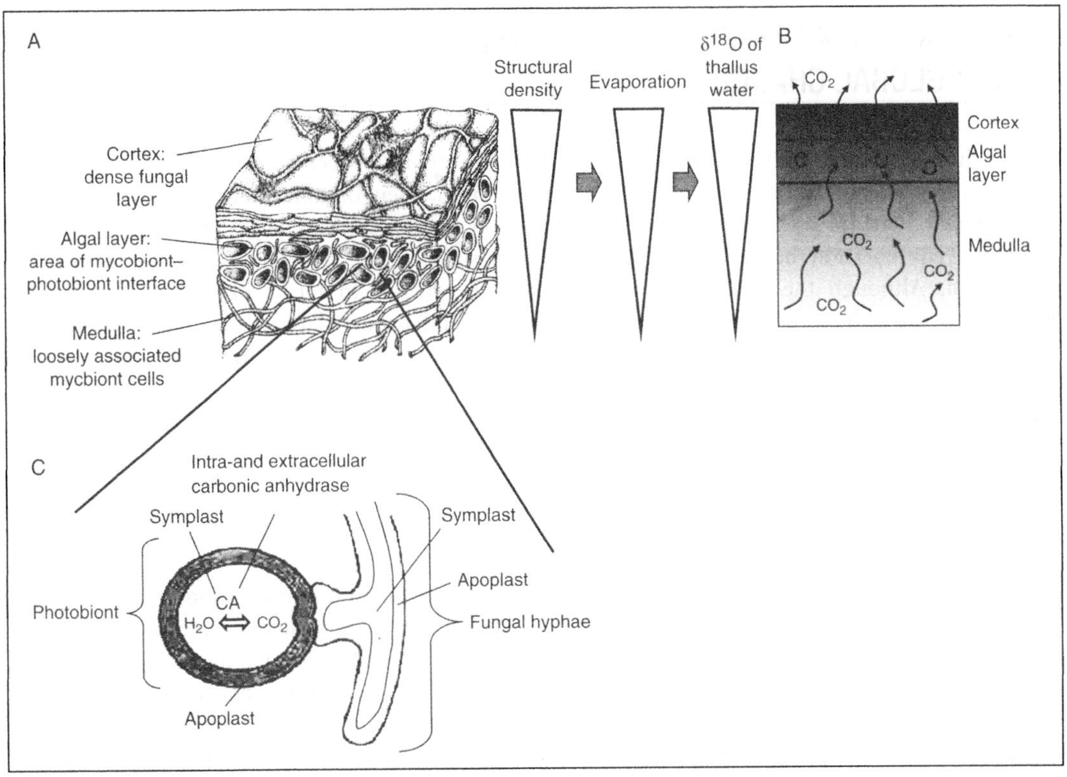

FIGURE 6.1 Schematic illustration of a stratified lichen thallus. (A) Illustrated cross section of a lichen thallus with distinctive cortex, algal layer, and medulla. Also shown are the causalities between structural density, evaporation rates, and expected thallus water $\delta^{18}O$; (B) suggested model of $\delta^{18}O$ of respiratory CO_2 diffusing through the thallus which is linked to thallus water $\delta^{18}O$ (gradual shading indicates increasing $\delta^{18}O$ enrichment along the thallus). Isotopic composition of upper thallus regions is reflected in $\delta^{18}O$ of respired CO_2 due to CA catalyzed $CO_2 \leftrightarrow H_2O$ equilibration; (C) detailed and enlarged scheme of the interface between photobiont and mycobiont within the algal layer with an example of catalyzing CA which exists in photobiont cells and fungal hyphae.

the side of Rubisco, which also decreases photorespiration. This mechanism is called CO_2-concentrating mechanism (CCM) and is associated with the presence of a specific cell structure, the pyrenoids in green algae, and the carboxysomes in cyanobacteria. However, some green algal photobionts lack a CCM and do not contain pyrenoids (Palmqvist *et al.* 1994a, Máguas *et al.* 1995).

Due to their unique characteristics as well as their cosmopolitan distribution (Kappen 1988, 1993), the use of lichens as biomonitoring organisms has become highly attractive. They are sensitive to a wide range of environmental changes, with both natural and human origin. Human activities such as urban and industrial pollution cause a reduction in lichen biodiversity. Numerous worldwide studies of lichen diversity confirm that sensitive species decline in polluted areas, while resistant and tolerant species remain (Nimis *et al.* 1991, Branquinho *et al.* 1999, Purvis *et al.* 2003). Moreover, lichens may also be used as good indicators of complex environmental changes that are normally associated with forestry and land-use changes (Asta *et al.* 2002, Ellis and Coppins 2006).

Our research evaluates the ecological value of poikilohydric lichens that substantially differ from homoiohydric vascular plants, as tracers of environmental factors associated with microhabitats and climate changes. For that, variations in carbon sources, water exchange, and carbon dioxide fluxes were assessed by $\delta^{13}C$ and $\delta^{18}O$ not only in dry material but also in respired CO_2 and thallus water.

II. LICHEN $\delta^{13}C$ AS TRACER FOR CARBON ACQUISITION, CARBON SOURCE, AND GLOBAL CHANGE

Since lichens grow slowly (0.5–5 mm $year^{-1}$) and are directly influenced by microclimatic conditions such as light, water, temperature, and CO_2 concentration, their organic material (OM) integrates environmental factors at their specific microhabitat over a long period. The OM is determined by an economic equilibration between carbon source and sink, which are mainly photosynthesis and respiration. Although the mycobiont predominates the thallus in terms of biomass, its carbon is acquired through photosynthesis in the photobiont and exported from the photobiont to the mycobiont. Thus, the carbon acquisition of the lichen depends on the WC, light intensity, and CO_2 fixation of the photobiont. Discrimination processes of ^{13}C can thus be related to CO_2 acquisition modes, CO_2 diffusion, and CO_2 sources.

A. Physiological Uptake Processes of Carbon

In general, the carbon isotope ratios ($\delta^{13}C$) of a plant's OM depend on fractionation processes during the photoautotrophic uptake of CO_2. The range of discrimination is related to the balance between carboxylation and diffusion limitation (*i.e.*, stomatal conductance) and through the ratio of internal to external partial carbon dioxide pressure. Uptake of CO_2 in lichens, on the other hand, may be more transient, as here it is a function of thallus WC and, thus, regulated by atmospheric humidity, dewfall, and precipitation. In contrast to homoiohydric plants, poikilohydric lichens, without the possibility of regulating WC through stomata, operate in a delicate equilibrium: sufficient water to reactivate the photobiont photosynthesis, and CO_2 uptake limited by diffusion in a water-saturated thalli.

Previous studies suggested that $\delta^{13}C$ of lichens is governed by the photobiont association, at which the variety of $\delta^{13}C$ within a photobiont group represents different degrees of diffusion limitation (Lange and Ziegler 1986, Lange *et al.* 1988). Later, this theory was complemented by the findings that $\delta^{13}C$ in lichens resulted from a combination of species-specific differences in resistances to inward CO_2 fluxes (Máguas and Brugnoli 1996, Máguas *et al.* 1997) and the CO_2-fixation mechanism of their photobiont (Máguas *et al.* 1993, 1995, Smith and Griffiths 1996, 1998, Smith *et al.* 1998). Recent studies confirm that basically two categories of lichen are identifiable (Lakatos *et al.* unpublished data), depending on the existence or absence of a CCM in the primary photobiont, revealing a significant difference of an average 10‰ between both groups (Figure 6.2; $p = 0.000$, $n = 230$). Photobionts such as cyanobacteria or green algae with CCM increase the internal carbon-pool near the carboxylation site of Rubisco. As a consequence of the increased substrate availability, the rate of carboxylation is increased while that of photorespiration is decreased. Moreover, the CCM tempers the effect of CO_2 diffusion resistance within the thallus, for example, at high WCs or supersaturation (Máguas *et al.* 1997). Thus, $\delta^{13}C$ of CCM containing lichens varies between −16‰ and −26‰ with a mean around −22‰ (Figure 6.2). In tripartite lichens with two photobionts, green alga without CCM and cyanobacteria with CCM—the latter organized in specialized thallus structures called cephalodia—the organic $\delta^{13}C$ is determined by the algal partner (Máguas *et al.* 1995, Green *et al.* 2002). Thus, OM of these tripartite lichens is characterized by $\delta^{13}C$ similar to C3 plants (mean of 18 species -32 ± 1.6‰; $n = 43$) since their green algae, such as *Myrmecia, Dictyochloropsis* (Smith and Griffiths 1996), and *Coccomyxa* (Palmqvist *et al.* 1994b), lack both pyrenoids and CCM (Figure 6.2). Similar values can be found in lichens with only one photobiont: the green algae *Trentepohlia,* which lack CCM (Lakatos *et al.* 2006), being the predominant photobiont in the tropics and subtropics (Sipman and Harris 1989). Thus, the presence or absence of a CCM explains basic $\delta^{13}C$ difference between two different functional groups (Figure 6.2).

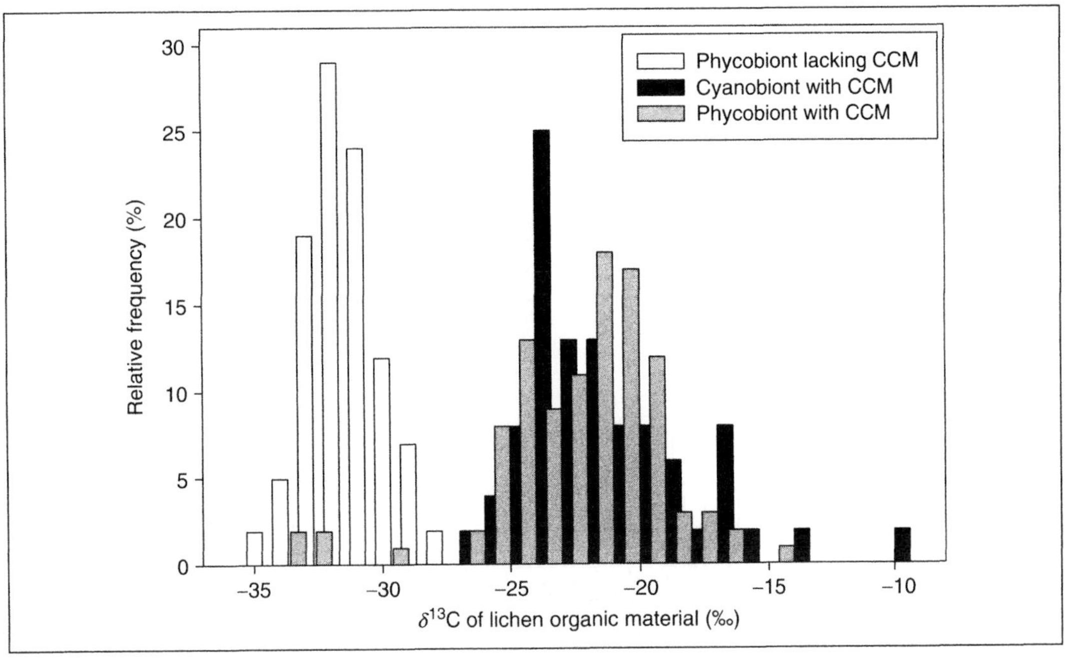

FIGURE 6.2 Relative frequency of $\delta^{13}C$ in OM of different lichens (species > 70; *n* > 250) classified with regard to presence or absence of a CCM in green algal and cyanobacterial photobionts. Outliers are the tropical corticolous *Phyllospora corallina* (exposed to bark respiration) and the marine cyanolichen *Lichina pygmaea* (often water saturated). Data from Lange *et al.* (1988), Máguas *et al.* (1995, 1997), Lakatos *et al.* (2006, unpublished data).

B. Morphology and Resistance to CO_2 Diffusion

Morphology, mass allocation, and physiology of the lichen thallus also vary with size (Larson 1987). As a foliose thallus increases in size, its surface to volume ratio decreases and its structural and physiological intrathalline variability amplifies. Moreover, it is generally accepted that foliose and fruticose lichens possess an apical or marginal growth and, therefore, the marginal tips are very active in comparison to the inner parts. This common growth pattern combined with the large structural and functional intrathalline differences found between marginal and central zones in large thalli led to the hypothesis of a certain division of functions within the same thallus: marginal zones maximize gas exchange, while central zones maximize water storage (Valladares *et al.* 1994, Máguas and Brugnoli 1996). Moreover, along the same thallus, CO_2 diffusion resistances could also be enhanced by morphological structures such as the conglutinate cortical layer, thallus thickness and density, and concomitant structural changes during water absorption. Hence, thinner thallus parts such as margins or tips, which are often rather loosely constructed, therefore displaying less CO_2 diffusion resistances, should also lead to less $\delta^{13}C$ discrimination (depleted $\delta^{13}C$) in comparison to central thallus structures. Our results confirmed that, provided a significant difference between marginal and central areas of the same thallus, the margins always present a higher $\delta^{13}C$ (more depleted in ^{13}C). Within the different functional lichen groups, it was possible to identify differences in 50% of the phycobiont lichen species (with green algae) with and without CCM (Figure 6.3).

However, the observed high heterogeneity of $\delta^{13}C$ of margins versus center parts within different growth forms and photobiont groups (from 0.25‰ to 2.5‰) did not give a clear indication of the expected discrimination factor associated with morphology and thallus structure, and no correlation with respect to growth form could be assessed (Lakatos *et al.* unpublished data). We may speculate that in lichens with a very active CCM (as in cyanobacteria lichens), $\delta^{13}C$ will be maintained throughout the thallus due to

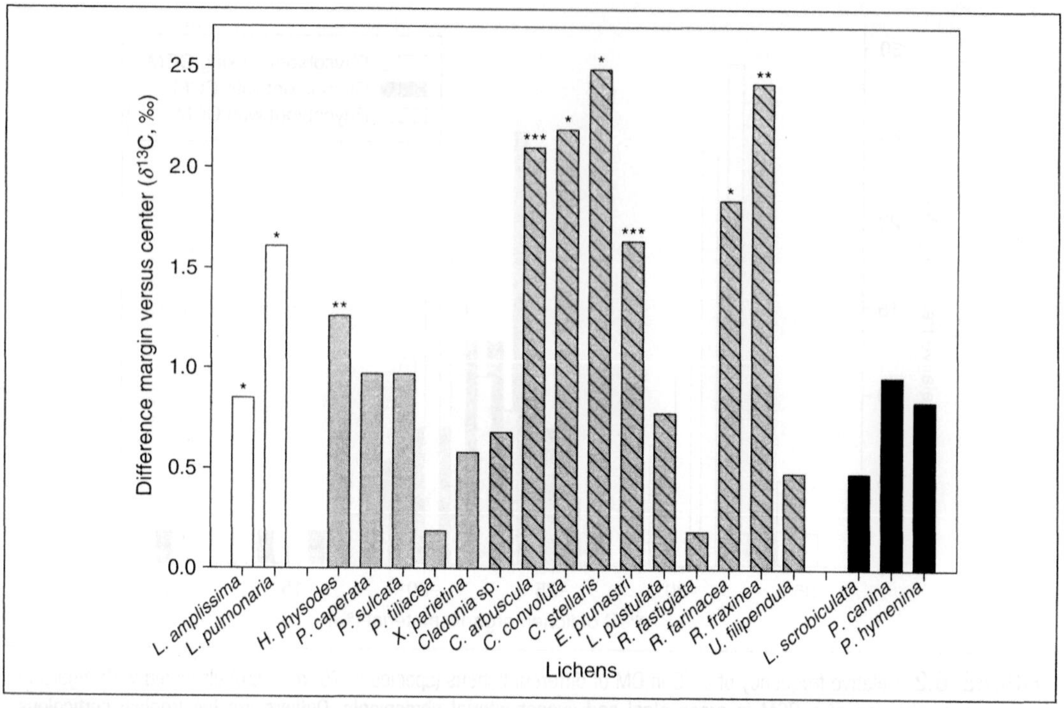

FIGURE 6.3 Difference in $\delta^{13}C$ of OM of spatial discrete lichen parts: thin margin versus thick center (n = 3–6; $*p < 0.05$, $**p < 0.01$, $***p < 0.001$). Classification into the three groups of photobionts, the first without (open bars) and two with CCM (phycobionts with gray and cyanobiont with black bars) regarding to foliose or fruticose (striped pattern) growth forms. Modified after Máguas and Brugnoli (1997, unpublished data).

relatively high and constant CO_2 concentrations. Thus, the $\delta^{13}C$ variation within a lichen thallus, by itself, does not seem to be a reliable tracer to predict gas diffusion resistances. Instead, we may just conclude that a clear decrease in $\delta^{13}C$ across the same lichen thallus may trace an increase in resistance to CO_2 diffusion, causing differences up to 2.5‰ (Figure 6.3).

C. Carbon Source Influences $\delta^{13}C$ in Microhabitats

Besides the major photosynthetic fractionations due to the transport and fixation of CO_2, one crucial factor influencing $\delta^{13}C$ is the origin of the carbon source used by lichens at a specific microenvironment. Several factors may contribute to this: (1) lichens can be attached to a substrate where the most direct CO_2 source is not atmospheric CO_2, and (2) respired CO_2 from different components within the ecosystem will be preferentially fixed by lichens, depending on where they are located. For example, it is well established that in macrohabitats such as closed forests, ambient CO_2 gradually changes (CO_2 profile) from the forest floor to the canopy (Sternberg *et al.* 1989, Buchmann *et al.* 1997). Additionally on a microscale, CO_2 deriving from differently respiring substrates (*e.g.*, soil, tree bark, leaf) can process $\delta^{13}C$ signals more depleted than that of ambient air.

To evaluate whether lichen $\delta^{13}C$ is influenced by these processes, lichens from different substrates, such as terricolous (soil), corticolous (bark), or pendulous growing (ambient air exposed) species, were studied in an open Mediterranean coastal sand dune habitat with substantial lichen ground cover. In addition, to evaluate the impact of natural CO_2 profiles, lichens growing on differently respiring trees (phorophytes) were studied at a tropical rain forest in French Guiana and Panama (Lakatos *et al.* 2006, submitted for publication).

The latter studies showed that the $\delta^{13}C$ of OM of adjacent corticolous lichens differed significantly, depending on whether they were more air exposed or closely attached to the bark. Attached lichen thalli utilized bark-respired CO_2, which is more depleted than ambient air $^{13}CO_2$. Additionally, corticolous lichens growing on trees within the forest differ in $\delta^{13}C$ with respect to their vertical distribution (middle story versus understory), thus confirming that the exposition to a pronounced ambient CO_2 and $\delta^{13}C$ profile within tropical forests significantly influences lichen's $\delta^{13}C$.

The CO_2 profile observed within the first 10 cm above the soil surface of Mediterranean sand dune vegetation (Portugal) also showed a highly pronounced gradient and $\delta^{13}C$ signals that were strongly influenced by substrate structure and air turbulence. In this ecosystem, the carbon source of terricolous lichens is mainly influenced by depleted soil respiration while pendulous growing species fix the enriched ambient air CO_2. Hence, $\delta^{13}C$ of OM of the former lichens was significantly more depleted compared to the latter species. Briefly, corticolous lichens are primarily exposed to bark-respired CO_2 ($\delta^{13}C$: $\sim -30‰$), terricolous to soil-respired CO_2 ($\sim -25‰$ to $-16‰$), and epilithic as well as pendulous species to ambient CO_2 ($\sim -8‰$). Moreover, the general trend exists that lichens with CCM are less masked in $\delta^{13}C$ by this source effect compared to those without CCM (Figures 6.4 and 6.5).

To summarize, in the studied microhabitats respired, $\delta^{13}C$-depleted CO_2 serves as carbon source for photosynthesizing lichens, thus biasing their characteristic isotopic signatures, which otherwise are determined by physiological processes. Thus, lichens can be used as tracers to point out the prevailing CO_2 sources of microhabitats. Therefore, lichens' OM would also indicate—especially in the younger thallus parts—an alteration of ambient $^{13}CO_2$ by changes of urban–rural and land-use boundaries.

D. Atmospheric Carbon Source Influences Lichen $\delta^{13}C$ over Decades

In order to investigate the effects of the decreasing $\delta^{13}C$ of atmospheric CO_2 caused by anthropogenic activities following the industrial revolution, herbarium samples of the tripartite foliose lichen *Lobaria pulmonaria* collected during 1846, 1923, 1945, 1953, and 1989 were analyzed (Table 6.1) (Máguas and Brugnoli 1996). The isotope composition of the lichens becomes progressively more depleted in ^{13}C from 1846 to 1989. The similarly observed changes in both the $\delta^{13}C$ of lichen samples and depletion in the atmospheric source CO_2 clearly indicate that the observed variations correlated with fossil fuel combustion and the consequent changes in $\delta^{13}C$ of the atmospheric CO_2. Additionally, irrespective of the collection date, a clear distinction was observed between marginal and central parts of this lichen

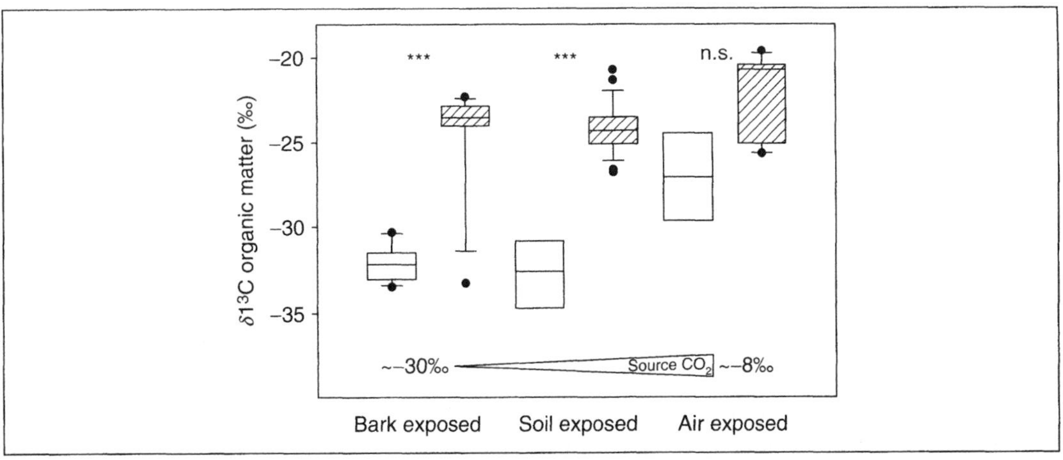

FIGURE 6.4 Box-Whisker plot: $\delta^{13}C$ of lichen bulk OM from different regions (Median, Quartile, Range, Outlier, $n = 3$–22; asterisks indicate $p < 0.001$). Classification regarding the exposition to CO_2 source into corticolous, terricolous, and ambient lichen groups with respect to CCM presence (striped pattern).

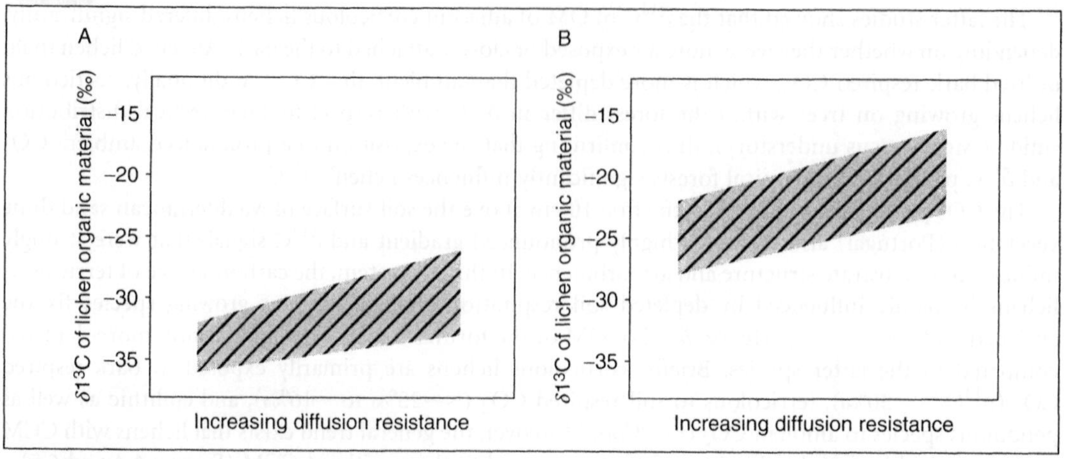

FIGURE 6.5 General model for $\delta^{13}C$ discrimination of poikilohydric organisms. The three discriminating factors are CCM absence (A) or presence (B) with $\delta^{13}C$ limit around −28‰, CO_2 diffusion limitation, and sources of CO_2— atmospheric CO_2 enriches while respired CO_2 depletes $\delta^{13}C$ from lichen organic matter (filled areas indicate the main $\delta^{13}C$ ranges of lichen organic matter).

TABLE 6.1 Variation in carbon isotope composition ($\delta^{13}C$) during the period 1846–present

Date	$\delta^{13}C$ source air (‰)	$\delta^{13}C$ whole thallus (‰)	Lichen thallus zone	Δ (‰)
1846	−6.5	−29.3	Margin	24.1 ± 0.2
			Center	22.3 ± 0.3
1923	−6.8	−29.4	Margin	24.8 ± 0.2
			Center	22.7 ± 0.2
1945	−6.8	−29.3	Margin	24.0 ± 0.4
			Center	22.9 ± 0.1
1953	−6.8	−29.7	Margin	25.5 ± 0.4
			Center	22.1 ± 0.3
1989	−8.0	−30.6	Margin	24.1 ± 0.1
			Center	22.6 ± 0.3

Source air CO_2 data from Friedli and coworker (1986) and lichen thallus data (herbarium samples of *Lobaria pulmonaria*) from Máguas and Brugnoli (1996).

species (Table 6.1). Carbon isotopic composition (Δ) was always higher (more depleted in ^{13}C) in marginal than in central lichen regions. These observations demonstrate that variations in $\delta^{13}C$ of herbarium thalli do not reflect physiological response to increasing CO_2 concentration, with variations in $\delta^{13}C$ among different collection periods entirely dependent on changes in $\delta^{13}C$ of the atmospheric source air CO_2. This evidence may be relevant for the use of lichens in studies related to global change, in conjunction with other studies, such as tree rings.

In conclusion, the $\delta^{13}C$ of lichen material demonstrates high heterogeneity and clear interactions between physiological, morphological, and sources effects. Lichen $\delta^{13}C$ can therefore be used as a tracer for carbon acquisition, environmental change of CO_2 sources, and global change. The presence or absence of a CCM in the primary photobiont accounts for a general difference of ~10‰. The highest natural variance observed within one species was up to 4‰. Outliers demonstrate that high CO_2 limitation by diffusion resistance can lead to a discrimination of 6–8‰. The influence of CO_2 source can create variances between free living and lichenized photobionts up to 13‰. As an orientation for tracing effects, a general model for $\delta^{13}C$ discrimination in poikilohydric lichens taking demonstrated factors into account is suggested. The three main discriminating factors for the carbon stable isotopic composition ($\delta^{13}C$) of OM of lichens are

(1) CCM presence ($\delta^{13}C$ frontier around $-28 \pm 2‰$), (2) CO_2 limitation by diffusion resistance, and (3) utilization of a CO_2 source different than well-mixed atmospheric CO_2 (Figure 6.5).

III. OXYGEN ISOTOPIC COMPOSITION OF THALLUS WATER AND RESPIRED CO_2: A TRACER FOR VARYING WATER SOURCES?

As mentioned earlier, poikilohydric cryptogams mainly influence water and CO_2 fluxes of an ecosystem but, compared to higher plants, they exhibit different mechanisms of water exchange. Their water status, for example, varies passively with surrounding environmental conditions, and they have neither a continuous influx of water nor stomata to control water deficit. Hence, during evaporation no isotopic steady state can be achieved and the $\delta^{18}O$ composition of both the thallus water and the evaporated water is expected to show progressive enrichment similar to the Rayleigh distillation process. As the water also transduces its oxygen isotopic signal to CO_2 via hydration of dissolved CO_2 (Amundson *et al.* 1998, Tans 1998, Stern *et al.* 1999), respired CO_2 should also reflect the oxygen isotopic composition of the thallus water.

To reveal the yet unknown basic fractionation processes taking place in lichens which lead to temporal and spatial $\delta^{18}O$ variations, the following two paragraphs focus on the fruticose (shrub-like) terricolous lichen *Cladina arbuscula* (reindeer lichen) as a representation, studied under controlled and natural conditions during desiccation periods (Hartard *et al.* 2007). In the genus *Cladina*, the lower medulla is glutinated, building a cylindric structure which then forms shrub-like cushions: the thicker stems close to the ground branch out into several thinning tips which become increasingly air exposed. In its natural habitat, these extremely dense cushions may cover up to 80% of the soil surface, for example, at Mediterranean coastal sand dune and boreal habitats. Due to its ecological importance as ground cover and nutrition resource (*e.g.*, for reindeer), field studies were conducted showing how their oxygen isotopic signal is influenced by varying environmental conditions at their natural habitat and how this ground cover influence soil–atmosphere water exchange processes (Hartard *et al.* 2007).

A. $\delta^{18}O$ Enrichment of Thallus Water and CO_2 During Desiccation

Laboratory experiments on *C. arbuscula* showed that the $\delta^{18}O$ of lichen thallus water became progressively enriched by about 7‰ over the course of a desiccation period (Figure 6.6). In general, the course of this enrichment showed two phases. At high WCs, the increase of enrichment appeared to be faster, whereas this enrichment diminished when proceeding dehydration slowed down (Figure 6.6A). This is probably mainly due to the progressive decline in lichen water potential ($\psi_{thallus}$). At water saturation, $\psi_{thallus}$ is much larger compared to ψ of the surrounding air. This large difference also causes high evaporative fractionation. Concurrently with the decrease in lichen WC, $\psi_{thallus}$ also declines and approaches that of the surrounding atmosphere. Consequently, the evaporative enrichment of thallus water also becomes less pronounced. At the same time, the isotopic exchange between the isotopically lighter surrounding water vapor and the remaining thallus water starts to become more apparent. This impact also contributes to the observed decline of thallus water enrichment.

The respired lichen CO_2 showed a similar enrichment pattern: revealing more pronounced $\delta^{18}O$ enrichment at high WCs (Figure 6.6B). Still, the comparison of $\delta^{18}O$ of respired CO_2 and of thallus water indicates that, down to a WC of 60–70% DW, respired CO_2 was about 1‰ more enriched than would be expected from overall thallus water (Figure 6.6C). At higher WC, water movement in lichens predominantly occurs in the liquid state, passively within the apoplast (Honegger and Hugelshofer 2000). The characteristic morphology of lichens (increasing structural density from the inner medulla to the

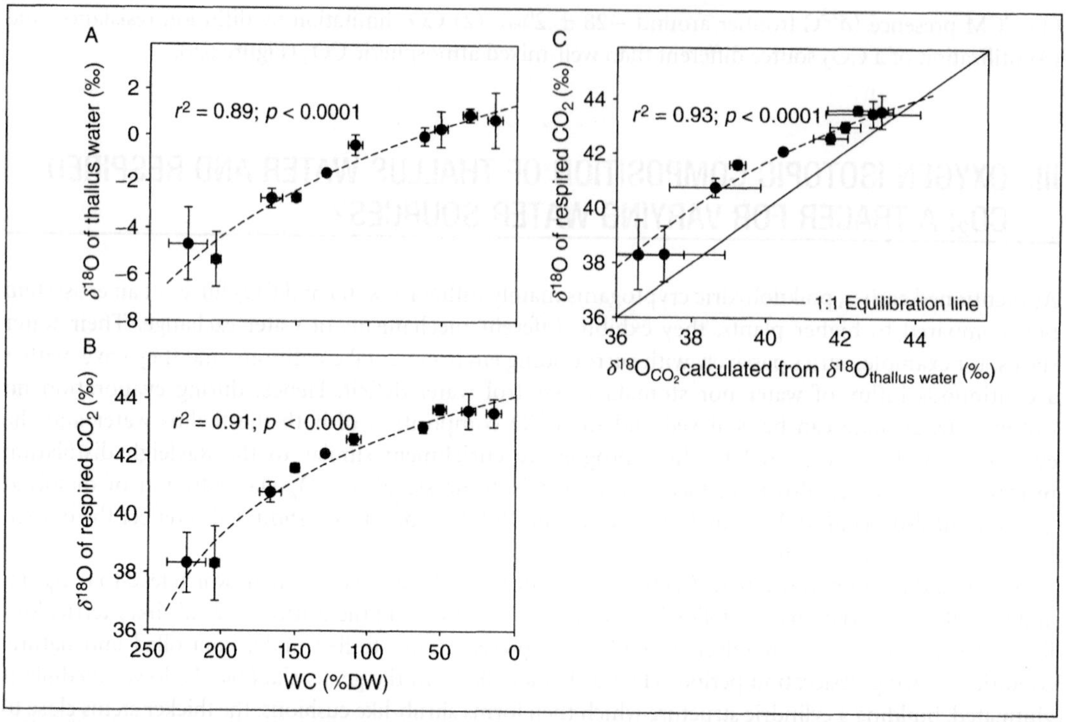

FIGURE 6.6 δ^{18}O enrichment of the lichen *C. arbuscula* during desiccation (changed after Hartard *et al.* 2007): (A) correlation between δ^{18}O of thallus water and (B) δ^{18}O of respired CO_2 and relative thallus WC; (C) relationship between measured δ^{18}O of respired CO_2 and CO_2 *calculated* to be in isotopic equilibrium with thallus water. Equilibrium calculations made using prevailing temperatures and isotopic fractionations from Brenninkmeijer *et al.* (1983).

outer algal layer and cortex) easily creates a gradient in vapor pressure deficit (VPD) from the inside to the outside of the thallus (Figure 6.1A and B). Considering this evaporation and, hence, isotopic enrichment should mainly take place at outer thallus parts, that is, at algal and cortical cells. Moreover, in contrast to higher plants, fully water-saturated lichens may hold about 50–60% of their total WC in the apoplast and other extracellular regions (Beckett 1997). During desiccation, these water pools are lost before intracellular water (Honegger 1997). It is therefore expected that outer thallus parts reveal stronger enrichment compared to inner parts and that extracellular water shows stronger enrichment compared to intracellular water. However, respiratory CO_2 released by lichens passes through the outer thallus parts where its δ^{18}O instantly may equilibrate with the more enriched water of this region (Figure 6.1B). This is at least due to the presence of both extracellular carbonic anhydrase (CA) in the photobiont and intracellular CA in the photobiont and mycobiont (Raven *et al.* 1990, Palmqvist and Badger 1996, Amoroso *et al.* 2005; Figure 6.1C).

In conclusion, it seems that, at higher WC, the oxygen isotopic composition of the overall thallus water resembles both less enriched water from inner thallus compartments as well as more enriched water from more exposed compartments. In contrast, the oxygen isotopic ratio of the respired CO_2 only represents the water pool equilibrated with the thallus' outer regions, which are assumed to be more enriched compared to overall thallus water. Below a WC of 60–70% (DW), thallus density strongly decreases because of loss of turgescence and the occurrence of air pores resulting from desiccation. Thus, VPD inside the thallus also increases, enhancing evaporation within the lichen thallus. Consequently, δ^{18}O of CO_2 converges with that of total thallus water due to the progressive alignment of the various water pools. This finally causes the relationship of respired CO_2 and thallus water to approach the 1:1 equilibration line indicating a gradual alignment of δ^{18}O of thallus water of the different compartments (Figure 6.6C).

Very similar enrichment patterns were found for a variety of different lichen species with distinct growth forms and morphological features (Lakatos *et al.* unpublished data). The results of these further studies indicate that the actual degree of enrichment also depends on the lichen's intrathalline resistance. Species with thick outer cortices and dense thallus structures, *i.e.*, great resistance toward diffusion of water vapor, revealed stronger overall enrichment compared to species with no outer cortex and rather loosely associated structures.

B. Lichens as Tracers for Water Sources and Their Effect on Soil: Atmosphere Water Exchange

As previously mentioned, lichens are not able to actively control their water deficit. Thus, they are inevitably always in a stage of approaching equilibrium with their surroundings. This special characteristic makes it particularly interesting to evaluate their applicability as tracers for water sources in the field, especially regarding the increasing interest in water vapor. Within field studies conducted at a coastal sand dune ecosystem (a natural habitat of *C. arbuscula*) in Portugal, different parts of *C. arbuscula* cushions ("stem" parts from the protected interior close to the ground and "tip" parts from the air-exposed top part of the cushions) were collected, and $\delta^{18}O$ of the thallus water was analyzed. Preliminary data of this study revealed that the different thallus parts exhibited distinct isotopic signals. As expected, the exposed top parts always showed lower WCs compared to the interior regions (Hartard *et al.* 2007). However, while the cushion's interior thallus water appeared to resemble the isotopic composition of the enriched soil surface water, the air-exposed tip parts revealed rather depleted $\delta^{18}O$ values (Figure 6.7A and B), especially at very low WCs. Only in the course of the first day, did the lichen WC, which had been increased by a short rain event the night before, experience a strong decline due to the pronounced change in microclimatic conditions (Figure 6.7B). This was accompanied by strong thallus water enrichment due to the high evaporative fractionation caused by the large water potential difference between the thallus and the surrounding air. However, in the afternoon, after the lichens had achieved equilibrium with atmospheric conditions, $\delta^{18}O$ of the thallus water became more depleted again, suggesting a simultaneous approach of an isotopic equilibration with the surrounding humidity. In addition, particularly green algal lichens exhibit the special ability to utilize different water sources. Especially after desiccation periods, when liquid water is no longer available, they take up water vapor in order to remain metabolically active. Consistent with this, it was observed throughout the remaining study period that the thallus water always seemed to resemble the more depleted water vapor which the lichen took up during times of high ambient humidity, that is early in the morning or in the evening.

In addition to the lichen sampling, microclimatic measurements of relative humidity (RH) and temperature (T) were carried out (Figure 6.7C and D). Data was collected both at ambient sites as well as from inside *C. arbuscula* cushions. The measurements showed that inside their cushion structure these cryptogams create their own favorable microhabitat by maintaining high relative humidity as well as more constant temperatures compared to ambient conditions (Figure 6.7C and D).

In conclusion, the isotopic composition of the thallus water of lichens growing in their natural habitat appears to depend mainly on two separate factors: the isotopic signal of the predominantly available water source and the water potential difference between the thallus and the surrounding air. From this, it follows that lichen thallus water $\delta^{18}O$ in the process of achieving equilibration with its surroundings experiences isotopic enrichment depending on the water potential difference between the thallus and the surrounding atmosphere. Thus, large evaporative forces, that is rapidly desiccating lichens, also cause strong evaporative enrichments, whereas lichens close to physical equilibration with their surroundings will approach isotopic equilibration with the surrounding isotopically lighter water vapor. However, $\delta^{18}O$ of lichens being already in "steady state," that is in continuous equilibrium with its surroundings, reflects the isotopic signal of the atmospheric vapor.

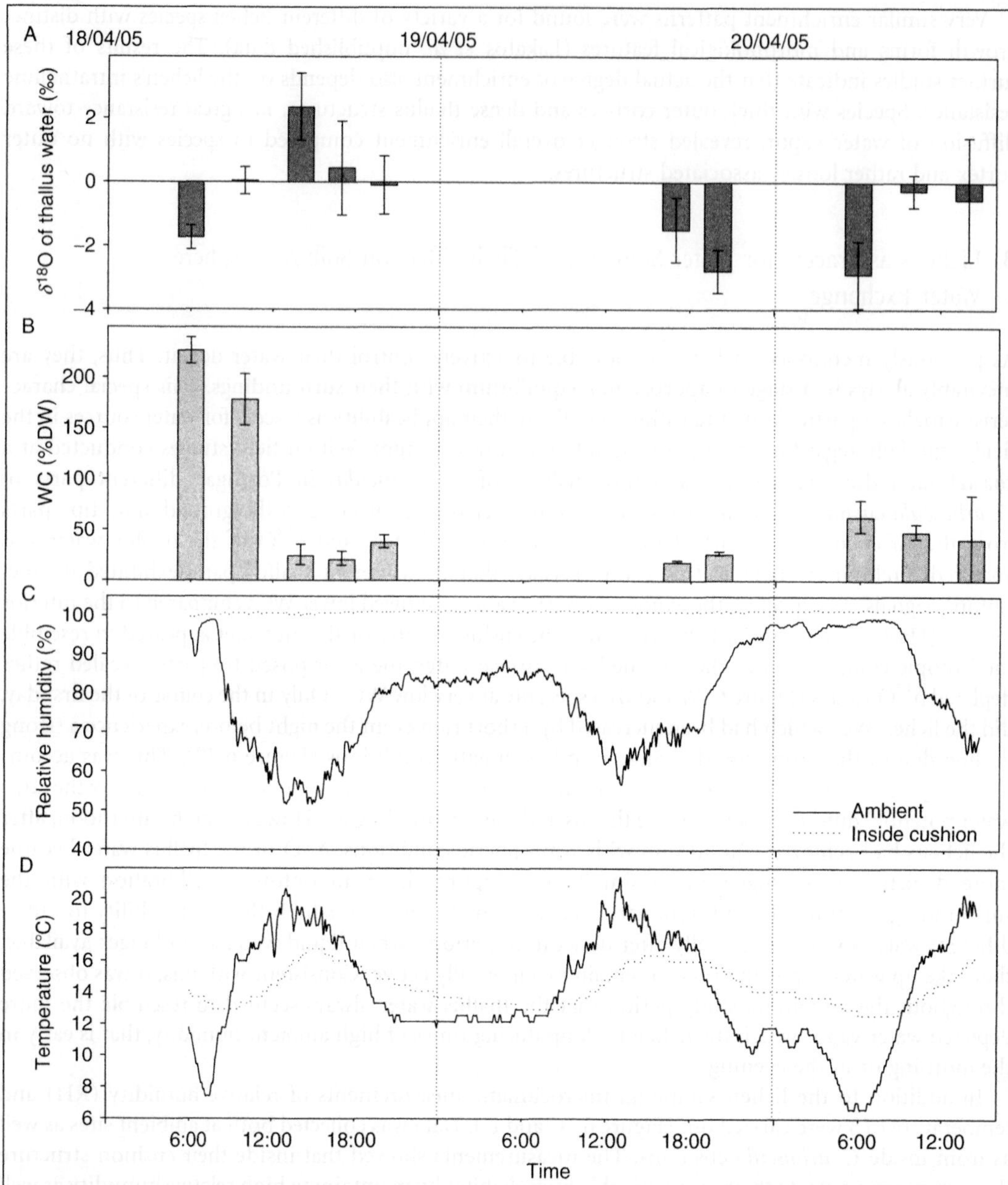

FIGURE 6.7 (A) $\delta^{18}O$ and (B) relative WC of the air exposed, upper thallus part of *C. arbuscula* cushions over a course of a 3 day field study in a Mediterranean sand dune ecosystem (Portugal). Also shown are prevailing microclimatic conditions of (C) relative humidity (RH) and (D) temperature (T) measured at ambient sites as well as inside the lichen cushions.

Thus, the data indicates strong influence of poikilohydric ground cover on soil evaporative fluxes, especially during drier periods without rain. During these periods, the organisms predominantly utilize the more depleted air moisture and, hence, also evaporate more depleted water vapor into the atmosphere. At the same time, the cushions build up an enlarged water-saturated boundary layer above the soil surface that reduces the highly enriched soil evaporation flux. As a consequence, the relative contribution of the lichen's depleted evaporation to overall ecosystem evaporative fluxes should also be increased.

Whether this hypothesis really holds true and to what extent this might become apparent in the isotopic composition of ecosystem fluxes still needs to be evaluated in further studies. However, with respect to the widely known and still existing problems in water vapor sampling, and especially in the context of the currently increasing interest and demand on isotopic data of water vapor, the presented findings may open up a new and easily accessible opportunity to use lichens as tracers for different water sources.

IV. OUTLOOK AND CONCLUSIONS

Lichens are highly diverse organisms offering a number of particular physiological and morphological properties which enable the assessment of several environmental and ecological factors by the use of stable isotope techniques. Lichens can be used as tracers of general trends as well as specific environmental changes—possibly to the same extent as they are already used as bioindicators for biomonitoring. They integrate both timescales (when considering different thallus parts) and spatial scales when considering available sources (CO_2 sources such as soil-respired, bark-respired, and ambient CO_2 as well as water sources such as rain, soil surface water, dew, and vapor). Moreover, lichens can integrate and reflect crucial environmental variables, such as water availability, temperature, and light, on a microscale (morphology, individuals, microhabitat, and so on) as well as on a macro scale (population, macrohabitat, ecosystem, and so on). Many environments with limited nutrition or/and water supply are predominated by these poikilohydric cryptogams, which may constitute more than half of the aboveground biomass, and at higher latitudes up to 100% ground cover by lichens. In these ecosystems, lichens may significantly influence both $\delta^{13}C$ and $\delta^{18}O$ from CO_2 and water fluxes. A comparison of the globally measured and modeled $\delta^{18}O$ of atmospheric CO_2—the latter being based *inter alia* on parameters of higher plant functional types (Cuntz *et al.* 2003a,b)—reveals that measured values are generally more depleted compared to modeled data, especially in higher latitudes. The aforementioned equilibration of lichen-respired CO_2 with the absorbed $\delta^{18}O$-depleted water vapor may possibly explain these discrepancies.

In conclusion, lichens fundamentally differ from higher plants in many processes and hence, they influence ecosystems in different ways. Thus, they can be used as tracers for complementary information. However, their importance within an ecosystem or even globally is still unknown, and the use of stable isotopes may offer us a better understanding. In the future, we could use lichens as sensitive monitors for land-use change rather than just community composition changes in biodiversity studies.

V. ACKNOWLEDGMENT

This research was supported within the EC-program NETCARB (HPRN-CT-1999-59), the ESF-program SIBAE (62561 EXGC EX03), the German and Portuguese Academic Exchange Service DAAD/GRICES (PPP-DAAD/GRICES; D/04/42019), and the Portuguese Science Foundation (POCTI/BIA-BDE/60140/2004). The authors would like to thank to Rodrigo Maia for technical assistance regarding the stable isotopic analysis.

VI. APPENDIX

The poikilohydric nature of lichens and their small growth size necessitate the performance of a special protocol for lichens to determine water and CO_2 fluxes for $\delta^{13}C$ and $\delta^{18}O$ analysis. The advantage of their size is the convenient handling inside of small vials, so-called exetainer.

A. Dehydration Experiment

To ensure initial water saturation, all lichen samples were submersed in water for 30 min prior to the experiment. Subsequently, the samples were slightly shaken to remove any excess water, and fresh weight at maximum WC (FW_{max}) was determined using an analytical balance. Except for the first set of samples to be analyzed [100% relative WC (WC_{rel})], all samples were then placed into a plexiglass container which was closed with a perforated lid to allow adequate dehydration. For the following 7 h, on the hour, a set of at least three samples of each lichen species was removed from the container, current FW was determined, and each sample was placed into an individual exetainer vial (Labco Limited, Buckinghamshire, United Kingdom). Each exetainer was wrapped in aluminum foil to ensure dark respiration and was then flushed with CO_2-free air. This was done by penetrating the exetainer's Teflon septum with two needles and flushing it with atmospheric air initially directed through a soda lime column. The air going out was directed into a CO_2 infrared gas analyser (BINOS, Hereus, Hanau, Germany). When the air coming out was CO_2-free, the needles were removed and the lichen samples were left to respire for at least 10 min. A preexperiment was conducted to investigate the lichens' respiration rates which were found to be highly species specific and mainly dependent on the lichens' prevailing water status. On the basis of this survey, specific respiration times were adjusted according to the expected respiration rates of the individual samples. That way it could be timed that, at the moment of isotopic analysis of the respired CO_2, the amount of CO_2 in the exetainers was similar for all samples.

B. Isotopic Analysis of Respired CO_2

To analyse the isotopic composition of the respired CO_2, sample exetainers were placed in a multiflow autosampler (Micromass) coupled to an IsoPrime mass spectrometer (Micromass). A gas sample of the respired CO_2 in each exetainer was entrained in the carrier gas, He_2, CO_2 was isolated via GC separation and, subsequently, carried to the mass spectrometer. Following the analysis, all exetainers with lichen samples were stored in a freezer until thallus water extractions took place.

C. Thallus Water Extractions

The thallus water of all lichen samples was extracted using cryogenic vacuum distillation. For this, each sample in the exetainer was taken from the freezer, opened, and instantly placed into the extraction vessel of a vacuum line. The vessel was then immersed in liquid nitrogen (LN_2) to keep the sample frozen, attached to the vacuum line, and evacuated. Subsequently, it was heated up to 100°C and the vaporized water was collected in a cold finger. When all moisture had been extracted from the sample, the cold finger was sealed under vacuum using a torch.

D. Isotopic Analysis of Extracted Thallus Water

Oxygen isotope ratios of the extracted water samples were determined by equilibration of water with CO_2. Three hundred microliters of the water sample was placed into a gastight vial and an aliquot of pure CO_2 was added. The vials were then placed into a constant temperature water bath of 17°C and CO_2 was allowed to equilibrate for 16 h. The equilibrated CO_2 was then analyzed for its oxygen isotopic composition using a multiflow autosampler (Micromass) coupled with an IsoPrime mass spectrometer (Micromass).

VII. REFERENCES

Amoroso, G., L. Morell-Avrachov, D. Müller, K. Kluge, and D. Sültemeyer. 2005. Upregulation of yeast carbonic anhydrase (nce103) by low CO_2. *Molecular Microbiology* **56:**549–558.

Amundson, R., L. Stern, T. Baisden, and Y. Wang. 1998. The isotopic composition of soil and soil-respired CO_2. *Geoderma* **82:**83–114.

Asta, J., W. Erhardt, M. Ferretti, F. Fornasier, U. Kirschbaum, P. L. Nimis, O. W. Purvis, S. Pirintsos, C. Scheidegger, C. van Haluwyn, and V. Wirth. 2002. Mapping lichen diversity as an indicator of environmental quality. Pages 273–279 *in* P. L. Nimis, C. Scheidegger, and P. A. Wolseley (Eds.) *Monitoring with Lichens—Monitoring Lichens.* Kluwer Academic Publisher, Dordrecht.

Beckett, R. P. 1997. Pressure–volume analysis of a range of poikilohydric plants implies the existence of negative turgor in vegetative cells. *Annals of Botany* **79:**145–152.

Bilger, W., S. Rimke, U. Schreiber, and O. L. Lange. 1989. Inhibition of energy-transfer to photosystem II in lichens by dehydration: Different properties of reversibility with green and blue-green phycobionts. *Journal of Plant Physiology* **134:**261–268.

Branquinho, C., F. Catarino, D. H. Brown, M. J. Pereira, and A. Soares. 1999. Improving the use of lichens as biomonitors of atmospheric metal pollution. *The Science of the Total Environment* **232:**67–77.

Brenninkmeijer, C. A. M., P. Kraft, and W. G. Mook. 1983. Oxygen isotope fractionation between CO_2 and H_2O. *Isotope Geoscience* **1:**181–190.

Brugnoli, E., and G. D. Farquhar. 2000. Photosynthetic fractionation of carbon isotopes. Pages 399–434 *in* R. C. Leegood, T. D. Sarkey, and S. von Caemmerer (Eds.) *Photosynthesis: Physiology and Metabolism.* Kluwer Academic Publisher, Netherlands.

Buchmann, N., J.-M. Guehl, T. S. Barigah, and J. R. Ehleringer. 1997. Interseasonal comparison of CO_2 concentrations, isotopic composition, and carbon dynamics in an Amazonian rainforest (French Guiana). *Oecologia* **110:**120–131.

Cowan, I. R., O. L. Lange, and T. G. A. Green. 1992. Carbon-dioxide exchange in lichens: Determination of transport and carboxylation characteristics. *Planta* **187:**282–294.

Cuntz, M., P. Ciais, G. Hoffmann, and W. Knorr. 2003a. A comprehensive global three-dimensional model of $\delta^{18}O$. *Journal of Geophysical Research* **108:**1–24.

Cuntz, M., P. Ciais, G. Hoffmann, C. E. Allison, R. J. Francey, W. Knorr, P. P. Tans, J. W. C. White, and I. Levin. 2003b. A comprehensive global three-dimensional model of $\delta^{18}O$ in atmospheric CO_2: 2. Mapping the atmospheric signal. *Journal of Geophysical Research* **108:**1–19.

Dawson, T. E., S. Mambelli, A. H. Plamboeck, P. H. Templer, and K. P. Tu. 2002. Stable isotopes in plant ecology. *Annual Review of Ecology and Systematics* **33:**507–559.

Ellis, C. J., and B. J. Coppins. 2006. Contrasting functional traits maintain lichen epiphyte diversity in response to climate and autogenic succession. *Journal of Biogeography* **33:**1643–1656.

Green, T. G. A., M. Schlensog, L. G. Sancho, J. B. Winkler, F. D. Broom, and B. Schroeter. 2002. The photobiont determines the pattern of photosynthetic activity within a single lichen thallus containing cyanobacterial and green algal sectors (photosymbiodeme). *Oecologia* **130:**191–198.

Hartard, B., C. Máguas, and M. Lakatos. 2007. $\delta^{18}O$ characteristics of poikilohydrous lichens and their effects on evaporative processes of the subjacent soil. *Isotopes in Environmental and Health Studies,* in press.

Honegger, R. 1997. Metabolic interaction at the mycobiont-photobiont interface in lichens. Pages 209–221 *in* G. C. Carroll and P. Tudzynski (Eds.) *The Mycota.* Springer Press, Berlin.

Honegger, R., and G. Hugelshofer. 2000. Water relations in the *Peltigera aptosa* group visualized with LTSEM techniques. *Bibliotheca Lichenologica* **75:**113–126.

Kappen, L. 1988. Ecophysiological relationships in different climatic regions. Pages 37–100 *in* M. Galun (Ed.) *Handbook of Lichenology,* Vol. 2. CRC Press, Boca Raton.

Kappen, L. 1993. Lichen in the Antarctic region. Pages 433–490 *in* E. I. Friedman (Ed.) *Antarctic Microbiology.* Wiley-Liss, New York.

Kappen, L., and F. Valladares. 1999. Opportunistic growth and desiccation tolerance: The ecological success of poikilohydrious autotrophs. Pages 9–80 *in* F. I. Pugnaire and F. Valladares (Eds.) *Handbook of Functional Plant Ecology.* Marcel Dekker, New York.

Lakatos, M., U. Rascher, and B. Büdel. 2006. Functional characteristics of corticolous lichens in the understory of a tropical lowland rain forest. *New Phytologist* **172:**679–695.

Lange, O. L., E. Kilian, and H. Ziegler. 1986. Water vapor uptake and photosynthesis of lichens: Performance differences in species with green and blue-green algae as phycobionts. *Oecologia* **71:**104–110.

Lange, O. L., T. G. A. Green, and H. Ziegler. 1988. Water status related photosynthesis and carbon isotope discrimination in species of the lichen genus *Pseudocyphellaria* with green or blue-green photobionts and in photosymbiodemes. *Oecologia* **75:**494–501.

Lange, O. L., T. G. A. Green, and U. Heber. 2001. Hydration-dependent photosynthetic production of lichens: What do laboratory studies tell us about field performance? *Journal of Experimental Botany* **52:**2033–2042.

Larson, D. W. 1987. The absorption and release of water by lichens. *Bibliotheca Lichenologica* **25:**351–360.

Máguas, C., and E. Brugnoli. 1996. Spatial variation in carbon isotope discrimination across the thalli of several lichen species. *Plant, Cell and Environment* **19:**437–446.

Máguas, C., H. Griffiths, J. Ehleringer, and J. Serodio. 1993. Characterization of photobiont associations in lichens using carbon isotope discrimination techniques. Pages 201–212 *in* J. Ehleringer, J. Hall, and G. Farquahr (Eds.) *Stable Isotopes and Plant Carbon–Water Relations.* Academic Press, New York.

Máguas, C., H. Griffiths, and M. S. J. Broadmeadow. 1995. Gas exchange and carbon isotope discrimination in lichens—evidence for interactions between CO_2-concentrating mechanisms and diffusion limitation. *Planta* **196:**95–102.

Máguas, C., F. Valladares, E. Brugnoli, and F. Catarino. 1997. Carbon isotope discrimination, chlorophyll fluorescence and qualitative structure in the assessment of gas diffusion resistances of lichens. Pages 119–135 *in* L. Kappen (Ed.) *New Species and Novel Aspects in Ecology and Physiology of Lichens. In honour of O.L. Lange.* J. Cramer, Stuttgart, Berlin.

Nimis, P. L., G. Lazzarin, and D. Gasparo. 1991. Lichens as bioindicators of air pollution by SO_2 in the Veneto region (NE Italy). *Studia Geobotanica* **11:**3–76.

Palmqvist, K. 2000. Tansley Review No. 117: Carbon economy in lichens. *New Phytologist* **148:**11–36.

Palmqvist, K., and M. Badger. 1996. Carbonic anhydrase(s) associated with lichen: *In vivo* activities, possible locations and putative roles. *New Phytologist* **132:**627–639.

Palmqvist, K., C. Maguas, M. R. Badger, and H. Griffiths. 1994a. Assimilation, accumulation and isotope discrimination of inorganic carbon in lichens: Further evidence for the operation of a CO_2 concentrating mechanism in cyanobacterial lichens. *Cryptogamic Botany* **4:**218–226.

Palmqvist, K., E. Ögren, and U. Lernmark. 1994b. The CO_2 concentrating mechanism is absent in the green alga lichen *Coccomyxa*: A comparative study of photosynthetic CO_2 and light responses of *Coccomyxa, Chlamydomonas reinhardtii* and barley protoplasts. *Plant, Cell and Environment* **17:**65–72.

Purvis, O. W., J. Chimonides, V. Din, L. Erotokritou, T. Jeffries, G. C. Jones, S. Louwhoff, H. Read, and B. Spiro. 2003. Which factors are responsible for the changing lichen floras of London? *The Science of the Total Environment* **310:**179–189.

Raven, J. A., A. M. Johnston, L. L. Handley, and S. G. McInroy. 1990. Transport and assimilation of inorganic carbon by *Lichina pygmaea* under emersed and submersed conditions. *New Phytologist* **114:**407–417.

Schlensog, M., B. Schroeter, and T. G. A. Green. 2000. Water dependent photosynthetic activity of lichens from New Zealand: Differences in the green algal and the cyanobacterial thallus parts of photosymbiodemes. *Bibliotheca Lichenologica* **75:**149–160.

Sipman, H. J. M., and R. C. Harris. 1989. Lichens. Pages 303–309 *in* H. Lieth and M. J. A. Werger (Eds.) *Tropical Rain Forest Ecosystems.* Elsevier Science Publishers, Amsterdam.

Smith, E. C., and H. Griffiths. 1996. The occurrence of the chloroplast pyrenoid is correlated with the activity of a CO_2-concentrating mechanism and carbon isotope discrimination in lichens and bryophytes. *Planta* **198:**6–16.

Smith, E. C., and H. Griffiths. 1998. Intraspecific variation in photosynthetic responses of trebouxioid lichens with reference to the activity of a carbon-concentrating mechanism. *Oecologia* **113:**360–369.

Smith, E. C., H. Griffiths, L. Wood, and J. Gillon. 1998. Intra-specific variation in the photosynthetic responses of cyanobiont lichens from contrasting habitats. *New Phytologist* **138:**213–224.

Stern, L., W. T. Baisden, and R. Amunson. 1999. Processes controlling the oxygen isotopic ratio of soil CO_2: Analytic and numerical modeling. *Geochimica et cosmochimica acta* **63:**799–814.

Sternberg, L. S. L., S. S. Mulkey, and S. J. Wright. 1989. Ecological interpretation of leaf carbon isotope ratios: Influence of respired carbon dioxide. *Ecology* **70:**1317–1324.

Tans, P. P. 1998. Oxygen isotopic equilibrium between carbon dioxide and water in soils. *Tellus* **50B:**163–178.

Valladares, F., C. Ascaso, and L. G. Sancho. 1994. Intrathalline patterns of some structural and physical parameters in the lichen genus *Lasallia. Canadian Journal of Botany* **72:**415–428.

Yakir, D., and L. S. L. Sternberg. 2000. The use of stable isotopes to study ecosystem gas exchange. *Oecologia* **123:**297–311.

CHAPTER 7

Foliar $\delta^{15}N$ Values as Indicators of Foliar Uptake of Atmospheric Nitrogen Pollution

Dena M. Vallano and Jed P. Sparks

Department of Ecology and Evolutionary Biology, Cornell University

Contents

I. INTRODUCTION

Among the cycles of elements important to life, the nitrogen (N) cycle has been the most perturbed beyond its natural state by human activities (Vitousek 1994, Vitousek *et al.* 1997, Hopkins *et al.* 1998). One manifestation of this perturbation is a dramatic increase in the amount of reactive N in the atmosphere. Anthropogenic emissions of atmospheric reactive N now exceed natural emissions and are the predominant source of atmospheric reactive N deposition to terrestrial ecosystems. Asia, Europe, and North America account for the majority of the current anthropogenic emissions and the highest levels of deposition. However, substantial increases in atmospheric reactive N are projected over the next 50 years for several other regions of the world as industrialization increases (Galloway *et al.* 2004, Phoenix *et al.* 2006). It is crucial that we improve our understanding of the potential sinks for reactive N as global inputs continue to increase.

Vegetation appears to be a significant sink for atmospheric reactive N, and the direct foliar incorporation of atmospheric N pollution may represent a considerable source of N to plants (Hill 1971, Hanson and Lindberg 1991, Rondon *et al.* 1993, Rondon and Granat 1994). Plants are capable of assimilating the N in gaseous N compounds, including the reactive nitrogen oxides ("NO_y," including NO, NO_2, HNO_3, and organic nitrates) and ammonia gases ("NH_x," including NH_3 and particulate NH_4^+), and incorporating this N into organic compounds for growth (Morikawa *et al.* 1998, Gessler *et al.* 2002). The capacity for foliar uptake of reactive N is likely influenced by plant species characteristics and reactive N concentrations in the atmosphere (Morikawa *et al.* 1998). In this chapter, we review the potential of N stable isotopes, specifically foliar $\delta^{15}N$ values, as an innovative and important tool for determining the magnitude and ultimate fate of atmospheric reactive N deposition to foliage globally.

Reactive N generation and redistribution globally is altering ecosystem function and the composition of natural plant assemblages both through growth-stimulating and phytotoxic effects (Wellburn 1990, Bobbink and Roelofs 1995, Pearson and Soares 1995, Vitousek *et al.* 1997, Lea 1998). Increasing the total reactive N input to an ecosystem can lead to several negative effects on ecosystem function such as soil and water acidification, increased loss of nutrients from the soil, changes to competitive interactions among plants, reductions in biodiversity, increased vulnerability of invasion, and nutrient imbalances in vegetation (Asner *et al.* 1997, Bobbink *et al.* 1998, Gilliam 2006). Moreover, plant performance can also be deleteriously affected through reductions in net photosynthesis, respiration, stomatal conductance, enzyme activities, and growth (Takahashi *et al.* 2005). Positive effects of increased reactive N deposition may include fertilization of N-limited ecosystems promoting increased carbon sequestration and plant growth (Beedlow *et al.* 2004). Although numerous studies have shown the impacts of total N deposition on ecosystem processes (Lovett 1994, Fahey *et al.* 1999, Wesely and Hicks 2000, Aber *et al.* 2003, Galloway *et al.* 2003), few have investigated the magnitude or influence of atmospheric N pollution assimilated directly by foliage (Sparks *et al.* 2001).

The ability of vegetation to incorporate N into plant metabolism from both leaves and roots suggests the fate of N deposited to an ecosystem may be fundamentally different depending on the entry pathway. For example, the chemical products of foliar N assimilation (most commonly NO_3^-) are plant nutrients and are directly incorporated from the leaf apoplast and used for growth (Morikawa *et al.* 1998, Gessler *et al.* 2000). In contrast, N additions to the soil surface during deposition may be incorporated into microbial biomass, revolatilized, or leached. Furthermore, additional N to the soil surface may have other effects (*e.g.*, changes in soil fertility, soil acidification, stripping of base cations). Thus, the direct entry of N through leaves is intriguing in that it could fundamentally change plant productivity through a direct influence on photosynthesis without the concomitant changes in soil chemistry often observed under high levels of soil N additions. However, it is still unclear if, and how, plant performance differs under N deposition to the soil (*i.e.*, fundamentally altering the reactive N supply to root systems) versus changes in N additions from the atmosphere (*i.e.*, changes in the magnitude of foliar incorporation). Both forms of N entry into an ecosystem will likely increase together under most scenarios of intensified industrialization. However, various forms of gaseous reactive N have significantly different atmospheric lifetimes, suggesting plant communities arrayed across the landscape experience not only unequal total N deposition but also different proportions of N through the root and foliar pathways.

There is some evidence to suggest the proportion of N incorporated through the foliar versus soil pathway may vary by environment. For instance, atmospheric reactive N is dominated by relatively reduced forms (*e.g.*, NO_2) in industrialized areas. These compounds deposit more readily through the leaf stomata than to other surfaces in the environment (Sparks *et al.* 2001). At distances further from sources of N pollution, much of the reactive N is further oxidized to organic forms such as peroxyacetyl and isoprene nitrate. These compounds also readily enter plant stomata (Sparks *et al.* 2003) at a rate higher than deposition to other surfaces. In regions still further from sources of reactive N, the air mass is often dominated by nitric acid (HNO_3). Nitric acid is very soluble in water and tends to deposit readily to many surfaces in the environment at a rate much faster than entry through the stomata (Byterowicz *et al.* 1998). Since the composition of atmospheric reactive N gases can vary across the landscape and are more or less likely to enter plant stomata, terrestrial ecosystems may receive differential amounts of N through the foliar incorporation pathway.

N stable isotope ratios ($^{15}N/^{14}N$) are a useful tool for estimating the amount of anthropogenic N in the system and quantifying the incorporation pathway into the plant [*e.g.*, through soil NO_3^- or NH_4^+, dissolved organic N, foliar uptake via stomata, cuticular diffusion (Robinson *et al.* 1998, 2000)] because forms of N available to plants often have varying isotopic compositions (Evans 2001, Kolb and Evans 2002). This may be especially relevant to industrialized, urban regions where pollutant compounds contribute a significant amount of N input via dry and wet deposition to ecosystems. Consequently, the difference in source isotopic composition can provide a natural tracer for the assessment of foliar uptake and assimilation of reactive N pollution in terrestrial ecosystems (Handley and Raven 1992, Handley and Scrimgeour 1997). Specific to foliar uptake, if the foliar incorporated source is significantly different from the soil source, then the foliar $\delta^{15}N$ value will be representative of both sources and can be used to quantify the proportion of each N source incorporated into plant biomass. If plants are able to assimilate significant quantities of gaseous N pollutants and incorporate them into organic compounds, foliar $\delta^{15}N$ will be an important tool to examine the role foliar uptake of atmospheric N pollution plays in the global N and carbon cycles through the alteration of photosynthesis and the removal of N pollutants from the atmosphere.

In this chapter, we discuss the use of foliar $\delta^{15}N$ values as a tool for determining the magnitude and impact of gaseous reactive N directly entering the foliage of vegetation. Although not an exhaustive treatment, we place this discussion within the context of the causative factors defining foliar $\delta^{15}N$ values. We also describe the current understanding of foliar $\delta^{15}N$ values to provide a background on which foliar uptake of reactive N can be considered.

II. NITROGEN SOURCES TO TERRESTRIAL VASCULAR PLANTS

The N sources assimilated by plants often have different isotopic compositions and many studies have used foliar $\delta^{15}N$ values to indicate the $\delta^{15}N$ of the dominant N source used by a plant and as a measure of N sources from both soil and atmospheric deposition (Evans 2001). However, we now know that foliar $\delta^{15}N$ values represent the integration of a range of processes (reviewed by

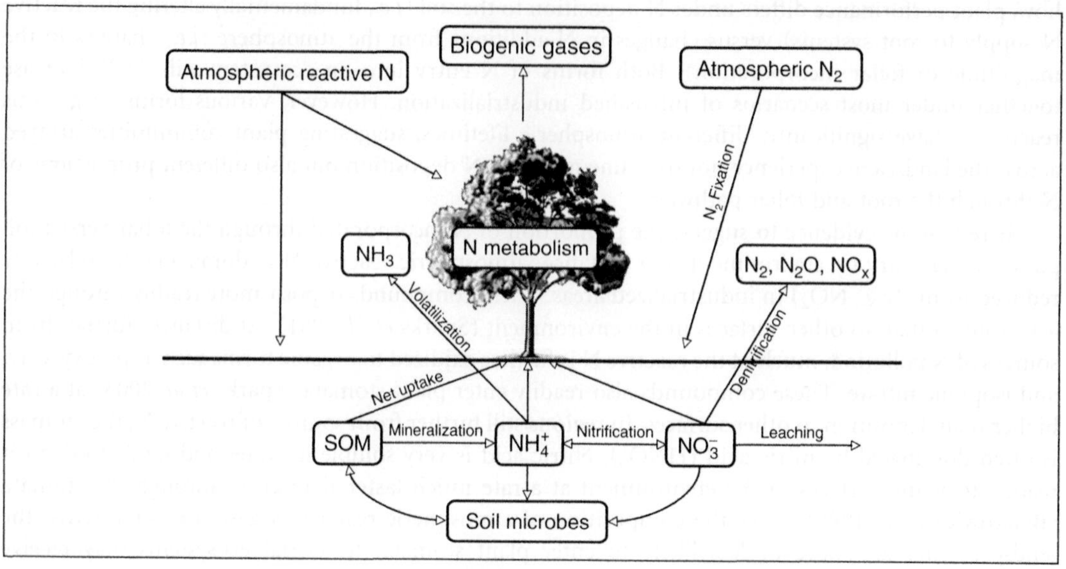

FIGURE 7.1 Primary source pools and uptake pathways in the N cycle. Boxes represent pools and arrows represent fluxes. Various N transformations in the soil and atmosphere lead to varying amounts of ^{15}N enrichment or depletion in natural ecosystems, which can vary greatly with geographical location and environmental conditions. SOM represents soil organic matter derived from plant, microbial, and faunal organic tissues (adapted from Dawson *et al.* 2002).

Handley and Scrimgeour 1997, Evans 2001, Robinson 2001, Stewart *et al.* 2002) and can vary by as much as ±10‰ in coexisting species (Handley and Scrimgeour 1997). The presence of multiple N sources with distinct isotopic values, differential processing of soil N pools (*e.g.*, temporal and spatial patterns of net mineralization and nitrification), temporal and spatial variation in N availability, mycorrhizal interactions, and alterations in plant N demand can all influence foliar $\delta^{15}N$ (Figure 7.1).

Identifying and differentiating the sources of N available to plants is necessary for a variety of ecological questions, including accurately interpreting foliar $\delta^{15}N$ values. Further, additional understanding of how specific N species deposited to terrestrial ecosystems impact foliar $\delta^{15}N$ values will provide us a tool to link plant physiological responses, terrestrial ecosystem function, and atmospheric chemistry. In this section, we briefly discuss various sources of plant-available N and their potential influences on foliar $\delta^{15}N$.

A. Soil Nitrogen Sources

Plants derive N primarily from the soil in the form of ammonium (NH_4^+), nitrate (NO_3^-), and, secondarily, from organic N. Plant-available soil N originates from decomposition of organic matter, biological N fixation, additions of N in organic or inorganic fertilizers, and deposition originating from fossil fuel combustion and animal production (Hopkins *et al.* 1998). There is enormous variability in the isotopic composition of the soil N pool available to plants. For example, evidence suggests that bulk soil $\delta^{15}N$ and the $\delta^{15}N$ of soil NO_3^- and NH_4^+ diverge from each other at various timescales. This suggests that foliar $\delta^{15}N$ may be interpreted as the integrated signal of $\delta^{15}N$ of all plant-available N forms over the growth history of a particular leaf. However, the leaf signal may not represent the current status of soil $\delta^{15}N$ at the time of sampling. Symbiotic interactions with fungi and N_2-fixing bacteria may also obscure the interpretation of the foliar $\delta^{15}N$ value (Evans 2001).

Further, observations have suggested that much of the variation in foliar $\delta^{15}N$ among coexisting species may be due to the internal metabolism of fungi participating in the mycorrhizal association (Handley *et al.* 1998). For example, under N-limited conditions, plants interacting with mycorrhizae will have a lower $\delta^{15}N$ compared to the fungus due to fractionation during N transfer from the symbiont to the host plant (Evans 2001).

Differential uptake of multiple N sources with different $\delta^{15}N$ values, differences in fractionation during uptake and assimilation, and physiological differences between plants and mycorrhizal symbionts are significant obstacles to the use of foliar $\delta^{15}N$ values as a tool to assess the foliar uptake of reactive N. Currently, there is no adequate method for isolating and analyzing $\delta^{15}N$ in soil N pools that are available for plant uptake. Consequently, the influence of foliar uptake on foliar $\delta^{15}N$ is confounded by variations in the soil system. Variability in substrate supply, abiotic conditions, organism assemblages, and their demand for N influence foliar $\delta^{15}N$ values to a degree beyond that of pollution additions to the soil surface or foliage, making most field-based studies challenging. Ultimately, enhancing our understanding of the fractionations and transformations occurring in and between soil N pools and whether there are systematic correlations among pools will dramatically improve our ability to use foliar $\delta^{15}N$ as an indicator and allow us to examine foliar N incorporation in natural systems.

B. Atmospheric Nitrogen Sources

Atmospheric N deposition can affect plant functioning via both foliar and soil uptake, with different effects on plant growth depending on the mode of entry and the type of N compound and plant species (Stulen *et al.* 1998). Atmospheric reactive N can enter terrestrial ecosystems via wet, dry, and occult deposition. Most forms of atmospheric reactive N are generated from the partial oxidation of N_2 at high temperatures and via the volatilization of animal waste. Total oxidized reactive N (NO_y) and ammonia gases (NH_x) are the most prevalent atmospheric reactive N sources, can be transported long distances, and are eventually deposited on terrestrial ecosystems as NO_3^- and NH_4^+ in precipitation, as gases, or as particles. Major sources of NO_y typically include power production, transportation sources, off-road mobile sources, and other industrial sources (EPA 2000). Sources of NH_x include both natural plant and soil emissions and anthropogenic emissions from agricultural and animal waste products (Srivastava 1992). In addition to inorganic N sources, there are several sources of naturally occurring organic N, including plant pollen and sea-spray droplets (Driscoll *et al.* 2003). Atmospheric organic N may also be derived from anthropogenic inorganic N compounds reacting with hydrocarbons or organic particles in the atmosphere (Prospero *et al.* 1996) and generally comprises 30% of total atmospheric reactive N deposition (Neff *et al.* 2002). All of these compounds have distinct deposition velocities, travel different distances from point sources of N pollution, and have different probabilities of being deposited to the soil or directly to the vegetation. For example, work by Sparks *et al.* (2003) and Turnipseed *et al.* (2006) has shown that peroxyacetylnitrate (PAN), an organic N pollutant and the dominant form of reactive N transported long distances in the atmosphere, is readily taken up through plant foliage, primarily deposits directly to vegetation via foliar incorporation and is less likely to be deposited to the soil N pool, making it a potential source of N to the plant that predominantly enters through the foliar incorporation pathway.

1. Wet Deposition

Most forms of reactive gas-phase N will eventually be oxidized to a form [most commonly nitric acid (HNO_3)] that is highly soluble and easily scavenged by particles or water droplets. These compounds are deposited to ecosystems within precipitation, low clouds, and fog. This process is collectively termed wet deposition and refers generally to the transfer of soluble compounds from the atmosphere to the earth's surface. Wet reactive N deposition contains a variety of N compounds, most of which are available for biological utilization, including inorganic (NO_3^-, NO_2, NH_4^+) and organic (amino acids, PAN, urea) chemical species (Peierls and Paerl 1997). Cloud deposition that occurs through impaction of fog droplets on exposed surfaces can contribute between 25% and 50% of total N deposition in high-elevation areas of the Northeastern United States (Anderson *et al.* 1999). These compounds can be deposited directly onto leaf surfaces and later incorporated into the plant tissue or deposited to the soil surface and added to the soil pool.

2. Dry Deposition

In the absence of an aqueous phase, gases and dry particles can be transported to and deposited into terrestrial ecosystems in the form of dry deposition. Dry deposition of gaseous N pollutants has been shown to be of a similar magnitude (or potentially greater) to wet deposition in many regions (Holland *et al.* 2004). Monitoring efforts in North America and Western Europe report that reactive N deposition flux to terrestrial ecosystems via dry deposition is dominated by NO_2, NH_3, HNO_3, and a relatively small amount of particle-nitrate. However, most monitoring efforts only assess inorganic forms of N, and Day *et al.* (2003) suggest that the dry deposition of organic N could be comparable to that of HNO_3. Dry deposition estimates are still highly uncertain in most areas, and methodology is still an active area of research. It is desirable for ecologists to not only evaluate the magnitude of gaseous N deposition on terrestrial ecosystems but to also monitor the isotopic composition of the deposited N and the resulting impact on foliar $\delta^{15}N$ values.

III. FOLIAR ASSIMILATION OF ATMOSPHERIC NITROGEN POLLUTANTS

The impact of atmospheric reactive N deposition to leaves depends on the physical and chemical characteristics of the pollutant and the plant species, including adsorption of the compound to the leaf surface, compound solubility within the internal plant cell wall and its reactivity with cellular components, and physiological factors such as stomatal conductance and the rate of N metabolism. Foliar uptake occurs through two routes: stomatal diffusion and uptake via the cuticle, although the latter is negligible for most chemical species. Cuticular uptake of N compounds (most commonly HNO_3) refers to the active or passive transport of a molecule through the lipid bilayer of the plasmalemma into the cytosol. Gaseous N pollutants enter the leaf through one of these two routes and are assimilated by different biochemical pathways (Figure 7.2). For example, ammonia gas diffuses through the stomata and is dissolved in the apoplastic space to form NH_4^+. Ammonium is transported into the cytoplasm where it is either partitioned into storage or assimilated by the glutamine synthetase/glutamate synthase cycle (Stulen *et al.* 1998). In contrast, after uptake via diffusion through the stomata, NO_2 enters the substomatal cavity of the leaf and disproportionates to form NO_3^- and nitrite (NO_2^-). Eventually, NO_2^- is reduced to NH_2 via nitrate reductase (NR) in the cytosol and nitrite reductase in the chloroplast, used in amino acid synthesis, and incorporated into primary N metabolism (Wellburn 1994, Weber *et al.* 1998).

A. Contribution of Foliar Uptake to Total Plant Nitrogen Assimilation

Vegetation has the capacity to assimilate atmospheric sources of N both through root and foliar pathways. Through modeling, direct foliar incorporation of reactive N has been estimated to contribute 3–16% of the plant N demand for new growth tissue (Holland *et al.* 2004).

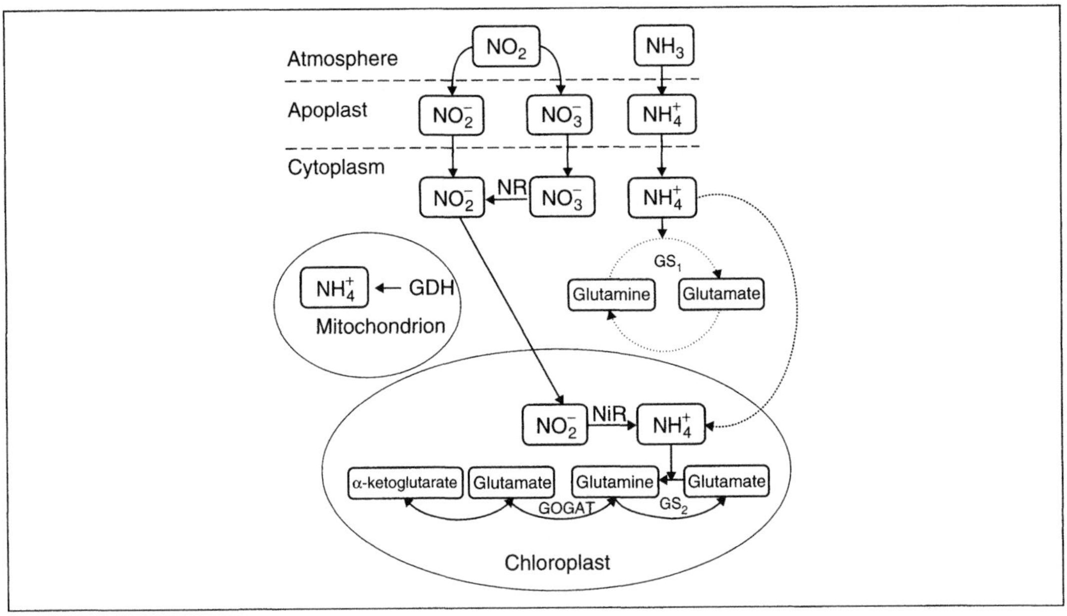

FIGURE 7.2 Primary metabolic processes involved in the foliar uptake and assimilation of atmospheric ammonia (NH_3) and nitrogen dioxide (NO_2). NH_3 and NO_2 enter the apoplast via stomatal or cuticular uptake, are assimilated into primary N metabolism by various biochemical pathways. Dashed lines indicate the possible role of cytosolic GS_1 or chloroplastic GS_2 in NH_4^+ assimilation derived from atmospheric NH_3 (adapted from Stulen *et al.* 1998).

Although this is a small proportion of the total N required for growth, N directly assimilated via foliar uptake is a direct addition to plant metabolism, would not necessarily influence soil acidity, and would likely influence plant productivity and carbon assimilation in N-limited terrestrial ecosystems. This additional source of N to plants could also increase plant growth under increased atmospheric carbon dioxide (CO_2). Carbon dioxide fertilization effects, when they are observed, are often not sustained as the plant community becomes limited by some secondary limitation. This limitation is often hypothesized to be N availability in many ecosystems (Beedlow *et al.* 2004). Plants become N deficient because of the relatively higher availability of carbon substrate for photosynthesis. Under such a scenario, N available through foliar incorporation would enhance or at least sustain more rapid growth under elevated CO_2. Experiments directly quantifying the proportion of N assimilated through foliage or demonstrating a positive relationship between foliar uptake of N and elevated CO_2 have not yet been forthcoming.

B. Variation in Foliar Nitrogen Assimilation Via the Foliar Uptake Pathway

The ability of vegetation to incorporate atmospheric reactive N via foliar uptake has been demonstrated, but few studies have examined the variation in leaf N assimilation among plant species. For example, Morikawa *et al.* (1998) documented a 600-fold variation in leaf N assimilation among 217 plant taxa. However, most studies, including this one, have been limited by the use of unnaturally high fumigation levels and limited mechanistic explanations as to why some species have a higher capacity for foliar uptake than others.

Reactive N uptake by leaves is strongly controlled by diffusion processes (Thoene *et al.* 1991, Weber and Rennenberg 1996) and variation in stomatal dynamics among species likely explains some of the observed variation in foliar uptake capacity. However, variation in foliar incorporation of reactive N among species is likely controlled by factors in addition to diffusional resistance. For example, if stomatal conductance remains fairly constant, variations in the biochemical capacity for NO_2 assimilation would alter the substomatal air cavity NO_2 concentration, and subsequently influence the NO_2 concentration gradient into the leaf and the rate of foliar NO_2 uptake (Hereid and Monson 2001). This "internal" resistance to uptake is often termed the mesophyll resistance and has been suggested to be controlled by a number of factors including: (1) the permeability of NO_3^-/NO_2^- ions through cell membranes and walls (Ammann *et al.* 1995), (2) the site of the activity in the primary nitrate assimilation pathway through which NO_y is metabolized (Wellburn 1990), and (3) the activity of ascorbic acid and other antioxidant compounds in the leaf apoplast (Teklemariam and Sparks 2004).

Work by Eller and Sparks (2006) has simultaneously investigated the hypothesized controlling factors over reactive N flux into leaves. In this study, the investigators measured stomatal conductance as an estimate of the diffusional resistance to gases entering the leaf, the concentration of apoplastic ascorbate to examine the reaction rates between gases and chemicals within cell walls of leaf tissue, and the activity of the enzyme NR as a proxy for the transport rate of NO_3^- out of the apoplast and then incorporated the strengths of each of these controlling factors into a multiple regression model describing leaf uptake of NO_2. As in previous studies, stomatal conductance was found to be the primary control over the foliar incorporation of NO_2. However, NR activity and leaf ascorbate concentrations explained significant variance in NO_2 leaf flux, suggesting that enzyme activities and antioxidant concentrations in the cell wall likely underlie the differences we observe across species in mesophyllic resistance to NO_2 uptake. For example, NR activity is variable across species and such differences could account for the broad variation in rates of NO_2 incorporation (Morikawa *et al.* 1998) because plants expressing higher NR activities may have higher NO_2 uptake capacity at elevated NO_2 levels where the cell wall may be saturated with NO_3^-/NO_2^- (Eller and Sparks 2006). Furthermore, antioxidants such as ascorbate could also contribute to uptake capacity depending on their concentration and associated ability to react with oxidants such as NO_2. Ultimately, the concentration of ascorbate may control the steady state flux of NO_2 into a leaf. Increasing our knowledge of these leaf biochemical processes will aid in our understanding of the existing variation in foliar uptake capacity among species.

IV. FOLIAR $\delta^{15}N$ AS A TOOL TO DETERMINE THE MAGNITUDE OF FOLIAR NITROGEN INCORPORATION

The differences in $\delta^{15}N$ between atmospheric N, plant-available soil N, and N derived from biological fixation may be used as a natural tracer in plant systems. Further, the $\delta^{15}N$ of atmospheric N compounds produced from anthropogenic activities may be significantly different from the natural, background N sources in the soil (Freyer 1991). This anthropogenic signal in foliar $\delta^{15}N$ has been reported in several studies (Qiao and Murray 1997, Ammann *et al.* 1999, Siegwolf *et al.* 2001) and may be a reliable indicator of N pollution addition. For example, the $\delta^{15}N$ of atmospheric NO_y pollution from traffic exhausts and industrial combustion processes is mostly positive (–1% to +5%) and atmospheric NH_3 originating from the volatilization of animal waste is usually very depleted (Heaton 1986, Macko and Ostrom 1994) compared to natural sources of inorganic N (*e.g.*, NO_3^- from mineralization) that are usually negative [between –5‰ and –2‰ (Nadelhoffer and Fry 1994)]. Because of these differences, the relative magnitude of different N sources can be observed at the leaf level (Ammann *et al.* 1999, Siegwolf *et al.* 2001). Only if there are differences in $\delta^{15}N$ among potential N sources can $\delta^{15}N$ be used as an indicator for any of these sources (Robinson 2001).

Assuming the $\delta^{15}N$ of the various sources are known, the amount of N assimilated by the plant can be quantified. The final isotopic composition of the leaf tissue is a combination of the various source values and masses. Take, for example, a plant that receives N from two sources: root uptake of soil NO_3^- and foliar uptake of atmospheric NO_2. If the isotopic ratios of both sources are known, we can use a two-ended mixing model and the proportion of N obtained from foliar uptake (X_{foliar}) is:

$$X_{foliar} = \frac{\delta_{plant} - \delta_{soil}}{\delta_{atmosphere} - \delta_{soil}} \quad (1)$$

where δ_{plant}, δ_{soil}, and $\delta_{atmosphere}$ are the N isotope compositions of the leaf, soil, and atmospheric N sources, respectively.

V. CURRENT CHALLENGES FOR USING FOLIAR δ^{15}N AS A TOOL TO DETERMINE THE MAGNITUDE OF FOLIAR NITROGEN INCORPORATION

The obstacles for using mixing models like the one described above and interpreting the foliar δ^{15}N value lie in the presence of simultaneously available multiple N sources and currently unknown fractionation events that occur during N transformation, incorporation, and assimilation. As discussed in Section II.A, this is a significant issue in the soil system allowing the generation of various foliar δ^{15}N values driven by differences in soil N cycling. Although the mechanisms are largely unknown, differences between the δ^{15}N of plant material versus soil δ^{15}N suggest some within-plant fractionation of N (Robinson *et al.* 1998, Comstock 2001). In addition, foliar incorporation of reactive N presumably exhibits an array of largely unknown fractionation events as well. For example, N gases assimilated through the stomata both have variable values in the atmosphere and are likely influenced by the same fractionation events caused by diffusion that influence the incorporation of CO_2 during photosynthesis. Therefore, a large fractionation is expected to occur during the diffusion of NO_2 into the leaf. Fractionations associated with transport into plant cells and N reduction enzymes, although not yet quantified, are also likely to occur. Further fractionation likely occurs during diffusion and biochemical assimilation pathways by a fine-scale separation of pools. As an example, in the substomatal cavity, NO_2 rapidly disproportionates in water within the cell wall and is converted to NO_3^-. Accurately defining the difference in isotopic composition between the gaseous NO_2 in the air and NO_3^- in the cell wall would describe the fractionation due to stomatal diffusion and disproportionation of NO_2. However, for foliar incorporation of reactive N, none of these fractionation events have been adequately resolved.

Developing a more complete understanding of the source values and fractionation events associated with foliar δ^{15}N in the plant–soil system is necessary to predict the importance of various N sources on the plant N budget (Handley *et al.* 1998). Promising techniques are being developed and used to elucidate some soil processes, such as dual-isotope techniques [*e.g.*, δ^{15}N and δ^{18}O in NO_3^- (Durka *et al.* 1994)]. Ultimately, a solid understanding of these processes will be necessary for the production of a mechanistic model describing how δ^{15}N changes during plant N acquisition and assimilation from both foliar and soil uptake pathways (Dawson *et al.* 2002).

VI. PROMISING METHODS USING FOLIAR δ^{15}N AS AN INDICATOR OF DIRECT LEAF NITROGEN ASSIMILATION

A possible strategy for quantifying the foliar incorporation of atmospheric reactive N pollution using foliar δ^{15}N values is to enhance the magnitude of differences in δ^{15}N in the various N pools against a background level such that the difference in signal may provide a quantifiable tracer.

In the laboratory, the external chemical and isotopic composition of the N source(s) can be identified and controlled. We describe this type of laboratory experiment in Section VI.A. In the field, a significant ^{15}N-enrichment of the gas-phase N source would be necessary to create a large enough signal separation. The naturally occurring range of $\delta^{15}N$ found in nature is generally no more than ±20‰ (Handley and Scrimgeour 1997). Therefore, the ^{15}N-enrichment of the gas source would need to be large enough to overcome this natural variation. Systems enriched in ^{15}N ameliorate many of the problems of interpretation brought about by fractionation among N pools and are useful for following the movement of N through a plant and for determining rates of N assimilation within the plant (Nadelhoffer and Fry 1994, Dawson *et al.* 2002). In the following sections, we discuss several of the more promising laboratory and field efforts using ^{15}N to quantify foliar N incorporation.

A. Hydroponics

Hydroponics systems allow for the control of the isotopic composition of the nutrient solution N and, if adequately replenished, remain constant through time. Traditional hydroponics systems have been used to quantify foliar $\delta^{15}N$ values (Yoneyama and Kaneko 1989, Evans *et al.* 1996, Qiao and Murray 1997, 1998, Pritchard and Guy 2005) and to explore the impact of NO_2 fumigation on root uptake of N (Muller *et al.* 1996). However, these studies did not directly address the magnitude nor the consequences of foliar uptake of reactive N.

We have used combined hydroponics–fumigation systems to address several of the unresolved questions associated with the foliar uptake of reactive N (Figure 7.3). Plants are transplanted into the system with the root systems suspended in nutrient solution and the shoot systems enclosed in a fumigation chamber. Using this system, we can simultaneously and independently control the forms, external concentrations, and isotopic compositions of N sources available to plant root and shoot systems. After specific exposure protocols (*i.e.*, different combinations of N gas concentrations to the leaves and N fertilizer to roots), plant biomass, and isotopic composition measurements are used to quantify the proportion of each N source utilized and the overall influence of the exposure protocol

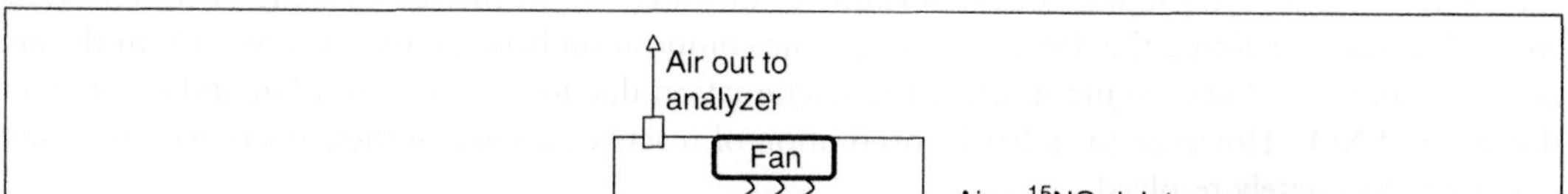

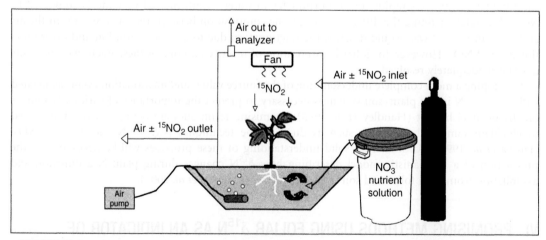

FIGURE 7.3 Design of an individual hydroponics–fumigation system used with seedlings for long-term (weeks) experiments. The root system is suspended in the nutrient solution while the shoot system is contained within the fumigation chamber. The seedling is sealed at the root–shoot junction with modeling clay to isolate the N sources. The nutrient solution is refreshed on a weekly basis and aerated by an aquarium pump. The NO_2 concentration of the incoming air to the fumigation chamber (Plexiglas) is fixed by mixing concentrated $^{15}NO_2$ gas with clean air by using mass flow controllers. It is vigorously mixed in the fumigation chamber by a fan.

on plant growth. Using this system, we have determined the partitioning of plant N among sources and the potential magnitude of NO_2 assimilation in tomato (*Lycopersicon esculentum*). Depending on the root availability of NO_3^-, NO_2 derived N in leaf tissue varied between 1% and 7% at the end of the fumigation (Figure 7.4).

We have also begun to address the variation in foliar N assimilation among plant species at realistic concentrations and how this assimilation may influence foliar $\delta^{15}N$ values. Using the same hydroponics–fumigation system to quantify species differences in foliar NO_2 assimilation, we have examined differences in growth and foliar $\delta^{15}N$ values in two species that contrast in their N reduction strategies: tomato (*L. esculentum*) and tobacco (*Nicotiana tabacum*). In natural and cultivated systems, *L. esculentum* primarily reduces N in the roots, while *N. tabacum* primarily reduces N in the leaves, suggesting that *N. tabacum* may have an increased capacity to incorporate and assimilate atmospheric N sources via the leaves. Seed-grown plants were transplanted into the system and exposed to low or high (50 and 500 μM) root NO_3^- supply and filtered or $^{15}NO_2$-enriched air (0, 20, and 40 ppb) for 4 weeks. A key finding from this work is that both species are capable of incorporating up to 15% of their total assimilated N using the foliar uptake pathway, although no significant variation in foliar uptake capacity between species was observed at the whole-plant level (Figure 7.5). This result suggests that the foliar incorporation of NO_2 contributes a significant proportion of N to plant metabolism and is consistent with modeled values of foliar uptake in the literature (Muller *et al.* 1996, Ammann *et al.* 1999). We also observed a NO_2-induced stimulation in whole-plant biomass production in both species similar to those observed in some fumigation studies (Muller *et al.* 1996, Siegwolf *et al.* 2001). These results suggest that NO_2 is potentially beneficial to biomass production and can change whole-plant allocation patterns under certain conditions. Further analysis of NR activity is necessary to highlight any species differences in foliar uptake capacity.

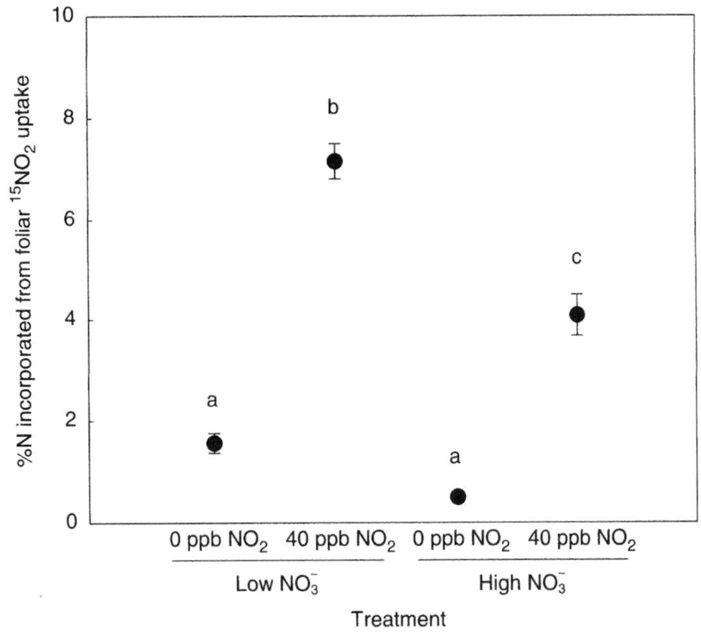

FIGURE 7.4 Proportion (%) of total N content attributable to the direct foliar incorporation of $^{15}NO_2$-N in *L. esculentum*. Data are means ±1 SE. Data with distinct lowercase letters indicate significant differences. From Vallano and Sparks (unpublished).

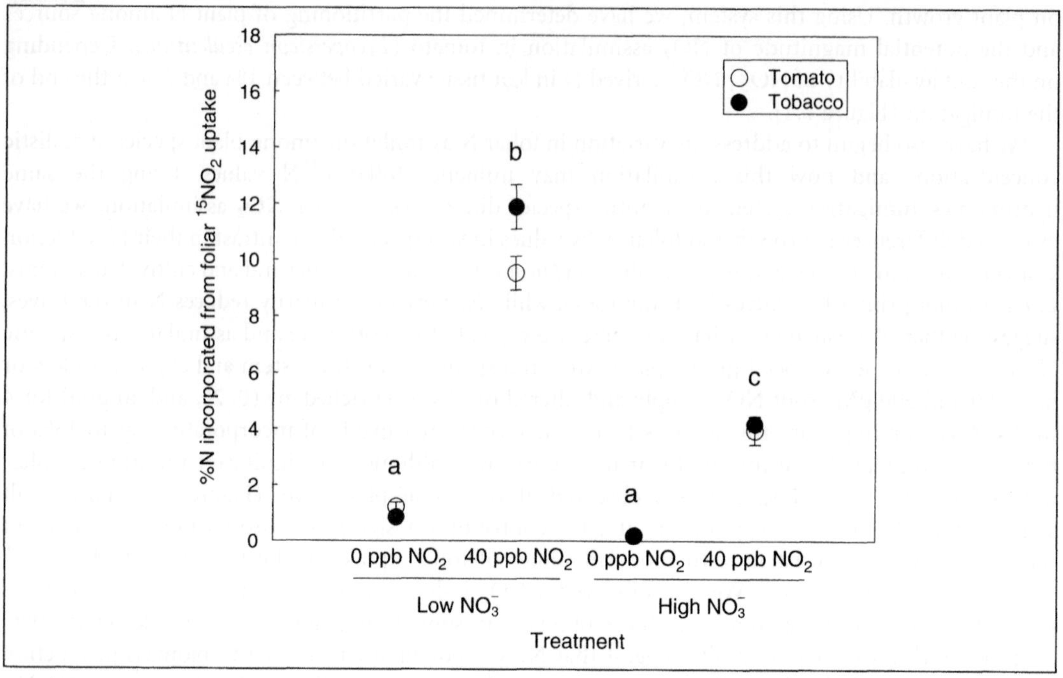

FIGURE 7.5 Proportion (%) of total N content attributable to the direct foliar incorporation of $^{15}NO_2$-N in *N. tabacum* (●) *and L. esculentum* (○). Data are means ±1 SE. Data with distinct lowercase letters indicate significant differences for both species. From Vallano and Sparks (unpublished).

Obviously, hydroponics–fumigation experiments like the ones we describe here are not an adequate simulation of natural conditions. However, they do lend significant insight into the transport of N pollutants into plants, and the physiological mechanisms that influence the capacity of various plant species to incorporate atmospheric reactive N pollution via foliar uptake. Our future work using these systems will include multiple species comparisons of foliar uptake capacity, combined applications of multiple atmospheric N pollution sources, and quantification of the fractionation events associated with the foliar N incorporation and how this influences foliar $\delta^{15}N$ values.

B. Field Studies

Controlled laboratory experiments have yielded promising results for using foliar $\delta^{15}N$ measurements as indicators of reactive N incorporation through leaves. However, can we use these, or similar, methods to estimate the proportion of foliar uptake of anthropogenic reactive N in natural systems? Multiple field studies have used foliar $\delta^{15}N$ values to estimate the contribution of total atmospheric N pollution in terrestrial ecosystems (Ammann *et al.* 1999, Siegwolf *et al.* 2001, Gessler *et al.* 2002, Wania *et al.* 2002). However, this work did not attempt to separate incorporation pathways and most were likely complicated by many of the soil factors discussed earlier (*e.g.*, mycorrhizal effects, N_2 fixation, heterogeneity of soil N sources, and transformations). Currently, no consistent mechanistic relationship has been reported between atmospheric N pollution, foliar uptake of atmospheric N pollution, and foliar $\delta^{15}N$ values.

In temperate forest ecosystems, high concentrations of total N pollution addition to ecosystems have resulted in the ^{15}N enrichment of soil and plants (Stewart *et al.* 2002). Measurements of foliar $\delta^{15}N$ in pine needles representing both the soil and foliar incorporation of N deposition in heavily polluted areas of Germany varied by an order of magnitude depending on the plants proximity to various emission sources (Jung *et al.* 1997). It has been suggested these trends are the result of the loss of ^{15}N depleted N from soil and plants due to volatilization causing enrichment of the residual N fraction (Hogberg *et al.* 1999). This is likely true, but the role and importance of foliar incorporation in determining the foliar value and what this means in terms of plant function is still unknown and likely an important factor in these measurements.

Ammann *et al.* (1999) attempted to quantify the influences of soil and foliar uptake pathways on pollutant N assimilation by sampling whole plant communities over a growing season and examining the resulting patterns of foliar $\delta^{15}N$. They estimated NO_2 uptake in *Picea abies* by analyzing atmospheric NO_2, foliar $\delta^{15}N$, and soil $\delta^{15}N$ values along a highway in Switzerland. This study showed that leaves of plants growing in proximity to an emission source reflected the ^{15}N values of the N pollution (Ammann *et al.* 1999). Specifically, the foliar $\delta^{15}N$ values increased from –3.7‰ to –1.5‰ within one growing season. Total anthropogenic inputs of both soil-derived N (–3.7‰) and gaseous NO_2 (5.7‰) contributed about 25% to the total N budget of vegetation adjacent to the emissions source. However, the experimental methodology did not allow for the separation of soil versus foliar uptake sources of anthropogenic N. Additionally, whether the differences in $\delta^{15}N$ are due to differential soil cycling, fractionations within the plant, or foliar uptake is unknown. It is critical that future research address the importance and resulting impacts of these fundamentally different modes of anthropogenic N incorporation to terrestrial ecosystems.

VII. FUTURE RESEARCH

Foliar $\delta^{15}N$ measurements have the potential to be powerful indicators of the relative contribution of soil versus foliar uptake of atmospheric reactive N inputs to terrestrial ecosystems in both rural and urban environments. Natural abundance foliar $\delta^{15}N$ measurements provide us with an integration of the $\delta^{15}N$ of the N sources and fractionation events that occur during foliar N incorporation, allocation, and emissions of N compounds from vegetation. However, the understanding of sources and fractionation events is incomplete and using foliar $\delta^{15}N$ as a tool to quantify a single process, like foliar uptake, is challenging. Although modeling efforts have made significant headway in identifying processes that influence foliar $\delta^{15}N$ (Comstock 2001), no work has examined the fractionation events explicitly associated with foliar N incorporation. For foliar $\delta^{15}N$ measurements to become an effective research tool in determining foliar uptake of N pollution, we must further develop models for ^{15}N uptake and discrimination by leaves to a point analogous to those developed for plant carbon isotope composition where all sources and fractionation events are considered (Farquhar *et al.* 1982, Evans 2001).

Additionally, the use of dual isotope methods promises to be a useful tool to investigate the impact of reactive N exposure; not only foliar uptake of atmospheric N pollution, but also as a way to analyze changes in physiology, morphology, and carbon partitioning in vegetation exposed to atmospheric reactive N sources (Durka *et al.* 1994, Siegwolf *et al.* 2001). Advances in methodology using $^{13}C/^{15}N$ and $^{18}O/^{15}N$ dual isotope approaches (Cernusak *et al.* 2002, Gaudinski *et al.* 2005, McIlvin and Casciotti 2006) provide great potential to expand our knowledge of N source partitioning and fractionation events along multiple spatial and temporal scales. In particular, natural abundance ratios of $^{18}O/^{16}O$ in atmospheric NO_3^- sources differ from those of nitrification derived NO_3^- and may be a powerful diagnostic tool. Together with $^{15}N/^{14}N$, stable oxygen isotope analysis of leaf water could be very useful in determining the fate of anthropogenic N sources to vegetation. This method has successfully been applied to identifying anthropogenic N sources to stream water (Piatek *et al.* 2005), but has not yet been applied to plant N dynamics, an area ripe for further investigation.

Laboratory-based hydroponics–fumigation systems provide a foundation to test and verify our current understanding of the foliar uptake of pollutant N and how this additional source of reactive N to plants will affect the plant N budget. It may be possible to use a parallel strategy to that applied to hydroponics–fumigation systems in the field. Examining the divergence between natural abundance foliar $\delta^{15}N$ and the $\delta^{15}N$ of extractable NO_3^-/NH_4^+ (plant-available soil N) along a gradient of anthropogenic N deposition will assess whether the difference in magnitude between these measurements reflects the incorporation of atmospheric N pollution directly via the leaves. Any such effort will need to account for the influence of mycorrhizae and verify the methodology of accurately quantifying soil solution $\delta^{15}N$.

Currently, there are no published experiments that have directly quantified the proportion of N assimilated through foliage or quantitatively demonstrated a positive relationship between foliar N uptake and anthropogenic N pollution in natural systems. Clearly, as atmospheric reactive N deposition increases across the globe, we need to develop effective methods to identify and trace the resulting ecological change due to human-mediated additions of reactive N. Future efforts must include both estimates of total N deposition and archives of foliar $\delta^{15}N$ values across a variety of ecosystems.

VIII. CONCLUSIONS

This chapter highlights the importance and utility of the foliar $\delta^{15}N$ value as a tool for determining the magnitude and impact of atmospheric N pollution directly entering the foliage of vegetation. Substantial progress has been made in furthering our understanding of the mechanisms and environmental factors that control foliar $\delta^{15}N$ and the contribution of foliar uptake of anthropogenic N pollution to individual plant N budgets. Recent hydroponics–fumigation studies have successfully used foliar $\delta^{15}N$ as a tool for investigating and quantifying the proportion of foliar incorporation of atmospheric N pollution. *In situ* coupled measurements of natural abundance foliar and soil solution $\delta^{15}N$ across a range of ecosystems and species will shed further light on the validity of foliar $\delta^{15}N$ as an accurate indicator of ecological change in plant N dynamics. A firm understanding of both the magnitude of foliar N incorporation and whether foliar ^{15}N values may be used as a tracer to quantify pollutant N assimilation is imperative for understanding the future global N cycle.

IX. REFERENCES

Aber, J. D., C. L. Goodale, S. V. Ollinger, M. Smith, A. H. Magill, M. E. Martin, R. A. Hallett, and J. L. Stoddard. 2003. Is nitrogen deposition altering the nitrogen status of northeastern forests? *Bioscience* **53**:375–389.

Ammann, M., P. vonBallmoos, M. Stalder, M. Suter, and C. Brunold. 1995. Uptake and assimilation of atmospheric NO_2-N by spruce needles (*Picea abies*): A field study. *Water, Air, and Soil Pollution* **85**:1497–1502.

Ammann, M., R. T. W. Siegwolf, F. Pichlmayer, M. Suter, M. Saurer, and C. Brunold. 1999. Estimating the uptake of traffic-derived NO_2 from ^{15}N abundance in Norway spruce needles. *Oecologia* **118**:124–131.

Anderson, J. B., R. E. Baumgardner, V. A. Mohnen, and J. J. Bowser. 1999. Cloud chemistry in the eastern United States, as sampled from three high-elevation sites along the Appalachian Mountains. *Atmospheric Environment* **33**:5105–5114.

Asner, G. P., T. R. Seastedt, and A. R. Townsend. 1997. The decoupling of terrestrial carbon and nitrogen cycles. *Bioscience* **47**:226–234.

Beedlow, P. A., D. T. Tingey, D. L. Phillips, W. E. Hogsett, and D. M. Olszyk. 2004. Rising atmospheric CO_2 and carbon sequestration in forests. *Frontiers in Ecology and Environment* **2**:315–322.

Bobbink, R., and J. G. M. Roelofs. 1995. Nitrogen critical loads for natural and semi-natural ecosystems: The empirical approach. *Water, Air, and Soil Pollution* **85**:2413–2418.

Bobbink, R., M. Hornung, and J. G. M. Roelofs. 1998. The effects of air-borne nitrogen pollutants on species diversity in natural and semi-natural European vegetation. *Journal of Ecology* **86:**717–738.

Byterowicz, A., K. Percy, G. Riechers, P. Padgett, and M. Krywult. 1998. Nitric acid vapor effects on forest trees—deposition and cuticular changes. *Chemosphere* **36:**697–702.

Cernusak, L. A., J. S. Pate, and G. D. Farquhar. 2002. Diurnal variation in the stable isotope composition of water and dry matter in fruiting Lupinus angustifolius under field conditions. *Plant Cell Environment* **25:**893–907.

Comstock, J. P. 2001. Steady-state isotopic fractionation in branched pathways using plant uptake of NO_3^- as an example. *Planta* **214:**220–234.

Dawson, T. E., S. Mambelli, A. H. Plamboeck, P. H. Templer, and K. P. Tu. 2002. Stable isotopes in plant ecology. *Annual Review of Ecology and Systematics* **33:**507–559.

Day, D. A., M. B. Dillon, P. J. Wooldridge, J. A. Thornton, R. S. Rosen, E. C. Wood, and R. C. Cohen. 2003. On alkyl nitrates, O-3, and the "missing NOy." *Journal of Geophysical Research (D)* **108.**

Driscoll, C. T., D. Whitall, J. Aber, E. Boyer, M. Castro, C. Cronan, C. L. Goodale, P. Groffman, C. Hopkinson, K. Lambert, G. Lawrence, and S. Ollinger. 2003. Nitrogen pollution in the northeastern United States: Sources, effects, and management options. *Bioscience* **53:**357–374.

Durka, W., E. D. Schulze, G. Gebauer, and S. Voerkelius. 1994. Effects of forest decline on uptake and leaching of deposited nitrate determined from ^{15}N and ^{18}O measurements. *Nature* **372:**765–767.

Eller, A., and J. P. Sparks. 2006. Predicting leaf-level fluxes of ozone and nitrogen dioxide: The relative roles of diffusion and biochemical processes. *Plant, Cell & Environment* **29:**1742–1750.

EPA, U. E. P. A. 2000. National emissions inventory: Air pollutant emissions trends. [Available online from www.epa.gov/ttn/chief/trends/trends98/index.html].

Evans, R. D. 2001. Physiological mechanisms influencing plant nitrogen isotope composition. *Trends in Plant Science* **6:**121–126.

Evans, R. D., A. J. Bloom, S. S. Sukrapanna, and J. R. Ehleringer. 1996. Nitrogen isotope composition of tomato (*Lycopersicon esculentum* Mill. cv. T-5) grown under ammonium or nitrate nutrition. *Plant, Cell & Environment* **19:**1317–1323.

Fahey, T. J., C. J. Williams, J. N. Rooney-Varga, C. C. Cleveland, K. M. Postek, S. D. Smith, and D. R. Bouldin. 1999. Nitrogen deposition in and around an intensive agricultural district in central New York. *Journal of Environmental Quality* **28:**1585–1600.

Farquhar, G. D., M. H. Oleary, and J. A. Berry. 1982. On the relationship between carbon isotope discrimination and the intercellular carbon-dioxide concentration in leaves. *Australian Journal of Plant Physiology* **9:**121–137.

Galloway, J. N., J. D. Aber, J. W. Erisman, S. P. Seitzinger, R. W. Howarth, E. B. Cowling, and B. J. Cosby. 2003. The nitrogen cascade. *Bioscience* **53:**341–356.

Galloway, J. N., F. J. Dentener, D. G. Capone, E. W. Boyer, R. W. Howarth, S. P. Seitzinger, G. P. Asner, C. C. Cleveland, P. A. Green, E. A. Holland, D. M. Karl, A. F. Michaels, *et al.* 2004. Nitrogen cycles: Past, present, and future. *Biogeochemistry* **70:**153–226.

Gaudinski, J. B., T. E. Dawson, S. Quideau, E. A. G. Schuur, J. S. Roden, S. E. Trumbore, D. R. Sandquist, S. W. Oh, and R. E. Wasylishen. 2005. Comparative analysis of cellulose preparation techniques for use with C-13, C-14, and O-18 isotopic measurements. *Analytical Chemistry* **77:**7212–7224.

Gessler, A., M. Rienks, and H. Rennenberg. 2000. NH_3 and NO_2 fluxes between beech trees and the atmosphere—correlation with climatic and physiological parameters. *New Phytology* **147:**539–560.

Gessler, A., M. Rienks, and H. Rennenberg. 2002. Stomatal uptake and cuticular adsorption contribute to dry deposition of NH_3 and NO_2 to needles of adult spruce (*Picea abies*) trees. *New Phytology* **156:**179–194.

Gilliam, F. S. 2006. Response of the herbaceous layer of forest ecosystems to excess nitrogen deposition. *Journal of Ecology* **94:**1176–1191.

Handley, L. L., and C. M. Scrimgeour. 1997. Terrestrial plant ecology and the ^{15}N natural abundance: The present limits to interpretation for uncultivated systems with original data for a Scottish old field. *Advances in Ecology Research* **27:**133–212.

Handley, L. L., and J. A. Raven. 1992. The use of natural abundance of nitrogen isotopes in plant physiology and ecology. *Plant, Cell & Environment* **15:**965–985.

Handley, L. L., C. M. Scrimgeour, and J. A. Raven. 1998. ^{15}N at natural abundance levels in terrestrial vascular plants: A precis. Pages 89–98 *in* H. Griffiths (Ed.) *Stable Isotopes: Integration of Biological, Ecological and Geochemical Processes.* Bios Scientific Publishers. Oxford, UK.

Hanson, P. J., and S. E. Lindberg. 1991. Dry deposition of reactive nitrogen-compounds—a review of leaf, canopy and non-foliar measurements. *Atmospheric Environment* **25A:**1615–1634.

Heaton, T. H. E. 1986. Isotopic studies of nitrogen pollution in the hydrosphere and atmosphere: A review. *Chemical Geology* **59:**87–102.

Hereid, D. P., and R. K. Monson. 2001. Nitrogen oxide fluxes between corn (*Zea mays* L.) leaves and the atmosphere. *Atmospheric Environment* **35:**975–983.

Hill, A. C. 1971. Vegetation: A sink for atmospheric air pollutants. *Journal of the Air Pollution Control Association* **21:**341.

Hogberg, P., M. N. Hogberg, M. E. Quist, A. Ekblad, and T. Nasholm. 1999. Nitrogen isotope fractionation during nitrogen uptake by ectomycorrhizal and non-mycorrhizal *Pinus sylvestris. New Phytology* **142:**569–576.

Holland, E. A., S. B. Bertman, M. A. Carroll, A. B. Guenther, P. B. Shepson, J. P. Sparks, K. Barney, and J. Lee-Taylor. 2004. A U.S. Nitrogen Science Plan: Atmospheric-Terrestrial Exchange of Reactive Nitrogen. UCAR, Boulder, CO. 38 p.

Hopkins, D. W., R. E. Wheatley, and D. Robinson. 1998. Stable isotope studies of soil nitrogen. Pages 75–88 *in* H. Griffiths (Ed.) *Stable Isotopes: Integration of Biological, Ecological, and Geochemical Processes.* Bios Scientific Publishers, Oxford, UK.

Jung, K., G. Gebauer, M. Gehre, D. Hofmann, L. Weissflog, and G. Schuurmann. 1997. Anthropogenic impacts on natural nitrogen isotope variations in *Pinus sylvestris* stands in an industrially polluted area. *Environtal Pollution* **97:**175–181.

Kolb, K. J., and R. D. Evans. 2002. Implications of leaf nitrogen recycling on the nitrogen isotope composition of deciduous plant tissues. *New Phytology* **156:**57–64.

Lea, P. J. 1998. Oxides of nitrogen and ozone: Can our plants survive? *New Phytoogy* **139:**25–26.

Lovett, G. M. 1994. Atmospheric deposition of nutrients and pollutants in North America: An ecological perspective. *Ecological Applications* **4:**629–650.

Macko, S. A., and N. E. Ostrom. 1994. Pollution studies using stable isotopes. Pages 45–62 *in* K. Lajtha and R. Michner (Eds.) *Stable Isotopes in Ecology.* Blackwell Scientific Publishers, Oxford, UK.

McIlvin, M. R., and K. L. Casciotti. 2006. Method for the analysis of delta O-18 in water. *Analytical Chemistry* **78:**2377–2381.

Morikawa, H., A. Higaki, M. Nohno, M. Takahasi, M. Kamada, M. Nakata, G. Toyohara, Y. Okamura, K. Matsui, S. Kitani, K. Fujita, K. Irifune, *et al.* 1998. More than a 600-fold variation in nitrogen dioxide assimilation among 217 plant taxa. *Plant, Cell & Environment* **21:**180–190.

Muller, B., B. Touraine, and H. Rennenberg. 1996. Interaction between atmospheric and pedospheric nitrogen nutrition in spruce (*Picea abies* L Karst) seedlings. *Plant, Cell & Environment* **19:**345–355.

Nadelhoffer, K. J., and B. Fry. 1994. Nitrogen isotopes in forest ecosystems. Pages 22–44 *in* K. Lajtha and R. H. Michener (Eds.) *Stable Isotopes in Ecology and Environmental Science.* Blackwell, Oxford, UK.

Neff, J. C., E. A. Holland, F. J. Dentener, W. H. McDowell, and K. M. Russell. 2002. The origin, composition and rates of organic nitrogen deposition: A missing piece of the nitrogen cycle? *Biogeochemistry* **57:**99–136.

Pearson, J., and A. Soares. 1995. A hypothesis of plant susceptibility to atmospheric pollution based on intrinsic nitrogen metabolism: Why acidity really is the problem. *Water, Air, and Soil Pollution* **85:**1227–1232.

Peierls, B. L., and H. W. Paerl. 1997. Bioavailability of atmospheric organic nitrogen deposition to coastal phytoplankton. *Limnology and Oceanography* **42:**1819–1823.

Phoenix, G. K., W. K. Hicks, S. Cinderby, J. C. I. Kuylenstierna, W. D. Stock, F. J. Dentener, K. E. Giller, A. T. Austin, R. D. B. Lefroy, B. S. Gimeno, M. R. Ashmore, and P. Ineson. 2006. Atmospheric nitrogen deposition in world biodiversity hotspots: The need for a greater global perspective in assessing N deposition impacts. *Global Change Biology* **12:**470–476.

Piatek, K. B., M. J. Mitchell, S. R. Silva, and C. Kendall. 2005. Sources of nitrate in snowmelt discharge: Evidence from water chemistry and stable isotopes of nitrate. *Water, Air, and Soil Pollution* **165:**13–35.

Pritchard, E. S., and R. D. Guy. 2005. Nitrogen isotope discrimination in white spruce fed with low concentrations of ammonium and nitrate. *Trees—Structure and Function* **19:**89–98.

Prospero, J. M., K. Barrett, T. Church, F. Dentener, R. A. Duce, J. N. Galloway, H. Levy, J. Moody, and P. Quinn. 1996. Atmospheric deposition of nutrients to the North Atlantic Basin. *Biogeochemistry* **35:**27–73.

Qiao, Z., and F. Murray. 1997. The effects of root nitrogen supplies on the absorption of atmospheric NO_2 by soybean leaves. *New Phytology* **136:**239–243.

Qiao, Z., and F. Murray. 1998. The effects of NO_2 on the uptake and assimilation of nitrate by soybean plants. *Environmental and Experimental Botany* **39:**33–40.

Robinson, D. 2001. δ^{15}N as an integrator of the nitrogen cycle. *Trends in Ecology and Evolution* **16:**153–162.

Robinson, D., L. L. Handley, and C. M. Scrimgeour. 1998. A theory for $^{15}N/^{14}N$ fractionation in nitrate-grown vascular plants. *Planta* **205:**397–406.

Robinson, D., L. L. Handley, C. M. Scrimgeour, D. C. Gordon, B. P. Forster, and R. P. Ellis. 2000. Using stable isotope natural abundances (δ^{15}N and δ^{13}C) to integrate the stress responses of wild barley (*Hordeum spontaneum* C. Koch.) genotypes. *Journal of Experimental Botany* **51:**41–50.

Rondon, A., and L. Granat. 1994. Studies on the dry deposition of NO_2 to coniferous species at low NO_2 concentrations. *Tellus. Series B, Chemical and Physical Meteorology* **46:**339–352.

Rondon, A., C. Johansson, and L. Granat. 1993. Dry deposition of nitrogen-dioxide and ozone to coniferous forests. *Journal of Geophysical Research (D)* **98:**5159–5172.

Siegwolf, R. T. W., R. Matyssek, M. Saurer, S. Maurer, M. S. Gunthardt-Goerg, P. Schmutz, and J. B. Bucher. 2001. Stable isotope analysis reveals differential effects of soil nitrogen and nitrogen dioxide on the water use efficiency in hybrid poplar leaves. *New Phytology* **149:**233–246.

Sparks, J. P., R. K. Monson, K. L. Sparks, and M. Lerdau. 2001. Leaf uptake of nitrogen dioxide (NO_2) in a tropical wet forest: Implications for tropospheric chemistry. *Oecologia* **127:**214–221.

Sparks, J. P., J. M. Roberts, and R. K. Monson. 2003. The uptake of gaseous organic nitrogen by leaves: A significant global nitrogen transfer process. *Geophysical Research Letters* **30:**2189–2193.

Srivastava, H. S. 1992. Nitrogenous pollutants in the atmosphere: Their assimilation and phytotoxicity. *Current Science* **63:**310–317.

Stewart, G. R., M. P. M. Aidar, C. A. Joly, and S. Schmidt. 2002. Impact of point source pollution on nitrogen isotope signatures ($\delta^{15}N$) of vegetation in SE Brazil. *Oecologia* **131**:468–472.

Stulen, I., M. Perez-Soba, L. J. De Kok, and L. Van der Eerden. 1998. Impact of gaseous nitrogen deposition on plant functioning. *New Phytology* **139**:61–70.

Takahashi, M., A. Higaki, M. Nohno, M. Kamada, Y. Okamura, K. Matsui, S. Kitani, and H. Morikawa. 2005. Differential assimilation of nitrogen dioxide by 70 taxa of roadside trees at an urban pollution level. *Chemosphere* **61**:633–639.

Teklemariam, T. A., and J. P. Sparks. 2004. Gaseous fluxes of peroxyacetyl nitrate (PAN) into plant leaves. *Plant, Cell & Environment* **27**:1149–1158.

Thoene, B., P. Schroder, H. Papen, A. Egger, and H. Rennenberg. 1991. Absorption of atmospheric NO_2 by spruce (*Picea Abies* L. Karst) trees.1. NO_2 influx and its correlation with nitrate reduction. *New Phytology* **117**:575–585.

Turnipseed, A. A., L. G. Huey, E. Nemitz, R. Stickel, J. Higgs, D. J. Tanner, D. L. Slusher, J. P. Sparks, F. Flocke, and A. Guenther. 2006. Eddy covariance fluxes of peroxyacetyl nitrates (PANs) and NO_y to a coniferous forest. *Journal of Geophysical Research* **111**: doi:10.1029/2005JD006631, D09304.

Vitousek, P. M. 1994. Beyond global warming—ecology and global change. *Ecology* **75**:1861–1876.

Vitousek, P. M., J. D. Aber, R. W. Howarth, G. E. Likens, P. A. Matson, D. W. Schindler, W. H. Schlesinger, and D. G. Tilman. 1997. Human alteration of the global nitrogen cycle: Sources and consequences. *Ecological Applications* **7**:737–750.

Wania, R., P. Hietz, and W. Wanek. 2002. Natural N-15 abundance of epiphytes depends on the position within the forest canopy: Source signals and isotope fractionation. *Plant, Cell & Environment* **25**:581–589.

Weber, P., and H. Rennenberg. 1996. Exchange of NO and NO_2 between wheat canopy monoliths and the atmosphere. *Plant Soil* **180**:197–208.

Weber, P., B. Thoene, and H. Rennenberg. 1998. Absorption of atmospheric NO_2 by spruce (*Picea abies*) trees. III. Interaction with nitrate reductase activity in the needles and phloem transport. *Botanica Acta* **111**:377–382.

Wellburn, A. R. 1990. Why are atmospheric oxides of nitrogen usually phytotoxic and not alternative fertilizers? *New Phytology* **115**:395–429.

Wellburn, A. R. 1994. *Air Pollution and Climate Change: The Biological Impact.* 2nd ed. Longman Scientific and Technical, New York.

Wesely, M. L., and B. B. Hicks. 2000. A review of the current status of knowledge on dry deposition. *Atmospheric Environment* **34**:2261–2282.

Yoneyama, T., and A. Kaneko. 1989. Variations in the natural abundance of ^{15}N in nitrogenous fractions of komatsuna plants supplied with nitrate. *Plant Cell Physiology* **30**:957–962.

CHAPTER 8

The Triple Isotopic Composition of Oxygen in Leaf Water and Its Implications for Quantifying Biosphere Productivity

A. Landais,[*] D. Yakir,[†] E. Barkan,[*] and B. Luz[*]

[*] *Institute of Earth Sciences, Hebrew University*
[†] *Department of Environmental Sciences and Energy Research*
The Weizmann Institute of Science

Contents

I. INTRODUCTION

Among the main processes that affect global climate changes are the interactions between the atmosphere and biosphere. Climatic conditions control the biosphere productivity and, in turn, vegetation strongly influences climate through the emission and consumption of greenhouse gases and through the terrestrial albedo. The only effective way to understand these interactions is based on examination of the past climate changes and the associated biosphere evolution. While many different proxies exist that allow reconstruction of past climates, our knowledge of the past biosphere is still very sketchy.

For the most part, past biosphere evolution is inferred from local studies of terrestrial and marine sediments. An important representative property of the past biosphere that is extensively measured in the ocean is paleoproductivity (Kohfeld *et al.* 2005). However, these studies give information only on local variations in biosphere productivity, and much effort in data compilation is needed to infer the evolution of biosphere productivity on global scale.

The triple isotopic composition of atmospheric oxygen is a new tracer that reflects the global oxygen-based biosphere productivity (Luz *et al.* 1999). Indeed, the isotopic composition of oxygen in the lower atmosphere results from a balance between (1) the flux of oxygen between the troposphere and the biosphere associated with mass-dependent fractionations and (2) the flux of oxygen between the troposphere and the stratosphere characterized by mass-independent fractionations. Thus, variations in the triple isotopic composition of oxygen in the troposphere should reflect the relative proportions of the stratospheric and biospheric exchange fluxes.

In the present chapter, we first describe some of basic principles underlying the mass-dependent and mass-independent fractionation in the oxygen cycle. We then detail how the change in the global oxygen-based biosphere productivity can be inferred from the past changes in the triple isotopic composition of O_2. We make a particular emphasize in this context of the necessity to know precisely the relationship between $\delta^{17}O$ and $\delta^{18}O$ during leaf transpiration. We report the results of our experimental studies on transpiration isotope effects that include variations along a leaf, daily variations, the influence of plant species, and the effects of environmental and climatic conditions. Finally, we use these results to perform a global budget of the three isotopes of oxygen in the atmosphere and show the implications for the estimate of the ratio between the last glacial maximum (LGM) and the present-day oxygen biosphere productivities.

II. MASS-DEPENDENT AND MASS-INDEPENDENT FRACTIONATIONS IN THE OXYGEN CYCLE

Oxygen exists as a mixture of three different stable isotopes (^{16}O, ^{17}O, and ^{18}O). These isotopes are all incorporated into molecules (H_2O, O_2, ...), but the isotopic composition of a molecular species does not always reflect the background isotopic composition characterized by $\delta^{17}O/\delta^{18}O = 0.5$ (where $\delta^{x}O = [^{x}O/^{16}O]_{sample}/[^{x}O/^{16}O]_{reference} - 1$, and x is 17 or 18 and the reference is usually atmospheric oxygen). Equilibrium chemical processes such as isotopic exchange, diffusion, evaporation, condensation are sensitive to isotopic mass, and different isotopes may thus be incorporated into the molecules at different rates and with different bond strengths. The resulting "conventional" mass-dependent fractionation processes produce small variations around the ratio $\delta^{17}O/\delta^{18}O = 0.5$ because the mass difference between ^{18}O and ^{16}O is 2 atomic mass units (amu), whereas for ^{17}O it is 1 amu. When such relationships between ^{17}O and ^{18}O compositions are observed the isotope fractionations are termed mass dependent.

The first exception to these mass-dependent relationships was found by Clayton *et al.* (1973), who observed that in the calcium–aluminum-rich inclusions in the Allende meteorite, $\delta^{17}O/\delta^{18}O = 1$.

A decade later, Thiemens and Heidenreich (1983) showed that ozone formation yields a product enriched in both ^{17}O and ^{18}O, with $\delta^{17}O/\delta^{18}O = 1$.

On a three-isotope plot with $\delta^{17}O$ on the ordinate and $\delta^{18}O$ on the abscissa, a mass-dependent enrichment is indicated by $\delta^{17}O \approx 0.5 \times \delta^{18}O$ and a mass-independent one by $\delta^{17}O \neq 0.5 \times \delta^{18}O$ (Thiemens and Heidenreich 1983).

The ^{17}O excess (or ^{17}O anomaly, $^{17}\Delta$) is classically defined as the difference between measured and expected ^{17}O values from mass-dependent fractionation $^{17}\Delta = \delta^{17}O - \lambda \times \delta^{18}O$ (Thiemens and Heidenreich 1983), with λ the slope of the mass-dependent fractionation. Until recently, a general slope of 0.52 has been used for such purposes. This value is a compromise value representing terrestrial mass-dependent fractionation and is referred to as the "terrestrial fractionation line" (Young *et al.* 2002).

As for the global O_2 cycle, O_2 in the lower atmosphere is exchanged with the terrestrial and oceanic biosphere through photosynthesis and respiration, and with the stratosphere through mixing of tropospheric and stratospheric air. The oxygen fluxes associated with the biosphere productivity modify the isotopic composition of atmospheric oxygen through mass-dependent fractionation in respiration (O_2 uptake into water) and photosynthesis (O_2 production from water). In contrast, photochemical reactions in the stratosphere give rise to a mass-independent isotope fractionation, producing approximately equal ^{17}O and ^{18}O enrichments in stratospheric ozone and carbon dioxide. Since O_2 is the ultimate reservoir of oxygen from ozone and carbon dioxide, such reactions drive a stratospheric O_2 isotopic anomaly (Thiemens *et al.* 1991, Bender *et al.* 1994, Luz *et al.* 1999). The result is a competition between (1) the biosphere–atmosphere cycle that erases any O_2 isotopic anomaly by cycling it through water with mass-dependent isotopic fractionation and (2) the troposphere–stratosphere cycle that restocks the troposphere with isotopically anomalous stratosphere-derived O_2. The magnitude of the observed tropospheric ^{17}O anomaly should, therefore, give the balance between the two cycles, and knowledge of the rate and signal of the cycle (2) permits estimates of the rates of the global O_2 cycle through the biosphere and, therefore, its productivity (Figure 8.1, Luz *et al.* 1999, Blunier *et al.* 2002).

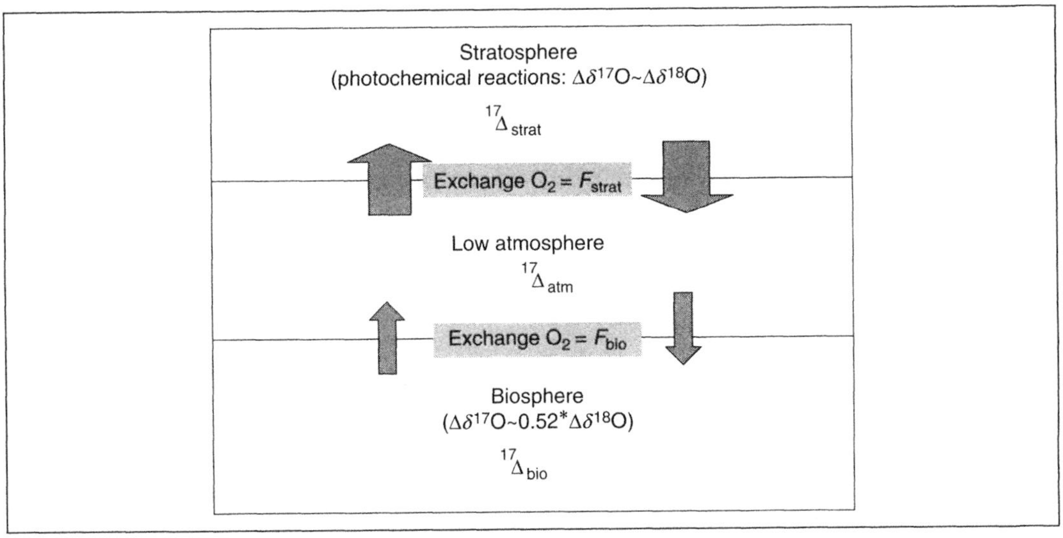

FIGURE 8.1 Simplified scheme for the exchange of O_2 between the biosphere, the stratosphere, and the low atmosphere with the associated triple isotopic composition. The isotopic balance corresponds to: $F_{strat}{*}^{17}\Delta_{strat} + F_{bio}{*}^{17}\Delta_{bio} = (F_{strat} + F_{bio}){*}^{17}\Delta_{atm}$ with F_{strat} and F_{bio} the O_2 fluxes of exchange between stratosphere and low atmosphere and between biosphere and low atmosphere. $^{17}\Delta_{strat}$, $^{17}\Delta_{bio}$, and $^{17}\Delta_{atm}$ are the isotopic anomalies associated with stratospheric O_2, biospheric O_2 (*i.e.*, O_2 consumed by respiration and produced by photosynthesis) and atmospheric O_2.

TABLE 8.1 Summary of the measured slopes (λ) for the different mass-dependent processes associated with biospheric productivity

Process	Slope (λ)
Photorespiration	0.509 ± 0.001^a
Dark respiration	$0.516 \pm 0.001^{a,b}$
Mehler reaction	0.525 ± 0.002^a
Meteoric water line (liquid precipitation)	0.528 ± 0.001^c
Evapotranspiration (h = relative humidity)	$(-0.0078 \pm 0.0026) \times h + 0.5216 \pm 0.0008^d$

[a] Helman *et al.* 2005.
[b] Angert *et al.* 2003.
[c] Meijer and Li 1998.
[d] Landais *et al.* 2006.

Variations in the triple isotopic composition of atmospheric oxygen are quite small (even when associated with major climatic changes in the past, ~40 permeg for the glacial–interglacial transition). Moreover, the advent of new analytical capabilities has only recently made it possible to determine the mass-dependent fractionation slopes with the required precision ($\sim\times 10^3$ that for conventional stable isotope methodology). Such improvements now enable a distinction between slopes of various physical and biologic processes. For example, Meijer and Li (1998) found a general slope of 0.528 for meteoric water while Helman *et al.* (2005) suggest a value of 0.525 for the Mehler reaction (O_2 photoreduction) and 0.509 for photorespiration (Table 8.1). Although the biological processes show only small variations around the expected value, it became apparent that part of the variations in the triple isotopic composition of atmospheric oxygen in the past is simply related to changes in the relative proportions of the different biospheric processes of oxygen uptake or production. It follows that in order to meaningfully interpret the past variations in terms of past oxygen productivity, it is necessary to know the exact relationships among the ratios $^{17}O/^{16}O$ and $^{18}O/^{16}O$ in all the relevant processes on Earth.

In recent years these relationships have been studied extensively in various biological processes (Angert *et al.* 2003, Helman *et al.* 2005). However, the $^{17}O/^{16}O$ versus $^{18}O/^{16}O$ trends in leaf water—the substrate of all terrestrial photosynthetic O_2 production, which is associated with large ^{18}O evaporative enrichment (Yakir 1997)—remained unknown. To this end, we studied $\delta^{17}O$ and $\delta^{18}O$ of water and the triple isotope effects associated with transpiration in leaves in field experiments in Israel, and in a European survey. This is the first study of $\delta^{17}O$ in leaf water and was reported in detail in Landais *et al.* (2006, 2007). These studies were made possible by the development of high precision methods that allow the analysis of small water samples by fluorination (Barkan and Luz 2005).

III. DEFINITIONS

On the basis of our better knowledge of the relationships between $\delta^{17}O$ and $\delta^{18}O$ in the different mass-dependent processes, the constant value of 0.52 for λ and the above definition of the ^{17}O anomaly were revised. Miller (2002) and Luz and Barkan (2005) used the δ notations and the natural log transformation to express $^{17}\Delta$ in a general form as:

$$^{17}\Delta = \ln(\delta^{17}O + 1) - \lambda \times \ln(\delta^{18}O + 1) \quad (8.1)$$

where $\delta^{*}O$ values are reported in ‰, and λ is the slope of the linear best fit line in a $\ln(\delta^{17}O + 1) - \ln(\delta^{18}O + 1)$ plot of isotopic ratios generated by mass-dependent fractionation. As stated earlier, this slope varies slightly depending on the isotope fractionation processes (kinetic or steady state) and among the different biological processes (Meijer and Li 1998, Angert *et al.* 2003, Helman *et al.* 2005, Luz and Barkan 2005). In a kinetic one-way process (substrate → product), λ is the slope of the trend line representing the reaction evolution path due to mass-dependent fractionation and thus depends on the reaction rates. Steady state corresponds to equilibrium fractionation between the substrate and the product.

In this chapter, we will only deal with steady state fractionation. Indeed, for the biospheric O_2 system, it is usually assumed that we are at or very near a steady state between photosynthetic production and respiratory consumption of O_2 gas (Bender *et al.* 1994). Near isotopic steady state is also generally assumed for evapotranspiration of leaf water (Dongmann *et al.* 1974, Flanagan *et al.* 1991). In this case, λ is the slope of the line connecting the source and the evaporating water in the leaf in the $\ln(\delta^{17}O + 1) - \ln(\delta^{18}O + 1)$ space (product and substrate) and does not necessarily represent a single fractionating process.

As in previous studies (Luz and Barkan, 2000, 2005, Angert *et al.* 2003), the variations in the triple isotopic composition of atmospheric oxygen were reported as the extent to which an O_2 sample is enriched in ^{17}O with respect to the present atmosphere (atm, PST). By definition, the present atmosphere anomaly, $^{17}\Delta_{\text{atm,PST}}$, is therefore nil. To describe the variations in the ^{17}O anomaly, $^{17}\Delta$, we use here Eq. (8.1) with the value of 0.516 for λ for the dominant mass-dependent process of dark respiration in the biological cycle (Angert *et al.* 2003, Luz and Barkan 2005).

IV. BUDGET OF TRIPLE ISOTOPES OF OXYGEN IN THE ATMOSPHERE AND PALEOPRODUCTIVITY

Oxygen associated with mass-dependent fractionation is produced in the lower atmosphere through photosynthesis. We note F_{bio} the O_2 exchange flux from the global biosphere productivity and $^{17}\Delta_{\text{bio}}$ the associated O_2 isotopic anomaly. The other source of oxygen in the lower atmosphere is the stratosphere–troposphere exchange of oxygen involving mass-independent stratospheric photochemistry between O_2, O_3, and CO_2. We note F_{strat} the oxygen O_2 flux of stratosphere–troposphere exchange and $^{17}\Delta_{\text{strat}}$ the isotopic anomaly of the stratospheric O_2. At steady state, oxygen is removed from the atmosphere by a similar stratosphere–troposphere exchange of oxygen with a flux F_{strat} and by the biospheric oxygen uptake with a flux F_{bio} and in both cases the fluxes carry the ambient ^{17}O anomaly, $^{17}\Delta_{\text{atm}}$. The resulting ^{17}O anomaly of oxygen in the atmosphere thus results from an oxygen isotopic balance of the two aforementioned fluxes, F_{bio} and F_{strat} (Figure 8.1; Luz *et al.* 1999):

$$F_{\text{bio}} \times (^{17}\Delta_{\text{bio}} - {}^{17}\Delta_{\text{atm}}) = F_{\text{strat}} \times (^{17}\Delta_{\text{strat}} - {}^{17}\Delta_{\text{atm}}) \qquad (8.2)$$

We are interested in describing the temporal variations in the biosphere productivity. On the basis of Eq. (8.2), it is possible to express the ratio of the biosphere productivity between two periods, t1 and t2 as:

$$\frac{F_{\text{bio,t1}}}{F_{\text{bio,t2}}} = \frac{F_{\text{strat,t1}}(^{17}\Delta_{\text{strat,t1}} - {}^{17}\Delta_{\text{atm,t1}})}{F_{\text{strat,t2}}(^{17}\Delta_{\text{strat,t2}} - {}^{17}\Delta_{\text{atm,t2}})} \times \frac{(^{17}\Delta_{\text{bio,t2}} - {}^{17}\Delta_{\text{atm,t2}})}{(^{17}\Delta_{\text{bio,t1}} - {}^{17}\Delta_{\text{atm,t1}})} \qquad (8.3)$$

Measurements of $^{17}\Delta_{\text{atm}}$ have been performed over the last 60 kyrs (Blunier *et al.* 2002) using the air trapped in a Greenland ice core. While $^{17}\Delta_{\text{atm}}$ is rather constant over the glacial period (around 40 permeg), a large change of $^{17}\Delta_{\text{atm}}$ occurred over the last deglaciation. As a consequence, we restrict our use of the $^{17}\Delta_{\text{atm}}$ to the quantification of the ratio of the biosphere productivity fluxes between the last glacial maximum (LGM) and today. Such ratio is expressed as:

$$\frac{F_{bio,LGM}}{F_{bio,PST}} = \frac{F_{strat,LGM}({}^{17}\Delta_{strat,LGM} - {}^{17}\Delta_{atm,LGM})}{F_{strat,PST}({}^{17}\Delta_{strat,PST} - {}^{17}\Delta_{atm,PST})} \times \frac{({}^{17}\Delta_{bio,PST} - {}^{17}\Delta_{atm,PST})}{({}^{17}\Delta_{bio,LGM} - {}^{17}\Delta_{atm,LGM})} \quad (8.4)$$

By definition, ${}^{17}\Delta_{atm,PST} = 0$, whereas we take +43 permeg for ${}^{17}\Delta_{atm,LGM}$ [note that the original value given by Blunier *et al.* (2002) is +38 permeg since it was calculated with a slope, λ, of 0.521 instead of 0.516 accepted in the present study as a reference slope]. As for ${}^{17}\Delta_{strat}$, Luz *et al.* (1999) suggests that it could be estimated from an isotopic budget between CO_2 and O_2. Indeed, in the stratosphere, the ozone recombination reaction ($O + O_2 \Rightarrow O_3$) causes O_3 to be mass independently fractionated and this anomalous fractionation is transmitted to CO_2 (Wen and Thiemens 1993, Yung *et al.* 1997). The ultimate source of oxygen in ozone is O_2 so that the isotopic anomaly of CO_2 is directly linked to the stratospheric isotopic anomaly of O_2. Unfortunately, up to now the stratospheric isotopic anomaly of CO_2 is not sufficiently constrained to provide the value of ${}^{17}\Delta_{strat}$ with the necessary accuracy for our purpose. For these reasons, previous works have assumed that the ratio of the production rates of anomalously depleted O_2 in the stratosphere, $F_{strat} \times ({}^{17}\Delta_{strat} - {}^{17}\Delta_{atm})$, between the LGM and present-day, is proportional to the ratio of atmospheric CO_2 concentrations between the LGM and the pre-industrial Holocene (Luz *et al.* 1999, Blunier *et al.* 2002). Then, using CO_2 concentrations of 280 ppmv for the preindustrial Holocene and 190 ppmv for the LGM (Barnola *et al.* 1987), Eq. (8.4) becomes:

$$\frac{F_{bio,LGM}}{F_{bio,PST}} = \frac{190}{280} \times \frac{{}^{17}\Delta_{bio,PST}}{{}^{17}\Delta_{bio,LGM} - 43} \quad (8.5)$$

The quantification of the past productivity therefore depends on quantifying the triple isotopic composition of oxygen produced by the biosphere at present-day and for the LGM. Because both ocean and land biospheres contribute to the global oxygen productivity, ${}^{17}\Delta_{bio}$ is the weighted average of ${}^{17}\Delta$ produced by the oceanic biosphere, ${}^{17}\Delta_{ocean}$, and by the terrestrial biosphere, ${}^{17}\Delta_{terr}$:

$${}^{17}\Delta_{bio} = \frac{F_O \times {}^{17}\Delta_{ocean} + F_T \times {}^{17}\Delta_{terr}}{F_O + F_T} \quad (8.6)$$

where F_O and F_T are the fluxes of oxygen associated with the oceanic and terrestrial productivities.

For present-day conditions the different ocean and land biosphere models give F_O/F_T ratio in the range of 0.45–0.59 (Bender *et al.* 1994, Blunier *et al.* 2002, Hoffmann *et al.* 2004). A global value for ${}^{17}\Delta_{ocean}$ was determined by Luz and Barkan (2000) as 249 permeg, but no global estimate of ${}^{17}\Delta_{terr}$ is available. Therefore, one of the last critical uncertainty in using the triple isotopic composition of oxygen to quantify temporal variations and past biosphere productivity is a correct determination of the variations with time of ${}^{17}\Delta_{terr}$.

${}^{17}\Delta_{terr}$ results from fractionation associated with the biologic consumption and production of oxygen for the terrestrial biosphere. As for the biologic oxygen consumption, Angert *et al.* (2003) and Helman *et al.* (2005) have determined the relationships between $\delta^{17}O$ and $\delta^{18}O$ during all the main oxygen uptake processes. Summarizing, they found slopes λ of 0.525, 0.509, and 0.516 for Mehler reaction, photorespiration, and dark reaction, respectively (Table 8.1). Taking into account the proportions of each process, these experimental results enable one to calculate an average slope for the global terrestrial oxygen uptake, λ_{resp_terr}, of 0.514 for present-day (Landais *et al.* 2007).

Concerning the oxygen production, it has been shown that photosynthesis transmits the isotopic composition of the substrate water to the oxygen produced without fractionation (Guy *et al.* 1993, Helman *et al.* 2005). Hence, the isotopic composition of O_2 produced by photosynthesis on land is that of leaf water. While large number of measurements and modeling studies have permitted to determine leaf water $\delta^{18}O$ for various climates and environments, no measurements of leaf water $\delta^{17}O$ have ever been performed. In order to evaluate the global effect of leaf transpiration on the triple isotopic composition of oxygen, we need to quantify the relationship between $\delta^{17}O$ and $\delta^{18}O$ in water during the process of leaf transpiration. The slope of this relationship, λ_{transp}, is defined as:

$$\lambda_{transp} = \frac{\ln(\delta^{17}O + 1)_{lw} - \ln(\delta^{17}O + 1)_{mw}}{\ln(\delta^{18}O + 1)_{lw} - \ln(\delta^{18}O + 1)_{mw}} \tag{8.7}$$

where the subscript "lw" stands for leaf water and "mw" for meteoric water, that is the source water for the plants.

The critical need for precise estimates of λ_{transp} is illustrated in Figure 8.2. The figure shows the stepwise derivation of two possible global ^{17}O anomalies, $^{17}\Delta_{bio1}$ and $^{17}\Delta_{bio2}$, based on the maximum range of λ_{transp} estimates available prior to our study. It demonstrates that such range precluded the use of the ^{17}O anomaly in estimating changes in biosphere productivity. For this illustration we depart from the source water for the plant, which lies on the meteoric water line of slope, $\lambda_{meteoric} = 0.528$ (Meijer and Li 1998). From this point, we can locate the position of the isotopic composition of leaf water only if we can combine the readily available measurements of average $\delta^{18}O$ enrichment of leaf water above that of source water (~14‰ today; Gillon and Yakir 2001), taking us along the x axis

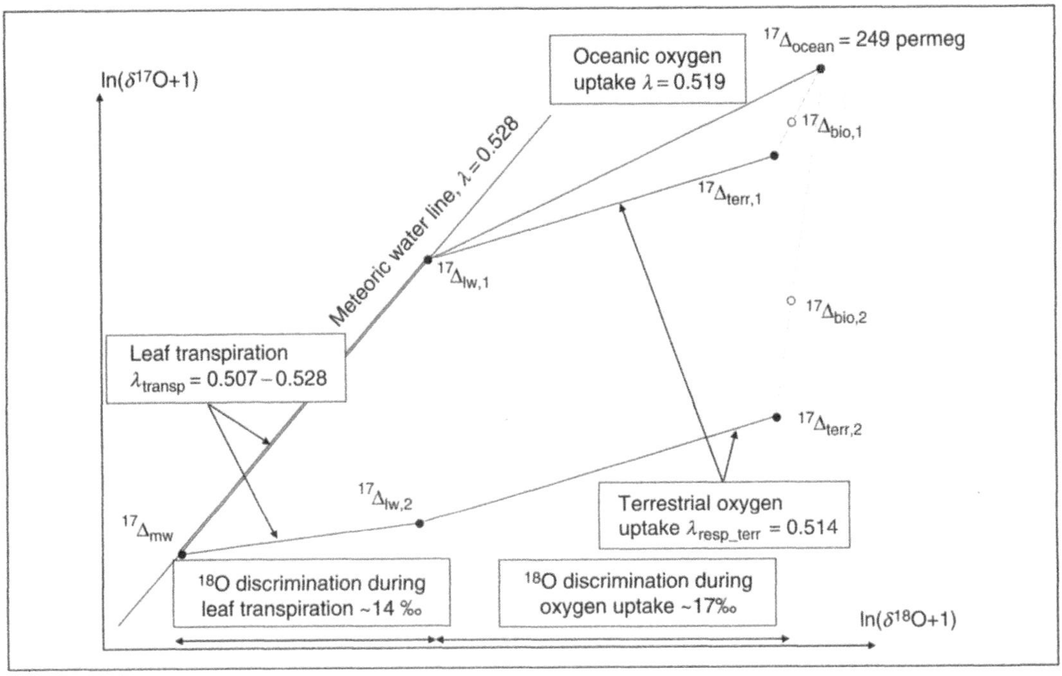

FIGURE 8.2 Scheme showing the determination of $^{17}\Delta_{terr}$ from the triple isotopic composition of meteoric water (characterized by $^{17}\Delta_{mw}$), the fractionation during leaf transpiration (characterized by λ_{transp} and leading to the triple isotopic composition of leaf water, $^{17}\Delta_{lw}$) and the fractionation associated with oxygen uptake (characterized by λ_{resp_terr}). We depict the influence of uncertainties on the slope of leaf transpiration (scenario 1: $\lambda_{transp} = 0.528$; scenario 2: $l_{transp} = 0.507$, see text) for: (1) the determination of the global triple isotopic composition of oxygen of leaf water (characterized by $^{17}\Delta_{lw,1}$ and $^{17}\Delta_{lw,2}$ depending on the scenario), (2) the determination of the global triple isotopic composition of oxygen associated with terrestrial biosphere productivity (characterized by $^{17}\Delta_{terr,1}$ and $^{17}\Delta_{terr,2}$ depending on the scenario), and (3) the determination of the global triple isotopic composition of oxygen associated with the global biosphere productivity (characterized by $^{17}\Delta_{bio,1}$ and $^{17}\Delta_{bio,2}$ depending on the scenario). (2) $^{17}\Delta_{bio}$ (empty circle) is deduced from $^{17}\Delta_{ocean}$ and $^{17}\Delta_{terr}$ according to Eq. (8.6). The triple isotopic composition of oxygen associated with global biosphere productivity should therefore lie on the gray line linking the triple isotopic composition of oxygen associated with terrestrial biosphere productivity and the one associated with oceanic biosphere productivity.

of the plot in Figure 8.2) with a reliable estimate of λ_{transp}. In the absence of any experimental determination of λ_{transp} we could only suggest a maximal value of 0.528, corresponding to the slope of the meteoric water line, or use the estimate of Angert *et al.* (2004), based on crude global budget, of 0.507 (from their estimated mean value of 0.512 ± 0.005). Figure 8.2 show the leaf water points, $^{17}\Delta_{lw1}$ and $^{17}\Delta_{lw2}$, derived from these two extremes. Once $^{17}\Delta_{lw}$ is established we can complete the estimate of $^{17}\Delta_{terr}$ and $^{17}\Delta_{bio}$ as we know [Landais *et al.* (2007) and references therein] that the effect of oxygen consumption on the terrestrial biosphere is along a slope λ_{resp_terr} of 0.514 for present-day (or 0.515 for LGM) with an associated shift in $\delta^{18}O$ by ~17‰ (along the *x* axis in Figure 8.2). The ocean component can be estimated simply from repeating the analysis of Luz and Barkan (2000). We can now infer the location of the triple isotopic composition of oxygen produced by the biosphere, $^{17}\Delta_{bio1}$ and $^{17}\Delta_{bio2}$ in Figure 8.2.

This illustration demonstrates the rough estimate of λ_{transp} that were available using the results for estimating the uncertainty of $^{17}\Delta_{terr}$, by ~294 permeg (=0.021*14*1000), *i.e.*, ~200 permeg in estimating $^{17}\Delta_{bio}$. Such variations in $^{17}\Delta_{bio}$ are much greater than the overall difference in $^{17}\Delta_{atm}$ of 43 permeg measured between the LGM and present-day. Clearly, in order to better estimate past biosphere productivity it was necessary to obtain accurate estimates of $\delta^{17}O$–$\delta^{18}O$ relationship associated with leaf transpiration. In this chapter, we describe the effort to provide a global value for λ_{transp}, representative of the different plants and conditions over the Earth.

V. EXPERIMENTAL

Extraction of leaf water was done following Wang and Yakir (2000). Leaves and stems or branches were collected in the field in 15-mL gastight vials after removing the petiole and the central vein from the leaves. Water was then quantitatively extracted by vacuum distillation at 60 °C directly from the vials for 3 h. In order to remove volatiles from the extracted water, we added few granules of activated charcoal and slowly stirred the water for 12 h. Tests showed that the distillation period was sufficient and the addition of charcoal did not modify the isotopic ratio of the water (for more detail, see Landais *et al.* 2006).

The analytical method for determination of the oxygen isotopic ratios of water is detailed in Barkan and Luz (2005). Briefly, 2 μL from the extracted water samples are converted into oxygen by fluorination using CoF_3 reagent. The produced oxygen is transferred to a stainless steel tube on a collection manifold immersed in liquid helium. After all samples are processed and sealed, the manifold is warmed to room temperature and connected to a Finnigan-MAT Delta^{plus} mass spectrometer. $\delta^{17}O$ and $\delta^{18}O$ of O_2 were measured simultaneously in dual inlet mode by multicollector mass spectrometry. All measurements were run against a working O_2 standard calibrated against V-SMOW.

VI. RESULTS

Here, we discuss the results of investigating the possible variations in λ_{transp}, the slope of the line connecting the isotopic composition of leaf water to the isotopic composition of source water in a $\ln(\delta^{17}O + 1) - \ln(\delta^{18}O + 1)$ plot. The uncertainty on λ_{transp} is 0.001, and more detail of experimental results and uncertainty analysis are given in Landais *et al.* (2006). Note that Landais *et al.* (2006) also confirmed previous conclusions that there are no changes in the isotopic composition of oxygen of water during uptake and transport in roots and stems (Gonfiantini *et al.* 1965, Wershaw *et al.* 1970). They also confirm that the evaporation from soil is a second order effect in the water isotopic enrichment between source and leaf (Saurer *et al.* 1997, Yakir and Sternberg, 2000). As a consequence, we use in the following the irrigation or long-term meteoric water isotopic composition for the calculation of λ_{transp} when the stem water isotopic composition was not measured.

A. Internal Leaf Variations

For the study of the variations of isotopic composition along a leaf, we used a long leaf of maize (*Zea mays* sp.) sampled at the beginning of the afternoon at the end of spring 2004 in a commercial cornfield in central Israel. The leaf was divided into eight sections along its length. Table 8.2 shows that the variations in λ_{transp} remained very small from one leaf to the other and the average λ_{transp} was 0.519 ± 0.001.

B. Variations During the Diurnal Cycle

To investigate the influence of short-term changes in environmental conditions, we sampled leaves from trees or large bushes during the diurnal cycle (6 a.m. to 7 p.m., May 2005, encompassing large variations in temperature and relative humidity). The selected plants were: bougainvillea (*Bungainvillea x buttiana* Holt. & Stand.), usually found in tropical environments and which typically shows high stomatal conductance; and coral tree (*Erythrina corallodendron* L.), a deciduous tree usually found in temperate climates. While the majority of the leaves were exposed to sunlight, some leaves were in the shade. For each plant, we sampled one branch in order to provide source water.

Table 8.2 displays the results associated with the diurnal cycle experiment. The leaf exposition (direct sunlight or shade) had no significant influence on λ_{transp}. The two plants showed similar evolution of λ_{transp}, and it remained roughly constant during the day with an average value of 0.517 ± 0.001. However, the slope associated with samples taken early in the morning or late in the afternoon, with higher relative humidity, appeared to have somewhat lower slopes than at midday with lower relative humidity.

C. Variations Among Different Species in the Same Site

To study the influence of plant species on the transpiration slope between $^{18}O/^{16}O$ and $^{17}O/^{16}O$, we selected common trees in the Mediterranean region that have clear differences in leaf morphology and phenology. (1) Three different oaks: one deciduous—*Quercus pedunculiflora* originating from

TABLE 8.2 Temporal and spatial variations in the transpiration slopes

			Bougainvillea	Coral tree	Maize	
Time	T (°C)	RH (%)	λ_{transp}	λ_{transp}	Leaf segment	λ_{transp}
					1 (base)	0.520
6:15	21	75	0.515	0.516	2	0.520
7:15	22	60	0.517	0.517	3	0.519
8:15	26	40	0.519	0.518	4	0.519
10:45	26	41	0.519	0.518	5	0.519
13:45	31	33	0.518	0.518	6	0.520
16:30	27	50	0.517	0.517	7	0.519
18:45	23	60	0.517	0.517	8 (top)	0.518
Average			0.517 ± 0.001	0.517 ± 0.001		0.519 ± 0.001

Whole leaves were sampled during the daily cycle in temperature and humidity from two plant species growing in the same site. Leaf segments (leaves were divided into eight equal segments from leaf base to leaf tip) from a maize plant growing in a commercial field were sampled at the samples at the same time (mid-morning).

TABLE 8.3 Influence of tree species on the transpiration slope

Tree species	λ_{transp}
Quercus pedunculiflora	0.517
Quercus alnifolia	0.520
Quercus ilex	0.519
Cedrus libani	0.518
Cedrus atlantica	0.520
Castanea sative	0.518
Average	0.519 ± 0.001

All samples were collected at the same site (Jerusalem Botanical Garden), on the same day at temperature 25 °C and relative humidity 32% in April 2005.

Turkey—and two evergreen species—*Q. alnifolia* originating from Cyprus and *Q. ilex* found from Southeast France to Turkey, (2) two different cedar trees: *Cedrus libani* originating from Lebanon and *Cedrus atlantica* from the Atlas mountains, and (3) one chestnut tree (*Castanea sative* sp.). All leaves and stems were sampled around midday in the Jerusalem Botanical Garden in April 2005.

The results for the survey over different tree species are given in Table 8.3. Again, λ_{transp} varied only between 0.517 and 0.520 and the average λ_{transp} was 0.519 ± 0.001.

D. Different Geographic Locations

The effect of environmental and climatic conditions was studied by midday sampling of branches and leaves of chestnut trees (*Castanea sative* sp.), at the same season (flowering season), in Jerusalem (April 2005), in the Budapest hills (Hungary, in May 2005) and in the western French countryside (Bouère, Mayenne, in May 2005). In addition, we used stem and leaf water obtained from a range of European sites participating in the WP5 of the Carboeuroflux project (www.weizmann.ac.il/ESER/wp5/; Hemming *et al.* 2005) that covered a large range of environmental and climatic conditions.

In Table 8.4, we give the λ_{transp} for chestnut trees sampled at three different sites. As in the previous studies, the slope is similar in the three sites and the average λ_{transp} was 0.518 ± 0.001.

The transpiration slopes, derived from measured $\delta^{17}O$ and $\delta^{18}O$ in leaf and stem water, for the large European survey is given in Table 8.4 and show generally similar results of a robust transpiration slope, but with a notable caveat: The results show a general tendency of lower λ_{transp} when relative humidity increases (rainy conditions or sampling in the evening). Our lowest value was 0.513 for relative humidity around 90–100%.

E. The Dependency of λ_{transp} on Relative Humidity

Our results show that λ_{transp} remains stable within a leaf and among different species and environments. However, variations in the ambient relative humidity seem to influence λ_{transp}. Using the entire dataset, the empirical relationship between λ_{transp} and relative humidity for relative humidity, h, between 32% and 100% λ_{transp} we observed a linear decrease with a best fit line according to Eq. (8.8):

$$\lambda_{transp} = (-0.0078 \pm 0.0026) \times h + 0.5216 \pm 0.0008 \quad (8.8)$$

Note that Eq. (8.8) was obtained by using Isoplot 3.00 (Ludwig 2003), which takes into account uncertainties in both relative humidity and λ_{transp} for the linear regression (details in Landais *et al.* 2006).

TABLE 8.4 Influence of environmental conditions on the slope of transpiration in the same tree species (chestnut leaves in the cities indicated at the top of the table) and of different tree species in different geographical locations and times (in countries indicated at the lower part of the table)

Place	T (°C)	RH (%)	λ_{transp}
Jerusalem	25	32	0.518
Budapest	22	45	0.518
Bouère	18	50	0.519
Denmark	–	95	0.513
Germany (1)	–	Sunny-20:30	0.513
Germany (2)	–	Sunny-18:30	0.516
Germany	–	Sunny-18:00	0.518
Germany	–	Sunny-18:30	0.516
Germany	–	Rainy-17:00	0.515
Germany	–	Sunny-16:00	0.518
Italy (1)	15	88	0.516
Italy (2)	15	88	0.515
Italy	–	Rainy	0.515
Italy	14	60	0.517
Netherlands	20	74	0.516
Portugal	21	40	0.518
Scotland	21	93	0.513
Scotland	8	88	0.513
Scotland	15	97	0.516
Scotland	10	90	0.514

Figures within parentheses indicate plant species sampled on the same place and date. When relative humidity was not measured, the regional weather in the day and the sampling time was indicated if known. When stem or irrigation water were not available, we used the $\delta^{18}O$ of meteoric water and estimated $\delta^{17}O$ the slope of the meteoric water line [0.528, Meijer and Li (1998)]. Samples were collected during summer 2004 for the chestnut and July–September 2001–2002 for the rest.

VII. DISCUSSION

A. Implications for the Budget of Triple Isotopes of Oxygen in the Atmosphere

The determination of the plant transpiration slope provided the missing link in the ^{17}O anomaly system, and it is now possible to evaluate the global triple isotopic composition of oxygen issued from the terrestrial biosphere. As discussed in Section IV earlier, we want to apply the ^{17}O approach to assess changes in global biosphere productivity between the LGM and the present. For this, we use our experimental determination of λ_{transp} to evaluate the triple isotopic composition of leaf water for these two periods. Because $\delta^{18}O$ of the meteoric water ($\delta^{18}O_{mw}$) and λ_{transp} varies with latitude, we estimate first the worldwide distribution of leaf water $\delta^{17}O$ and $\delta^{18}O$ and their global mean.

This is done using the global $\delta^{18}O_{mw}$ dataset of the IAEA network (Edwards *et al.* 2002) and computing the corresponding $\delta^{17}O_{mw}$ from the known relationships between these two values (Meijer and Li 1998). Leaf water $\delta^{18}O$ is then estimated, globally, using the Craig and Gordon (1965) expression of evaporation applied to leaf water (Dongmann *et al.* 1974, Flanagan *et al.* 1991), using $\delta^{18}O_{mw}$, and the equilibrium water vapor isotopic composition, together with available data on relative humidity and temperature. From the latitudinal and seasonal variability of these input parameters, we therefore produced monthly maps (2°×2°) of leaf water $\delta^{18}O$. Finally, from the global distribution of relative humidity, we calculated the corresponding distribution of λ_{transp} and inferred the leaf water $\delta^{17}O$, $\delta^{17}O_{lw}$, on each grid point as:

$$\ln(\delta^{17}O_{lw}+1)=\ln(\delta^{17}O_{mw}+1)+\lambda_{transpi}\times[\ln(\delta^{18}O_{lw}+1)-\ln(\delta^{18}O_{mw}+1)] \quad (8.9)$$

We now need to weight the relative contribution of each grid-point to the global mean $\delta^{17}O_{lw}$ and $\delta^{18}O_{lw}$ values. We weight the leaf water values by the photosynthetic flux from each grid-point using the dynamic global vegetation model Organizing Carbon and Hydrology in Dynamic Ecosystems (ORCHIDEE; Krinner *et al.* 2005). The weighted values are then integrated to provide the global mean values of $\delta^{17}O_{lw}$ = 3.5‰ and of $\delta^{18}O_{lw}$ = 7‰, which is consistent with the constraints such as of Gillon and Yakir (2001). And when expressed with respect to present-day atmospheric oxygen, we infer the ^{17}O anomaly imposed by the terrestrial photosynthesis from Eq. (8.1) as $^{17}\Delta_{lw}$ = 144 permeg.

We repeated the same exercise for the LGM. However, for this time period, we inferred both the global mean $\delta^{18}O_{lw}$ and $\delta^{17}O_{lw}$ since no estimate is available for the global LGM $\delta^{18}O_{lw}$. In this case, we obtained estimated values of 4.12‰, 8.03‰, and 157 permeg, respectively, for $\delta^{17}O_{lw}$, $\delta^{18}O_{lw}$, and $^{17}\Delta_{lw}$: The different values between the present-day and LGM values are directly related to the larger extent of the ice sheets during the LGM. Indeed, with large ice sheets, the vegetation of the LGM is very sparse in the high latitudes and the associated photosynthetic production is accordingly decreased. The high latitudes are associated with low $\delta^{18}O$ of meteoric water so that with less contribution of the high latitudes, the global meteoric water $\delta^{18}O$ and hence the global $\delta^{18}O_{lw}$ increases. The scheme in Figure 8.3 explains the associated change in $\delta^{17}O$ and hence in $^{17}\Delta_{lw}$. It shows that with the decrease in the mean meteoric water $\delta^{18}O$ over the deglaciation, the associated decrease in $\delta^{17}O$ is along the meteoric water line of slope 0.528. This slope is higher than the slope of 0.516 used to calculate the ^{17}O anomaly on Earth and therefore the anomaly $^{17}\Delta_{lw}$ is higher during the LGM than present-day.

This difference in $^{17}\Delta_{lw}$ over the deglaciation has important implications for the quantification of the past biosphere productivity, since it implies a decrease of $^{17}\Delta_{bio}$ between the LGM and present-day of about 13 permeg. Such a change is not negligible compared to the 43 permeg change in the $^{17}\Delta_{atm}$.

B. Quantification of LGM Productivity

The change in $^{17}\Delta_{lw}$ is not the only factor that modifies $^{17}\Delta_{bio}$ over the deglaciation. First, changes are expected in the oxygen consumption component. A change in the relative proportions of the dark respiration, the photorespiration and the Mehler reaction alters the global λ_{resp_terr} because of the slightly different slopes associated with these processes (Angert *et al.* 2003, Helman *et al.* 2005). The fraction of photorespiration varies with plant species (Lloyd and Farquhar 1994, von Caemmerer 2000). For example, whereas C3 plants photorespire, C4 plants do not. Using the ORCHIDEE model (Krinner *et al.* 2005), we estimated that the large decrease in C4 plants over the deglaciation leads to an increase of the fraction of photorespiration from 0.15 to 0.3 between the LGM and present-day (Landais *et al.* 2007). Such change results in a decrease of λ_{resp_terr} from 0.515 to 0.514, and then a decrease of $^{17}\Delta_{terr}$ by 17 permeg over the deglaciation. Combining the fractionations in the hydrological cycle (meteoric water and leaf transpiration discussed earlier), and the fractionations associated with the oxygen uptake processes, $^{17}\Delta_{terr}$ decreases from 140 to 110 permeg over the deglaciation (see Figure 8.3 for a schematic representation).

Second, $^{17}\Delta_{ocean}$ is also modified over the deglaciation. Indeed, the $\delta^{18}O$ of the mean ocean decreases by 1‰ between the LGM and present-day as a consequence of the decrease of the polar ice sheets (Waelbroeck *et al.* 2002). Because the isotopic composition of the ice sheets lies on the meteoric water line (Barkan and Luz 2005), the triple isotopic composition of the mean ocean is modified along the meteoric water line of slope 0.528. This slope is larger than the slope λ = 0.516 used to express the ^{17}O anomaly [Eq. (8.1)] so that the ^{17}O anomaly of the mean oceanic water is larger by 12 permeg

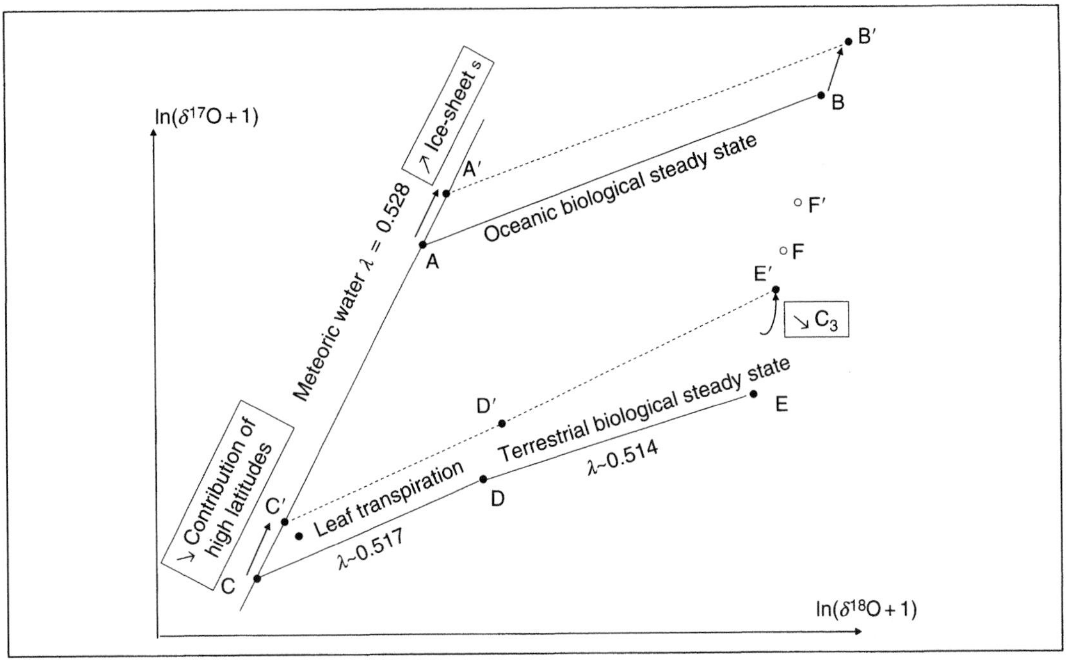

FIGURE 8.3 Scheme showing the relative positions of the isotopic composition of:

- Global ocean (A for present-day, A′ for the LGM)
- Oxygen issued from oceanic productivity (B for present-day, B′ for the LGM)
- Mean meteoric water (C for present-day, C′ for the LGM)
- Mean leaf water (D for present-day, D′ for the LGM)
- Oxygen issued from terrestrial productivity (E for present-day, E′ for the LGM)

From this scheme, it is possible to obtain the triple isotopic composition of oxygen issued from the global biosphere at present and for the LGM and characterized by $^{17}\Delta_{bio}$ as depicted in Figure 8.2. This result is indicated with empty circles labeled F (present) and F′ (LGM). The products (F and F′) can then be used together with observed ^{17}O anomaly of atmospheric O_2 in global isotopic mass balance [Eq. (8.5)] to estimate changes in biospheric productivity.

during the LGM than present-day. We assumed that the slope associated with the oceanic oxygen uptake processes remained the same between the LGM and present-day so that the shift by 12 permeg of the mean ocean, that is, the substrate for oceanic photosynthesis, is directly transmitted to the isotopic composition of the oxygen produced from the oceanic biosphere and $^{17}\Delta_{ocean}$ decreases from 261 to 249 permeg over the deglaciation (see Figure 8.3 for a schematic representation).

Summarizing, both $^{17}\Delta_{ocean}$ and $^{17}\Delta_{terr}$ decrease over the deglaciation. It results in an associated decrease of $^{17}\Delta_{bio}$. From Eq. (8.6) and estimates of the ratio F_O/F_T [0.45–0.59 for present-day and 0.56–1.4 for the LGM (Landais *et al.* 2007)], we calculated that $^{17}\Delta_{bio}$ decreased from 198 ± 9 permeg to 157 ± 4 between the LGM and present-day. Finally, Eq. (8.5) enables one to obtain 69 ± 6% for the ratio of the biosphere oxygen productivity between the LGM and present-day.

VIII. CONCLUSIONS

In order to close a major gap in the global biogeochemical cycle of oxygen on Earth, we have estimated the relationship between $\delta^{17}O$ and $\delta^{18}O$ during leaf transpiration. We have especially studied numerous factors that can influence the isotopic composition of water during transpiration: the microscale effects along a leaf, the local environmental conditions (relative humidity and temperature), differences among plant species, and the influence of different environmental conditions through a European survey. We have found that the mean transpiration slope, λ_{transp}, is surprisingly robust and that the only significant effect is associated with changes in relative humidity, *h*, so that $\lambda_{transp} = 0.522–0.008$ h.

We then combined our new estimates of λ_{transp}, previous λ estimates for other processes, global model estimates of plant distribution, and fluxes and the measured ^{17}O anomaly during the LGM and at present to apply the ^{17}O approach and infer the overall change in the biosphere productivity between the two periods. We estimated that the ^{17}O anomaly in oxygen exchanged between biosphere and atmosphere changed between 157 permeg for present-day and 198 permeg for the LGM. On the basis of these estimates in the context of global oxygen budget, we propose that biospheric productivity of the LGM was 69 $\pm$ 6% of that at the present.

Finally, we note that while the ^{17}O approach is a rather difficult way to assess the past biospheric productivity, requiring accurate measurements and model outputs, this approach is, nevertheless, the only one available to us at present. Next challenges would be first, to link such global estimates both to the local estimates of paleoproductivity (Kohfeld *et al.* 2005); and second, to use the ^{17}O approach to link oxygen cycle studies with those of the carbon cycle. This will help improve our understanding of the interactions between climate change and the evolution of the carbon cycle.

IX. REFERENCES

Angert, A., S. Rachmilevitch, E. Barkan, and B. Luz. 2003. Effect of photorespiration, the cytochrome pathway, and the alternative pathway on the triple isotopic composition of atmospheric O_2. *Global Biogeochemical Cycles* **17**: doi:10.1029/2002GB001933.

Angert, A., C. D. Cappa, and D. J. DePaolo. 2004. Kinetic 17O effects in the hydrologic cycle: Indirect evidence and implications. *Geochimica et Cosmochimica Acta* **68**(17):3487–3495.

Barkan, E., and B. Luz. 2005. High precision measurements of O-17/O-16 and O-18/O-16 of O2 in H_2O. *Rapid Communications in Mass Spectrometry* **19**:3737–3742.

Barnola, J. M., D. Raynaud, Y. S. Korotkevich, and C. Lorius. 1987. Vostok ice core provides 160,000-year record of atmospheric CO_2. *Nature* **329**:408–414.

Bender, M., T. Sowers, and L. D. Labeyrie. 1994. The Dole effect and its variation during the last 130,000 years as measured in the Vostok core. *Global Biogeochemical Cycles* **8**(3):363–376.

Blunier, T., B. Barnett, M. L. Bender, and M. B. Hendricks. 2002. Biological oxygen productivity during the last 60,000 years from triple oxygen isotope measurements. *Global Biogeochemical cycles* **16**(3), doi:10.1029/2001GB001460.

Clayton, R. N., L. Grossman, and T. K. Mayeda. 1973. A component of primitive nuclear composition in carbonaceous meteorites. *Science* **182**:485–488.

Craig, H., and L. Gordon. 1965. Deuterium and oxygen-18 in the ocean and the marine atmosphere. Pages 9–130 *in* E. Tongiorgi (Ed.) *Stable Isotopes in Oceanographic Studies and Paleotemperatures.* V. Lishi E Figli, Pisa.

Dongmann, G., H. W. Nurnberg, H. Forstel, and K. Wagener. 1974. On the enrichment of H218O in the leaves of transpiring plants. *Radiation and Environmental Biophysics* **11**:41–52.

Edwards, T. W. D., S. J. Birks, and J. J. Gibson. 2002. Isotope tracers in global water and climate studies of the past and present. *In* International Conference on the Study of Environmental Change Using Isotope Techniques, edited by I.A.E. Agency, Vienna.

Flanagan, L. B., J. D. Marshall, and J. R. Ehleringer. 1991. Comparison of modeled and observed environmental influences on the stable oxygen and hydrogen isotope composition of leaf water in *Phaseolus vulgaris* L. *Plant Physiology* **96**:623–631.

Gonfiantini, R., S. Gratziu, and E. Tongiorgi. 1965. *Oxygen isotope composition of water in leaves.* Pages 405–410, International Atomic Energy Agency, Technical Report Series, Vienna.

Gillon, J., and D. Yakir. 2001. Influence of carbonic anhydrase activity in terrestrial vegetation on the 18O content of atmospheric CO_2. *Science* **291**:2584–2587.

Guy, R. D., M. L. Fogel, and J. A. Berry. 1993. Photosynthetic fractionation of the stable isotopes of oxygen and carbon. *Plant Physiology* **101**:37–47.

Helman, Y., E. Barkan, D. Eisenstadt, B. Luz, and A. Kaplan. 2005. Fractionation of the three stable oxygen isotopes by oxygen producing and consuming reactions in photosynthetic organisms. *Plant Physiology* **138**:2292–2298.

Hemming, D., D. Yakir, and WP5 members. 2005. Pan-European $\delta^{13}C$ signatures from forest ecosystems. *Global Change Biology* **11**:1065–1072.

Hoffmann, G., M. Cuntz, C. Weber, P. Ciais, P. Friedlingstein, M. Heimann, J. Jouzel, J. Kaduk, E. Maier-Reimer, U. Seibt, and K. Six. 2004. A model of the Earth's Dole effect. *Global Biogeochemical Cycles* **18**:GB1008, doi:10.1029/2003GB002059.

Kohfeld, K. E., C. Le Quere, S. P. Harrison, and R. F. Anderson. 2005. Role of Marine Biology in Glacial-Interglacial CO_2 cycles. *Science* **308:**74–78.

Krinner, G., N. Viovy, N. de Noblet-Ducoudre, J. Ogee, J. Polcher, P. Friedlingstein, P. Ciais, S. Sitch, and I. C. Prentice. 2005. A dynamic global vegetation model for studies of the coupled atmosphere-biosphere system. *Global Biogeochemical Cycle* **19:**GB1015, doi:10.1029/2003GB002199.

Landais, A., E. Barkan, D. Yakir, and B. Luz. 2006. The triple isotopic composition of oxygen in leaf water. *Geochimica et Cosmochimica Acta* **70:**4105–4115.

Landais, A., J. Lathiere, E. Barkan, and B. Luz. 2007. Reconsidering the interpretation of triple oxygen isotopic measurements in the air trapped in the ice core as a marker of paleoproductivity. *Global Biogeochemical Cycles* **21:**GB1025, doi:10.1029/2006GB002739.

Lloyd, J., and G. D. Farquhar. 1994. 13C discrimination during CO_2 assimilation by the terrestrial biosphere. *Oecologia* **99:**201–215.

Ludwig, K. 2003. A Geochronological Toolkit for Microsoft Excel, Berkeley Geochronology Center, Berkeley, p. 71, http://www.bgc.org/klprogrammenu.html.

Luz, B., and E. Barkan. 2000. Assessment of oceanic productivity with the triple-isotope composition of dissolved oxygen. *Science* **288:**2028–2031.

Luz, B., and E. Barkan. 2005. The isotopic ratios O-17/O-16 and O-18/O-16 in molecular oxygen and their significance in biogeochemistry. *Geochimica et Cosmochimica Acta* **69**(5):1099–1110.

Luz, B., E. Barkan, M. L. Bender, M. H. Thiemens, and K. A. Boering. 1999. Triple-isotopic composition of atmospheric oxygen as a tracer of biosphere productivity. *Nature* **400:**547–550.

Meijer, H. A. J., and W. J. Li. 1998. The use of electrolysis for accurate d17O and d18O isotope measurements in water. *Isotopes in Environmental and Health Studies* **34:**349–369.

Miller, M. F. 2002. Isotopic fractionation and the quantification of 17O anomalies in the oxygen three-isotopes system: An appraisal and geochemical significance. *Geochimica et Cosmochimica Acta* **66**(11):1881–1889.

Saurer, M., S. Borella, and M. Leuenberger. 1997. $\delta^{18}O$ of tree rings of beech (*Fagus sylvatica*) as a record of $\delta^{18}O$ of the growing season precipitation. *Tellus* **49B:**80–92.

Thiemens, M. H., and J. E. Heidenreich. 1983. The mass-independent fractionation of oxygen: A novel isotope effect and its possible cosmochemical implications. *Science* **219:**1073–1075.

Thiemens, M. H., T. Jackson, K. Mauersberger, B. Schueler, and J. Morton. 1991. Oxygen isotope fractionation in stratospheric CO_2. *Geophysical Research Letters* **18:**669–672.

von Caemmerer, S. 2000. Biochemical models of leaf photosynthesis, Commonwealth Science and Industrial Research Organization, Collingwood, Australia.

Waelbroeck, C., L. Labeyrie, E. Michel, J.-C. Duplessy, J. F. McManus, K. Lambeck, E. Balbon, and M. Labracherie. 2002. Sea level and deep temperature changes derived from benthic foraminifera benthic records. *Quaternary Science Reviews* **21:**295–306.

Wershaw, R. L., I. Freidman, S. L. Heller, and P. A. Frank. 1970. Hydrogen isotopic fractionation of water passing through trees. Pages 55–67 *in* G. D. Hobson (Ed.) *Advances in Organic Geochemistry.* Pergamon Press, Oxford.

Wang, X. F., and D. Yakir. 2000. Using stable isotope of water in evapotranspiration studies. *Hydrological Processes* **14:**1407–1421.

Wen, J., and M. H. Thiemens. 1993. First multi-isotope study of the O(D) + CO_2 exchange and stratospheric consequences. *Journal of Geophysical Research* **98:**12801–12808.

Yakir, D. 1997. Oxygen-18 of leaf water: A crossroad for plant-associated isotopic signals. Pages 147–168 *in* H. Griffith (Ed.) *Stable Isotopes and the Integration of Biological, Ecological and Geochemical Processes.* Bios, Oxford.

Yakir, D., and L. da S. L. Sternberg. 2000. The use of stable isotopes to study ecosystem gas exchange. *Oecologia* **123:**297–311.

Young, E. D., A. Galy, and H. Nagahara. 2002. Kinetic and equilibrium mass-dependent isotope fractionation laws in nature and their geochemical and cosmochemical significance. *Geochimca et Cosmochimica Acta* **66:**1094–1104.

Yung, Y. L., A. Y. T. Lee, F. W. Irion, W. B. DeMore, and J. Wen. 1997. Carbon dioxide in the atmosphere: Isotopic exchange with ozone and its use as a tracer in the middle atmosphere. *Journal of Geophysical Research* **102:**10857–10866.

Section 3

Animal-Based Isotope Data as Indicators of Ecological Change

CHAPTER 9

An Isotopic Exploration of the Potential of Avian Tissues to Track Changes in Terrestrial and Marine Ecosystems

Keith A. Hobson

Environment Canada

Contents

I. INTRODUCTION

Birds are among the most successful of all vertebrates and currently occupy all of the earth's major ecosystems. Birds have evolved to exploit an impressive array of niches within habitats of both terrestrial and marine biomes where they range from extreme dietary specialists to generalists. This broad trophic diversity includes frugivory and nectarivory to piscivory and carnivory. Birds are also among the most volant of all life forms and several species migrate annually between breeding and wintering sites that may be many thousands of kilometers apart. Other species are entirely sedentary and range little more than a few kilometers or less during their annual cycles. For those species that do migrate, some are highly philopatric to their breeding and wintering areas whereas others disperse or are nomadic. Birds are also conspicuous, accessible, and amenable to scientific investigation and there is a good deal of interest in using birds as indicators of ecological change at various spatial and temporal scales (Furness and Camphuysen 1997, Burger and Gochfeld 2004). Birds also produce a range of tissues that are amenable to dietary reconstruction using stable isotopes and so there is growing interest in using isotopic measurements of birds to monitor larger-scale ecological processes that are associated with characteristic isotopic abundance (Thompson *et al.* 1995, Stapp *et al.* 1999, Ainley *et al.* 2006). This chapter will review how the isotopic measurement of avian tissues can be used to monitor ecological change. Here, ecological change will be considered to be both changes in baseline food web isotopic composition (*i.e.*, using birds as true isotopic monitors of their environment) and ecological changes involving the birds themselves (*i.e.*, using stable isotopes to investigate how birds respond to environmental change).

II. THE TISSUE ISSUE

Animals are related isotopically to their environment by means of an isotopic diet-tissue discrimination factor. This factor, in turn, is the net change resulting from several processes involving isotopic fractionation in rate-limiting chemical reactions and kinetic processes. Researchers interested in using birds or other animals to monitor isotopic abundance in the environment are, to a large extent, at the mercy of the tissues available and ultimately on how well such discrimination factors are known. Fortunately, birds are amenable to captive study where they have been raised on known isotopic diets and so data on discrimination values are available for several species and tissue types, primarily for stable isotopes of carbon and nitrogen (Table 9.1). However, these studies have shown considerable variance that might ultimately be related to issues of nutrition or species-specific metabolic processes. For example, it has become clear that the elemental C:N ratios in diets can influence the diet-tissue discrimination factor related to both ^{15}N and ^{13}C (Pearson *et al.* 2003, Robbins *et al.* 2005).

In addition to providing a range of metabolically active tissues common to other taxa (*e.g.*, blood and muscle), birds produce metabolically inactive keratinous tissues like feathers and claws. Birds also produce eggs which reflect the diet of the female at the time of laying or her mobilization of stored nutrients. Eggs are particularly interesting because they contain such a range of individual tissue types reflecting different metabolic pathways. Egg yolk contains both protein and lipid fractions that are deposited during the follicular growth period which can take several days. Egg albumen and shell membranes are largely protein and reflect a shorter period of dietary integration since they are formed rapidly just before shell deposition. Shells contain an inorganic calcium carbonate matrix imbued with collagen-like protein that reflects most recent diet of the laying female. The carbonate fraction of eggshells can be assayed for $\delta^{18}O$, $\delta^{13}C$, and $\delta^{44}Ca$ measurements (Schaffner and Swart 1991), whereas the protein fraction can be assayed for $\delta^{13}C$, $\delta^{15}N$, δD, $\delta^{18}O$, and $\delta^{34}S$ measurements. Hobson (1995) conducted captive rearing studies of quail (*Coturnix japonica*) and Peregrine Falcons (*Falco peregrina*) and established estimates of discrimination factors associated with a variety of organic and inorganic components of eggs. Interestingly, that study established different diet-tissue discrimination factors associated with the herbivore versus the carnivore model that were in turn associated with differential mobilization of lipids versus carbohydrates from the diet. Such studies underline the importance of establishing diet-tissue discrimination factors as accurately as possible using experimental techniques in order to decipher just how reliably stable isotope values in avian tissues can be used to monitor isotopic changes in the environment.

Another fundamental consideration in isotopic sampling of tissues as a measure of environmental isotope values is the temporal period of integration of tissues. Again, birds have been used by several researchers to evaluate elemental tissue turnover rates by switching the isotopic composition of diets. As expected, tissue turnover rate conforms largely with our understanding of tissue metabolism or cell

TABLE 9.1 Estimates of $\delta^{13}C$ and $\delta^{15}N$ discrimination factors between food and birds' blood and feathers from published studies in which diet was known or controlled

		Discrimination factors		
Species	**Food items**	**$\delta^{13}C$**	**$\delta^{15}N$**	**References**
Whole Blood				
King penguin *Aptenodytes patagonica*	Herring	−0.8	2.1	Cherel *et al.* 2005
Rockhopper penguin *Eudyptes chrysocome*	Capelin	0.0	2.7	Cherel *et al.* 2005
Ring-billed gull *Larus delawarensis*	Perch	−0.3	3.1	Hobson and Clark 1992
Great skua *Catharacta skua*	Sprat	1.1	2.8	Bearhop *et al.* 2002
	Beef	2.3	4.2	Bearhop *et al.* 2002
Peregrine falcon *Falco peregrinus*	Quail	0.2	3.3	Hobson and Clark 1992
Canvasback *Aythia valisineria*	Commercial diet	1.5	3.0	Haramis *et al.* 2001
Japanese quail *Coturnix japonica*	Commercial diet	1.2	2.2	Hobson and Clark 1992
Dunlin *Calidris alpina pacifica*	Mixed diet	1.3	2.9	Evans-Ogden *et al.* 2004
Garden warbler *Sylvia borin*	Control diet	1.7	2.4	Hobson and Bairlein 2003
Yellow-rumped Warbler *Dendroica coronata*	Mealworms and bananas	−1.2 to 2.2	1.7 to 2.7	Pearson *et al.* 2003
Feathers				
Humboldt's penguin *Spheniscus humboldti*	Anchovy	–	4.8	Mizutani *et al.* 1992
King penguin *Aptenodytes patagonica*	Herring	0.1	3.5	Cherel *et al.* 2005
Rockhopper penguin *Eudyptes chrysocome*	Capelin	0.1	4.4	Cherel *et al.* 2005
Common cormorant *Phalacrocorax carbo*	Mackerel	–	3.7	Mizutani *et al.* 1992
	Sprat	2.6	4.9	Bearhop *et al.* 1999
European shag *Phalacrocorax aristotelis*	Sprat	2.0	3.6	Bearhop *et al.* 1999
Ring-billed gull *Larus delawarensis*	Perch	0.2	3.0	Hobson and Clark 1992
Black-tailed gull *Larus crassirostris*	Saurel	–	5.3	Mizutani *et al.* 1992
Great skua *Catharacta skua*	Sprat	2.1	4.6	Bearhop *et al.* 2002
	Beef	2.2	5.0	Bearhop *et al.* 2002
Nankeen night heron *Nycticorax caledonicus*	Saurel	–	4.2	Mizutani *et al.* 1992
Great white egret *Egretta alba*	Saurel	–	3.9	Mizutani *et al.* 1992
Grey heron *Ardea cinerea*	Saurel	–	4.3	Mizutani *et al.* 1992
Scarlet ibis *Eudocimus ruber*	Mixed diet	–	4.5	Mizutani *et al.* 1992
White ibis *Eudocimus albus*	Mixed diet	–	4.3	Mizutani *et al.* 1992
Flamingo *Phoenicopterus* spp.	Mixed diet	–	5.6	Mizutani *et al.* 1992
Peregrine falcon *Falco peregrinus*	Quail	2.1	2.7	Hobson and Clark 1992
Chicken *Gallus gallus*	Commercial diet	−0.4	1.1	Hobson and Clark 1992
Japanese quail *Coturnix japonica*	Commercial diet	1.4	1.6	Hobson and Clark 1992
Garden warbler *Sylvia borin*	Control diet	2.7	4.0	Hobson and Bairlein 2003

Food, but not blood, was lipid-extracted for isotope measurements, which were performed on the whole food items, not on a given prey tissue.

replacement. Blood plasma and liver turnover rapidly (half-lives of the order of 2–3 days) whereas muscle and blood cellular fractions turnover more slowly (half-lives of 12–20 days), and bone collagen dietary integration is slow enough in adults to represent the bird's lifetime. At the other extreme, stable isotope values of breath carbon dioxide are expected to represent diet over a period of hours since the last meal (Podlesak *et al.* 2005). Importantly, we expect allometric scaling relationships between body size and turnover rates to hold, thus allowing for estimates of isotopic turnover rates in metabolically active tissues for species that have not been investigated experimentally.

For metabolically inert tissues, dietary integration has previously been expected to reflect diet only during the period of tissue growth. However, Cerling (Chapter 11) presents evidence in mammals that hair may be formed from more than one endogenous nutrient pool and similar studies using bird feathers are now warranted. Nonetheless, there is reasonable evidence that birds mobilize primarily local nutrients during feather formation since feathers provide a very convenient latitudinal marker for hydrogen isotope studies (see below). For feathers, this provides an important application to avian studies since the sequence, location, and duration of molt is well known for several species. Feather and claw growth has yet to be well established for most wild birds (Bearhop *et al.* 2003), but claw isotope signatures have been shown to correlate well with feather values in cases where the growth periods coincide (Hobson *et al.* 2006). Another avian keratinous material that has yet to be used for isotopic analysis is the ramphotheca or horny plate covering the bone structure of the beak.

III. BIRDS AS INDICATORS

Ecological change may involve changes in fundamental ecological processes that result in baseline food web isotope changes. In such cases, birds and other organisms may be used as proxies to sample the environment. There are several advantages to this approach since it may be otherwise impossible to sample baseline inorganic or lower trophic levels due to logistical constraints and isotopic spatial and temporal variability is typical of complex systems. In such cases, birds or other upper trophic-level organisms are expected to integrate this complexity and thus may be better isotopic samplers of the environment. However, here, it will be important to know the ecology of the species or population of interest and their spatial use of the environment. It may also be necessary to focus on dietary specialists so that isotopic changes through time may be unequivocally associated with changes in food web labels rather than changes in diet per se. Fortunately, some stable isotopes, such as ^{34}S, ^{2}H, and ^{13}C, show little change with trophic level and so their measurement will reflect baseline food web changes rather than changes in diet. On the other hand, δ^{15}N values change significantly with trophic level and so interpretation of changes in avian δ^{15}N values will be more complicated. The measurement of several stable isotopes in avian tissues may allow much more effective use of birds as monitors of ecological change (Lott *et al.* 2003, Hebert and Wassenaar 2005).

Ecological change may also result in dietary changes in individuals or populations rather than actual baseline isotopic changes. For example, global climate change may result in changes in species composition and abundance of insects and so drive dietary responses in birds. In marine and freshwater systems, commercial fisheries or the introduction of invasive species can result in fundamental changes in the trophic structure of aquatic food webs and these changes can result in trophic changes in waterbird communities (Becker and Beissinger 2006). The monitoring of avian δ^{15}N values is an effective means of tracking trophic changes that may in turn be driven by larger-scale environmental change. Moreover, environmental change may be subtle and go undetected whereas routine avian monitoring is feasible and may provide the proverbial canary in the coalmine.

Migratory birds provide an interesting means of detecting the influence of global climate change since their phenology of movement and their wintering latitude can be dictated by continental weather conditions (reviewed by Murphy-Klassen *et al.* 2005). Use of the measurement of feather deuterium (δD) values as a proxy for latitude where feathers were grown has provided a convenient means of monitoring both breeding and wintering latitude of birds, especially in North America (Hobson 2005). The phenology of movement of birds sampled during migration has also been investigated using feather δD measurements (Dunn *et al.* 2006). The isotopic analysis of bird tissues can thus function to monitor both fundamental baseline isotopic changes in the environment as well as avian responses to biotic conditions that may or may not be related to these changes. The following sections will deal with examples of the kinds of environmental information that can be gleaned from isotopic measurements of avian tissues in terrestrial and marine systems.

IV. TERRESTRIAL ISOSCAPES

Stable isotope measurements of consumers indicate both source and trophic information of individuals and, in this sense, birds are no different from any other components of the food web. Thus, $\delta^{13}C$ values can typically distinguish between C_3 and C_4 or between C_3 and CAM linkages of birds to primary production and δD measurements are also useful for distinguishing between CAM and C4-based food webs (Alisauskas *et al.* 1998, Wolf and Martinez del Rio 2000).

In addition to photosynthetic pathway, other mechanisms influence plant isotopic composition. C_3 plants can become enriched in ^{13}C as a result of water-use efficiency mechanisms. Marra, Hobson and Holmes (1998) made use of this phenomenon to track winter habitat use by American Redstarts (*Setophaga ruticilla*) by examining muscle tissue of newly arriving birds on their breeding grounds some weeks later. Birds that occupied poorer quality scrub habitat were enriched in ^{13}C compared to those that occupied cooler and moister forest habitat that tended to arrive on the breeding grounds later. That study was the first to infer consequences of habitat use at one point in the annual cycle with subsequent events at another, an accomplishment virtually impossible without the use of intrinsic tracers such as stable isotopes. Although it was not the purpose of their investigation, these authors also provided an interesting example of how the isotopic measurement of tissues of birds arriving on their breeding grounds in North America could be used to indicate the relative abundance of good (moist forest) versus poor (xeric scrub) habitat availability on the wintering grounds. For birds that are able to use secondary habitats, the decreasing cover of moist forest in the tropics will presumably result in a positive shift in the distribution of tissues of arriving migrants. Chamberlain *et al.* (2000) relied on $\delta^{13}C$ and $\delta^{15}N$ measurements of feathers of Scandinavian Willow Warblers (*Phylloscopus trochilis*) grown on African winter quarters to infer climate and subsequent geographic location of African origin. While such measurements may be generally poor at defining origins of migrants, they may be a more useful means of tracking use of xeric versus mesic habitats through time.

Factors other than climate per se can influence terrestrial plant isotopic composition. In the case of ^{13}C, agricultural land-use practices can alter the exposure of plants to sunlight by removing canopy forest. This involves potential enrichment due to water-use efficiency mechanisms but also, in the case of previously dense forest, the removal of the canopy effect that tends to lower plant $\delta^{13}C$ values (Schleser and Jayasekera 1985, Brooks *et al.* 1997). All things being equal, we might thus expect animal food webs to be more enriched in ^{13}C in agricultural areas versus forested areas. Land-use practices and anthropogenic factors also influence plant $\delta^{15}N$ values. The use of animal-based and chemical fertilizers can cause enrichment in ^{15}N in food webs either directly or through ammonification which favors the evaporative loss of isotopically lighter nitrogenous compounds (Nadelhoffer and Fry 1994). Tilling soils also exposes them to the atmosphere leading to enrichment through the same process. As a result, agricultural landscapes tend to be more enriched in ^{15}N and their use by birds can be traced (Hobson 1999, Hebert and Wassenaar 2001).

Progress is being made with GIS models to predict global distributions of C_3 and C_4 plants using remote sensing information on climate and algorithms of how plants are expected to respond to these conditions (Still *et al.* 2003). These continental-scale depictions of plant photosynthetic pathway and expected isotopic distributions in plants could be augmented by GIS layers corresponding to distributions of C4 agricultural crops such as corn for those cases where species are known to depend on them (Wassenaar and Hobson 2000). Recently, Powell *et al.* (unpublished) modeled the potential African wintering origins of Hoopoes (*Upupa epops*) breeding in Europe. That study was based on models of expected distributions of plant $\delta^{13}C$ and $\delta^{15}N$ values and of annual precipitation δD that were converted to expected isotopic distributions of feathers using known isotopic discrimination factors between plants and feathers. This was the first study to combine three isoscapes to delineate origins of a migratory organism. Impressively, the region of overlap of all three isotopes in sub-Saharan Africa corresponded well with the known wintering grounds of Hoopoes.

The measurement of stable isotopes of other elements, such as Sr, holds some promise and Beard and Johnson (2000) provided a depiction of the expected distribution of $\delta^{87}Sr$ values in geologic substrate across the United States. Similar depictions are required for other parts of North America and other continents. A useful first step in evaluating the promise of this isotope in animal movement studies will be to measure $\delta^{87}Sr$ signatures in animal tissues grown across known isotopic gradients.

A. Deuterium Patterns in Precipitation

The relative abundance of deuterium (δD) in environmental waters is strongly influenced by kinetic processes associated with temperature, evaporation, and condensation (Craig 1961). Schiegl (1970) summarized several effects that influence the deuterium content of precipitation. The *latitudinal effect* refers to the general decrease in precipitation D with increasing distance from the equator due to the associated temperature gradient. The *altitude effect* similarly refers to a depletion of D in precipitation with altitude. The altitude effect is of the order of 1.6‰/100 m at low latitudes and 4.8‰/100 m at high latitudes (Dansgaard 1964, Ziegler 1988). Deuterium content in precipitation also undergoes a *seasonal effect* at higher latitudes whereby summer precipitation tends to be more enriched in D than in winter. The *continental effect* refers to the phenomenon of air masses becoming more depleted in D as they move from the coast to interior locations on a continent. Finally, the *total precipitation* effect refers to the general depletion in precipitation D with increasing amount of rain fallout. The combination of these factors results in relatively robust patterns of deuterium in precipitation across continents (IAEA/WMO 2001).

For food webs involving birds, all hydrogen originates from plant water, in turn derived ultimately from precipitation or groundwater and from drinking water. As plants fix hydrogen in food webs, deuterium content of plant organic matter becomes depleted relative to precipitation. Although continental patterns of monthly or annual δD in precipitation have been known for decades (Sheppard *et al.* 1969, Taylor 1974), it was the investigations of Miller *et al.* (1988) on beetle exoskeletons and Cormie *et al.* (1994) on deer bone collagen that first demonstrated how growing season average precipitation values could be transferred to higher-order herbivores, in turn allowing past climatic data to be derived from the isotopic analysis chitin and bones. Hobson and Wassenaar (1997) and Chamberlain *et al.* (1997) demonstrated that continental patterns in precipitation δD were transferred through food webs to feathers grown at known locations in North America. Importantly, a strong relationship was found between feather δD (δD_f) and the mean growing season δD precipitation value (δD_p). The intercept of these relationships further provided an estimate of the discrimination factor between precipitation and feather formation, a value typically of the order of −25‰ to −30‰ for insectivorous passerines largely reflecting initial precipitation to plant isotopic discrimination

(Wassenaar and Hobson 2000). Further studies have largely confirmed that long-term average continental patterns of δD_p can be used to link tissue δD values of consumers with approximate latitude in North America (Meehan *et al.* 2001, Kelly *et al.* 2002).

Bowen *et al.* (2005) examined δD_p patterns in other continents. That study revealed several other systems where the use of tissue δD values of migratory organisms could be used to infer origins. In particular, the European continent shows a general depletion in δD_p with latitude along a southwest to northeast gradient (Hobson *et al.* 2004a). However, unlike North America, Hobson *et al.* (2004a) found that there was equally good agreement between δD_f and both annual mean δD_p and mean growing season δD_p. South America shows vast regions of the Amazon basin with relatively constant δD_p values but the more temperate southern regions of Argentina and Chile show a useful stepwise depletion in δD_p with latitude (Bowen *et al.* 2005). The δD_p situation in Africa appears to be highly dynamic from month to month. However, some attempts have been made to use measurements of δD_f to infer origins of sub-Saharan migrants from Europe (Møller and Hobson 2004, Pain *et al.* 2004, Yohannes *et al.* 2005) and of medium distance migrants within Europe (Bearhop *et al.* 2005).

Plant δD_p values decrease and $\delta^{13}C$ values increase with altitude (Körner *et al.* 1988, Körner *et al.* 1991, Graves *et al.* 2002). Hobson *et al.* (2003) demonstrated that δD and $\delta^{13}C$ values of hummingbird feathers reflected altitude along a 4000-m gradient in the Ecuadorean Andes clearly demonstrating the utility of stable isotope measurements of avian tissues to associate birds with altitude. There are several species of tropical birds that are known to move altitudinally during their annual cycles. On the Hawaiian islands, viable populations of native avifauna are restricted to elevations above which they are not susceptible to avian malaria and there is great interest in tracking altitudinal movements of these birds to monitor their exposure to malaria and their response to global climate change (van Riper *et al.* 1986).

Of considerable interest to conservation biologists is the development of tools that better allow them to interpret trends in wildlife populations through time. For birds, population trends for species breeding in remote areas can be monitored through migration monitoring stations where birds are captured en route and relative capture rates quantified. This is the case for the Canadian migration monitoring network that intercepts birds breeding in the vast boreal forest. The problem in the past has been that it is extremely difficult to interpret these trends since the catchment area for any given station is largely undefined. Unless regions of origins of birds are described, it is impossible to link ecological and anthropogenic changes on the landscape with any subsequent changes in populations. Following on from Wassenaar and Hobson (2001), Dunn *et al.* (2006) used feather δD measurements to define catchment areas for five species of migrant songbirds moving through five stations across the southern region of the country. For each species, that study showed considerable variation in catchment region across the country suggesting that population trends at each cluster of stations can be interpreted independently (Figure 9.1). Dunn *et al.* (2006) also determined that later arriving birds in the autumn were typically from more northern origins and this indicates that population trends could also be stratified within stations by time periods representing different geographic regions.

B. Eggs

Egg tissues are formed from nutrients acquired directly from diet of the laying female or from her endogenous sources or both. If these sources differ isotopically, it is possible to quantify relative contributions to egg formation, providing isotopic discrimination factors between diet and egg tissues and between endogenous sources and egg tissues are known. For macromolecules, such as yolk protein, albumen, and yolk lipid, stable isotope analyses provide proximate and ultimate links between the egg components and female diet. This is also true for measurements of the organic, collagen-like component of eggshells since this is formed quickly from diet at the time of shell formation. The inorganic (largely calcium carbonate) fraction of eggshells provides yet another useful substrate for isotopic analyses since C in the carbonate is

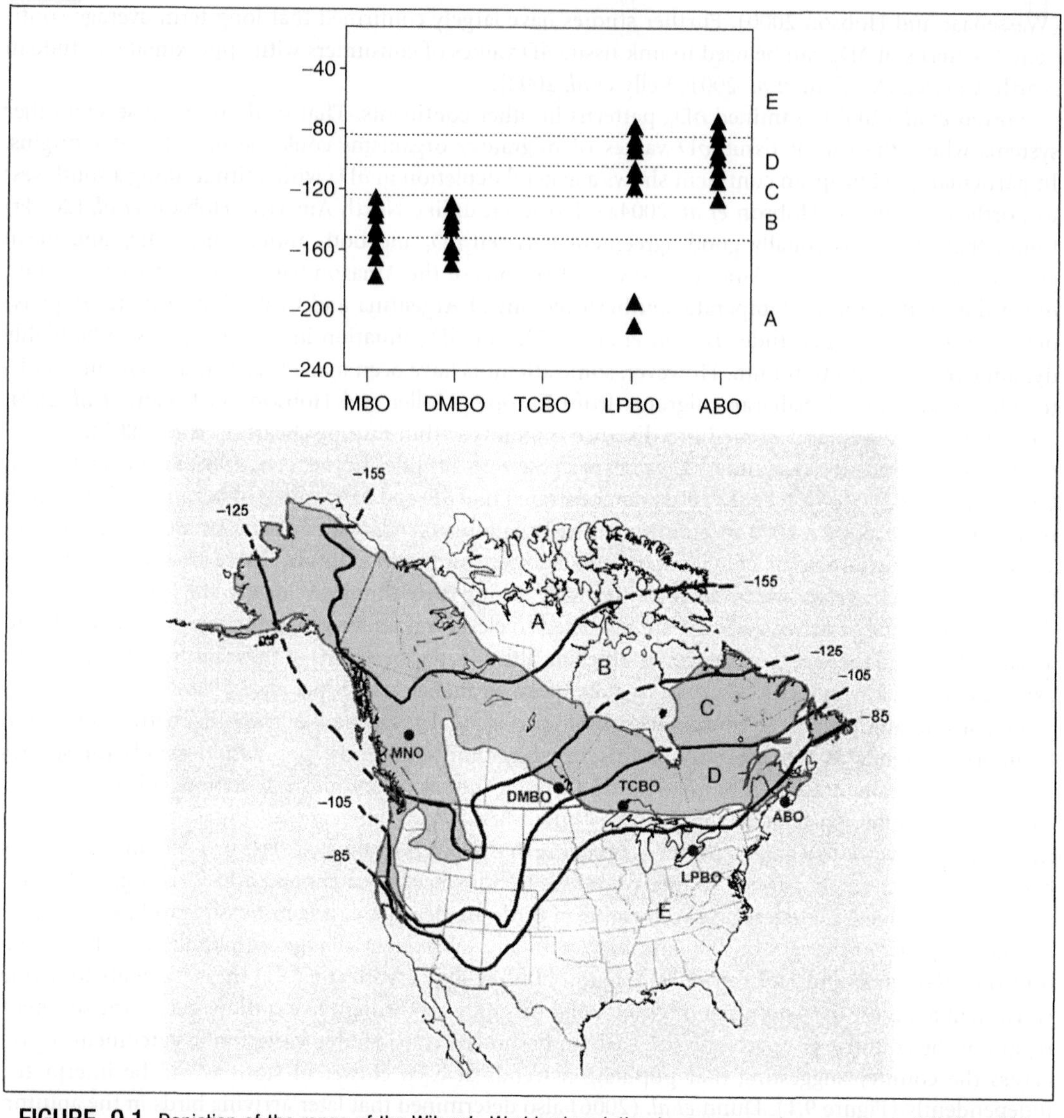

FIGURE 9.1 Depiction of the range of the Wilson's Warbler in relation to mean growing season precipitation δD in North America and zones considered by Dunn *et al.* (2006) to describe the catchment area of this and other species moving through five migration monitoring stations. The inset shows the results of feather δD analyses from these stations illustrating the different catchment areas for western and eastern locations. This information can be used to better assist in the determination of long-term population changes of this species.

derived directly from diet and the O is derived from drinking water. Von Schirnding *et al.* (1982) demonstrated this by relating $\delta^{13}C$ values in eggshells of African Ostrich (*Struthio camelus*) to those in local diet. Since drinking water largely reflects local meteoric water, oxygen isotope values in eggshell carbonates can reflect continental patterns in the same way that was demonstrated for feather δD values. Folinsbee *et al.* (1970) conducted a controlled feeding experiment in which chickens were supplied with drinking water of known $\delta^{18}O$ values and their egg carbonates analyzed. They found an approximate linear relationship between $\delta^{18}O$ values in drinking water and egg carbonates. Erben *et al.* (1979) confirmed this effect also on modern avian eggshells, and Sarkar *et al.* (1991) extended this kind of analysis to the investigation of dietary reconstructions based on dinosaur eggshells.

V. MARINE ISOSCAPES

Seabirds can be extremely useful samplers of their marine environments and the isotopic analysis of seabird tissues can provide valuable information on current and past environmental conditions. Stable nitrogen isotope analyses provide trophic-level estimates (Hobson and Welch 1992) and $\delta^{13}C$ and $\delta^{34}S$ analyses provide information on relative dependence on benthic versus pelagic food webs and on marsh plants versus pelagic producers in estuarine situations (Peterson and Howarth 1987, France 1995, Hobson *et al.* 1997). However, the world's oceans are not isotopically homogeneous and considerable isotopic structure can be found regionally. Stable isotopic signatures characterizing primary production in marine food webs are determined by the isotopic composition of inorganic substrates and biogeochemical processes associated with elemental uptake by the algal cell (reviewed by Michener and Schell 1994). In marine waters, strong opposing gradients in CO_2 concentration and planktonic $\delta^{13}C$ have been observed (Rau *et al.* 1989, Popp *et al.* 1999, Lourey *et al.* 2004). However, unless clear empirical evidence is available that marine organic $\delta^{13}C$ consistently responds to variations in a single or small set of parameters in a given region, interpreting past variations in $\delta^{13}C$ will be problematic. Perhaps the best demonstration of this was the study of Schell *et al.* (1998) who isotopically characterized the Bering, Chukchi, and western Beaufort sea food webs. That study showed the pronounced isotopic ($\delta^{15}N$, $\delta^{13}C$) influence of the upwelling over the Bering Sea shelf on downstream food webs. Other studies have demonstrated latitudinal gradients in zooplankton $\delta^{13}C$ values in both hemispheres (Rau *et al.* 1982) and the occurrence of strong isotopic anomalies such as the Southern Ocean Convergence (Best and Schell 1996).

Variation in marine food web $\delta^{15}N$ values can be related to sources of inorganic nitrogen important to phytoplankton including N_2 gas, ammonia, and nitrate. High $\delta^{15}N$ nitrate is often associated with oxygen-depleted water and significant isotopic variation in particulate organic nitrogen occurs around the world (Michener and Schell 1994). Less is known about spatial variation in marine food web δD values. However, marine δD will be influenced by mixing of seawaters of different isotopic values ultimately associated with inputs from the terrestrial hydrologic cycle. Ocean water δD values are positively correlated with salinity and temperature (Jouzel 1999, Masson *et al.* 2000). Knowledge of relevant marine isoscapes can be useful in interpreting stable isotope values of seabird tissues since some species are wide ranging and may encounter new isotopic regions throughout their annual cycles (Cherel *et al.* 2000). In other cases, such variation will be confounding.

A. Tracing Past Changes in Marine Productivity and Trophic Position

Schell (2000) investigated the long-term productivity of the Bering and Chukchi seas using Bowhead Whale (*Balaena mysticetus*) baleen as a proxy for marine baseline isotopic signatures. That study showed a significant decline in baleen $\delta^{13}C$ values over the last 40 years strongly suggesting reduction in food web productivity. Hobson *et al.* (2004b) used a similar approach to investigate a possible ecological regime shift in the Gulf of Alaska by examining isotopically $\delta^{13}C$ and $\delta^{15}N$ values in archived Steller sea lion (*Eumatopias jubatus*) teeth and in Tufted Puffin (*Fratercula cirrhata*) and Crested Auklet (*Aethia cristatella*) feathers. Despite extreme variation in feather $\delta^{15}N$ values through time indicating trophic plasticity in puffin and auklet diet, that study similarly showed a steady and significant decrease in feather $\delta^{13}C$ values. Archived feathers from museum specimens or other materials of known provenance can thus provide a powerful means of examining historical patterns in stable isotope values in marine food webs (Ainley *et al.* 2006, Becker and Beissinger 2006).

Seabird eggshells are often preserved in archaeological sites and can be used to infer information on past diets and the isotopic nature of food webs. Emslie and Patterson (2007) recently examined eggshells from current and extinct colonies of Adelie penguins (*Pygoscelis adeliae*) in the Ross Sea region of Antarctica. Their examination of the $\delta^{13}C$ and $\delta^{15}N$ values in the organic portion, and $\delta^{13}C$ values in the inorganic (carbonate) fraction, of eggs reflected short-term diet of the laying female and showed a distinct dietary shift in penguin diet within the last 200–300 years (Figure 9.2). During most of the Holocene, penguins fed on high-trophic foods such as fish but most recently switched to lower trophic level krill. The timing of the abrupt shift coincides with the rapid depletion of whales and seals from the eighteenth to twentieth centuries in the Southern Ocean. The massive slaughter of marine mammals in the Southern Ocean is thought to have caused an excess of krill that is thought to have benefited penguins (Fraiser *et al.* 1992).

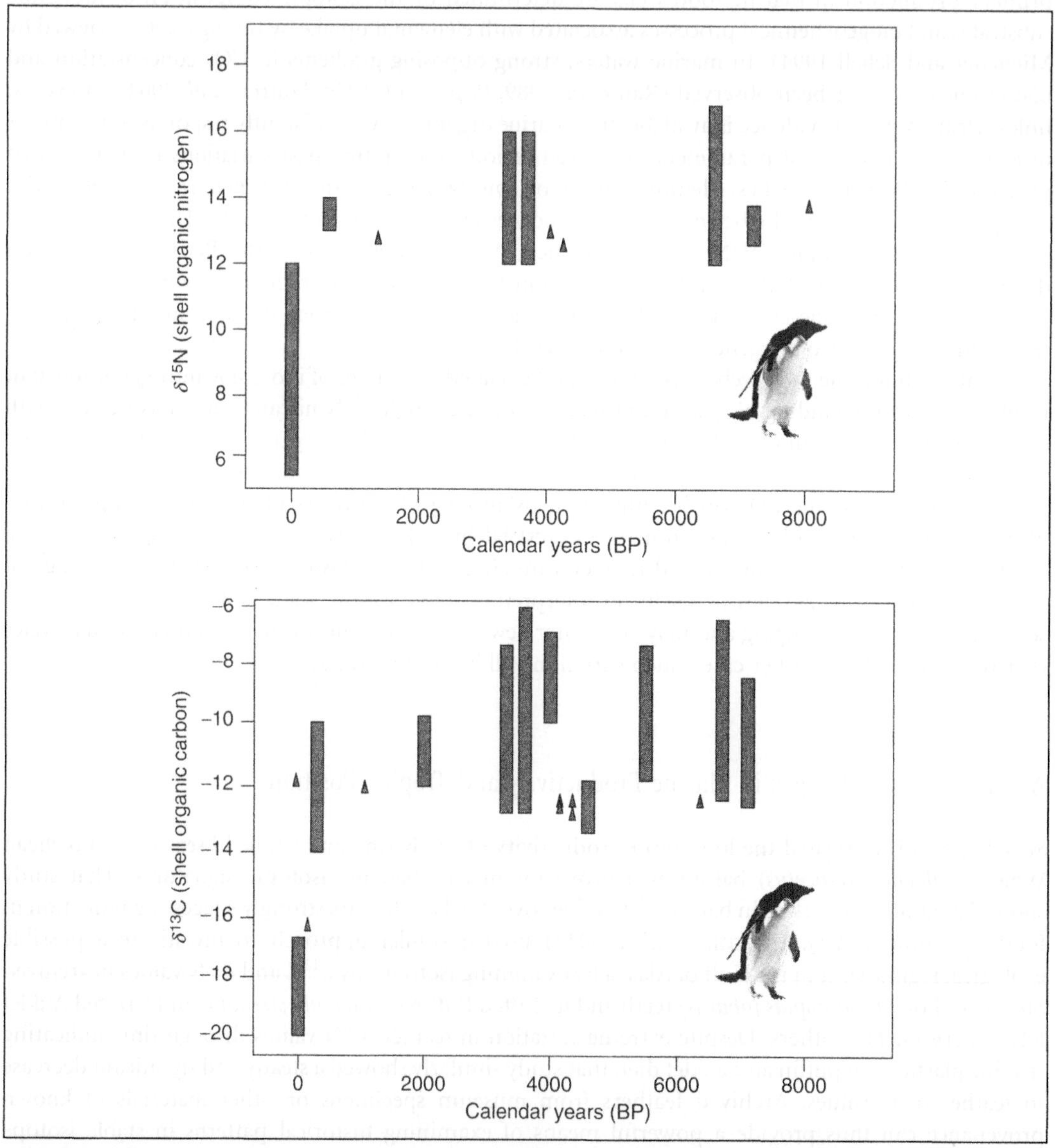

FIGURE 9.2 Stable isotope measurements (ranges) of modern and past Adelie penguin eggshells from the Ross Sea region of Antarctica. Depicted are the $\delta^{15}N$ and $\delta^{13}C$ values for the organic (protein) matrix of the eggshells. A similar pattern was also found for $\delta^{13}C$ values of the inorganic (carbonate) fraction. From Emslie and Patterson (2007).

In a similar study to that performed by Emslie and Patterson on penguin eggshells, Thompson *et al.* (1995) examined long-term changes in $\delta^{15}N$ and $\delta^{13}C$ values in feathers of Northern Fulmars (*Fulmarus glacialis*) from two northeast Atlantic colonies. This species is a seabird known to have scavenged offal associated with whaling activities at the turn of the twentieth century. Since much of the whaling in the area was of relatively high-trophic-level, toothed whales (Odontoceti), the authors reasoned that feathers from birds of that era should be relatively enriched in ^{15}N and ^{13}C compared to modern samples. This was indeed the case but it is still unclear whether this effect was primarily due to the reduction in whaling or trophic reduction in fisheries generally. Fisheries have generally shifted from a concentration on older, larger individuals to younger, smaller classes during that time period.

VI. ENVIRONMENTAL CONTAMINANTS

The advent of stable isotope analysis of consumer tissues has provided a valuable tool to assist in the interpretation of levels of contaminants in food webs (Kidd 1998). This is primarily due to the dual use of stable isotopes as indicators of trophic level and source of feeding. Many environmental contaminants are known to either magnify or bioaccumulate in food webs whereas others decrease or biodepurate with trophic level. Since $\delta^{15}N$ values provide a proxy for trophic level and represent a continuous variable, the slope of the relationship between contaminant concentration and tissue $\delta^{15}N$ values provides a quantitative measure of the tendency for bioaccumulation of that contaminant. The first demonstration of this tool using birds was provided by Jarman *et al.* (1996) and Sydeman *et al.* (1997) who examined organochlorine concentrations and $\delta^{15}N$ values in eggs of a seabird community off the California coast. Those authors quantified the expected bioaccumulation of these contaminants but, more importantly, suggested that the examination of the residuals of such a relationship could indicate whether those contaminants were mobilized from endogenous (lipid) stores or were more directly related to immediate diet of the laying female. Species lying above the contaminant versus tissue $\delta^{15}N$ line had higher organochlorine levels than explained by the diet of the female during laying and so represented a situation of routing of lipids to eggs that were accumulated when the birds earlier fed at higher trophic level. Thus, the careful interpretation of the contaminant versus $\delta^{15}N$ relationship can reveal more information than rates of contaminant accumulation alone.

Bird feathers are an extremely valuable tool for the current and retrospective analysis of heavy metals, especially mercury (reviewed by Monteiro and Furness 1995). Since almost all the mercury present in bird feathers is in the form of methyl mercury, problems of contamination from atmospheric deposition of elemental mercury are generally avoided. Several studies have also shown that mercury concentration in feathers reflects those in body burdens in a dose-dependent fashion. In addition, because birds void body burdens of mercury during annual molt and growth of new feathers, their feather mercury concentrations are independent of the age of the individual (Furness *et al.* 1990). Bird feather $\delta^{15}N$ values also reflect diet during the feather formation period (Mizutani *et al.* 1991) and so the combined measurement of mercury and $\delta^{15}N$ values can be used to retrospectively examine body burden of mercury and diet. However, Thompson *et al.* (1998) found that for two species of seabird, feather mercury and $\delta^{15}N$ were uncoupled, a consequence of the fact that temporal periods of integration of contaminant and diet differed.

One area of contaminant research that will benefit from the use of stable isotope measurements in marine systems is the interpretation of contaminant trends through time. Braune *et al.* (2002) have monitored a variety of contaminants in the eggs of seabirds at various high Arctic colonies. At Prince Leopold Island, their record of egg contaminant represents three species of seabirds and goes back to 1975. Since several of these contaminants (DDT, PCB) bioaccumulate, it is important to take into account any changes in trophic level of laying females through time. Thus, while contaminant loads may show a decreasing trend, this may be due to birds switching to lower trophic-level prey and not necessarily due to decreasing levels of that contaminant in the environment. Using $\delta^{15}N$ measurements of these eggs, Braune *et al.* (2002) were able to demonstrate that the contaminant trends were real and not the result of diet shifts. Future environmental contaminant studies should use tissue $\delta^{15}N$ values as a covariate in trend analyses.

VII. FUTURE PROSPECTS

Despite exponential increase in the application of stable isotope methods to study animal ecology in recent years (Hobson 2005), the formal development of how isotopic measurement of animals can be used to trace ecological change is lacking. Nonetheless, as this chapter demonstrates, there is considerable potential to use tissues of birds and other higher trophic-level organisms to trace current and past isotopic patterns related to fundamental ecological change. This will be both in terms of birds as proxies to record baseline isotopic changes and in terms of using stable isotope measurements as a tool to investigate how birds and other wildlife contend with global change. The archives of avian tissue available in museums will be particularly valuable as a resource for isotopic investigations (Smith *et al.* 2003) as will archaeological collections of avian faunal material. This will implicitly involve the careful development of protocols to use absolute minimal amounts of material for analysis and to choose those materials that can best answer the question of interest. Coordination among the scientific community will be especially important in order to prevent a situation whereby museum curators are forced to contend with too many requests, especially for rare materials. These materials are valuable and we are already seeing a reluctance among some curators to use avian specimens for stable isotope analyses.

For the specific association of avian tissue isotopic measurements to track ecological change, there are two general areas of research that will be important. First, we require more controlled captive studies of birds to help us understand the scope of variation in diet-tissue isotopic discrimination factors and elemental tissue turnover rates. These discrimination factors really are the currency that ultimately links bird isotope data with more fundamental baseline food web and abiotic isotope patterns. This area of research will also benefit from close collaborations between isotope ecologists and nutritional physiologists and biochemists. Second, isotope ecologists interested in using birds or other organisms as isotopic indicators of ecological change are limited by our current knowledge of terrestrial and marine isoscapes.

Refinement in our understanding of terrestrial isoscapes will involve the continual development and ground truthing of algorithms that predict the distribution of plant photosynthetic types on the landscape based on climate and elevation models. For deuterium patterns in rainfall, more refined datasets than those available through the long-term Global Network for Isotopes in Precipitation (GNIP) dataset are almost certainly required. A better understanding of spatial variance in the kriged contour maps available for the major continents will further allow statistical confidence limits to be applied to feather δD datasets. More coverage of precipitation δD is required for just about all regions. One intriguing possibility will be to exploit tap water δD measurements for those areas of the continents where it can be demonstrated that tap water closely follows meteoric water (Chapter 18).

Knowledge of marine isoscapes throughout the world really is at a primitive level. If these were developed, our understanding of the response of marine organisms, including seabirds, to oceanic ecological changes would be greatly enhanced. One possibility will be to develop algorithms to predict baseline phytoplankton isotope values that are analogous to those developed for terrestrial plant distributions. A valuable archive to test any theoretical marine isoscape will be the long-term Continuous Plankton Recorder survey (Warner and Hays 1994) in the North Atlantic.

VIII. REFERENCES

Ainley, D. G., K. A. Hobson, G. H. Rau, L. I. Wassenaar, and P. C. Augustinus. 2006. A 10,000 year record of isotopic variability in snow petrel mumiyo: Further evidence for a mid-Holocene climate shift affecting coastal Antarctic. *Marine Ecology Progress Series* **306:**31–40.

Alisauskas, R. T., E. E. Klaas, K. A. Hobson, and C. D. Ankney. 1998. Stable-carbon isotopes support use of adventitious color to discern winter origins of lesser snow geese. *Journal of Field Ornithology* **69:**262–268.

Beard, B. L., and C. M. Johnson. 2000. Strontium isotope composition of skeletal material can determine the birth place and geographic mobility of humans and animals. *Journal of Forensic Science* **45:**1049–1061.

Bearhop, S., D. R. Thompson, S. Waldron, I. C. Russell, G. Alexander, and R. W. Furness. 1999. Stable isotopes indicate the extent of freshwater feeding by cormorants *Phalacrocorax carbo* shot at inland fisheries in England. *Journal of Applied Ecology* **36:**75–84.

Bearhop, S., S. Waldron, S. C. Votier, and R. W. Furness. 2002. Factors that influence assimilation rates and fractionation of nitrogen and carbon stable isotopes in avian blood and feathers. *Physiological and Biochemical Zoology* **75:**451–458.

Bearhop, S., R. W. Furness, G. M. Hilton, S. C. Votier, and S. Waldron. 2003. A forensic approach to understanding diet and habitat use from stable isotope analysis of (avian) claw material. *Functional Ecology* **17:**270–275.

Bearhop, S., W. Fiedler, R. W. Furness, S. C. Votier, S. Waldron, J. Newton, G. Bowen, P. Berthold, and K. Farnsworth. 2005. Assortative mating as a mechanism for rapid evolution of a migratory divide. *Science* **310:**502–504.

Becker, B. H., and S. R. Beissinger. 2006. Centennial decline in the trophic level of an endangered seabird after fisheries decline. *Conservation Biology* **20:**470–479.

Best, P. B., and D. M. Schell. 1996. Stable isotopes in southern right whale (*Eubalaena australis*) baleen as indicators of seasonal movements, feeding and growth. *Marine Biology* **124:**483–494.

Bowen, G. J., L. I. Wassenaar, and K. A. Hobson. 2005. Application of stable hydrogen and oxygen isotopes to wildlife forensic investigations at global scales. *Oecologia* **143:**337–348.

Braune, B. M., G. M. Donaldson, and K. A. Hobson. 2002. Contaminant residues in seabird eggs from the Canadian Arctic. II. Spatial trends and evidence from stable isotopes for intercolony differences. *Environmental Pollution* **117:**133–145.

Brooks, J. R., L. B. Flanagan, N. Buchmann, and J. R. Ehleringer. 1997. Carbon isotope composition of boreal plants: Functional grouping of life forms. *Oecologia* **110:**301–311.

Burger, J., and M. Gochfeld. 2004. Marine birds as sentinels of environmental pollution. *EcoHealth* **1:**263–274.

Chamberlain, C. P., J. D. Blum, R. T. Holmes, X. Feng, T. W. Sherry, and G. R. Graves. 1997. The use of isotope tracers for identifying populations of migratory birds. *Oceologia* **109:**132–141.

Chamberlain, C. P., S. Bensch, X. Feng, S. Akesson, and T. Andersson. 2000. Stable isotopes examined across a migratory divide in Scandinavian willow warblers (*Phylloscopus trochilus trochilus* and *Phlloscopus trochilus acredula*) reflect African winter quarters. *Proceedings of the Royal Society (London) B* **267:**43–48.

Cherel, Y., K. A. Hobson, and H. Weimerskirch. 2000. Using stable-isotope analysis of feathers to distinguish moulting and breeding origins of seabirds. *Oecologia* **122:**155–162.

Cherel, Y., K. A. Hobson, and S. Hassani. 2005. Isotopic discrimination between diet and blood and feathers of captive penguins: Implications for dietary studies in the wild. *Physiological and Biochemical Zoology* **78:**106–115.

Cormie, A. B., H. P. Schwarcz, and J. Gray. 1994. Relation between hydrogen isotopic ratios of bone collagen and rain. *Geochimica et Cosmochimica Acta* **58:**377–391.

Craig, H. 1961. Isotopic variations in meteoric water. *Science* **133:**1702–1703.

Dansgaard, W. 1964. Stable isotopes in precipitation. *Tellus* **16:**436–468.

Dunn, E. H., K. A. Hobson, L. I. Wassenaar, D. Hussell, and M. L. Allen. 2006. Identification of summer origins of songbirds migrating through southern Canada in autumn. *Avian Conservation and Ecology* **1:**4 http://www.ace-eco.org/vol1/iss2/art4/

Emslie, S. D., and W. P. Patterson. 2007. Abrupt recent shift in $\delta^{13}C$ and $\delta^{15}N$ values in Adelie Penguin eggshell in Antarctica. *Proceedings of the National Academy of Science* (in press).

Erben, H. K., J. Hoefs, and K. H. Wedepohl. 1979. Palaeobiological and isotopic studies of eggshells from a declining dinosaur species. *Palaeobiology* **5:**380–414.

Evans-Ogden, L. J., K. A. Hobson, and D. B. Lank. 2004. Blood isotopic ($\delta^{13}C$ and $\delta^{15}N$) turnover and diet-tissue fractionation factors in captive Dunlin. *Auk* **121:**170–177.

Folinsbee, R. E., P. Fritz, H. R. Krouse, and A. R. Robblee. 1970. Carbon-13 and oxygen-18 in dinosaur, crocodile and bird eggshells indicate environmental conditions. *Science* **168:**1353–1356.

Fraiser, W. R., W. Z. Trivelpiece, D. G. Ainley, and S. G. Trivelpiece. 1992. Increases in Antarctic penguin populations: Reduced competition with whales or a loss of sea ices due to environmental warming. *Polar Biology* **11:**525–531.

France, R. L. 1995. Carbon-13 enrichment in benthic compared to plantonic algae: Foodweb implications. *Marine Ecology Progress Series* **124:**307–312.

Furness, R. W., and L. C. J. Camphuysen. 1997. Seabirds as monitors of the marine environment. *ICES Journal of Marine Sciences* **54:**723–726.

Furness, R. W., S. A. Lewis, and J. A. Mills. 1990. Mercury levels in the plumage of red-billed gulls Larus novehollandiae scopulinus of known sex and age. *Environmental Pollution* **63:**33–39.

Graves, G. R., C. S. Romanek, and A. R. Navarro. 2002. Stable isotope signature of philopatry and dispersal in a migratory songbird. *Proceedings of the National Academy of Science* **99:**8096–8100.

Haramis, G. M., D. G. Jorde, S. A. Macko, and J. L. Walker. 2001. Stable-isotope analysis of canvasback winter diet in upper Chesapeake Bay. *Auk* **118:**1008–1017.

Hebert, C., and L. I. Wassenaar. 2001. Stable nitrogen isotopes in waterfowl feathers reflect agricultural land use in western Canada. *Environmental Science and Technology* **35:**3482–3487.

Hebert, C., and L. I. Wassenaar. 2005. Stable isotopes provide evidence for poor northern pintail production on the Canadian prairies. *Journal of Wildlife Management* **69:**101–109.

Hobson, K. A. 1995. Reconstructing avian diets using stable-carbon and nitrogen isotope analysis of egg components: Patterns of isotopic fractionation and turnover. *Condor* **97:**752–762.

Hobson, K. A. 1999. Stable-carbon and nitrogen isotope ratios of songbird feathers grown in two terrestrial biomes: Implications for evaluating trophic relationships and breeding origins. *Condor* **101:**799–805.

Hobson, K. A. 2005. Stable isotopes and the determination of avian migratory connectivity and seasonal interactions. *Auk* **122:**1037–1048.

Hobson, K. A., and F. Bairlein. 2003. Isotopic fractionation and turnover in captive garden warblers (*Sylvia borin*): Implications for delineating dietary and migratory associations in wild passerines. *Canadian Journal of Zoology* **81:**1630–1635.

Hobson, K. A., and H. E. Welch. 1992. Determination of trophic relationships within a high Arctic marine food web using δ^{13}C and δ ^{15}N analysis. *Marine Ecology Progress Series* **84:**9–18.

Hobson, K. A., and L. I. Wassenaar. 1997. Linking breeding and wintering grounds of neotropical migrant songbirds using stable hydrogen isotopic analysis of feathers. *Oecologia* **109:**142–148.

Hobson, K. A., and R. G. Clark. 1992. Assessing avian diets using stable isotopes. II. Factors influencing diet-tissue fractionation. *Condor* **94:**189–197.

Hobson, K. A., K. D. Hughes, and P. J. Ewins. 1997. Using stable-isotope analysis to identify endogenous and exogenous sources of nutrients in eggs of migratory birds: Applications to Great Lakes contaminants research. *Auk* **114:**467–478.

Hobson, K. A., L. I. Wassenaar, B. Milá, I. Lovette, C. Dingle, and T. B. Smith. 2003. Stable isotopes as indicators of altitudinal distributions and movements in an Ecuadorean hummingbird community. *Oecologia* **136:**302–308.

Hobson, K. A., G. Bowen, L. I. Wassenaar, Y. Ferrand, and H. Lormee. 2004a. Using stable hydrogen isotope measurements of feathers to infer geographical origins of migrating European birds. *Oecologia* **141:**477–488.

Hobson, K. A., E. H. Sinclair, A. E. York, J. Thomason, and R. E. Merrick. 2004b. Retrospective isotopic analyses of Steller's sea lion tooth annuli and seabird feathers: A cross-taxa approach to investigating regime and dietary shifts in the Gulf of Alaska. *Marine Mammal Science* **20:**621–638.

Hobson, K. A., S. Van Wilgenburg, L. I. Wassenaar, H. Hands, W. Johnson, M. O'Melia, and P. Taylor. 2006. Using stable-hydrogen isotopes to delineate origins of sandhill cranes harvested in the Central Flyway of North America. *Waterbirds* **29:**137–147.

IAEA/WMO. 2001. Global network for isotopes in precipitation, the GNIP database.

Jarman, W. M., K. A. Hobson, W. J. Sydeman, C. E. Bacon, and E. B. McLaren. 1996. Influence of trophic position and feeding location on contaminant levels in the Gulf of the Farallones food web revealed by stable isotope analysis. *Environmental Science and Technology* **30:**654–660.

Jouzel, J. 1999. Calibrating the isotopic paleothermometer. *Science* **286:**910–911.

Kelly, J. F., V. Atudorei, Z. D. Sharp, and D. M. Finch. 2002. Insights into Wilson's Warbler migration from analyses of hydrogen stable-isotope ratios. *Oecologia* **130:**216–221.

Kidd, K. A. 1998. Use of stable isotope ratios in freshwater and marine biomagnification studies. Pages 357–376 *in* J. Rose (Ed.) *Environmental Toxicology: Current Developments.* Gordon and Breach Science Publishers, New York.

Körner, C., G. D. Farquhar, and Z. Roksandic. 1988. A global survey of carbon isotope discrimination plants from high altitude. *Oecologia* **74:**623–632.

Körner, C., G. D. Farquhar, and S. C. Wong. 1991. Carbon isotope discrimination by plants follows latitudinal and altitudinal trends. *Oecologia* **74:**623–632.

Lott, C. A., T. A. Meehan, and J. Smith. 2003. Estimating the latitudinal origins of migratory birds using hydrogen and sulfur stable isotopes in feathers: Influence of marine prey base. *Oecologia* **134:**505–510.

Lourey, M. J., T. W. Trull, and B. Tillbrook. 2004. Sensitivity of δ^{13}C of Southern Ocean suspended and sinking organic matter to temperature, nutrient utilization, and atmospheric CO_2. *Deep-Sea Research* 1, **151:**281–305.

Marra, P. P., K. A. Hobson, and R. T. Holmes. 1998. Linking winter and summer events in a migratory bird using stable carbon isotopes. *Science* **282:**1884–1886.

Masson, V., F. Vimeus, J. Jouzel, V. Morgan, M. Delmotte, P. Ciais, C. Hammer, S. Johnsen, V. Y. Lipenkov, E. Mosley-Thompson, J.-R. Petit, E. J. Steig, *et al.* 2000. Holocene climate variability in Antarctica based on 11 ice-core isotopic records. *Quarternary Research* **54:**348–358.

Meehan, T. D., C. A. Lott, Z. D. Sharp, R. B. Smith, R. N. Rosenfield, A. C. Stewart, and R. K. Murphy. 2001. Using hydrogen isotope geochemistry to estimate the natal latitudes of immature Cooper's hawks migrating through the Florida Keys. *Condor* **103:**11–20.

Miller, R. F., P. Fritz, and A. V. Morgan. 1988. Climatic implications of D/H ratios in beetle chitin. *Palaeogeography, Palaeoclimatology and Plaeoecology* **66:**277–288.

MØller, A. P., and K. A. Hobson. 2004. Heterogeneity in stable isotope profiles predicts coexistence of two populations of barn swallows *Hirundo rustica* differing in morphology and reproductive performance. *Proceedings of the Royal Society (London)* **271:**1355–1362.

Monteiro, L. R., and R. W. Furness. 1995. Seabirds as monitors of mercury in the marine environment. *Water Air and Soil Pollution* **80:**851–870.

Murphy-Klassen, H. M., T. J. Underwood, S. G. Sealy, and A. A. Czyrnyj. 2005. Lon-term trends in spring arrival dates of migrant birds at Delta Marsh, Manitoba, in relation to climate change. *Auk* **122:**1130–1148.

Michener, R. H., and D. M. Schell. 1994. Stable isotope ratios as tracers in marine aquatic food webs. Pages 138–158 *in* K. Lajtha and R. H. Michener (Eds.) *Stable Isotopes in Ecology and Environmental Science.* Blackwell Scientific, London.

Mizutani, H., M. Fukuda, Y. Kabaya, and E. Wada. 1991. Carbon isotope ratio of feathers reveals feeding behavior of cormorants. *Auk* **107:**400–403.

Mizutani, H., M. Fukuda, and Y. Kabaya. 1992. ^{13}C and ^{15}N enrichment factors of feathers of 11 species of adult birds. *Ecology* **73:**1391–1395.

Nadelhoffer, K. J., and B. Fry. 1994. Nitrogen isotopes studies in forest ecosystems. Pages 22–44 *in* K. Lajtha and R. H. Michener (Eds.) *Stable Isotopes in Ecology and Environmental Science.* Blackwell Scientific, London.

Pain, D. J., R. E. Green, B. Giebing, A. Kozulin, A. Poluda, U. Ottosson, M. Flade, and G. M. Hilton. 2004. Using stable isotopes to investigate migratory connectivity of the globally threatened aquatic warbler *Acrocephalus paludicola. Oecologia* **138:**168–174.

Pearson, S. F., D. J. Levey, C. H. Greenberg, and C. Martinez del Rio. 2003. Effects of elemental composition on the incorporation of dietary nitrogen and carbon isotopic signatures in an omnivorous songbird. *Oecologia* **135:**516–523.

Peterson, B. J., and R. W. Howarth. 1987. Sulfur, carbon and nitrogen isotopes used to trace organic matter flow in the salt marsh estuaries of Sapelo Island, Georgia. *Limnology and Oceanography* **32:**1195–1213.

Podlesak, D. W., S. R. McWilliams, and K. A. Hatch. 2005. Stable isotopes in breath, blood, feces and feathers can indicate intra-individual changes in the diet of migratory songbirds. *Oecologia* **142:**501–510.

Popp, B. N., T. Trull, F. Kenig, S. G. Wakeham, T. M. Rust, B. Tilbrook, F. B. Griffiths, S. W. Wright, H. J. Marchant, R. R. Bidigare, and E. A. Laws. 1999. Controls on the isotopic composition of Southern Ocean phytoplankton. *Global Biogeochemical Cycles* **13:**827–843.

Rau, G. H., R. E. Sweeney, and L. R. Kaplan. 1982. Plankton $^{13}C/^{12}C$ ratio changes with latitude: Differences between northern and southern oceans. *Deep Sea Research I* **29:**1035–1039.

Rau, G. H., T. Takahashi, and D. J. Marais. 1989. Latitudinal variations in planktonic δ^{13}C: Implicaitons for CO_2 and productivity of past oceans. *Nature* **341:**516–518.

Robbins, C. T., L. A. Felicetti, and M. Sponheimer. 2005. The effect of dietary protein quality on nitrogen isotope discrimination in mammals and birds. *Oecologia* **144:**534–540.

Sarkar, A., S. K. Bhattachyra, and D. M. Mohabey. 1991. Stable-isotope analysis of dinosaur eggshells: Paleoenvironmental implications. *Geology* **19:**1068–1071.

Schaffner, F. C., and P. K. Swart. 1991. Influence of diet and environmental water on the carbon and oxygen isotopic signatures of seabird eggshell carbonate. *Bulletin of Marine Science* **48:**23–38.

Schell, D. M. 2000. Declining carrying capacity of the Bering Sea: Isotopic evidence from whale baleen. *Limnology and Oceanography* **45:**459–462.

Schell, D. M., B. A. Barnett, and K. Vinette. 1998. Carbon and nitrogen isotope ratios in zooplankton of the Bering, Chukchi and Beaufort seas. *Marine Ecology Progress Series* **162:**11–23.

Schiegl, W. G. 1970. Natural deuterium in biogenic materials. Influence of environment and geophysical applications Ph.D. thesis, University of South Africa, Pretoria.

Schleser, G. H., and R. Jayasekera. 1985. δ^{13}C-variations of leaves in forests as an indication of reassimilated CO_2 from the soil. *Oecologia* **65:**536–542.

Sheppard, S. M. F., R. L. Nielsen, and H. P. Taylor. 1969. O and H isotope ratios of clay minerals from porphyry copper deposits. *Economic Geology* **64:**755–777.

Smith, T. B., P. Marra, M. S. Webster, I. Lovette, L. Gibbs, R. T. Holmes, K. A. Hobson, and S. Rohwer. 2003. A call for feather sampling. *Auk* **120:**218–221.

Sydeman, W. J., K. A. Hobson, P. Pyle, and E. B. McLaren. 1997. Trophic relationships among seabirds in central California: Combined stable isotope and conventional dietary approach. *Condor* **99:**327–336.

Stapp, P., G. A. Polis, and F. Sanchez-Pinero. 1999. Stable isotopes reveal strong marine and El Nino effects on island food webs. *Nature* **401:**467–469.

Still, C. J., J. A. Berry, G. J. Collatz, and R. S. DeFries. 2003. Global distribution of C3 and C4 vegetation: Carbon cycle implications. *Global Biogeochemical Cycles* **17**:1006–1029.

Taylor, H. P., Jr. 1974. An application of oxygen and hydrogen isotope studies to problems of hydrothermal alteration and ore deposition. *Economic Geology* **69**:843–883.

Thompson, D. R., R. W. Furness, and S. A. Lewis. 1995. Diets and long-term changes in $\delta^{15}N$ and $\delta^{13}C$ values in Norhern Fulmars *Fulmarus glacialis* from two northeast Atlantic colonies. *Marine Ecology Progress Series* **125**:3–11.

Thompson, D. R., S. Bearhop, J. R. Speakman, and R. W. Furness. 1998. Feathers as a means of monitoring mercury in seabirds: Insights from stable isotope analysis. *Environmental Pollution* **101**:193–200.

van Riper, C., S. G. van Riper, M. L. Goff, and M. Laird. 1986. The epizootiology and ecological significance of malaria in Hawaiian landbirds. *Ecological Monographs* **56**:327–344.

Von Schirnding, Y., N. J. van der Merwe, and J. C. Vogel. 1982. Influence of diet and age on carbon isotope ratios in ostrich eggehsell. *Archaeometry* **24**:3–20.

Warner, A. J., and G. C. Hays. 1994. Sampling by the continuous plankton recorder survey. *Progress in Oceanography* **34**:237–256.

Wassenaar, L. I., and K. A. Hobson. 2000. Stable-carbon and hydrogen isotope ratios reveal breeding origins of red-winged blackbirds. *Ecological Applications* **10**:911–916.

Wassenaar, L. I., and K. A. Hobson. 2001. A stable-isotope approach to delineate geographical catchment areas of avian migration monitoring stations in North America. *Environmental Science and Technology* **35**:1845–1850.

Wolf, B., and C. Martinez del Rio. 2000. Use of saguaro fruit by white-winged doves: Isotopic evidence of a tight ecological association. *Oecologia* **124**:536–543.

Yohannes, E., K. A. Hobson, D. Pearson, L. I. Wassenaar, and H. Biebach. 2005. Stable isotope analyses of feathers help identify autumn stopover sites of three long-distance migrants in northeastern Africa. *Journal of Avian Biology* **36**:235–241.

Ziegler, H. 1988. Hydrogen isotope fractionation in plant tissues. Pages 105–123 *in* P. W. Rundel, J. R. Ehleringer, and K. A. Nagy (Eds.) *Stable Isotopes in Ecological Research.* Springer-Verlag, New York.

CHAPTER 10

Use of the Stable Isotope Composition of Fish Scales for Monitoring Aquatic Ecosystems

Clive N. Trueman* and Andy Moore†

*School of Ocean and Earth Science, National Oceanography Centre, Southampton University of Southampton Waterfront Campus
†Lowestoft Laboratory, Centre for Environment, Fisheries and Aquaculture Science

Contents

I. INTRODUCTION

A. Ocean Climate and Ecosystems

There is widespread interest in natural and human-induced variability in environmental and climate systems and their effects on ecosystem structure and function. The open ocean is the largest ecological system on the planet, and ecosystem processes within the open ocean profoundly influence global biogeochemical cycles. The ocean also provides food sustaining an increasing proportion of the world's population. The structure and function of ocean ecosystems are clearly linked to climate, and it is likely that changes in climate due to increased warming or perturbations of natural climate forcing cycles will similarly impact marine ecosystem function. The El Niño Southern Oscillation, for instance, exerts strong control on relative abundances of Pacific sardine (*Sardinops sagax*) and anchovy (*Englauris ringens*) stocks (Chavez *et al.* 2003). In the North Atlantic both phytoplankton and zooplankton communities have been associated with variations in the North Atlantic Oscillation (NAO) and Northern Hemisphere temperature indices (Fromentin and Planque 1996, Beaugrand *et al.* 2002, 2003). Observed increases in marine temperatures in the North Atlantic have triggered major reorganizations in plankton community structure and have led to movements of plankton over 10° latitude (Beaugrand *et al.* 2003). The effects of changing distributions and seasonal timing (phenology) of primary production cascade through the food web to produce a wide range of effects on the marine ecosystem at all trophic levels (Beaugrand and Reid 2003, Beaugrand *et al.* 2003, Brander and Mohn 2004, Edwards and Richardson 2004). Understanding the mechanisms linking climate, environmental change, and ecosystem function would have major implications for the prediction of biogeochemical cycling and management of aquatic resources.

Identifying the influence of climate on ecosystem structure and function requires datasets of population structure and/or behavior extending over several decades. Direct, physical collection programs such as the Continuous Plankton Recorder (CPR) survey in the North Atlantic continue to provide a record of long-term response of marine planktonic communities to climate change, but long-term monitoring experiments are difficult to establish and maintain, particularly in the marine environment (Duarte *et al.* 1992, Brander *et al.* 2003). Consequently, there are few existing multi-decadal records of marine ecosystem structure and function. Even if it were possible to establish a new series of direct monitoring programmes, we cannot afford to wait another ~30 years to recover fundamental information. Some method of retrospective analysis is therefore necessary to investigate the relationship between marine ecosystem function and climate forcing.

The isotopic composition of carbon and nitrogen in marine organic materials provides a possibility to investigate marine ecosystem response to environmental change. $\delta^{13}C$ values are influenced by critical ecological variables such as the source of carbon, the extent of primary production, and phytoplankton taxonomy. $\delta^{15}N$ values in organic particles are influenced by inorganic nitrogen concentrations and phytoplankton taxonomy, and particularly by the trophic structure of the ecosystem. Stable isotope analyses could be performed directly on phytoplankton, using for instance samples obtained during CPR survey programs. Marine environments are extremely dynamic, however, and short-term temporal and spatial variations in environmental conditions may be translated rapidly to plankton, obscuring longer term trends. To avoid the effects of short-term variation, some method of integrating isotope values over temporal scales of weeks or months (while ideally limiting spatial averaging) is required. All organisms within a marine food web ultimately derive carbon and nitrogen from primary production, and the isotopic composition of a consumer's tissues will be averaged over the time taken to grow the sampled tissue. Fish represent attractive isotopic integrators as commercial fish are routinely sampled in huge numbers from all of the world's oceans. It is important to note that fish are mobile organisms, and the interpretation of fish isotope signals is limited by the knowledge of the area sampled. Scales provide a particularly attractive target tissue for isotopic analysis as they are easily sampled and stored. Most importantly, the scales of commercially exploited fish are routinely sampled to establish age and population structure, and in many cases these scales are kept in archived collections, often covering more than 60 years. Retrospective analysis of the isotopic composition of archived fish scales can therefore provide information concerning temporal changes in ecosystem structure and function (Wainwright *et al.* 1993, Satterfield and Finney 2002, Perga and Gerdeaux 2003, Pruell *et al.* 2003, Gerdeaux and Perga 2006). Scales also present an attractive option for future long-term monitoring of aquatic ecosystems.

The purpose of this chapter is to briefly review the potential for using the stable isotope composition of archived fish scales to provide decadal records of variability in ecosystem structure in both freshwater and marine ecosystems.

II. ARCHIVED SCALES AS RECORDS OF ISOTOPIC CHANGE

A. Composition and Isotope Chemistry of Scales

A typical teleost scale consists of two portions; a hard upper well-mineralized layer composed of calcium phosphate similar to the mineral apatite (the external layer or EL) overlying a poorly mineralized layer composed largely of collagen known as the basal or fibrillar plate (Figure 10.1; Fouda 1979, Zylberberg and Nicols 1982, Zylberberg 2004). The upper well-mineralized external layer

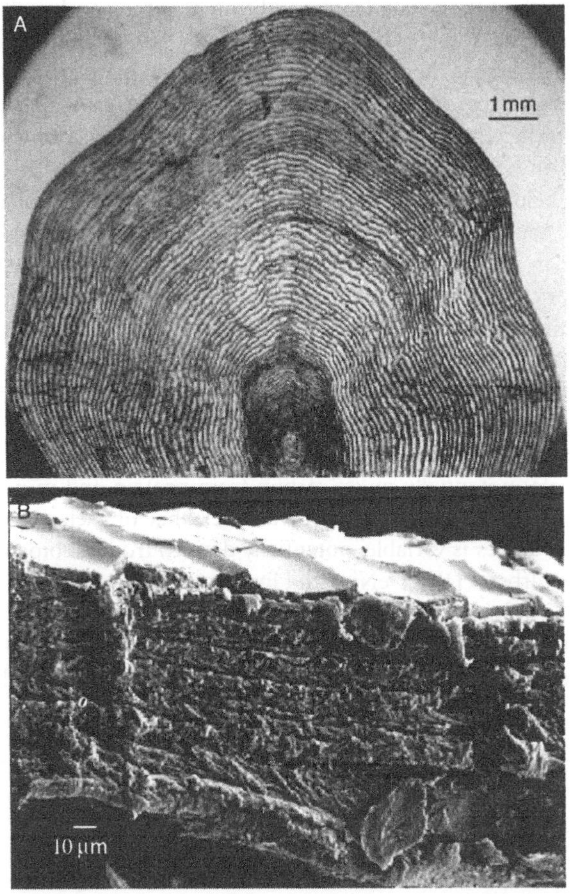

FIGURE 10.1 Scales from Atlantic salmon (*S. salar*). (A) Ventral view of an adult salmon scale showing incremental growth lines formed from concentric mineral ridges (circuli). (B) SEM image of a transverse section through an adult scale showing the mineralized external layer (EL) and successive collagen laminae forming the basal or fibrillar plate (CL).

of teleost scales is analogous to bone, consisting of a matrix of thin collagen fibrils and mineral crystals. Unlike bone, in the EL of scales, crystals do not appear to be associated with collagen fibrils, but form aggregated clusters which increase in size until they fuse and the EL becomes mineralized (Zylberberg and Nicols 1982, Zylberberg 2004). The EL comprises around 30% of the scale by mass and increases in area but not in thickness during the growth of the scale as formation occurs through increments of centrifugally grown rings (circuli), taking the form of concentric parallel ridges (Neave 1936) (Figure 10.1).

The underlying basal or fibrillary plate varies in thickness, being thin at the anterior growing portion of the scale and increasing in thickness toward the centre (focus) of the scale (Fouda 1979). The basal plate is composed of successive layers of collagen arranged in a lamellar fashion. Thick (c. 100 nm in diameter, Zylberberg 2004) collagen fibrils are arranged in superimposed layers, with the fiber direction changing from one layer to the next, forming a plywood structure (Zylberberg *et al.* 1988, Weiner *et al.* 1997).

B. Growth of Scales and Sampling

The growth of teleost scales begins by formation and mineralization of the overlying EL. Formation of the EL is followed by growth of the basal plate by addition of successive lamellae of collagen fibers and finally mineralization of the basal plate (Neave 1936, Fouda 1979, Zylberberg *et al.* 1988, 2004). Thus, the EL grows centrifugally whereas the collagen-containing basal plate grows by successive addition of collagen lamellae. Each successive collagen layer extends from the growing tip posterially toward the focus, and underplates earlier formed collagen. A vertically cut section of a scale, guided by reading apatite circuli, will contain both collagen formed contemporaneously with the growth of the incremental line immediately below the apatite layer, and also younger-formed collagen in vertically stacked sequences (Figure 10.2). Sampling such vertical sections in the inner (juvenile) portions of the scale would therefore include an isotopic component from collagen formed in later life, and so would not truly reflect the diet history solely for the intended period of growth (Wainwright *et al.* 1993, Satterfield and Finney 2002, Hutchinson and Trueman 2006). Similarly, sampling whole scales would provide a record of diet biased toward the last season of growth rather than an integrated record of whole life diet. Scale sampling is therefore ideally limited to the last-formed portion of scale, representing the last growth season (Figure 10.2), although this may not be practical when dealing with small scales in fish such as cod (*Gadus morhua*). Whole scales are composed of protein and carbonated apatite, thus sampling of complete scales potentially mixes carbon from protein and mineral (carbonate) sources. While the isotopic composition of scale protein will track that of dietary protein, the origin of the carbon in biomineral carbonates is variable, potentially derived from all biochemical types in diet and from dissolved inorganic carbon (DIC; DeNiro and Epstein 1978, Lee-Thorp and van der Merwe 1989, Kalish 1991). To avoid the influence of mineral carbon in scale analyses some authors suggest removal of the inorganic fraction of scales via acid dissolution (Perga and Gerdeaux 2003). Decalcification may alter the isotopic composition of protein (Pinnegar and Polunin 1999) possibly through preferential

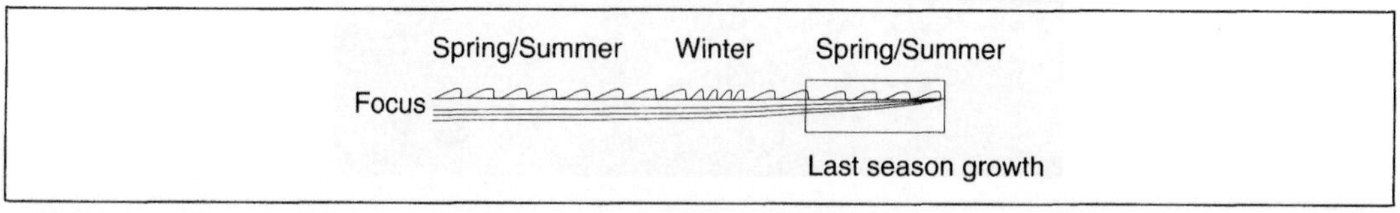

FIGURE 10.2 Schematic section through a fish scale indicating the spatial relationship between collagen laminae and mineral circuli. A vertical section through the scale will contain collagen of varying ages at all points except the last season of growth.

loss of acid-mobile amino acids. In scales of fish such as Atlantic salmon (*Salmo salar*), the apatite component is <30% by mass of the scale, resulting in a ratio of protein:minersal-derived carbon ratio of >100:1, and in these cases acidic decalcification is unnecessary and may be avoided.

C. Isotope Chemistry of Scales

A close relationship between the isotopic composition of scale and muscle tissue has been demonstrated in several fish species (Estep and Vigg 1985, Satterfield and Finney 2002, Perga and Gerdaux 2003, Pruell *et al.* 2003, Estrada *et al.* 2005, Figure 10.3). Scale $\delta^{13}C$ values are typically enriched over muscle values by c. 2–4‰, similar to the isotopic offset between muscle and bone collagen (Lee-Thorp and van der Merwe 1989). This is consistent with the suggestion that carbon in scales is almost

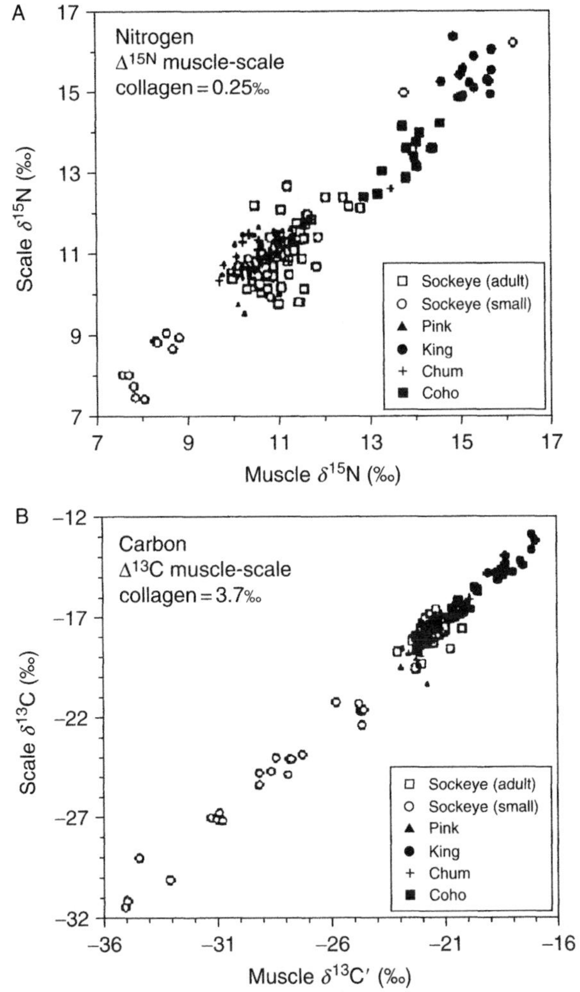

FIGURE 10.3 Relationship between the stable isotope composition of scales and muscle in five species of Pacific salmon. Reprinted from Progress in Oceanography 53, Satterfield, F. R. and B. P. Finney, Stable isotope analysis of Pacific salmon: insight into trophic status and oceanographic conditions over the last 30 years, 231–246 (2002), with permission from Elsevier.

exclusively derived from the protein component (dominantly collagen) and that the carbon contained in the carbonated apatite mineral component is quantitatively insignificant. Scale $\delta^{15}N$ values do not differ significantly from muscle values in most cases. Separation between scale and muscle $\delta^{15}N$ values may indicate differences in isotopic assimilation rates between the two tissues with a rapidly changing diet. Scales are therefore a reliable record of integrated dietary $\delta^{13}C$ and $\delta^{15}N$ values and archived scales provide a suitable target for retrospective analyses of tissue stable isotope composition.

Estimating the period of time represented by scale sampling is an essential aspect of stable isotope analyses (Satterfield and Finney 2002). The isotopic composition of tissues in a growing animal will respond rapidly to changes in diet as the new isotope values will be incorporated directly into new tissue. As growth rates slow, the rate of isotopic equilibration between diet and tissue is reduced as any change in diet composition is diluted by the metabolism of existing tissues (*i.e.*, metabolic turnover) (Hesslein *et al.* 1993, Jardine *et al.* 2004, Trueman *et al.* 2005). In fast growing, active fish such as salmon the isotopic turnover time in adult fish is likely on the order of months (Hesslein *et al.* 1993, Jardine *et al.* 2004, Trueman *et al.* 2005) and sampling the last season's growth of a scale will provide an estimate of the isotopic composition of the fish integrated over the last 3–4 months of life. This is supported by the consistent relationship between the isotopic composition of muscle and scale samples from single fish (Estep and Vigg 1985, Satterfield and Finney 2002, Estrada *et al.* 2005). Sampling striped bass, however, Pruell *et al.* (2003) obtain a relatively poor correlation between scale and muscle $\delta^{13}C$ values, suggesting integration of $\delta^{13}C$ values in scale collagen over longer timescales in whole bass scales. Clearly, it is important to conduct experimental "proof of concept" studies with the species of choice before utilizing tissues in monitoring programmes. Such studies should involve controlled diet experiments to establish isotopic discrimination factors and also diet-switching studies to determine isotopic integration time and metabolic turnover rates.

III. POTENTIAL CAUSES OF VARIATION IN $\delta^{13}C$ AND $\delta^{15}N$ VALUES IN SCALE COLLAGEN

Variations in the isotopic composition of animal tissues rise from variations in the isotopic composition at the base of the food web and from the consumer's trophic level due to the preferential excretion of isotopically light carbon and nitrogen (Petersen and Fry 1987, Cabana and Rasmussen 1996). Hence, the stable isotope composition of a consumer's tissues may potentially yield insights into variations in both the planktonic ecosystem and the food web trophic structure.

A. Controls on $\delta^{13}C$ Values in Marine Ecosystems

The isotopic composition of carbon in phytoplankton varies spatially and temporally within marine ecosystems (Michener and Schell 1994 and references therein). $\delta^{13}C$ values in phytoplankton are influenced by growth rates, taxonomy, and $\delta^{13}C$ values in DIC. $\delta^{13}C$ values in DIC are themselves potentially influenced by the balance between export of carbon from upper ocean waters via sinking of particulate organic matter (POM), and inputs to surface waters via oxidation (remineralization) of POM, upwelling, and atmospheric drawdown (Figure 10.4). $\delta^{13}C$ values in POM are typically 20–27‰ lighter than atmospheric carbon. Thus export of organic carbon via particle settling may lead to net increases in $\delta^{13}C$ values of DIC. Recycling of organic carbon via oxidation releases ^{12}C-rich C to DIC, lowering bulk $\delta^{13}C$ values (Post 2002). Recycled C may also be returned to surface waters through upwelling. Where export of carbon exceeds remineralization and/or upwelling (for instance at times of high surface productivity), $\delta^{13}C$ values in surface DIC may increase. Where recycling of organic carbon

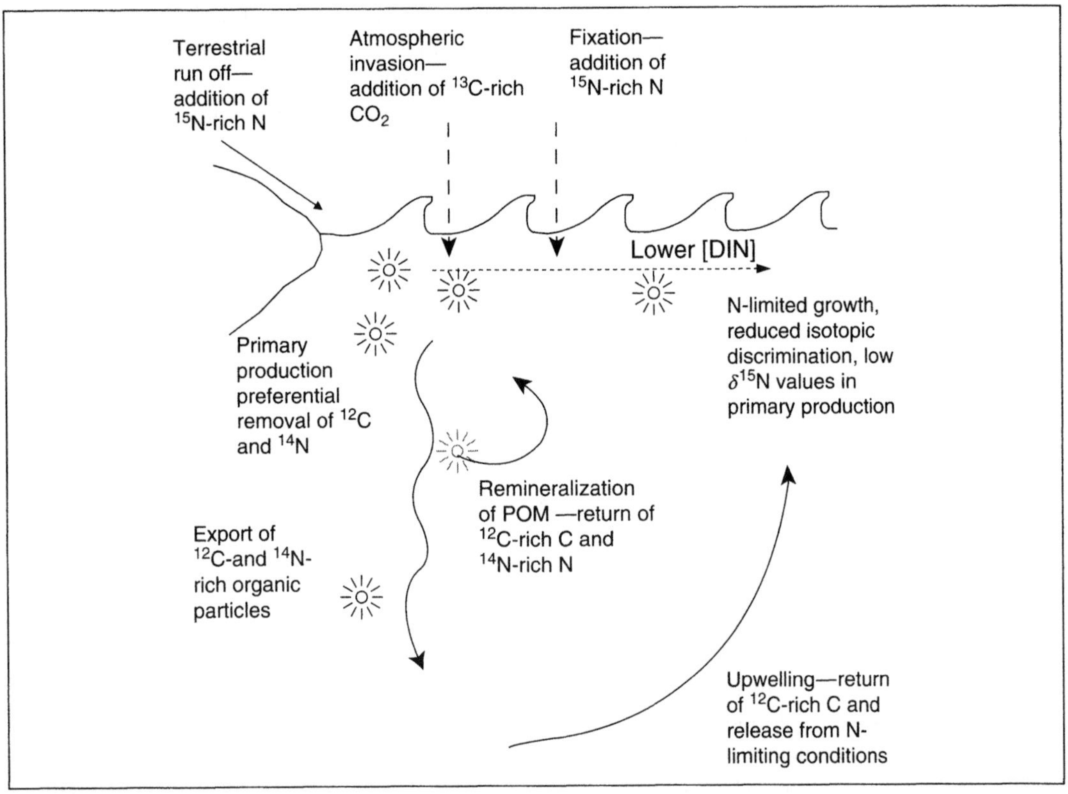

FIGURE 10.4 Simplified diagram illustrating some processes influencing the isotopic composition of carbon and nitrogen in phytoplankton. Photosynthesis discriminates between carbon isotopes to varying extents depending on a number of variables including growth rate, DIC concentration and taxon, with ^{12}C preferentially removed into organic tissue. The isotopic composition of nitrogen in primary production varies depending on phytoplankton taxon and the availability of dissolved inorganic nitrate (DIN). Terrestrial runoff introduces DIN to surface waters. Fast settling of POM preferentially removes ^{12}C and ^{14}N from surface waters, and may reduce DIC and particularly DIN concentrations. As N concentrations become limiting for phytoplankton growth, isotopic discrimination is reduced and $\delta^{15}N$ values of phytoplankton are lowered. Remineralization of POM and upwelling returns ^{12}C-rich carbonate and DIN to surface waters, releasing phytoplankton from nitrogen limitation. Invasion of ^{13}C-rich atmospheric CO_2 may occur where surface waters are unsaturated with respect to CO_2.

exceeds export (for instance in upwelling regions), $\delta^{13}C$ values in DIC may be relatively low. Where C is removed from surface waters via increased export, atmospheric invasion of relatively ^{13}C-rich CO_2 into surface water may also be increased, raising bulk $\delta^{13}C$ values in DIC (Hayes 1993, Schindler *et al.* 1997, Bade *et al.* 2004, Perga and Gerdeaux 2004).

Fractionation of carbon isotopes during photosynthesis is also dependent on the relative concentrations of intracellular and extracellular CO_2 and thus carbon demand (Rau *et al.* 1992). Theoretical considerations and experimental data indicate that maximum isotopic discrimination between tissue and extracellular carbon will be expressed when growth rates are low and extracellular carbon concentrations are high (Laws *et al.* 1995). Where carbon is abundant fractionation will be most pronounced, whereas at low DIC concentrations fractionation is reduced. Tissue-DIC spacing in phytoplankton may vary by as much as 25‰ indicating the importance of growth rate and carbon demand in controlling bulk $\delta^{13}C$ levels in POM (Raven *et al.* 1994, Perga and Gerdeaux 2003).

Increased export of carbon may therefore lead to increased $\delta^{13}C$ values in DIC, increased carbon demand during algal cell growth and reduced fractionation associated with photosynthesis. High levels of C export are typically associated with raised marine fertility and associated high levels of primary productivity, and high $\delta^{13}C$ values in POM are frequently interpreted as reflecting increased levels of primary production (Schelske and Hodell 1991, 1995, Hayes 1993, Meyers and Bernasconi 2005).

The taxonomic composition of phytoplankton, however, also significantly influences the isotopic composition of bulk suspended POM as the physiology of isotope fractionation varies between different phytoplankton taxa (Goericke *et al.* 1994). In the Georges Bank ecosystem, for instance, phytoplankton $\delta^{13}C$ values vary by at least 8‰ with the highest values of $\delta^{13}C$ corresponding to large, fast growing diatoms (Fry and Wainwright 1991). ^{13}C enrichment in diatoms relative to other phytoplankton has been observed in several other studies (Berry 1989, Hinga *et al.* 1994, Pancost *et al.* 1999) and may be related to different mechanisms for transport of carbon into the cell, however the degree of isotopic enrichment between diatoms and other phytoplankton taxa is inconsistent (Vuorio *et al.* 2006). The isotopic composition of carbon in animal tissues also varies with trophic level, and the typical enrichment in ^{13}C from one trophic level to the next is estimated at c. 0.4‰ (Post 2002, McCutchan *et al.* 2003). Variations in both $\delta^{13}C$ and $\delta^{15}N$ values related to plankton cell size were noted by Rau *et al.* (1990), with smaller size fractions associated with lower $\delta^{13}C$ and $\delta^{15}N$ values. The isotopic composition of carbon in fish scales may thus be influenced by relative levels of primary productivity, phytoplankton taxonomy, and plankton trophic structure.

B. Controls on $\delta^{15}N$ Values Within Marine Ecosystems

As with carbon, variations in the isotopic composition of nitrogen in consumer tissue may reflect processes at the base of the food web and the consumer's trophic level. Phytoplankton preferentially incorporate ^{14}N, consequently the $\delta^{15}N$ values of dissolved inorganic nitrogen (DIN) (and primary production) increase as long as DIN concentrations remain high (Figure 10.4). Nitrogen is frequently a limiting nutrient in marine ecosystems, however, and in such cases isotopic discrimination is reduced and $\delta^{15}N$ values in phytoplankton are relatively low. Several studies have demonstrated that $\delta^{15}N$ values in DIN and suspended POM vary inversely with DIN concentration (Miyake and Wada 1967, Altabet and Francois 1994, Wu *et al.* 1997, Rau *et al.* 1998). Upward diffusion or upwelling of deep water brings new DIN with relatively high $\delta^{15}N$ values potentially releasing phytoplankton growth from nitrogen limitation and allowing increased isotopic discrimination. Surface water $\delta^{15}N$ values thus reflect the balance between the overall DIN concentrations, removal of DIN into primary production and its replacement via upwelling or terrestrial runoff. If production exceeds replenishment, DIN becomes limiting, isotopic discrimination is reduced and $\delta^{15}N$ values in phytoplankton will be reduced and vice versa. The relationship between $\delta^{15}N$ values of primary production and the concentration of DIN results in spatial variation in $\delta^{15}N$ values of phytoplankton, particularly evident as gradients of decreasing $\delta^{15}N$ values with increasing distances from sources of DIN. For instance, suspended POM and zooplankton collected in the Central North Pacific were depleted in ^{15}N by c. 7‰ compared to materials collected in coastal sites in British Columbia and central California (Mullin *et al.* 1984, Wu *et al.* 1997).

Finally, taxonomic composition and trophic structure also influence $\delta^{15}N$ values of bulk primary productivity, as $\delta^{15}N$ values increase with increasing size fraction (Rau *et al.* 1990, Wu *et al.* 1997) reflecting the step-wise enrichment in ^{15}N in tissues associated with increasing trophic levels (Minagawa and Wada 1984). Estimating absolute trophic levels from isotope data requires a thorough understanding of the isotopic composition of prey species, which is frequently absent. Furthermore, estimating absolute trophic level for any species within an ecosystem from stable isotope analysis requires that the diet-tissue isotopic fractionation factor is known and constant. There is increasing evidence that such assumptions are not valid and that the often quoted value of c. 3.4‰ increase in $\delta^{15}N$ values per trophic level cannot be confidently applied across wide ranges of species and ecological guilds (McCutchan *et al.* 2003, Vanderklift and Ponsard 2003, Focken 2004). In marine ecosystems, average values are likely to lie between 2‰ and 4‰ (McCutchan *et al.* 2003), but estimates of absolute trophic level based on tissue $\delta^{15}N$ values may be suspect. New approaches based on analyses of single amino acids hold much promise to overcome this problem (Chapter 12).

Thus, in open marine (offshore) ecosystems, DIN supply, community structure, and particularly food chain length potentially exert stronger influences on consumer $\delta^{15}N$ values than $\delta^{13}C$ values, whereas changes in the carbon export and taxonomic composition of phytoplankton are likely to be more strongly expressed in $\delta^{13}C$ values. As the isotopic composition of both carbon and nitrogen in animal tissues is influenced by trophic level, $\delta^{15}N$ and $\delta^{13}C$ values determined in a range of species inhabiting the same ecosystem are expected to covary. Decoupling of $\delta^{15}N$ and $\delta^{13}C$ values indicates that influences other than the trophic structure of the ecosystem (*e.g.*, differences in carbon export rates and /or spatial variation in the isotopic composition of DIN) affect either carbon and/or nitrogen isotopic composition.

The stable carbon and nitrogen isotope composition of fish scales may therefore provide insight into many topical questions in marine ecosystem science as they reflect both base level and consumer level fluctuations. Fish are mobile data recorders, however, and spatial variations in base $\delta^{13}C$ and $\delta^{15}N$ values will also contribute to the total variation in scale collagen measured over any time series. Given the large number of variables potentially affecting $\delta^{13}C$ and $\delta^{15}N$ values in scale collagen, attributing measured variations in the isotopic composition of scale collagen to one or more specific cause may be problematic.

Some key questions in aquatic ecosystem science which could be addressed in part through stable isotope analysis of archived scales are outlined below.

1. Does aquatic ecosystem structure experience significant annual and decadal fluctuations? If so, are they related to climate forcing?
2. If climate forcing does exert control on ecosystem structure and/or abundance is this influence at the level of primary production or at higher trophic levels?
3. Do any recognized ecosystem/environment fluctuations correlate with target fish population variables such as mortality, population size, growth, or recruitment?

IV. APPLICATION OF STABLE ISOTOPE RECORDS OF ARCHIVED FISH SCALES TO STUDY ENVIRONMENTAL CHANGE

A. Eutrophication in Lakes

The potential for scales to record long-term changes in the environmental conditions of lakes was investigated by Perga and Gerdeaux (2003) and Gerdeaux and Perga (2006) through the analysis of archived collections of scales from whitefish (*Coregonus lavaretus*) from subalpine lakes. Scale archives covering >50 years were analysed, including periods where lakes suffered both eutrophic and oligotrophic conditions. Whitefish are zooplanktiverous consumers in the lake ecosystems studied, and variations in isotopic composition of scale collagen were interpreted as reflecting integrated records of the isotopic composition of zooplankton.

Scales were recovered from fish from three lakes; Geneva, Constance, and Annecy. In the period from the 1960s to the 1980s Lakes Geneva and Constance underwent eutrophication mainly due to anthropogenic phosphorus loading. Remediation measures taken in the 1980s lead to a reduction in P loading and both lakes returned to oligotrophic conditions. In contrast, P concentrations in Lake Annecy remained relatively low throughout the sampled period.

Significant interannual variations in both $\delta^{15}N$ and $\delta^{13}C$ values of scales were found, but the two isotope systems did not covary, suggesting that variations in $\delta^{13}C$ reflect changes in the isotopic composition of zooplankton rather than changes in diet. This interpretation is supported by analyses of stomach contents of whitefish which show no significant change in diet over the sampled period. $\delta^{13}C$ values in scales varied more extensively in Lakes Geneva and Constance, and correlated strongly with annual phosphorus concentrations, with more positive $\delta^{13}C$ values associated with high P concentrations and eutrophic conditions (Figure 10.5). This trend is consistent with increasing export of organic carbon and/or atmospheric invasion during eutrophic periods (Figure 10.4). In either case, the observed relationship between scale $\delta^{13}C$ values and annual P concentrations indicates that whitefish scales represent effective integrators of plankton $\delta^{13}C$ values and can be used to monitor long-term ecosystem change.

B. Changes in Terrestrial Versus Marine Carbon Sources in Estuaries

An 18-year record of $\delta^{13}C$ values in collagen of archived striped bass scales from Chesapaeke Bay was presented by Pruell *et al.* (2003). Over the sampling period, $\delta^{13}C$ values in scale collagen increased systematically, which could reflect either changes in the source of carbon for primary productivity (particularly increases in isotopically negative terrestrially derived carbon) or differences in trophic structure. The $\delta^{13}C$ composition of prey species of striped bass in Chesapeake Bay varied from low values for Atlantic menhaden; average $\delta^{13}C$ = c. −19.5‰, to high values for crustaceans; average $\delta^{13}C$ = c. −16‰. Over the sampled time period, the population of Atlantic menhaden declined significantly, and invertebrates were increasingly noted in stomach contents of striped bass, strongly suggesting that the observed temporal changes in scale collagen $\delta^{13}C$ values reflected long-term changes in ecosystem structure.

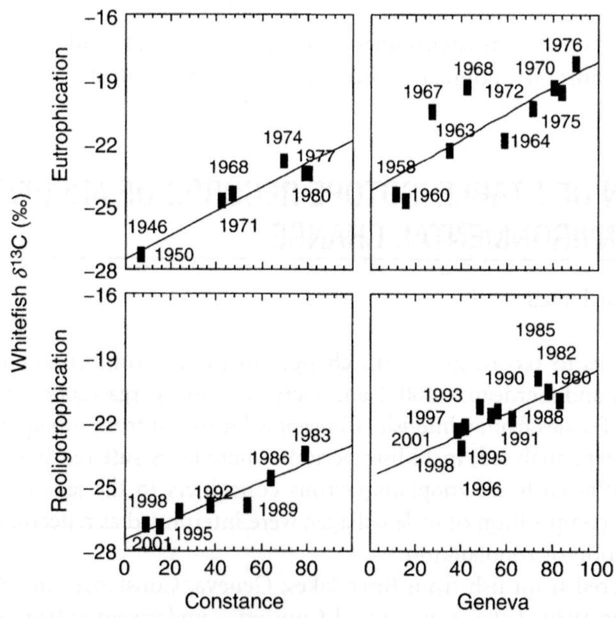

FIGURE 10.5 Relationship between mean annual $\delta^{13}C$ values recorded in scales of whitefish and total phosphorus concentration (P_{TOT}) during eutrophication and reoligotrophication of lakes Constance and Geneva. Changes in P_{TOT} stimulated changes in phytoplankton, resulting in variations in stable-carbon compositions of phytoplankton that were in turn transferred to fish. Reprinted from Gerdeaux and Perga (2006) with permission from the American Society for Limnology and Oceanography.

C. Long-Term Changes in Marine Fisheries

Temporal variations in the stable isotope composition of archived fish scales from seven commercially important demersal fish species in the Georges Bank ecosystem were studied by Wainwright *et al.* (1993). $\delta^{15}N$ values in scale collagen from all species varied from c. 10‰ to 14‰, indicating dietary separation with considerable overlap between species. The inferred trophic levels broadly agreed with gut content analyses for five of the seven species but suggested less trophic separation than expected. Discrepancies between trophic relationships inferred from stable isotope data and gut content analyses are not surprising as isotope data reflect the composition of assimilated diet averaged over the period of tissue growth, but require knowledge of tissue-diet isotope fractionation whereas gut content analyses provide a "snapshot" of ingested diet and may be biased by the proportion of identifiable remains found.

Significant interannual variation in both $\delta^{13}C$ and $\delta^{15}N$ values was found in scales from all species sampled from the Georges Bank, and species varied in a concerted manner with respect to $\delta^{13}C$ values for at least part of the temporal record (Figure 10.6). As $\delta^{13}C$ values are suggested to be a sensitive indicator of changes in the isotopic composition of primary production, the concerted changes in $\delta^{13}C$ values in scales from different species suggest that fish derived energy from the same phytoplankton source. The observed variations in $\delta^{13}C$ values in scales may reflect changes in primary productivity (and thus may correspond to climatic variables) or may reflect changes in the proportions of diatoms in net phytoplankton.

The study of Wainwright *et al.* (1993) also included an archive of haddock (*Melanogrammus aeglefinnus*) scales covering years 1929–1987. $\delta^{15}N$ values in haddock scales declined by 2.45‰ over the study period, suggesting a progressive decline in trophic level (Figure 10.6). $\delta^{13}C$ values declined between 1929 and 1960, then increased to 1987, suggesting long-term changes in the food web at the level of primary production. A long-term decline in $\delta^{13}C$ values between 1930 and 1950 is consistent with a gradual reduction in the importance of diatoms in primary productivity on the Georges Bank. Smaller, cyclical isotopic variations were superimposed on the longer-term trends and the isotopic composition of fish scales was significantly correlated with a combination of environmental and anthropogenic variables including the NAO and indices of fishing intensity. The observed reduction in $\delta^{15}N$ values was coincident with an increase in weight at age 2 interpreted as reflecting release from intraspecific competition caused by over-fishing. It is likely, therefore, that the long-term trend in $\delta^{15}N$ values in haddock scales indicates a decline in trophic level in response to intense fishing pressure, and long-term changes in $\delta^{13}C$ values reflect major changes in plankton composition. Shorter term, decadal variations in $\delta^{15}N$ and $\delta^{13}C$ values appear to reflect the impact of climate forcing cycles on ecosystem structure.

Satterfield and Finney (2002) provide a 33-year record of $\delta^{13}C$ and $\delta^{15}N$ values in scales of sockeye salmon from a single Alaskan stock. The $\delta^{13}C$ record exhibited minor variability around the mean with no long-term trend, suggesting limited changes in either carbon export rates or phytoplankton assemblage. By contrast, $\delta^{15}N$ values varied more extensively with longer-term interdecadal trends. A lack of clear correspondence between $\delta^{13}C$ and $\delta^{15}N$ values in this data series suggests that changes in trophic level over time do not adequately explain the observed variation, however at present insufficient data are available to explain either longer-term trends or interannual variability.

As part on an ongoing study into the long-term behavior of Atlantic salmon (*S. salar*) from a single stock from the North East English coast by the authors, the isotopic composition of collagen in the last season of growth was determined in an archived series of scales initially covering years 1985–1997. Between 20 and 40 individual fish were sampled per year, including fish returning after one winter at sea (grilse) and after multiple (usually 2) winters at sea (MSW).

While the time series is incomplete, the $\delta^{13}C$ record shows large inter-annual variations of >2‰, with no suggestion of long-term change over the limited time series (Figure 10.7). Over the same time

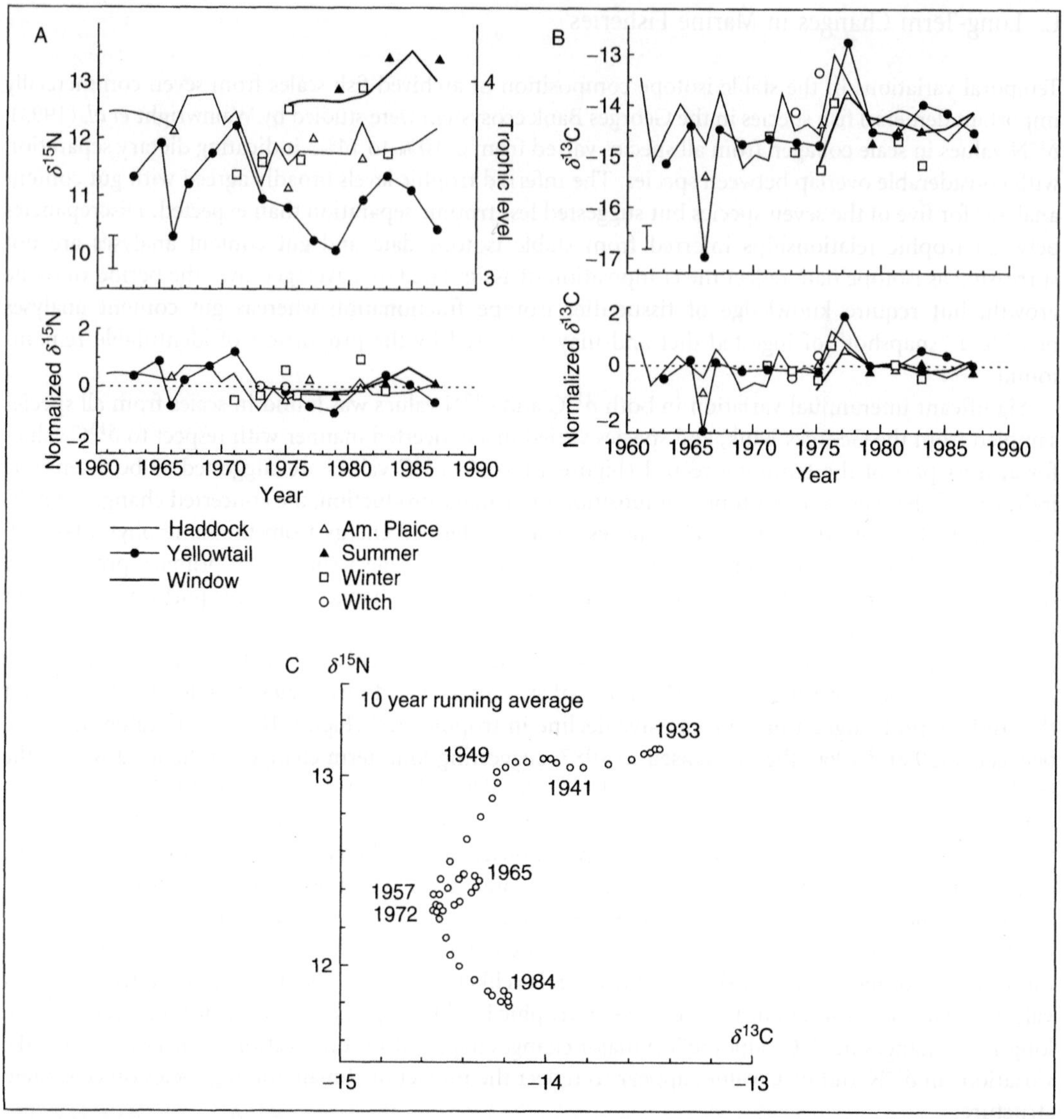

FIGURE 10.6 Long-term variations in (A) $\delta^{13}C$ and (B) $\delta^{15}N$ values in scales from seven fish species from the Georges Bank and (C) Dual isotope plot of 10-year running averages of $\delta^{13}C$ and $\delta^{15}N$ values in scales of *M. aeglefinus.* Upper plots in (A) and (B) show composite samples of four to seven individuals, lower plots are the same values normalized by subtracting the mean for that species. All fish are of equivalent age, or corrected to reflect age by applying length versus isotope regressions. Error bar represents the maximum difference between replicate composites. Reprinted from Wainwright *et al.* (1992) with kind permission from Springer Science and Business Media.

series, $\delta^{15}N$ values vary interannually by >2‰ and also show no significant long-term trend. Within the whole dataset, $\delta^{13}C$ and $\delta^{15}N$ values are not correlated, suggesting that interannual changes at both base level (carbon export rates, location and/or plankton composition) and consumer level (trophic level) influence the isotope record.

$\delta^{13}C$ values in grilse and MSW fish covary, suggesting that both cohorts derive carbon from the same primary source, and that changes in carbon export rates and/or plankton composition occur over the sampled time series. While the causes of the observed changes in $\delta^{13}C$ are uncertain, the $\delta^{13}C$

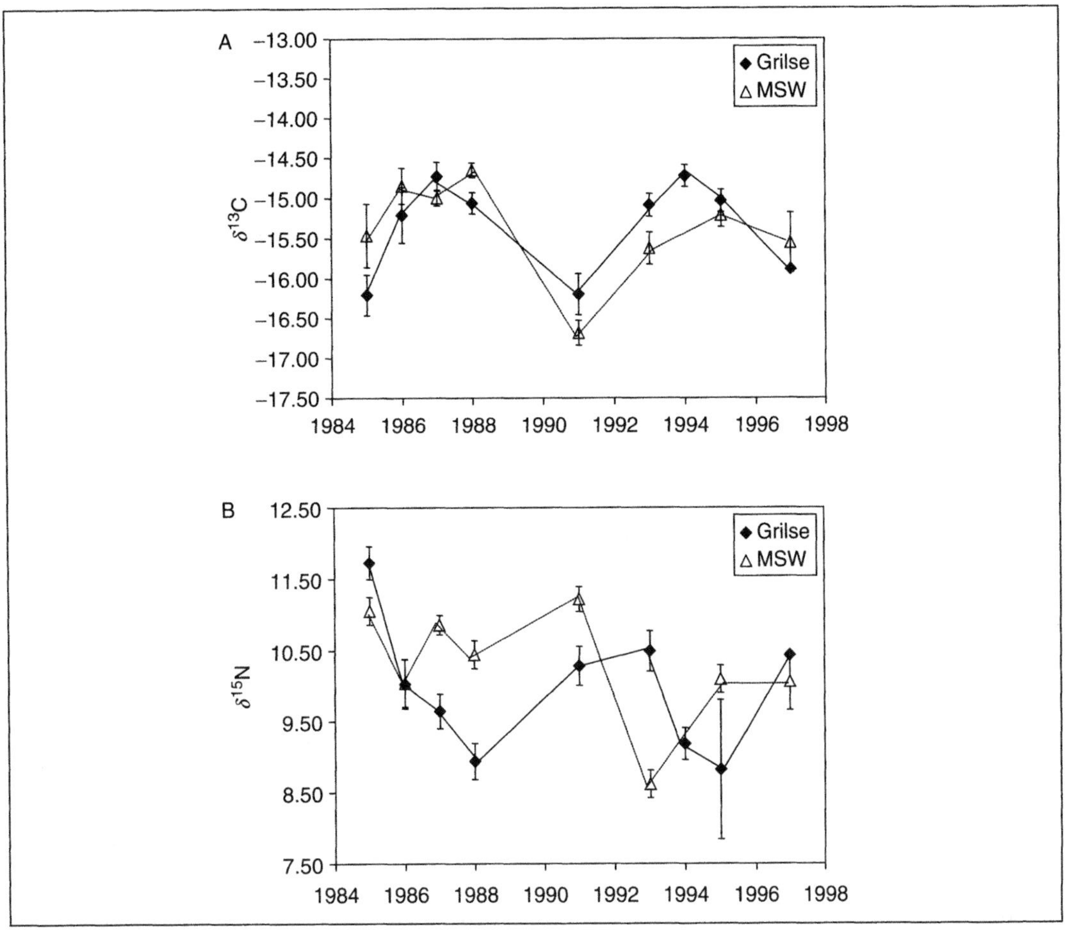

FIGURE 10.7 Variations in (A) $\delta^{13}C$ and (B) $\delta^{15}N$ values in scales from Atlantic salmon representing the last season of growth. Fish were sampled on their return to home waters, returning either as one sea-winter (1SW, closed symbols) or multi sea-winter (MSW, open symbols) fish. $\delta^{13}C$ values in 1SW and MSW fish vary in concert indicating temporal changes in the isotopic composition of carbon at the base of the food web. $\delta^{15}N$ values in 1SW and MSW fish do not covary, indicating separation in either trophic level and/or location between the two cohorts.

record for both grilse and MSW fish are significantly correlated with estimates of total salmon returns. This correlation suggests that climate influences on primary production significantly impact marine mortality for Atlantic salmon, and is in agreement with the results of a large-scale study comparing long-term trends in plankton abundance and diversity with climate variables and salmon population estimates (Beaugrand and Reid, 2003).

Within the sampled population of *S. salar*, scale $\delta^{15}N$ values are typically higher in MSW fish suggesting a higher trophic occupancy (typically around 0.5 trophic level separation assuming a tissue-diet separation of c. 2.3‰, Trueman *et al.* 2005). This is expected as salmon returning after multiple winters at sea are larger than those returning after 1 year at sea. Alternatively, the N data may suggest that the two cohorts of salmon feed in differing areas. $\delta^{15}N$ values in grilse are more variable than those in MSW fish (ANOVA $F_{1,155} = 4.24$, $p < 0.05$), suggesting that grilse are either more sensitive to changes in ecosystem structure (such as changes in the relative proportions of prey items of differing trophic level), or occupy regions with greater inter-annual variability in basal $\delta^{15}N$ values.

V. CONCLUSIONS AND FUTURE DIRECTIONS

The case studies presented above demonstrate that the stable isotope composition of fish scales record significant temporal variations which can be interpreted in the context of environmental and ecological change.

Isotopic changes caused by consumer level effects (*e.g.*, changes in diet or trophic structure) should be manifested in both $\delta^{13}C$ and $\delta^{15}N$ values with greater variation in $\delta^{15}N$ values. $\delta^{13}C$ values in scales from different species within a single ecosystem would be expected to covary if all are deriving carbon from the same phytoplankton source. Decoupling of $\delta^{13}C$ and $\delta^{15}N$ values in scales suggests changes in the isotopic composition of primary production which may in turn reflect changing environmental conditions or spatial separation.

The relatively few studies conducted to date demonstrate that large temporal changes in both $\delta^{13}C$ and $\delta^{15}N$ values are found in fish scales, but frequently additional information is required to interpret these changes. It is generally easier to interpret variations in $\delta^{13}C$ values as these are certainly linked to changes at the base of the food web. Interpreting scale isotope data in marine ecosystems is further complicated by the fact that fish are mobile data recorders and distinguishing geographic effects from temporal effects may be extremely difficult. This problem is likely to be more important in coastal/shelf settings where offshore gradients in $\delta^{15}N$ values are most pronounced, and variations in plankton may be marked over small regional areas. The development of compound specific isotope analyses will provide a wealth of new opportunities as a measure of the isotopic composition of the basal food web and the trophic level can be obtained directly from any single sample of consumer tissue. See Chapter 12 for an application of this methodology.

Some significant research hurdles must be cleared to fully exploit the potential of fish scales for long-term monitoring of marine ecosystems.

1. Assessing the relationship between scale $\delta^{13}C$ values and phytoplankton ecology.

Coupled records of the isotopic composition of phytoplankton and scale collagen would provide further information about the link between climate variables and fish isotope composition. Large-scale records of variation in phytoplankton isotopic composition related to total phytoplankton abundance and diversity would further aid interpretation of scale isotope records.

2. Development of predictive isotope distribution ("isoscape") models for marine ecosystems.

Many dynamic ocean climate models exist, some coupled with biological models predicting plankton abundance. Extension of these models to predict spatial variations in the isotopic composition of primary productivity would greatly assist interpretation of the isotopic composition of fish scales.

3. Exploration of stable isotope systems other than C and N (*e.g.*, $\delta^{34}S$ and δD).

Multiple isotope tracers are increasingly used in aquatic ecology, particularly to resolve questions of spatial origin, movement, and migration (Hesslein *et al.* 1992, Birchall *et al.* 2005, Dube *et al.* 2005, Whitledge *et al.* 2006). As yet, these systems have not been studied in fish scales and the relationships between tissue composition on the one hand and water composition, diet composition, and metabolic processes on the other are frequently uncertain. With more experimentation and process-driven research the use of multiple isotope (and elemental) tracers must help to improve our understanding of the fundamental ecosystem processes responsible for causing isotopic variations in consumer tissues.

With these developments, long-term monitoring of many aquatic ecosystems could be achieved with relatively little additional sampling effort, exploiting commercial fishery landings. A vast amount of potentially useful scale tissue is collected every day, new archived collections could be initiated and maintained with relatively little effort and cost. Existing scale archives represent an extremely valuable resource and their curation and archiving should be supported and maintained.

VI. REFERENCES

Altabet, M. A., and R. Francois. 1994. Sedimentary nitrogen isotopic ratio as a recorder for surface ocean nitrogen utilization. *Global Biogeochemical Cycles* **8:**103–116.

Bade, D. L., S. R. Carpenter, J. J. Cole, P. C. Hanson, and R. H. Hesslein. 2004. Controls of delta C-13-DIC in lakes: Geochemistry, lake metabolism and morphology. *Limnology and Oceanography* **49:**1161–1172.

Beaugrand, G., and P. C. Reid. 2003. Long-term changes in phytoplankton, zooplankton and salmon related to climate. *Global Change Biology* **9:**801–817.

Beaugrand, G., P. C. Reid, F. Ibanez, J. A. Lindley, and M. Edwards. 2002. Reorganisation of North Atlantic marine copepod biodiversity and climate. *Science* **296:**1692–1694.

Beaugrand, G., K. M. Brander, J. A. Lindley, S. Souissi, and P. C. Reid. 2003. Plankton effect on cod recruitment in the North Sea. *Nature* **426:**661–664.

Berry, J. A. 1989. Studies of mechanisms affecting the fractionation of carbon isotopes in photosynthesis. Pages 82–94 *in* P. W. Rundel, J. R. Ehleringer, and K. A. Nagy (Eds.) *Stable Isotopes in Ecological Research.* Springer-Verlag, New York.

Birchall, J., T. C. O'Connell, T. H. E. Heaton, and R. E. M. Hedges. 2005. Hydrogen isotope ratios in animal body protein reflect trophic level. *Journal of Animal Ecology* **74:**877–881.

Brander, K. M., and R. Mohn. 2004. Effect of the North Atlantic Oscillation on recruitment of Atlantic cod (*Gadus morhua*). *Canadian Journal of Fisheries and Aquatic Sciences* **61:**1558–1564.

Brander, K. M., R. R. Dickson, and M. Edwards. 2003. Use of Continuous Plankton Recorder information in support of marine management: Applications in fisheries, environmental protection, and in the study of ecosystem response to environmental change. *Progress in Oceanography* **58:**175–191.

Cabana, G., and J. B. Rasmussen. 1996. Comparison of aquatic food chains using nitrogen isotopes. *Proceedings of the National Academy of Sciences of the United States of America* **93:**10844–10847.

Chavez, F. P., J. Ryan, S. E. Lluch-Cota, and M. Niguen. 2003. From anchovies to sardines and back: Multidecadal change in the Pacific Ocean. *Science* **299:**217–221.

DeNiro, M. J., and S. Epstein. 1978. Influence of diet on the distribution of carbon isotopes in animals. *Geochimica et Cosmochimica Acta* **42:**495–506.

Duarte, C. M., J. Cebrian, and N. Marba. 1992. Uncertainty of detecting sea change. *Nature* **356:**190.

Dube, M. G., G. A. Benoy, S. Blenkinsopp, J. M. Ferone, R. B. Brua, and L. I. Wassenaar. 2005. Application of multi-stable isotope (C-13, N-15, S-34, Cl-37) assays to assess spatial separation of fish (longnose sucker Catostomus catostomus) in an area receiving complex effluents. *Water Quality Research Journal of Canada* **40:**275–287.

Edwards, M., and A. J. Richardson. 2004. Impact of climate change on marine pelagic phenology and trophic mismatch. *Nature* **430:**881–884.

Estep, M. L. F., and S. Vigg. 1985. Stable carbon and nitrogen isotope tracers of trophic dynamics in natural populations and fisheries of the Lahontan Lake system, Nevada. *Canadian Journal of Fisheries and Aquatic Sciences* **42:**1712–1719.

Estrada, J. A., M. Lutcavage, and S. R. Thorrold. 2005. Diet and trophic position of Atlantic bluefin tuna (*Thunnus thynnus*) inferred from stable carbon and nitrogen isotope analysis. *Marine Biology* **147:**37–45.

Focken, U. 2004. Feeding fish with diets of different ratios of C-3- and C-4-plant-derived ingredients: A laboratory analysis with implications for the back-calculation of diet from stable isotope data. *Rapid Communications in Mass Spectrometry* **18:**2087–2092.

Fouda, M. M. 1979. Studies on scale structure in the common goby *Pomatoschistus microps* Krøyer. *Journal of Fish Biology* **15:**173–183.

Fromentin, J.-M., and B. Planque. 1996. *Calanus* and environment in the eastern North Atlantic. II. Influence of the North Atlantic Oscillation on *C. finmarchicus* and *C. heligolandicus. Marine Ecology Progress Series* **134:**111–118.

Fry, B., and S. C. Wainwright. 1991. Diatom sources of δ^{13}C-rich carbon in marine food webs. *Marine Ecology Progress Series* **76:**149–157.

Gerdeaux, D., and M.-E. Perga. 2006. Changes in whitefish scales δ^{13}C during eutrophication and reoligotrophication of subalpine lakes. *Limnology and Oceanography* **51:**772–780.

Goericke, R., J. P. Montoya, and B. Fry. 1994. Physiology of isotopic fractionation in algae and cyanobacteria. Pages 187–222 *in* K. Lajtha and R. H. Mitchener (Eds.) *Stable Isotopes in Ecology and Environmental Science.* Blackwell Scientific, Oxford.

Hayes, J. M. 1993. Factors controlling ^{13}C contents of sedimentary organic compounds: Principles and evidence. *Marine Geology* **113:**111–125.

Hesslein, R. H., M. D. Capel, D. E. Fox, and K. A. Hallard. 1992. Stable isotopes of sulphur, carbon and nitrogen as indicators of trophic level and fish migration in the lower MacKenzie River Basin, Canada. *Canadian Journal of Fisheries and Aquatic Sciences* **48:**2258–2265.

Hesslein, R. H., K. A. Hallard, and P. Ramlal. 1993. Replacement of sulphur, carbon, and nitrogen in tissue of growing broad whitefish (*Coregonus-nasus*) in response to a change in diet traced by delta-S-34, delta-C-13 and delta-N-15. *Canadian Journal of Fisheries and Aquatic Sciences* **50:**2071–2076.

Hinga, K. R., M. A. Arthur, M. E. Q. Pilson, and D. Whitaker. 1994. Carbon isotope fractionation by marine phytoplankton in culture: The effects of CO_2 concentration, pH, temperature and species. *Global Biogeochemical Cycles* **8:**91–102.

Hutchinson, J., and C. N. Trueman. 2006. Stable isotope analyses of collagen in fish scales: Limitations set by scale architecture. *Journal of Fish Biology* **69:**1874–1880.

Jardine, T. D., D. L. MacLatchy, W. L. Fairchild, R. A. Cunjak, and S. B. Brown. 2004. Rapid carbon turnover during growth of Atlantic salmon (*Salmo salar*) smolts in sea water, and evidence for reduced food consumption by growth-stunts. *Hydrobiologia* **527:**63–75.

Kalish, J. M. 1991. C-13 and O-18 isotopic disequilibria in fish otoliths: Metabolic and kinetic effects. *Marine Ecology Progress Series* **75:**191–203.

Laws, E. A., B. N. Popp, R. R. Bidigare, M. C. Kennicutt, and S. A. Macko. 1995. Dependence of phytoplankton carbon isotopic composition on growth rate and $[CO_2]$aq: Theoretical considerations and experimental results. *Geochimica et Cosmochimica Acta* **59:**1131–1138.

Lee-Thorp, J., and N. J. van der Merwe. 1989. Stable carbon isotope ratio differences between bone collagen and bone apatite, and their relationship to diet. *Journal of Archaeological Science* **16:**585–599.

McCutchan, J. H., Jr., W. M. Lewis, Jr., C. Kendall, and C. C. McGrath. 2003. Variation in trophic shift for stable isotope ratios of carbon, nitrogen, and sulphur. *Oikos* **102:**378–390.

Meyers, P. A., and S. M. Bernasconi. 2005. Carbon and nitrogen isotope excursions in mid-Pleistocene sapropels form the Tyrrhenian Basin: Evidence for climate-induced increase in microbial primary production. *Marine Geology* **220:**41–58.

Michener, R. H., and D. M. Schell. 1994. Stable isotope ratios as tracers in marine aquatic food webs. Pages 138–157 *in* K. Lajtha and R. H. Mitchener (Eds.) *Stable Isotopes in Ecology and Environmental Science.* Blackwell Scientific, Oxford.

Minagawa, M., and E. Wada. 1984. Stepwise enrichment of N-15 along food chains – further evidence and the relation between delta-N-15 and animal age. *Geochimica et Cosmochimica Acta* **48:**1135–1140.

Miyake, Y., and E. Wada. 1967. The abundance ratio of $^{15}N/^{14}N$ in marine environments. *Records of Oceanographic Works in Japan* **9:**1135–1140.

Mullin, M. M., G. H. Rau, and R. W. Eppley. 1984. Stable nitrogen isotopes in zooplankton: Some geographic and temporal variations in the north Pacific. *Limnology and Oceanography* **29:**1267–1273.

Neave, F. 1936. The development of the scales of salmon. *Transactions of the Royal Society of Canada Section 5* **3:**55–74.

Pancost, R. D., K. H. Feeman, and S. G. Wakeham. 1999. Controls on the carbon-isotope compositions of compounds in Peru surface waters. *Organic Geochemistry* **30:**319–340.

Perga, M.-E., and D. Gerdeaux. 2003. Using the $\delta^{13}C$ and $\delta^{15}N$ of whitefish scales for retrospective ecological studies: Changes in isotope signatures during restoration of Lake Geneva, 1980–2001. *Journal of Fish Biology* **63:**1197–1207.

Perga, M.-E., and D. Gerdeaux. 2004. Changes in the $\delta^{13}C$ of pelagic food webs: The influence of lake area and trophic status on the isotopic signature of whitefish (*Coregonus lavaretus*). *Canadian Journal of Fisheries and Aquatic Sciences* **61:**1485–1492.

Petersen, B. J., and B. Fry. 1987. Stable isotopes in ecosystem studies. *Annual Review of Ecology and Systematics* **18:**293–320.

Pinnegar, J. K., and N. V. C. Polunin. 1999. Differential fractionation of d13C and d15N among fish tissues: Implications for the study of trophic interactions. *Functional Ecology* **13:**225–231.

Post, D. M. 2002. Using stable isotopes to estimate trophic position: Models, methods, and assumptions. *Ecology* **83:**703–718.

Pruell, R. J., B. K. Taplin, and K. Cicchelli. 2003. Stable isotope ratios in archived striped bass scales suggest changes in trophic structure. *Fisheries Management and Ecology* **10:**329–336.

Rau, G. H., J.-L. Teyssie, F. Rassoulzadegan, and S. W. Fowler. 1990. $^{13}C/^{12}C$ and $^{15}N/^{14}N$ variations among size-fractionated marine particles: Implications for their origin and trophic relationships. *Marine Ecology Progress Series* **59:**33–38.

Rau, G. H., T. Takahashi, D. J. Desmarais, D. J. Repeta, and J. H. Martin. 1992. The relationship between delta-C-13 of organic matter and $[CO_2(AQ)]$ in ocean surface-water: data from a JGOFS site in the northeast Atlantic Ocean and a model. *Geochimica et Cosmochimica Acta* **56:**1413–1419.

Rau, G. H., C. Low, J. T. Pennington, K. R. Buck, and F. P. Chavez. 1998. Suspended particulate nitrogen $\delta^{15}N$ versus nitrate utilization: Observations in Monterey Bay, CA. *Deep-Sea Research II* **45:**1603–1616.

Raven, J. A., A. M. Johnston, J. R. Newman, and C. M. Scrimgeour. 1994. Inorganic carbon acquisition by aquatic photolithotrophs of the Dighty Burn, Angua, UK: Uses and limitations of natural abundance measurements of carbon isotopes. *New Phytologist* **127:**271–286.

Satterfield, F. R., and B. P. Finney. 2002. Stable isotope analysis of Pacific salmon: Insight into trophic status and oceanographic conditions over the last 30 years. *Progress in Oceanography* **53:**231–246.

Schelske, C. L., and D. A. Hodell. 1991. Recent changes in productivity and climate of Lake Ontario detected by isotopic analysis of sediments. *Limnology and Oceanography* **36:**961–975.

Schelske, C. L., and D. A. Hodell. 1995. Using carbon isotopes of bulk sedimentary organic matter to reconstruct the history of nutrient loading and eutrophication in Lake Erie. *Limnology and Oceanography* **40:**918–929.

Schindler, D. E., S. R. Carpenter, J. J. Cole, J. F. Kitchell, and M. L. Pace. 1997. Influence of food web structure on carbon exchange between lakes and the atmosphere. *Science* **277:**248–251.

Trueman, C. N., R. A. R. McGill, and P. H. Guyard. 2005. The effect of growth rate on tissue-diet isotopic spacing in rapidly growing animals. An experimental study with Atlantic salmon (*Salmo salar*). *Rapid Communications in Mass Spectrometry* **19:**3239–3247.

Vanderklift, M. A., and S. Ponsard. 2003. Sources of variation in consumer-diet delta N-15 enrichment: A meta-analysis. *Oecologia* **136:**169–182.

Vuorio, K., M. Meili, and J. Sarvala. 2006. Taxon-specific variation in the stable isotopic signatures ($\delta^{13}C$ and $\delta^{15}N$) of lake phytoplankton. *Freshwater Biology* **51:**807–822.

Wainwright, S. C., M. J. Fogarty, R. C. Greenfield, and B. Fry. 1993. Long-term changes in the Georges Bank food web: Trends in isotopic compositions of fish scales. *Marine Biology* **115:**481–493.

Weiner, S., T. Arad, I. Sabanay, and W. Traub. 1997. Rotated plywood structure of primary lamellar bone in the rat: Orientations of the collagen fibril arrays. *Bone* **20:**509–514.

Whitledge, G. W., B. M. Johnson, and P. J. Martinez. 2006. Stable hydrogen isotope composition of fishes reflects that of their environment. *Canadian Journal of Fisheries and Aquatic Sciences* **63:**1746–1751.

Wu, J., S. C. Calvert, and C. S. Wong. 1997. Nitrogen isotope variations in the subarctic northeast Pacific: Relationships to nitrate utilization and trophic structure. *Deep-Sea Research I* **44:**287–314.

Zylberberg, L. 2004. New data on bone matrix and its proteins. *Comptes Rendus Palevol* **3:**591–604.

Zylberberg, L., and G. Nicols. 1982. Ultrastructure of scales in a teleost (*Carassius auratus* L.) after use of rapid freeze-fixation and freeze-substitution. *Cell and Tissue Research* **223:**349–367.

Zylberberg, L., J. Bereiter-Hahn, and J.-Y. Sire. 1988. Cytoskeletal organisation and collagen orientation in the fish scales. *Cell and Tissue Research* **253:**597–607.

CHAPTER 11

The Reaction Progress Variable and Isotope Turnover in Biological Systems

Thure E. Cerling,*,† Gabriel J. Bowen,*,‡ James R. Ehleringer,* and Matt Sponheimer*,§

*Department of Biology, University of Utah
†Department of Geology and Geophysics, University of Utah
‡Department of Earth and Planetary Sciences, Purdue University
§Department of Anthropology, University of Colorado

Contents

I. INTRODUCTION

Biological systems under natural field conditions are rarely at steady state, but are frequently responding to changes in the abundance of external resources, environmental conditions, and phenology and development. This poses a challenge for interpreting the stable isotope ratios sequentially recorded in biological materials because those "recorder" tissues may be responding to external resource pools as well as to pools within the organism or ecological reservoir itself. The reaction progress model describes stable isotope turnover in ecological systems; it allows the determination of half-lives, multiple turnover pools, and their relative contribution to the whole, and can identify transit times not controlled by first-order reaction kinetics. In addition, it is useful in determining sampling intervals specific to isotope turnover studies. Thus, the reaction progress model provides a powerful tool for quantitatively reconstructing discrete resource change events (such as a change in dietary food source of an animal) from the effects of other pools that contribute to the isotope ratio of the "recorder."

Stable isotope analyses have been used as tracers of ecological processes for many different sets of conditions (see chapters in this volume). As a "tracer" or "recorder" of an ecological change, it is the sequential changes in the isotope composition of a material or tissue that serves as a recorder. Examples of a linear ecological recorder include the isotope ratios along the segment of an animal hair, the sequential carbonate deposition in soil caliche, corals, or in mollusk shells, and the xylem cells that form tree rings. In some cases, the stable isotope ratio of that recorder reflects the immediate state of the environment, as would be the case for carbonates in corals laid down in equilibrium with the surrounding ocean water. In other cases, the isotope ratio of a segment of biological tissue reflects a mixing of different pools such as the amino acids in animal hair that can be derived from the most recent food source as well as from internal pools associated with metabolism and decomposition of existing protein pools within the body.

There are various ways in which previous studies have attempted to resolve patterns that involve multiple pools that can contribute to a single ecological recorder. Turnover in biological systems has been described using exponential curves to fit data. This includes animal tissue studies (Fry and Arnold 1982, Hobson and Clark 1992), soil studies (Balesdent and Mariotti 1996), and other studies. A typical turnover experiment involves an instantaneous change and measuring the change in stable isotope ratio over time.

The exponential fit method has the implicit assumption that the system is behaving according to first-order kinetics. In this chapter, we discuss the reaction progress model to describe isotope turnover in biological systems. This concept has been used for isotope exchange reactions in geochemical systems (Criss *et al.* 1987, Criss 1999) and has been applied to isotope turnover in animal tissues (Ayliffe *et al.* 2004, West *et al.* 2004, Cerling *et al.* 2007) and in laboratory scale exchange reactions (Bowen *et al.* 2005). Here we use the reaction progress variable to examine other cases where first-order kinetics is assumed to play the dominant role in isotope turnover.

In this chapter, we describe first-order kinetics and show that it is a more useful way than a simple exponential fit to understanding isotope turnover in biological systems. We then use a number of examples, primarily from existing literature, to demonstrate how rate constants are better understood using this approach. We then discuss implications for changes at the ecosystem scale.

II. METHODS

Stable isotope ratios in natural systems are described by:

$$\delta^k M_A = \left(\frac{R_A}{R_{\text{standard}}} - 1\right) \times 1000 \tag{11.1}$$

where k is the rare isotope species of interest, M is the element of interest, A is the phase being of interest (*e.g.*, water, diet, and tissue), R_A is the ratio of the heavy to light isotope (*e.g.*, $^{13}C/^{12}C$) in the sample, and R_{standard} the isotope standard.

A first-order rate constant is:

$$\frac{dN}{dt} = -\lambda N \tag{11.2}$$

where N is the number of atoms or molecules in the system being described, t is time (sec), and λ is the rate constant (sec^{-1}). Thus, the reaction rate is directly proportional to the amount of material (N) in the system. This is integrated as:

$$N = N_0 e^{-\lambda t} \tag{11.3}$$

This equation is linearized by taking the natural logarithm:

$$\ln\left(\frac{N}{N_0}\right) = -\lambda t \tag{11.4}$$

where N_0 is the initial number of atoms or molecules. The half-life for the system is:

$$t_{1/2} = \frac{\ln(2)}{\lambda} \tag{11.5}$$

and the mean life is:

$$t_{1/2} = \frac{1}{\lambda} \tag{11.6}$$

Many stable isotope turnover experiments are modeled according to an exponential fit model. This assumes first-order rate kinetics, as we show below, and we accept that assumption. The exponential fit is of the form:

$$\delta^t = ae^{-\lambda t} + c \tag{11.7}$$

where δ^t is the stable isotope of interest undergoing isotope exchange with time *t*, *a* and *c* are parameters derived from the "best fit," and λ is a first-order rate constant. Three important parameters result from this equation: λ is the data-derived rate constant from which the half-life is derived as in Eq. (11.5), "*c*" is the data-derived isotope equilibrium factor δ^{eq}, "*a*" is the isotope difference between the initial and final equilibrium states (Tieszen *et al.* 1983), so that "$c + a$" is the data-derived initial isotope value δ^{init}. In many cases, the initial and final isotope equilibrium conditions are known independently and should be compared to those derived from the exponential fit (Cerling *et al.* 2007).

Isotope turnover or exchange reactions begin under some initial isotope ratio conditions and proceed to the final state. The progress along this path can be normalized to the nominal fraction 1.0 at $t = 0$ by (Criss 1999):

$$\frac{\delta_A^t - \delta_A^{\text{eq}}}{\delta_A^{\text{init}} - \delta_A^{\text{eq}}} = 1 - F \tag{11.8}$$

where δ_A^{init}, δ_A^t, and δ_A^{eq} are the isotopic δ-values of "*A*" under the initial conditions, at time "*t*" after exchange begins, and for the final equilibrium conditions, respectively and where $F = 0$ at the beginning of the exchange reaction and $F = 1$ in the final state (*i.e.*, $t \gg 1/\lambda$). This term $(1 - F)$ describes the reaction progress. Criss (1999) pointed out that for trace isotope changes in an open system that:

$$\frac{R_A^t - R_A^{\text{eq}}}{R_A^{\text{init}} - R_A^{\text{eq}}} = e^{-\lambda t} \tag{11.9}$$

This equation is to be converted to the common "δ" notation as:

$$\frac{\delta_A^t - \delta_A^{\text{eq}}}{\delta_A^{\text{init}} - \delta_A^{\text{eq}}} = e^{-\lambda t} \tag{11.10}$$

Equations (4), (8), and (10) are combined to give:

$$\ln(1 - F) = -\lambda t \tag{11.11}$$

This expression describes various biological and geological "turnover", exchange, and reaction progress experiments that are commonly treated as exponential functions. Cerling *et al.* (2007) show that tissue turnover experiments, when treated in this fashion, can yield far more information about the system than an exponential description; for example, multiple rate constants can be derived, delayed release of red blood cells can be determined along with a more accurate determination of the rate constants for turnover, and this leads to a forward model to calculate dietary input from sequential measurements of animal tissues such as hair.

III. RESULTS

In this section, we provide examples of biological systems where rate constants can readily be determined using the reaction progress variable. The first example is of a "dry-down" experiment of hair. The second example concerns turnover of soil carbon during an agricultural change from C_3 to C_4 plants. The third example involves elimination of feces by an animal that changed from a C_3-based to a C_4-based diet. The fourth example discussed is in the case for multiple turnover pools in ecological systems. Cerling *et al.* (2007) have already discussed examples of tissue turnover during diet-switch experiments and the determination of multiple rate constants. These examples illustrate the variety of applications of considering biological turnover experiments in the reaction progress model context.

1. Bowen *et al.* (2005) carried out an isotope exchange experiment where hair samples were equilibrated with water vapor-enriched D and ^{18}O (δD and $\delta^{18}O$ of the liquid water was +192‰ and +4.9‰, respectively) and then freeze dried for up to 10 days. This represents a "classical" reaction progress experiment as described by Criss (1999) for geochemical systems. Figure 11.1 shows that the rate constant for isotope exchange for both D and ^{18}O is similar, with both having half-lives of 1.4 days for isotope exchange.
2. The amount of soil carbon is approximately twice the amount of carbon in the terrestrial biosphere and therefore has a significant role in the global carbon budget. In this section, we use the data of Balesdent and Mariotti (1996) to illustrate turnover of carbon in a modified ecosystem where a C_4 crop was maintained on a field for 13 years; previously this field had a C_3 vegetation. Figure 11.2 shows the original data of Balesdent and Mariotti, along with data plotted as the reaction progress variable. Although the data is sparse, the reaction progress variable gives half-lives of 62, 9.9, and 5.6 years for the <10-μm, 20- to 200-μm, and 200- to 2000-μm size fractions. In addition, the nonzero intercept of the 200- to 2000-μm size fraction plot indicates that a very short-lived turnover pool may be present (<1 year).
3. The third example illustrates the delayed release of a substance and the simultaneous determination of a first-order rate constant. Sponheimer *et al.* (2003) conducted a controlled diet experiment where alpacas (*Llama pacos*) and horses (*Equus caballus*) were switched from a C_3 to a C_4 diet. Feces were collected and analyzed during this experiment. Figure 11.3 shows the original data along with the data plotted as the reaction progress variable for two different alpaca. After about 3 days, the $\ln(1 - F)$ versus time plot is linear for two different individuals, consistent with a first-order rate constant for fecal carbon turnover. However, this analysis also gives a "transit" time of about 35 h from ingestion to expulsion. The turnover rate of fecal carbon for these two individuals had an average value of 46.3 h (Figure 11.3B). Application of this model to the equids of this study shows that the two animals had slightly different physiological behaviors, with "transit" times of 17 and 20 h with half-lives of 4 and 9 h, respectively. The traditional method for describing such an experiment would include the transit time where such a time represents the first appearance of a marker; in our case, this is an equivalent definition. Mean retention time is the time to which half the material has been passed; in this case the mean retention time does not

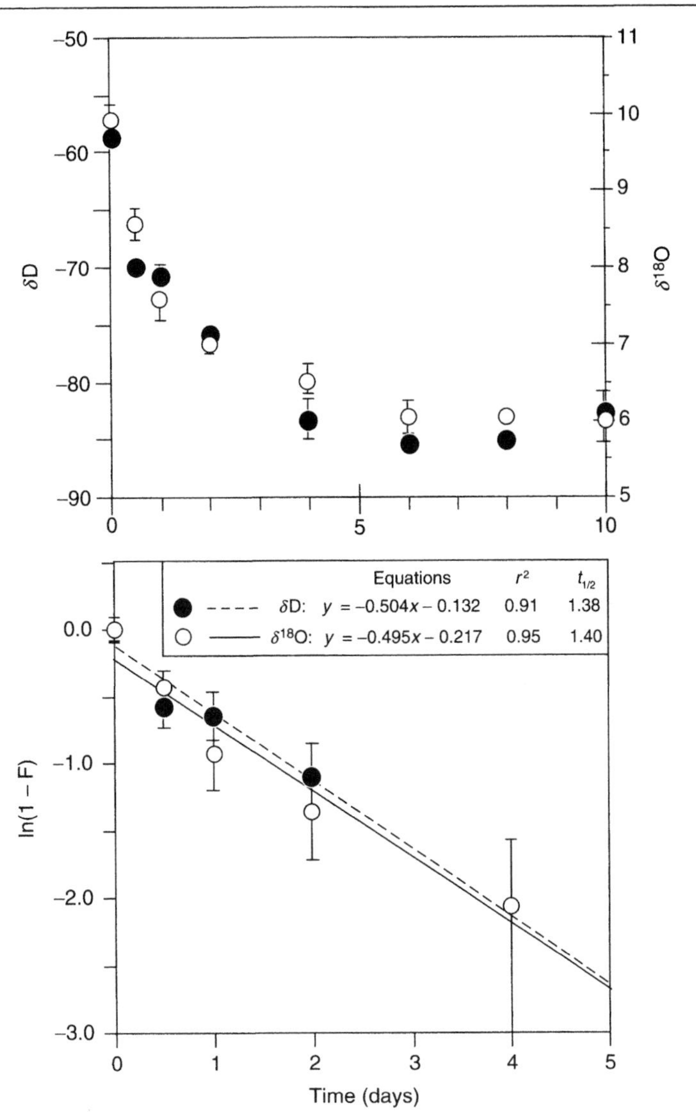

FIGURE 11.1 Dry-down experiment showing δD and $\delta^{18}O$ values for hair that was first equilibrated with water enriched in heavy isotopes and then dried in a freeze drier. (A) Measured δD and $\delta^{18}O$ values of hair dried in a freeze drier for varying lengths of time. (B) The same data plotted as the reaction progress variable, with rate constants for water loss derived from the δD and $\delta^{18}O$ data, respectively.

correspond to the half-life which refers to the physiological process only where half the material is replaced by other material by physiological processes and does not include transit. Thus, the traditional "mean retention time" would be the sum of the transit time and the half-life for turnover of fecal carbon.

4. There are several examples in animal physiological studies that have identified multiple turnover pools (Ayliffe *et al.* 2004). Such behavior is expected in the soil carbon system, but most experiments have not been of sufficient sampling detail to determine multiple carbon turnover pools. Experimental design should allow the multiple turnover pool concept to be examined.

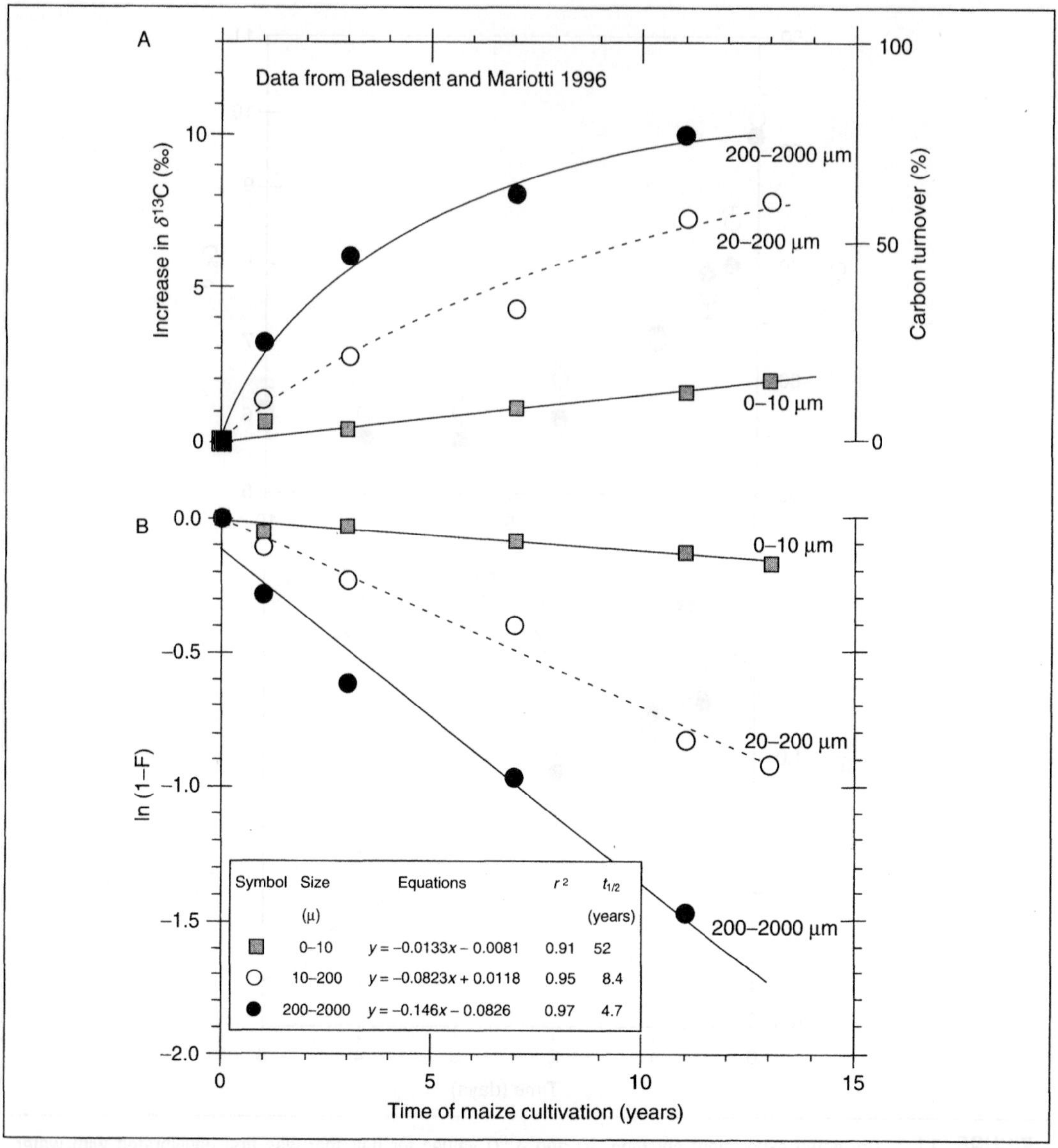

FIGURE 11.2 Stable isotope values and the reaction progress variable for soils converted from C_3 to C_4 biomass. Panel (A) is the original data of Balesdent and Mariotti (1996). Panel (B) shows the same data plotted as the reaction progress variable, with rate constants derived from first-order reaction kinetics.

IV. DISCUSSION

The purpose of this chapter is to illustrate the possibilities of using the reaction progress variable to stable isotope studies. We provide four examples of the reaction progress model to ecological applications. The reaction progress model has advantages over the traditional exponential fit of data, principally because of the linearization of the variable:

$$\ln(1 - F) = -\lambda t$$

This makes the recognition of multiple pools (Cerling *et al.* 2007) easier than the conventional method. Data from related experiments, such as approaching an equilibrium state from two directions or using different stable isotopes (Figure 11.1), can be used together to calculate rate constants. This is particularly important when data sampling intervals are limited. Very short turnover pools can often be inferred using the reaction progress model: the nonzero intercept of the 200- to 2000-mm size fraction in the example 2 (Figure 11.2B) suggests that a very short turnover pool may be present, but that the sampling interval is too sparse to capture it. Delay, or transit, times can also be determined from this visualization, and can be separated from turnover times (Figure 11.3B).

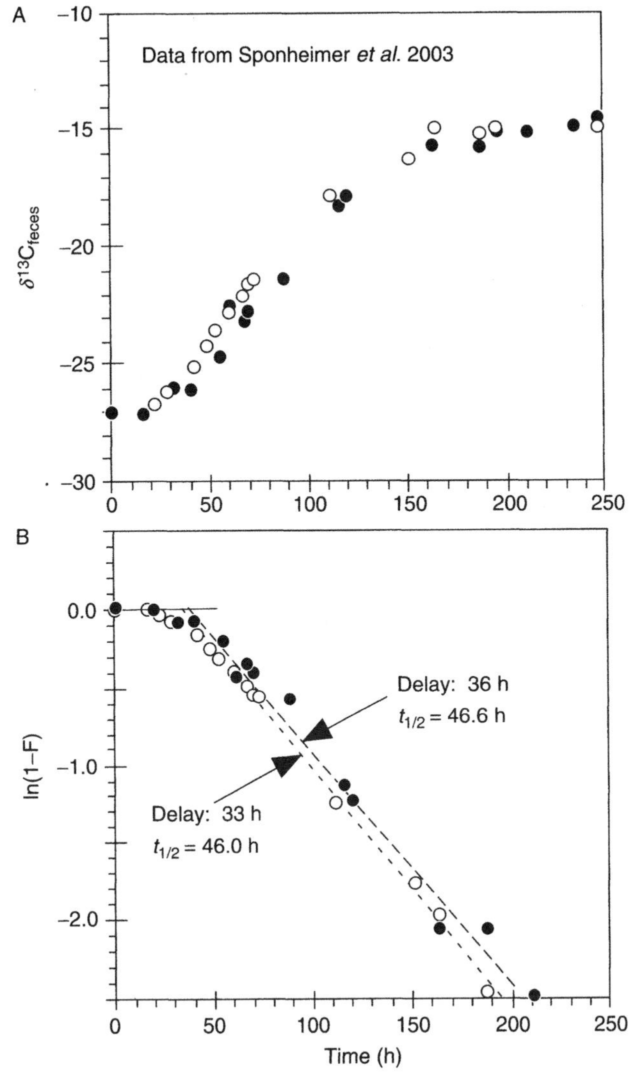

FIGURE 11.3 Stable isotope values of feces of the alpaca and the reaction progress variable for a diet-switch experiment. The two different animals of this experiment are shown as solid or closed symbols, respectively. Panel (A) is the original data of Sponheimer *et al.* (2003). Panel (B) shows the same data plotted as the reaction progress variable, with the turnover rates shown for each individual.

The reaction progress model has important implications for ecological change. It shows how multiple rate constants can be derived from isotope turnover experiments and how multiple experiments can be combined to derive rate constants of ecological turnover processes. For example, this is likely to be important in examining soil carbon turnover in ecosystems under different climatic conditions. With multiple rate constants identified, other experiments can be designed to identify processes responsible for different turnover pools. The example cited in the chapter, the study by Balesdent and Mariotti (1996), is only one example of many that is needed to understand carbon turnover in soils as it is related to climate variables such as precipitation and temperature.

The delay, or transit, time in biological systems can be readily separated from turnover. Previously, these have been treated together and the computed turnover times have been the sum of the transit and true turnover reactions: this has the result of giving an erroneously long turnover time and also giving no insight into the processes related to the transit process. For example, it has been reported that short-term changes in the carbon isotope ratios of ecosystem respiration and soil respiration are closely correlated with atmospheric vapor pressure deficits (Ekblad and Högberg 2001, Bowling *et al.* 2002, McDowell *et al.* 2004). Mechanistically, these ecosystem-scale changes are thought to be mediated by reductions in leaf stomatal conductance in response to increased water vapor-deficit gradients. A decreased stomatal conductance affects photosynthesis and is translated into changes in the isotope ratios of leaf carbohydrates that are formed and then translocated to the roots where the carbohydrates are subsequently respired. Application of the progress reaction model can help resolve time lag and mixed carbohydrate pools that contribute to the observed changes in the carbon isotope ratio of ecosystem-scale respiration.

Finally, this method is useful in the planning of sample intervals: the simple geometric progression 1, 2, 4, 8, 16, 32, 64, 128, and 256 is too sparse to adequately describe isotope turnover if more than one isotope pool is present. We recommend a sample frequency that increases as $(2)^{1/2}$, this captures more of the detail needed to determine multiple half-lives. The reaction progress model is particularly useful in planning experiments; models can be run to see what sample intervals are important. For example, a system with a significant transit time should include a sampling protocol that will capture both the transit time and the initial isotope turnover that is measured after the transit time interval. A simple geometric sampling protocol could under sample the turnover portion of the pool (*e.g.*, in Figure 11.3, single samples at 1, 2, 4, 8, 16, 32, 64, and 128 h would under sample the turnover portion of the experiment).

V. CONCLUSIONS

The reaction progress model is better suited than exponential fit models for deriving rate constants from stable isotope turnover experiments. This makes it possible to determine multiple turnover pools and identify transit times through the ecological system being studied.

VI. REFERENCES

Ayliffe, L. K., T. E. Cerling, T. Robinson, A. West, M. Sponheimer, B. Passey, J. Hammer, B. Roeder, M. D. Dearing, and J. R. Ehleringer. 2004. Turnover of carbon isotopes in tail hair and breath CO_2 of horses fed an isotopically varied diet. *Oecologia* **139:**11–22.

Balesdent, J., and A. Mariotti. 1996. Measurement of soil organic matter turnover using ^{13}C natural abundance. Pages 83–111 *in* T. W. Boutton and S.-I. Yamasaki (Eds.) *Mass Spectrometry of Soils.* Marcel Dekker, New York.

Bowling, D. R., N. G. McDowell, B. J. Bond, B. E. Law, and J. R. Ehleringer. 2002. ^{13}C content of ecosystem respiration is linked to precipitation and vapor pressure deficit. *Oecologia* **131:**113–124.

Bowen, G. J., L. Chesson, K. Nielson, T. E. Cerling, and J. R. Ehleringer. 2005. Treatment methods for the determination of δ^2H and $\delta^{18}O$ of hair keratin by continuous-flow isotope ratio mass spectrometry. *Rapid Communications in Mass Spectrometry* **19:**2371–2378.

Cerling, T. E., L. K. Ayliffe, M. D. Dearing, J. R. Ehleringer, B. H. Passey, D. W. Podlesak, A.-M. Torregrossa, and A. G. West. 2007. Determining biological tissue turnover using stable isotopes: The reaction progress variable. *Oecologia* **151:**175–189.

Criss, R. E. 1999. *Principles of Stable Isotope Distribution,* Oxford Press, Oxford. 254 p.

Criss, R. E., R. T. Gregory, and H. P. Taylor. 1987. Kinetic theory of oxygen isotope exchange between minerals and water. *Geochimica et Cosmochimica Acta* **51:**1099–1108.

Ekblad, A., and P. Högberg. 2001. Natural abundance of ^{13}C in CO_2 respired from forest soils reveal speed of link between tree photosynthesis and root respiration. *Oecologia* **127:**305–308.

Fry, B., and C. Arnold. 1982. Rapid $^{13}C/^{12}C$ turnover during growth of brown shrimp (*Penaeus aztecus*). *Oecologia* **54:**200–204.

Hobson, K. A., and R. G. Clark. 1992. Assessing avian diets using stable isotopes. I. Turnover of ^{13}C in tissues. *Condor* **94:**181–188.

McDowell, N. G., D. R. Bowling, A. Schauer, J. Irvine, B. J. Bond, B. E. Law, and J. R. Ehleringer. 2004. Associations between carbon isotope ratios of ecosystem respiration, water availability, and canopy conductance. *Global Change Biology* **10:**1767–1784, doi:10.1111/j.1365-2486.2004.00837.

Sponheimer, J., T. Robinson, L. Ayliffe, B. Roeder, L. Shipley, E. Lopez, A. West, J. Hammer, B. Passey, T. Cerling, D. Dearing, and J. Ehleringer. 2003. An experimental study of carbon isotopes in the diets, feces and hair of mammalian herbivores. *Canadian Journal of Zoology* **81:**871–876.

Tieszen, L. L., T. W. Boutton, K. G. Tesdahl, and N. A. Slade. 1983. Fractionation and turnover of stable carbon isotopes in animal tissues: Implications for $\delta^{13}C$ analysis of diet. *Oecologia* **57:**32–37.

West, A. G., L. K. Ayliffe, T. E. Cerling, T. F. Robinson, B. Karren, M. D. Dearing, and J. R. Ehleringer. 2004. Short-term diet changes revealed using stable carbon isotopes in horse tail-hair. *Functional Ecology* **18:**616–624.

CHAPTER 12

Insight into the Trophic Ecology of Yellowfin Tuna, *Thunnus albacares*, from Compound-Specific Nitrogen Isotope Analysis of Proteinaceous Amino Acids

Brian N. Popp,* Brittany S. Graham,† Robert J. Olson,‡ Cecelia C. S. Hannides,† Michael J. Lott,§ Gladis A. López-Ibarra,¶ Felipe Galván-Magaña,¶ and Brian Fry||

**Department of Geology and Geophysics, University of Hawaii*
†Department of Oceanography, University of Hawaii
‡Inter-American Tropical Tuna Commission
§Department of Biology, University of Utah
¶Centro Interdisciplinario de Ciencias Marinas-Instituto Politécnico Nacional La Paz
||Department of Oceanography & Coastal Sciences and Coastal Ecology Institute, School of the Coast and Environment Louisiana State University

Contents

I. INTRODUCTION

There is widespread concern and debate on the extent that commercial fisheries are altering the structure and function of marine ecosystems (Pauly *et al.* 1998, NRC 1999, Myers and Worm 2003, Hampton *et al.* 2005, Sibert *et al.* 2006). Selective removal of large predatory fishes from food webs can impart changes in trophic structure and stability via trophic cascades, defined as inverse patterns in abundance or biomass across more than one trophic level in a food web (Carpenter *et al.* 1985, Pace *et al.* 1999). Recent calls for policy makers to adopt an ecologically based approach to fisheries management (Botsford *et al.* 1997, Pikitch *et al.* 2004) places renewed emphasis on achieving accurate depictions of trophic links and biomass flows through the food web in exploited systems. Such an approach would take into consideration the indirect effects of fishing, such as declines in diversity, changes in the species composition of the prey community, and changes in trophic-level structure (*e.g.*, aggregate removals at various trophic levels; Gislason *et al.* 2000).

There is general agreement of the importance of measuring changes in trophic structure as a means to evaluate fishery impacts on ecosystems, and ecosystem indicators that take trophic level into consideration are desirable (Gislason *et al.* 2000, Murawski 2000, Rice 2000). Monitoring the trophic level of key food web components and functional groups instead of the mean trophic level of the fisheries catch (Pauly *et al.* 2001) serves as a useful fisheries-independent metric of ecosystem change and sustainability because it integrates an array of biological and ecological relationships and processes. In addition to adopting ecosystem metrics, ecosystem-based fisheries management is facilitated through the development of multispecies models that represent indirect ecological interactions among species or guilds (Latour *et al.* 2003). Among these models, mass-balance models of food webs (Cox *et al.* 2002, Olson and Watters 2003) explicitly represent trophic links between biomass pools based on diet relations determined from stomach contents analysis.

Stable isotope ratios have been used extensively in ecosystems research and are a valuable compliment to traditional methods used to study food webs (Peterson and Fry 1987). In particular, nitrogen isotopic ratios have been frequently used to examine trophic dynamics (Peterson and Fry 1987, Lajtha and Michener 1994). At each trophic level, an increase of ~3‰ has been observed in the bulk tissue $\delta^{15}N$ values of many consumers (Deniro and Epstein 1981, Minagawa and Wada 1984, Post 2002). However, the $\delta^{15}N$ value of any consumer is predominantly a function of both the trophic level of that consumer and the $\delta^{15}N$ value of the primary producers at the base of the food web. In marine environments, the microalgae that support marine food webs typically have $\delta^{15}N$ values that change spatially and seasonally due to incomplete utilization of nitrogenous nutrients (Altabet 2001, Lourey *et al.* 2003), uptake of partly denitrified nitrate (Cline and Kaplan 1975, Voss *et al.* 2001, Sigman *et al.* 2005), and because primary producers can use different sources of nitrogen (nitrate, ammonium, N_2) in different areas and seasons (Dugdale and Goering 1967, Owens 1987, Dugdale and Wilkerson 1991, Dore *et al.* 2002). When determining the relative trophic level of top predators, characterizing the $\delta^{15}N$ values of the base of marine food webs can be challenging because marine microalge have very short life spans and can be difficult to isolate from other organic suspended particulate material. An alternative approach is to use primary consumers (*e.g.*, zooplankton or bivalve mollusks), which may integrate short-term and spatial variability in the $\delta^{15}N$ values of their diet, to represent trophic level 2 or slightly higher (Jennings *et al.* 2002, Post 2002). Unfortunately, zooplankton are also not ideal for this purpose, since they too have short life spans and many are omnivorous (Rolff 2000).

Compound-specific isotopic analyses (CSIA) can compliment whole-tissue or whole-animal isotopic results and can distinguish metabolic and trophic-level relationships in a food web from changes in isotopic compositions at the base of the food web (Uhle *et al.* 1997, Fantle *et al.* 1999, McClelland and Montoya 2002). For example, Uhle *et al.* (1997) used the $\delta^{13}C$ of individual fatty and amino acids to elucidate the sources of metabolic carbon used for synthesis of these compounds in foraminifera.

Previous research has also shown that $\delta^{13}C$ values of essential and nonessential amino acids (NAA) distinguished between the basal carbon sources and diet of a consumer (Fantle *et al.* 1999). These researchers showed that essential amino acids (EAA), which are produced only by plants and bacteria, were not heavily fractionated by juvenile blue crabs whereas NAA were fractionated to a greater extent. Jim *et al.* (2006), using laboratory rats grown on diets of isotopically and nutritionally manipulated purified C_3 and/or C_4 macronutrients, found that EAA and conditionally indispensable amino acids were routed from diet to collagen with little isotopic fractionation, whereas NAA differed by up to 20‰. The EAA and NAA designation is based on the flow of carbon through biochemical systems, and although it is a convenient way to organize our thoughts, it unfortunately does not necessarily provide an accurate picture of the origins of amino nitrogen. For example, McClelland and Montoya (2002) indicated that a mixture of EAA and NAA were incorporated with little alteration in $\delta^{15}N$ values from dietary sources into herbivorous zooplankton fed a known algal diet. These authors found $\delta^{15}N$ values of glycine, lysine, phenylalanine, serine, threonine, and tyrosine were nearly identical in producer and consumer. Of these amino acids, only lysine, phenylalanine, and threonine are considered EAA. On the other hand, the amino acids alanine, aspartic acid, glutamic acid, isoleucine, leucine, proline, and valine were enriched in ^{15}N by ~5–7‰ in the consumer relative to those in the producer. Leucine, isoleucine, and valine are considered EAA. The work of McClelland and Montoya (2002) clearly showed that the classic essential and nonessential grouping of amino acids did not correlate well with either $\delta^{15}N$ values or trophic position in their simple laboratory food web study. However, the principle finding of McClelland and Montoya (2002) for applications of isotope ecology is that the $\delta^{15}N$ values of some amino acids in consumers apparently can provide accurate determination of the isotopic composition of the base of the food web. We consider this group of compounds the "source" amino acids. On the other hand, other amino acids are either synthesized by animal consumers *de novo* or undergo significant transamination and deamination reaction, are enriched in ^{15}N by ~5–7‰ relative to the first group of amino acids, and appear to reflect the trophic level of the consumer (McClelland and Montoya 2002). We consider these the "trophic" amino acids.

In this chapter, we tested the premise that a single sample from an upper trophic level pelagic predator fish could provide information on both the trophic level of the fish and the $\delta^{15}N$ value at the base of the food web. We analyzed the nitrogen isotopic composition of individual amino acids in white muscle tissue (WMT) of yellowfin tuna (*Thunnus albacares*) from the eastern tropical Pacific (ETP) to determine if the observed variations in the $\delta^{15}N$ values of WMT are primarily controlled by the nitrogen isotopic composition at the base of the food web or the trophic level of the tuna. We used the difference between the $\delta^{15}N$ values of source and trophic amino acids to estimate the trophic level of yellowfin, and we compare this to estimates based on two independent methods, stomach content analysis and the isotopic difference between the $\delta^{15}N$ values of mesozooplankton and yellowfin tuna in the ETP. We found that nitrogen isotopic analyses of individual amino acids in tuna can be used to distinguish between nutrient and trophic dynamics in pelagic ecosystems, and we discuss implications of these results for investigating the long-term impact of commercial fishing on the food web structure of marine ecosystems.

II. OCEANOGRAPHIC SETTING

The ETP contains some of the most biologically productive waters of the world's oceans. In the ETP, the northeast and southeast trade winds converge north of the equator along the intertropical convergence zone. The northeast trade winds drive the north equatorial current to the west at ~10°N, and the southeast trade winds drive the south equatorial current to the west at ~3°S (Wyrtki 1966). The equatorial countercurrent flows eastward between the north and south equatorial currents in the region where the trade winds are the weakest. This circulation pattern results in a band of cold, nutrient-rich water near the equator that extends west from South America far into the central equatorial Pacific. The intensity and spatial extent of this "cold tongue" can vary seasonally and interannually (Chelton *et al.* 2001). Ekman drift in the most eastern part of the ETP carries these nutrient-rich waters poleward along the coasts of Baja California and of Ecuador and Peru (Philander *et al.* 1987).

Coastal upwelling along the eastern boundary of the ETP and oceanic upwelling along offshore divergences bring new macronutrients (nitrogen, phosphorous, and silicon) to the euphotic zone (Wyrtki 1981), and can account for the high biological productivity of this region (Fiedler *et al.* 1991). Oxic respiration associated with the sinking of organic matter produced by high biological productivity can result in oxygen conditions low enough ($\sim$0.1 mL liter^{-1}) that bacteria can use nitrate as an alternative electron acceptor in the respiratory process (*i.e.*, denitrification). Although these low-oxygen oceanic regions represent only 0.1% of the total ocean volume, half of the global denitrification occurs in pelagic oxygen minimum zones with O_2 levels less than 0.05 mL liter^{-1} (Codispoti and Christensen 1985). The importance of pelagic denitrification in the ETP has been recognized for many years on the basis of N–O–P stoichiometric relationships, the existence of a nitrite maximum within the O_2 minimum zone (Brandhorst 1959, Cline and Richards 1972, Thomas 1966), and observations of apparent N_2O consumption in the O_2 minimum zone (Cohen and Gordon 1978). The ETP is the largest region of the world's oceans with low oxygen and high rates of water column denitrification, which results in $\sim$35–45% of global pelagic denitrification (Cline and Richards 1972, Codispoti and Richards 1976).

Denitrification preferentially consumes $^{14}NO_3^-$, which leads to a marked increase in nitrate $\delta^{15}N$ values in oceanic regions with strong oxygen minimum zones (Cline and Kaplan 1975, Liu and Kaplan 1989, Voss *et al.* 2001). Nitrogen isotopic compositions of nitrate coupled with consideration of the nitrate deficit in the water column indicate that denitrification in the ETP strongly affects the $\delta^{15}N$ values of NO_3^- (Sigman *et al.* 2005), as well as that of suspended and sinking particulate matter (Voss *et al.* 2001). Consequently, variations in the $\delta^{15}N$ value of primary producers, and in turn, consumers in the ETP can be affected by the spatial and temporal intensity of algal production and denitrification.

III. SAMPLE COLLECTION AND ANALYTICAL METHODS

A. Samples

Yellowfin tuna were captured by purse seine fishing vessels in the eastern Pacific Ocean between August 16, 2003 and December 6, 2004, and were sampled on board the vessels by observers of the Inter-American Tropical Tuna Commission (IATTC 2004). The fish were measured [fork length (FL), mm] and the sex determined if the fish were mature enough to do so. Samples of WMT were removed from the dorsal region adjacent to the second dorsal fin and stored at -20°C until processed further. Stomachs and liver samples were also collected, but the data reported here are only for white muscle. Subsamples of WMT from up to six individuals per purse seine set and size class ($<$900- and $\geq$900-mm FL) were combined into one sample for stable isotope analysis. Purse seine set locations are shown in Figure 12.1.

Samples of zooplankton were collected by a 0.6-m diameter bongo net (Smith and Richardson 1977) on board the NOAA research ships *David Starr Jordan* and *McArthur II* in the ETP from August 5–December 5, 2003 (Figure 12.1). The bongo net frame with two 333-μm mesh cylindrical–conical nets was towed obliquely from 200 m for 15 min, and the material collected by the inboard net was stored at -20°C until processed further. A flow meter was used on the outboard net, where an average of about 400 m^3 of water was filtered per tow. In the laboratory, the zooplankton samples were thawed slowly,

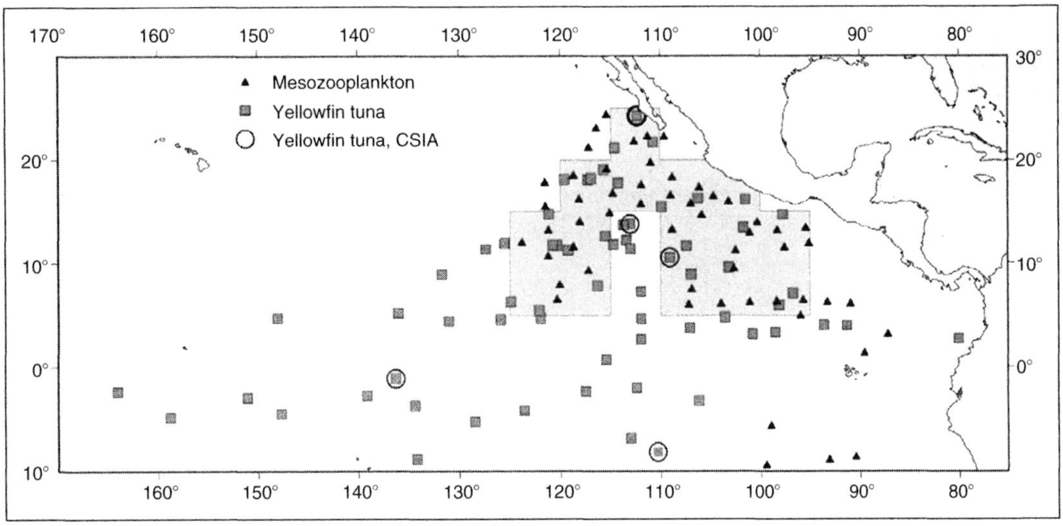

FIGURE 12.1 Purse seine set locations (squares) and zooplankton collection sites (triangles) for yellowfin tuna and mesozooplankton samples, respectively, used for bulk stable isotope analysis. Circles are locations of the yellowfin tuna samples used for compound-specific stable isotope analysis. Fifteen 5 $\times$ 5-degree areas used for trophic level estimates based on comparisons of yellowfin and mesozooplankton δ^{15}N are indicated by shading.

sorted for copepods (to species level), amphipods (to order), euphausiids (to order), and chaetognaths (to phylum) using a stereoscopic microscope, and refrozen. Collectively, we define these taxa as components of the mesozooplankton guild, after Chai *et al.* (2002) and Olson and Watters (2003).

Samples of WMT for CSIA were chosen from similar-size yellowfin tuna along a broad latitudinal gradient in the δ^{15}N values. The latitudinal gradient was defined based on nitrogen isotopic analysis of 95 composite samples of tuna bulk WMT. The samples selected for CSIA were from fish that ranged from about 600 to 800 mm (Table 12.1). Although our sample from the equatorial region was chosen from fish caught further to the west than the other four samples (Figure 12.1), the δ^{15}N value of bulk WMT is representative of yellowfin tuna caught along the equator in the ETP (Figure 12.2).

B. Bulk Isotope Analyses

The tuna tissue samples were lyophilized or oven dried (60 °C, ~24 h) and homogenized to a fine powder using a mortar and pestle. An average of about 70 individuals per species (for the copepods), per order (for the amphipods and euphausiids), or per phylum (for the chaetognaths) per sample was combined into a single sample for stable isotope analysis. One-hundred forty-nine mesozooplankton samples were analyzed. Bulk carbon and nitrogen isotopic compositions of tuna and mesozooplankton were determined using an online carbon–nitrogen analyzer coupled with an isotope ratio mass spectrometer (Finnigan ConFlo II/Delta-Plus). Isotope values are reported in standard δ-notation relative to the international V-PDB and atmospheric N_2 for carbon and nitrogen, respectively. A glycine standard was analyzed approximately every 10 samples to ensure accuracy of all isotope measurements. Furthermore, several samples were measured in duplicate or triplicate, and the analytical error associated with these measurements was typically $\leq$0.2‰.

TABLE 12.1 Sample location in the ETP, number of individuals included in the composite sample, mean FL (±SD), C/N ratio, and stable isotopic data for bulk WMT from the yellowfin tuna used for amino acid compound-specific isotope analysis

Sample location						
Latitude	Longitude	Number of individuals	Mean FL (mm)	C/N	$\delta^{13}C$ (‰)	$\delta^{15}N$ (‰)
8°13.8′S	110°19.2′W	6	630 (±86)	3.6	−15.9	10.4
1°10.8′S	136°16.2′W	6	595 (±13)	3.6	−16.0	10.9
10°31.8′N	109° 1.2′W	6	682 (±80)	3.7	−16.0	13.5
13°45.0′N	113° 4.2′W	6	801 (±148)	3.8	−16.0	14.7
24°15.0′N	112°19.2′W	2	643 (±5)	3.8	−16.7	15.6

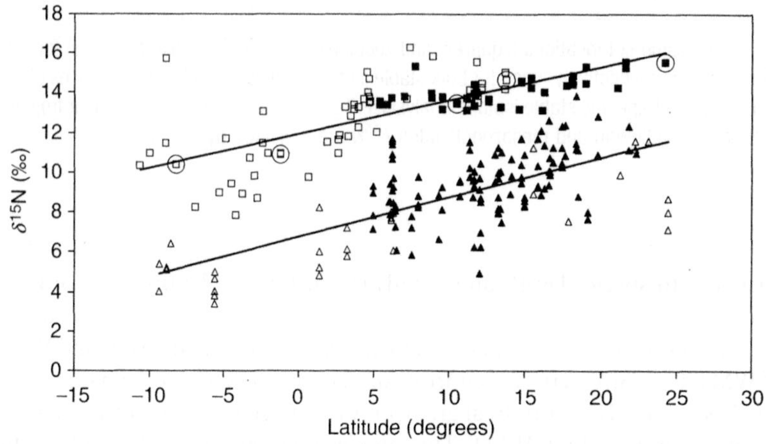

FIGURE 12.2 Bulk $\delta^{15}N$ values for WMT of yellowfin tuna (squares) and for whole mesozooplankton (triangles) versus latitude in the ETP. The samples used to estimate yellowfin trophic level by comparison with mesozooplankton bulk $\delta^{15}N$ values are represented by filled squares and triangles (see shaded region in Figure 12.1). The five yellowfin tuna samples used for CSIA determinations are shown by circles. Equations describing these lines are: $\delta^{15}N_{bulk} = 0.17$ (±0.03)*Latitude + 11.9 (±0.3), $\delta^{15}N_{mesozoo} = 0.20$ (±0.03)*Latitude + 6.8 (±0.5). The slopes of the regressions are similar and not different at the 95% confidence interval.

C. Acid Hydrolysis

The samples were prepared for compound-specific nitrogen isotope analysis of amino acids by acid hydrolysis followed by derivatization to produce trifluoroacetic amino acid esters (Macko *et al.* 1997). Approximately 10 mg homogenized sample was hydrolyzed following procedures modified from Cowie and Hedges (1992). The homogenized sample and norleucine (0.43-nmol mg^{-1} sample) were transferred to 5 mL reaction vials (Reacti-Vial, Pierce Scientific), 1-mL 6 N HCl added (Sequanal Grade, Pierce Scientific), the vial flushed with N_2 and capped using a Teflon/silicone liner (18-mm Tuf-Bond, Pierce Scientific) before heating at 150°C for 70 min. Norleucine was used as an internal recovery standard. The hydrolysate was evaporated to dryness at 55°C under a gentle stream of N_2 and the residue redissolved in 1-mL 0.01-N HCl. This solution was purified by filtration (0.22-μm Millex-GP, Millipore Corporation) followed by a rinse with 1-mL 0.01-N HCl. The solution was further purified by cation-exchange chromatography (Dowex 50W8-400, Sigma-Aldrich) following the method of Metges *et al.* (1996). Briefly, a 5-cm column of cation-exchange resin was prepared in a Pasteur pipette and the amino acids eluted with 4-mL 2-N ammonium hydroxide. The eluant was evaporated to dryness under a stream of N_2 at 80°C. Prior to derivatization, samples were reacidified with 1-mL 0.01-N HCl and then evaporated to dryness under a stream of N_2 at 55°C. The samples were split and one-half was archived.

D. Derivatization

The samples were first reacted with acidified isopropanol to esterify the carboxyl terminus of the amino acids. Approximately 2 mL of 4:1 isopropanol:acetyl chloride were added to each sample, the vial was flushed with N_2 and sealed with a Teflon-lined cap, and then the sample was heated (110°C) for 60 min. The samples were then dried under a stream of N_2 at 60°C. The resultant amino acid esters were acylated by the addition of 1 mL 3:1 methylene chloride:trifluoroacetic anhydride (TFAA, 99+%, Pierce Scientific). The vials were flushed with N_2, sealed with Teflon-lined caps and heated (100°C) for 15 min. Samples were then further purified using solvent extraction (Ueda *et al.* 1989). The acylated amino acid esters were evaporated at room temperature under a stream of N_2 and then redissolved in 3 mL 1:2 chloroform:P-buffer ($KH_2PO_4 + Na_2HPO_4$ in Milli-Q water, pH 7). Vigorous shaking caused the acylated amino acid esters to partition into the chloroform and high boiling point contamination ended up in the buffer. The solvents were separated by centrifugation (10 min at 600 *g*) the chloroform was transferred to a clean vial, and the solvent extraction process repeated. Veuger *et al.* (2005) showed full recovery of the acylated amino acid esters using this technique. Finally, to ensure complete derivatization, the acylation step was repeated. The samples were stored in 3:1 methylene chloride: TFAA at 4°C and analyzed within 1 month.

E. Compound-Specific Isotopic Analyses

The nitrogen isotopic composition of TFA derivatives of amino acids were analyzed by isotope ratio monitoring gas chromatography-mass spectrometry using a Finnigan MAT 252 mass spectrometer interfaced to a Trace GC gas chromatograph through a GC-C III combustion furnace (980°C), reduction furnace (650°C), and liquid nitrogen cold trap. L-2-Aminoadipic acid for which the $\delta^{15}N$ value had been previously independently determined was coinjected as an internal standard. Prior to analysis, the TFA derivatives were dried at room temperature under a stream of N_2 and redissolved in 100 μL of ethyl acetate. The samples (1 μL) and L-2-aminoadipic acid (1 μL, ~20 nmole) were injected (split/splitless, 5:1 split ratio) onto a 50-m HP Ultra-2 column (0.32-mm i.d., 0.5-μm film thickness) at an injector temperature of 180°C and a constant helium flow rate of 2 mL min^{-1}. The column oven was initially held for 2 min at 52°C and increased, in stages, to temperatures of 190°C at a rate of 8.0°C min^{-1}, 300°C at 10.0°C min^{-1}, then finally held at 300°C for 8 min. The samples were analyzed in triplicate, and the measured nitrogen isotope compositions were corrected relative to the $\delta^{15}N$ value of the amino acid internal standard. Reproducibility associated with these isotopic measurements averaged 1.4‰ and ranged from 0.1‰ to 4.4‰.

F. Tuna-Mesozooplankon Comparisons

Bulk $\delta^{15}N$ values of yellowfin tuna and mesozooplankton were compared to derive independent estimates of trophic level over a range of latitudes in the ETP. The data were compared in fifteen 5 × 5-degree areas (Figure 12.1) where samples of both taxa were collected. The trophic level of the yellowfin tuna (TL_{YFT}) sampled from each purse seine set in the 5 × 5-degree areas was calculated as:

$$TL_{YFT} = \frac{\delta^{15}N_{YFT} - \delta^{15}N_{Mesozoo}}{TEF} + TL_{Mesozoo} \qquad (12.1)$$

where $TL_{Mesozoo}$ is the estimated trophic level of mesozooplankton in the ETP, $\delta^{15}N_{Mesozoo}$ is the mean $\delta^{15}N$ value of mesozooplankton in each 5-degree area, and TEF represents the ‰ trophic enrichment factor = $\delta^{15}N_{Consumer} - \delta^{15}N_{Diet}$. $TL_{Mesozoo}$ was estimated as 2.7 by Olson and Watters (2003), based on the nutrient–phytoplankton–zooplankton-detritus model of Chai *et al.* (2002) for the eastern equatorial Pacific. The $TL_{Mesozoo}$ depends on the relative proportions of mesozooplankton predation on microzooplankton and grazing on diatoms, derived from the nitrogen balance of Chai *et al.* (2002). The trophic level of a food web component is 1.0 plus the weighted average of the trophic levels of its prey. That is,

$$TL = 1.0 + \sum_{i}(P_i \times TL_i) \qquad (12.2)$$

where P_i is the diet proportion of the *i*th prey group and TL_i is the trophic level of the *i*th prey group. Several workers have adopted an average TEF value of 3.4‰ for many taxa (Minagawa and Wada 1984, Vander Zanden and Rasmussen 2001, Post 2002); however, there is acknowledged variance in TEF values (Gannes *et al.* 1997). For example, compilations of data for laboratory-grown ammonotelic fish show lower TEF values (2.3‰ McCutchan *et al.* 2003; 2.0‰ Vanderklift and Ponsard 2003). Field studies of the nitrogen isotopic compositions of juvenile yellowfin tuna and their prey from Hawaii yielded an average TEF of 2.1‰ based on ~85% characterization of the tuna diet (Graham *et al.* 2007).

IV. RESULTS AND DISCUSSION

A. Variation in $\delta^{15}N$ Values of WMT

The bulk WMT nitrogen isotopic composition of the five yellowfin tuna samples chosen for CSIA varied from 10.4‰ at ~10°S to 15.6‰ at ~25°N near the tip of Baja California (Table 12.1). Variation in $\delta^{15}N$ values with latitude in this small number of samples is representative of the variation in the $\delta^{15}N$ values of WMT in our much larger dataset for the ETP (Figure 12.2). This consistent spatial variation in the $\delta^{15}N$ values of an upper-level predator could be a result of several factors, including variation in the organism's trophic level due to dietary differences, in the organism's physiology, or in the nutrient dynamics at the base of the food web. Assuming a TEF of 3‰ for each trophic level in this ecosystem, if the observed gradient in $\delta^{15}N$ values with latitude were due to dietary differences alone, then it would represent a gradient of ~1.7 trophic levels occupied by yellowfin tuna. This observation is not explained by an ontogenetic gradient in foraging behavior or food habits (M. Bocanegra-Castillo and F. Galván-Magaña, unpublished data) or by correlation between tuna size and $\delta^{15}N$ values of bulk WMT (Table 12.1, and unpublished data for the ETP). Additional factors can affect the $\delta^{15}N$ values of organisms, such as diet quality and quantity, and even protein catabolism can affect a consumer's $\delta^{15}N$ value (see review by Gannes *et al.* 1997). However, the diets of yellowfin tuna are protein-rich, and we have observed no obvious relationship between $\delta^{15}N$ values and stomach fullness.

As described earlier, changes in the intensity of denitrification can affect the nitrogen isotopic composition of the nitrate pool in the ETP. For example, Voss *et al.* (2001) showed that the δ^{15}N values of sinking particles in the ETP, which most likely includes organic matter derived from algae, bacteria, and zooplankton, change from ~9‰ at 14°N to 11.2‰ at 24°N. If the gradient in δ^{15}N values with latitude originates at the base of the food web in the ETP and that signal is propagated up the food web to the upper-level pelagic predators, it provides another viable explanation for the observed latitudinal shift in the δ^{15}N values of bulk WMT. However, the degree to which isotopic variability at the base of the food web affects a consumer's δ^{15}N value depends on the animal's movement patterns, life span, tissue turnover rates, and foraging behavior. Highly mobile organisms will integrate the δ^{15}N values of the prey consumed over large spatial scales, whereas less mobile consumers will reflect local spatial and shorter temporal trends in nutrient dynamics at the base of the food web.

B. Variation in the δ^{15}N Values of Amino Acids

To help interpret the coherent spatial variation in the δ^{15}N values of bulk WMT, we analyzed the isotopic composition of individual amino acids in ETP yellowfin tuna. The nitrogen isotopic composition of eight individual amino acids and two combinations of chemically related amino acids were determined (Table 12.2). However, before considering the results of the CSIA, it is informative to review sources of amino nitrogen for metabolism. Nitrogen in the body and the diet of tunas is predominantly protein and the amino acids from which protein is synthesized. Proteins are synthesized from ~20 common amino acids, which are divided into two classes, EAA and NAA (Young and El-Khoury 1995). Essentiality of amino acids is not the same in all organisms, but is often species-specific (NRC 1994, 1995) and, as discussed above, does not provide an accurate picture of the origins of amino nitrogen. Protein ingested by higher organisms is denatured in the stomach and hydrolyzed to amino acids and short polypeptides. Most of these compounds are absorbed and transported to the liver, where about 75% of the amino acids are incorporated into the organism (NRC 1994, 1995). The excess amino acids are catabolized, producing ammonia, which is eliminated mainly through the gills as ammonium. The amino acids incorporated via the liver may undergo deamination and transamination to provide the precursors for gluconeogenesis, lipogenesis, and protein synthesis (Smutna *et al.* 2002).

TABLE 12.2 Bulk and amino acid δ^{15}N values from the WMT and estimated trophic level of yellowfin tuna as a function of latitude in the ETP

	8°13.8′S	1°10.8′S	10°31.8′N	13°45.0′N	24°15.0′N
Bulk WMT	10.4	10.9	13.5	14.7	15.6
Amino acids					
Alanine	25.4 [a](±0.2)	18.0 (±1.7)	24.6 (±2.6)	28.5 (±2.0)	29.3 (±1.9)
Glycine	1.7 (±0.7)	−2.3 (±1.5)	2.8 (±1.5)	7.2 (±1.8)	7.5 (±0.3)
Leucine + isoleucine	29.5 (±1.3)	25.3 (±0.7)	29.0 (±2.0)	29.7 (±1.8)	33.5 (±2.9)
Proline	19.8 (±0.3)	17.8 (±0.6)	22.3 (±4.4)	22.0 (±3.8)	26.5 (±0.7)
Aspartic acid	29.7 (±0.1)	27.7 (±0.5)	29.2 (±1.3)	31.1 (±2.1)	32.5 (±2.8)
Glutamic acid	27.3 (±0.7)	23.4 (±0.3)	24.7 (±1.1)	29.2 (±1.2)	30.6 (±2.2)
Phenylalanine	2.7 (±1.6)	2.7 (±1.0)	3.2 (±3.1)	5.3 (±0.9)	7.5 (±1.4)
Tyrosine + lysine	8.3 (±0.4)	0.3 (±0.4)	5.0 (±1.1)	8.8 (±1.5)	11.9 (±0.6)
Argine	2.2 (±1.5)	−1.1 (±0.3)	2.1 (±1.4)		6.6 (±1.0)
Histidine	2.9 (±0.9)	−2.4 (±1.4)	2.1 (±2.3)	5.7 (±0.5)	6.8 (±1.1)
[b]Trophic level	4.7 (±0.2)	4.7 (±0.3)	4.1 (±0.4)	4.2 (±0.4)	4.3 (±0.4)
[c]Trophic level	4.8 (±0.1)	4.3 (±0.2)	4.4 (±0.3)	4.4 (±0.3)	4.3 (±0.2)

[a] Values in parentheses are 1 SD.
[b] Trophic level = $1 + ((\delta^{15}N_{Glutamic\ acid} - \delta^{15}N_{Glycine})/7)$.
[c] Trophic level = $1 + ((\delta^{15}N_{\Sigma trophic} - \delta^{15}N_{\Sigma source})/7)$.

Transamination reactions do not occur in some amino acids, whereas in others there is complete equilibration and these reactions can change the ^{15}N content of the amino acids (Hare and Estep 1983, Macko *et al.* 1987, Hare *et al.* 1991). For example, Hare and Estep (1983) found a 19‰ range in $\delta^{15}N$ values in amino acids from bovine tendon collagen. Tissues can vary in their protein composition and therefore in their amino acid distribution (Wilson and Poe 1985, Gunasekera *et al.* 1997), which in turn, can affect the bulk tissue $\delta^{15}N$ values.

The distribution of $\delta^{15}N$ values of individual amino acids in yellowfin tuna caught in the ETP are strongly bimodal, comprising a group of "high" $\delta^{15}N$ amino acids and a group of "low" $\delta^{15}N$ amino acids (Table 12.2). Enrichment in ^{15}N is not related to the class of amino acid essentiality but rather follows the patterns originally observed by McClelland and Montoya (2002). For example, within the high $\delta^{15}N$ group (Table 12.2), alanine, aspartic, and glutamic acid are NAA whereas leucine and isoleucine are considered EAA (Schepartz 1973). It should be noted that leucine and isoleucine have similar metabolic origins and belong to the pyruvate family of amino acids (Stryer 1988). In the low $\delta^{15}N$ group (Table 12.2), phenylalanine is an EAA, whereas glycine is usually considered to be a NAA (Schepartz 1973).

The $\delta^{15}N$ values of individual amino acids have not commonly been measured in marine organisms (Schmidt *et al.* 2004). However, nitrogen isotopic compositions of individual amino acids have been determined for laboratory cultures of rotifers (McClelland and Montoya 2002), size-fractionated zooplankton from the tropical Atlantic (McClelland *et al.* 2003) and postlarval euphausiids from the Southern Ocean (Schmidt *et al.* 2004). These authors found that alanine, leucine, isoleucine, aspartic acid, and glutamic acid were strongly fractionated in food–web relationships, whereas the ^{15}N content of glycine, lysine, phenylalanine, serine, and tyrosine did not appear to change within the food web. These results imply that the $\delta^{15}N$ values of some amino acids, such as alanine, aspartic acid, and glutamic acid, should reflect the relative trophic position of yellowfin tuna in the food web and that the $\delta^{15}N$ values of glycine and phenylalanine should record the $\delta^{15}N$ value of the source of nitrogen supporting production. Although glycine is not typically considered an EAA, results suggest that its carbon skeleton is derived from an EAA without alteration of the ^{15}N content of the amino nitrogen in the molecule, or that in the marine organisms studied, glycine is derived from dietary sources (McClelland and Montoya 2002, McClelland *et al.* 2003, Schmidt *et al.* 2004). Glycine is thought to derive mainly from serine, which is produced from 3-phosphoglycerate, an intermediate in glycolysis (Stryer 1988). It is possible that glycine is a conditionally EAA (Reeds 2000) in yellowfin tuna. The degree to which glycine may be regarded as essential or indispensable could be a function of the quantity of serine in the diet of pelagic marine organisms. McClelland and Montoya (2002) also noted little ^{15}N enrichment of these amino acids in a consumer relative to its controlled diet, suggesting that the $\delta^{15}N$ value of glycine and serine appear to record the $\delta^{15}N$ value of the source of nitrogen supporting production. One scenario that would give a conserved $\delta^{15}N$ value of glycine would be that glycine travels through the food web primarily as part of an amino acid dimer, with the other part of the dimer being a rare amino acid. In this scenario, the glycine-containing dimers would be conserved and shunted from prey protein into predator protein and not undergo expected metabolic fractionations associated with deaminations and transaminations. Of course, this is speculation at this time. In reality, we do not know why the nitrogen isotopic composition of glycine reflects the $\delta^{15}N$ value at the base of the food web, but existing laboratory (McClelland and Montoya 2002) and field research (McClelland *et al.* 2003, Schmidt *et al.* 2004) suggests that glycine is conservative and records the $\delta^{15}N$ values at the base of the food web in large and small marine organism. On the other hand, the amino acids alanine, aspartic acid, and glutamic acid are mainly derived from intermediates in the citric acid cycle (Stryer 1988) and all show ^{15}N enrichment in consumer organisms relative to glycine, lysine, phenylalanine, serine, and tyrosine in the same sample (McClelland and Montoya 2002, McClelland *et al.* 2003, Schmidt *et al.* 2004).

Given the observed isotopic behavior of source and trophic amino acids, two general scenarios can explain an increase in the $\delta^{15}N$ values of yellowfin tuna with latitude in the ETP. If the trophic level of

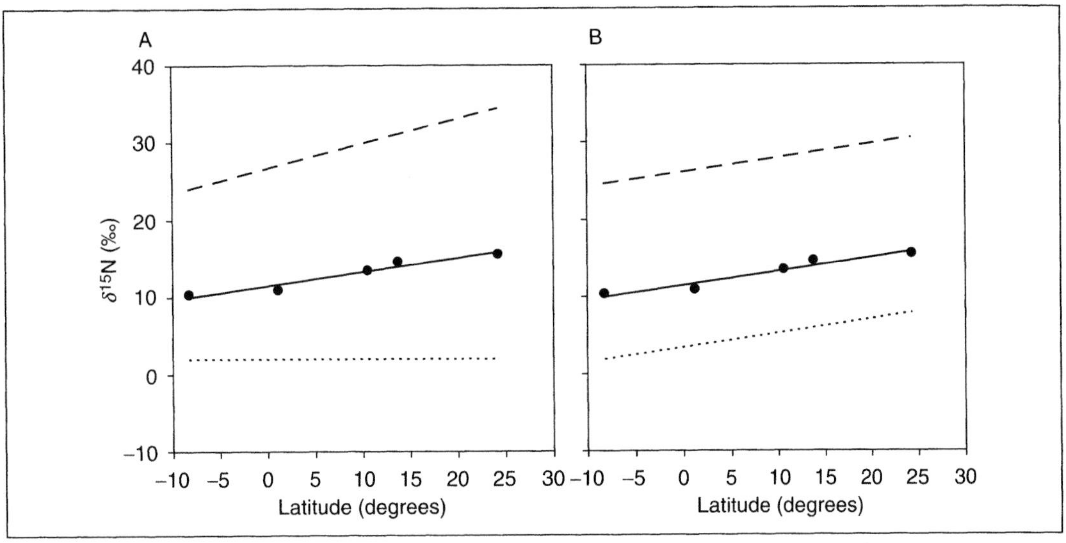

FIGURE 12.3 Conceptual diagrams showing changes with latitude in the δ^{15}N values of ETP yellowfin tuna bulk WMT and of source (dotted line) and trophic (dash line) amino acids with (A) no change in the δ^{15}N value at the base of the food web and a gradient with latitude in the trophic level of tuna and (B) change with latitude in the δ^{15}N values of phytoplankton at the base of the food web and no gradient in tuna trophic level. The filled circles are the bulk δ^{15}N values for WMT of the yellowfin tuna used for compound-specific isotope analysis.

yellowfin tuna increased with latitude, then one would expect the δ^{15}N values of source amino acids (glycine and phenylalanine) to remain constant at all latitudes, and differences between the δ^{15}N values of source and trophic (alanine, aspartic acid, or glutamic acid) amino acids to increase as the trophic level increases to the north (Figure 12.3A). If the trophic level of the yellowfin tuna remained constant over the region, then the δ^{15}N values the source amino acids should parallel those of bulk WMT and the δ^{15}N values of the trophic amino acids (Figure 12.3B). Our results reveal that the δ^{15}N values of glycine, phenylalanine, alanine, aspartic acid, and glutamic acid in tuna show latitudinal trends similar to the δ^{15}N values of bulk WMT (Figure 12.4), indicating that the ^{15}N enrichment in the north is due to changes in δ^{15}N values at the base of the food web. Increasing δ^{15}N values to the north is consistent with the effects of denitrification on the δ^{15}N values of nitrate and the transfer of this isotopic composition to phytoplankton (see also Fig. 7A in Voss *et al.* 2001). It is remarkable that the δ^{15}N values of a highly mobile and metabolically active predator would track processes at the base of the food web, and our results suggest that, even though yellowfin tuna are capable of basin-wide migrations, in the ETP they may have a relatively high level of regional residency.

C. Trophic Level of ETP Yellowfin Tuna

We used the difference between the δ^{15}N values of glutamic acid and glycine to estimate the trophic level of yellowfin tuna in the ETP, assuming the difference between the δ^{15}N values of glutamic acid and glycine is 7‰ per trophic level. We chose 7‰ based on the work of McClelland and Montoya (2002). These authors cultured the marine rotifer *Brachionus plicatilis* on a diet of the alga *Tetraselmis suecica* to examine changes in the nitrogen isotopic composition of individual amino acids between a plankton consumer and their food source. They found a ~2‰ increase in bulk δ^{15}N value with trophic position, which resulted from averaging large increases in the δ^{15}N values of some amino acids and little or no change in the δ^{15}N values of others. McClelland and Montoya (2002) proposed that the amino acids

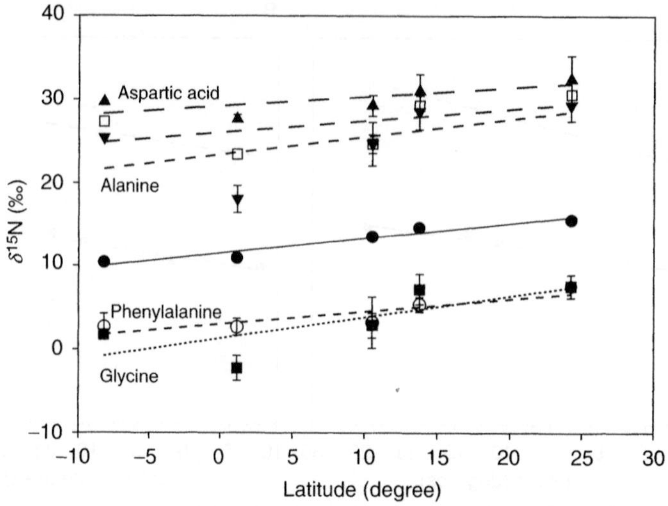

FIGURE 12.4 $\delta^{15}N$ values as a function of latitude for bulk WMT and the source (glycine, phenylalanine) and trophic (alanine, aspartic acid, and glutamic acid shown by dash line) amino acids from the WMT of ETP yellowfin tuna. The parallel trends in the $\delta^{15}N$ values of source and trophic indicate that the ^{15}N enrichment in the north is due to a trend in $\delta^{15}N$ values at the base of the food web. Equations describing these lines are: $\delta^{15}N_{bulk} = 0.18\ (\pm0.07)*\text{Latitude} + 11.6\ (\pm0.1)$, $\delta^{15}N_{glycine} = 0.26\ (\pm0.35)*\text{Latitude} + 1.4\ (\pm0.7)$, $\delta^{15}N_{phenylalanine} = 0.15\ (\pm0.14)*\text{Latitude} + 3.1\ (\pm0.3)$, $\delta^{15}N_{alanine} = 0.23\ (\pm0.49)*\text{Latitude} + 23.4\ (\pm0.9)$, $\delta^{15}N_{aspartic} = 0.11\ (\pm0.17)*\text{Latitude} + 29.2\ (\pm0.3)$, $\delta^{15}N_{glutamic} = 0.14\ (\pm0.34)*\text{Latitude} + 25.9\ (\pm0.6)$. The slopes of all regressions are similar and not different at the 95% confidence interval.

showing consistently large increases in $\delta^{15}N$ values provided a more robust estimate of the trophic level of a consumer than bulk tissue. In particular, glutamic acid was enriched in ^{15}N by ~7‰ in the consumer relative to the food (McClelland and Montoya 2002). The trophic level estimated from the weighted mean difference between the $\delta^{15}N$ values of glutamic acid and glycine, assuming an amino acid TEF of 7‰, is 4.5 ± 0.1 (1 SD). Weighting was based on the analytical uncertainty in the $\delta^{15}N$ values of the amino acids (Table 12.2). In contrast, the trophic level calculated from the weighted mean difference between the $\delta^{15}N$ values of glutamic acid and phenylalanine is 4.2 ± 0.1 (1 SD). We prefer to use the $\delta^{15}N$ value of glycine, rather than phenyalanine for trophic level estimates because phenylalanine can be used in large amounts to form tyrosine if the latter is not adequately supplied in the diet (Schepartz 1973) and using our CSIA methods, the $\delta^{15}N$ value of glycine was easier to determine relative to phenylalanine because close elution between phenylalanine and glutamic acid made quantification of the $\delta^{15}N$ value of phenylalanine more difficult. For comparison, we also calculated trophic level based on the average difference between trophic (alanine + aspartic acid + glutamic acid) and source (glycine + phenylalanine) amino acids ($=\delta^{15}N_{\sum trophic} - \delta^{15}N_{\sum source}$). The trophic level calculated from the weighted mean difference between the $\delta^{15}N$ values of the trophic and source amino acids, assuming a TEF of 7‰, is 4.6 ± 0.1 (1 SD) (Table 12.2).

Bulk $\delta^{15}N$ values of mesozooplankton exhibited the same geographical trend as the bulk WMT and amino acid $\delta^{15}N$ values of yellowfin tuna in the ETP, increasing from ~10°S to ~25°N (Figure 12.2). These short-lived zooplankton at low trophic levels are more likely to track spatial and temporal changes in nutrient dynamics than long-lived, highly active tuna predators. The similar spatial pattern in tuna and mesozooplankton bulk $\delta^{15}N$ values provides further evidence that the ^{15}N enrichment in the north is due to changes in the $\delta^{15}N$ values at the base of the food web. Applying Eq. (12.1) to the $\delta^{15}N$ values of both taxa (Figure 12.2, filled symbols) in the fifteen 5 × 5 degree areas where both taxa were collected (Figure 12.1) yielded a yellowfin tuna trophic level estimate ranging from 4.1 (±0.3 SD) assuming a TEF value of 3.4‰, to 4.9 (±0.5 SD) assuming the field-based TEF value of 2.1‰. These estimates are in close agreement with the trophic-level estimates derived from the compound-specific stable isotope data. In addition, trophic level estimates for yellowfin tuna using diet data in a mass balance ecosystem model for the ETP were 4.6–4.7 (Olson and Watters 2003). Thus, results of CSIA of amino acids in yellowfin tuna appear to be a sensitive indicator of the trophic level of this upper-level pelagic predator, a conclusion in broad agreement with that of McClelland and Montoya (2002).

V. IMPLICATIONS

Our results set the stage for the application of compound-specific stable isotope techniques to support ecosystem-based approaches for the management of pelagic tuna fisheries. Fisheries that target specific components of the food web act as potential agents of ecological change, in some cases profoundly restructuring marine food webs (Estes *et al.* 1998, Jackson *et al.* 2001, Worm and Myers 2003). One way that fishing may alter exploited ecosystems is termed "fishing down the food web" (Pauly *et al.* 1998). The commonly held interpretation of fishing down the food web is a gradual reduction in the mean trophic level of fisheries landings caused by serial depletion of high-trophic-level species and replacement by lower-trophic-level species, although there is another more-tenable interpretation (Essington *et al.* 2006). High-seas purse seine and longline fisheries target tuna and billfish species that are dominant, high-level predators in pelagic ecosystems. A decrease in the biomass of top predators could impart a top–down trophic cascade, which could affect the overall structure and function of the ecosystem (Carpenter *et al.* 1985, Pace *et al.* 1999). Fisheries-induced restructuring of food webs has not been demonstrated empirically for high-seas pelagic ecosystems of the Pacific Ocean, although modeling studies of the central north Pacific (Cox *et al.* 2002) and the ETP (Olson and Watters 2003) have shown strong evidence for top-down effects of harvesting predators on the productivity of their prey (Walters *et al.* 2005). An analysis of all available data from Pacific tuna fisheries for 1950–2004 indicated substantial, though not catastrophic, impacts of fisheries on top-level predators and minor impacts on the ecosystem in the Pacific Ocean (Sibert *et al.* 2006). We propose that trophic level estimates derived from amino acid nitrogen isotopic analysis of archived samples of pelagic fishes will provide valuable insight into the historic effects of fishing on pelagic marine ecosystems.

To examine the potential historical effects of commercial fishing on pelagic fisheries, using stable isotope data, both the trophic level and isotope baseline of a single archived sample must be estimated. Previous workers (Thompson *et al.* 1995, Bearhop *et al.* 2001, Jennings *et al.* 2002, Becker and Beissinger 2006) have used bulk isotopic compositions to examine short- and long-term effects of commercial fishing on the trophic level of marine predators. These studies required extensive characterization of isotopic variability at the base of marine food webs, which as we noted above can be notoriously difficult. The principle advantage of $\delta^{15}N$ analyses of individual amino acids is that using only a sample of the consumer, the trophic level of the consumer and the $\delta^{15}N$ value at the base of the food web can be estimated. That is, additional sampling of basal food resources and prey is not required, since predator trophic levels and the basal isotope values can be obtained simply by comparing $\delta^{15}N$ values of trophic and source amino acids extracted from the tissues of the predator. Time series of trophic-level and baseline isotope estimates for a given predator, using CSIA studies, could provide insight into the relative importance of fisheries and physical forcing in structuring marine ecosystems.

VI. SUMMARY AND FUTURE RESEARCH

Bulk $\delta^{15}N$ values of yellowfin tuna WMT increased by ~5‰ from 10°S to 25°N, and followed spatial trends documented for hundreds of other $\delta^{15}N$ analyses of tuna and mesozooplankton. We observed parallel latitudinal trends in the $\delta^{15}N$ values of bulk WMT, source and trophic amino acids, indicating that the ^{15}N enrichment in the north was due to changes in the $\delta^{15}N$ values at the base of the food web. The increase in $\delta^{15}N$ values to the north was consistent with the effects of denitrification on the $\delta^{15}N$ value of nitrate and the transfer of this isotopic composition to phytoplankton.

We used the difference between the $\delta^{15}N$ values of trophic (alanine, aspartic acid, and glutamic acid) and source (glycine and phenylalanine) amino acids to estimate the trophic level of ETP yellowfin tuna. Assuming the difference between the $\delta^{15}N$ values of source and trophic amino acids changed by 7‰ per trophic level (McClelland and Montoya 2002), we estimated that the trophic level of ETP tuna ranges from 4.2 to 4.6. This amino acid-derived estimate matched well the estimate of 4.6–4.7 derived from diet analysis (Olson and Watters 2003) and an estimate of 4.1–4.9 calculated from a model based on the difference between the $\delta^{15}N$ values of bulk mesozooplankton and yellowfin tuna in the ETP. The implication of our results is that $\delta^{15}N$ analyses of individual amino acids in tuna can be used to estimate the $\delta^{15}N$ values at the base of the food web and the trophic level from a single sample. Other results of compound-specific nitrogen isotopic analysis of amino acids in marine food webs suggest that this generalization holds for marine organisms at a variety of trophic levels (McClelland and Montoya 2002, McClelland *et al.* 2003, Schmidt *et al.* 2004). Compound-specific nitrogen isotopic analysis of amino acids can offer a unique opportunity to elucidate the nitrogen dynamics in a variety of food webs if these generalizations apply to other ecosystems.

We propose that differences between the $\delta^{15}N$ values of source and trophic amino acids can be used to examine possible historical changes in the trophic level of archived samples of fishes to investigate potential effects of fisheries removal on the trophic dynamics of pelagic ecosystems. However, before CSIA of amino acids is broadly applied to ecological studies, we must first determine tissue-specific turnover rates of amino acids and test several critical assumptions. The primary assumption that must be evaluated is the constancy and the mechanisms underlying the 7‰ per trophic level difference between the $\delta^{15}N$ values of the source and trophic amino acids. The agreement between trophic level estimates based on diet analysis (4.6–4.7) and those from differences between the $\delta^{15}N$ values of glutamic acid − glycine (4.5 ± 0.1) and Σtrophic − Σsource (4.6 ± 0.1) amino acids provides some level of confidence in the TEF of 7‰; however, this TEF value must be further evaluated using laboratory and additional field studies. Second, we need to better investigate the origins and metabolic cycling of amino nitrogen in the amino acids in organisms, and if these origins and metabolic cycling change at the ecosystem level, for example, as a response to nitrogen availability. Third, it is known that EAA are produced by plants and bacteria. However, little is known about controls on the $\delta^{15}N$ values of bacterially produced amino acids (Veuger *et al.* 2005). We must evaluate amino acid production and cycling in the microbial-loop (*sensu* Azam *et al.* 1983) on the $\delta^{15}N$ values at the base of the food web and of marine organisms that feed especially in mesopelagic environments.

To examine historical changes in trophic level, we must also consider the effects of preservation on archived specimens. Archived samples of fish are typically preserved with alcohol or formalin. Preservation of fish tissues with alcohol and formalin can have a small, uniform affect on bulk nitrogen stable isotope values (Arrington and Winemiller 2002). Bulk nitrogen isotope values generally increase by less than 1‰ in animal tissues treated with preservatives, which is small relative to the observed shifts in $\delta^{15}N$ values associated with trophic dynamics (Hobson *et al.* 1997, Bosley and Wainright 1999). Preliminary results on CSIA of amino acids in subtropical Pacific zooplankton (Hannides and Popp, unpublished data) indicated little change in the $\delta^{15}N$ values of amino acids in samples preserved in formalin as compared with similar frozen samples. We therefore suspect that affects of preservation of animal tissues in alcohol or formalin to be small for individual amino acids, but this supposition must be thoroughly tested. Consequently, several assumptions should be critically evaluated before the $\delta^{15}N$ values of individual amino acids in organisms can be used to examine animal physiology, foraging behavior, movement patterns, and trophic-level estimates in archived specimens to examine natural and anthropogenic changes to ecosystem structure and function. If these assumptions can be evaluated, the future application of CSIA of amino acids may have great potential for many biological fields, from animal physiology to conservation biology.

VII. ACKNOWLEDGMENTS

We thank H. Kreuzer-Martin, N. C. Popp, J. Tanimoto, and T. Rust for assistance and advice in the laboratory, J. Ehleringer for access to equipment and supplies at the University of Utah, where we began our CSIA research on amino acids in tuna, and J. Sibert for thoughtful comments and encouragement throughout our research on tuna. We also thank S. Hernandez-Trujillo, CICIMAR Project CGPI:20060472, for his support. We are grateful to NOAA Fisheries, Southwest Fisheries Science Center, United States, especially L. Ballance, V. Andreassi, K. Kopitsky, M. Kelley, C. Hall, R. Dotson, D. Griffiths, and N. Bowlin, for kindly collecting zooplankton samples on the STAR2003 cruises. Samples of tuna were collected by a team of observers in Ecuador and Mexico, with the valuable assistance of E. Largacha, H. Pérez, K. Loor, V. Fuentes, C. de la A, A. Basante, W. Paladines, F. Cruz, C. Maldonado, and the captains and crew of several purse seine vessels. The chapter was improved by reviews of W. Bayliff and two anonymous reviewers. This research was funded by Cooperative Agreement NA17RJ1230 between the Joint Institute for Marine and Atmospheric Research (JIMAR) and the National Oceanic and Atmospheric Administration (NOAA). The views expressed herein are those of the authors and do not necessarily reflect the views of NOAA of any of its subdivisions. BSG was supported by a Pelagic Fisheries Research Program graduate assistantship. CCSH was supported by an EPA STAR Graduate Fellowship. FGM and GALI were supported by CONACyT and COFAA-PIFI-IPN. BF was supported in part by NOAA Coastal Ocean grant NA16OP2670. This is SOEST contribution number 7119.

VIII. REFERENCES

Altabet, M. A. 2001. Nitrogen isotopic evidence for micronutrient control of fractional NO_3^- utilization in the equatorial Pacific. *Limnology and Oceanography* **46:**368–380.

Arrington, D. A., and K. O. Winemiller. 2002. Preservation effects on stable isotope analysis of fish muscle. *Transactions of the American Fisheries Society* **131:**337–342.

Azam, F., T. Fenchel, J. G. Field, J. S. Gray, and L. A. T. F. Meyer-Reil. 1983. The ecological role of water-column microbes in the sea. *Marine Ecology Progress Series* **10:**257–263.

Bearhop, S., D. R. Thompson, R. A. Phillips, S. Waldron, K. C. Hamer, C. M. Gray, S. C. Votier, B. P. Ross, and R. W. Furness. 2001. Annual variation in great skua diets: The importance of commercial fisheries and predation on seabirds revealed by combining dietary analyses. *The Condor* **103:**802–809.

Becker, B. H., and S. R. Beissinger. 2006. Centennial decline in the trophic level of an endangered seabird after fisheries decline. *Conservation Biology* **20:**470–479.

Bosley, K. L., and S. C. Wainright. 1999. Effects of preservatives and acidification on the stable isotope ratios (^{15}N:^{14}N, ^{13}C:^{12}C) of two species of marine animals. *Canadian Journal of Fisheries and Aquatic Sciences* **56:**2181–2185.

Botsford, L. W., J. C. Castilla, and C. H. Peterson. 1997. The management of fisheries and marine ecosystems. *Science* **277:**509–515.

Brandhorst, W. 1959. Nitrification and denitrification in the eastern tropical North Pacific. *Journal du Conseil International pour l'Exploration de la Mer* **25:**3–20.

Carpenter, S. R., J. F. Kitchell, and J. R. Hodgson. 1985. Cascading trophic interactions and lake productivity. *BioScience* **35:**634–639.

Chai, F., R. C. Dugdale, T.-H. Peng, F. P. Wilkerson, and R. T. Barber. 2002. One-dimensional ecosystem model of the equatorial Pacific upwelling system. Part I: Model development and silicon and nitrogen cycle. *Deep-Sea Research II* **49:**2713–2745.

Chelton, D. B., S. K. Esbensen, M. G. Schlax, N. Thum, M. H. Freilich, F. J. Wentz, C. L. Gentemann, M. J. McPhaden, and P. S. Schopf. 2001. Observations of coupling between surface wind stress and sea surface temperature in the eastern tropical Pacific. *Journal of Climate* **14:**1479–1498.

Cline, J. D., and F. A. Richards. 1972. Oxygen deficient conditions and nitrate reduction in the eastern tropical North Pacific Ocean. *Limnology and Oceanography* **17:**885–900.

Cline, J. D., and I. R. Kaplan. 1975. Isotopic fractionation of dissolved nitrate during denitrification in the eastern tropical North Pacific Ocean. *Marine Chemistry* **3:**271–299.

Codispoti, L. A., and F. A. Richards. 1976. An analysis of the horizontal regime of denitrification in the eastern tropical North Pacific. *Limnology and Oceanography* **21:**379–388.

Codispoti, L. A., and J. C. Christensen. 1985. Nitrification, denitrification and nitrous oxide cycling in the eastern tropical South Pacific. *Marine Chemistry* **16:**277–300.

Cohen, Y., and L. I. Gordon. 1978. Nitrous oxide production in the oxygen minimum of the eastern tropical North Pacific: Evidence for its consumption during denitrification and possible mechanisms for its production. *Deep Sea Research* **25:**509–524.

Cowie, G. L., and J. I. Hedges. 1992. Sources and reactivities of amino acids in a coastal marine environment. *Limnology and Oceanography* **37:**703–724.

Cox, S. P., T. E. Essington, J. F. Kitchell, S. J. D. Martell, C. J. Walters, C. Boggs, and I. Kaplan. 2002. Reconstructing ecosystem dynamics in the central Pacific Ocean, 1952–1998. II. A preliminary assessment of the trophic impacts of fishing and effects on tuna dynamics. *Canadian Journal of Fisheries and Aquatic Sciences* **59:**1736–1747.

Deniro, M. J., and S. Epstein. 1981. Influence of diet on the distribution of nitrogen isotopes in animals. *Geochimica et Cosmochimica Acta* **45:**341–351.

Dore, J. E., J. R. Brum, L. M. Tupas, and D. M. Karl. 2002. Seasonal and interannual variability in sources of nitrogen supporting export in the oligotrophic subtropical North Pacific Ocean. *Limnology and Oceanography* **47:**1595–1607.

Dugdale, R. C., and J. J. Goering. 1967. Uptake of new and regenerated forms of nitrogen in primary production. *Limnology and Oceanography* **12:**196–206.

Dugdale, R. C., and F. P. Wilkerson. 1991. Low specific nitrate uptake rate: A common feature of high-nutrient, low-chlorophyll marine ecosystems. *Limnology and Oceanography* **36:**1678–1688.

Essington, T. E., A. H. Beaudreau, and J. Wiedenmann. 2006. Fishing through marine food webs. *Proceedings of the National Academy of Sciences of the United States of America* **103:**3171–3175.

Estes, J. A., M. T. Tinker, T. M. Williams, and D. F. Doak. 1998. Killer whale predation on sea otters linking oceanic and nearshore ecosystems. *Science* **282:**473–476.

Fantle, M. S., A. I. Dittel, S. M. Schwalm, C. E. Epifanio, and M. L. Fogel. 1999. A food web analysis of the juvenile blue crab, *Callinectes sapidus*, using stable isotopes in whole animals and individual amino acids. *Oecologia* **120:**416–426.

Fiedler, P. C., V. Philbrick, and F. P. Chavez. 1991. Oceanic upwelling and productivity in the eastern tropical Pacific. *Limnology and Oceanography* **36:**1834–1850.

Gannes, L. Z., D. M. O'Brian, and C. Martinez del Rio. 1997. Stable isotopes in animal ecology: Assumptions, caveats, and a call for more laboratory studies. *Ecology* **78:**1271–1276.

Gislason, H., M. Sinclair, K. Sainsbury, and R. O'Boyle. 2000. Symposium overview: Incorporating ecosystem objectives within fisheries management. *ICES Journal of Marine Science* **57:**468–475.

Graham, B. S., D. Grubbs, K. Holland, and B. N. Popp. 2007. A rapid ontogenetic shift in the diet of juvenile yellowfin tuna from Hawaii. *Marine Biology* **150:**647–658, doi: 10.1007/s00227-006-0360-y.

Gunasekera, R. M., K. F. Shim, and T. J. Lam. 1997. Influence of dietary protein content on the distribution of amino acids in oocytes, serum and muscle of Nile tilapia, *Oreochromis niloticus* (L). *Aquaculture* **152:**205–221.

Hampton, J., J. R. Sibert, P. Kleiber, M. N. Maunder, and S. H. Harley. 2005. Decline of Pacific tuna populations exaggerated? *Nature* **434:**E1–E2.

Hare, P. E., and M. L. F. Estep. 1983. Carbon and nitrogen isotopic composition of amino acids in modern and fossil collagens. *Carnegie Institution of Washington Year Book* **82:**410–414.

Hare, P. E., M. L. Fogel, T. W. Stafford, A. D. Mitchel, and T. C. Hoering. 1991. The isotopic composition of carbon and nitrogen in individual amino acids isolated from modern and fossil proteins. *Journal of Archaeological Science* **18:**277–292.

Hobson, K. A., H. L. Gibbs, and M. L. Gloutney. 1997. Preservation of blood and tissue samples for stable-carbon and stable-nitrogen isotope analysis. *Canadian Journal of Zoology* **75:**1720–1723.

IATTC. 2004. Annual report of the Inter-American Tropical Tuna Commission, 2003. Inter-American Tropical Tuna Commission, 98 p.

Jackson, J. B. C., M. X. Kirby, W. H. Berger, K. A. Bjorndal, L. W. Botsford, B. J. Bourque, R. H. Bradbury, R. Cooke, J. Erlandson, J. A. Estes, T. P. Hughes, S. Kidwell, *et al.* 2001. Historical overfishing and the recent collapse of coastal ecosystems. *Science* **293:**629–638.

Jennings, S., S. P. R. Greenstreet, L. Hill, G. J. Piet, J. K. Pinnegar, and K. J. Warr. 2002. Long-term trends in the trophic structure of the North Sea fish community: Evidence from stable-isotope analysis, size-spectra and community metrics. *Marine Biology* **141:**1085–1097.

Jim, S., V. Jones, S. H. Ambrose, and R. P. Evershed. 2006. Quantifying dietary macronutrient sources of carbon for bone collagen biosynthesis using natural abundance stable carbon isotope analysis. *British Journal of Nutrition* **95:**1055–1062.

Lajtha, K., and R. H. Michener. 1994. *Stable Isotopes in Ecology and Environmental Science.* Blackwell Scientific Publications, Oxford.

Latour, R. J., M. J. Brush, and C. F. Bonzek. 2003. Toward ecosystem-based fisheries management: Strategies for multispecies modeling and associated data requirements. *Fisheries* **28:**10–22.

Liu, K.-K., and I. R. Kaplan. 1989. The eastern tropical Pacific as a source of ^{15}N-enriched nitrate in seawater off southern California. *Limnology and Oceanography* **34:**820–830.

Lourey, M. J., T. W. Trull, and D. M. Sigman. 2003. Sensitivity of δ^{15}N of nitrate, surface suspended and deep sinking particulate nitrogen to seasonal nitrate depletion in the Southern Ocean. *Global Biogeochemical Cycles* **17:**1081, doi:10.1029/2002GB001973.

Macko, S. A., M. L. Fogel, P. E. Hare, and T. C. Hoering. 1987. Isotopic fractionation of nitrogen and carbon in the synthesis of amino acids by microorganisms. *Chemical Geology* **65:**79–92.

Macko, S. A., M. E. Uhle, M. H. Engel, and V. Andrusevich. 1997. Stable nitrogen isotope analysis of amino acid enantiomers by gas chromatography/combustion/isotope ratio mass spectrometry. *Analytical Chemistry* **69:**926–929.

McClelland, J. W., and J. P. Montoya. 2002. Trophic relationships and the nitrogen isotopic composition of amino acids in phytoplankton. *Ecology* **83:**2173–2180.

McClelland, J. W., C. M. Holl, and J. P. Montoya. 2003. Relating low $\delta^{15}N$ values of zooplankton to N_2 fixation in the tropical North Atlantic: Insights provided by stable isotope ratios of amino acids. *Deep-Sea Research I* **50:**849–861.

McCutchan, J. H., Jr., W. M. Lewis, C. Kendall, and C. C. McGrath. 2003. Variation in trophic shift for stable isotope ratios of carbon, nitrogen and sulfur. *OIKOS* **102:**378–390.

Metges, C. C., K. Petzke, and U. Hennig. 1996. Gas chromatography/combustion/isotope ratio mass spectrometric comparison of N-acetyl- and N-pivaloyl amino acid esters to measure ^{15}N isotopic abundances in physiological samples: A pilot study on amino acid synthesis in the upper gastrointestinal tract of minipigs. *Journal of Mass Spectrometry* **31:**367–376.

Minagawa, M., and E. Wada. 1984. Stepwise enrichment of ^{15}N along food chains: Further evidence and the relation between ^{15}N and animal age. *Geochimica et Cosmochimica Acta* **48:**1135–1140.

Murawski, S. A. 2000. Definitions of overfishing from an ecosystem perspective. *ICES Journal of Marine Science* **57:**649–658.

Myers, R. A., and B. Worm. 2003. Rapid worldwide depletion of predatory fish communities. *Nature* **423:**280–283.

NRC. 1994. *Nutrient Requirements of Poultry.* 9th edn. National Academic Press, Washington, DC.

NRC. 1995. *Nutrient Requirements of Laboratory Animals.* 4th edn. National Academic Press, Washington, DC.

NRC. 1999. *Sustaining Marine Fisheries.* National Academy Press, Washington, DC.

Olson, R. J., and G. M. Watters. 2003. A model of the pelagic ecosystem in the eastern tropical Pacific Ocean. *Inter-American Tropical Tuna Commission, Bulletin* **22:**135–218.

Owens, N. J. P. 1987. Natural variations in $\delta^{15}N$ in the marine environment. *Advances in Marine Biology* **24:**389–451.

Pace, M. L., J. J. Cole, S. R. Carpenter, and J. F. Kitchell. 1999. Trophic cascades revealed in diverse ecosystems. *Trends in Ecology and Evolution* **14:**483–488.

Pauly, D., V. Christensen, J. Dalsgaard, R. Froese, and F. Torres, Jr. 1998. Fishing down marine food webs. *Science* **279:**860–863.

Pauly, D., M. Lourdes Palomares, R. Froese, P. Saa, M. Vakily, D. Preikshot, and S. Wallace. 2001. Fishing down Canadian aquatic food webs. *Canadian Journal of Fisheries and Aquatic Sciences* **58:**51–62.

Peterson, B. J., and B. Fry. 1987. Stable isotopes in ecosystem studies. *Annual Review of Ecology and Systematics* **18:**293–320.

Philander, S. G. H., W. J. Hurlin, and A. D. Seigel. 1987. Simulation of the seasonal cycle of the tropical Pacific Ocean. *Journal of Physical Oceanography* **17:**1986–2002.

Pikitch, E. K., C. Santora, E. A. Babcock, A. Bakun, R. Bonfil, D. O. Conover, P. Dayton, P. Doukakis, D. Fluharty, B. Heneman, E. D. Houde, J. Link, *et al.* 2004. Ecosystem-based fishery management. *Science* **305:**346–347.

Post, D. M. 2002. Using stable isotopes to estimate trophic position: Models, methods and assumptions. *Ecology* **83:**703–718.

Reeds, P. J. 2000. Criteria and significance of dietary protein sources in humans. *Journal of Nutrition* **130:**1835S–1840S.

Rice, J. C. 2000. Evaluating fishery impacts using metrics of community structure. *ICES Journal of Marine Science* **57:**682–688.

Rolff, C. 2000. Seasonal variation in $\delta^{13}C$ and $\delta^{15}N$ of size-fractionated plankton at a coastal station in the northern Baltic proper. *Marine Ecology Progress Series* **203:**47–65.

Schepartz, B. 1973. *Regulation of Amino Acid Metabolism in Mammals* W.B. Saunders Company, London, 205 p.

Schmidt, K., J. W. McClelland, E. Mente, J. P. Montoya, A. Atkinson, and M. Voss. 2004. Trophic-level interpretation based on $\delta^{15}N$ values: Implications of tissue-specific fractionation and amino acid composition. *Marine Ecology Progress Series* **266:**43–58.

Sibert, J., J. Hampton, P. Kleiber, and M. Maunder. 2006. Biomass, size, and trophic status of top predators in the Pacific Ocean. *Science* **314:**1773–1776.

Sigman, D. M., J. Granger, P. J. DiFiore, M. M. Lehmann, R. Ho, G. Cane, and A. van Geen. 2005. Coupled nitrogen and oxygen isotope measurements of nitrate along the eastern North Pacific margin. *Global Biogeochemical Cycles* **19:**GB4022, doi:10.1029/2005GB002458.

Smith, P. E., and S. L. Richardson. 1977. Standard techniques for pelagic fish egg and larva surveys. *FAO Fisheries Technical Paper* **175:**1–100.

Smutna, M., L. Vorlova, and Z. Svobodova. 2002. Pathobiochemistry of ammonia in the internal environment of fish (review). *Acta Veterinaria Brno* **71:**169–181.

Stryer, L. 1988. *Biochemistry,* 3rd edn. W. H. Freeman and Company, New York.

Thomas, W. H. 1966. On denitrification in the northeastern tropical Pacific Ocean. *Deep-Sea Research* **13:**1109–1114.

Thompson, D. R., R. W. Furness, and S. A. Lewis. 1995. Diets and long-term changes in $\delta^{15}N$ and $\delta^{13}C$ values in Northern fulmars *Fulmarus glacialis* from two northwest Atlantic colonies. *Marine Ecology Progress Series* **125:**3–11.

Ueda, K., S. L. Morgan, A. Fox, J. Gilbart, A. Sonesson, L. Larsson, and G. Odham. 1989. D-Alanine as a chemical marker for the determination of streptococcal cell wall levels in mammalian tissues by gas chromatography/negative ion chemical ionisation mass spectrometry. *Analytical Chemistry* **61:**265–270.

Uhle, M. E., S. A. Macko, H. J. Spero, M. H. Engel, and D. W. Lea. 1997. Sources of carbon and nitrogen in modern planktonic foraminifera: The role of algal symbionts as determined by bulk and compound specific stable isotopic analyses. *Organic Geochemistry* **27**:103–113.

Vanderklift, M. A., and S. Ponsard. 2003. Sources of variation in consumer-diet $\delta^{15}N$ enrichment: A meta-analysis. *Oecologia* **136**:169–182.

Vander Zanden, M. J., and J. B. Rasmussen. 2001. Variation in $\delta^{15}N$ and $\delta^{13}C$ trophic fractionation: Implications for aquatic food web studies. *Limnology and Oceanography* **46**:2061–2066.

Veuger, B., J. J. Middleburgh, T. S. Boschker, and M. Houtekamer. 2005. Analysis of ^{15}N incorporation into D-alanine: A new method for tracing nitrogen uptake by bacteria. *Limnology and Oceanography: Methods* **3**:230–240.

Voss, M., J. W. Dippner, and J. P. Montoya. 2001. Nitrogen isotope patterns in the oxygen-deficient waters of the Eastern Tropical North Pacific Ocean. *Deep-Sea Research I* **48**:1905–1921.

Walters, C. J., V. Christensen, S. J. Martell, and J. F. Kitchell. 2005. Possible ecosystem impacts of applying MSY policies from single-species assessment. *ICES Journal of Marine Science* **62**(3):558–568.

Wilson, R. P., and W. E. Poe. 1985. Relationship of whole body and egg essential amino acid patterns to amino acid requirement patterns in channel catfish, *Ictalurus punctatus*. *Comparative Biochemistry and Physiology* **80B**:385–388.

Worm, B., and R. A. Myers. 2003. Meta-analysis of cod-shrimp interactions reveals top-down control in oceanic food webs. *Ecology* **84**:162–173.

Wyrtki, K. 1966. Oceanography of the eastern equatorial Pacific Ocean. *Oceanography and Marine Biology Annual Reviews* **4**:33–68.

Wyrtki, K. 1981. An estimate of equatorial upwelling in the Pacific. *Journal of Physical Oceanography* **11**:1205–1214.

Young, V. R., and E. El-Khoury. 1995. The notion of the nutritional essentiality of amino acids, revisited, with a note on the indispensable amino acid requirements in adults. Pages 191–232 *in* L. A. Cynober (Ed.) *Amino Acid Metabolism and Therapy in Health and Nutritional Disease*. CRC Press, Boca Raton, FL.

Section 4

Isotope Composition of Trace Gasses, Sediments and Biomarkers as Indicators of Change

CHAPTER 13

Temporal Dynamics in $\delta^{13}C$ of Ecosystem Respiration in Response to Environmental Changes

Christiane Werner,* Stephan Unger,* João S. Pereira,† Jaleh Ghashghaie,‡ and Cristina Máguas§

*Experimental and Systems Ecology, University of Bielefeld
†Instituto Superior de Agronomia, Universidade Técnica de Lisboa
‡Laboratoire d'Écologie, Systématique et Evolution, CNRS-UMR 8079, IFR 87, Bâtiment 362 Université de Paris XI
§Centro de Ecologia e Biologia Vegetal, Faculdade de Ciências, Universidade Lisboa

Contents

I. INTRODUCTION

Global atmospheric carbon dioxide concentration, which is a primary greenhouse gas, has been rapidly increasing as a result of fossil fuel combustion, land use change, and biomass burning (Keeling *et al.* 1995). The rising CO_2 concentration has now been identified as a primary cause for climate change (IPCC 2007). Facing the current threats to the world's climate, a better understanding of the terrestrial carbon cycle namely the processes of CO_2 exchange between the biosphere and the atmosphere is of utmost importance.

Stable isotopes are sensible tracers for the human impact on the atmosphere, as the $\delta^{13}CO_2$ released from fossil fuel sources is highly depleted in the heavy carbon isotope. Hence, the isotopic composition of carbon in the atmosphere and biosphere parallels current environmental changes. Model approaches based on these alterations of the isotopic composition of the atmosphere have indicated that the terrestrial ecosystems, particularly of the Northern Hemisphere, are a significant sink for the CO_2 released by human activities (Tans *et al.* 1990, Ciais *et al.* 1995, Valentini *et al.* 2000). However, this effect may be counterbalanced by increased ecosystem respiration in response to global warming (Canadell *et al.* 2000, Schulze *et al.* 2000), which constitutes a major component of ecosystem carbon balance (Valentini *et al.* 2000, Reichstein *et al.* 2002).

The changes in ecosystem carbon stocks and fluxes may be traced by their stable isotopic compositions, which provide an independent way to partition photosynthetic carbon fixation from respiratory carbon released (Yakir and Wang 1996, Bowling *et al.* 2001, Ogée *et al.* 2003). Photosynthesis enriches the air in and near terrestrial ecosystems in $^{13}CO_2$ (Farquhar *et al.* 1989), whereas respiration tends to dilute the air of the heavy isotope (Buchmann *et al.* 2002). The large difference between the isotope composition of respiratory and tropospheric CO_2 is now frequently used to estimate the isotopic composition of ecosystem respiration ($\delta^{13}C_R$, Flanagan and Ehleringer 1998, Yakir and Sternberg 2000; for definitions of terms see Table 13.1). However, whereas a solid foundation exists for our understanding of photosynthetic carbon isotope discrimination (Δ) at the leaf-scale (Farquhar *et al.* 1982, for a review see Brugnoli and Farquhar 2000), our theoretical and empirical understanding of isotope fractionation during dark respiration as well as temporal variability of $\delta^{13}C_R$ at ecosystem level is comparatively weak. Modeling approaches used for ecosystem studies only consider leaf photosynthetic discrimination and ignore any discrimination which could occur during dark respiration. The finding of Lin and Ehleringer (1997) on isolated protoplasts reinforced the assumption that fractionation during dark respiration is negligible.

Recent works, however, support the view that "apparent" carbon isotope discrimination occurs during dark respiration in leaves of many C_3 species. Respired CO_2 is ^{13}C-enriched compared to leaf major metabolites (Duranceau *et al.* 1999, Ghashghaie *et al.* 2001, Tcherkez *et al.* 2003, Xu *et al.* 2004, Hymus *et al.* 2005, Klumpp *et al.* 2005, Mortazavi *et al.* 2006, Prater *et al.* 2006). This response is highly variable, changing with species and environmental factors, for example drought and temperature (for a review, see Ghashghaie *et al.* 2003). Further, leaf respiration is not the only component of ecosystem respiration. The contribution of other plant parts or ecosystem compartments (*e.g.*, trunk and heterotrophic and autotrophic soil respiration) can be substantial (Damesin and Lelarge 2003, Badeck *et al.* 2005, Damesin *et al.* 2005, Klumpp *et al.* 2005). Any temporal changes in the relative contribution of respiration of each organ should induce changes in ecosystem $\delta^{13}C_R$ (Bowling *et al.* 2003a,b). Therefore, under natural conditions, the $\delta^{13}C$ of leaf respired CO_2, and thus the ecosystem $\delta^{13}C_R$ can be expected to vary substantially if environmental conditions change. We will present evidence that these changes can be important even on very short-term scales of hours to days.

In this chapter, we will give an overview of the dynamics of ecosystem productivity (Section II) and $\delta^{13}C_R$ at annual, seasonal, and diurnal timescales (Section III) for a case study of a Mediterranean evergreen oak woodland. These ecosystems are vulnerable to climate change (Giorgi 2006) and are

TABLE 13.1 Definition of terms

Ecosystem respiration (R_{eco})—amount of CO_2 respired by the ecosystem, *i.e.*, the sum of autotrophic respiration (R_A) and heterotrophic respiration (R_H)

Eddy-covariance—statistical tool to analyze high frequency wind and scalar atmospheric data series, yielding values of fluxes (*e.g.*, of CO_2) at the ecosystem level

Gross primary productivity (GPP)—gross rate of photosynthetic CO_2 uptake of the plant canopy

Net ecosystem exchange (NEE)—net rate of carbon exchange between the atmosphere and the ecosystem: NEE = GPP − R_{eco}. In the micro-meteorological notation, negative values indicate the ecosystem as a carbon sink (*i.e.*, CO_2 net uptake by the ecosystem) and positive values indicate the ecosystem as a carbon source (*i.e.*, CO_2 release by the ecosystem)

Net primary productivity (NPP)—net rate of biomass (or energy) accumulation of the ecosystem. It is equal to GPP excluding plant respiration (autotrophic respiration, R_A), (NPP = GPP − R_A)

$\delta^{13}C_R$—integrated signal of carbon isotope composition of ecosystem respired CO_2, estimated in general by Keeling plot analysis (linear regression between $\delta^{13}C$ and inverse of CO_2 concentrations, *i.e.* $1/[CO_2]$)

$\delta^{13}C_{res}$—carbon isotope composition of respired CO_2 by different compartments (leaves, soil, trunks, and so on)

Time lag analysis—correlation analysis between $\delta^{13}C_R$ and environmental parameters of preceding days, using shifted time series to determine the time lag with the highest correlation between both factors. It is generally interpreted as the delay between photosynthetic carbon fixation, its transport time from leaf to roots, and subsequent release by respiration (Ekblad and Högberg 2001, Bowling *et al.* 2002)

Dark respiration—mitochondrial respiration releasing CO_2 in darkness

Apparent fractionation during dark respiration—$\delta^{13}C$ difference between potential substrates (or plant organic matter) as source carbon for respiration and CO_2 produced in the dark

Respiratory quotient (RQ)—ratio of CO_2 produced to O_2 consumed by dark respiration. RQ depends on the oxygenation level of the substrates used for respiration: when carbohydrates are degraded the RQ = 1; when less oxygenated metabolites (*e.g.*, fatty acids) are degraded, the RQ is lower (around 0.7)

endangered by desertification processes. Furthermore, they are particularly useful to evaluate the temporal dynamics in $\delta^{13}C_R$ (at timescales of hours to days) due to the pronounced seasonal environmental changes and its link to net ecosystem productivity. Particularly, we will focus on the impact of the vegetation on ecosystem respiration. We will present a survey on the knowledge of the mechanisms of carbon isotope fractionation during dark respiration (Section IV) and discuss the implications of rapid dynamics in $\delta^{13}C_R$ at larger temporal and spatial scales (Section V).

II. THE DYNAMICS OF MEDITERRANEAN ECOSYSTEM PRODUCTIVITY AND RESPIRATION

The marked seasonality of the Mediterranean type climate, with hot and dry summers and mild, rainy winters bares good conditions for the study of dynamics in $\delta^{13}C$ of respired CO_2 to assess ecosystem functioning in response to environmental factors. Our experiments were conducted near *Herdade da Mitra* (38°32′26″N, 8°00′01″W, 220–250 m a.s.l.), near Évora (Portugal) henceforth called Mitra (David *et al.* 2004, Carreiras *et al.* 2006, Figure 13.1). The vegetation is of the savanna-type characterized by a sparse tree cover consisting of holm oak (*Quercus ilex* ssp. *rotundifolia*) and cork oak (*Quercus suber*), associated with an understorey of herbs and shrubs, mainly *Cistus salvifolius* L. The average stand density and tree crown cover were about 30 trees ha^{-1} and 21%, respectively (Carreiras *et al.* 2006).

The surface fluxes of CO_2, water vapor, and energy fluxes were continuously measured using an eddy covariance system (Aubinet *et al.* 2000) on a tower at a height of 28 m, that is, about 20 m above

FIGURE 13.1 Map of Europe indicating sites of the Carboeuroflux-Project and location of study site (Mitra, Portugal, Section II), adapted from http://www.bgc-jena.mpg.de/public/carboeur/sites/ind_sites.html

the canopy. The system consisted of a three-dimensional sonic anemometer (Solent Model R2, Gill Instruments Ltd., Lymington, United Kingdom) and a closed-path dual CO_2/H_2O infrared gas analyzer (IRGA, model LI-7000, LI-COR, Inc., Lincoln, NE, US). Real-time data were acquired at 20-Hz sampling rate using the software by Kolle (Eddysoft). The quality of all primary flux data was assured by a routine of equipment calibration and non-representative 30-min data were excluded using standard procedures (Papale *et al.* 2006). The gap-filling and flux-partitioning methods proposed by Reichstein *et al.* (2005) were used to fill data gaps and to separate the observed net primary productivity (NEE) into gross primary productivity (GPP) and ecosystem respiration (R_{eco}). All measurements, including Keeling plots, were done in the footprint area of the flux tower.

The annual R_{eco} varied with total rainfall (Figure 13.2A and B), as well as with the frequency and temporal distribution of rain events. The seasonal variation in GPP followed the usual pattern of the regional climate with a maximum in spring and minimum during summer. Although R_{eco} was high when GPP was high, it had a maximum impact on NEE at the end of summer or autumn, following the first rains when photosynthesis was still low but pulses of CO_2 efflux followed each rainfall event (Pereira *et al.* 2004, Jarvis *et al.* 2007). During the prolonged rainless season, soil biological activity declines. On rewetting there is a sudden "burst" in mineralization and CO_2 release—the Birch effect (Rey *et al.* 2005, Jarvis *et al.* 2007). The amount of carbon returned to the atmosphere in this way can reduce significantly the annual net carbon gain by the forests (Pereira *et al.* 2004, Jarvis *et al.* 2007).

The rain events at the end of summer in Mediterranean ecosystems may provide an excess of mineralization in the soil during the early phases of the wet cycle (Rey *et al.* 2005). At that time, there are no herbaceous plants utilizing the nutrients released, thus leading to the loss of carbon and nitrogen from the soil pools (Pereira *et al.* 2004, Jarvis *et al.* 2007). In such cases, summer rains may often have a negative effect on plant productivity (Snyder *et al.* 2004). The deep-rooted perennials, in particular, cannot use rainfall until the water reaches deeper soil horizons, and can be partly decoupled from soil

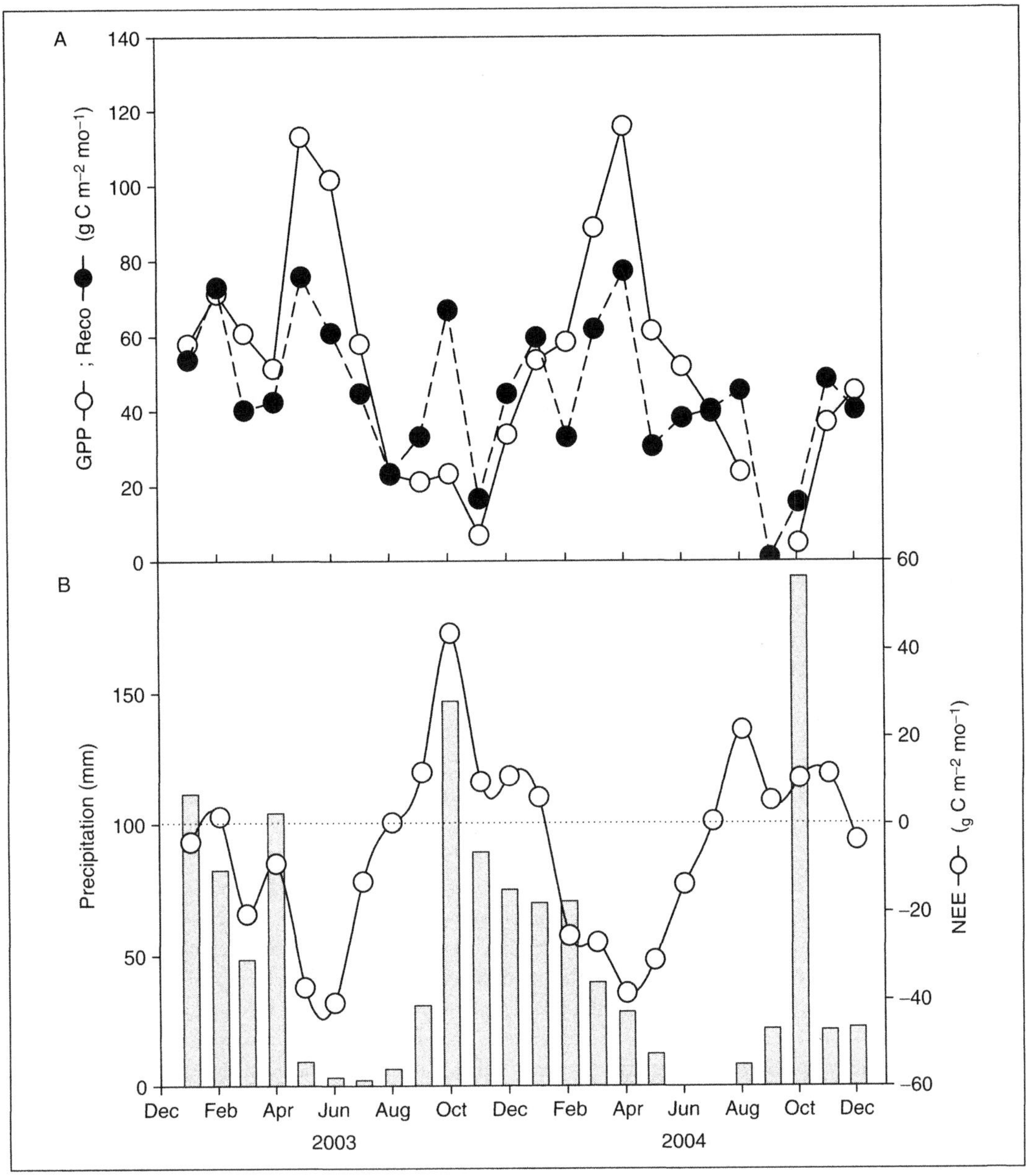

FIGURE 13.2 Annual dynamics of ecosystem respiration (R_{eco}, black circles), gross primary production (GPP, white circles) (A), rainfall (bars) and net ecosystem exchange (NEE, white circles) (B) for a Mediterranean oak forest (Mitra site, Portugal) for the years 2003 and 2004.

water content. Although the oak-savanna trees (*e.g.*, the oak woodland at Mitra) were shown to perform a substantial hydraulic lift (Kurz-Besson *et al.* 2006, Otieno *et al.* 2006) through their deep roots, allowing the redistribution of ground water to the shallower root system during the night in summer, this was not enough to change the seasonal pattern in metabolism, especially in R_{eco}.

Another consequence of drought is that ecosystem respiration only responds to temperature during the wet period of the year (Pereira *et al.* 2004, Rey *et al.* 2005, Jarvis *et al.* 2007), soil moisture being the main limiting factor. Moreover, the impact of high summer temperatures on soil respiration might be limited by the known inhibition of microbial respiration by low soil moisture (reviewed in Jarvis *et al.* 2007) and a decline in root respiration due to lack of recent photosynthates, tissue death, and seasonal acclimation to high temperatures (Loveys *et al.* 2003). For example, in 2003 the heat waves of the summer, which affected severely ecosystem metabolism in central Europe (Ciais *et al.* 2005), were not effective in changing ecosystem metabolism at Mitra because it was already down-regulated by drought (unpublished data).

The long rainless season will decrease GPP and have an important impact on ecosystem productivity, NPP. Moreover, one of the traits of future climate scenarios for western Mediterranean is the lengthening of the dry season (Miranda *et al.* 2002) and a trend toward increasing drought caused by warmer air temperatures and dryer winter–spring periods during the last decades of the twentieth century, namely in March (Miranda *et al.* 2002, Luterbacher and Xoplaki 2003).

III. VARIATIONS IN ECOSYSTEM RESPIRATION ($\delta^{13}C_R$) AT DIFFERENT TEMPORAL SCALES

The different components contributing to the large seasonal changes in ecosystem productivity and, particularly, in ecosystem respiration presented in Section II can be analyzed using stable isotopes. Hence, spatial and temporal variations in the $\delta^{13}C$ of ecosystem respired CO_2 ($\delta^{13}C_R$) are a critical aspect when studying terrestrial carbon fluxes in response to environmental change. While $\delta^{13}C_R$ has been considered to remain relatively constant on the short-term scale, this section will present evidence for pronounced variation at annual, seasonal and diurnal timescales in a case study for the Mediterranean semi-arid oak woodland at Mitra.

A. Analysis of $\delta^{13}C$ of Ecosystem Respired CO_2 Through Keeling Plot Approach

The isotopic composition of ecosystem respired CO_2 can be assessed by a Keeling plot approach, which is based on a two-source mixing model. An important assumption is that both source CO_2 and background CO_2 remain constant during the sampling period, which can be achieved by restricting the sampling period to 30 min (Bowling *et al.* 2001), for example, by sampling vertical profiles in CO_2 concentrations. However, this assumption is not met when nocturnal time series are used for Keeling plot sampling (Section V). The isotopic signature of ecosystem respiration ($\delta^{13}C_R$) can be calculated as the *y*-intercept of a linear regression of $\delta^{13}C$ versus the inverse of the CO_2 mixing ratio obtained from vertical profiles solving the following isotopic mass balance equation (Keeling 1958):

$$\delta^{13}C_a = c_T(\delta^{13}C_T - \delta^{13}C_R)\left(\frac{1}{c_a}\right) + \delta^{13}C_R$$

where $\delta^{13}C$ denotes the isotopic composition and c denotes CO_2 concentrations [CO_2] of the mixing ratios. The subscripts indicate sample air from several heights above and within the canopy (a), tropospheric air (T), and air respired from the ecosystem (R). Keeling plot intercepts are calculated from geometric mean (Model II) regressions and uncertainties can be expressed as standard errors of the intercept estimated from ordinary least square (OLS Model I) regressions. To remove outliers, residual analyses were performed. Data points were removed from the regression when the residual of an individual data point was higher than three times the standard deviation (as proposed by Pataki *et al.* 2003). Regressions were rejected when not significant ($\alpha = 0.01$). R^2 of obtained Keeling plot regressions ranged between 0.92 and 0.99.

To process large sample numbers, we used small volume (12 mL) soda glass vials (Exetainer, Labco, High Wycombe, United Kingdom) capped with pierceable septa for atmospheric air sampling.

Air was collected at nine different heights from 0.5 to 24 m at the Mitra site (Section II). The exetainers were flushed for 1 min, and CO_2 concentrations were measured with an IRGA at the outlet (BINOS 100 4P, Rosemount Analytical, Hanau, Germany). Samples were repeatedly collected from the top to the bottom, resulting in 18–25 samples (10 for diurnal Keeling plots). Sample collection was completed within 30 min. Samples were analyzed within 12–20 h after sample collection (LIE, ICAT, Lisbon) with a continuous-flow isotope ratio mass spectrometer (ISOPRIME, GV, Manchester, United Kingdom) connected to a Multiflow automatic sampler (GV, Manchester, United Kingdom).

B. Annual and Seasonal Variation in $\delta^{13}C_R$

Recent work showed that $\delta^{13}C_R$ can undergo significant seasonal variation (*e.g.*, up to 8‰, McDowell *et al.* 2004a, see also Lai *et al.* 2005, Ponton *et al.* 2006). However, variation does not only occur along the annual cycle, but pronounced changes in $\delta^{13}C_R$ can occur at much shorter timescales (Knohl *et al.* 2005, Mortazavi *et al.* 2005). This is clearly visible by comparing variations in $\delta^{13}C_R$ at timescales ranging from monthly, daily to hourly records of $\delta^{13}C_R$, as shown for the case study in a Mediterranean type ecosystem in Figure 13.3.

$\delta^{13}C_R$ exhibited a high annual variability of >7‰, which is similar or slightly lower than the maximum variation found in other ecosystems, for example, in forests with a marked seasonality in precipitation (Bowling *et al.* 2002, McDowell *et al.* 2004a). Much less variation in $\delta^{13}C_R$ was found in other ecosystems (Fessenden and Ehleringer 2003, Knohl *et al.* 2005, Mortazavi *et al.* 2005). During the Mediterranean dry season (Figures 13.2B and 13.3A), no clear enrichment in $\delta^{13}C_R$ was detectable. Such enrichment in $\delta^{13}C_R$ during drought has been found in some other ecosystems (Pataki *et al.* 2003, McDowell *et al.* 2004a, Hemming *et al.* 2005, Lai *et al.* 2005).

Periods of particularly high variability in $\delta^{13}C_R$ were observed during seasonal changes, and a high sampling frequency is required to capture these rapid dynamics (Figure 13.3B). In the Mediterranean climate regions, the transition from a productive period in spring into summer drought or the release from drought after the first rain falls can occur within a few weeks. During these rapid changes in environmental conditions, pronounced isotopic disequilibrium, for example, through marked changes in carbon isotope discrimination and/or soil respiration (Section III.D) induced large shifts in $\delta^{13}C_R$. A rapid increase in air temperature and vapor pressure deficit (VPD) in May resulted in a pronounced enrichment in $\delta^{13}C_R$ (Figure 13.3B). The inverse pattern was observed after the first rainfall at the end of summer in September, which markedly increased soil respiration (data not shown), subsequently followed by a decline in $\delta^{13}C_R$ (Figure 13.3B). The variation in $\delta^{13}C_R$ (<5‰) exceeded two-third of the total range observed at this site. This indicates that short-term changes in environmental conditions are driving factors stimulating changes in $\delta^{13}C_R$. These changes in ecosystem activity during pronounced and rapid environmental changes provide excellent case studies to explore the response of $\delta^{13}C_R$ to environmental factors, as shown below (Section III. D). However, $\delta^{13}C_R$ is difficult to assess during periods of low rates of ecosystem metabolism. Insufficient buildup of a CO_2 gradient (<30 ppm) at night did not meet the criteria required for reliable Keeling plot analysis in most sampling occasions during the summer drought, which is a major difficulty in all ecosystems with periods of low metabolic activity, for example, during drought.

C. Short-Term Dynamics in $\delta^{13}C_R$ During Diurnal Cycles

While pronounced seasonal or weekly variations in $\delta^{13}C_R$ have been reported elsewhere (McDowell *et al.* 2004a, Knohl *et al.* 2005), until recently, $\delta^{13}C_R$ was considered to remain constant during the night (Pataki *et al.* 2003). The high sampling frequency during 24-h cycles revealed large nocturnal

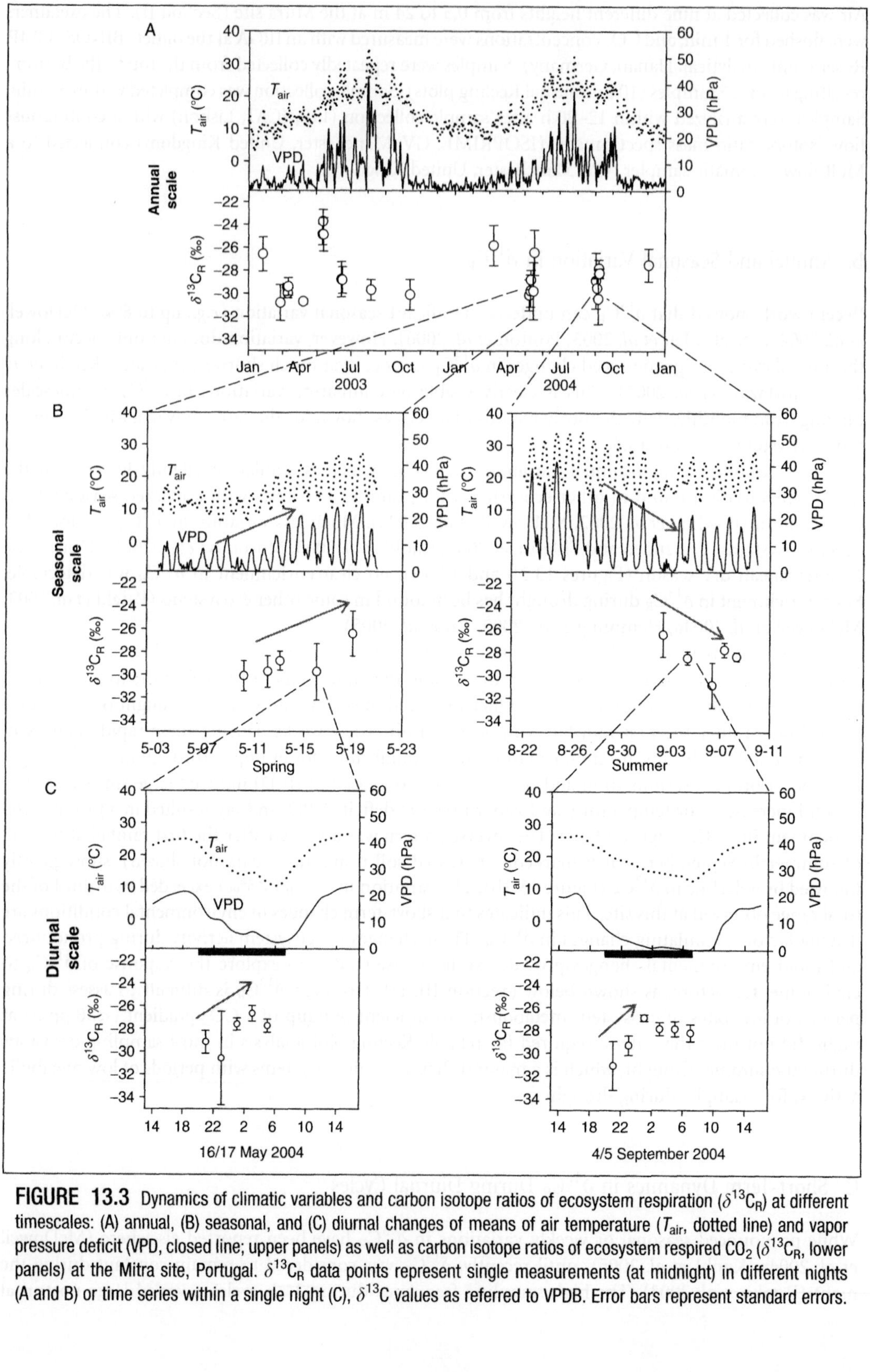

FIGURE 13.3 Dynamics of climatic variables and carbon isotope ratios of ecosystem respiration ($\delta^{13}C_R$) at different timescales: (A) annual, (B) seasonal, and (C) diurnal changes of means of air temperature (T_{air}, dotted line) and vapor pressure deficit (VPD, closed line; upper panels) as well as carbon isotope ratios of ecosystem respired CO_2 ($\delta^{13}C_R$, lower panels) at the Mitra site, Portugal. $\delta^{13}C_R$ data points represent single measurements (at midnight) in different nights (A and B) or time series within a single night (C), $\delta^{13}C$ values as referred to VPDB. Error bars represent standard errors.

shifts in $\delta^{13}C_R$, both in May and September (Figure 13.3C). During both periods a nocturnal shift of about 4‰ was observed. This nocturnal shift reached nearly the same magnitude as the total $\delta^{13}C_R$ variation during the seasonal changes in both periods (compare Figure 13.3B and C).

As far as we are aware, the only study in which substantial nocturnal changes in $\delta^{13}C_R$ were published beside of ours is that of Bowling *et al.* (2003b). Ogée *et al.* (2003) did not find significant changes in $\delta^{13}C_R$ during one nocturnal cycle, similarly to Schnyder *et al.* (2004), who found nocturnal variations of $\delta^{13}C_R$ <2‰, concluding that $\delta^{13}C_R$ remained relatively constant. Still *et al.* (2003) explained similar small nocturnal variation with changing wind and footprint conditions at the tower. Mortazavi *et al.* (2006) found $\delta^{13}C_R$ being 1‰ more enriched for sampling the Keeling plot during a nocturnal time series in comparison to a height profile. However, at most sites nocturnal changes in $\delta^{13}C_R$ are not assessed. Bowling *et al.* (2003b) reported even larger shifts (>6‰) within a single night. Contrary to our results they reported an increasing nocturnal depletion in $\delta^{13}C_R$.

There are several possible reasons for these nocturnal shifts in $\delta^{13}C_R$, which may occur through changes in respiratory substrate; if diurnal changes in photosynthetic discrimination will translate into nocturnal variation of $\delta^{13}C$ of respired CO_2; if the relative contribution of different respiratory fluxes (*e.g.*, foliage, soil, roots) changes; or through changes in fractionation in the respiratory pathways. Changes in respiratory substrate may occur if recent photosynthate is respired early in the evening and slowly the substrate changes to stored carbon as the night progresses. Additionally, changes in $\delta^{13}C$ of the assimilates could be reflected in root respiratory signals, mediated by the transport rate and amount of metabolites reaching the roots, even though Göttlicher *et al.* (2006) did not found marked variation in $\delta^{13}C$ of root starch during one diurnal course, in spite of significant variation in leaf starch at two of three heights in the canopy. Another aspect may arise from changes in relative proportions of the different respiratory fluxes, for example, leaf and soil respiration, to total ecosystem respiration (Bowling *et al.* 2003b), which could be triggered by different changes in air and soil temperature. Changes in soil temperature and moisture might also affect heterotrophic and autotrophic respiration differently (Boone *et al.* 1998, Bhupinderpal *et al.* 2003). Finally, there is increasing evidence that $\delta^{13}C$ of leaf dark respiration ($\delta^{13}C_{resp}$) can undergo quite dramatic diurnal changes of 5–10‰ under natural conditions (Hymus *et al.* 2005, Prater *et al.* 2006). We found similar diurnal enrichment in *Q. ilex* of 8‰ under laboratory conditions and rapid depletion upon darkness (Werner *et al.* 2007). This has been explained by apparent ^{13}C fractionation during leaf dark respiration (Ghashghaie *et al.* 2003, Tcherkez *et al.* 2003) which will be discussed in detail in Section IV.

In summary, for the above mentioned reasons nocturnal variations in $\delta^{13}C_R$ might be a common phenomenon, which should be examined in more ecosystems. The implications of short-term dynamics in $\delta^{13}C_R$ on larger temporal and spatial scales will be discussed in Section V.

D. Responsiveness of $\delta^{13}C_R$ to Environmental Changes

Variations in $\delta^{13}C_R$ have been linked to photosynthetic discrimination that depends on a variety of environmental factors, such as VPD, precipitation, irradiance, and temperature through their effect on the ratio of internal versus external CO_2 concentration (c_i/c_a). The linkage between assimilation and ecosystem respiration is not immediate, but occurs with a certain delay (time lag) of several days. This is generally interpreted as the time between carbon fixation, assimilate transport to roots, and subsequent release through root respiration and exudates (Ekblad and Högberg 2001, Bowling *et al.* 2002). Several studies have shown time-lagged relationships between environmental drivers of photosynthetic discrimination and isotopic composition of respired CO_2 from soils (Ekblad and Högberg 2001, McDowell *et al.* 2004a,b) or ecosystems (Bowling *et al.* 2002, Knohl *et al.* 2005, Mortazavi *et al.* 2005, Werner *et al.* 2006). Hence, these relationships provide information on the responsiveness of $\delta^{13}C_R$ to changes in environmental parameters. For the Mediterranean oak forest however, strong time-lagged

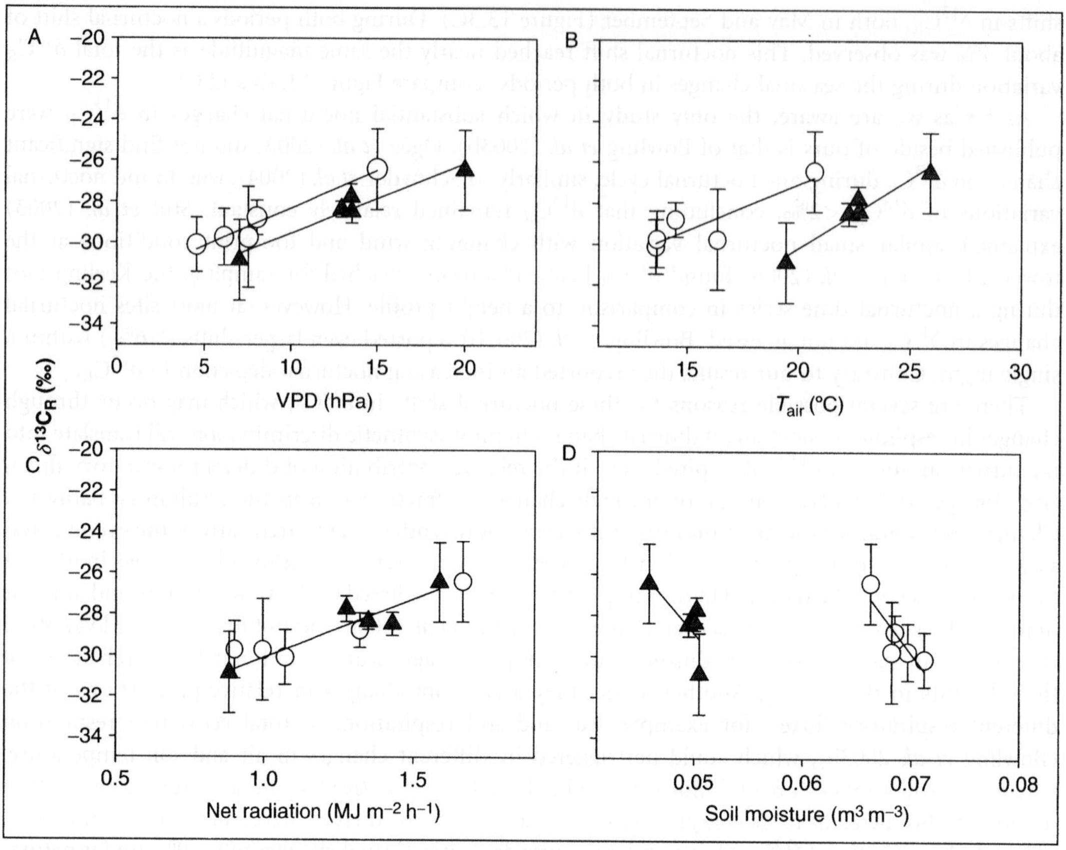

FIGURE 13.4 Relationships between carbon isotope ratios of ecosystem respired CO_2 ($\delta^{13}C_R$) and time-lagged daytime averages of climatic variables: (A) vapor pressure deficit (VPD), (B) air temperature (T_{air}), (C) net radiation, and (D) soil moisture during May (open circles) and September (shaded triangles) in 2004. Shifted time series (up to 10 days) with different averaging periods (0–3 days) were tested, and best correlations were found for time lags of 5 and 3 days for May and September, respectively. $\delta^{13}C$ values as referred to VPDB. Error bars represent standard errors. Redrawn from Werner *et al.* 2006.

relationships occurred only during the seasonal transition periods from spring to drought (May) and after the first rain events (September), whereas no correlation was found on an annual basis. The reasons can be interfered from Figure 13.4: $\delta^{13}C_R$ was highly correlated to atmospheric drivers of photosynthetic discrimination (VPD), air temperature (T_{air}), net radiation (R_n) (Figure 13.4A–C), inducing a 4‰ shift in both periods. However, the regressions were offset between May and September indicating a different responsiveness of $\delta^{13}C_R$ to changes in for example temperature in different seasons. This can be explained by the well documented seasonal temperature acclimation of maximum photosynthetic capacity during Mediterranean summer (Larcher 2000). It indicates that the influence of most environmental factors on $\delta^{13}C_R$ is not constant, but varies throughout the year, probably with changes, e.g., in carbon allocation, tissue metabolism, drought adaptations. Only the correlation of $\delta^{13}C_R$ with net radiation was equal for both seasons (Figure 13.4C). Therefore, in ecosystems where the vegetation expresses marked physiological acclimation to seasonal changes in environmental conditions a correlation of $\delta^{13}C_R$ for the whole annual cycle cannot be expected.

Furthermore, there are indications that heterotrophic soil respiration can markedly influence $\delta^{13}C_R$ (McDowell *et al.* 2004a), indicating belowground respiration as a second important component of $\delta^{13}C_R$ in these systems (Figure 13.4D, for a thorough discussion on different relationships with above and belowground components and seasonal shifts in time lags, see Werner *et al.* 2006). For example during September 2004, a rain event caused a sudden increase in soil respiration (Birch effect, see Section II). Time lags were 2–3 days shorter in September compared to May, reflecting the immediate response to the rainfall. Ecophysiological measurements indicated that changes in $\delta^{13}C_R$ were controlled by soil respiration rather than changes in photosynthetic discrimination, as trees responded slowly to the summer rainfall event (Section II, Pereira *et al.* 2004). Nevertheless, the time-lagged correlation between $\delta^{13}C_R$ with environmental factors was high in September. Hence, this indicates that these time-lagged relationships need to be interpreted with care, as all driving environmental parameters are autocorrelated: in this case the rain event was associated with a large drop in VPD, net radiation, and temperature.

Correlations of $\delta^{13}C_R$ with environmental drivers were also found on shorter timescales with nocturnal changes of $\delta^{13}C_R$ (data not shown). However, the time frame of the nocturnal enrichment in $\delta^{13}C_R$ is similar to the diurnal changes in environmental parameters, and therefore these correlations might not indicate a causal relationship.

The respiratory signal is influenced by large post-photosynthetic fractionation processes as will be discussed in the next Section (IV). Hence, not only the isotopic signature of recent assimilates but also the carbohydrate pools and their use in different metabolic pathways may be the primary cause of changes in the respiratory $\delta^{13}C$. The time-lagged relationship with environmental parameters could be based on the differences in pool sizes (*e.g.*, high accumulation of photosynthates under high light conditions) and its subsequent use in different metabolic pathways as discussed below. Future research is required to fill the gap in our understanding of processes controlling the respiratory signals and dynamics in $\delta^{13}C_R$.

In summary, relationships between $\delta^{13}C_{res}$ and environmental drivers can contain valuable information on ecosystem responsiveness if physiological processes are taken into account.

IV. CARBON ISOTOPE FRACTIONATION DURING DARK RESPIRATION OF PLANTS

"Apparent" carbon isotope fractionation during dark respiration is determined as the difference in $\delta^{13}C$ of bulk organic matter (or potential respiratory substrates, mainly carbohydrates) and that of respired CO_2. This was mainly investigated in leaves showing, despite huge variability, that the leaf respired CO_2 is generally ^{13}C-enriched compared to leaf material (Section I). Mass balance calculations showed that this "apparent" fractionation could lead to ^{13}C-depletion in the remaining leaf organic matter at the end of the night compared to photosynthetic products synthesized during the day (Ghashghaie *et al.* 2003). Few works on heterotrophic organs showed that CO_2 respired by tree trunks is in general ^{13}C-enriched, while that of roots is ^{13}C-depleted compared to bulk organic matter or carbohydrates (for a review, see Badeck *et al.* 2005). The opposite respiratory fractionations of aboveground and belowground compartment could partly counterbalance the effects of each other on $\delta^{13}C$ of organic matter at whole plant level. The only published data supporting this assumption were obtained by Klumpp *et al.* (2005) on three herbaceous plants under controlled conditions. More mass balance studies on other species and under different environmental conditions are needed to evaluate these effects under controlled and natural conditions.

A. Mechanisms of Carbon Isotope Fractionation During Dark Respiration

Metabolic origin of the carbon isotope signature of dark respired CO_2 by plants and its variability has been discussed by Ghashghaie *et al.* (2003) and Tcherkez *et al.* (2003): ^{13}C-enrichment in dark respired CO_2 by leaves is mainly attributed to the heterogeneous ^{13}C-distribution within natural hexose molecules, so called "fragmentation fractionation" by Tcherkez *et al.* (2004). Indeed, Rossmann *et al.* (1991) experimentally demonstrated that C-3 and C-4 of glucose molecules (extracted from sugar beet syrup and from maize flour) are ^{13}C-enriched compared to other positions (*i.e.*, C-1, C-2, C-5, and C-6). During glycolysis, C_1 of pyruvate coming from C_3 and C_4 of glucose molecules being decarboxylated by pyruvate dehydrogenase (PDH) reaction, CO_2 released by PDH reaction is expected to be ^{13}C-enriched compared to hexose molecules, while the other carbon atoms (lighter ones) of hexoses incorporated into acetyl-CoA molecules are decarboxylated in the Krebs cycle (Figure 13.5). Acetyl-CoA molecules are partially deviated to the biosynthesis of metabolites, for example, fatty acids and isoprenoids, well known to be ^{13}C-depleted compared to carbohydrates (Park and Epstein 1961). Accordingly, Ghashghaie *et al.* (2001) proposed that if all the carbon coming from carbohydrates is consumed during dark respiration (*i.e.*, heavy carbon being decarboxylated by PDH and light one by Krebs cycle), no "apparent" fractionation will be observed, that is, the overall CO_2 released by dark respiration will have the signature of the carbohydrates used. This is probably the case of isolated protoplasts, which consume all the carbon of the given sugars without any biosynthesis (Lin and Ehleringer 1997). By contrast, in case of partly deviation of light carbon (acetyl-CoA) to biosynthetic pathways, the overall CO_2 respired in the dark is expected to be ^{13}C-enriched. This is probably the case of intact leaves (see above). Therefore, the variability in the relative metabolic pathways (glycolysis versus Krebs cycle) could induce variability in the signature of respired CO_2 (Figure 13.5).

Tcherkez *et al.* (2003) showed that changing leaf temperature changes the isotope signature of CO_2 respired in the dark by *Phaseolus* leaves, from −18‰ at low temperature (10°C) to −24‰ at high

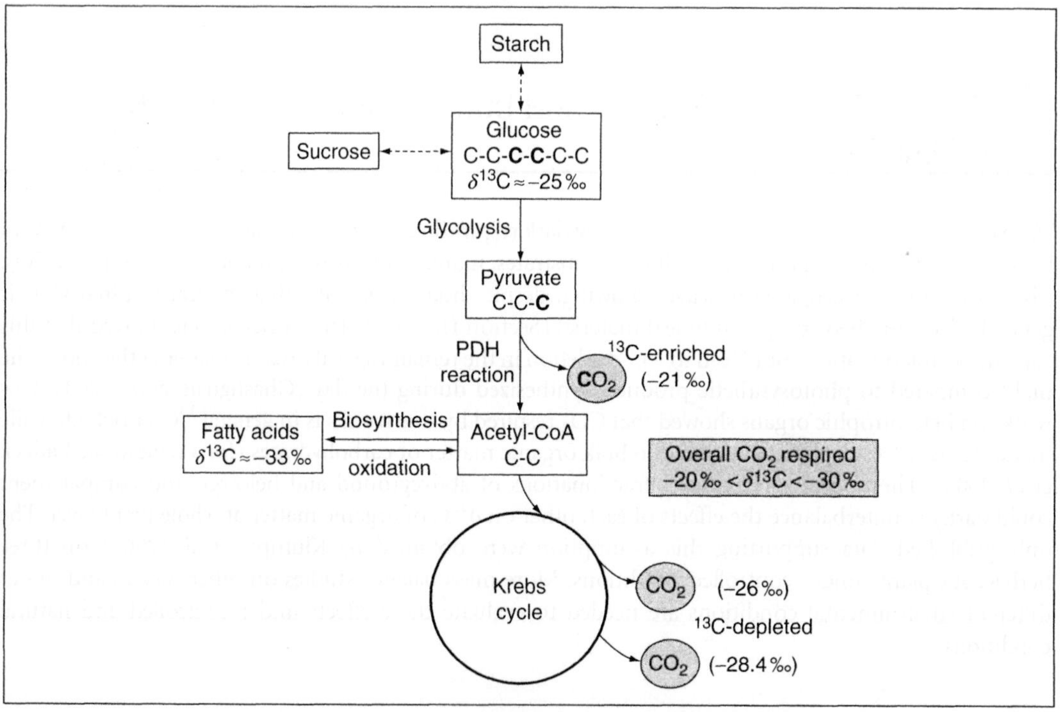

FIGURE 13.5 Respiratory metabolic "pattern" explaining $\delta^{13}C$ of respired CO_2 in the dark. $\delta^{13}C$ values are those measured in French bean leaves by Tcherkez *et al.* (2003) except those of CO_2 in parentheses, which are derived from positional $\delta^{13}C$ values in natural glucose as given by Rossmann *et al.* (1991). Carbon atoms C-3 and C-4 of glucose and thus C_1 of pyruvate which is decarboxylated during PDH reaction are ^{13}C-enriched (bold letters: **C**). Other carbon atoms of glucose forming acetyl-CoA, which are then decarboxylated in the Krebs cycle, are ^{13}C-depleted. Redrawn and simplified from Tcherkez *et al.* (2003).

temperature (32 °C). They suggested that at low temperature and, consequently low respiration rates, the biosynthetic deviation is high and thus the respired CO_2 is ^{13}C-enriched. In contrast, at high temperature and dark respiration rates, the biosynthetic deviation is low and thus the signature of respired CO_2 is very close to that of the sugars. They also showed that leaf-respired CO_2 becomes more and more ^{13}C-depleted during continuous darkness ranging from −20‰ at the beginning of a dark period to −30‰ after a few days of darkness (more slowly at low temperatures). They suggested that the decrease in carbohydrate pool size during continuous darkness decreases the rate of decarboxylation of heavy carbon atoms by PDH reaction. Eventually, when the carbohydrate pool is exhausted and fatty acids are being oxidized instead, supplying the Krebs cycle with more depleted carbon, the signature of respired CO_2 becomes closer to that of fatty acids (around −33‰). Additionally, they showed a positive linear correlation between the isotopic signature of dark respired CO_2 and the respiratory quotient (RQ): the ratio of CO_2 produced to O_2 consumed by dark respiration. As expected, RQ was high (RQ = 1) when carbohydrates were used as respiratory substrates (*i.e.*, at low temperature and at the beginning of the darkness) and subsequently respiratory CO_2 was ^{13}C-enriched. RQ decreased when carbohydrate pool size was reduced during the darkness and/or when temperature was higher and less oxygenated substrates like fatty acids were oxidized.

Accordingly, huge changes in the signature of dark respired CO_2 by leaves can occur which could reflect on $\delta^{13}C_R$ at the ecosystem level. Indeed, both temperature and carbohydrate pool size could change under natural conditions leading to the changes in the rate of pyruvate decarboxylation via glycolysis relative to the oxidation of highly reduced metabolites, for example, fatty acids, that is the shift in the substrate used for dark respiration. This may explain the diurnal variation in $\delta^{13}C$ of leaf respired CO_2 (about 7–10‰) reported in the literature (Hymus *et al.* 2005, Prater *et al.* 2006, Werner *et al.* 2007). When growth has ceased and there is a low metabolic demand for respiratory products of the Krebs cycle, the synthesis of secondary compounds derived from ^{13}C-depleted acetyl-CoA could be favored relative to the oxidation of these compounds (Hymus *et al.* 2005) with increasing assimilate pool during the day. This is supported by a linear relationship between the ^{13}C-enrichment in leaf respired CO_2 and the cumulative fixed carbon by photosynthesis (Hymus *et al.* 2005). These results support the "metabolic pattern" proposed by Tcherkez *et al.* (2003) relating the changes in $\delta^{13}C$ of leaf respired CO_2 to changes in carbohydrate pool size as explained earlier. The short-term variation in ecosystem $\delta^{13}C_R$ reported in Section III could therefore be partly explained by changes in plant metabolic and biosynthetic pathways.

B. Contribution of Multiple Pools to Nighttime Respiration

The carbon isotope signature of ecosystem respired CO_2 is considered to be similar to that of the photosynthetic products of the preceding light period. However, Nogués *et al.* (2004, 2006a,b) showed in labeling experiments that recent photoassimilates contributed only partly to leaf dark respiration. The contribution of old carbon (reserves) was substantial. The relative contribution of new carbon to leaf dark respiration was shown to be variable depending on species, for example, about 50% in *Phaseolus vulgaris* and around 10% in *Ranunculus glacialis* (Nogués *et al.* 2004, 2006b, respectively) and changed seasonally, for example, about 60% (June) and around 30% (October) in *Fagus sylvatica* (Nogués *et al.* 2006a). They showed the existence of at least two respiratory pools at leaf level in *P. vulgaris*, one with a rapid turnover rate of a few hours and one with a slow turnover rate of a few days (Nogués *et al.* 2004). Similar results were reported by Schnyder *et al.* (2003) at whole plant level in sunflower, suggesting that any change in environmental conditions in the field changing photosynthetic discrimination will be reflected on ecosystem $\delta^{13}C_R$ but with a certain time lag, as discussed in Section III.D.

The theory of "apparent" discrimination during dark respiration of plants (mainly leaves) is now largely accepted by the stable isotope community working at both plant and ecosystem level. However, ecosystems (*e.g.*, forests) are complex systems composed of autotrophic and heterotrophic compartments with different respiratory signatures that can change with environmental conditions. There are time lags between assimilation of carbon in the leaves and its transfer to (and use in) different compartments and also changes in relative contribution of the recently fixed carbon compared to old reserves. Thus, the contribution of each compartment to ecosystem $\delta^{13}C_R$ cannot be easily evaluated. More data on different plants/organs and on different ecosystems/compartments are needed in order to better understand the origin of dynamics in the signature of ecosystem respired CO_2 and hence to be able to incorporate them into the carbon cycle models.

V. IMPLICATIONS AT LARGER TEMPORAL AND SPATIAL SCALES

Stable isotopes are important tracers to monitor the impact of environmental changes on ecosystem functioning, particularly for understanding and quantification of changes in respiratory and photosynthetic processes involved in ecosystem carbon fluxes. However, in spite of recent insights, we still lack a full mechanistic understanding of the temporal and spatial variability of post-photosynthetic fractionation during dark respiration and, particularly in $\delta^{13}C_R$. As shown here, there is now evidence that variation in $\delta^{13}C$ of respired CO_2 can be substantial at spatial scales ranging from the leaf (Ghashghaie *et al.* 2001, Tcherkez *et al.* 2003, Hymus *et al.* 2005, Prater *et al.* 2006, Werner *et al.* 2007) and the trunk (Damesin and Lelarge 2003) to the ecosystem (Bowling *et al.* 2002, Mortazavi *et al.* 2005, Werner *et al.* 2006) and in the short temporal scales, from hours to days (Figure 13.3).

Given our knowledge on the strong relationship of $\delta^{13}C_R$ and environmental drivers, we suggest that it is reasonable to expect nocturnal variations in $\delta^{13}C_R$ in many ecosystems. Indeed, based on this knowledge it seems difficult to justify the constancy in nocturnal $\delta^{13}C_R$ (Section III.D) even though periods of stable $\delta^{13}C_R$ may occur, similarly to what was found on coarser timescales (McDowell *et al.* 2004b).

This has large implications for the sampling protocols used to collect nocturnal Keeling plots, since timing of data collection will be decisive. The Keeling plot method assumes either one respiratory source with a single isotopic composition, or that the relative contributions of component fluxes that might differ in isotopic composition (such as foliar and soil respiration) do not change over the sampling period (Bowling *et al.* 2003a). However, in many ecosystems large CO_2 gradients required for reliable Keeling plots are difficult to capture in short time periods due to low activity of the systems, which is commonly overcome by extending the time of sampling over several hours until a sufficiently large gradient is reached (commonly 2–8 h, see Pataki *et al.* 2003). Considering high dynamics in respiratory $\delta^{13}C$ both at the leaf and ecosystem scale, we highlight the importance of standardized sampling protocols for nocturnal Keeling plots (Mortazavi *et al.* 2006). It still needs to be evaluated which time at night is the most appropriate to collect samples for a representative value of $\delta^{13}C_R$.

The quality of $\delta^{13}C_R$ measurements has important implications for the reliability of our estimates in ecosystem response to global change, both in the past and, particularly, for future predictions. Most recent models assume constant $\delta^{13}C_R$ even across different biomes. Partitioning studies use one nocturnal measurement of $\delta^{13}C_R$ to separate net ecosystem exchange into respiratory and photosynthetic fluxes (Bowling *et al.* 2001, Knohl and Buchmann 2005). Sensitivity analyses are required to evaluate the effect of nonconstant nocturnal $\delta^{13}C_R$ in these model predictions.

These recent insights on temporal variation in respired $\delta^{13}C$ are partially attributed to advances in methodology. Formerly, atmospheric air samples had to be collected in large volumes; purified and concentrated for analysis. Nowadays automatic sampling of small, low-cost septum capped vials allows a high sample throughput and better replication of measurements. Still, capturing the rapid variations of respired $\delta^{13}C$ of different ecosystem components might require "online" isotope measurements in the field. New technical development; like the outdoor continuous-flow IRMS developed by Schnyder *et al.* (2004) or optical systems like the tuneable diode laser (Bowling *et al.* 2003b) might provide a breakthrough for high-time resolved measurements of $\delta^{13}C_R$. A better mechanistic understanding of the fractionation processes during dark respiration and the dynamics controlling $\delta^{13}C_R$ will strongly improve the reliability of our records of ecological changes in ecosystem functioning and, particularly the prognostic power of models to evaluate future scenarios.

In conclusion, there is now clear evidence that:

- Substantial changes of ecosystem respired $^{13}CO_2$ occur at different timescales and with different responsiveness to changes in environmental factors (*i.e.*, with variable time lags).
- Short term changes may at some extent be explained by changes in the respiratory metabolism.
- Apparent fractionation during dark respiration is important and highly variable with conditions and species.

Future research is required:

- To evaluate the dynamics in respiratory fractionation in different organs (roots, shoots), plant species (functional groups), and different ecosystem compartments (heterotrophic, autotrophic respiration) and their feedback at the ecosystem level
- To assess the implications of short-term dynamics in $\delta^{13}C_R$ for partitioning studies at the ecosystem level and evaluation of environmental changes
- For technological advances in high-time resolved measurements of $\delta^{13}C_{res}$ of different ecosystem compounds and common sampling protocols for $\delta^{13}C_R$ in global networks
- A better process-based understanding of the use of stable isotopes as tracers of the respiratory fluxes in the carbon cycle to monitor ecological change in ecosystem functioning

VI. ACKNOWLEDGMENTS

This work was financed by the German DFG (WE 2681/2-1; ISOFLUX project). We kindly acknowledge that climate data were collected by T. S. David in the PIDDAC project (216/2001; Ministério da Agricultura; Portugal). Further, financial support from the European MIND project (EVK2-CT-2002-00158) to J. S. Pereira; and through the European Community's Human Potential Programme under contract HPRN-CT-1999-00059 (NETCARB) to J. Ghashghaie and C. Máguas is acknowledged. The authors want to thank R. Maia for technical assistance with isotopic analysis and J. Banza for technical support at the tower.

VII. REFERENCES

Aubinet, M., A. Grelle, A. Ibrom, Ü. Rannik, J. Moncrieff, T. Foken, A. S. Kowalski, P. H. Martin, P. Berbigier, C. Bernhofer, R. Clement, J. Elbers, *et al.* 2000. Estimates of the annual net carbon and water exchange of forests: The EUROFLUX-Methodology. *Advances in Ecological Research* **30:**113–175.

Badeck, F. W., G. Tcherkez, S. Nogués, C. Piel, and J. Ghashghaie. 2005. Post-photosynthetic fractionation of stable carbon isotopes between plant organs: A widespread phenomenon. *Rapid Communications in Mass Spectrometry* **19:**1381–1391.

Bhupinderpal, S., A. Nordgren, M. Ottosson Löfvenius, M. N. Högberg, P.-E. Mellander, and P. Högberg. 2003. Tree root and soil heterotophic respiration as revealed by girdling of boreal Scots pine forest: Extending observations beyond the first year. *Plant, Cell and Environment* **26:**1287–1296.

Boone, R. D., K. J. Nadelhoffer, J. D. Canary, and J. P. Kaye. 1998. Roots exert a strong influence on the temperature sensitivity of soil respiration. *Nature* **396:**570–572.

Bowling, D. R., P. P. Tans, and R. K. Monson. 2001. Partitioning net ecosystem carbon exchange with isotopic fluxes of CO_2. *Global Change Biology* **7:**127–145.

Bowling, D. R., N. G. McDowell, B. J. Bond, B. E. Law, and J. R. Ehleringer. 2002. ^{13}C content of ecosystem respiration is linked to precipitation and vapor pressure deficit. *Oecologia* **131:**113–124.

Bowling, D. R., N. G. McDowell, J. M. Welker, B. J. Bond, B. E. Law, and J. R. Ehleringer. 2003a. Oxygen isotope content of CO_2 in nocturnal ecosystem respiration: 1. Observations in forests along a precipitation transect in Oregon, USA. *Global Biogeochemical Cycles* **17:**31-1–31-14.

Bowling, D. R., S. D. Sargent, B. D. Tanner, and J. R. Ehleringer. 2003b. Tunable diode laser absorption spectroscopy for stable isotope studies of ecosystem-atmosphere CO_2 exchange. *Agricultural and Forest Meteorology* **118:**1–19.

Brugnoli, E., and G. D. Farquhar. 2000. Photosynthetic fractionation of carbon isotopes. Pages 399–434 *in* R. C. Leegood, T. D. Sharkey, and S. von Caemmerer (Eds.) *Photosynthesis: Physiology and Mechanisms.* Kluver Akademic Publisher, The Netherlands.

Buchmann, N., J. R. Brooks, and J. R. Ehleringer. 2002. Predicting daytime carbon isotope ratios of atmospheric CO_2 within forest canopies. *Functional Ecology* **16:**49–57.

Canadell, J. G., H. A. Mooney, D. Baldocchi, J. A. Berry, J. R. Ehleringer, C. B. Field, S. T. Gower, D. Y. Hollinger, J. E. Hunt, R. B. Jackson, S. W. Running, G. R. Shaver, *et al.* 2000. Carbon Metabolism of the Terrestrial Biosphere: A multi-technique approach for improved understanding. *Ecosystems* **3:**115–130.

Carreiras, J. M. B., J. M. C. Pereira, and J. S. Pereira. 2006. Estimation of tree canopy cover in evergreen oak woodlands using remote sensing. *Forest Ecology and Management* **223:**45–53.

Ciais, P., P. P. Tans, M. Trolier, J. W. C. White, and R. J. Francey. 1995. A large northern hemisphere terrestrial CO_2 sink indicated by the $^{13}C/^{12}C$ ratio of atmospheric CO_2. *Science* **269:**1098–1102.

Ciais, P., M. Reichstein, N. Viovy, A. Granier, J. Ogee, V. Allard, M. Aubinet, N. Buchmann, C. Bernhofer, A. Carrara, F. Chevallier, N. De Noblet, *et al.* 2005. Europe-wide reduction in primary productivity caused by the heat and drought in 2003. *Nature* **437:**529–533.

Damesin, C., and C. Lelarge. 2003. Carbon isotope composition of current-year shoots from *Fagus sylvatica* in relation to growth, respiration and use of reserves. *Plant, Cell and Environment* **26:**207–219.

Damesin, C., C. Barbaroux, D. Berveiller, C. Lelarge, M. Chaves, C. Maguas, R. Maia, and J. Y. Pontailler. 2005. The carbon isotope composition of CO_2 respired by trunks: Comparison of four sampling methods. *Rapid Communications in Mass Spectrometry* **19:**369–374.

David, T. S., M. I. Ferreira, S. Cohen, J. S. Pereira, and J. S. David. 2004. Constraints on transpiration from an evergreen oak tree in southern Portugal. *Agricultural and Forest Meteorology* **122:**193–205.

Duranceau, M., J. Ghashghaie, F. Badeck, E. Deléens, and G. Cornic. 1999. $\delta^{13}C$ of leaf carbohydrates in relation to dark respiration in *Phaseolus vulgaris* L. under progressive drought. *Plant, Cell and Environment* **22:**515–523.

Ekblad, A., and P. Högberg. 2001. Natural abundance of ^{13}C in CO_2 respired from forest soils reveals speed of link between photosynthesis and root respiration. *Oecologia* **127:**305–308.

Farquhar, G. D., M. H. O'Leary, and J. A. Berry. 1982. On the relationship between carbon isotope discrimination and the intercellular carbon dioxide concentration in leaves. *Australian Journal of Plant Physiology* **9:**121–137.

Farquhar, G. D., J. R. Ehleringer, and K. T. Hubick. 1989. Carbon isotope discrimination and photosynthesis. *Annual Review of Plant Physiology and Plant Molecular Biology* **40:**503–537.

Fessenden, J. E., and J. R. Ehleringer. 2003. Temporal variation in $\delta^{13}C$ of ecosystem respiration in the Pacific Northwest: Links to moisture stress. *Oecologia* **136:**129–136.

Flanagan, L. B., and J. Ehleringer. 1998. Ecosystem-atmosphere CO_2 exchange: Interpreting signals of change using stable isotope ratios. *Trends in Ecology and Evolution* **13:**10–14.

Ghashghaie, J., M. Duranceau, F. Badeck, G. Cornic, M. T. Adeline, and E. Deléens. 2001. $\delta^{13}C$ of CO_2 respired in the dark in relation to ^{13}C of leaf metabolites: Comparison between *Nicotiana sylvestris* and *Helianthus annuus* under drought. *Plant, Cell and Environment* **24:**145–515.

Ghashghaie, J., F. Badeck, G. Lanigan, S. Nogúes, G. Tcherkez, E. Deléens, G. Cornic, and H. Griffiths. 2003. Carbon isotope fractionation during dark respiration and photorespiration in C_3 plants. *Phytochemistry Reviews* **2:**145–161.

Giorgi, F. 2006. Climate change hot-spots. *Geophysical Research Letters* **33:** doi:10.1029/2006GL025734.

Göttlicher, S., A. Knohl, W. Wanek, N. Buchmann, and A. Richter. 2006. Short-term changes in carbon isotope composition of soluble carbohydrates and starch: From canopy leaves to the roots. *Rapid Communications in Mass Spectrometry* **20:**653–660.

Hemming, D., D. Yakir, P. Ambus, M. Aurela, C. Besson, K. Black, N. Buchmann, R. Burlett, A. Cescatti, R. Clement, P. Gross, A. Granier, *et al.* 2005. Pan-European delta C-13 values of air and organic matter from forest ecosystems. *Global Change Biology* **11:**1065–1093.

Hymus, G. J., K. Maseyk, R. Valentini, and D. Yakir. 2005. Large daily variation in ^{13}C-enrichment of leaf-respired CO_2 in two Quercus forest canopies. *New Phytologist* **167:**377–384.

Jarvis, P. G., A. Rey, C. Petsikos, M. Rayment, J. S. Pereira, J. Banza, J. S. David, F. Miglietta, and R. Valentini. 2007. Drying and wetting of soils stimulates decomposition and carbon dioxide emission: The "Birch Effect." *Tree Physiology:* **27:**929–940.

Keeling, C. D. 1958. The concentration and isotopic abundance of atmospheric carbon dioxide in rural areas. *Geochimica et Cosmochimica Acta* **13:**322–334.

Keeling, C. D., T. P. Whorf, M. Wahlen, and J. van der Plicht. 1995. Interannual extremes in the rate of rise of atmospheric carbon dioxide since 1980. *Nature* **375**:666–670.

Klumpp, K., R. Schäufele, M. Lötscher, F. A. Lattanzi, W. Feneis, and H. Schnyder. 2005. C-isotope composition of CO_2 respired by shoots and roots: Fractionation during dark respiration. *Plant, Cell and Environment* **28**:241–250.

Knohl, A., and N. Buchmann. 2005. Partitioning the net CO_2 fluxes of a deciduous forest into respiration and assimilation using stable carbon isotopes. *Global Biogeochemical Cycles* **19**:GB4008, 1–14.

Knohl, A., R. A. Werner, W. A. Brand, and N. Buchmann. 2005. Short-term variations in $\delta^{13}C$ of ecosystem respiration reveals link between assimilation and respiration in a deciduous forest. *Oecologia* **142**:70–82.

Kurz-Besson, C., D. Otieno, R. L. Vale, R. T. W. Siegwolf, M. Schmidt, A. Herd, C. Nogueira, T. S. David, J. S. David, J. Tenhunen, J. S. Pereira, and M. Chaves. 2006. Hydraulic lift in cork oak trees in a savannah-type Mediterranean ecosystem and its contribution to the local water balance. *Plant and Soil* **282**:361–378.

Lai, C. T., J. R. Ehleringer, A. J. Schauer, P. P. Tans, D. Y. Hollinger, U. Paw, J. W. Munger, and S. C. Wofsy. 2005. Canopy-scale delta ^{13}C of photosynthetic and respiratory CO_2 fluxes: Observations in forest biomes across the United States. *Global Change Biology* **11**:633–643.

Larcher, W. 2000. Temperature stress and survival ability of Mediterranean sclerophyllous plants. *Plant Biosystems* **134**:279–295.

Lin, G., and J. E. Ehleringer. 1997. Carbon isotopic fractionation does not occur during dark respiration in C3 abd C4 plants. *Plant Physiology* **114**:391–394.

Loveys, B. R., L. J. Atkinson, D. J. Sherlock, R. L. Roberts, A. H. Fitter, and O. K. Atkin. 2003. Thermal acclimation of leaf and root respiration: An investigation comparing inherently fast- and slow-growing plant species. *Global Change Biology* **9**:895–910.

Luterbacher, J., and E. Xoplaki. 2003. 500 Year winter temperature and precipitation variability over the Mediterranean area and its connection to the large-scale atmospheric circulation. *In* H.-J. Bolle (Ed.) *Mediterranean Climate. Variability and Trends.* Springer Verlag, Berlin Heidelberg.

McDowell, N. G., D. R. Bowling, A. Schauer, J. Irvine, B. J. Bond, B. E. Law, and J. R. Ehleringer. 2004a. Associations between carbon isotope ratios of ecosystem respiration, water availability and canopy conductance. *Global Change Biology* **10**:1767–1784.

McDowell, N. G., D. R. Bowling, B. J. Bond, J. Irvine, B. E. Law, P. Anthoni, and J. R. Ehleringer. 2004b. Response of the carbon isotopic content of ecosystem, leaf, and soil respiration to meteorological driving factors in a *Pinus ponderosa* ecosystem. *Global Biogeochemical Cycles* **18**(GB1013):1–12.

Miranda, P., F. E. S. Coelho, A. R. Tomé, and M. A. Valente. 2002. 20th Century Portuguese Climate and Climate Scenarios. Pages 25–83 *in* F. D. Santos, K. Forbes, and R. Moita (Eds.) *Climate Change in Portugal. Scenarios, Impacts, and Adaptation Measures.* Lisbon, Gradiva.

Mortazavi, B., J. P. Chanton, J. L. Prater, A. C. Oishi, R. Oren, and G. Katul. 2005. Temporal variability in ^{13}C of respired CO_2 in a pine and a hardwood forest subject to similar climatic conditions. *Oecologia* **142**:57–69.

Mortazavi, B., J. P. Chanton, and M. C. Smith. 2006. Influence of C-13-enriched foliage respired CO_2 on delta C-13 of ecosystem-respired CO_2. *Global Biogeochemical Cycles* **20**:1–9.

Nogués, S., G. Tcherkez, G. Cornic, and J. Ghashghaie. 2004. Respiratory carbon metabolism following illumination in intact French bean leaves using $^{13}C/^{12}C$ isotope labelling. *Plant Physiology* **136**:3245–3254.

Nogués, S., C. Damesin, G. Tcherkez, F. Maunoury, G. Cornic, and J. Ghashghaie. 2006a. $^{13}C/^{12}C$ isotope labelling to study leaf carbon respiration and allocation in twigs of field grown beech trees. *Rapid Communications in Mass Spectrometry* **20**:219–226.

Nogués, S., G. Tcherkez, P. Streb, A. Pardo, F. Baptist, R. Bligny, J. Ghashghaie, and G. Cornic. 2006b. Respiratory carbon metabolism in the high mountain plant species *Ranunculus glacialis. Journal of Experimental Botany* **57**(14):3837–3845.

Ogée, J., P. Peylin, P. Ciais, T. Bariac, Y. Brunet, P. Berbigier, C. Roche, P. Richard, G. Bardoux, and J. M. Bonnefond. 2003. Partitioning net ecosystem carbon exchange into net assimilation and respiration using $^{13}CO_2$ measurements: A cost-effective sampling strategy. *Global Biogeochemical Cycles* **17**(2):39-1–18.

Otieno, D., C. Kurz-Besson, J. Liu, M. Schmidt, R. D. Vale, T. David, R. T. W. Siegwolf, J. Pereira, and J. Tenhunen. 2006. Seasonal variations in soil and plant water status in a *Quercus suber* L. stand: Roots as determinants of tree productivity and survival in the mediterranean-type ecosystem. *Plant and Soil* **283**:119–135.

Papale, D., M. Reichstein, M. Aubinet, E. Canfora, C. Bernhofer, W. Kutsch, B. Longdoz, S. Rambal, R. Valentini, T. Vesala, and D. Yakir. 2006. Towards a standardized processing of Net Ecosystem Exchange measured with eddy covariance technique: Algorithms and uncertainty estimation. *Biogeosciences* **3**:571–583.

Park, R., and S. Epstein. 1961. Metabolic fractionation of ^{13}C and ^{12}C in plants. *Plant Physiology* **36**:133–138.

Pataki, D. E., J. R. Ehleringer, L. B. Flanagan, D. Yakir, D. R. Bowling, C. J. Still, N. Buchmann, J. O. Kaplan, and J. A. Berry. 2003. The application and interpretation of Keeling plots in terrestrial carbon cycle research. *Global Biogeochemical Cycles* **17**:22-1–22-4.

Pereira, J. S., J. S. David, T. S. David, M. C. Caldeira, and M. M. Chaves. 2004. Carbon and water fluxes in mediterranean-type ecosystems: Constraints and adaptations. Pages 467–498 *in* K. Esser, U. Lüttge, W. Beyschlag, and J. Murata (Eds.) *Progress in Botany.* Springer-Verlag, Berlin Heidelberg.

Ponton, S., L. B. Flanagan, K. P. Alstad, B. G. Johnson, K. Morgenstern, N. Kljun, T. A. Black, and A. G. Barr. 2006. Comparison of ecosystem water-use efficiency among Douglas-fir forest, aspen forest and grassland using eddy covariance and carbon isotope techniques. *Global Change Biology* **12:**294–310.

Prater, J. L., B. Mortazavi, and J. P. Chanton. 2006. Diurnal variation of the $\delta^{13}C$ of pine needle respired CO_2 evolved in darkness. *Plant, Cell and Environment* **29:**202–211.

Reichstein, M., J. D. Tenhunen, O. Roupsard, J.-M. Ourcival, S. Rambal, S. Dore, and R. Valentini. 2002. Ecosystem respiration in two Mediterranean evergreen Holm Oak forests: Drought effects and decomposition dynamics. *Functional Ecology* **16:**27–39.

Reichstein, M., E. Falge, D. Baldocchi, D. Papale, M. Aubinet, P. Berbigier, C. Bernhofer, N. Buchmann, T. Gilmanov, A. Granier, T. Grunwald, K. Havrankova, *et al.* 2005. On the separation of net ecosystem exchange into assimilation and ecosystem respiration: Review and improved algorithm. *Global Change Biology* **11:**1424–1439.

Rey, A., C. Petsikos, P. G. Jarvis, and J. Grace. 2005. Effect of temperature and moisture on rates of carbon mineralization in a Mediterranean oak forest soil under controlled and field conditions. *European Journal of Soil Science* **56:**589–599.

Rossmann, A., M. Butzenlechner, and H. L. Schmidt. 1991. Evidence for a nonstatistical carbon isotope distribution in natural glucose. *Plant Physiology* **96:**609–614.

Schnyder, H., R. Schäufele, M. Lötscher, and T. Gebbing. 2003. Disentangling CO_2 fluxes: Direct measurements of mesocosm-scale natural abundance $^{13}CO_2/^{12}CO_2$ gas exchange, ^{13}C discrimination, and labelling of CO_2 exchange flux components in controlled environments. *Plant, Cell and Environment* **26:**1863–1874.

Schnyder, H., R. Schäufele, and R. Wenzel. 2004. Mobile, outdoor continuous-flow isotope-ratio mass spectrometer system for automated high-frequency C-13- and O-18-CO_2 analysis for Keeling plot applications. *Rapid Communications in Mass Spectrometry* **18:**3068–3074.

Schulze, E.-D., C. Wirth, and M. Heimann. 2000. Managing forests after Kyoto. *Science* **289:**2058–2059.

Snyder, K. A., L. A. Donovan, J. J. James, R. L. Tiller, and J. H. Richards. 2004. Extensive summer water pulses do not necessarily lead to canopy growth of Great Basin and northern Mojave Desert shrubs. *Oecologia* **141:**325–334.

Still, C. J., J. A. Berry, M. Ribas-Carbó, and B. R. Helliker. 2003. The contribution of C_3 and C_4 plants to the carbon cycle of tallgrass prairie: An isotopic approach. *Oecologia* **136:**347–359.

Tans, P. P., I. Y. Fung, and T. Takahashi. 1990. Observational constraints on the global atmospheric CO_2 buget. *Science* **247:**1431–1438.

Tcherkez, G., S. Nogués, J. Bleton, G. Cornic, F. Badeck, and J. Ghashghaie. 2003. Metabolic origin of carbon isotope composition of leaf dark-respired CO_2 in French bean. *Plant Physiology* **131:**237–244.

Tcherkez, G., G. Farquhar, F. Badeck, and J. Ghashghaie. 2004. Theoretical considerations about carbon isotope distribution in glucose of C3 plants. *Functional Plant Biology* **31:**857–877.

Valentini, R., G. Matteucci, A. J. Dolman, E.-D. Schulze, C. Rebmann, E. J. Moors, A. Granier, P. Gross, N. O. Jensen, and K. Pilegaard. 2000. Respiration as the main determinant of carbon balance in European forests. *Nature* **404:**861–865.

Werner, C., S. Unger, J. S. Pereira, R. Maia, J. S. David, C. Kurz-Besson, T. S. David, and C. Maguas. 2006. Importance of short-term dynamics in carbon isotope ratios of ecosystem respiration ($\delta^{13}C_R$) in a Mediterranean oak woodland and linkage to environmental factors. *New Phytologist* **172:**330–346.

Werner, C., N. Hasenbein, R. Maia, W. Beyschlag, and C. Máguas. 2007. Evaluating high time-resolved changes in carbon isotope ratio of respired CO_2 by a rapid in-tube incubation technique. *Rapid Communications in Mass Spectrometry* **21:**1352–1360

Xu, C. Y., G. H. Lin, K. L. Griffin, and R. N. Sambrotto. 2004. Leaf respiratory CO_2 is ^{13}C-enriched relative to leaf organic components in five species of C_3 Plants. *New Phytologist* **163:**499–505.

Yakir, D., and X.-F. Wang. 1996. Fluxes of CO_2 and water between terrestrial vegetation and the atmosphere estimated from isotope measurements. *Nature* **380:**515–517.

Yakir, D., and L. Sternberg. 2000. The use of stable isotopes to study ecosystem gas exchange. *Oecologia* **123:**297–311.

CHAPTER 14

To What Extent Can Ice Core Data Contribute to the Understanding of Plant Ecological Developments of the Past?

Markus Leuenberger

Climate and Environmental Physics, Physics Institute, University of Bern

Contents

I. INTRODUCTION

Ecological changes are subject to many different influences. One of those is a change of the atmospheric gas composition, namely of trace gases. Such changes can be traced by either direct atmospheric concentration measurements or determination of the isotopic composition of those gas species.

However, those records are rather short since they have commenced from the late fifties of the last century only. Therefore, it is important and helpful to explore archives where such information is available. For changes of the atmospheric gas composition and its isotopic signature, polar ice cores are best suited since they exhibit the most direct access to this information. Observed changes in these long-term records are fundamental to learn more about the sensitivity of ecological changes on the atmospheric gas composition. Particular focus is given to the $\delta^{13}C$ change of atmospheric carbon dioxide during the industrialization. A linear correction table is given based on a composite record from air derived from ice core, firn, and archived samples as well as direct atmospheric air. This correction is thought to be applied to temporal carbon isotope series from plant material as obtained for instance tree-ring studies. Several effects influencing this correction are briefly discussed.

Since the early measurements by Keeling (1958), we know that terrestrial biological activity has a strong influence on the CO_2 of the atmosphere. He clearly could detect the breathing of the large biological systems by precisely monitoring the CO_2 concentration throughout the season at Mauna Loa, Hawaii. On top of this seasonal cycling he found a long-term increase that was discovered to document the human emissions of fossil fuel burning into the atmosphere (Keeling 1968, 1970, 1973). Due to his continuous effort this record became one of the worldwide most recognized records. A large part of our community was influenced by the early findings others have redirected their research.

By setting up records at different sites over the globe, Keeling was able to clearly demonstrate that a gradient exists between the two hemispheres representing the unequal distribution of land masses and therefore terrestrial biomes. The seasonal amplitudes on the Northern Hemisphere proved to be much larger than in the Southern Hemisphere where at some of the locations hardly any seasonal signal could be detected, but still retaining the long-term trend (Mook *et al.* 1983, Keeling *et al.* 1984, Bacastow *et al.* 1985). This shows another interesting and important fact that the atmosphere is a well-mixed reservoir with a mixing time of 1–2 years between the hemispheres (Keeling *et al.* 1989a,b).

By a careful examination of small differences between years within and across different records, he was actually able to extract information of interannual as well as intra-annual breathing signal changes (Keeling *et al.* 1976a,b). As already mentioned above, his early work has influenced many scientists and lead for instance to pioneering excursions of Jet Langway and others to Polar sites, where they wanted to look for the possibility to extend Keeling's record back in time by using ice cores. That was back in the late fifties. A first deep drilling was undertaken in 1961 at Camp Century. There were many efforts to extract the air from the ice and to analyze its CO_2 concentration to a similar precision as Keeling did in the atmosphere. However, the community had to deal with many failures and drawbacks. It took another 20 years to really get reliable, similarly precise, CO_2 concentration values from ice cores (Neftel *et al.* 1982).

Since that time, roughly 20 years have passed during which a wealth of information from ice cores from several Artic and Antarctic sites was retrieved (Figure 14.1). Ice cores are among the best archives for paleoclimate reconstruction. This is due to their direct recordings of climate change in the different phases of ice (ice matrix, air bubbles, contaminants) through, for example, greenhouse gases (CO_2, CH_4, N_2O), temperature proxies (δ^{18}Oice, $\delta^{15}N$, $\delta^{36}Ar$), chemical proxies (CO_3^-, Na^+, and so on), volcanic eruption proxies (electrical conductivity, sulfur), and many more. We know from detailed analyses of different ice cores from Greenland and Antarctica that there have been significant changes in the greenhouse gas concentrations in former times (Figure 14.2; Raynaud *et al.* 1993, Petit *et al.* 1999).

Due to the higher dust content in Greenland ice cores and potential implications on the gas compositions of entrapped bubble air such as CO_2 production through decomposition of organic compounds or from carbonates, reconstructions of greenhouse gas concentrations and their isotope compositions are generally restricted to Antarctic ice cores. This is especially the case for CO_2 (Anklin *et al.* 1995), since Antarctic ice cores have lower dust concentrations of at least a factor of 10 (Fuhrer *et al.* 1993, 1996, Mayewski *et al.* 1996, Röthlisberger *et al.* 2000) due to their far more distant locations from industrial places compared to Greenlandic sites. However, analyses of ice cores from both, the northern and southern, ice cores allow us to trace hemispheric gradients of any kind of parameter back in time.

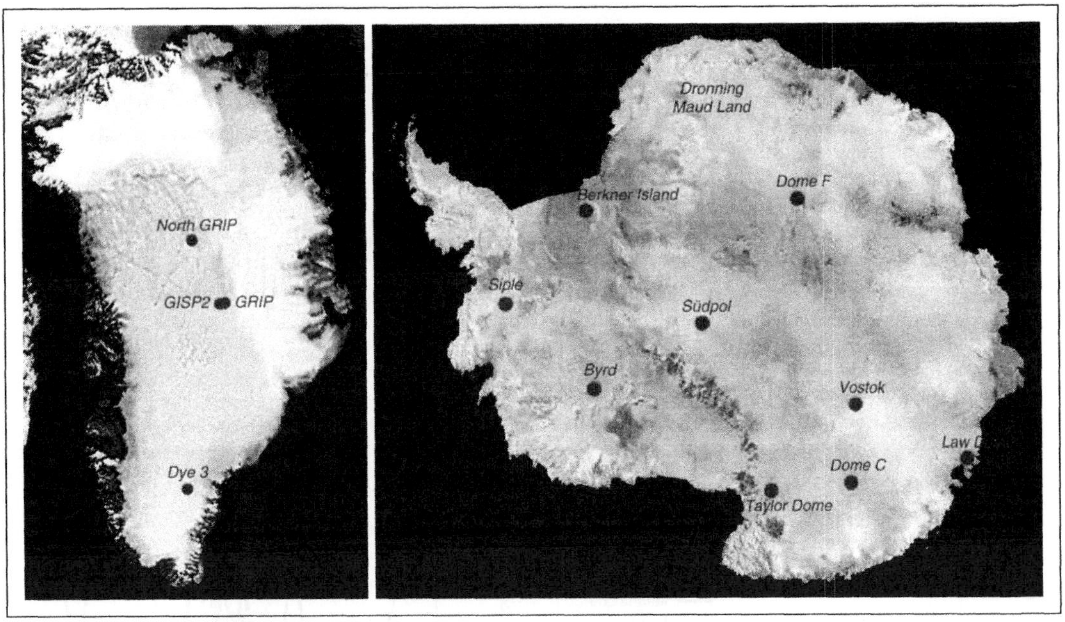

FIGURE 14.1 Ice core drilling sites in Greenland as well as in Antarctica.

This study deals with the influence of biosphere–atmosphere exchange on the atmospheric gas composition. Focus is given to CO_2 and its isotopic carbon signature since (1) issues regarding methane are discussed by White *et al.* (Chapter 15) and (2) information regarding nitrous oxide and its isotopes are still very vaguely understood. However, it should be noted that studies on the nitrogen cycle are important since it carries information about the sources and sinks of plant nutrients and links of the nitrogen cycle to the carbon cycle.

II. CONCENTRATION RECORDS

In the following sections, the gas records for different time intervals extracted from ice cores as well as records from direct atmospheric measurements are discussed and partly shown in Figures 14.2–14.7. Most of the records are available either from the NOAA Paleoclimate archive server (http://www.ncdc.noaa.gov/paleo/data.html) or from the corresponding atmospheric networks, see below.

A. The Last 650,000 Years

Already in the early eighties of the former century it was observed that the CO_2 concentration of the last glacial period was mostly below 220 parts per million (ppm). During the glacial to interglacial transitions it rose generally to about 280 ppm. The variability during glacial periods was rather small with about 20 ppm (Neftel *et al.* 1982). A breakthrough was then obtained by analyzing the Vostok ice core extending back to 400,000 years (Barnola *et al.* 1987, Petit *et al.* 1999). Another published record covers the last 650,000 years (Siegenthaler *et al.* 2005). Around 430,000 years ago, the climate system underwent a significant change (Brunhes transition) in that the ice water isotope values became significantly more enriched during interglacial periods (VI–VII) (still very negative compared to

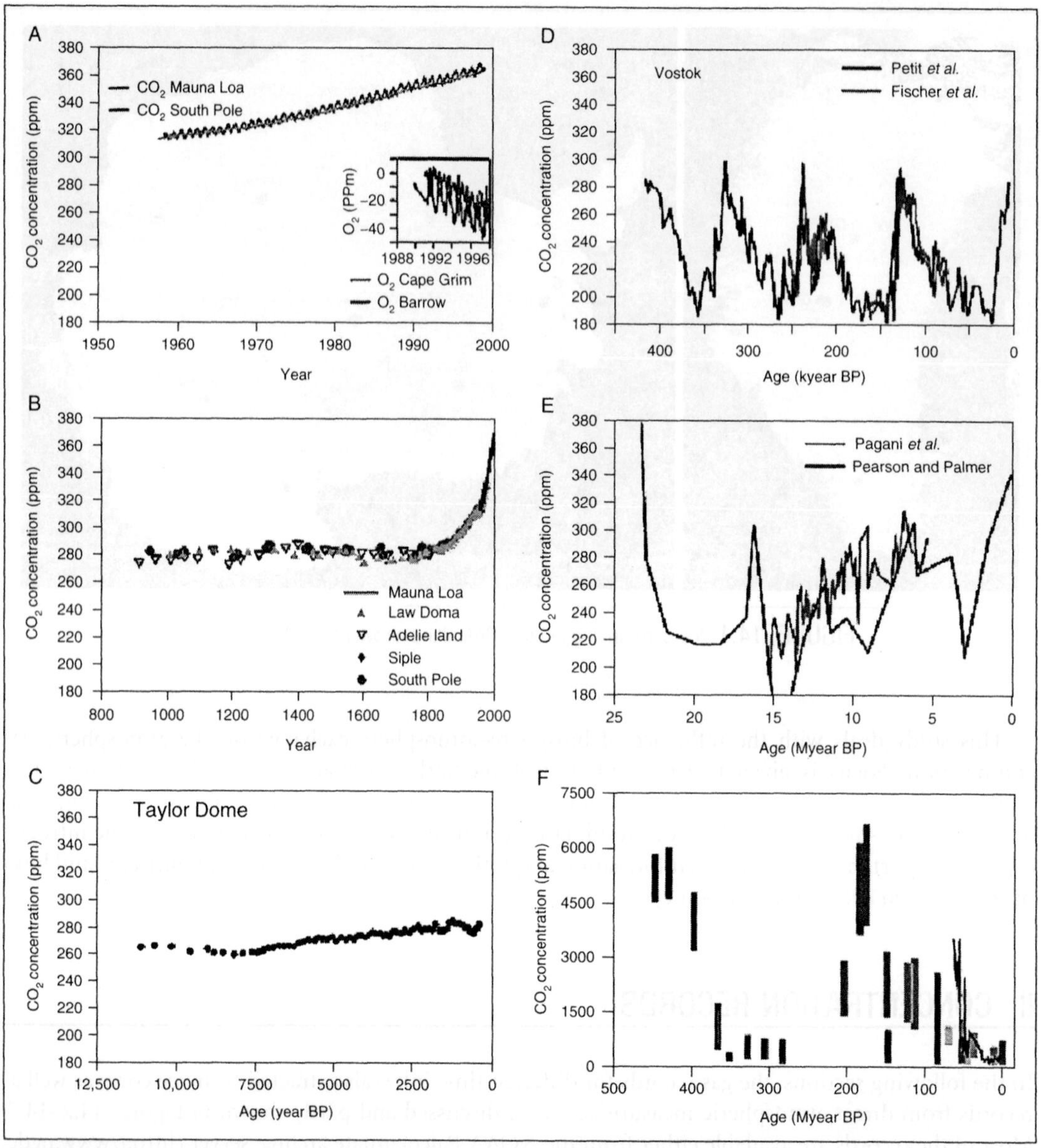

FIGURE 14.2 Reprinted from IPCC (Houghton *et al.* 2001): Variations in atmospheric CO_2 concentration on different timescales. (A) Direct measurements of atmospheric CO_2 (Keeling and Whorf 2000) and O_2 concentration from 1990 onward (Battle *et al.* 2000). O_2 concentration is expressed as the change from an arbitrary standard. (B) CO_2 concentration in Antarctic ice cores for the past millennium (Siegenthaler *et al.* 1988, Neftel *et al.* 1994, Barnola *et al.* 1995, Etheridge *et al.* 1996). Atmospheric measurements at Mauna Loa (Keeling and Whorf 2000) are shown for comparison. (C) CO_2 concentration in the Taylor Dome Antarctic ice core (Indermühle *et al.* 1999). (D) CO_2 concentration in the Vostok Antarctic ice core (Fischer *et al.* 1999, Petit *et al.* 1999). (E) Geochemically inferred CO_2 concentrations from Pagani *et al.* (1999) and Pearson and Palmer (2000). (F) Geochemically inferred CO_2 concentrations: colored bars represent different published studies cited by Berner (1997). The data from Pearson and Palmer (2000) are shown by a black line (BP = before present). **(See Color Plate.)**

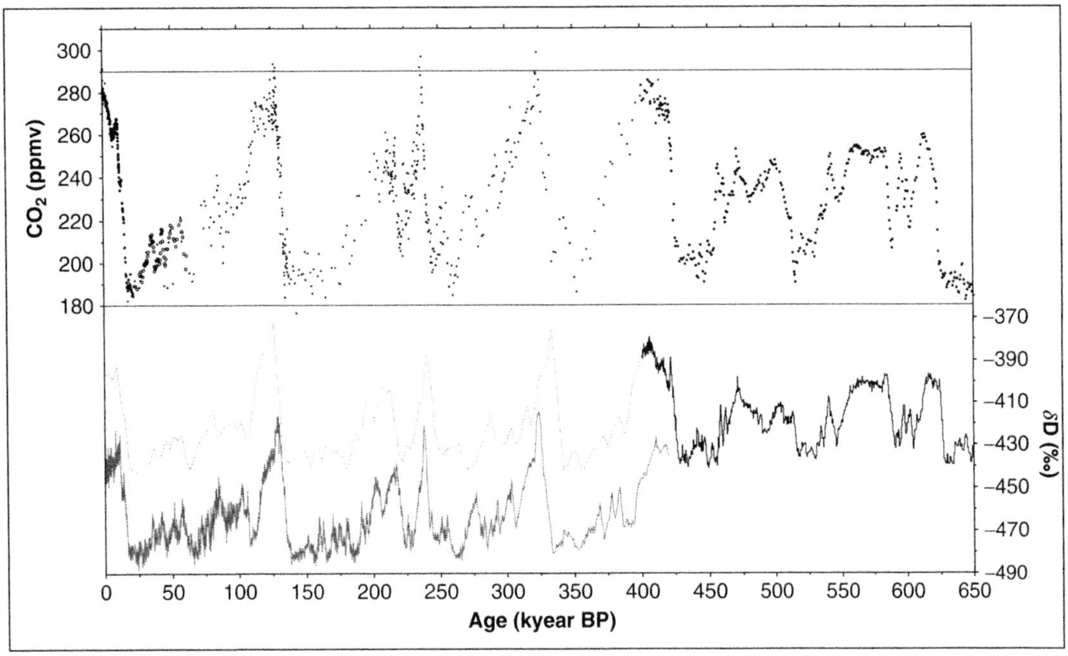

FIGURE 14.3 Reprinted from Science (Siegenthaler *et al.* 2005): A composite CO_2 record over 6½ ice age cycles back to 650,000 years BP. The record results from the combination of CO_2 data from three Antarctic ice cores. Black dots: Dome C, 0–22 kyear BP (Monnin *et al.* 2001, 2004) and 390–650 kyear BP [this work including data from 31 depth intervals over termination V of (EPICA Community Members 2004)]. Rhomboids: Vostok, 0–420 kyear BP (Fischer *et al.* 1999, Petit *et al.* 1999). Open dots: Taylor Dome, 22–62 year BP (Indermühle *et al.* 2000). Light grey line: δD from Dome C, 0–400 kyear BP (EPICA Community Members 2004) and 400–650 kyear BP (black line). Gray line: δD from Vostok, 0–420 kyear BP (Petit *et al.* 1999).

today's ocean values) compared to interglacials (I–V). Figure 14.3 summarizes the evolution over the last 650,000 years. Methane on the other hand did show dramatic variations also during glacial periods as documented in Figure 14.4 (Spahni *et al.* 2005). Nitrous oxide showed variations in range of 200–260 ppb during glacial periods associated with Dansgaard–Oeschger (DO) events (Flückiger *et al.* 2004).

We know that the size of the biospheric reservoir did change by 400–1200 GtC during the last glacial to interglacial transition associated with the large change in CO_2 (Duplessy and Labeyrie 1988, Adams *et al.* 1990). Despite the fact that the exact size of the inventory change is still under debate, it requires an additional CO_2 transfer from the ocean via the atmosphere to the terrestrial biosphere in addition to explain the observed atmospheric CO_2 increase.

B. The Last Transition and the Holocene

Of interest here is certainly the change from a glacial to an interglacial condition. As documented in Figure 14.5, CO_2 rose from 180 ppm at 17,000 years BP stepwise to 265 ppm at 11,000 years BP (Monnin *et al.* 2001). For the Holocene, the last 10,000 years, very detailed investigations were performed on Antarctic ice cores (Indermühle *et al.* 1999, Monnin *et al.* 2004). Variations of the CO_2 concentration were in the order of less than 20 ppm. Between 10,000 and 8500 years BP the CO_2 concentration was decreasing by about 5 ppm to a level of 260 ppm followed by 1500 years of more or less constant values. At around 7000 years BP, the CO_2 concentration began to steadily rise again from

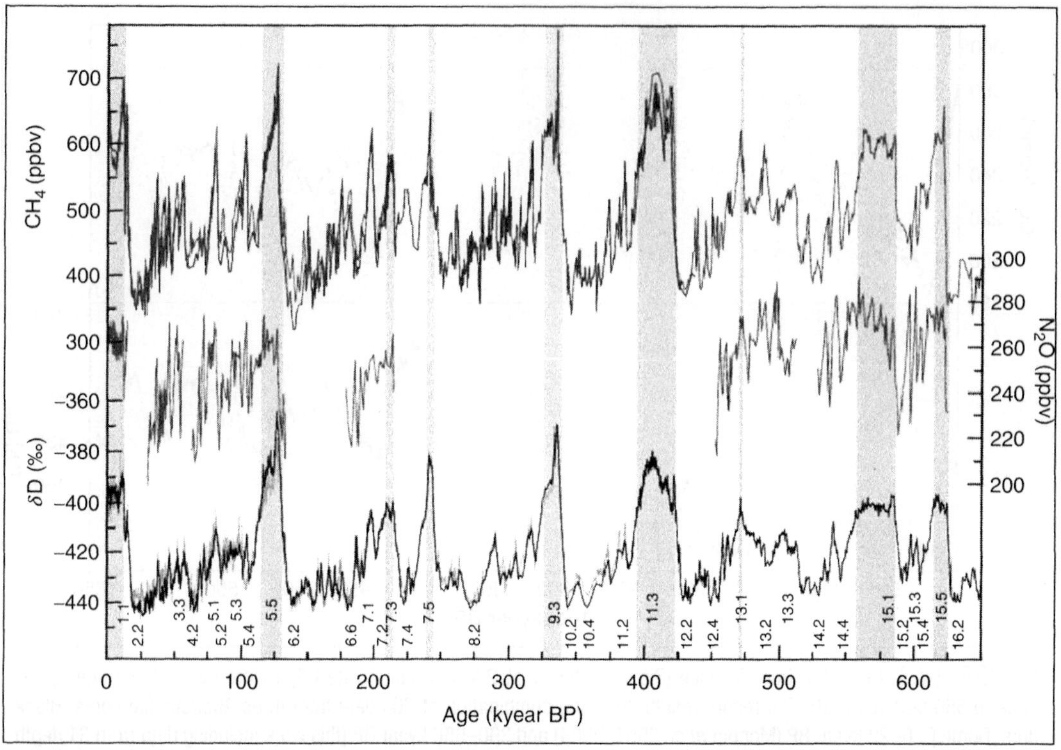

FIGURE 14.4 Reprinted from Science (Spahni *et al.* 2005). CH_4 record over the last 650 kyears composed of Dome C CH_4 (purple line) (Flückiger *et al.* 2002, Monnin *et al.* 2001, and new data) and Vostok CH_4 (blue line) (Delmotte *et al.* 2004, Petit *et al.* 1999). Also shown are the N_2O data measured along the Dome C ice cores (red line) (Flückiger *et al.* 2002, Stauffer *et al.* 2002, 2003, and new data) and δD records from Dome C (black line) as well as from Vostok (gray line, +42 permil) (Petit *et al.* 1999). N_2O artifacts are not shown in this figure. Gray shaded areas highlight interglacial periods with a δD value >-403 permil as defined in EPICA Community Members (2004). Numbers of MIS are given at the bottom of the figure (Sarnthein and Tiedemann 1990). All data are shown on the EDC2 timescale (EPICA Community Members 2004). (**See Color Plate.**)

260 to 280 ppm at the middle of the nineteenth century. Methane increased significantly during the transition from about 350 ppb to about 700 ppb followed by a decrease to 550 ppb at around 5500 years BP and an increase back to 700 ppb during the last millennium. Nitrous oxide increased during the transition from about 200 to 270 ppb followed by a decrease during the early Holocene to 255 ppb and with a recovery to 270 ppb toward the last millennium (Flückiger *et al.* 2002).

C. The Last Millennium

During most of the last millennium only moderate variations with no long-term trend were observed with a mean concentration of ≅280 ppm for CO_2, ≅270 ppb for N_2O, and ≅700 ppb for CH_4. The records from different ice cores and measured by several laboratories do agree quite nicely. The period of 1200–1400 years AD seem to be about 5 ppm higher in CO_2 than the centuries before and the values around 1600–1750 AD. Those differences are most probably the result of an enhanced terrestrial storage during the latter period compared to the previous period (Joos *et al.* 1999). At around 1750 BP when industrialization started, CO_2 began to rise at an unprecedented rate (Figure 14.6).

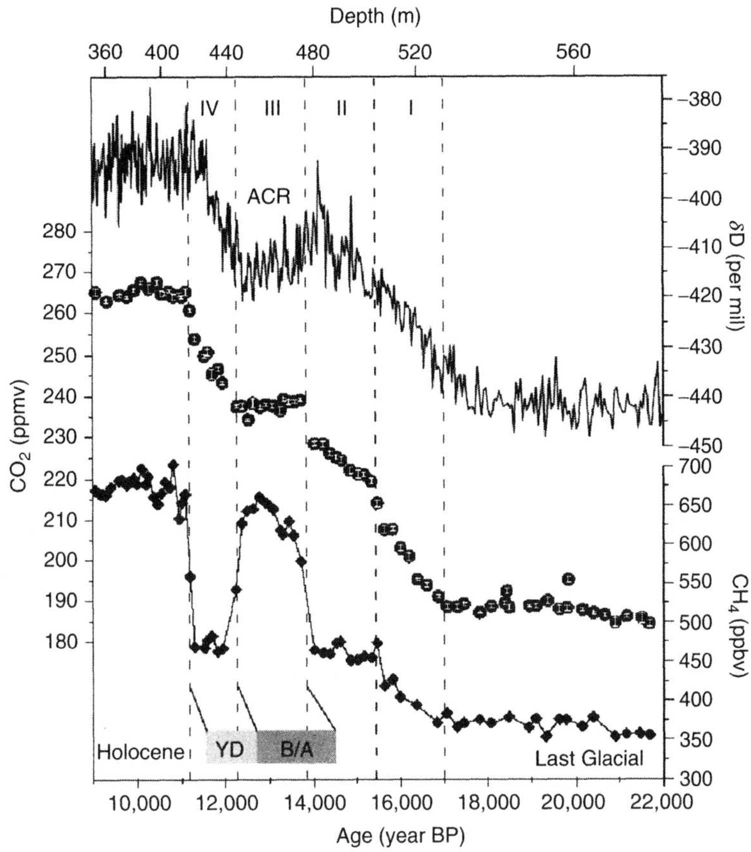

FIGURE 14.5 Reprinted from Science (Monnin *et al.* 2001): The solid curve indicates the Dome C δD in the ice as a proxy for local temperature (Jouzel *et al.* 2001). Solid circles represent CO_2 data from Dome C (mean of six samples; error bars, 1σ of the mean). Diamonds show methane data from Dome C (the 1s uncertainty is 10 ppbv). The timescale used for the gas-ice age is from work by Schwander *et al.* (2001) (the depth at the top of the figure is only valid for the CO_2 and methane records). In the CO_2 and methane records, four intervals (I through IV) can be distinguished during the transition. The δD record is highly correlated with the CO_2 record, with the exception that the increased rates during intervals I and II are not significantly different in the deuterium record. The YD and the B/A events recorded in Greenland ice cores are indicated by shaded bars according to the GRIP timescale. Comparisons of the methane record with that of GRIP demonstrate that the YD corresponds to interval IV and the B/A event corresponds to interval III.

D. The Industrial Period

During the last 250 years the increase of CO_2 was steadily speeding up. This increase is still continuing due to the burning of fossil fuels as well as land-use changes. Up to the year 1920 AD, the CO_2 evolution can be understood by a terrestrial source mainly dominated through land-use changes. After 1920, model results suggest a significant terrestrial sink in contrast to continued growth emissions from land-use changes (Joos *et al.* 1999). This strongly calls for an increasing CO_2 fertilization effect parallel to the CO_2 increase. This will be investigated further below. Within this short time the CO_2 concentration did increase by about 100 ppm, that is, from 280 to 380 ppm today (global mean, NOAA CMDL CCGG cooperative air sampling network). This corresponds to an increase of more than 35% from the preindustrial value. This dramatic change certainly had and still has an influence on the biosphere.

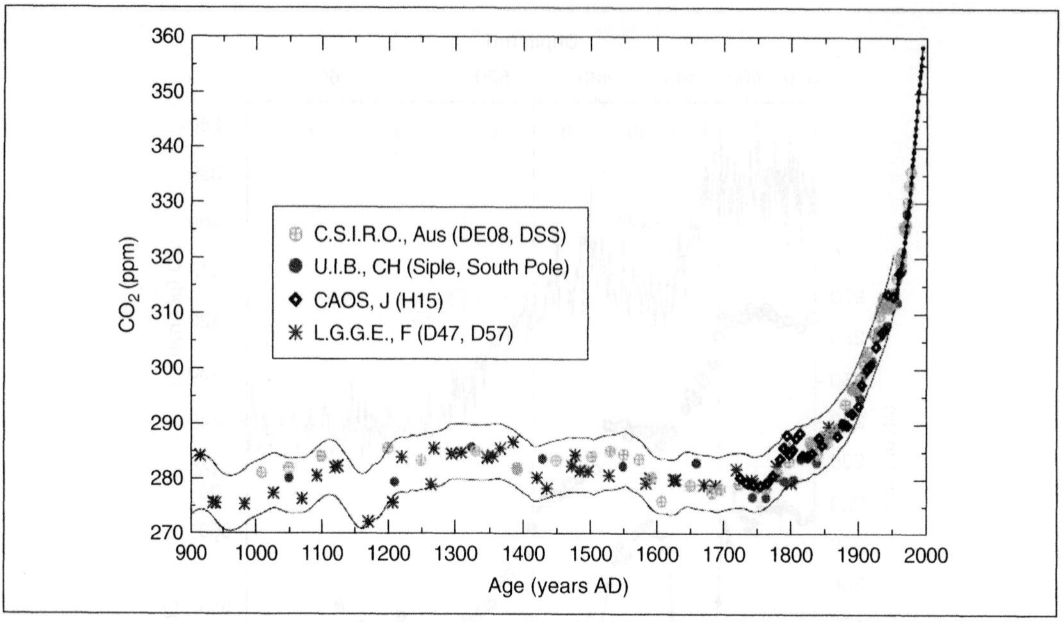

FIGURE 14.6 CO_2 data available for the last millennium from the different Antarctic cores (modified after Barnola 1999). The data are from Neftel *et al.* (1985), Friedli *et al.* (1986), and Siegenthaler *et al.* (1988) for Siple and South Pole data; Barnola *et al.* (1995) for D47 and D57; Nakazawa *et al.* (1993) and Kawamura *et al.* (1997) for H15; Etheridge *et al.* (1996) for DE08 and DSS. The continuous lines are the limits of the ±5-ppmv envelope around a 50 years running mean through the yearly interpolated data.

E. Direct Atmospheric Measurements

Since the late seventies trace gases are measured at many locations. They are networked as for instance the NOAA/CMDL/CCGG network (http://www.cmdl.noaa.gov/ccgg/iadv/index.php). Similar, but smaller networks are run by CSIRO/CAR (http://www.bom.gov.au/inside/cgbaps/) and Scripps institution of oceanography and the Canadian Meteorological Service. These networks are based on flask sampling as well as continuous measurements by a number of observatories. During the last couple of years the Europeans have initiated a network within the framework of EU projects such as AEROCARB and the follow-up CARBOEUROPE-IP (http://www.carboeurope.org/).

III. RELATIONS OF CO_2 AND CH_4 CONCENTRATION RECORDS TO TEMPERATURE OR TEMPERATURE PROXIES

It is obvious that there should be a dependence of the atmospheric gas records on temperature changes due to several causal relationships of involved processes to temperature, for example photosynthesis/respiration cycles or gas solubility depend on temperature. Therefore, it is logical to correlate temperature estimates with changes in greenhouse gas compositions. In Figure 14.7, CO_2 concentrations are correlated to the ice δD values representing temperature variations. The correlation coefficients are rather good with $r^2 = 0.57$ for marine isotope stage (MIS) 12–16, $r^2 = 0.70$ for MIS 1–12, and $r^2 = 0.84$ for the last transition. The sensitivity of the CO_2 concentration to temperature based on ice water isotopes cannot directly be evaluated from this correlation since the sensitivity of the isotope values to

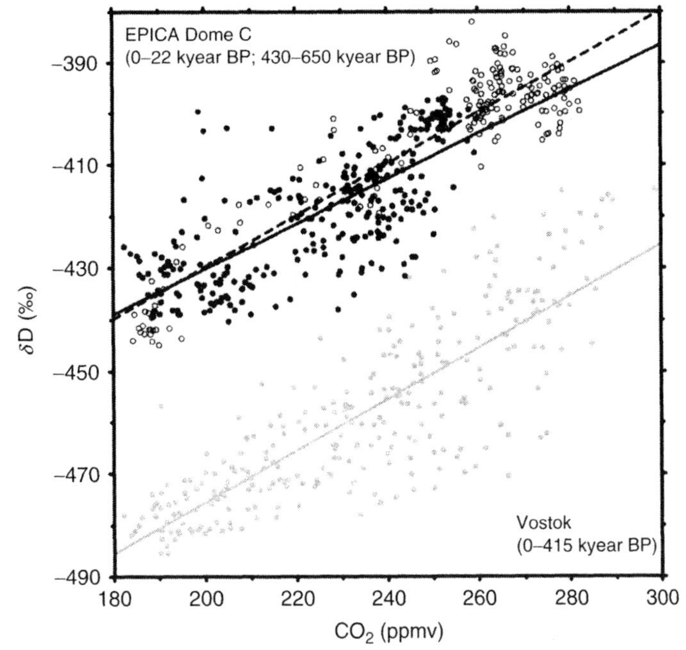

FIGURE 14.7 Reprinted from Science (Siegenthaler *et al.* 2005): Correlation between δD, a proxy for Antarctic temperature, and CO_2 for three datasets. The new data from Dome C cover the beginning of MIS 12–16 (black dots; black line: linear fit $\delta D = 0.44‰\ ppmv^{-1} \times CO_2 - 517.80‰$, $r^2 = 0.57$), and the period from MIS 1 to 11 is covered by data from the Vostok ice core [gray dots (Petit *et al.* 1999); gray line: linear fit ($\delta D = 0.44‰\ ppmv^{-1} \times CO_2 - 517.86‰$, $r^2 = 0.70$)] and Dome C Holocene and termination I [black circles (Monnin *et al.* 2001, 2004); black dashed line: linear fit ($\delta D = 0.50‰\ ppmv^{-1} \times CO_2 - 529.87‰$, $r^2 = 0.84$)]. The offset in the δD values from these two cores is due to the different distance to the open ocean, elevation, and surface temperature of the two sites (Masson *et al.* 2000).

temperature may have changed over time. It is believed that this change is mainly due to a change in the seasonal distribution of precipitation and/or a change of the source area of precipitation (Huber *et al.* 2006). This change is more significant for the Greenland ice sheet than for the large and much more remote Antarctic ice sheet. An impression of such changes of sensitivity can be obtained by a comparison of ice water isotopes from Greenland and an independent temperature estimate using atmospheric nitrogen isotopes.

Temperature estimates are one of the most important goals in paleoclimate research. Ice water isotopes are certainly a very good means to extract temperature information. But it was shown over the last decade by two different approaches that it is difficult to scale the variations based on the spatial correlation sensitivity as observed by Dansgaard (1964). On the one hand, it was shown that estimates of temperature shifts based on direct borehole temperature measurements during the last termination (change from glacial to interglacial conditions between 20,000 and 10,000 years BP) were significantly larger than calculated via the spatial dependence for Greenland (Dahl-Jensen *et al.* 1998). On the other hand, it was noticed that rapid temperature changes are recorded in ice cores due to the physical effect occurring in the uppermost part of an ice sheet, the firn, by altering the nitrogen isotope composition of molecular nitrogen (Severinghaus *et al.* 1996, Lang *et al.* 1999, Landais *et al.* 2004, Huber *et al.* 2006). All these studies showed that the sensitivity of ice water isotopes is less than the present spatially derived sensitivity. However, these studies are restricted to abrupt temperature variations and are, therefore, mainly applied to Greenland ice samples. This leads to much larger temperature changes of up to a factor of two as one would obtain using the spatial sensitivity.

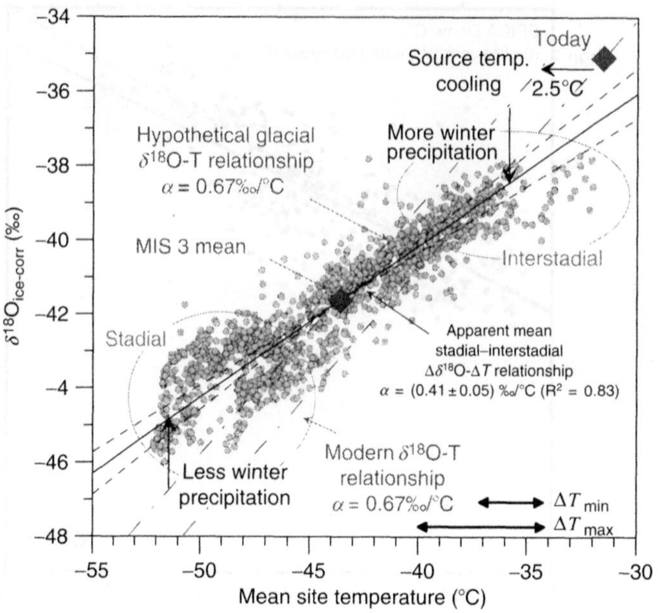

FIGURE 14.8 Relationship between ^{18}O of precipitation and temperature estimates based on nitrogen isotope variations (after Huber *et al.* 2006). δ^{18}O has been corrected ($\delta^{18}O_{ice\text{-}corr}$) for the oceanic change in oxygen isotopes due to changes in land ice extent according to benthic foraminifera measurements.

The results by Huber *et al.* (2006), who investigated MIS 3 in detail, support the idea of a change in the seasonal precipitation distribution. During DO events (warm phases), precipitation was driven towards more winter precipitation whereas the opposite, less winter precipitation, seem to be the case during cold phases. This leads to the apparent smaller sensitivity (slopes of around 0.4 permil/°C compared to the spatially derived slope of 0.67 permil/°C) (Figure 14.8). However, as pointed out by Huber *et al.* (2006), another explanation would be a shift of the source region and associated source temperature changes. Further investigations are needed to clarify this issue.

These temperature estimates allow us to investigate the correlation between the methane concentration and temperature changes. The records, as obtained on NorthGRIP ice, match very well even in small details for certain periods (Figure 14.9). However, the temperature sensitivity for methane did change between DO events 15 and 12 by nearly a factor of two from about 13.3 to about 7.0 ppb/°C (Figure 14.10). This must go along with significant changes in the source strengths, potentially due to changes in the insolation in tropical to mid-northern latitudes (Brook *et al.* 1996, Flückiger *et al.* 2004). In this respect, the results of Keppler *et al.* (2006) are interesting since they propose a significant direct methane source from the biosphere. If such a methane source exists it is very likely that it also would scale with temperature. Hence, the good temperature relation with methane could be the result of such a coupling. If this is the case, the biosphere changes could be estimated in relation to the methane variations as already used by Leuenberger (1997). Depending on the size of the methane source it may be that it could influence the isotopic carbon budget on short timescales of decades (due to the slow response of the deep ocean).

To which extent the glacial–interglacial transitions can act as an example for the industrial situation has to be investigated carefully since they are fundamentally different. The industrial situation that we experience now is driven by emissions of fossil fuels leading to the observed increase in CO_2 that itself leads to a temperature increase through its radiative forcing potential. Hence, the temperature follows the CO_2. The opposite is observed during the natural glacial–interglacial changes for which temperature leads the CO_2 signal (Siegenthaler *et al.* 2005). We certainly learn much about the processes by investigating the past but we have to be careful in applying the findings to our present

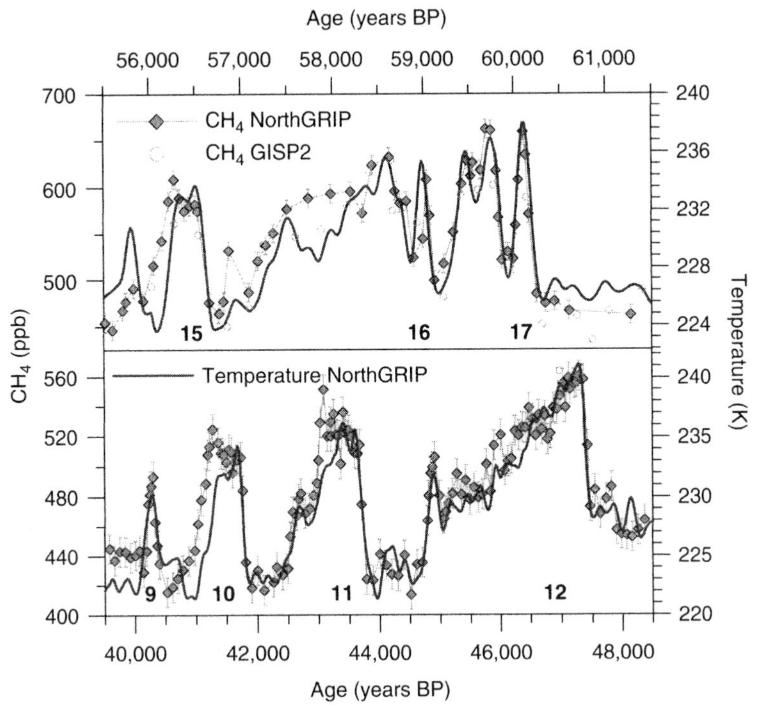

FIGURE 14.9 Comparison of the temporal evolution of methane and estimated temperatures based on nitrogen isotopes for MIS 3 (after Huber *et al.* 2006).

situation. Nevertheless, one lesson to be learnt from the past is that today's rate of CO_2 increase was not experienced before and it leads to a positive feedback mechanism that further enhances the temperature increase.

IV. CARBON ISOTOPE RECORDS

There are only limited records available for atmospheric carbon isotope composition ($\delta^{13}C$) (Friedli *et al.* 1986, Leuenberger *et al.* 1992, Smith *et al.* 1997a,b, Fischer *et al.* 1999, Francey *et al.* 1999). As mentioned above Greenland ice samples could be affected by excess CO_2, that is CO_2 produced *in situ* by acid–carbonate reaction or decomposition of biogenic remains transported to Greenland. This is also influencing the $\delta^{13}C$ and therefore, the ice core community focused to measure CO_2 and other affected tracers on Antarctic ice. Since in this study we will focus on the influence of atmospheric $\delta^{13}C$ variation of CO_2 on living plant material, we restrict our discussion to the last millennium during which significant variations occurred mainly during the industrial period. This further restricts the anyway low number of available $\delta^{13}C$ records.

A. The Industrial Period

For the industrial period only two records for $\delta^{13}C$ are available from Antarctica: the Law Dome record (Francey *et al.* 1999) and the Siple Dome record (Friedli *et al.* 1986). The precision of the Siple Dome record is not as good as for the Law Dome record and the original values were corrected by Leuenberger *et al.* (1992) by −0.10 permil due to the meanwhile discovered gravitational effect and a gas extraction effect that has not been taken into account by Friedli *et al.* (1986). Therefore, in the following

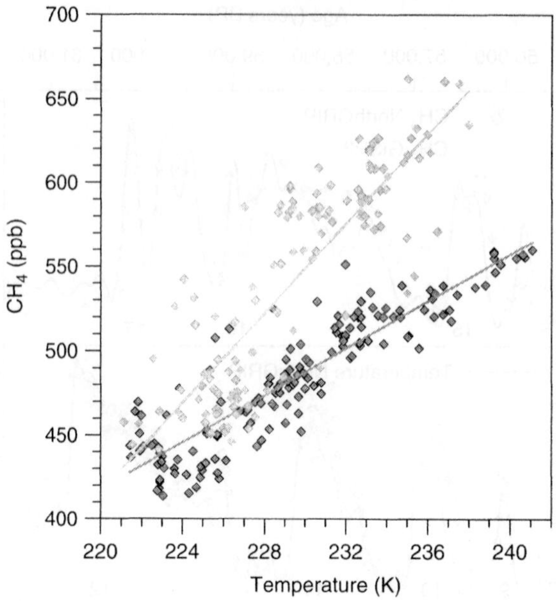

FIGURE 14.10 Temperature sensitivity of methane for MIS 3. Note the significant sensitivity shift during MIS 3, from about 13.6 ppb K^{-1} during DO 17-15 to about 7.7 ppb K^{-1} during DO 9–12.

calculations, we have assumed a $\delta^{13}C$ uncertainty of 0.15 permil for the Siple Dome record. For the Law Dome record we did use an uncertainty of 0.05 permil, slightly higher than the uncertainty stated in the original paper. The reason is that we compare two records which were measured in different laboratories having their own calibration scheme and using different techniques, therefore not the precision but the accuracy is needed.

Francey *et al.* (1999) combined their ice core data with direct atmospheric measurements and they were overlapped with archived air covering the period of 1978–1996 and firn air from Law Dome covering the period of 1978–1993 (Figure 14.11).

This graph nicely documents the dilution of atmospheric CO_2 with isotopically depleted CO_2 originating either from fossil fuel combustion or from land-use changes and to a lesser extent from biomass burning events. The latter two CO_2 sources can be distinguished from the fossil fuel contribution via radiocarbon measurements before the start of bomb tests around 1960. The relative contributions of those effects are discussed for instance by Joos *et al.* (1999). Note that the CO_2 emissions are by far larger than the corresponding increase in the atmosphere mirrors, since about half of the emitted CO_2 has already been taken up by the ocean, assuming that the biosphere remained constant. Because plants are building their carbohydrates based on atmospheric CO_2 and water taken up by the roots, these carbohydrates will mirror the $\delta^{13}C$ variations in the atmosphere. In order to adequately account for this effect a corresponding correction has to be applied when analyzing plant series, for example tree-ring series for paleoclimatic studies. Therefore, the atmospheric $\delta^{13}C$ needs to be known. The following three methods were applied in order to produce a correction record for the observed atmospheric $\delta^{13}C$ trend.

1. Linear Interpolation Technique

Linear interpolation between points is the simplest way of filling gaps of data. If done section by section during which a linear behavior is present then any kind of dependence can be matched. In this case, the resolution of the data is crucial. Since the carbon-13 record shows such a linear behavior for rather

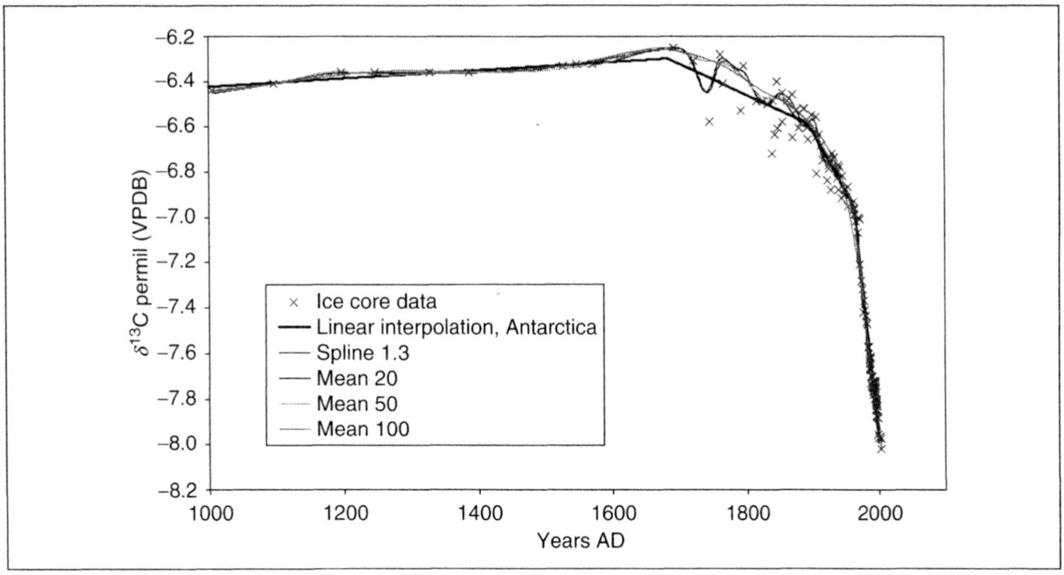

FIGURE 14.11 Compilation of $\delta^{13}C$ data (crosses) from ice cores, firn sampling, archived air tanks, and direct atmospheric air (see text for details). Lines document different data treatment that have been applied (see text for details).

long-time sections, linear interpolation is justified to be applied (Figure 14.11). In Table 14.1, we have summarized the linear functions for the corresponding time intervals that represent atmospheric carbon-13 trends to a high degree.

2. Spline Smoothing Interpolation Technique

Spline smoothing is another technique to treat data in order to extract a mean trend. When applying a spline smoothing algorithm to the data you can choose a smoothing factor that adjusts the degree of agreement between the spline function and the data. Depending on the spread and resolution of the data, spline functions may have large excursions between single data points. In order to account for the reliability of data points they can be weighted by their uncertainty. We applied a spline smoothing as described in Reinsch (1967).

3. Cutoff Frequency Interpolation Technique

Cutoff frequency interpolation is another kind of gap-filling. As derived from its name it cuts off any variations below a certain frequency. A description of the mathematical formulation is given in Enting (1987). One can combine such a method with a Monte–Carlo simulation in order to obtain a band of uncertainty for the interpolation function. Cutoff frequency should be estimated from the underlying physical or chemical processes driving the data variations. In the case of ice core data, firn diffusion and the close-off procedure lead to a natural smoothing of the atmospheric record. The mean age distribution of a sample depends mainly on the accumulation and temperature at the site. Low accumulation sites integrate over a long period of up to several hundred years compared to high accumulation sites that integrate over a couple of decades with a minimal age distribution of two decades.

The results obtained with these three different techniques are illustrated in Figure 14.11. Spline smoothing is critically dependent on the smoothing factor applied. A spline smoothing factor of 1.3 leads to a similar spline function as with the cutoff frequency method using a cutoff of 20 years (Figure 14.12). This seems to be unrealistic because it captures variations in $\delta^{13}C$ that are unexpected

TABLE 14.1 Correction for the atmospheric trend in the form $\Delta\delta^{13}C = A*Time + B$, $\delta^{13}C_{corr} = \delta^{13}C_{plants} - \Delta\delta^{13}C$

Time periods (years BP)	Southern Hemisphere	North–South Gradient	Northern Hemisphere
1000–1682	$A = 0.00018459$	$A = 0.0000222$	$A = 0.00020674$
	$B = -0.18459$	$B = -0.0221526$	$B = -0.20674257$
1683–1891	$A = -0.00138829$	$A = -0.00016661$	$A = -0.00155490$
	$B = 2.46158732$	$B = 0.29541415$	$B = 2.75700147$
1892–1961	$A = -0.00522930$	$A = -0.00062757$	$A = -0.00585687$
	$B = 9.72796331$	$B = 1.16744913$	$B = 10.89541244$
1962–2005	$A = -0.02615944$	$A = -0.00313938$	$A = -0.02929882$
	$B = 50.78324932$	$B = 6.09447817$	$B = 56.87772749$

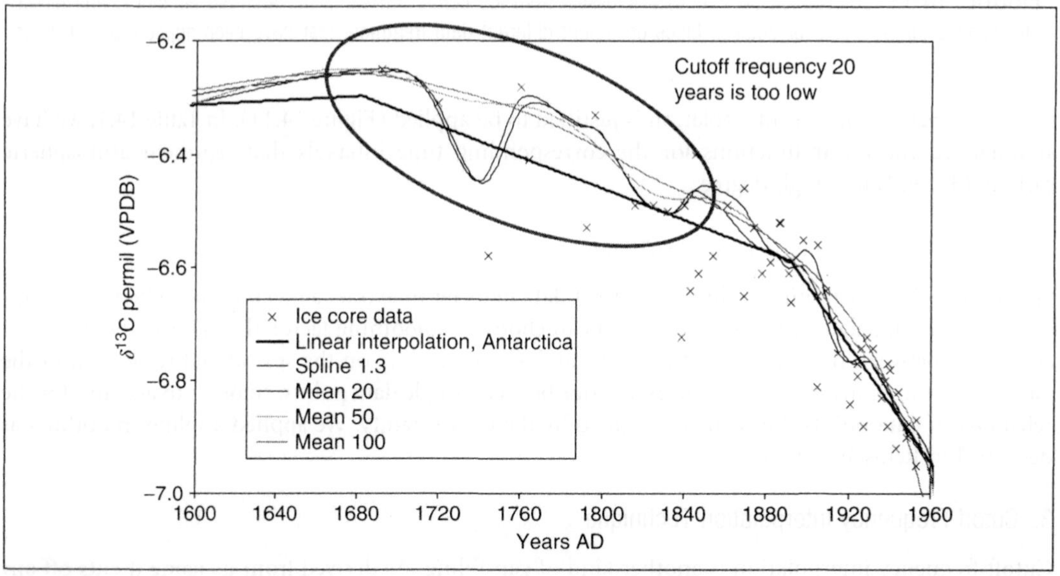

FIGURE 14.12 A low cutoff frequency of 20 years leads to large variations according to the scatter of the underlying data. Such variations are not supported by biogeochemical models. (**See Color Plate.**)

regarding the fast dampening of a Dirac Delta-Puls input (unit impulse function) to the atmosphere (Joos *et al.* 1996). On the other hand, a large smoothing factor leads, similarly as a high cutoff frequency, to a loss of information since smoothing becomes too strong (Figure 14.13). Due to the rather low resolution of the ice core record and its rather high associated uncertainty of 0.05 permil for the Law Dome record and 0.15 permil for the Siple record that are in the range of the variations seen up to 1900, we decided to base our final correction on the linear interpolation of the data. An alternative would be a smoothing function of intermediate cutoff frequency, maybe 50 years, as shown in Figures 14.12–14.14. The differences are generally below 0.10 permil.

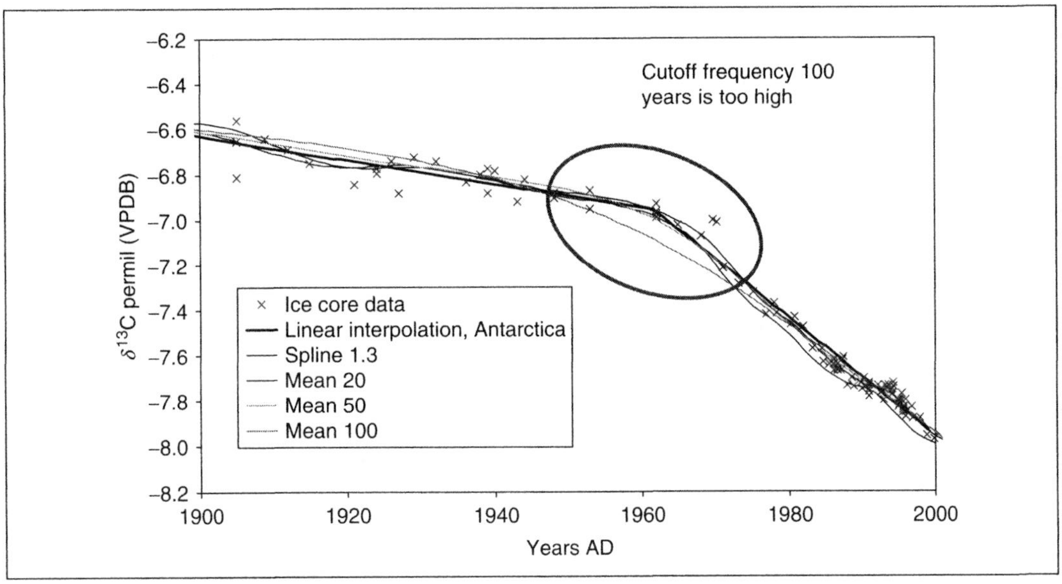

FIGURE 14.13 A high cutoff frequency of 100 years could lead to a loss of information as shown for the period from 1940 to 1970 (pink line). (**See Color Plate.**)

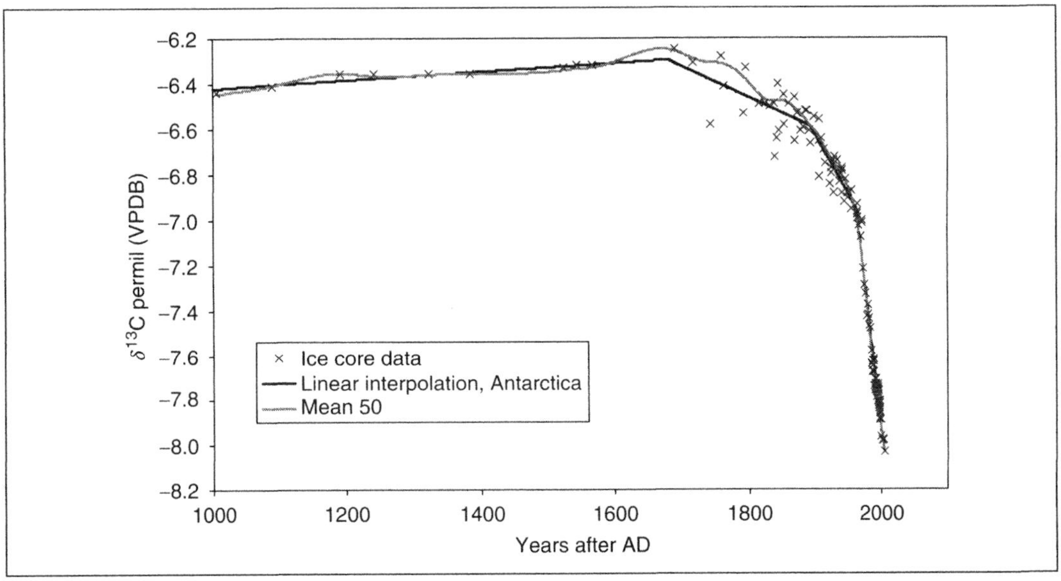

FIGURE 14.14 Approximation of measured $\delta^{13}C$ values as a basis for $\delta^{13}C$ correction of plant material as a function of time (Table 14.1).

V. CORRECTION FOR THE LAST 1000 YEARS

The linear approach leads to a correction value of 1.67 permil for today compared to the year 1000 AD for the Southern Hemisphere and 1.87 for the Northern Hemisphere. The value for the year 1000 AD of $-6.42 + -0.05$ permil has been obtained by linearly extrapolating the measured ice core values for the period 1000–1570. Included in the correction file for the Northern Hemisphere is the effect of the increasing hemispheric gradient during industrialization due to a higher fossil fuel output in the industrialized countries that are mainly concentrated on the Northern Hemisphere (Table 14.1).

VI. CHANGE IN HEMISPHERIC GRADIENT DURING INDUSTRIALIZATION

The change in atmospheric $\delta^{13}C$ is due to the emissions of fossil fuels and hence in relation to the CO_2 increase observed during industrialization as demonstrated in Figure 14.15A. The air originates from ice cores, firn, archived samples, and direct air from Law Dome and Cape Grim, respectively. Therefore,

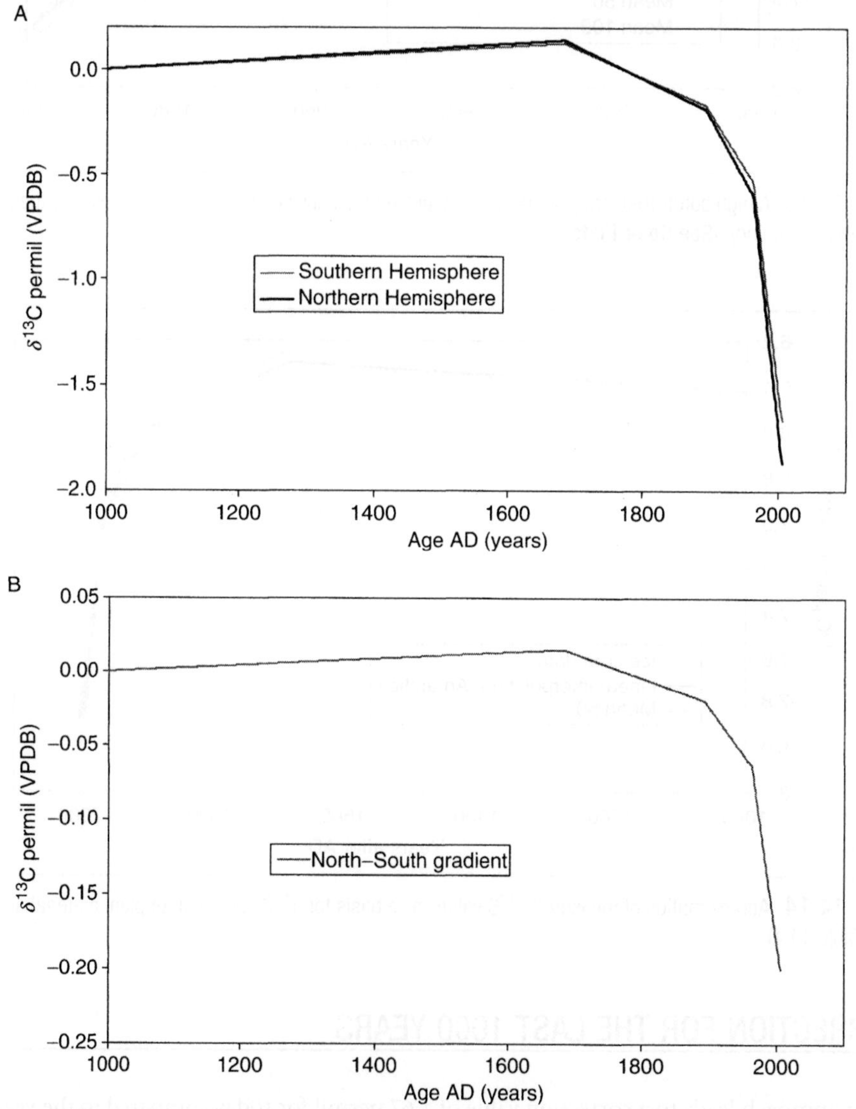

FIGURE 14.15 (A) $\delta^{13}C$ correction for the Northern and Southern Hemisphere. (B) The North–South gradient based on a scaled difference observed today (see text for details).

we safely can assume that also the gradient in $\delta^{13}C$ should scale to the corresponding CO_2 gradient between the hemispheres. Since the gradient is known neither for CO_2 nor for $\delta^{13}C$ except for the last 10–15 years, we make use of the $\delta^{13}C$ decrease itself. Scaling of today's observed gradient to the evolution of the observed $\delta^{13}C$ values leads to an increasing gradient with time. The influence of this estimated hemispheric gradient on the correction is documented in Figure 14.15B. The correction for the Northern Hemisphere compared to the Southern Hemisphere is in the order of 0.2 permil today.

VII. POTENTIAL IMPLICATIONS FROM GROWING SEASON LENGTHENING

Atmospheric CO_2 as well as $\delta^{13}C$ of CO_2 have seasonal signatures due to the exchange with the biosphere (photosynthesis and respiration processes). The mean seasonal amplitude for $\delta^{13}C$ is as high as 1 permil (maximum to minimum value) and the corresponding change in CO_2 is about 15 ppm. Figure 14.16 shows the mean seasonality at four stations, Mauna Loa, Alert, Point Barrow, and Hegyhatsal. The differences in the observed pattern and strength of the seasonality are mainly representing the different fingerprints (region of signal capture) for those stations. Hegyhatsal shows the strongest seasonality which is not surprising since it is strongly influenced by continental climate with a large biospheric imprint. In contrast, Mauna Loa exhibits only small seasonal variations due to its remoteness of extended land areas. Values have been taken from the NOAA/CMDL/CCGG network.

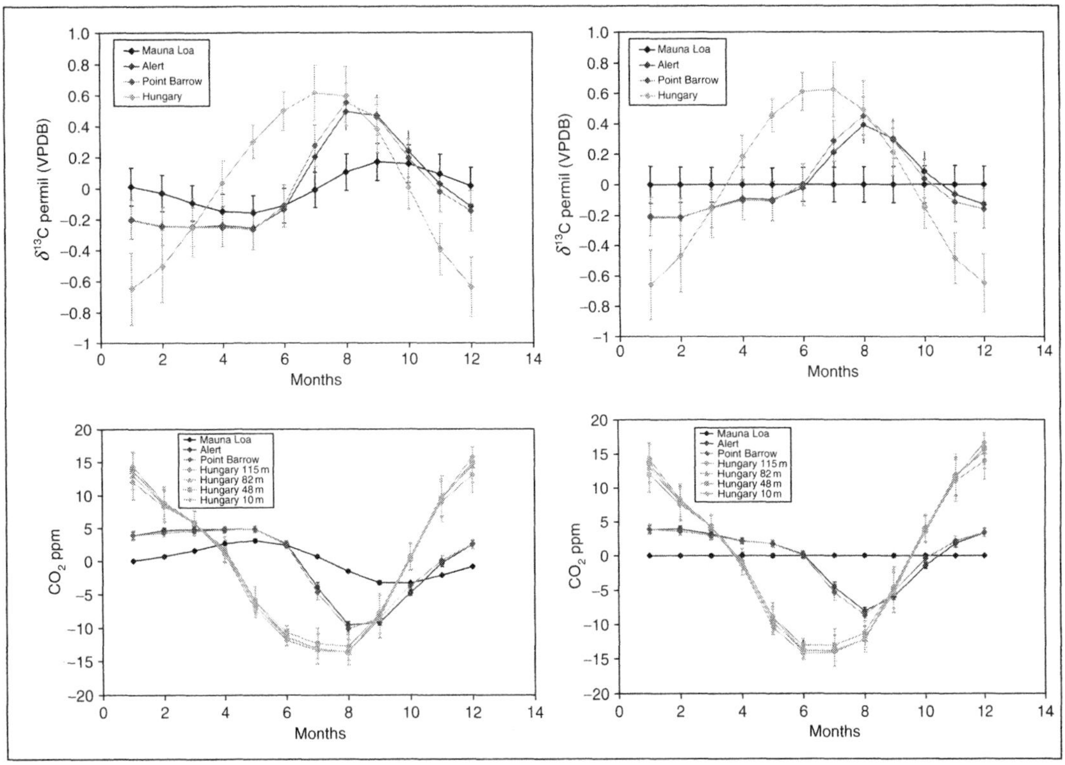

FIGURE 14.16 Left: Seasonal variations and corresponding errors in CO_2 and $\delta^{13}C$ of four stations from the NOAA/CMDL/CCGG cooperative air sampling network (data are from Mauna Loa, Alert, Point Barrow, and the Hegyhatsal site). Right: Same data, but plotted as differences from Mauna Loa taken as reference.

Assume for now, that CO_2 does influence climate and lead to a change in the start and length of the growing season, then two questions can be posed. What are the influences of this shift and widening of the growing season on the $\delta^{13}C$ correction? Can we potentially use this approach to investigate changes of growing season length and/or start? These questions were already addressed by Keeling *et al.* (1995, 1996) and Myneni *et al.* (1997). Here, this issue will be discussed only qualitatively. If the growing season of a particular site is indeed lengthened, then this would lead to lower $\delta^{13}C$ values in the plants if the start of the growing season at this particular site is equal or before the start of the northern hemispheric mean growing season, defined here as zero level passage (Figure 14.16). In the other case, it would lead to an increase of the $\delta^{13}C$ in plant material. If the growing season would be lengthened anywhere retaining productivity, it would lead to larger amplitudes but not necessarily to a shift in the isotopic composition in plants. Hence, changing gradients in atmospheric $\delta^{13}C$ between monitoring stations can be used to infer not only increases in assimilation but also lengthening of the growing season. Therefore, a network of tree-ring sites could give us information not only about temperature and precipitation changes but potentially also about growing season changes.

VIII. CO_2 FERTILIZATION EFFECT

As already mentioned above, the rising CO_2 concentration is likely to enhance the photosynthetic rate and therefore leads to sequestration of CO_2. Similarly, rising CO_2 concentration leads to an enhanced greenhouse effect resulting in a temperature increase. This will affect respiration, however, most probably to a lesser extent than photosynthesis due to lower temperature sensitivity. To what extent

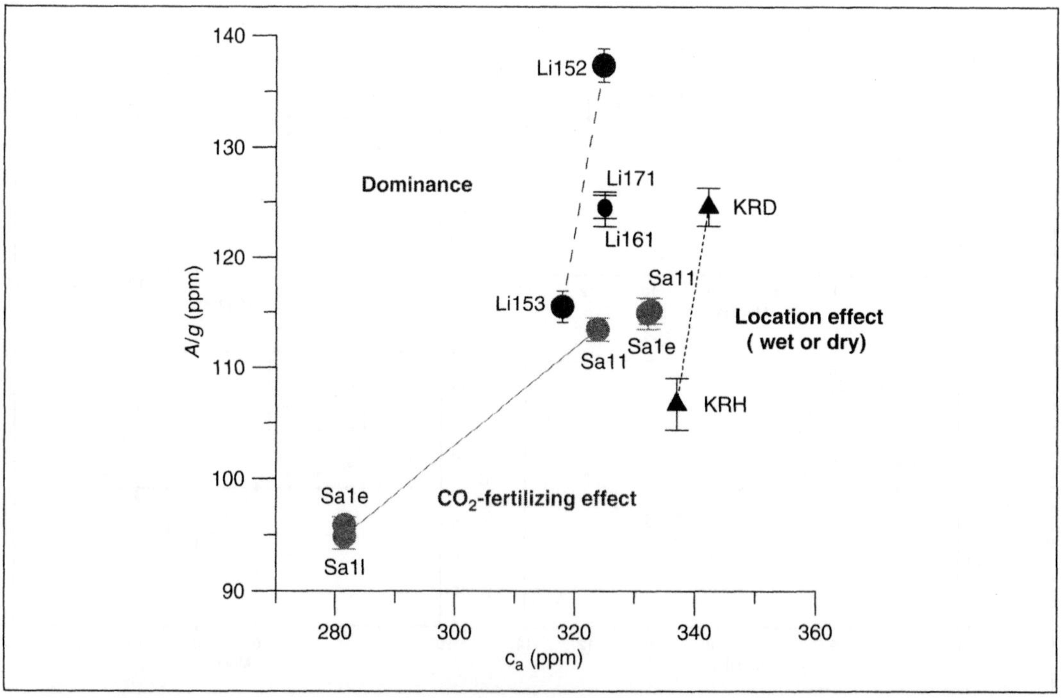

FIGURE 14.17 Interpretation of observed changes in the mean ratio for assimilation to lead conductance for two periods exhibiting different CO_2 concentrations. Data are based on combined tree-ring isotope measurements series from three locations: two Swiss sites (Salvenach and Krauchthal) and one German site (Limlingerode).

and where this is happening is still under debate. There is growing evidence that the terrestrial biosphere is taking up a significant amount of the released CO_2. Estimates vary around 1 ± 1 GtC, corresponding to about 0–25% of the annual emissions. At present 50–60% accumulates in the atmosphere and 20–30% is taken up by the oceans (Joos *et al.* 1999, Battle *et al.* 2000, Kohler *et al.* 2005). The major problem of uncertainty originates from the short monitoring period on which the estimates are based. In this respect, the terrestrial plants and in particular the trees can be of help, since they carry the history of the rising CO_2 availability as well as their reaction to it. Hence, there is a potential to extract information about the fertilization effect in a quantitative way by carefully looking at the isotope information in tree rings over time. A multiple isotope approach (*i.e.*, combination of carbon and oxygen isotope measurements on tree rings) may be the best way of tackling this problem as shown by Saurer and Siegwolf (Chapter 4). However, one has to be careful with the interpretation of the isotopes since the impacts on tree-ring isotopes are manifold. As documented by Figure 14.17, the ratio of assimilation to leaf conductance, A/g, as derived from $\delta^{13}C$ of cellulose can vary due to changes in site conditions, for example dominance-suppression behavior and fertilization effect. Excluding dominance-suppression effects requires a good site selection for which a long-term knowledge is available about the development of the investigated site. Furthermore, one should not combine isotope ratio measurements with different site characteristics if one aims to extract a fertilization effect. A dominant tree will behave differently as compared to a smaller dominated tree because it has not access to free atmospheric air but has to reassimilate a large part of respired air, partly from the soil. Additionally, it has to deal with far less light, mainly indirect illumination compared to the dominant tree. This is documented in Figure 14.17 with data from three different locations, Salvenach, Krauchthal in Switzerland, and Limlingerode in Germany (Borella *et al.* 1998). In case of the Swiss site Salvenach, we have found a location that should be well suited to investigate a fertilization effect since it is a large area that was planted with oak and maintained in an extensive way. Since the soil characteristics are very similar throughout the stand and each single oak has the same age, we can assume that dominance-suppression effects are rather low. This can actually be verified by inspection on site where these 270 years old oaks are of very similar size. Mean A/g values from a period around 1800 AD to a period of 1960 AD show a clear increase in A/g. This, however, does not tell us whether A or g has changed. As pointed out by Saurer and Siegwolf (Chapter 4), we can disentangle those two influences by adding oxygen results. We have done that and saw no clear trend in $\delta^{18}O$ indicating that leaf conductance did not change significantly. Therefore, it is very likely that A did increase documenting a CO_2 fertilization effect. The increase in A/g corresponds to about 20%. Whether the CO_2 fertilization effect amounts to the same percentage cannot be evaluated with confidence yet.

IX. CONCLUSIONS

In order to reliably reconstruct climate variations from biological systems, a thorough understanding of the processes involved is required. In this respect ice core data can help since variations of trace gases over several glacial–interglacial cycles document the basic relationships with several parameters, mainly temperature. Temperature reconstruction itself is difficult and most reconstructions are critically dependent on their calibration scheme. For studies based on living plants, it has to be considered that climate underwent a significant change starting early during the industrialization. This is nicely documented by the increasing CO_2 concentration and its associated isotopic carbon composition. The latter being very important when analyzing time series of living plant material, such as tree rings. A correction for the atmospheric $\delta^{13}C$ change is tabulated and several influencing processes are briefly discussed. Furthermore, it is discussed and documented in an example that the interpretation of isotopic composition in plants is very much dependent on site conditions. Therefore, site selection is a critical issue.

X. ACKNOWLEDGMENTS

The author would like to thank the organizers for the invitation to the Joint BASIN and SIBAE meeting at Tomar, Portugal in March 2006 about "Isotopes as Tracers of Ecological Change." The author is very grateful to two anonymous reviewer comments that significantly improved the chapter. The presented work was financially supported by the EU-project ISONET [contract number (EVK2-CT-20021-00147)] and the Bundesamt für Bildungswesen, Schweiz [contract number (01.0499-2)]. The author would like to thank Renato Spahni and Urs Siegenthaler for figure copies of their publications.

XI. REFERENCES

Adams, J. M., H. Faure, L. Fauredenard, J. M. McGlade, and F. I. Woodward. 1990. Increases in terrestrial carbon storage from the last glacial maximum to the present. *Nature* **348**(6303):711–714.

Anklin, M., J. M. Barnola, J. Schwander, B. Stauffer, and D. Raynaud. 1995. Processes affecting the CO_2 concentrations measured in Greenland ice. *Tellus* **47B**:461–470.

Bacastow, R. B., C. D. Keeling, and T. P. Whorf. 1985. Seasonal amplitude increase in atmospheric CO_2 concentration at Mauna Loa, Hawaii, 1959–1982. *Journal of Geophysical Research-Atmospehers* **90**(ND6):10529–10540.

Barnola, J. M. 1999. Status of the atmospheric CO_2 reconstruction from ice cores analyses. *Tellus Series B-Chemical and Physical Meteorology* **51**(2):151–155.

Barnola, J. M., D. Raynaud, Y. S. Korotkevich, and C. Lorius. 1987. Vostok ice core provides 160,000-year record of atmospheric CO_2. *Nature* **329**:408–413.

Barnola, J.-M., M. Anklin, J. Porcheron, D. Raynaud, J. Schwander, and B. Stauffer. 1995. CO_2 evolution during the last millenium as recorded by Antarctic and Greenland ice. *Tellus* **47B**:264–272.

Battle, M., M. L. Bender, P. P. Tans, J. W. C. White, J. T. Ellis, T. Conway, and R. J. Francey. 2000. Global carbon sinks and their variability inferred from atmospheric O_2 and $\delta^{13}C$. *Science* **287**:2467–2470.

Berner, R. A. 1997. The rise of plants and their effect on weathering and atmospheric CO_2. *Science* **276**:544–546.

Borella, S., M. Leuenberger, M. Saurer, and R. T. W. Siegwolf. 1998. Reducing uncertainties in $\delta^{13}C$ analysis of tree rings: Pooling, milling and cellulose extraction. *Journal of Geophysical Research* **103**(D16):19519–19526.

Brook, E. J., T. Sowers, and J. Orchardo. 1996. Rapid variations in atmospheric methane concentration during the past 110,000 years. *Science* **273**:1087–1091.

Dahl-Jensen, D., K. Mosegaard, N. Gundestrup, G. D. Clow, S. J. Johnsen, A. W. Hansen, and N. Balling. 1998. Past temperatures directly from the Greenland ice sheet. *Science* **282**:268–271.

Dansgaard, W. 1964. Stable isotopes in precipitation. *Tellus* **16**:436–468.

Delmotte, M., J. Chappellaz, E. Brook, P. Yiou, J. M. Barnola, C. Goujon, D. Raynaud, and V. I. Lipenkov. 2004. Atmospheric methane during the last four glacial-interglacial cycles: Rapid changes and their link with Antarctic temperature. *Journal of Geophysical Research-Atmospheres* **109**(D12), D12104, doi:10.1029/2003JD004417.

Duplessy, J. C., and L. Labeyrie. 1988. Norwegian sea deep water variations over the last climatic cycle: Paleo-oceanographical implications. Pages 83–116 *in* H. Wanner and U. Siegenthaler (Eds.) *Lecture Notes in Earth Science.* Springer-Verlag, New York.

Enting, I. G. 1987. On the use of smoothing splines to filter CO_2 data. *Journal of Geophysical Research* **92**(D9):10977–10984.

EPICA Community Members. 2004. Eight glacial cycles from an Antarctic ice core. *Nature* **429**:623–628.

Etheridge, D. M., L. P. Steele, R. L. Langenfields, R. J. Francey, J. -M. Barnola, and V. I. Morgan. 1996. Natural and anthropogenic changes in atmospheric CO_2 over the last 1000 years from air in Antarctic ice and firn. *Journal of Geophysical Research* **101**:4115–4128.

Fischer, H., M. Wahlen, J. Smith, D. Mastroianni, and B. Deck. 1999. Ice core records of atmospheric CO_2 around the last three glacial terminations. *Science* **283**:1712–1714.

Flückiger, J., E. Monnin, B. Stauffer, J. Schwander, T. F. Stocker, J. Chappellaz, D. Raynaud, and J.-M. Barnola. 2002. High resolution Holozene N_2O ice core record and its relationship with CH_4 and CO_2. *Global Biogeochemical Cycles* **16**(1), doi:10.1029/2001GB001417.

Flückiger, J., T. Blunier, B. Stauffer, J. Chappellaz, R. Spahni, K. Kawamura, J. Schwander, T. F. Stocker, and D. Dahl-Jensen. 2004. N_2O and CH_4 variations during the last glacial epoch: Insight into global processes. *Global Biogeochemical Cycles, GB* **1020**, doi:10.1029/2003GB002122.

Francey, R. J., M. R. Manning, C. E. Allison, S. A. Coram, D. Etheridge, R. L. Langenfelds, D. C. Lowe, and L. P. Steele. 1999. A history of delta C-13 in atmospheric CH_4 from the Cape Grim air archive and Antarctic firn air. *Journal of Geophysical Research* **104**:23631–23643.

Friedli, H., H. Lötscher, H. Oeschger, U. Siegenthaler, and B. Stauffer. 1986. Ice core record of the $^{13}C/^{12}C$ ratio of atmospheric CO_2 in the past two centuries. *Nature* **324:**237–238.

Fuhrer, K., A. Neftel, M. Anklin, and V. Maggi. 1993. Continuous measurements of hydrogen peroxide, formaldehyde, calcium and ammonium concentrations along the new GRIP ice core from Summit, Central Greenland. *Atmospheric Environment* **27A:**1873–1880.

Fuhrer, K., A. Neftel, M. Anklin, T. Staffelbach, and M. Legrand. 1996. High-resolution ammonium ice core record covering a complete glacial-interglacial cycle. *Journal of Geophysical Research* **101**(D2):4147–4164.

Houghton, J. T., Y. Ding, D. J. Griggs, M. Noguer, P. J. van der Linden, and D. Xiaosu (Eds.). Climate Change 2001: The Scientific Basis, Contribution of Working Group I to the Third Assessment Report of the Intergovernmental Panel on Climate Change (IPCC), pp. 944. Cambridge University Press, UK.

Huber, C., M. Leuenberger, R. Spahni, J. Flückiger, J. Schwander, T. F. Stocker, S. J. Johnsen, A. Landais, and J. Jouzel. 2006. Isotope calibrated Greenland temperature record over marine isotope stage 3 and its relation to CH_4. *Earth Planetary Science Letters* **243:**504–519.

Indermühle, A., T. F. Stocker, F. Joos, H. Fischer, H. J. Smith, M. Wahlen, B. Deck, D. Mastroianni, J. Tschumi, T. Blunier, and B. Stauffer. 1999. Holocene carbon-cycle dynamics based on CO_2 trapped in ice at Taylor Dome, Antarctica. *Nature* **398:**121–126.

Indermühle, A., E. Monnin, B. Stauffer, T. F. Stocker, and M. Wahlen. 2000. Atmospheric CO_2 concentration from 60 to 20 kyr BP from the Taylor Dome ice core, Antarctica. *Geophysical Research Letters* **27**(5):735–738.

Joos, F., M. Bruno, R. Fink, U. Siegenthaler, T. F. Stocker, and C. LeQuere. 1996. An efficient and accurate representation of complex oceanic and biospheric models of anthropogenic carbon uptake. *Tellus Series B-Chemical and Physical Meteorology* **48**(3):397–417.

Joos, F., R. Meyer, M. Bruno, and M. Leuenberger. 1999. The variability in the carbon sinks as reconstructed for the last 1000 years. *Geophysical Research Letters* **26**(10):1437–1440.

Jouzel, J., V. Masson, O. Cattani, S. Falourd, M. Stievenard, B. Stenni, A. Longinelli, S. J. Johnsen, J. P. Steffenssen, J. R. Petit, J. Schwander, R. Souchez, *et al.* 2001. A new 27 ky high resolution East Antarctic climate record. *Geophysical Research Letters* **28**(16):3199–3202.

Kawamura, K., T. Nakazawa, T. Machida, S. Morimoto, S. Aoki, Y. Fujii, and O. Watanabe. 1997. Precise estimates of the atmospheric CO_2 concentration and its carbon isotopic ratio during the last 250 years from an Antarctic ice core, H15, in *5th International Carbon dioxide Conference*, Cairns, Australia.

Keeling, C. D. 1958. The concentration and isotopic abundances of atmospheric carbon dioxide in rural areas. *Geochimica Et Cosmochimica Acta* **13**(4):322–334.

Keeling, C. D. 1968. Carbon dioxide from fossil fuel—its effect on natural carbon cycle and on global climate. *Transactions-American Geophysical Union* **49**(1):183.

Keeling, C. D. 1970. Is carbon dioxide from fossil fuel changing mans environment. *Proceedings of the American Philosophical Society* **114**(1):10–17.

Keeling, C. D. 1973. Industrial production of carbon-dioxide from fossil fuels and limestone. *Tellus* **25**(2):174–198.

Keeling, C. D., and T. P. Whorf. 2000. Atmospheric CO_2 records from sites in the SIO air sampling network, in *Trends: A compendium of data on global change.* Carbon Dioxide Information Analysis Center, Oak Ridge National Laboratory, US Department of Energy, Oak Ridge, Tennessee, USA.

Keeling, C. D., J. A. Adams, and C. A. Ekdahl. 1976a. Atmospheric carbon-dioxide variations at South Pole. *Tellus* **28**(6):553–564.

Keeling, C. D., R. B. Bacastow, A. E. Bainbridge, C. A. Ekdahl, P. R. Guenther, L. S. Waterman, and J. F. S. Chin. 1976b. Atmospheric carbon-dioxide variations at mauna-loa observatory, Hawaii. *Tellus* **28**(6):538–551.

Keeling, C. D., A. F. Carter, and W. G. Mook. 1984. Seasonal, latitudinal, and secular variations in the abundance and isotopic-ratios of atmospheric CO_2 2. Results from oceanographic cruises in the tropical pacific-ocean. *Journal of Geophysical Research-Atmospheres* **89**(ND3):4615–4628.

Keeling, C. D., R. B. Bacastow, A. F. Carter, S. C. Piper, T. P. Whorf, M. Heimann, W. G. Mook, and H. Roeloffzen. 1989a. A three-dimensional model of atmospheric CO_2 transport based on observed winds: 1. Analysis of observational data. *Geophysical Monographs of the American Geophysical Union* **55:**165–236.

Keeling, C. D., S. C. Piper, and M. Heimann. 1989b. A three-dimensional model of atmospheric CO_2 transport based on observed winds: 4. Mean annual gradients and interannual variations. Pages 305–363 *in* J. H. Peterson (Ed.) *Aspects of Climate Variability in the Pacific and Western Americas.* AGU, Washington.

Keeling, C. D., T. P. Whorf, M. Wahlen, and J. Vanderplicht. 1995. Interannual extremes in the rate of rise of atmospheric carbon-dioxide since 1980. *Nature* **375**(6533):666–670.

Keeling, C. D., J. F. S. Chin, and T. P. Whorf. 1996. Increased activity of northern vegetation inferred from atmospheric CO_2 measurements. *Nature* **382**(6587):146–149.

Keppler, F., J. T. G. Hamilton, M. Brass, and T. Rockmann. 2006. Methane emissions from terrestrial plants under aerobic conditions. *Nature* **439**(7073):187–191.

Kohler, P., F. Joos, S. Gerber, and R. Knutti. 2005. Simulated changes in vegetation distribution, land carbon storage, and atmospheric CO_2 in response to a collapse of the North Atlantic thermohaline circulation. *Climate Dynamics* **25**(7–8):689–708.

Landais, A., N. Caillon, C. Goujon, A. Grachev, J. M. Barnola, J. Chappellaz, J. Jouzel, V. Masson-Delmotte, and M. Leuenberger. 2004. Quantification of rapid temperature change during DO event 12 and phasing with methane inferred from air isotopic measurements. *Earth and Planetary Science Letters* **225**:221–232.

Lang, C., M. Leuenberger, J. Schwander, and S. Johnsen. 1999. 16°C rapid temperature variation in central Greenland 70000 years ago. *Science* **286**:934–937.

Leuenberger, M. C. 1997. Modeling the signal transfer of seawater $\delta^{18}O$ to the $\delta^{18}O$ of atmospheric oxygen using a diagnostic box model for the terrestrial and marine biosphere. *Journal of Geophysical Research-Atmospehers* **102**(C12):26841–26850.

Leuenberger, M., U. Siegenthaler, and C. C. Langway. 1992. Carbon isotope composition of atmospheric CO_2 during the last ice age from an Antarctic ice core. *Nature* **357**:488–490.

Masson, V., F. Vimeux, J. Jouzel, V. Morgan, M. Delmotte, P. Ciais, C. Hammer, S. Johnsen, V. Y. Lipenkov, E. Mosley-Thompson, J. R. Petit, E. J. Steig, M. Stievenard, and R. Vaikmae. 2000. Holocene climate variability in Antarctica based on 11 ice-core isotopic records. *Quaternary Research* **54**(3):348–358.

Mayewski, P. A., M. S. Twickler, S. I. Whitlow, L. D. Meeker, Q. Yang, J. Thomas, K. Kreutz, P. M. Grootes, D. L. Morse, E. J. Steig, E. D. Waddington, E. S. Saltzman, *et al.* 1996. Climate changes during the last deglaciation in Antarctica. *Science* **272**:1636–1638.

Monnin, E., A. Indermühle, A. Dällenbach, J. Flückiger, B. Stauffer, T. F. Stocker, D. Raynaud, and J.-M. Barnola. 2001. Atmospheric CO_2 concentrations over the last glacial termination. *Science* **291**:112–114.

Monnin, E., E. J. Steig, U. Siegenthaler, K. Kawamura, J. Schwander, B. Stauffer, T. F. Stocker, J.-M. Barnola, B. Bellier, D. Raynaud, and H. Fischer. 2004. Evidence for substantial accumulation rate variability in Antarctica during the Holocene through synchronization of CO_2 in the Taylor Dome, Dome C and DML ice cores. *Earth and Planetary Science Letters* **224**:45–54.

Mook, W. G., M. Koopmans, A. F. Carter, and C. D. Keeling. 1983. Seasonal, latitudinal, and secular variations in the abundance and isotopic-ratios of atmospheric carbon-dioxide.1. Results from land stations. *Journal of Geophysical Research-Oceans and Atmospheres* **88**(NC15):915–933.

Myneni, R. B., C. D. Keeling, C. J. Tucker, G. Asrar, and R. R. Nemani. 1997. Increased plant growth in the northern high latitudes from 1981 to 1991. *Nature* **386**(6626):698–702.

Nakazawa, T., T. Machida, K. Esumi, M. Tanaka, Y. Fujii, S. Aoki, and O. Watanabe. 1993. Measurements of CO_2 and CH_4 concentrations in air in a polar ice core. *Journal of Glaciology* **39**(132):209–215.

Neftel, A., H. Oeschger, J. Schwander, B. Stauffer, and R. Zumbrunn. 1982. Ice core sample measurements give atmospheric CO_2 content during the past 40,000 yr. *Nature* **295**:220–223.

Neftel, A., E. Moor, H. Oeschger, and B. Stauffer. 1985. Evidence from polar ice cores for the increase in atmospheric CO_2 in the past two centuries. *Nature* **315**(6014):45–47.

Neftel, A., H. Friedli, E. Moor, H. Lötscher, H. Oeschger, U. Siegenthaler, and B. Stauffer. 1994. Historical CO_2 record from the Siple station ice core. Pages 11–14 *in* T. A. Boden, D. P. Kaiser, R. J. Sepanski, and F. W. Stoss (Eds.) *Trends '93: A Compendium of Data on Global Change.* Carbon Dioxide Information Analysis Center, Oak Ridge.

Pagani, M., M. A. Arthur, and K. H. Freeman. 1999. Miocene evolution of atmospheric carbon dioxide. *Paleoceanography* **14**:273–292.

Pearson, P. N., and M. R. Palmer. 2000. Atmospheric carbon dioxide concentrations over the past 60 million years. *Nature* **406**:695–699.

Petit, J. R., J. Jouzel, D. Raynaud, N. I. Barkov, J.-M. Barmola, I. Basile, M. Bender, J. Chapppellaz, M. Davis, G. Delaygue, M. Demotte, V. M. Kotlyakov, *et al.* 1999. Climate and atmospheric history of the past 420000 years from the Vostok ice core, Antarctica. *Nature* **399**:429–436.

Raynaud, D., J. Jouzel, J. M. Barnola, J. Chappellaz, R. J. Delmas, and C. Lorius. 1993. The ice record of greenhouse gases. *Science* **259**:926–933.

Reinsch, C. H. 1967. Smoothing by spline functions. *Numerische Mathematik* **10**(3):177–183.

Röthlisberger, R., M. A. Hutterli, S. Sommer, E. W. Wolff, and R. Mulvaney. 2000. Factors controlling nitrate in ice cores: Evidence from the Dome C deep ice core. *Journal of Geophysical Research-Atmospehers* **105**(D16):20565–20572.

Sarnthein, M., and R. Tiedemann. 1990. Younger Dryas-style cooling events at glacial terminations I–VI at ODP site 658: Associated benthic $\delta^{13}C$ anomalies constrain meltwater hypothesis. *Paleoceanography* **5**(6):1041–1055.

Schwander, J., J. Jouzel, C. U. Hammer, J. R. Petit, R. Udisti, and E. Wolff. 2001. A tentative chronology for the EPICA Dome Concordia ice core. *Geophysical Research Letters* **28**(22):4243–4246.

Severinghaus, J. P., E. J. Brook, T. Sowers, and R. B. Alley. 1996. Gaseous thermal diffusion as a gas-phase stratigraphic marker of abrupt warmings in ice core climate records (abstract). *EOS Transaction American Geophysical Union* **77**(17, Spring Meet. Suppl.): S157.

Siegenthaler, U., H. Friedli, H. Loetscher, E. Moor, A. Neftel, H. Oeschger, and B. Stauffer. 1988. Stable-isotope ratios and concentration of CO_2 in air from polar ice cores. *Ann. Glaciol.* **10**:151–156.

Siegenthaler, U., T. F. Stocker, E. Monnin, D. Luthi, J. Schwander, B. Stauffer, D. Raynaud, J. M. Barnola, H. Fischer, V. Masson-Delmotte, and J. Jouzel. 2005. Stable carbon cycle-climate relationship during the late Pleistocene. *Science* **310** (5752):1313–1317.

Smith, H. J., M. Wahlen, D. Mastoianni, and K. C. Taylor. 1997a. The CO_2 concentration of air trapped in GISP2 ice from the Last Glacial Maximum-Holocene transition. *Geophys. Res. Lett.* **24**(1):1–4.

Smith, H. J., M. Wahlen, D. Mastroianni, K. C. Taylor, and P. Mayewski. 1997b. The CO_2 concentration of air trapped in Greenland Ice Sheet Project 2 ice formed during periods of rapid climate change. *Journal of Geophysical Research* **102**(C12):26577–26582.

Spahni, R., J. Chappellaz, T. F. Stocker, L. Loulergue, G. Hausammann, K. Kawamura, J. Fluckiger, J. Schwander, D. Raynaud, V. Masson-Delmotte, and J. Jouzel. 2005. Atmospheric methane and nitrous oxide of the late Pleistocene from Antarctic ice cores. *Science* **310**(5752):1317–1321.

Stauffer, B., J. Flückiger, E. Monnin, M. Schwander, J. M. Barnola, and J. Chappellaz. 2002. Atmospheric CO_2, CH_4 and N_2O records over the past 60,000 years based on the comparison of different polar ice cores. *Annals of Glaciology* **35**:202–208.

Stauffer, B., J. Flückiger, E. Monnin, T. Nakazawa, and S. Aoki. 2003. Discussion of the reliability of CO_2, CH_4 and N_2O records from polar ice cores. *Memoirs of National Institute of Polar Research* **57**(Special Issue):139–152.

CHAPTER 15

The Global Methane Budget over the Last 2000 Years: $^{13}CH_4$ Reveals Hidden Information

James W. C. White,* Dominic F. Ferretti,*,† John B. Miller,‡ David M. Etheridge,§ Keith R. Lassey,† David C. Lowe,† Cecelia M. MacFarling,§,¶ Mark F. Dreier,* Cathy M. Trudinger,§ and Tas van Ommen||

*Department of Geological Sciences and Environmental Studies Program, University of Colorado
†National Institute of Water and Atmospheric Research Ltd.
‡NOAA/ESRL, Global Monitoring Division, Institute of Arctic and Alpine Research
§CSIRO Atmospheric Research
¶School of Earth Sciences, University of Melbourne
||Department of the Environment and Heritage, Australian Antarctic Division, and Antarctic Climate & Ecosystems CRC

Contents

I. INTRODUCTION

Methane is arguably the most dynamic greenhouse gas in the atmosphere. During the last glacial period, the concentration of CH_4 in the atmosphere (pCH_4) rose and fell by about 50% in decades during rapid climate change events (Petit *et al.* 1999, Brook *et al.* 2000, Sowers 2006). From the mid-Holocene to about 300 years ago pCH_4 rose steadily by about 25% (Chappellaz *et al.* 1997, Brook *et al.* 2000). With human population increase and industrialization, pCH_4 is now about 250% higher than it was in the early 1800s (Blunier *et al.* 1993, Etheridge *et al.* 1998), a significantly larger percentage increase than that of carbon dioxide, the gas which is most often the focus of greenhouse gas mitigation and adaptation strategies. Surprisingly, in the last decade pCH_4 has stabilized at a global average of $\sim$1750 ppb. The reasons for these changes are not well understood (Lowe *et al.* 1997, Dlugokencky *et al.* 2003). Our incomplete understanding of the global methane budget is in part due to the difficulty in quantifying emissions from all of the diverse methane sources. As an example of this, work suggests that we may have been missing as much as 50% of modern methane sources (Keppler *et al.* 2006). Complicating the problem is that fluxes from most of these sources vary with climate (Chappellaz *et al.* 1997, Petit *et al.* 1999, Brook *et al.* 2000). Human impacts on methane are also important. Methane emissions from rice cultivation, cattle farming, and biomass burning clearly impact the atmospheric methane budget today, and may have been important long before the industrial era (Subak 1994, Ruddiman and Thomson 2001).

The baseline against which we have measured early human impacts on methane so far is pCH_4 levels in the past obtained from air trapped in ice cores (Rasmussen and Khalil 1984, Blunier *et al.* 1993, Etheridge *et al.* 1998). The story told by these measurements during the late preindustrial Holocene (LPIH: ca. 0–1700 AD) has generally been one of a stable, slowly increasing pCH_4 baseline prior to around 1700 AD (Figure 15.1). Our goal in this study is to use measurements of the stable carbon isotopes of methane ($\delta^{13}CH_4$) to help in source partitioning and thereby enhance our understanding of the LPIH methane budget. Carbon isotopes of methane are useful because the sources can be loosely binned into categories with distinct $\delta^{13}CH_4$ values. At this stage we exclude the not well-understood plant source ($\delta^{13}CH_4 \sim -50‰$) and use the following three groups for our source partitioning: biogenic (*e.g.*, anaerobic bacteria in wetlands or cattle rumen) at $\delta^{13}C \approx -60‰$; fossil (*e.g.*, natural gas seeps, fossil fuel mining) at $\approx -40‰$; and pyrogenic (biomass burning) at $\approx -25‰$ for C_3 vegetation or $\approx -12‰$ for C_4 vegetation. Although the range of individual measured values for these three classes of sources can be large, these source categories are useful distinctions for broadly interpreting changes in $\delta^{13}CH_4$ and the global methane budget.

Due to the small amount of air trapped in ice core bubbles and typical sample size requirements for $\delta^{13}CH_4$ analyses (Lowe *et al.* 1997), such measurements from ice cores have been scarce to date (Craig *et al.* 1988, Francey *et al.* 1999). The sample size requirement issue has been much improved with the advent of precise gas chromatography-isotope ratio mass spectrometry (GC-IRMS) analyses, opening the door for a high temporal resolution record of $\delta^{13}CH_4$. Here, we review 2000-year records of pCH_4 and $\delta^{13}CH_4$ at high temporal resolution and precision from the Law Dome ice cores in Antarctica (Ferretti *et al.* 2005). This record reveals that contrary to expectations, $\delta^{13}CH_4$, unlike pCH_4, was not slowly varying in the preindustrial period, and that its value was significantly different from that expected from previous estimates. We believe that these isotope analyses not only reveal the presence of a heavier human hand on the methane cycle than previously thought but also that methane history has been linked significantly to human history over the last two millennia.

II. METHODS

A. Materials and Procedures

Ice core samples were recovered from Law Dome, East Antarctica (66.7°S, 112.8°E). Law Dome is very useful for high-resolution records of gases in ice because of the relatively high snow accumulation rate. The age of the firn when it is compressed into ice, and the air bubbles trapped in that ice, can vary from

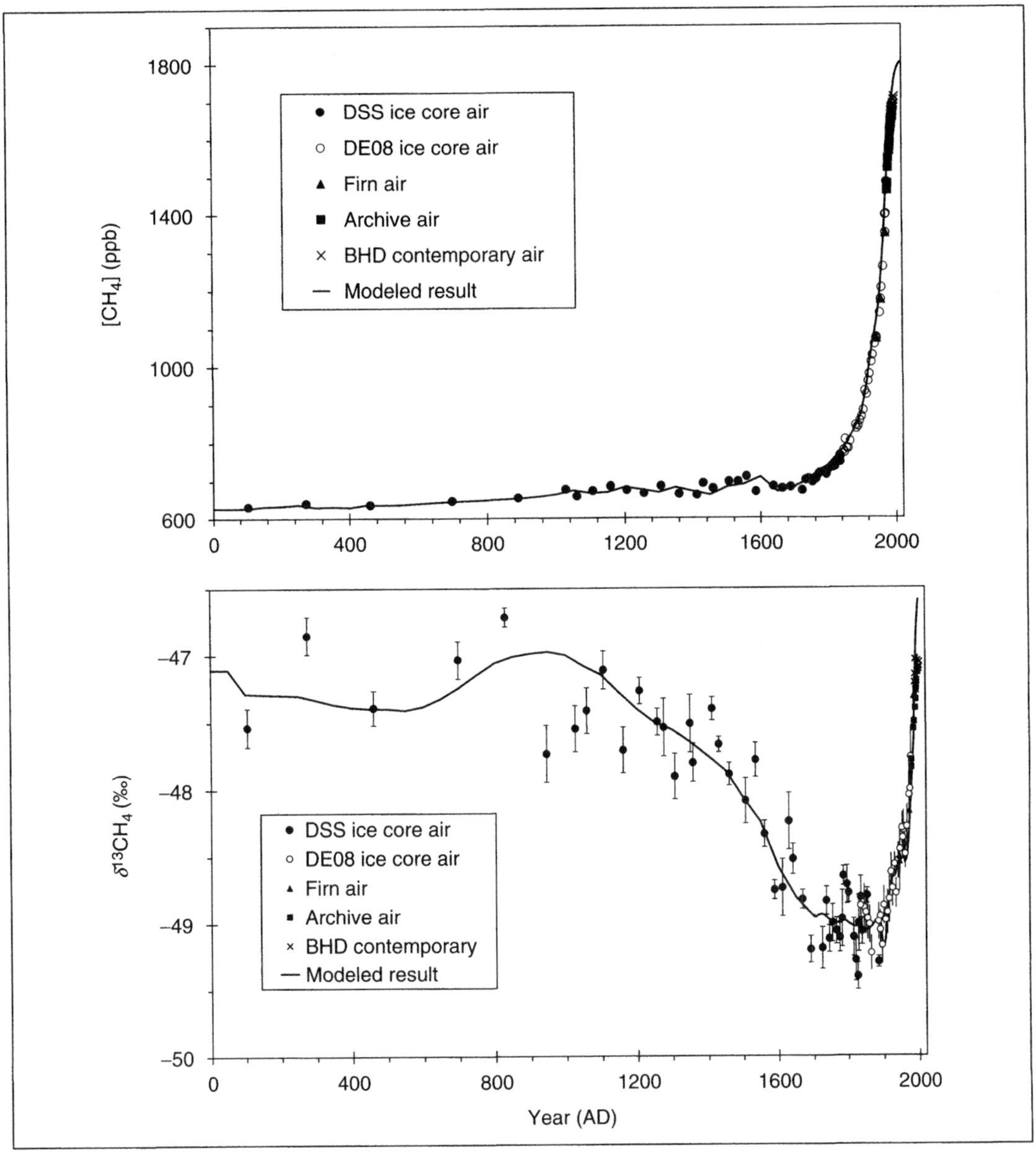

FIGURE 15.1 The 2000-year Law Dome pCH_4 (top panel) and $\delta^{13}CH_4$ (bottom panel) records. Between 0 and 1000 AD $\delta^{13}CH_4$ is at least 2‰ enriched from the expected value of about −50‰ (Miller 1999, Lassey *et al.* 2000). Between 1000 and 1700 AD $\delta^{13}CH_4$ undergoes a large 2‰ depletion, while pCH_4 varies by less than 55 ppb. The agreement between ice cores, firn and archive air, and contemporary air samples from Baring Head, New Zealand demonstrates the integrity of all samples types (Lowe *et al.* 1997, Francey *et al.* 1999).

decades to millennia, depending on the accumulation rate of snow. In addition, atmospheric gases mix, primarily by diffusion, in the porous, 40- to 80-m-thick snow and firn layer of an ice sheet. The higher the snow accumulation rate is, the less time air spends in the firn; the less the air is mixed, and the better the temporal resolution of gases in the ice. For studies such as this where we seek high temporal resolution over a relatively short time period, 2 thousand years, the Law Dome area is an excellent choice. When greater age is desired, areas with lower accumulation rates of snow are required. This will necessarily lower the temporal resolution, however, so choosing ice coring sites is always a trade-off between age and temporal resolution.

Two ice cores were used. The DE08 core has a higher snow accumulation rate and thus a slightly smaller age spread of gases per sample of about 10–11 years compared to age spread of 15 years for the DSS core. The DE08 core was drilled using a dry thermal drilling technique, and the DSS core was drilled using a wet drill with kerosene/perchloroethylene drilling fluid. Relatively shallow ice cores can be drilled dry, that is, without a fluid. Deeper holes require a fluid to maintain pressure in the hole and keep it from closing during the drilling season. The differences in the drilling techniques for these two cores allow us to check for the impact of different drilling techniques on the integrity of the gases trapped in the ice.

A dry extraction technique (Etheridge *et al.* 1988) was used to release the trapped air bubbles from the ice. Methane can be extracted from air in an ice core using both dry and wet techniques. In the dry technique, the ice is kept frozen and either crushed or shaved to release the air. In a wet technique, the ice is melted and the air that is released is captured. Gases such as methane that are not highly soluble in water can be extracted using either technique. Gases such as carbon dioxide that are highly soluble in water cannot be recovered quantitatively using the wet technique. As we also analyzed the air for CO_2, the dry technique was used. $\delta^{13}CH_4$ analyses were performed by GC-IRMS (Miller *et al.* 2002). Tests were preformed with helium only running through the system to monitor the integrity of the GC-IRMS system for leaks, which were found to account for errors of less than 0.005‰ in $\delta^{13}CH_4$. Tests were also performed with air-free ice to monitor the combined ice core extraction and GC-IRMS performance, and indicated that the air extraction process did not, within measurement uncertainty, shift the $\delta^{13}CH_4$.

Nonlinearity corrections were made to help account for the large concentration and isotopic range (~1100 ppb and ~2‰, respectively). Two corrections are applied. First, to correct for the nonlinearity arising from sample versus reference peak intensity differences, we analyzed a gas of known composition but varied the sample loop sizes to change the peak intensities. The amount linearity was 0.06‰/nA and applied to all ice core sample analyses to correct for any sample versus reference beam size differences that arose during analyses. At most this correction amounted to 0.17‰. Second, similarly, to correct for the nonlinearity arising from sample versus reference isotope ratio differences, we analyzed three working reference gases spanning about 4‰, while measuring the $\delta^{13}CH_4$ difference from their assigned values. The isotopic linearity result was −0.03‰/‰ and applied to all ice core sample analyses, correcting the sample versus reference isotopic differences. This correction was at most −0.06‰.

B. Ice Core Integrity

Deeper ice can exhibit regions of microfracturing from drilling and pressure release that either release air or allow contaminants to enter, or both. Contamination of this sort would be proportional to the amount of fracturing and thus expected to increase both CH_4 mixing and isotopic ratios. This was not observed, nor were abnormally elevated concentrations of CH_4, CO_2, CO, and N_2O. Alternatively, contamination from hydrocarbon-based drilling fluid into deeper/older ice core samples, driven by the high pressures at depth, could enrich the samples in $\delta^{13}C$. However, only heavily fractured and much deeper and older ice (8–30 ka BP) exhibited such contamination. These samples are not included in this study.

We also intercompared the $\delta^{13}CH_4$ measurements at similar ages between the two ice cores. The different drilling techniques, storage times, accumulation rates, temperatures, and trace element chemistries had no affect on the $\delta^{13}CH_4$. This agreement between samples from different cores is important as it discounts the possibility of *in situ* biological or chemical reactions in the ice core, reactions that tend to preferentially consume $^{12}CH_4$.

Diffusive methane loss increasing with age/depth/overpressure would discriminate against $^{13}CH_4$ and thus enrich the remaining $\delta^{13}CH_4$ value. Several factors minimize this possibility. First, the $\delta^{13}CH_4$

trend is relatively stable between 0 and 1000 AD, contrary to that expected if the diffusive loss was depth or overpressure driven. Second, other isotopes from the same ice core ($\delta^{13}CO_2$, $\delta^{15}N_2$, or Ar/N_2) do not show fractionations typical of diffusive loss. Third, ice samples from three cores with different age/depth/overpressure profiles and time since drilling (up to seven years difference) indicate that the age/depth/overpressure is not a problem and ^{12}C diffusive loss does not occur.

During extraction of ice core air, the headspace over the ice core sample is evacuated and pumped to remove impurities. At this stage, preferential removal of the lighter ^{12}C in older/more fractured ice could occur (Craig *et al.* 1988). Our data suggest that such a process may have occurred only during analyses of very old and fractured ice (>8 ka BP). As noted earlier, any such data were not used in this study.

Standard diffusion and gravity corrections are used (Trudinger *et al.* 1997). The gravity correction is constrained using $\delta^{15}N_2$ analyses. Nitrogen is the main component of the atmosphere, and any change in nitrogen isotopes of N_2 in air over the last 2000 years recovered from the ice core will be caused by gravitational separation in the firn as opposed to natural variability. As the pCH_4 is nearly constant for much of the early part of the record, and as the diffusion correction goes to zero as the change in methane concentration with time goes to zero, for all but the last 100 years or so of the record, the diffusion correction is very small. Thermal diffusion corrections are only required during abrupt, large climate changes, and as none are found in the last 2000 years at Law Dome, they are not important for this study.

Finally, we can compare the ice core $\delta^{13}CH_4$ measured here by GC-IRMS with $\delta^{13}CH_4$ measurements made from firn air, archived air, and contemporary air samples, all made using standard dual inlet mass spectrometry techniques. The agreement is excellent (Figure 15.1), as is the agreement between pCH_4 measurements from other ice cores in Antarctica and Greenland (Rasmussen and Khalil 1984, Blunier *et al.* 1993). Thus, while the record we will now present has some interesting surprises, all of the quality control evidence points to the conclusion that the measured trend is not an artifact of air stored in ice, but in fact the real atmospheric signal.

III. CARBON ISOTOPES AND THE METHANE BUDGET

As previously described, methane has numerous sources and sinks, and in terms of the $^{13}CH_4$ content we partition the sources into three main categories: biogenic, fossil, and pyrogenic. Table 15.1 gives approximate magnitudes and isotopic signatures of these source categories. The biogenic source is typified by wetlands, rice cultivation, ruminants, and termites. The ultimate source is anaerobic biota and thus is typically quite depleted in $^{13}CH_4$. The fossil source is leakage from natural gas sources, either naturally from sources such as methane clathrates or coal beds or human mediated, which are typically enhanced versions of these leakages. The pyrogenic source is burning of biomass. The methane produced by this mechanism is very close to the source carbon isotopic value (Chanton *et al.* 2000), and thus is lighter if C_3 plants are burned than if C_4 plants are burned. The sizes of the sinks of methane are not fully constrained, but it appears clear that destruction of methane by hydroxyl radical in the atmosphere is by far the dominant sink. Destruction in soils and the atmosphere by chlorine radical are thought to be relatively minor sinks. We note that the chlorine radical sink carries with it a very large isotopic fractionation, and thus needs to be carefully considered (Saueressig *et al.* 1995). In this study, we will assume that the sinks of methane have remained constant over time, as has the relative magnitude of the sinks. In the future, we will explore more fully the sinks of methane, which may have varied over time, but as of now, too little is known about the temporal variability in the sinks to do much more than speculate. Not included in Table 15.1 is a very new source of methane produced by plants and plant litter in the presence of oxygen, recently proposed by Keppler *et al.* (2006). This new source is thought to be large, perhaps as much as 50% of the modern budget. As this is a new source and

TABLE 15.1 Simplified methane budget

	Fluxes preindustrial (0–1700AD)	Fluxes today	Average $\delta^{13}CH_4$ fractionations	Average δD of CH_4 fractionations
Sources				
Biogenic	180	380	−60‰	−325‰
Fossil	20	120	−40‰	−175‰
Pyrogenic	25	30	−25‰(C_3) −12‰(C_4)	−210‰
Total	225	530		
Sinks				
Hydroxyl	200	500	−5.4	−215‰
Chlorine	10	10	66	510‰
Soils	10	10	−21	−80‰

Fluxes are in Tg year^{-1}. Isotopic signatures are averages, and are useful here to roughly partition the sources. As described in the text, the recently discovered plant source is not included here. Fluxes are approximate, and the total source and sink fluxes reflect this uncertainty and are not intended to balance exactly.

experiments to measure its strength have not been replicated as of the time of this writing, we will not include this source in the following discussion, but will address this new potential source separately later in this study.

IV. DATA

The pCH_4 and $\delta^{13}CH_4$ records are shown in Figure 15.1. On the basis of the relative stability of the methane concentrations prior to the mid-1800s, as well as the presumption of a minimal anthropogenic influence on methane prior to 1700 AD, we expected atmospheric $\delta^{13}CH_4$ to be equally stable in the LPIH. Furthermore, based on an assumed dominance of wetland sources ($\delta^{13}CH_4 \approx -60$‰) and assuming a sink fractionation of about −6‰ (primarily destruction by hydroxyl radical), earlier model simulations have suggested a stable atmospheric $\delta^{13}CH_4$ at −54% to −50‰ in the preindustrial period (Miller 1999, Lassey *et al.* 2000). As can be seen in Figure 15.1, neither of these expectations is born out by the observations. Two fundamental conundrums emerge from the data. First, in contrast to previous modeling studies, $\delta^{13}CH_4$ is several per mil more enriched in the LPIH than expected, especially during 0–1000 AD. Second, in contrast to the relatively stable pCH_4 in the LIPH, these first $\delta^{13}CH_4$ measurements reveal large and unexpected changes, notably a ~2‰ depletion during 1000–1700 AD.

V. DISCUSSION

Our focus in this chapter will be on the preindustrial period. Using the concentration and isotope data shown in Figure 15.1, we can partition the source fluxes into the three general categories discussed earlier. To do this two assumptions are required: (1) that methane sinks are constant over this time period, as discussed earlier; and (2) that the fossil sources are constant in the preindustrial period.

To partition the sources of methane we use the simple box model of Lassey *et al.* (2000) based on mass balance. The resulting source partitioning is shown in Figure 15.2. The biogenic source is dominant (Table 15.1), and cannot vary significantly in the LPIH because pCH_4 did not vary significantly in that period (see Figure 15.1). As LPIH $\delta^{13}CH_4$ did change significantly, however, the

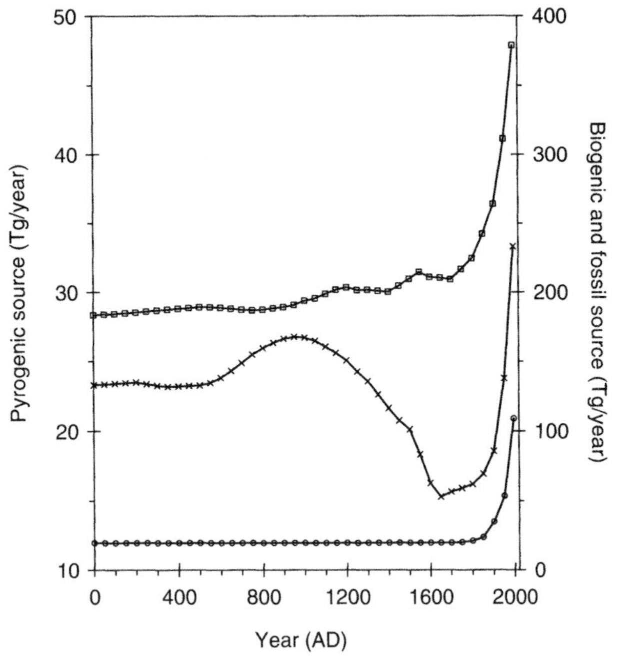

FIGURE 15.2 The partitioning of sources of methane using the constraints of the methane concentration data and the carbon isotope data. The lowest curve is the fossil sources, which are assumed to be constant prior to the industrial period. The top curve is the biogenic sources. The middle curve is the pyrogenic sources. Pyrogenic sources, while relatively small, exert a strong influence on the isotopes as they are significantly heavier than the atmosphere. Sinks are assumed constant in this partitioning of sources.

pyrogenic sources are likely to have varied. Pyrogenic sources, while only a small component of the total source, are significantly heavier than atmospheric $\delta^{13}CH_4$, and thus exert a relatively strong leverage on the isotopic composition of atmospheric methane. We calculate that a global pyrogenic emission of $\sim$25 $\pm$ 5 Tg year^{-1} is required to match measured $\delta^{13}CH_4$ and pCH_4 between 0 and 1000 AD. Here we assume a C_3/C_4 biomass burning ratio of 60/40. This gives more weight to the subtropics and tropics where C_4 plants are more abundant, and results in a more conservative estimate of biomass burning. We note that if only C_4 biomass burning occurred, then a lower emission of $\sim$22 $\pm$ 5 Tg year^{-1} is required, as C_4 biomass burns with a $\delta^{13}CH_4$ of about -12‰.

The next question then is what caused these changes, particularly in the pyrogenic sources. We investigate here two general causes of changes in methane source fluxes: natural climate changes and human activities. As noted earlier, we assume that the overall methane sink strength varies negligibly during the LPIH, as do the relative strengths of the sinks. Climate changes can alter the source strengths via a variety of mechanisms. As an example, changes in the ratio of precipitation to evaporation can alter wetland extent and distributions, and thus biogenic sources of methane, and warmer temperatures and droughts can lead to enhanced biomass burning. Human involvement in methane sources can also be significant. Examples include burning biomass to clear forests or drive game in grasslands, irrigating previously dry areas (especially for rice cultivation), and draining wetlands. We begin by looking at natural climate changes over the past 2000 years. Our strategy will be to first explain the observations of methane concentration and its carbon isotopes by invoking natural climate change, and if this proves inadequate, then to look for human influences.

The most pronounced natural climatic changes over the last 2000 years were the Medieval Warm Period ($\sim$900–1100 AD) and the cooling that followed and persisted for several centuries, the Little Ice

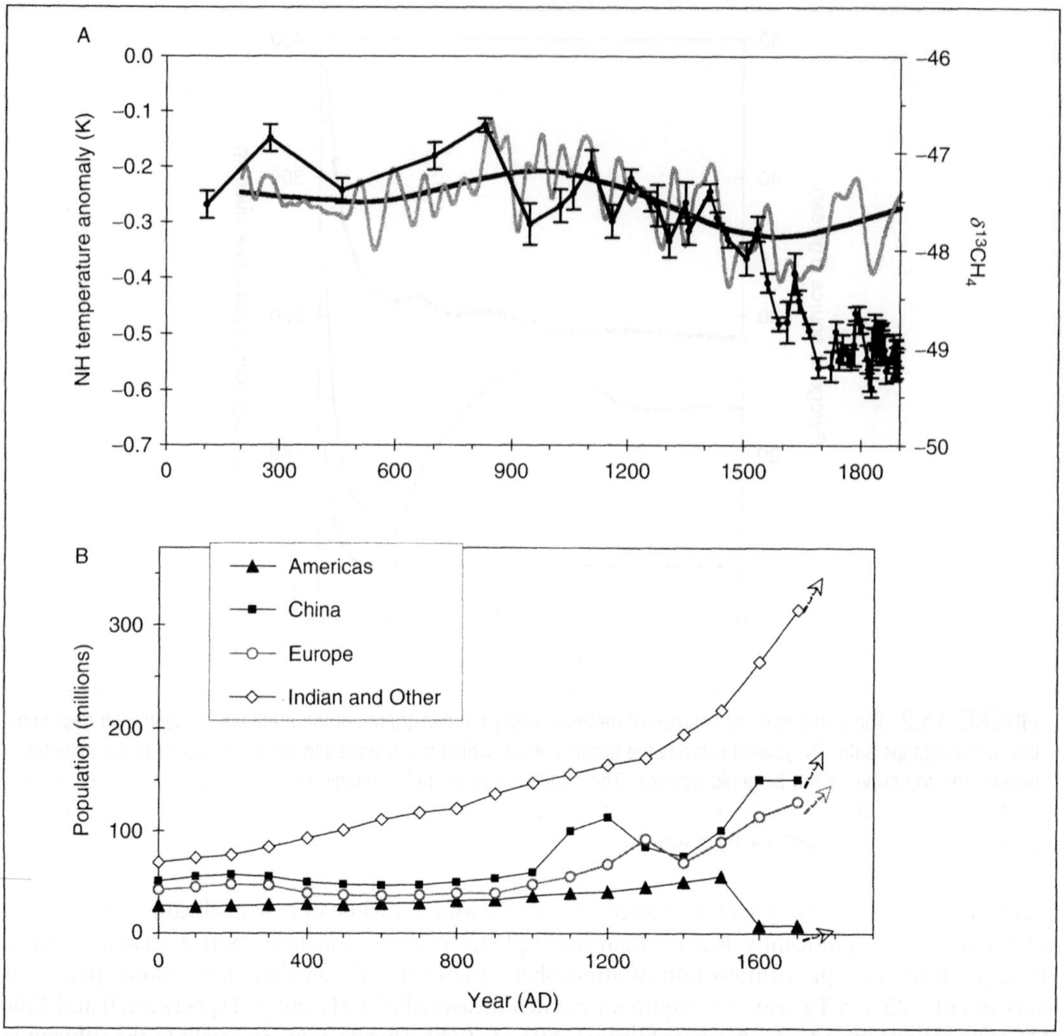

FIGURE 15.3 Natural climate and human population variations. (A) $\delta^{13}CH_4$ (dark line with data points) and Northern Hemisphere temperature anomaly (Jones and Mann 2004) (gray line without data points) relative to the 1961–1990 reference mean. Note how the similarity between the curves stops at ~1500 AD. Although, globally averaged temperature variations were relatively small (~0.2 K), regional temperature variations, especially in continental northern boreal wetland areas, were much greater (Sowers 2006) (up to 1–2°C). (B) Regional human population variations (Etheridge *et al.* 1988, Van Aardenne *et al.* 2001). The most important change occurs at ~1500 AD where the population of the Americas declines substantially causing a simultaneous decrease in biomass-burning emissions.

Age (~1500–1900 AD). A good correlation between Northern Hemispheric temperature anomalies (Jones and Mann 2004) and $\delta^{13}CH_4$ (Figure 15.3A) is observed during the time period of 0–1500 AD. The basic features in the curves appear to be common, such as the highest temperatures around 900 AD when the $\delta^{13}CH_4$ is also at its highest. The gradual cooling from 900 to 1500 AD is matched by a gradual decline in the $\delta^{13}CH_4$ values. Many of the decadal oscillations in both records are synchronous as well, although a lack of temporal detail in the $\delta^{13}CH_4$ record prevents us from making detailed comparisons. Lacking a hemispheric-wide precipitation record covering the past two millennia, we hypothesize that precipitation is itself correlated with temperature such that warm periods are also dryer and cool periods are also wetter. This general correlation has been noted in some records

(Verschuren *et al.* 2000, Hallett *et al.* 2003) and is consistent with evaporation demand. It is a broad generalization, however, a fact that should be kept in mind when evaluating our interpretation of the data. The lack of a precipitation record comparable in spatial and temporal resolution to the temperature record is a significant source of uncertainty in this study as precipitation amounts partly control wetland extents as well as the propensity of the biomass to burn. If one accepts the warm–dry and cool–wet climate correlation, then one can expect that both temperature and precipitation will influence pyrogenic and biogenic methane sources, such that during warm–dry climates, pyrogenic emissions are elevated and biogenic emissions are reduced when compared to cool-wet climates. This hypothesis is supported by independent evidence from lake and ice core records of precipitation, drought, and biomass burning (Swetnam 1993, Savarino and Legrand 1998, Verschuren *et al.* 2000, Carcaillet *et al.* 2002, Hallett *et al.* 2003). Although most available precipitation records are highly regional, and not all records or regions identically agree, a drier climate, with more biomass burning, is generally reported during 0–1000 AD compared to 1000–1500 AD.

Natural temperature variability, however, does not fully capture $\delta^{13}CH_4$ variations. While temperatures reached a minimum around 1500 AD, the $\delta^{13}CH_4$ record continued to decline significantly after this time. Between 1500 and 1600 AD, the $\delta^{13}CH_4$ drops by 1‰, half of the total depletion observed between 0 and 1700 AD. Between 1500 and 1700 AD, temperatures fluctuate, but do not get significantly colder than they were at 1500 AD. Assuming that the Northern Hemispheric temperature record (and implied precipitation) captures the climate impact on $\delta^{13}CH_4$, we must then seek another contribution to change in global methane composition from 1500 to 1700 AD.

Given the assumptions we have made thus far, particularly that sinks are constant, the most likely candidate for a nonclimatic change in source fluxes is changes in anthropogenic emissions. This would include agricultural practices like rice cultivation and cattle farming, but more likely is a change in biomass burning, as this source has the most leverage on the $\delta^{13}CH_4$. The pCH_4 does drop by about 40 ppb from the mid 1500s to 1600, but this is a small change relative to the change in the carbon isotopes. Work by Ruddiman (2003) has pointed to a relationship between large-scale population changes and the global carbon cycle during the LPIH, and we hypothesize that during this time early humans would have impacted the global methane cycle in a number of ways. Estimated trends in the LPIH of human population (McEvedy and Jones 1978, Denevan 1992) (Figure 15.3B) show a generally increasing population interrupted by periods of population decline due to pandemics. Of particular note is a tragically large decline of about 90% in the population of Native Americans between 1500 and 1600 AD (Figure 15.3B). This massive die-off is thought to be due to the introduction of diseases by European explorers (Denevan 1992). The native population of the Americas were thought to burn large amounts of grasslands, and some forests, annually (Anderson 1994, Woodcock and Wells 1994, Ruddiman and Thomson 2001, Mann 2005). Since forests in Europe and China were mostly cleared by around 0 AD (Ruddiman and Thomson 2001), it is therefore likely that Native Americans were key players in the LPIH pyrogenic emissions of methane to the atmosphere. We therefore postulate that the 1500 AD drop in $\delta^{13}CH_4$ is evidence for the substantial impact of humans on the preindustrial methane budget—in this case a rapid fall in human population in the Americas causing a decline in biomass-burning emissions, which in turn causes a trend toward more depleted global $\delta^{13}CH_4$.

To quantify the anthropogenic and natural pyrogenic and biogenic emissions, we make the following assumptions. First, we use a smoothed version of the Northern Hemispheric temperature anomaly record (Figure 15.3A) to drive the impact of climate on natural methane sources. We do not have sufficient confidence in all of the assumptions to interpret decadal changes. Second, we use population changes to estimate anthropogenic impact on methane emissions. The population of the Americas is taken to be a proxy for the biomass-burning component, and the combined populations of China and India are taken to be a proxy for the biogenic component. Third, we hold fossil methane sources constant. Although evidence (Etiope 2004) has suggested that natural fossil emissions of methane may be higher than thought, there is no evidence that these emissions would change significantly over time, and in particular would be elevated during 0–1000 AD relative to 1000–1700 AD.

To split the total pyrogenic emission into anthropogenic and natural components, we use the observation that the climate $\delta^{13}CH_4$ attribution accounts for roughly half of the $\delta^{13}CH_4$ depletion between 1000 and 1700 AD to postulate that the remaining 50% of the pyrogenic flux at 0 AD is anthropogenic. The modeled fluxes for the pyrogenic and biogenic sources separated into natural and human-caused fluxes are shown in Figure 15.4. During 1000–1700 AD, with the onset of the Little Ice Age, the climate impact can be seen in both the pyrogenic and biogenic natural fluxes, approximate changes of −5 Tg and +17 Tg, respectively. The impact of population changes on pyrogenic and biogenic human fluxes can be seen as well. During the 1000–1700 AD period, the biogenic human flux remains relatively constant, but the pyrogenic human flux decreases by ~5 Tg between 1500 and 1600 AD. While there are a number of assumptions that go into partitioning these fluxes, we note that the LPIH global $\delta^{13}CH_4$ can be reproduced as a simple combination of the climate and population curves in Figure 15.3, reinforcing the idea that climate and human impacts are the main drivers of change in $\delta^{13}CH_4$.

The modeled source evolution successfully reconciles the general features of $\delta^{13}CH_4$ and pCH_4 (Figure 15.1) to within 0.5‰ and 15 ppb, respectively. Key uncertainties are the amount of natural fossil emissions and their variability over time, and the ratio of C_3 to C_4 biomass that is burned, which we assumed to be 1.6 for anthropogenic biomass burning. If more C_4 grasslands burn, then the total pyrogenic emissions required to match the elevated $\delta^{13}CH_4$ from 0 to 1500 AD are lower, as would be the change in biomass burning required around 1500 AD. This would not, however, alter the shape of the inferred pyrogenic source curves, and thus does not alter our conclusion that anthropogenic burning appears to be a substantial part of the preindustrial methane budget.

A. A New Methane Source?

Keppler *et al.* (2006) reported finding a new source of methane from terrestrial plants under aerobic conditions. While the mechanism for this new source remains unknown, it is potentially a large source and if corroborated, it will need to be incorporated into our modeling. Their estimated magnitude of this source was 62–236 Tg year^{-1}, which is a large proportion of the modern methane budget, ~10–50%, and an even larger proportion of the methane budget in the late Holocene preindustrial

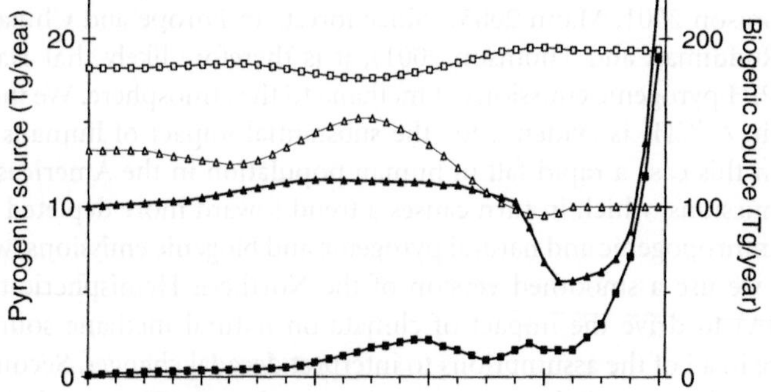

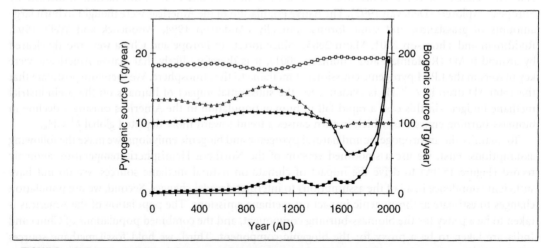

FIGURE 15.4 The partitioning of fluxes of methane into biogenic (top and bottom curves) and pyrogenic (middle) fluxes. The natural sources (thin lines with hollow symbols) are driven by the temperature changes shown in Figure 15.3. The anthropogenic sources (thick lines with solid symbols) are driven by the human population curves in Figure 15.3.

period, ~30–100%. In estimating the global magnitude of this source they used a "bottom up" approach, relying on their measured fluxes in controlled conditions, and then scaling up to the global level in proportion to net primary productivity. While we wait until confirmation of this surprising result before incorporating it fully into our modeling, we can use the isotopic and concentration data from the Law Dome ice core to provide a better constraint on the upper limit of the source.

This new source has different isotope signatures between C_3 and C_4 plants. Using a 60/40 C_3/C_4 global photosynthesis ratio, the estimated $\delta^{13}CH_4$ signature of this source is −50‰ (Keppler *et al.* 2006). As this value is isotopically heavier than the anaerobic plant source (Table 15.1), substituting this source for part of the anaerobic source will have the effect of making the atmospheric value heavier, thus lessening the need to call on biomass burning to balance the concentration and isotopic budgets in the preindustrial period. Using our simple model, we find that the upper ranges of the Keppler *et al.* are too high by about a factor of two, and are limited to a maximum of about 100 Tg $year^{-1}$ in the preindustrial and about 175 Tg $year^{-1}$ in the modern atmosphere. In making this estimate, we have maximized this new source by limiting biomass burning and fossil emissions to their lowest reported estimates of 10 Tg $year^{-1}$ in the preindustrial period (Judd *et al.* 1993, Scheehle and Kruger, 2006). While the plant source could, as suggested by Keppler *et al.*, help explain our finding that the preindustrial $\delta^{13}CH_4$ value was unexpectedly high at −47‰, we note that elevated levels of biomass burning still remain an attractive explanation for this finding. In particular, the drop in $\delta^{13}CH_4$ after 1500 AD is hard to reconcile without calling on a change in biomass burning. Because of its minimal isotope leverage (its $\delta^{13}CH_4$ value of −50‰), the strength of the putative aerobic source would have to change substantially in order for this to be the explanation. As of this writing, the Keppler results have just been published and we have just begun investigating the impacts. As of now, we suggest that the aerobic source was probably not as large as Kepper *et al.* calculated, and it remains to be seen how well it can explain the history of $\delta^{13}CH_4$ over the preindustrial period.

VI. CONCLUSIONS

Our measurements and modeling suggest that humans have long been important players in the atmospheric methane budget. This conclusion is entirely dependent on measurements of carbon isotopes of methane, as methane concentration changes are not large in the preindustrial period. This study thus points to the need to measure isotopes in greenhouse gases in the past along with concentrations if we are to fully reveal the variability in the cycles of these gases, whether that variability be "natural" or caused by humans.

Several issues remain unresolved, however. The $\delta^{13}CH_4$ value from 0 to 1000 AD is at least 2‰ more enriched than expected. If we attribute this enrichment to biomass burning, it requires almost twice as much biomass burning at 1000 AD than at 1800 AD. Here, we attribute this change partly to changes in human impacts on methane, and partly to climate changes. However, recent studies have questioned our understanding of the methane cycle and suggest that we have much to learn about the sources of methane. We note also that we need to better understand variability in methane sinks, especially sinks such as reactive chlorine that have large isotope fractionations and thus large isotopic leverage. Reducing these uncertainties is a particularly important task if humans were indeed the cause of the elevated levels of methane even as far back as 8000 years BP (Ruddiman and Thomson 2001, Ruddiman 2003). At some time in the Holocene, the true pre-Anthropocene is likely to have existed, when humans were not populous enough to have been a factor in the methane budget. It remains to be seen when that time was, and what was the "natural" methane cycle. What is clear is that stable isotopes have revealed important changes in the methane budget that are hidden in the concentration record. With the advent of new GC-IRMS techniques that require much less sample volume, we have shown that ice cores, with their limited amount of air, are excellent archives for these stable isotopes.

In future studies, the $\delta^{13}CH_4$ record should be extended back in time in Antarctic and Greenland ice cores. Pushing the record back in time beyond that when human influence was probable should help address the issue of the elevated $\delta^{13}CH_4$ values and whether biomass burning, or perhaps other causes are behind these enriched values. The interhemispheric gradient revealed from cores in both poles could help to constrain the location of the changes in the sources, such as high Northern Hemispheric sources of wetlands versus tropical wetland sources. Also, adding hydrogen isotopes of methane may also be helpful in understanding past and present changes in the methane budget. While the isotope sink fractionations and source isotopic signatures for the carbon and hydrogen isotopes of methane tend to covary (Table 15.1), there is enough difference in the $\delta^{13}C$ and δD signatures to warrant the use of both isotopes. Figure 15.5 shows a plot of $\delta^{13}C$ versus δD for the modern atmosphere, and lines expected for 10% and 20% increases and decreases in the sources and sinks given in Table 15.1. The slopes of the lines are not the same, are vary from about 15‰/‰ for a change in the hydroxyl destruction to about 3‰/‰ for a change in the biomass-burning source. This tells us that the hydrogen isotopes carry some unique information not in the carbon isotopes. While a test of this approach has yet to be done, and thus its utility has not been established, there appears to be potential in analyzing hydrogen isotopes of methane. There is clearly much work to do before we have a good understanding of the methane budget in the past, and the role of humans in shaping our environment in the preindustrial period. What is clear is that stable isotopes will play a key role in advancing this understanding.

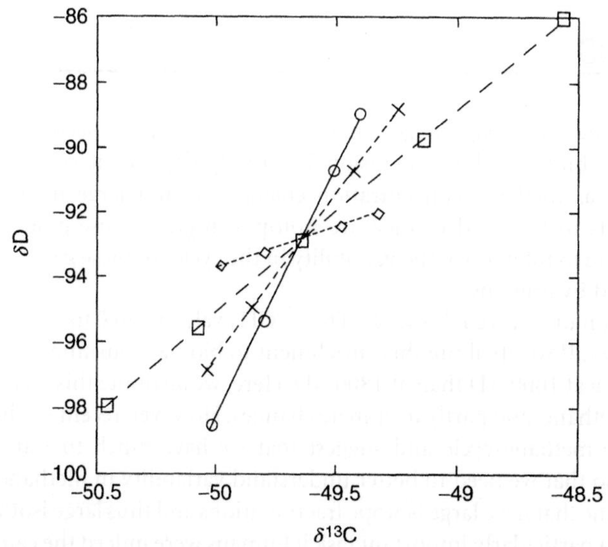

FIGURE 15.5 Plot of the carbon and hydrogen isotope ratios of methane expected for a 10% and 20% increase or decrease in (1) the hydroxyl sink (open circles), (2) the chlorine sink (×), (3) the anaerobic source (open squares), and (4) the biomass-burning source (open diamonds). The slopes of the lines are (1) hydroxyl sink: 15.6‰/‰, (2) chlorine sink: 10.0‰/‰, (3) the anaerobic source: 5.5‰/‰, and (4) the biomass-burning source: 2.7‰/‰. The values for the isotopic fractionations in the sinks and the average isotopic values for the sources used in calculating the lines are given in Table 15.1. The plot shows that the combination of carbon and hydrogen isotopes may be helpful in identifying changing sources and sinks of methane.

VII. REFERENCES

Anderson, M. K. 1994. Prehistoric anthropogenic burning for nonagricultural purposes in temperate regions: A net source, sink, or neutral to the global carbon budget? *Chemosphere* **29:**913–934.

Blunier, T., J. Chappellaz, J. Schwandae, J.-M. Barnola, T. Desperts, B. Stauffar, and D. Raynaud. 1993. Atmospheric methane record from a Greenland ice core over the last 1000 year. *Geophysical Research Letters* **20:**2219–2222.

Brook, E. J., S. Harder, J. P. Severinghaus, E. J. Steig, and C. M. Sucher. 2000. On the origin and timing of rapid changes in atmospheric methane during the last glacial period. *Global Biogeochemical Cycles* **14:**559–572.

Carcaillet, C., H. Almquist, H. Asnong, R. H. W. Bradshaw, J. S. Carrion, M.-J. Gaillard, K. Gajewski, J. N. Haas, S. G. Haberle, P. Hadorn, S. D. Muller, P. J. H. Richard, *et al.* 2002. Holocene biomass burning and global dynamics of the carbon cycle. *Chemosphere* **49:**845–863.

Chanton, J. P., C. M. Rutkowski, C. C. Schwartz, D. E. Ward, and L. Boring. 2000. Factors influencing the stable carbon isotopic signature of methane from combustion and biomass burning. *Journal of Geophysical Research* **105**(D2):1867–1877.

Chappellaz, J., *et al.* 1997. Changes in the atmospheric CH_4 gradient between Greenland and Antarctica during the Holocene. *Journal of Geophysical Research* **102:**15987–15997.

Craig, H., C. C. Chou, J. A. Welhan, C. M. Stevens, and A. Engelkemeir. 1988. The isotopic composition of methane in polar ice cores. *Science* **242:**1535–1539.

Denevan, W. 1992. *The Native population of the Americas in 1492.* University of Wisconsin Press, Madison.

Dlugokencky, E. J., S. Houweling, L. Bruhwiler, K. A. Masarie, P. M. Lang, J. B. Miller, and P. P. Tans. 2003. Atmospheric methane levels off: Temporary pause or a new steady-state? *Geophysical Research Letters* **30:**doi:10.1029/2003GL018126.

Etheridge, D., G. Pearman, and F. de Silva. 1988. Atmospheric trace gas variations as revealed by air trapped in an ice core from Law Dome, Antarctica. *Annals of Glaciology* **10:**28–33.

Etheridge, D. M., L. P. Steele, R. J. Francey, and R. L. Langenfelds. 1998. Atmospheric methane between 1000 AD and present: Evidence of anthropogenic emissions and climate variability. *Journal of Geophysical Research* **103:**15979–15993.

Etiope, G. 2004. New directions: GEM—geologic emissions of methane, the missing source in the atmospheric methane budget. *Atmospheric Environment* **38:**3099–3100.

Ferretti, D. F., J. B. Miller, J. W. C. White, D. M. Etheridge, K. R. Lassey, D. C. Lowe, C. M. MacFarling Meure, M. F. Dreier, C. M. Trudinger, T. D. van Ommen, and R. L. Langenfelds. 2005. Unexpected changes to the global methane budget over the past 2000 years. *Science* **309:**1714–1717.

Francey, R. J., M. R. Manning, C. E. Allison, S. A. Coram, D. M. Etheridge, R. L. Langenfelds, D. C. Lowe, and L. P. Steele. 1999. A history of $\delta^{13}C$ in atmospheric CH_4 from the Cape Grim Air Archive and Antarctic firn air. *Journal of Geophysical Research* **104:**23631–23643.

Hallett, D., R. Mathewes, and R. A. Walker. 2003. 1000-year record of forest fire, drought and lake level change in southeastern British Columbia. Canada. *The Holocene* **13:**751–761.

Jones, P. D., and M. E. Mann. 2004. Climate over past millennia. *Reviews of Goephysics* **42:**2003RG000143.

Judd, A. G., R. H. Charlier, A. Lacroix, G. Lambert, and C. Rouland. 1993. Pages 432–456 *in* M. A. K. Khalil (Ed.) *Atmospheric Methane: Sources, Sinks, and Role in Global Change.* Springer-Verlag, Berlin.

Keppler, F., J. T. G. Hamilton, M. Brass, and T. Rockmann. 2006. Methane emissions from terrestrial plants under aerobic conditions. *Nature* **439:**187–191.

Lassey, K. R., D. C. Lowe, and M. R. Manning. 2000. The trend in atmospheric methane $\delta^{13}C$ and implications for isotopic constraints on the global methane budget. *Global Biogeochemical Cycles* **14:**41–49.

Lowe, D. C., M. R. Manning, G. W. Brailsford, and A. M. Bromley. 1997. The 1991–1992 atmospheric methane anomaly: Southern hemisphere ^{13}C decrease and growth rate fluctuations. *Geophysical Research Letters* **24:**857–860.

Mann, C. C. 2005. *1491 New Revelations of the Americas Before Columbus.* 465 p., Alfred A. Knopf, New York.

McEvedy, C., and R. Jones. 1978. *Atlas of World Population History.* Penguin Books, New York.

Miller, J. 1999. *Application of Gas Chromatography Isotope Ratio Mass Spectrometry (GC-IRMS) to Atmospheric budgets of C18OO and 13CH4.* University of Colorado, Boulder.

Miller, J. B., K. A. Mack, R. Dissly, J. W. C. White, E. J. Dlugokencky, and P. P. Tans. 2002. Development of analytical methods and measurements of $^{13}C/^{12}C$ in atmospheric CH_4 from NOAA Climate Monitoring and Diagnostics Laboratory Global Air Sampling Network. *Journal of Geophysical Research* **107:**10.1029/2001JD000630.

Petit, J. R., J. Jouzel, D. Raynaud, N. I. Barkov, J.-M. Barnola, I. Basile, M. Benders, J. Chappellaz, M. Davis, G. Delayque, M. Delmotte, V. M. Kotlyakov, *et al.* 1999. Climate and atmospheric history of the past 420,000 years from the Vostok ice core, Antarctica. *Nature* **399:**429–436.

Rasmussen, R. A., and M. A. K. Khalil. 1984. Atmospheric methane in the recent and ancient atmospheres: Concentrations, trends, and interhemispheric gradient. *Journal of Geophysical Research* **89:**11599–11605.

Ruddiman, W. F. 2003. The anthropogenic greenhouse era began thousands of years ago. *Climatic Change* **61:**261–293.

Ruddiman, W. F., and J. S. Thomson. 2001. The case for human causes of increased atmospheric CH_4 over the last 5000 years. *Quaternary Science Reviews* **20:**1769–1777.

Saueressig, G., P. Bergamaschi, J. N. Crowley, H. Fischer, and G. W. Harris. 1995. Carbon kinetic isotope effect in the reaction of CH_4 with Cl atoms [CH_4]. *Geophysical Research Letters* **22:**1225–1228.
Savarino, J., and M. Legrand. 1998. High northern latitude forest fires and vegetation emissions over the last millennium inferred from the chemistry of a Greenland ice core. *Journal of Geophysical Research* **103:**8267–8279.
Scheehle, E. A., and D. Kruger. 2006. Global anthropogenic methane and nitrous oxide emissions. *Energy Journal* **28:**33–44
Sowers, T. 2006. Late quaternary atmospheric CH_4 isotope record suggests marine clathrates are stable. *Science* **311:**838–840.
Subak, S. 1994. Methane from the house of Tudor and the Ming Dynasty: Anthropogenic emissions in the sixteenth century. *Chemosphere* **29:**843–854.
Swetnam, T. W. 1993. Fire history and climate change in Giant Sequoia gorves. *Science* **262:**885–889.
Trudinger, C. M., I. G. Enting, R. J. Francey, D. M. Etheridge, R. J. Francey, V. A. Levchenko, L. P. Steele, D. Raynaud, and L. Arnaud. 1997. Modeling air movement and bubble trapping in firn. *Journal of Geophysical Research* **102D:**6747–6763.
Van Aardenne, J. A., F. J. Dentener, J. G. J. Olivier, C. Goldewijk, and J. Lelieveld. 2001. A 1 degrees ×1 degrees resolution data set of historical anthropogenic trace gas emissions for the period 1890–1990. *Global Biogeochemical Cycles* **15:**909–928.
Verschuren, D., R. Laird, and D. F. Cumming. 2000. Rainfall and drought in equatorial east Africa during the past 1100 years. *Nature* **403:**410–414.
Woodcock, D. W., and P. V. Wells. 1994. The burning of the new world: The extent and significance of broadcast burning by early humans. *Chemosphere* **29:**935–948.

CHAPTER 16

Compound-Specific Hydrogen Isotope Ratios of Biomarkers: Tracing Climatic Changes in the Past

Gerd Gleixner and Ines Mügler

Department of Biogeochemical Processes, Max-Planck Institute for Biogeochemistry

Contents

I. INTRODUCTION

Climatic factors like temperature or water availability are major drivers of terrestrial ecosystem processes. Their interactions strongly control the opening of leaf stomata and consequently net ecosystem productivity and water-use efficiency. These climatic effects are in turn preserved in the ^{13}C content of biomass. Additionally to carbon isotope discrimination, evapotranspiration enriches water

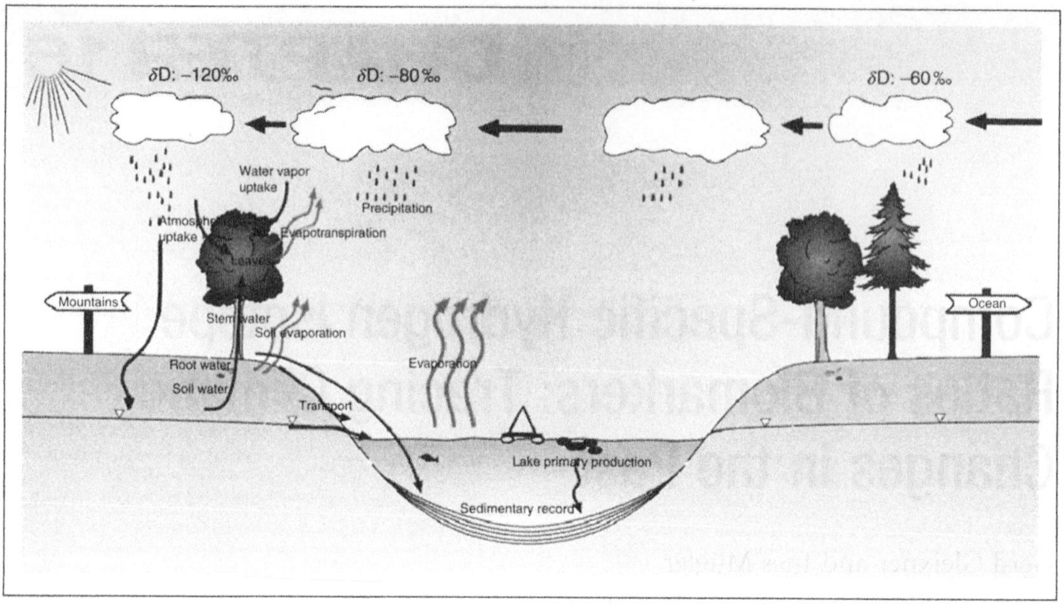

FIGURE 16.1 Fractionation processes for hydrogen isotopes on the ecosystem scale and contribution of organic matter from various sources to sedimentary organic matter.

isotopes in leaf or in lake water (Figure 16.1), which is also recorded in the deuterium content of biomass or sedimentary organic matter. In this chapter, it is demonstrated how the climate system controls the isotopic content of surface waters and why hydrogen isotopes are more suitable to monitor climatic changes than carbon isotopes. The current knowledge about the transfer of these isotopic signals into biomass and the benefit of using organic biomarkers to reconstruct past changes in ecoystems will be summarized. Finally, it will be evaluated if corresponding signals are preserved over geological timescales and if climate models are suitable to predict the isotopic content of past precipitation.

II. IMPORTANCE OF WATER AND THE WATER CYCLE FOR THE CLIMATE SYSTEM

For all known life forms on Earth, water is of essential importance. Living cells contain over 80% of water and water covers over 70% of Earth's surface. The estimated volume of 1.4 billion m^3 (UNEP 2002) is distributed in various forms and reservoirs. Due to its high effective heat capacity and mobility, water fundamentally influences the Earth's climate. In general, the Earth's climate system is driven by energy delivered from the sun, which equals to the solar constant of 1.367 W m^{-2}. About 30% of this incoming radiation are reflected by clouds (20%), the atmosphere (6%), or the surface (4%). The remaining 70% are absorbed by the Earth's surface (51%) and by clouds or the atmosphere (19%). Since more than two-thirds of the Earth's surface are covered with water, the oceans store significant amounts of the incoming energy resulting in regional temperature and salinity gradients. These gradients cause density differences that drive global ocean currents like the Gulf stream which is heating, for example, Western Europe with this energy.

The largest part from the Earth's absorbed energy is again emitted as long wave or thermal radiation which is then almost completely absorbed by atmospheric gases. This natural greenhouse effect leads to a rise of the average surface temperature by ~34°C and thus enables life on Earth. Greenhouse gases

TABLE 16.1 Contribution of greenhouse gases on the natural greenhouse effect

		Temperature effect	
Greenhouse gas		(°C)	(%)
Water vapor	H_2O	32.3	95
Carbon dioxide	CO_2	1.2	3.6
Methane	CH_4	0.1	0.4
Nitrous oxide	N_2O	0.3	1.0
CFC's and others		0.1	0.1

differentially contribute to the naturally occurring greenhouse effect (Table 16.1). Carbon dioxide (CO_2) contributes about 3.6% and methane (CH_4), nitrous oxide (N_2O), and CFC's and other gases together about 1.4% to the natural greenhouse effect. These gases are most important for the anthropogenic greenhouse effect as their absorption closes the open wavelength window for outgoing radiation. Water vapor (H_2O) is by far the most important natural greenhouse gas, which contributes over 95% to the Earth's natural greenhouse effect.

Water vapor is also an important part of the global hydrological cycle. Solar radiation leads to evaporation of water (latent heat flux) from marine and terrestrial ecosystems. Most important are oceanic surfaces that contribute to roughly 86% of global evaporation. About 90% of this vapor returns into the ocean by direct precipitation while wind and global circulations transport the remaining 10% across the continents. This oceanic water vapor contributes ca. 30% to the continental water flux (Gat 1996). The remaining 70% result from evaporation of terrestrial ecosystems. While traveling over the continents atmospheric vapor condensates and precipitates as rainfall, hail, or snow. The runoff returns directly via streams and rivers or delayed by reservoirs like glaciers, lakes or groundwater flow down the slope back to the ocean.

In summary, this suggests that the Earth's climate is directly reflected in the intensity of the hydrological cycle which feeds back to weathering rates and biomass growth. Hence, reconstructions of the hydrological cycle will provide information on past climates.

III. STABLE ISOTOPES OF WATER AND THEIR VARIATION IN THE HYDROLOGICAL CYCLE

Hydrogen and oxygen, the chemical constituents of water have two (^{1}H: protium and ^{2}H: deuterium) and three (^{16}O, ^{17}O, ^{18}O) stable isotopes, respectively. The isotope content of water samples is calculated as relative difference to an international standard, defined as δ value.

$$\delta\text{value}[‰] = \left[\frac{R_{\text{sample}} - R_{\text{standard}}}{R_{\text{standard}}}\right] \times 1000$$

where R_{sample} and R_{standard} are the isotope ratios (^{2}H/^{1}H and ^{18}O/^{16}O) of the sample and standard, respectively. Vienna Standard Mean Ocean Water is the internationally accepted standard for measurements of natural water samples (Coplen 1995). The isotope ratios are equal to the previous SMOW standard:

^{2}H/^{1}H $= 155.95 \pm 0.08 \times 10^{-6}$ (DeWit *et al.* 1980)
^{18}O/^{16}O of $2005.2 \pm 0.45 \times 10^{-6}$ (Baertschi 1976)

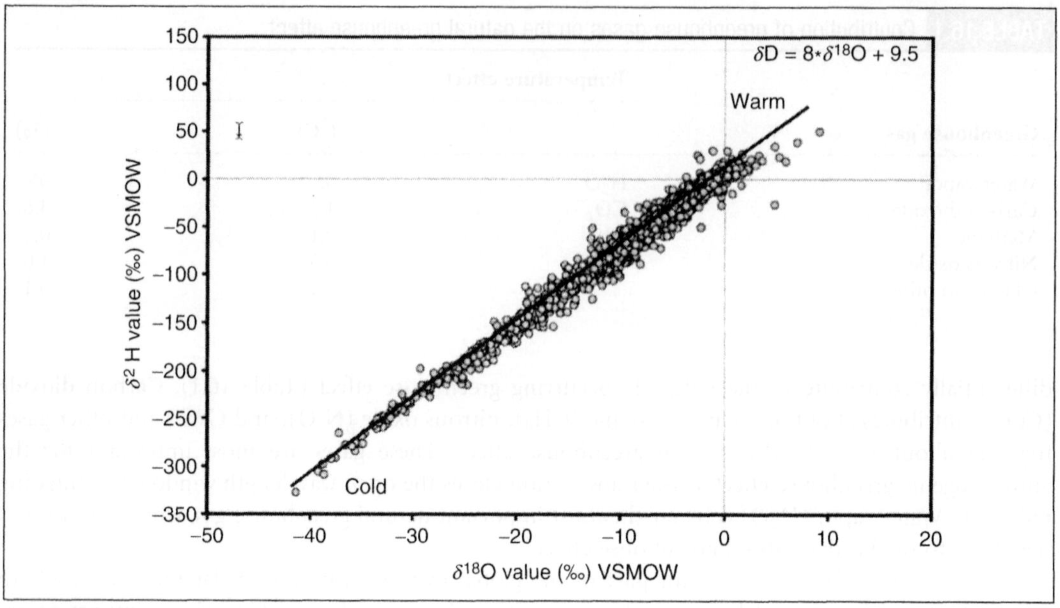

FIGURE 16.2 The relationship of $\delta^{18}O$ and $\delta^{2}H$ values in precipitation on a global scale (based on IAEA data http://www.iaea.org).

Craig (1961) first observed that deuterium and oxygen-18 of meteoric waters (precipitation and atmospheric water vapor) correlate on the global scale (δ^2H value = 8 × $\delta^{18}O$ value + 10‰) and δ^2H and $\delta^{18}O$ values can be calculated according the "global meteoric water line" (Figure 16.2).

Beneficially, in the hydrological cycle both the evaporation of water and the condensation of water vapor during atmospheric transport lead to an offset of isotope ratios of $^{18}O/^{16}O$ and $^2H/^1H$ from the global mean. In general, heavier isotopes (^2H, ^{17}O, and ^{18}O) remain in the liquid phase (Gonfiantini 1986). Thus, vapor is depleted in heavy isotopes relative to the water source and droplets formed from water vapor are heavier than this vapor. This is mostly caused by the lower amount of energy needed to evaporate light molecules (Craig 1961, Gat 1971, Merlivat 1978). This effect of isotopic discrimination is calculated by the fractionation factor α that describes the ratio of the heavy isotopes to the lighter isotope in the liquid (R_l) relative to the vapor (R_v) phase: $\alpha = R_l/R_v > 1$.

The fractionation is smaller nearby the boiling point of water than at lower temperatures (Figure 16.3), and the equilibrium fractionation factor α is highest at low temperatures. Additionally, a continuous removal of heavy raindrops from the pool of vapor results in an ongoing depletion of the remaining vapor or the evaporation of vapor from closed basins leads to ongoing enrichment of heavy isotopes in the remaining water (Figure 16.4). This process is described by an exponential function[1] $R = R_0 f^{(\alpha-1)}$ called Rayleigh distillation.

Evaporation and Rayleigh distillation cause isotopic signals of precipitation to vary in a predictable manner and correlate on the regional and global scale (Rozanski *et al.* 1992, Araguas-Araguas *et al.* 2000). Ongoing initiatives to characterize the isotopic variability in precipitation are taken by the Global Network for Isotopes in Precipitation (Araguas-Araguas *et al.* 2000) in cooperation with the International Atomic Energy Agency (IAEA) and the World Meteorological Organization (WMO) that launched a global isotope in precipitation monitoring network in 1962.

[1] In case of evaporation process, R describes the isotope ratios of the remaining fraction of water (f) after evaporation from initial water (R_0). α is the fractionation factor during evaporation.

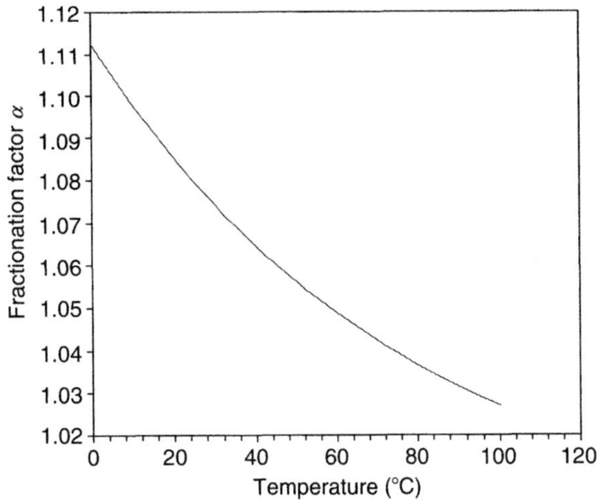

FIGURE 16.3 Dependence of fractionation factor α between vapor and water on temperature (modified after Majoube 1971).

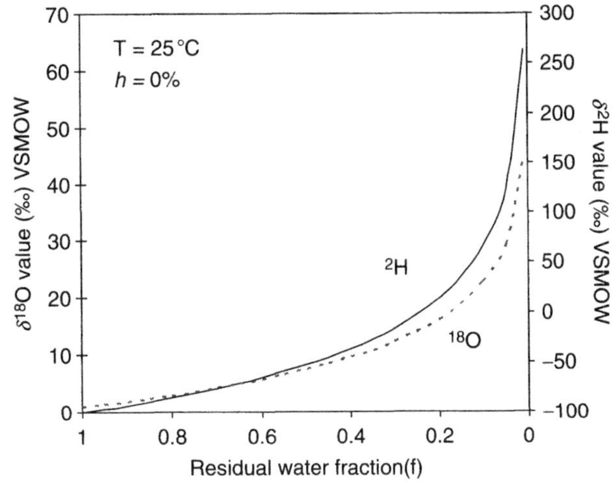

FIGURE 16.4 Isotopic enrichment of remaining water after evaporation of water vapor from a closed pool (temperature = 25°C; humidity 0%).

A first systematic analysis of these network data by Dansgaard (1964) revealed additional relationships between the isotopic signature of precipitation on the one hand and air temperature, precipitation amount, latitude, altitude, and distance to the coast on the other hand. Further studies confirmed these effects (Rozanski *et al.* 1982, Gonfiantini 1986, Araguas-Araguas *et al.* 2000) leading to an isotopically ordered world of precipitation (Figure 16.5). The isotopic content of precipitation nearby the coast or at low elevated continental areas is close to the δD and δ^{18}O values of the ocean. The progressive transport of moisture from intertropical regions toward the pole leads to a gradual depletion in deuterium and oxygen-18 of precipitation which is called the "*latitude effect*" (Figure 16.5). Additionally, progressive depletion in heavy isotopes occurs with increasing elevation.

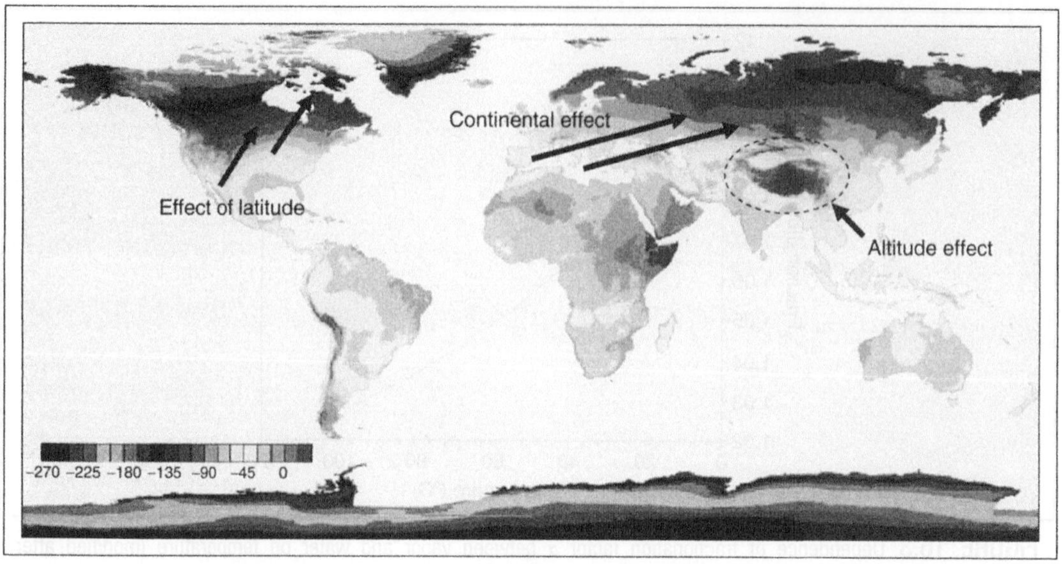

FIGURE 16.5 Global distribution of mean annual hydrogen isotope ratios in precipitation and isotope sensitive processes (modified after Bowen and Revenaugh 2003). (**See Color Plate.**)

Decreasing temperatures force enhanced condensation. This "*altitude effect*" accounts for -0.15‰ to −0.50‰ per 100-m height for oxygen-18 and −1‰ to −4‰ per 100-m for deuterium (Holdsworth *et al.* 1991), respectively. Predominantly in midlatitudes, an increasing distance from the coast to the inner continents leads to the gradual rainout of air masses corresponding with its depletion of heavy isotopes. In midlatitude Europe, the "*continental effect*" contributes to the depletion of heavy isotopes of around −2‰ per 1000 km for oxygen-18 (Rozanski *et al.* 1982). An apparent correlation between isotope content of precipitation and amount of rainfall is observed in regions with neglecting seasonal temperature variations. The "*amount effect*" is described as the impact of a rainout event generated by deep convective clouds. Thus, high precipitation amounts lead to the depletion of heavier isotopes overprinting the impact of temperature on the isotope signature of precipitation.

In conclusion, the major fluxes and reservoirs in the water cycle and the accompanying isotope effects are well known (Craig 1961, Dansgaard 1964, Huntington 2006). The isotopic signature of precipitation at a given location represents the history of its corresponding air mass and reflects the regional climate characteristics of the water source area. Thus, the use of stable isotopes as tracers to characterize the actual hydrological system (Epstein and Mayeda 1953, Craig 1961, Dansgaard 1964, Gonfiantini 1986) is well established and numerous basic empirical studies during the past decades lead to a solid understanding of processes affecting the isotopic signature of meteoric water at different spatial scales (Dansgaard 1964, Rozanski 1993, Gat 1996). Using long-term climatic archives this information can be transferred to the temporal scale.

IV. LONG-TERM WATER CYCLE PATTERN RECORDED BY INORGANIC MOLECULES

Isotope data that is retrieved from archives, such as ice cores from glaciers or polar ice caps, corals, or microfossils from lacustrine sediments, is assumed to be linked to the isotopic signal of their water source in the past and to the past climate signal as well as to the corresponding environmental parameters like air or water temperature, precipitation, or humidity. Each of these archives has certain benefits and disadvantages in conjunction with the temporal or spatial resolution of the isotope signal as well as its preservation.

Marine sediments belong to the climate archives that go back farthest in time. Best marine climatic records are known from biogenic carbonates of corals or laminated sediments deposited in anoxic basins or in accumulation regions at the continental margins. Unfortunately, continuous climate records are sparse mostly because these archives lie in shelf areas or shallow water, which are strongly influenced by varying continental signals and changes in water depth. Adequate temporal resolution can be achieved in open ocean areas but unfortunately here sedimentation rates are low and thermal response of the ocean to climate change is low. However, the isotopic composition of seawater was successfully reconstructed for the whole Phanerozoic that covers the last 600 million years (Veizer *et al.* 1999).

Glaciers and ice shields are the best available climate recorders as they directly store precipitation, atmospheric gases, aerosols, and dust. Analyses of annual layers of snow and ice from the Polar Regions provided the basis of our knowledge on variations of climatic factors like trace gas concentrations, humidity, or intensity of atmospheric circulation patterns (Petit *et al.* 1981, Mayewski *et al.* 1994, Indermuhle *et al.* 1999, Thompson *et al.* 2003, Raynaud *et al.* 2005). Ice core records are available only for limited locations worldwide. Most important are cores from Greenland and Antarctica. The 3000-m long GRIP and GISP cores were drilled on the summit of Greenland and provided paleoclimate information back to the last interglacial more than 100,000 years ago (Zielinski *et al.* 1995, Augustin *et al.* 2004). In Antarctica, the Vostok ice core reached back 420,000 years and covers four past glacial cycles (Watanabe *et al.* 2003). The by now longest ice core, also from the Antarctica, is the EPICA core that reveals eight previous glacial cycles and dates back 740,000 years (Augustin *et al.* 2004; Figure 16.6). In addition to these polar ice cores much shorter ice cores covering the time back to the LGM are collected from the glaciers of high mountain ranges, that is Mount Kilimanjaro in

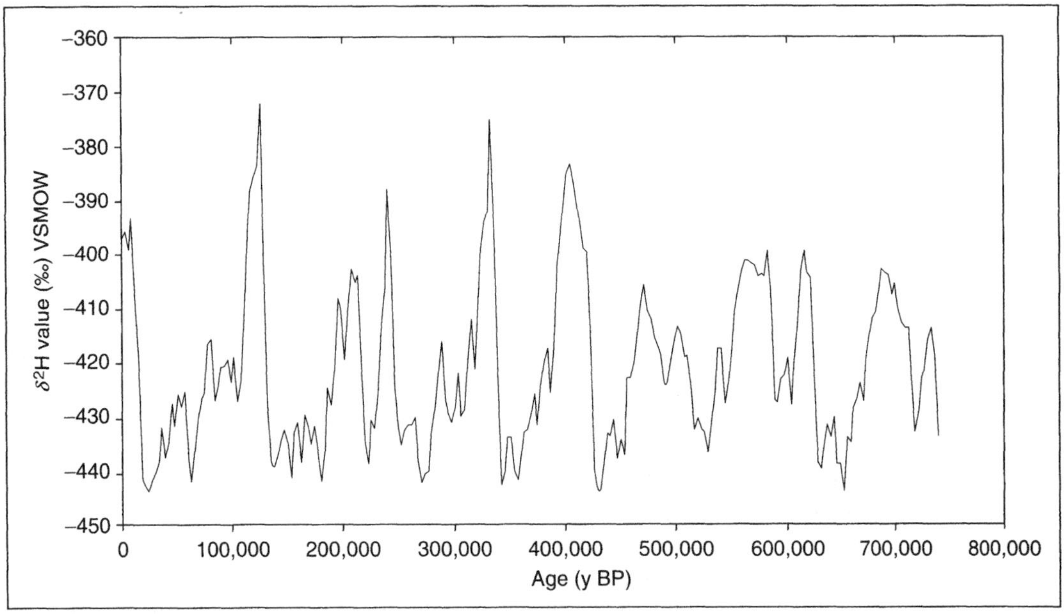

FIGURE 16.6 Deuterium values from the 740,000 years EPICA Dome C Ice core (modified after supplementary data (http://www.nature.com/nature/journal/v429/n6992/suppinfo/nature02599.html).

South Africa (Thompson *et al.* 2002), in Tibet and the Himalaya (Thompson 2000) or the Andes (Thompson *et al.* 2000). The shorter time covered by continental ice cores demonstrates the sensitivity of ice to elevated temperatures. However, beneficially the isotopic signature of ^2H and ^{18}O from precipitation is directly preserved in the ice itself and no further transfer function is needed to reconstruct the isotopic composition of precipitation and macroscale climate information can be reconstructed. Distinction between regional and global climate signals still is difficult because of site limitations either to high altitudes or Polar Regions. Finally, reconstructions based on ice core records are limited in time since maximum ages of polar ice caps does not exceed 1 million years.

On the continents lacustrine deposits archive the best paleoclimatic information. The impact of changing climate is here much stronger than in the oceanic sediments and lacustrine deposits additionally provide a continuous record and a high temporal and spatial resolution. Commonly they contain, like in the marine record, authigenic carbonates or siliceous remains from planctonic and benthic organisms. Their $\delta^{18}O$ and δD values are mainly controlled by the isotopic signal of lake water and temperature which can be successfully reconstructed (v. Grafenstein *et al.* 1999). Unfortunately, the relation of stable isotopes and ambient water for this well-established method is also dependant on further environmental factors. The transfer functions of water isotopes into the biomarkers depend on physiological effects for each species and on the salt content. Therefore, other proxies on environmental conditions are necessary to increase data liability for reconstructions. In addition primary mineral remains that have not undergone secondary exchange reactions are not available in all lake sediments.

V. LONG-TERM WATER CYCLE PATTERN RECORDED BY ORGANIC MOLECULES

Beside mineral remains, organic molecules record paleoclimatic information. Most well known are annually grown tree rings. Primarily analyses focused on the annual tree-ring growth reconstructing temperature and precipitation amount. Floating chronologies are available for the whole Holocene. Additionally, the nonexchangeable hydrogen and oxygen isotopes of tree-ring cellulose with well-established transfer functions record the isotope content of leaf water (Sternberg and DeNiro 1983, Yakir and DeNiro 1990, Dawson 1993, White *et al.* 1994). The isotopic composition of leaf water is mainly controlled by the isotopic composition of groundwater and the transpiration rate (Figure 16.1). The latter is mostly influenced by the relative humidity (Roden and Ehleringer 1999).

Organic remains are also found in the sedimentary record of marine and lacustrine deposits. Most interestingly for paleoclimate reconstructions, however, are lacustrine sediments, as climate shifts on the continents are larger and in lakes both terrestrial and aquatic organic matter is deposited (Figure 16.1). Continuous sedimentation and age control in laminated sediments enable detailed environmental records. Organic matter in lake sediments incorporates individual molecules that derive from distinct biotic sources. These molecular fossils or so-called biomarkers are deposited in almost all sediments. Their abundance and composition identify past biocenosis and environmental parameters of their formation (Meyers and Lallier-Verges 1999, Meyers 2003). Most of these biomarkers belong to the biochemical group of lipids which are rich in carbon-bound hydrogen. Comparing molecules with carbon-bound hydrogen atoms to heteroatoms like oxygen, nitrogen, or sulfur, carbon-bound hydrogen is nonexchangeable at temperatures up to 150°C (Schimmelmann *et al.* 1999). Consequently, biomarkers belonging to that group might be suitable indicator for the primary signature of the water source during biosynthesis, since they are stable even over geological timescales.

VI. COMPOUND-SPECIFIC ISOTOPE RATIOS OF BIOMARKERS RECORD RECENT CLIMATE

The variety of biomarker substances emerging as promising proxies in paleoclimate studies is constantly increasing over the last decades. Moreover, progress in analytical methods to measure isotope ratios on individual organic compounds for carbon (Hayes *et al.* 1990) and hydrogen (Burgoyne and Hayes 1998, Hilkert *et al.* 1999) improved the characterization of the carbon and hydrogen sources of individual biomarker compounds offering several benefits over measuring bulk organic fractions. Thus, since the specific compounds are measured in the archive material in different horizons identically, different biochemical pathways of lipid biosynthesis that leads to distinct isotopic fractionations can be excluded. The measurement of hydrogen isotopes is done specifically on the carbon-bound hydrogen which is nonexchangeable for temperatures up to 150°C (Schimmelmann *et al.* 1999) and therefore preserves the biological source information.

Of particular interest among all biomarkers are *n*-alkanes. They are recalcitrant and can be isolated even from 600 million-year-old Phanerozoic sediments. Three major sources for *n*-alkanes in lake sediments are known (Figure 16.1): aquatic organisms, like algae, submerged or floating vascular plants, and terrestrial vascular plants (Meyers 2003). Each of these organisms synthesize *n*-alkanes that differ in their molecular structures. Short-chain *n*-alkanes with 17 and 19 carbon atoms are derived from algae; submerged aquatic plants and Sphagnum species synthesize *n*-alkanes with 23–25 carbon atoms (Baas *et al.* 2000, Ficken *et al.* 2000). The leaf waxes of terrestrial higher plants contain large proportions of *n*-alkanes with 25–31 carbon atoms, broad leaf trees have high amounts of n-C_{27} and n-C_{29} (Eglinton and Hamilton 1967), and grasses mainly n-C_{31} (Maffei 1996).

Organisms and plants synthesizing organic compounds, such as lipids, use their ambient water as their primary source of hydrogen. Thus, aquatic organisms use the lake water to produce *n*-alkanes and the meteoric water serves as hydrogen source for terrestrial plants *n*-alkanes. The isotope composition of the particular *n*-alkanes reveals an isotopic difference relative to the source water. This fractionation was shown to be independent from temperature (Estep and Hoering 1980, DeNiro and Epstein 1981) but controlled by the isotope signature of their biosynthetic precursor and fractionation and hydrogenation during biosynthesis (Sessions *et al.* 1999). Furthermore, Sessions *et al.* (1999) demonstrated that the hydrogen isotopic composition of source water is recorded in biomarkers like *n*-alkanes. Nevertheless, transfer functions for each biomarker are needed to quantitatively relate the isotopic composition of source water with the isotope signatures that is preserved in the particular biomarker. Therefore, various surface sediment studies covering lake transects along climatic gradients in Europe and Northern America (Huang *et al.* 2004, Sachse *et al.* 2004a), analyses of lacustrine sedimentary *n*-alkanes in different climates (Sauer *et al.* 2001, Chikaraishi and Naraoka 2003), and vegetation-specific biomarker studies (Sessions *et al.* 1999, Ficken *et al.* 2000, Chikaraishi and Naraoka 2003, Liu and Huang 2005, Sachse *et al.* 2006, Smith and Freeman 2006) are performed.

Although the widespread data regarding fractionation factors are still controversial and lacks of accurate quantitative information, consistent knowledge is established concerning the general processes involved during biosynthetic hydrogen incorporation in *n*-alkanes. The fractionation of hydrogen during photosynthesis by aquatic organisms leading to a depletion of deuterium of the aquatic derived *n*-alkanes is independent from temperature and reveals more or less constant values over large climatic gradients (Figure 16.7) (Sessions *et al.* 1999, Chikaraishi and Naraoka 2003, Huang *et al.* 2004, Sachse *et al.* 2004b).

Terrestrial *n*-alkanes seem to be enriched in deuterium relative to the aquatic ones with ranges between 10‰ and 60‰ (Chikaraishi and Naraoka 2003, Sachse *et al.* 2004b). The terrestrial *n*-alkanes

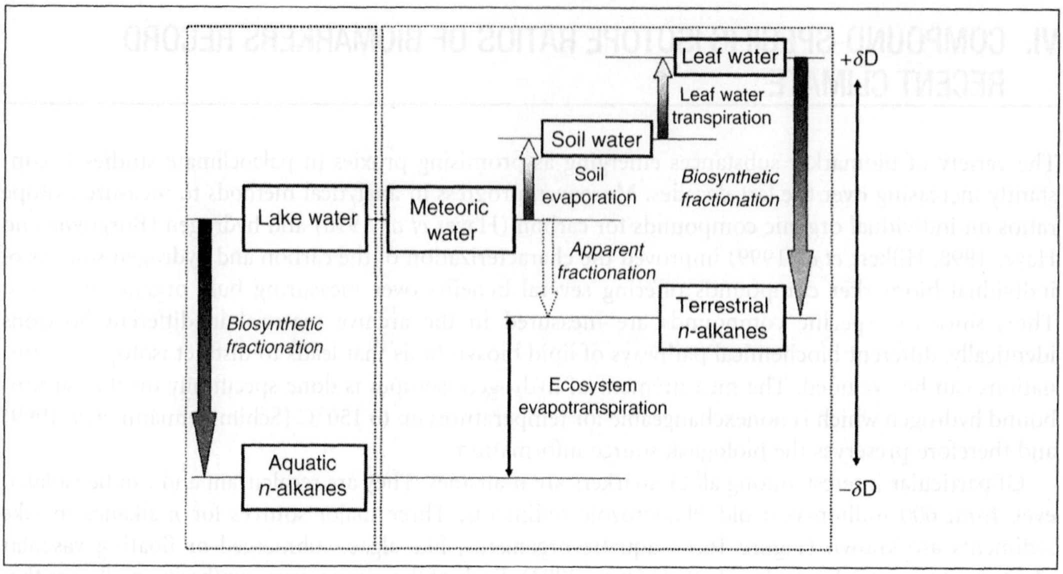

FIGURE 16.7 Isotopic relationships between δD values of source water and *n*-alkanes of terrestrial and aquatic origin in lake sediments (modified after Sachse *et al.* 2006).

isotope ratios are assumed to be affected by evapotranspiration effects first described by Yapp and Epstein (1982). Evaporation of soil water leads to an initial enrichment of meteoric water and transpiration effects at the plants leaf level amplify this enrichment (Leaney *et al.* 1985, Sachse *et al.* 2004b). Evapotranspiration of the source water from terrestrial organisms therefore causes an isotopic offset between hydrogen isotope ratios of terrestrial and aquatic derived *n*-alkanes. Quantitative assessment of this offset is of particular importance as it might serve as evapotranspiration proxy in ecosystems (Figure 16.7).

Whereas a strong linear relationship between the hydrogen isotope values of meteoric waters and *n*-alkanes from modern lacustrine sediments is shown, the specific impact of plant type, physiology, and climate that account for the δ^2H signature of *n*-alkanes in terrestrial plants is still controversially discussed. The hydrogen isotope ratio of *n*-alkanes in plants is supposed to be controlled by leaf architecture, photosynthetic pathway, and growth form (Bi *et al.* 2005, Krull *et al.* 2006, Liu *et al.* 2006, Smith and Freeman 2006). Since previous studies on δD values of plant lipids compared C_4 grasses to C_3 trees, shrubs, and herbs, the results are contradictory about the effects of growth forms, plant metabolism, and climate during incorporation of hydrogen (Sternberg *et al.* 1984, Ziegler 1989, Chikaraishi and Naraoka 2003). The establishment of consistent data on the biochemical fractionation of hydrogen in plant lipids has only begun focusing the plant-specific modifications that alter the isotope ratio of the precipitation water entering the soil–plant system through evapotranspiration.

VII. COMPOUND-SPECIFIC HYDROGEN ISOTOPE RATIOS IN CONTRASTING ECOSYSTEMS

To ensure the reliability established transfer functions have to be tested in distinct climate settings. The above-mentioned lake transect studies mainly focused on study sites in humid climate conditions (Huang *et al.* 2004, Sachse *et al.* 2004b). In order to figure out variations in isotopic fractionation of lacustrine *n*-alkanes and water source in different climates a comparison between δD values of sedimentary *n*-alkanes from lake sediments in Central European humid climate conditions and arid climate conditions at the Central Tibetan Plateau was conducted (Mügler *et al.*, submitted for publication). In general, this comparison revealed opposite directions of isotopic differences between aquatic and terrestrial *n*-alkanes in the two ecosystems (Figure 16.8). Under humid climate conditions terrestrial *n*-alkanes are enriched in deuterium relative to aquatic derived *n*-alkanes due to the evapotranspirative enrichment of soil and leaf water by ~30‰ (Sachse *et al.* 2004a, 2006). In contrast, in arid climate settings the aquatic *n*-alkanes record the enrichment in deuterium. This leads to an isotopic difference between aquatic and terrestrial biomarkers of about −55‰ (Figure 16.9). Thus, under arid climate conditions, aquatic organisms use lake water for biosynthesis that is isotopically enriched due to exceptional evaporation of lake water. Short-living terrestrial plants with their vegetation period during the strong monsoonal rain period use water directly from precipitation for biosynthesis (Figure 16.8) and are negligibly influenced by evapotranspiration. This opposite pattern of this isotopic difference therefore indicates the general hydrological characteristics of lake systems. Furthermore, considering its absolute values it might serve as the basis to quantify evapotranspiration rates and thus as a new proxy for paleoevapotranspiration.

The established transfer functions were also validated for the Neogene. Schefuss *et al.* (2005) successfully reconstructed past African rainfall variations during the last 20,000 years using the isotopic content of terrestrial *n*-alkanes. Abrupt climate changes such as the 8.2-ka event were reconstructed using δD values of biomarkers from lake sediments in New England and Massachusetts (Hou *et al.* 2006, Shuman *et al.* 2006). In general, δD values of individual compounds were successfully applied to reconstruct hydrologic conditions in the Miocene (Andersen *et al.* 2001), the late Quaternary (Huang *et al.* 2002, Liu and Huang 2005, Schefuss *et al.* 2005), or the Holocene (Xie *et al.* 2000). Consequently, early postsedimentary processes are not changing the hydrogen isotopic composition of sediments.

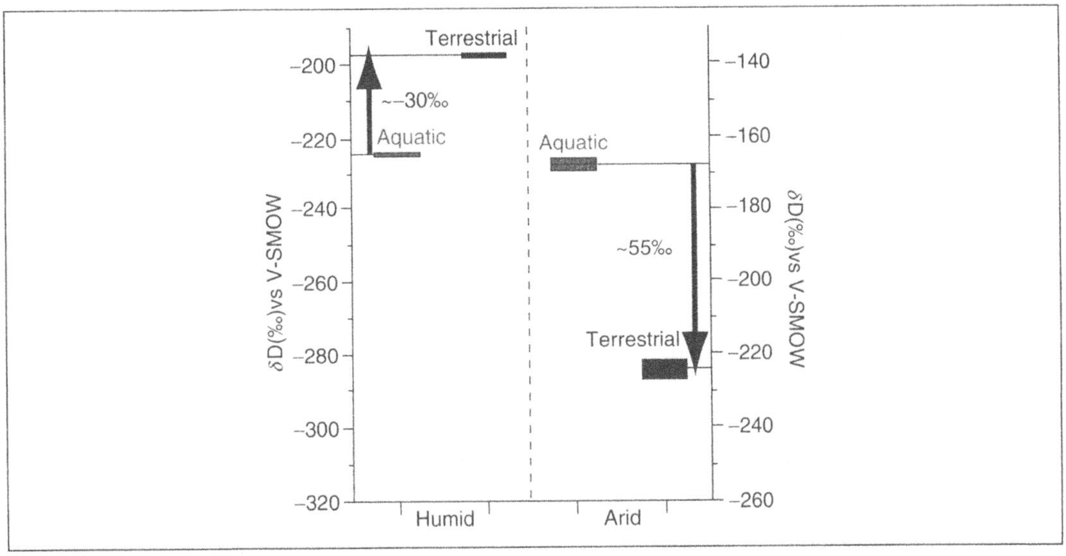

FIGURE 16.8 Isotopic difference of aquatic and terrestrial *n*-alkanes within the mid-European temperate climate Holzmaar (left) and within the highly evaporative environment of Nam Co, Tibetan Plateau (right) (modified after Mügler *et al.*, submitted for publication).

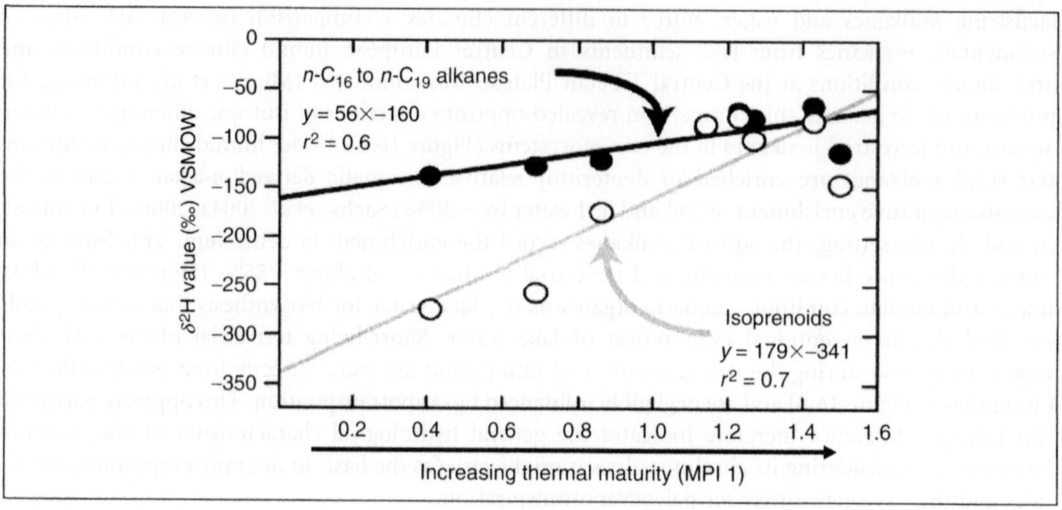

FIGURE 16.9 Increasing δD values of *n*-alkanes and isoprenoids with increasing degree of maturity (modified after Radke *et al.* 2005).

VIII. THE STABILITY OF COMPOUND-SPECIFIC HYDROGEN ISOTOPE RATIOS OVER THE GEOLOGICAL PAST

For geologic older sediments further analyses are required to test the reliability of compound-specific hydrogen isotope ratios since thermal maturity of bulk sediments is known to effect strongly the isotopic composition of organic matter (Schimmelmann *et al.* 1999). Radke *et al.* (2005) used δD values of *n*-alkanes extracted from different sections of the copper shale with increasing thermal maturity to investigate the effect of thermal maturity. The copper shale sediment was deposited due to an anoxic event in a short time period 258 My ago. Different burial depth caused increasing thermal maturity of this relative homogeneous deposit. The deuterium content of *n*-alkanes and isoprenoids linearly increased with increasing thermal maturity (Figure 16.9). With increasing maturity, the biosynthesis-based isotopic difference between *n*-alkanes and isoprenoids was lost. Several studies confirmed this lack of hydrogen isotopic difference between *n*-alkanes and isoprenoids in late mature sediments (Schimmelmann *et al.* 2004, 2006, Sessions *et al.* 2004, Dawson *et al.* 2005, Pedentchouk *et al.* 2006), and therefore suggest that climatic information in the δD signature is significantly altered in mature sediments. However, both, the linear relationship of isotopic change with thermal maturity as well as the isotopic offset between isoprenoids and *n*-alkanes can be used to correct for the influence of thermal maturity (Radke *et al.* 2005). This suggests that hydrogen isotope ratios of *n*-alkanes can be applied for paleoclimate reconstructions in the whole Phanerozoic.

IX. WATER ISOTOPES IN PALEOCLIMATE MODELS

Quantitative understanding of water isotope signatures in the water cycle is continuously growing with the development of isotope hydrology and paleoclimatology. First, very simplified models based only on Rayleigh distillation modeled the distribution of water isotopes in precipitation (Dansgaard 1964). These models were continuously improved implementing processes like evaporative fractionation

TABLE 16.2 Water isotope studies using general circulation models

Timescale	AGCM simulation
Seasonal cycle under modern conditions	Joussaume *et al.* 1984 Jouzel *et al.* 1987, 1991 Hoffmann and Heimann 1997 Hoffman *et al.* 1998 Mathieu *et al.* 1999 Yoshimura *et al.* 2003
Interannual variability	Cole *et al.* 1993 Hoffmann *et al.* 1998
Glacial–interglacial cycles	Joussaume and Jouzel 1993 Jouzel *et al.* 1994 Hoffmann and Heimann 1997 Werner *et al.* 2001

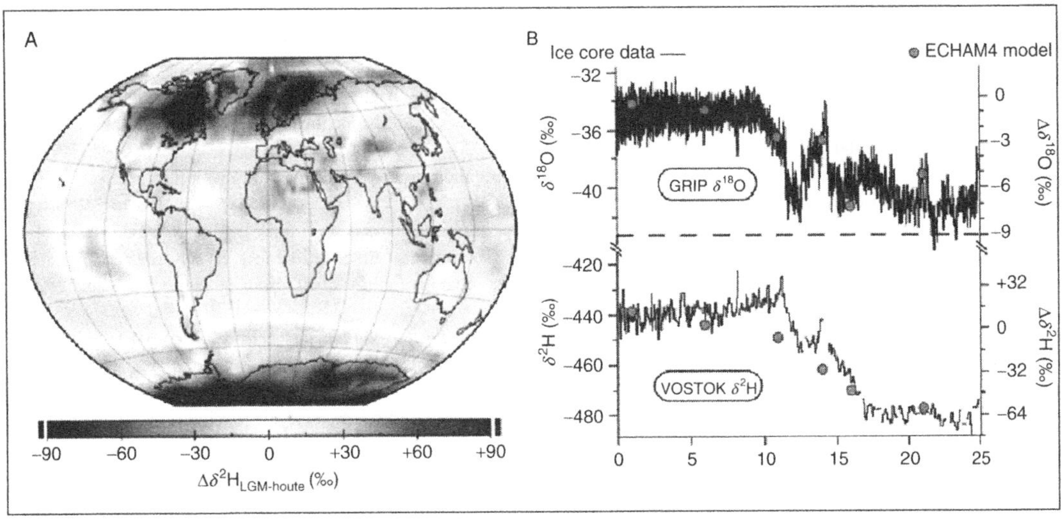

FIGURE 16.10 (A) ECHAM4-simulated changes of δD values in precipitation during the last glacial maximum compared to recent climate conditions. (B) Comparison of simulated and measured δD and $\delta^{18}O$ values in precipitation back to the last glacial maximum using GRIP and VOSTOK ice cores. (**See Color Plate.**)

(Merlivat and Jouzel 1979) or reevaporation of precipitation from the terrestrial surface (Rozanski *et al.* 1982). The full three-dimensional complexity of the fractionation processes in the hydrological cycle was started to be represented in atmospheric general circulation models (AGCM), that is the model of the Laboratoire de Meteorologie Dynamique (LDM; Joussaume *et al.* 1984) and the European Center Model Hamburg (ECHAM; Hoffmann and Heimann 1993). Until now a great variety of models working on different time and spatial scales exist (Table 16.2). However, direct comparison to water isotopes reconstructed from archives are still sparse. First, results using the global atmospheric circulation models ECHAM as well as the regional climate model REMO (developed by the Max-Planck Institute for Meteorology, Hamburg) gave good agreement between the simulated and experimental results from the polar ice core δD and $\delta^{18}O$ values (Figure 16.10A; Werner *et al.* 2001). Most interestingly predict these simulations large isotopic differences in the order of 120‰ between the LGM and today (Figure 16.10A).

Unfortunately, a direct comparison to compound-specific isotope ratios measured on biomarkers is still not possible, as biological-based modules incorporated in climate models need to be further developed. To extend the simulation of climate conditions in longer geological timescales additional changes in the distribution of continents and the ocean currents will be necessary.

X. CONCLUSIONS

This chapter suggests that compound-specific isotope ratios of biomarkers are most suitable to reconstruct past climate and environmental conditions. Further research will be needed to derive improved transfer functions for the isotopic signal of source water into biomarkers. The combination of experimental studies using lake sediments and modeling studies will help to evaluate the quality of climate predictions made by current circulation models.

XI. REFERENCES

Andersen, N., H. A. Paul, S. M. Bernasconi, J. A. McKenzie, A. Behrens, P. Schaeffer, and P. Albrecht. 2001. Large and rapid climate variability during the Messinian salinity crisis: Evidence from deuterium concentrations of individual biomarkers. *Geology* **29**(9):799–802.

Araguas-Araguas, L., K. Froehlich, and K. Rozanski. 2000. Deuterium and oxygen-18 isotope composition of precipitation and atmospheric moisture. *Hydrological Processes* **14**(8):1341–1355.

Augustin, L., C. Barbante, P. R. F. Barnes, J. M. Barnola, M. Bigler, E. Castellano, O. Cattani, J. Chappellaz, D. DahlJensen, B. Delmonte, G. Dreyfus, G. Durand, *et al.* 2004. Eight glacial cycles from an Antarctic ice core. *Nature* **429**(6992):623–628.

Baas, M., R. Pancost, B. van Geel, and J. S. S. Damste. 2000. A comparative study of lipids in Sphagnum species. *Organic Geochemistry* **31**(6):535–541.

Baertschi, P. 1976. Absolute O-18 content of standard mean ocean water. *Earth and Planetary Science Letters* **31**(3):341–344.

Bi, X., G. Sheng, X. Liu, C. Li, and J. Fu. 2005. Molecular and carbon and hydrogen isotopic composition of *n*-alkanes in plant leaf waxes. *Organic Geochemistry* **36**(10):1405–1417.

Bowen, G. J., and J. Revenaugh. 2003. Interpolating the isotopic composition of modern meteoric precipitation. *Water Resources Research* **39**(10):9-1–9-13.

Burgoyne, T. W., and J. M. Hayes. 1998. Quantitative production of H-2 by pyrolysis of gas chromatographic effluents. *Analytical Chemistry* **70**(24):5136–5141.

Chikaraishi, Y., and H. Naraoka. 2003. Compound-specific delta D-delta C-13 analyses of *n*-alkanes extracted from terrestrial and aquatic plants. *Phytochemistry* **63**(3):361–371.

Cole, J. E., D. Rind, and R. G. Fairbanks. 1993. Isotopic responses to interannual climate variability simulated by an atmospheric general-circulation model. *Quaternary Science Reviews* **12**(6):387–406.

Coplen, T. B. 1995. Reporting of stable hydrogen, carbon, and oxygen isotopic abundances (Technical report). *Geothermics* **24**(5–6):708–712.

Craig, H. 1961. Isotopic variations in meteoric waters. *Science* **133**(346):1702–1703.

Dansgaard, W. 1964. Stable isotopes in precipitation. *Tellus* **16**(4):436–468.

Dawson, D., K. Grice, and R. Alexander. 2005. Effect of maturation on the indigenous delta D signatures of individual hydrocarbons in sediments and crude oils from the Perth Basin (Western Australia). *Organic Geochemistry* **36**(1):95–104.

Dawson, T. E. 1993. Hydraulic lift and water-use by plants: Implications for water-balance, performance and plant-plant interactions. *Oecologia* **95**(4):565–574.

DeNiro, M. J., and S. Epstein. 1981. Isotopic composition of cellulose from aquatic organisms. *Geochimica et Cosmochimica Acta* **45**(10):1885–1894.

DeWit, J. C., C. M. Van der Straaten, and W. G. Mook. 1980. Determination of the absolute hydrogen isotope ratio of V-SMOW and SLAP. *Geostandards Newsletter* **4**:33–36.

Eglinton, G., and R. J. Hamilton. 1967. Leaf epicuticular waxes. *Science* **156**(3780):1322.

Epstein, S., and T. Mayeda. 1953. Variation of O-18 content of waters from natural sources. *Geochimica et Cosmochimica Acta* **4**(5):213–224.

Estep, M. F., and T. C. Hoering. 1980. Biogeochemistry of the stable hydrogen isotopes. *Geochimica et Cosmochimica Acta* **44**(8):1197–1206.

Ficken, K. J., B. Li, D. L. Swain, and G. Eglinton. 2000. An *n*-alkane proxy for the sedimentary input of submerged/floating freshwater aquatic macrophytes. *Organic Geochemistry* **31**(7–8):745–749.

Gat, J. R. 1971. Comments on stable isotope method in regional groundwater investigations. *Water Resources Research* **7**(4):980.

Gat, J. R. 1996. Oxygen and hydrogen isotopes in the hydrologic cycle. *Annual Review of Earth and Planetary Sciences* **24:**225–262.

Gonfiantini, R. 1986. Environmental isotopes in Lake studies. Pages 113–163 *in* P. Fritz and J. C. Fontes (Eds.) *Handbook of Environmental Isotope Geochemistry,* Vol. 2. Elsevier, Amsterdam.

Hayes, J. M., K. H. Freeman, B. N. Popp, and C. H. Hoham. 1990. Compound-specific isotopic analyses: A novel tool for reconstruction of ancient biogeochemical processes. *Organic Geochemistry* **16**(4–6):1115–1128.

Hilkert, A. W., C. B. Douthitt, H. J. Schluter, and W. A. Brand. 1999. Isotope ratio monitoring gas chromatography mass spectrometry of D H by high temperature conversion isotope ratio mass spectrometry. *Rapid Communications in Mass Spectrometry* **13**(13):1226–1230.

Hoffmann, G. H., and M. Heimann. 1993. Water tracers in the ECHAM GCM. *International Symposium and Isotope Techniques in the Study of the Past and Current Environmental Changes in the Hydrosphere and the Atmosphere.* International Atomic Agency, Vienna, pp. 3–14.

Hoffmann, G., and M. Heimann. 1997. Water isotope modeling in the Asian monsoon region. *Quaternary International* **37:**115–128.

Hoffmann, G., M. Werner, and M. Heimann. 1998. Water isotope module of the ECHAM atmospheric general circulation model: A study on timescales from days to several years. *Journal of Geophysical Research-Atmospheres* **103**(D14):16871–16896.

Holdsworth, G., S. Fogarasi, and H. R. Krouse. 1991. Variation of the stable isotopes of water with altitude in the Saint Elias Mountains of Canada. *Journal of Geophysical Research - Atmosphere* **96**(D4):7483–7494.

Hou, J. Z., Y. S. Huang, Y. Wang, B. Shuman, W. W. Oswald, E. Faison, and D. R. Foster. 2006. Postglacial climate reconstruction based on compound-specific D/H ratios of fatty acids from Blood Pond, New England. *Geochemistry, Geophysics, Geosystems* **7:**Q03008.

Huang, Y. S., B. Shuman, Y. Wang, and T. Webb. 2002. Hydrogen isotope ratios of palmitic acid in lacustrine sediments record late Quaternary climate variations. *Geology* **30**(12):1103–1106.

Huang, Y. S., B. Shuman, Y. Wang, and T. Webb. 2004. Hydrogen isotope ratios of individual lipids in lake sediments as novel tracers of climatic and environmental change: A surface sediment test. *Journal of Paleolimnology* **31**(3):363–375.

Huntington, T. G. 2006. Evidence for intensification of the global water cycle: Review and synthesis. *Journal of Hydrology* **319**(1–4):83–95.

Indermuhle, A., T. F. Stocker, F. Joos, H. Fischer, H. J. Smith, M. Wahlen, B. Deck, D. Mastroianni, J. Tschumi, T. Blunier, R. Meyer, and B. Stauffer. 1999. Holocene carbon-cycle dynamics based on CO_2 trapped in ice at Taylor Dome, Antarctica. *Nature* **398**(6723):121–126.

Joussaume, S., and J. Jouzel. 1993. Paleoclimatic tracers: An investigation using an atmospheric general-circulation model under ice-age conditions 2. water isotopes. *Journal of Geophysical Research-Atmospheres* **98**(D2):2807–2830.

Joussaume, S., R. Sadourny, and J. Jouzel. 1984. A general-circulation model of water isotope cycles in the atmosphere. *Nature* **311**(5981):24–29.

Jouzel, J., C. Lorius, J. R. Petit, C. Genthon, N. I. Barkov, V. M. Kotlyakov, and V. M. Petrov. 1987. Vostok ice core: A continuous isotope temperature record over the last climatic cycle (160,000 years). *Nature* **329**(6138):403–408.

Jouzel, J., R. D. Koster, R. J. Suozzo, G. L. Russell, J. W. C. White, and W. S. Broecker. 1991. Simulations of the Hdo and (H2O)-O-18 atmospheric cycles using the nasa giss general-circulation model: Sensitivity experiments for present-day conditions. *Journal of Geophysical Research-Atmospheres* **96**(D4):7495–7507.

Jouzel, J., R. D. Koster, R. J. Suozzo, and G. L. Russell. 1994. Stable water isotope behavior during the last glacial maximum: A general-circulation model analysis. *Journal of Geophysical Research-Atmospheres* **99**(D12):25791–25801.

Krull, E., D. Sachse, I. Mugler, A. Thiele, and G. Gleixner. 2006. Compound-specific delta C-13 and delta H-2 analyses of plant and soil organic matter: A preliminary assessment of the effects of vegetation change on ecosystem hydrology. *Soil Biology and Biochemistry* **38**(11):3211–3221.

Leaney, F. W., C. B. Osmond, G. B. Allison, and H. Ziegler. 1985. Hydrogen-isotope composition of leaf water in C-3 and C-4 plants: Its relationship to the hydrogen-isotope composition of dry-matter. *Planta* **164**(2):215–220.

Liu, W. G., and Y. S. Huang. 2005. Compound specific D/H ratios and molecular distributions of higher plant leaf waxes as novel paleoenvironmental indicators in the Chinese Loess Plateau. *Organic Geochemistry* **36**(6):851–860.

Liu, W. G., H. Yang, and L. W. Li. 2006. Hydrogen isotopic compositions of *n*-alkanes from terrestrial plants correlate with their ecological life forms. *Oecologia* **150**(2):330–338.

Maffei, M. 1996. Chemotaxonomic significance of leaf wax alkanes in the gramineae. *Biochemical Systematics and Ecology* **24**(1):53–64.

Majoube, M. 1971. Fractionnement en oxygene-18 et en deuterium entre l'eau et sa vapeure. *Journal de Chimie et Physique* **58:**1423–1436.

Mathieu, A., G. Seze, C. Guerin, H. Dupuis, and A. Weill. 1999. Mesoscale boundary layer clouds structures as observed during the semaphore campaign. *Physics and Chemistry of the Earth Part B-Hydrology Oceans and Atmosphere* **24**(8):933–938.

Mayewski, P. A., L. D. Meeker, S. Whitlow, M. S. Twickler, M. C. Morrison, P. Bloomfield, G. C. Bond, R. B. Alley, A. J. Gow, P. M. Grootes, D. A. Meese, M. Ram, *et al.* 1994. Changes in atmospheric circulation and ocean ice cover over the North-Atlantic during the last 41,000 years. *Science* **263**(5154):1747–1751.

Merlivat, L. 1978. Molecular diffusivities of (H2O)-O-1 6 Hd16O, and (H2O)-O-18 in gases. *Journal of Chemical Physics* **69**(6):2864–2871.

Merlivat, L., and J. Jouzel. 1979. Global climatic interpretation of the deuterium-oxygen-18 relationship for precipitation. *Journal of Geophysical Research—Oceans and Atmospheres* **84**(NC8):5029–5033.

Meyers, P. A. 2003. Applications of organic geochemistry to paleolimnological reconstructions: A summary of examples from the Laurentian Great Lakes. *Organic Geochemistry* **34**(2):261–289.

Meyers, P. A., and E. Lallier-Verges. 1999. Lacustrine sedimentary organic matter records of late Quaternary paleoclimates. *Journal of Paleolimnology* **21**(3):345–372.

Mügler, I. S., D. Werner, M. Xu, B. Wu, G. Yao, T. Gleixner, G. Lake Evaporation recorded in δD values of sedimentary *n*-alkanes of Nam Co, Tibetan Plateau, submitted for publication.

Pedentchouk, N., K. H. Freeman, and N. B. Harris. 2006. Different response of delta D values of *n*-alkanes, isoprenoids, and kerogen during thermal maturation. *Geochimica et Cosmochimica Acta* **70**(8):2063–2072.

Petit, J. R., M. Briat, and A. Royer. 1981. Ice-age aerosol content from east Antarctic ice core samples and past wind strength. *Nature* **293**(5831):391–394.

Radke, J., A. Bechtel, R. Gaupp, W. Puttmann, L. Schwark, D. Sachse, and G. Gleixner. 2005. Correlation between hydrogen isotope ratios of lipid biomarkers and sediment maturity. *Geochimica et Cosmochimica Acta* **69**(23):5517–5530.

Raynaud, D., J. M. Barnola, R. Souchez, R. Lorrain, J. R. Petit, P. Duval, and V. Y Lipenkov. 2005. Palaeoclimatology: The record for marine isotopic stage 11. *Nature* **436**(7047):39–40.

Roden, J. S., and J. R. Ehleringer. 1999. Observations of hydrogen and oxygen isotopes in leaf water confirm the Craig-Gordon model under wide-ranging environmental conditions. *Plant Physiology* **120**(4):1165–1173.

Rozanski, K., C. Sonntag, and K. O. Munnich. 1982. Factors controlling stable isotope composition of European precipitation. *Tellus* **34**(2):142–150.

Rozanski, K., L. Araguás-Araguás, and R. Gonfiantini. 1992. Relation between long-term trends of O-18 isotope composition of precipitation and climate. *Science* **258**(5084):981–985.

Rozanski, K., L. Araguás-Araguás, and R. Gonfiantini. 1993. Isotopic patterns in modern global precipitation. Pages 1–3 *in* L. K. Swart, P. K. McKenzie, and J. Savin S (Eds.) *Climate Change in Continental Isotopic Records.* American Geophysical Union, Washington.

Sachse, D., J. Radke, R. Gaupp, L. Schwark, G. Luniger, and G. Gleixner. 2004a. Reconstruction of palaeohydrological conditions in a lagoon during the 2nd Zechstein cycle through simultaneous use of delta D values of individual *n*-alkanes and delta O-18 and delta C-13 values of carbonates. *International Journal of Earth Sciences* **93**(4):554–564.

Sachse, D., J. Radke, and G. Gleixner. 2004b. Hydrogen isotope ratios of recent lacustrine sedimentary *n*-alkanes record modern climate variability. *Geochimica et Cosmochimica Acta* **68**(23):4877–4889.

Sachse, D., J. Radke, and G. Gleixner. 2006. Delta D values of individual n-alkanes from terrestrial plants along a climatic gradient: Implications for the sedimentary biomarker record. *Organic Geochemistry* **37**(4):469–483.

Sauer, P. E., T. I. Eglinton, J. M. Hayes, A. Schimmelmann, and A. L. Sessions. 2001. Compound-specific D/H ratios of lipid biomarkers from sediments as a proxy for environmental and climatic conditions. *Geochimica et Cosmochimica Acta* **65**(2):213–222.

Schefuss, E., S. Schouten, and R. R. Schneider. 2005. Climatic controls on central African hydrology during the past 20,000 years. *Nature* **437**(7061):1003–1006.

Schimmelmann, A., M. D. Lewan, and R. P. Wintsch. 1999. D/H isotope ratios of kerogen, bitumen, oil, and water in hydrous pyrolysis of source rocks containing kerogen types I, II, IIS, and III. *Geochimica et Cosmochimica Acta* **63**(22):3751–3766.

Schimmelmann, A., A. L. Sessions, C. J. Boreham, D. S. Edwards, G. A. Logan, and R. E. Summons. 2004. D/H ratios in terrestrially sourced petroleum systems. *Organic Geochemistry* **35**(10):1169–1195.

Schimmelmann, A., A. L. Sessions, and M. Mastalerz. 2006. Hydrogen isotopic (D/H) composition of organic matter during diagenesis and thermal maturation. *Annual Review of Earth and Planetary Sciences* **34**:501–533.

Sessions, A. L., T. W. Burgoyne, A. Schimmelmann, and J. M. Hayes. 1999. Fractionation of hydrogen isotopes in lipid biosynthesis. *Organic Geochemistry* **30**(9):1193–1200.

Sessions, A. L., S. P. Sylva, R. E. Summons, and J. M. Hayes. 2004. Isotopic exchange of carbon-bound hydrogen over geologic timescales. *Geochimica et Cosmochimica Acta* **68**(7):1545–1559.

Shuman, B., Y. S. Huang, P. Newby, and Y. Wang. 2006. Compound-specific isotopic analyses track changes in seasonal precipitation regimes in the Northeastern United States at ca 8200cal yr BP. *Quaternary Science Reviews* **25** (21–22): 2992–3002.

Smith, F. A., and K. H. Freeman. 2006. Influence of physiology and climate on delta D of leaf wax n-alkanes from C-3 and C-4 grasses. *Geochimica et Cosmochimica Acta* **70**(5):1172–1187.

Sternberg, L., M. J. DeNiro, and H. Ajie. 1984. Stable hydrogen isotope ratios of saponifiable lipids and cellulose nitrate from Cam, C-3 and C-4 Plants. *Phytochemistry* **23**(11):2475–2477.

Sternberg, L. D. L. O., and M. J. D. Deniro. 1983. Biogeochemical implications of the isotopic equilibrium fractionation factor between the oxygen-atoms of acetone and water. *Geochimica et Cosmochimica Acta* **47**(12):2271–2274.

Thompson, L. G. 2000. Ice core evidence for climate change in the Tropics: Implications for our future. *Quaternary Science Reviews* **19**(1–5):19–35.

Thompson, L. G., E. Mosley-Thompson, and K. A. Henderson. 2000. Ice-core palaeoclimate records in tropical South America since the last glacial maximum. *Journal of Quaternary Science* **15**(4):377–394.

Thompson, L. G., E. Mosley-Thompson, M. E. Davis, K. A. Henderson, H. H. Brecher, V. S. Zagorodnov, T. A. Mashiotta, P. N. Lin, V. N. Mikhalenko, D. R. Hardy, and J. Beer. 2002. Kilimanjaro ice core records: Evidence of Holocene climate change in tropical Africa. *Science* **298**(5593):589–593.

Thompson, L. G., E. Mosley-Thompson, M. E. Davis, P. N. Lin, K. Henderson, and T. A. Mashiotta. 2003. Tropical glacier and ice core evidence of climate change on annual to millennial time scales. *Climatic Change* **59**(1–2):137–155.

v. Grafenstein, U., H. Erlenkeuser, A. Bräuer, J. Jouzel, and S. J. Johnsen. 1999. A Mid-European decadal isotope-climate record from 15,500 to 5000 years B.P. *Science* **284:**1654–1657.

Veizer, J., D. Ala, K. Azmy, P. Bruckschen, D. Buhl, F. Bruhn, G. A. F. Carden, A. Diener, S. Ebneth, Y. Godderis, T. Jasper, G. Korte, *et al.* 1999. Sr-87/Sr-86, delta C-13 and delta O-18 evolution of Phanerozoic seawater. *Chemical Geology* **161**(1–3):59–88.

Watanabe, O., J. Jouzel, S. Johnsen, F. Parrenin, H. Shoji, and N. Yoshida. 2003. Homogeneous climate variability across East Antarctica over the past three glacial cycles. *Nature* **422**(6931):509–512.

Werner, M., M. Heimann, and G. Hoffmann. 2001. Isotopic composition and origin of polar precipitation in present and glacial climate simulations. *Tellus Series B - Chemical and Physical Meteorology* **53**(1):53–71.

White, J. W. C., J. R. Lawrence, and W. S. Broecker. 1994. Modeling and interpreting D/H ratios in tree-rings: A test-case of white-pine in the Northeastern United-States. *Geochimica et Cosmochimica Acta* **58**(2):851–862.

Xie, S., C. J. Nott, L. A. Avsejs, F. Volders, D. Maddy, F. M. Chambers, A. Gledhill, J. F. Carter, and R. P. Evershed. 2000. Palaeoclimate records in compound-specific delta D values of a lipid biomarker in ombrotrophic peat. *Organic Geochemistry* **31**(10):1053–1057.

Yakir, D., and M. J. DeNiro. 1990. Oxygen and hydrogen isotope fractionation during cellulose metabolism in Lemna-Gibba L. *Plant Physiology* **93**(1):325–332.

Yapp, C. J., and S. Epstein. 1982. A re-examination of cellulose carbon-bound hydrogen delta-D measurements and some factors affecting plant-water D/H relationships. *Geochimica et Cosmochimica Acta* **46**(6):955–965.

Yoshimura, K., T. Oki, N. Ohte, and S. Kanae. 2003. A quantitative analysis of short-term O-18 variability with a Rayleigh-type isotope circulation model. *Journal of Geophysical Research-Atmospheres* **108**(D20):13-1–13-15.

Ziegler, H. 1989. Hydrogen isotope fractionation in plant tissue. Pages 105–123 *in* P. W. Rundel, J. R. Ehleringer, and K. A. Nagy (Eds.) *Stable Isotopes in Ecological Research.* Springer Verlag, New York.

Zielinski, G. A., P. A. Mayewski, L. D. Meeker, S. Whitlow, M. S. Twickler, M. Morrison, D. A. Meese, A. J. Gow, and R. B. Alley. 1995. The GISP ice core record of volcanism since 7000 BC—reply. *Science* **267**(5195):257–258.

CHAPTER 17

Stable Carbon and Oxygen Isotopes in Recent Sediments of Lake Wigry, NE Poland: Implications for Lake Morphometry and Environmental Changes

Anna Paprocka

Institute of Geological Sciences, Polish Academy of Sciences

Contents

I. INTRODUCTION

Lake sediments preserve various biological and inorganic components which provide valuable informations about lake history, climatic changes in the past, and the current state of a lake and its catchment environment. One of the sources of environmental information in sediments that are commonly studied using stable carbon and oxygen isotope analyses are authigenic calcium carbonates, mostly formed as calcite.

The formation of calcium carbonate depends on the concentration of bicarbonate and calcium ions in water and corresponds to the following reaction:

$$Ca^{2+} + 2HCO_3^- \leftrightarrow CaCO_3(s) + H_2O + CO_2$$

The major factor inducing precipitation of carbonates is the CO_2 assimilation during photosynthesis by aquatic macrophytes and phytoplankton. In the lakes of temperate and high-latitude regions, carbonates are produced mainly in the summer months during periods of maximum primary productivity. Carbonate minerals are also formed biogenically as ostracode and mollusk shells or by charophyte algae. In addition, temperature increase and diffusive removal of carbon dioxide from water can generate precipitation of calcium carbonate.

Among the variety of geochemical proxies, stable carbon and oxygen isotope analyses of carbonate sediments have been considered to be a useful tool for paleoenvironmental and paleoclimatic reconstructions (Siegenthaler and Oeschger 1980, Edwards and Wolfe 1996, von Grafenstein *et al.* 1996, Schwalb and Dean 1998, Mayer and Schwark 1999, Makhnach *et al.* 2000). On account of increasing lakes degradation and cultural eutrophication, which are now a significant problems in many parts of the world, the isotope analyses of sediments and lake water are more frequently used to trace the recent changes related to natural processes or human impacts in the catchment area (Schelske and Hodell 1991, Neumann *et al.* 2002, Ekdahl *et al.* 2004, Jinglu *et al.* 2004). During the last decade stable carbon isotope analyses are also applied to calculate the carbon budget in lake systems (Wachniew and Różański 1997, Ogrinc *et al.* 2002). The aim of the carbon balance studies is to look at the lakes as a sink of carbon or a source of CO_2 interacting with the atmosphere in the global carbon cycle.

Although there are numerous stable isotope studies of lake sediments the suitable interpretation of the isotopic C, O data is still difficult or impossible to decipher without some additional assumptions. This is due to the various factors and complexity of processes that influence stable carbon and oxygen isotope record in sediments.

The $\delta^{18}O$ of lacustrine carbonates depends mostly on the isotopic composition of lake water and the temperature of water during carbonates precipitation. Both of these factors are closely related to the local climate. Likewise, the oxygen ratio ($^{18}O/^{16}O$) of lake water is determined by the isotopic composition of meteoric water which is also a function of climate, especially of air temperature. The $\delta^{18}O$ variations in lacustrine carbonates are also interpreted as changes in precipitation/evaporation ratio (Siegenthaler and Oeschger 1980, Edwards and Wolfe 1996) since the ^{16}O isotopes are preferentially transferred into the vapor phase leaving the water reservoir enriched in heavier oxygen isotopes (^{18}O).

The temperature relation with $\delta^{18}O$ of carbonates is commonly used to reconstruct past climate changes. But paleotemperature reconstructions based on the oxygen isotope records require an assumption that carbonate minerals precipitated in isotopic equilibrium with ambient water of the studied lake. The equilibrium oxygen isotope fractionation between carbonates and water temperature varies by $-0.24‰/°C$ (Friedman and O'Neil 1977). For the middle and high latitude regions, the oxygen isotope composition of the mean annual precipitation follows closely long-term changes of mean surface air temperatures with a gradient of ca. $+0.6‰/°C$ (Dansgaard 1964, Różański *et al.* 1992). Combination of these two fractionation factors leads to an estimated coefficient of ca. $+0.36‰/°C$ for lacustrine carbonates (Yu and Eicher 1998). The value of this coefficient can be reasonable if we assume that $\delta^{18}O$ of precipitation has not been affected by additional factors like changes in the isotopic composition of the moisture source, air mass trajectory, and disequilibrium isotope fractionation (Leng and Marshall 2004). It should be noticed that carbonate minerals do not always precipitate in isotopic equilibrium with their environments, which can be particularly associated with seasonally rapid carbonate precipitation. In lakes where rates of mineral precipitation are relatively high, the carbonate may have $\delta^{18}O$ values 2–3‰ below those expected for equilibrium (Fronval *et al.* 1995).

Variations in $\delta^{13}C$ values have been shown to reflect the rate of the lake productivity and changes in the trophic state as well as the redox conditions in the bottom water and early diagenetic processes in the surface sediment layer (Hollander and Smith 2001, Neumann *et al.* 2002). Carbon isotope composition of precipitating carbonates is only little affected by temperature and fractionation between carbonate minerals and dissolved inorganic carbon in lake water (DIC = $CO_{2(aq)} + HCO_3^- + CO_3^{2-}$). Thus, $\delta^{13}C$ values of carbonates in lake sediments mainly reflect the carbon isotope evolution of DIC during various carbon cycle transformations within the lake system. The dominant controls on the carbon isotope composition of DIC are CO_2 exchange between atmosphere and lake water, photosynthesis/respiration of aquatic organisms (Wachniew and Różański 1997, Vreča 2003), and lake stratification and circulation (Myrbo and Shapley 2006). Fractionation of carbon isotopes during photosynthesis is caused by the kinetic effects induced by phytoplankton and aquatic plants assimilation of ^{12}C-enriched CO_2. Subsequently, assimilation leads to precipitation of the enriched in ^{13}C carbonate minerals. Therefore, the increase of $^{13}C/^{12}C$ ratios in lake carbonates is often interpreted as a result of an intensive primary productivity in lake water.

The stable carbon and oxygen isotope proxies are also useful for identifying changes in the hydrological balance of lake by comparison of isotopic variations of both $\delta^{18}O$ and $\delta^{13}C$ (Hammarlund *et al.* 2003). The common relation for lacustrine carbonates is a positive covariance between $\delta^{13}C$ and $\delta^{18}O$ values (Li and Ku 1997). This effect is usually interpreted as a consequence of carbonates formation in a closed-basin lakes with long residence time of water. In the contrary, the lack or weak correlation is typical for hydrologically open lake systems (Talbot 1990). However, the isotopic covariance in such lakes at temperate latitudes can also be determined. An alternative explanation is that the $\delta^{13}C$-$\delta^{18}O$ covariance in open-basins is a direct record of change in regional climate (Drummond *et al.* 1995). The factors that simultaneously control the $\delta^{18}O$ of lake water and $\delta^{13}C$-DIC and further the isotopic record of sediments are changes in magnitude of water balance elements, especially precipitation and evaporation. All these patterns depend on temperature and air masses circulation. It is thought that if water volume declines, the intense evaporation will result in the increasing $\delta^{18}O$ values of lake water. Such conditions will also increase the $\delta^{13}C$ value of dissolved inorganic carbon in lake water as an effect of $^{12}CO_2$ diffusion to the atmosphere. According to Li and Ku (1997), the increase of the $\delta^{13}C$ during period of water level decrease can also be attributed to the vertical mixing of water which supplies nutrients from hypolimnion to the euphotic zone. This can result in enhancement of primary productivity in epilimnetic water which raises the $\delta^{13}C$-DIC values. In contrast, carbonate depletion in heavy carbon and oxygen isotopes is usually the effect of a positive water balance of closed-basin lake. But it must be noticed that the correlation between the two isotopic proxies can be also weak when the lake level is in steady state or during periods of higher $\sum CO_2$ of the lake (Li and Ku 1997). Likewise, the lack of the $\delta^{13}C$-$\delta^{18}O$ covariance can indicate the occurrence of the postdepositional processes, such as methanogenesis or precipitation, and dissolution of carbonate minerals in sediments.

Although the major processes that control fractionation of carbon and oxygen isotopes are relatively well known, the knowledge about all the factors which influence changes in isotopic composition of sediments are still insufficient. It should be considered there is a wide range of indirect factors and in-lake processes that can modify the isotopic composition and alter the isotopic signals of ecological changes recorded in lacustrine sediments. Therefore, the formation of isotopic composition of carbonates is more complex that was described above. The current work was undertaken on the assumption that the changes in isotopic signal in sediments are caused by the factors related to the lake morphometry such as the circulation of water column, rate of water exchange, and lake resistance to changes in trophic state. For example, in lakes with short time of water exchange and limited catchment area variations in the $\delta^{18}O$ of carbonates reflect changes in the isotopic composition of mean annual precipitation. Closed-basin lakes located at the same region as hydrologically open lakes will lose water through evaporation and will produce carbonates enriched in ^{18}O isotopes—the isotopic signal of temperature and precipitation changes will then be altered. In addition, in lakes with different hydrodynamic conditions (mixed or stratified lakes) the carbonate sediments will record the different carbon isotope signal of environmental changes. The difference can be caused by the circulation of water column as a valid mechanism which controls the distribution of carbonate species with different isotopic signature. Likewise, the stable isotope composition of lacustrine carbonates is, to some extent, the derivative of the $\delta^{18}O$ values of lake and inflow waters and their $\delta^{13}C$ of dissolved inorganic carbon. Thus, the contribution of surface and groundwater supplying the lake should also be considered as an important factor that reduces the isotopic signal of climate changes and the rate of lake productivity preserved in carbonate sediments. Moreover, it was observed (Leng *et al.* 1999, Dean 2002, Schwalb and Dean 2002) that even the lakes that are located at the same region and which are exposed to the same climate conditions can produce carbonates with different stable carbon and oxygen isotope records. This indicates that each lake reflects an individual isotopic signature in response to different hydrological condition of lake, their morphology, and sensitivity to trophic state changes and human impacts in the catchment. Therefore, the wide range of indirect factors and complexity of in-lake processes that influence stable isotope record in lacustrine sediments often leads to the interpretative objections and equivocal conclusions.

Another problem is that the stable carbon and oxygen isotope data of lake sediments are frequently interpreted on the basis of the one sediment core analyses. While, the representative site for the core sampling can be easily determined in small lakes, in case of large and complex basins the choice of a one place for research can be impossible. Therefore, the isotopic measurements restricted to the one site of a lake are less reliable as they preclude from relating the results to the variety limnological and ecological factors which are not necessarily uniform for a whole reservoir.

This chapter presents stable carbon and oxygen isotope records in recent sediments sampled in different parts of Lake Wigry, NE Poland. It provides an attempt to relate the isotopic data of lake sediments to the diverse character of the main Lake Wigry subbasins with respect to their bathymetry, hydrological conditions, trophic state, and pollution. The aim of presented work is also to detect how temporal variations of stable carbon and oxygen compositions of carbonates in sediment profiles of a complex basin reflect the recent environmental changes in lake watershed.

II. STUDY SITE

Lake Wigry is located in the macroregion of the Lithuanian Lake District, NE Poland (54°00′N, 23°01′E) and is one of the 42 lakes of Wigry National Park (Figure 17.1). The lake is also one of the biggest and deepest lakes in Poland. The main physical features of Lake Wigry are presented in Table 17.1. In limnological nomenclature it is of the melt-out furrow type which was formed by exaration processes during the retreat of the Baltic glacier phase of the last glaciation. The postglacial bedrock is rich in calcareous pebbles which results in a high concentration of calcium carbonate in Lake Wigry water. As a consequence, the most common sediments are lacustrine chalk in the coastal and central shallows, and carbonate gyttja accumulated in deeper parts of the basins. Typical organic sediments do not exist in the lake, whereas clastic sediments are restricted to narrow zones below the cliffs (Rutkowski 2004).

A. Spatial Variability

The strongly diversified bottom topography of Lake Wigry and its complex morphometry divide the lake into several subbasins (called ploso) (Figure17.1) which are different with respect to bathymetry, the type of water supply, and the intensity of water exchange (Table 17.2; Bajkiewicz-Grabowska 2002). The subbasins also differ in trophic state (Table 17.2) and thermal and oxygen conditions in water (Niewolak 1999, Górniak 2006).

Wigry is a dimictic lake with two stages of thermal stratification and circulation of water column. The major surface inlet is Czarna Hańcza River and Wiatrołuża River which enter the lake in its northern subbasin where the only surface outflow takes place. This part of the lake receives about 60% of water of the total surface inflow to Lake Wigry. The groundwater contribution in feeding the lake is assumed to range from 4% (30% of total groundwater inlet) in the southwestern as well as in the north

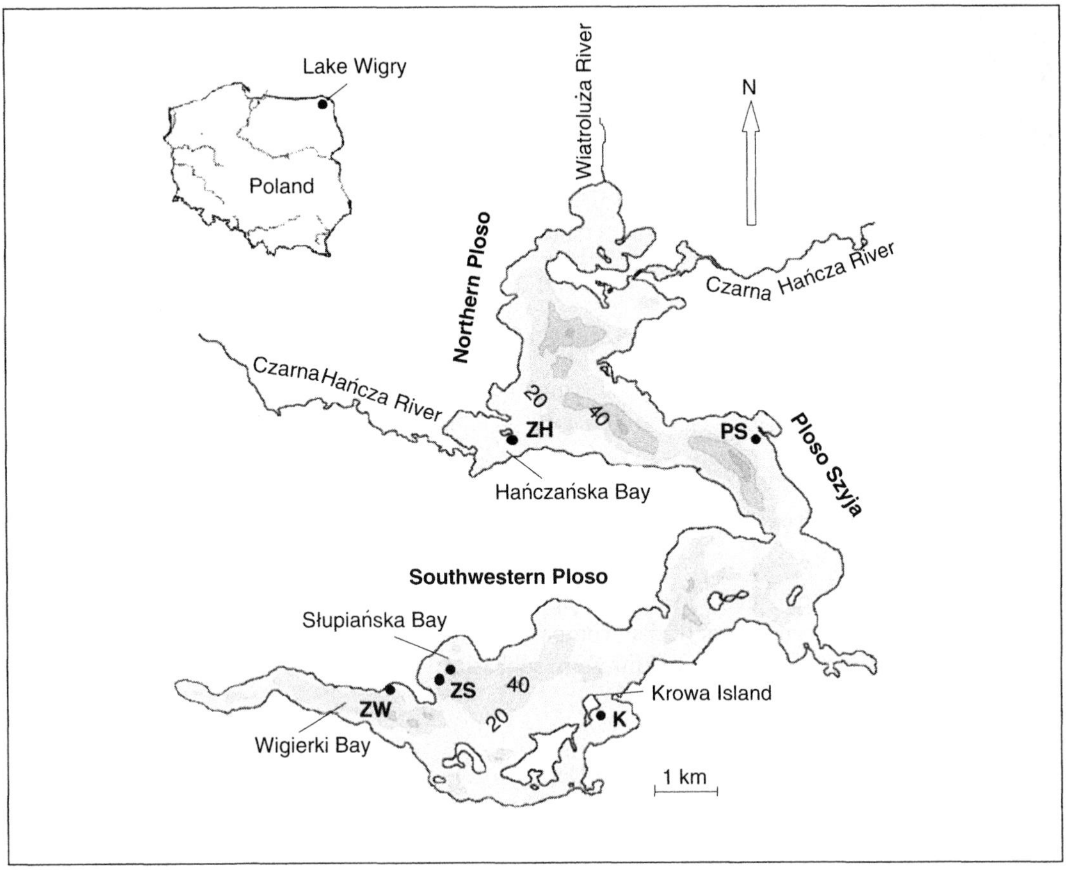

FIGURE 17.1 Location of Lake Wigry and sampling sites.

TABLE 17.1 Main physical features of Lake Wigry

Water surface	2118 ha
Total catchment	453.7 km^2
Maximum length	17.5 km
Maximum width	3.5 km
Maximum depth	73 m
Mean depth	15.8 m

to the 1% in the Ploso Szyja. According to Bajkiewicz-Grabowska (1999, 2002), the time of water exchange in the Northern Ploso is two- and threefold greater than in the southwestern subbasin due to the flow-through character of the northern part of the lake. The shape of the lake and spatial variation of the rate of water exchange is conductive to diverse habitat condition and trophic state (Górniak 2006). At present, Wigry is described as a mesotrophic lake, but in some places, particularly in the northern subbasin, signs of eutrophy are observed (Zdanowski 1992, Niewolak 1999, Górniak 2006). In the Northern Ploso and Szyja, the Carlson's trophic state index (TSI) is calculated to range from 50 to 61 (Table 17.2) which correspond to moderate eutrophy. For the southern parts of Lake Wigry, the TSI values are lower (40–49) and indicate mesotrophy conditions. The higher Secchi disk depth and the

TABLE 17.2 Spatial variation of Trophic State (TS) and Trophic State Index (TSI) in Lake Wigry Water in 1986–2002 and Water Exchange Coefficient (WE), Secchi Disc Depth (SD), Total Phosphorus (TP), Chlorophyll a (Chl-a), and Dissolved Organic Carbon (DOC) concentration in surface water (Epilimnion) of Lake Wigry in 2002 (Bajkiewicz-Grabowska 2002,* Górniak 2006)

Subbasin	WE* (%)	SD (m)	TP (μgP liter^{-1})	Chl-a (μg liter^{-1})	DOC (mgC liter^{-1})	Carlson's TSI 1986	1996	2002	TS
Northern Ploso	96	3.0	42	6.77	6.40	53.2	51.2	50.5	E
Hańczańska Bay	–	4.0	120	2.2	7.35	53.9	62.1	50.5	E
Ploso Szyja	82	3.0	36	10.61	10.12	52.7	55.7	51.3	E
Southwestern Ploso	33	4.6	23	2.34	6.78	53.8	56.4	42.1	M
Wigierki Bay	48	4.5	15	2.73	5.26	53.6	54.5	40.5	M

E and M indicate eutrophy and mesotrophy, respectively.

lower concentration of total phosphorus, chlorophyll a, and dissolved organic carbon in epilimnion of the Southwestern Ploso and Wigierki Bay compared to the other parts of the lake (Table 17.2) also indicate a lower trophic state in the southwestern subbasins.

B. Ecological Changes Based on Archival Data

Although Lake Wigry is well located, surrounded by a forested region, and far from industrial centers, during the last century the lake has undergone abrupt changes in productivity and trophic state due to nutrient-enrichment and impact of tourism. Environmental changes in the catchment began about 350 years ago together with settlement expansion, which was particularly strong in the nineteenth century. Distinct reduction of water quality and an elevated trophic state of the lake were identified in the 1970s and 1980s especially in the northern, flow-through subbasin (Table 17.2) (Górniak 2006). Over the last decades the major sources of pollution have been tourism development and mostly municipal sewage derived from the nearby city of Suwałki by the Czarna Hańcza River. Since the Wigry National Park was created in 1989, many steps have been undertaken to protect the lake. But the most effective factor in reducing degradation of Lake Wigry environment was thorough modernization of the Suwałki sewage treatment plant in the years 1993–1995. As a result, the efficiency of removing nitrogen and phosphorus from sewage has greatly increased and the lake is now in the state of relative balance between nutrient loading and their use within the lake (Kamiński 1999). Although biogenic substances and their concentrations have recently decreased, the input of other pollutants still present a threat to Lake Wigry.

III. SAMPLING SITES AND METHODS

Carbonate sediment cores were sampled at three different subbasins of Lake Wigry (Figure 17.1). The first core was taken at the Hańczanska Bay of the Northern Ploso—the relatively most polluted and the only flow-through part of the lake (profile ZH at water depth of 11 m). The next one was sampled at the deepest basin, Ploso Szyja, in its littoral zone where lacustrine chalk is the prevailing sediment (profile PS at water depth of 1.5 m). Another two cores were taken at the Southwestern Ploso with low-nutrient loading and longer time of water exchange: one of these two sediment cores at the southwestern subbasin was drilled nearby the Krowa island (profile K at water depth of 10 m), and the second at wind-sheltered Słupiańska Bay (profile ZS at water depth of 20 m). The fifth core was sampled in the narrow Wigierki Bay at the western part of Lake Wigry (profile ZW at water depth of 0.85 m). Additional sediments were also sampled in the Słupiańska Bay for control analysis and to compare the $\delta^{13}C$ and $\delta^{18}O$ values with those of previous sediment profile ZS. Undisturbed sediment cores to a depth of 34 cm were taken in 2003 and 2005 at all six stations using a gravity core sampler and were sectioned into 1-cm intervals for isotope analysis.

Stable carbon and oxygen analysis of the CO_2 gas extracted from $CaCO_3$ were performed at the Institute of Geological Sciences of the Polish Academy of Sciences. CO_2 from the dried and powdered carbonate sediments was liberated by reacting samples with 100% phosphoric acid under vacuum conditions, at 25 °C. The evolved CO_2 was then purified in a vacuum line. Isotope ratios are given as δ (‰) relative to the international standard (V-PDB). The weight percentage of $CaCO_3$ in all sediment samples was calculated on the basis of CO_2 volume expanded on a calibrated mercury manometer. The weight percentage of organic matter in additional samples was estimated by removing of OM with hydrogen peroxide and reweighting. Uncertainties of preparation and measurements of carbon and oxygen isotopes were ±0.1‰ and ±0.2‰, respectively, and ±1% for $CaCO_3$ and OM.

The dating of sediments by the ^{210}Pb method was performed in the Institute of Geological Sciences of the Polish Academy of Sciences and in the Institute of Oceanology of the Polish Academy of Sciences. ^{210}Pb was determined indirectly by measuring ^{210}Po using α-spectrometry method. A constant rate of unsupported ^{210}Pb supply model (CRS) was used to calculate the sediments age. This chapter presents the age model results established for three sediment profiles—PS, ZS, and K. At present, the ^{210}Pb dating procedure requires more sediment material from the Hańczańska and Wigierki Bay, thus the preliminary age data of the profile ZH and ZW are not included in the chapter.

IV. RESULTS

The age model results reported in the calendar year AD show that upper 9-cm fragment of profile PS and 15 cm of profile ZS and K represent the last ca.100 years (Figure 17.2). The highest calcium carbonate concentration (90–99% $CaCO_3$) and the lowest organic matter content (1–8% OM) are observed in sediments sampled at Ploso Szyja and the southwestern subbasins (ZS, K, and ZW) (67–99% $CaCO_3$ and 1–31% OM) (Table 17.3). The sediments derived from the Northern Ploso (ZH) are carbonate-depleted (32–36% $CaCO_3$) with higher concentration of organic matter (61–65% OM) (Table 17.3).

The isotopic analyses of the recent sediments displayed considerable differences in $\delta^{13}C$ and $\delta^{18}O$ values between the upper fragments (15 cm) of all studied cores (Table 17.4). The lowest average $\delta^{13}C$ values (−4.8‰) are recorded for sediment samples of the profile ZH. The farther from the northern part of the lake, the more enrichment sediments in ^{13}C are observed (Table 17.4). As a result, the highest average $\delta^{13}C$ values of −3.3‰ and −3.4‰ are determined for the profile K (Southwestern Ploso) and ZW (Wigierki Bay), respectively. The study also reveals the lowest ^{18}O content in sediments from the northern subbasin (average −9.7‰) and more enriched carbonates in ^{18}O from the southern part of the lake (−8.9‰ for profile K and −9.2‰ for ZS). However, the $\delta^{18}O$ values do not rise toward the southwestern parts of the lake and the highest average $\delta^{18}O$ values were analyzed for the profile PS (−8.3‰) as well as for the profile ZW (−8.0‰) (Table 17.4).

The variability of isotopic data between studied sediments is also shown in the vertical profiles. The characteristic trend observed for the lacustrine chalk sampled at Ploso Szyja is a positive $\delta^{13}C$-$\delta^{18}O$ covariance (Figure 17.3). The isotopic record of these carbonates displays the five alternating periods of enrichment and depletion of sediments in ^{13}C and ^{18}O (Figure 17.2). The data obtained from

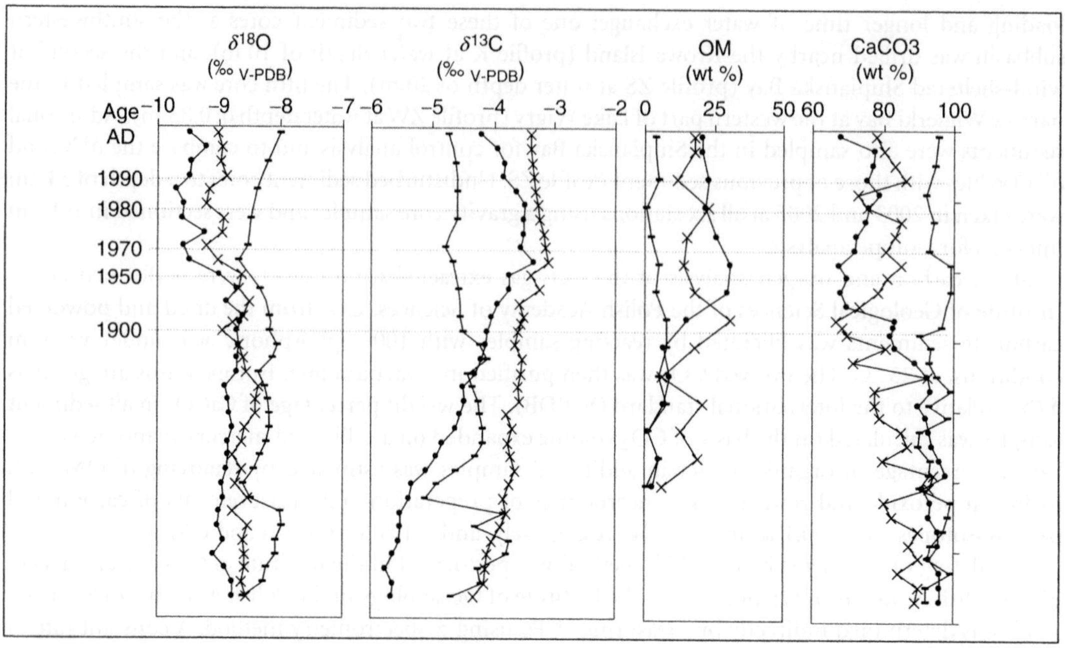

FIGURE 17.2 Temporal variation of $\delta^{18}O$, $\delta^{13}C$ values, and calcium carbonate and organic matter content in the recent sediments of Lake Wigry. Symbol • represents sediment profile ZS, x-profile K, – - profile PS.

TABLE 17.3 Calcium carbonate and organic matter content of the upper fragment (15 cm) of five sediment cores from Lake Wigry

	Upper 1–15 cm of sediment core					
	$\%CaCO_3$			%OM		
Core name	Minimum	Maximum	Average	Minimum	Maximum	Average
ZH	32	36	34	61	65	63
PS	90	99	95	1	8	4
ZS	68	84	74	15	31	24
K	67	86	79	14	31	19
ZW	87	98	94	1	12	5

measurements of profile ZS and K distinguish two sections in sediment cores which differ in isotopic records and calcium carbonate and organic matter concentration. The lower section (12–34 cm) is characterized by the increase in $\delta^{13}C$ values toward the upper fragments of profile ZS (from −5.7‰ to −3.7‰) and K (from −4.5‰ to −3.2‰) and almost uniform oxygen isotope composition (Figure 17.2). A positive $\delta^{13}C$-$\delta^{18}O$ and $\delta^{13}C$-%OM covariance (Figures 17.3 and 17.4) and a negative correlation between $\delta^{13}C$-$\%CaCO_3$ are also detected (Figure 17.4). The $^{13}C/^{12}C$ ratios of carbonates in the first 11 cm of the profile ZS and K are less variable than in the lower section and range from −4.0‰ and −3.4‰ to −3.3‰ and −3.2‰, respectively (Figure 17.2). In addition, the slight decreasing of $\delta^{13}C$ values is observed in the surface part of the cores. However, the $\delta^{13}C$ values do not correlate with the other proxies ($\delta^{18}O$, $\%CaCO_3$). Carbonates from the upper section are also $\delta^{18}O$-depleted by 0.5–1.0‰ compared to the lower fragment of these two profiles.

TABLE 17.4 Stable carbon and oxygen isotope compositions of the upper fragment (15 cm) of five sediment cores from Lake Wigry

	Upper 1–15 cm of sediment core					
Core name	**1-5**	**6–10**	**11–15**	**Minimum**	**Maximum**	**Average**
$\delta^{18}O$ (‰$_{V\text{-}PDB}$)						
ZH	−9.4	−9.9	−9.8	−10.1	−9.0	−9.7
PS	−8.3	−8.3	−8.2	−8.7	−8.0	−8.3
ZS	−9.5	−9.5	−8.8	−9.7	−8.6	−9.2
K	−9.0	−9.1	−8.7	−9.2	−8.5	−8.9
ZW	−7.9	−7.7	−8.2	−8.5	−7.6	−8.0
$\delta^{13}C$ (‰$_{V\text{-}PDB}$)						
ZH	−4.9	−4.8	−4.7	−5.0	−4.7	−4.8
PS	−4.7	−4.5	−4.3	−4.8	−4.1	−4.5
ZS	−3.9	−3.5	−3.9	−4.2	−3.3	−3.8
K	−3.4	−3.2	−3.5	−3.6	−3.1	−3.3
ZW	−3.6	−3.4	−3.2	−3.6	−3.1	−3.4

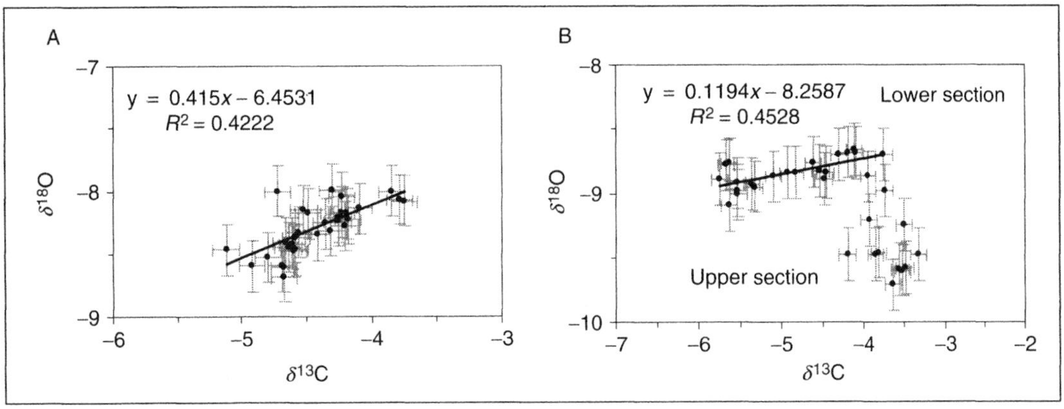

FIGURE 17.3 Relationship between $^{18}O/^{16}O$ and $^{13}C/^{12}C$ ratios in sediments from (A) Ploso Szyja-profile PS and (B) the southwestern subbasin—profile ZS.

The results of the control analysis of two sediment cores derived from the same part of the lake (Słupiańska Bay) revealed similar $\delta^{13}C$ and $\delta^{18}O$ data which confirms the reliability of performed isotopic measurements (Figure 17.5).

V. DISCUSSION

The stable carbon and oxygen isotope analyses of sediments derived from diverse subbasins of Lake Wigry give the opportunity to compare the isotopic records with regard to lake morphology and spatial differences of its trophic state, hydrological conditions, and the time of water exchange. Of great interest is the investigation of the isotopic record response to trophic state variety and temporal changes in biological productivity.

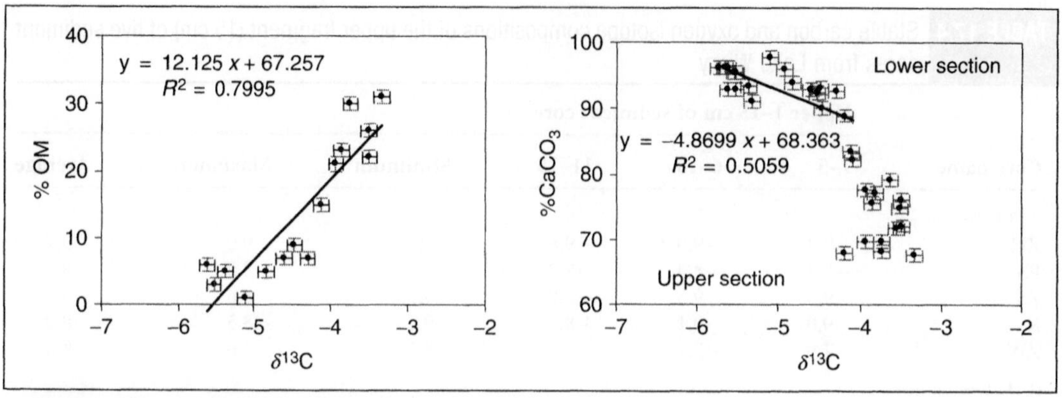

FIGURE 17.4 Relationship between $\delta^{13}C$ values and OM and $CaCO_3$ contents in the upper and lower sections of sediment profile (ZS) from the southwestern subbasin.

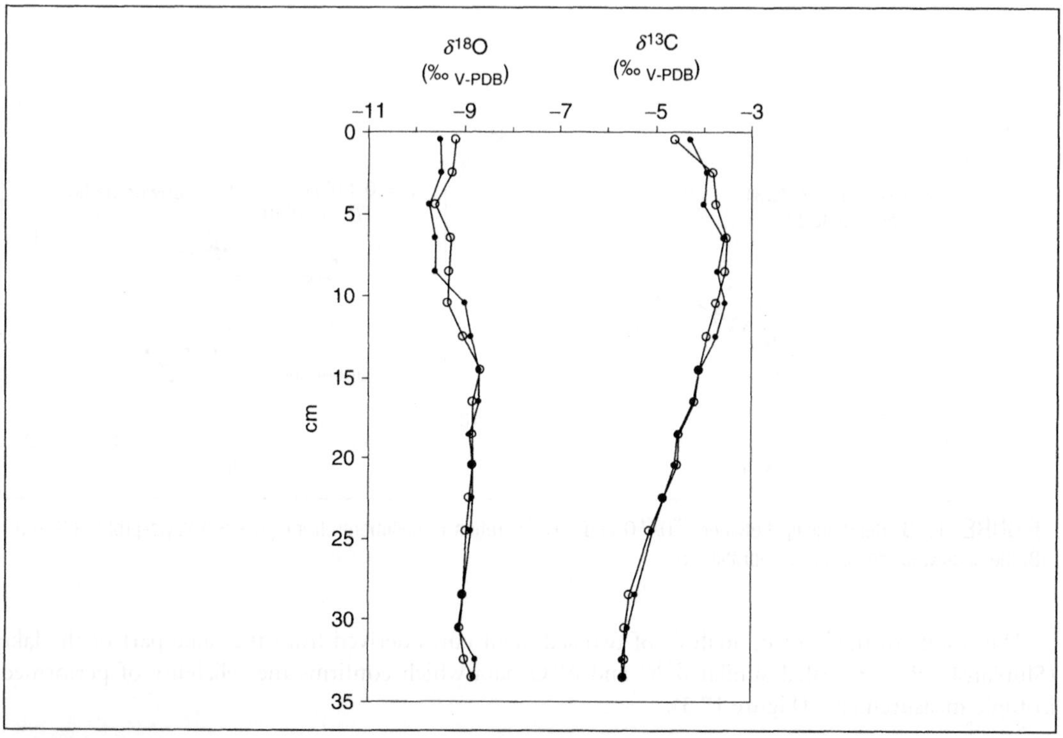

FIGURE 17.5 Comparison of $\delta^{18}O$ and $\delta^{13}C$ values of sediments sampled at the Słupiańska Bay (southwestern subbasin): • profile ZS, ○ additional profile.

An increasing eutrophication is generally assumed to be reflected by increasing $^{13}C/^{12}C$ ratio in lacustrine sediments. However, the $\delta^{13}C$ values of organic-rich sediments in the northern subbasin, which is now recognized as a moderate eutrophic, are lower compared to the $\delta^{13}C$ of sediments in mesotrophic southwestern subbasins. In fact, Hollander and McKenzie (1991) showed that, in some conditions, intense primary productivity of a lake may result in a slow decrease of the $\delta^{13}C$ values of dissolved inorganic carbon and precipitated carbonates. It is possible when enhance bioproductivity occurs with the accompanying limitation of CO_2 in water which leads to assimilation of $^{13}CO_2$ by aquatic plants. As a consequence, the carbon isotope fractionation decreases during photosynthesis. It was also found (Hollander and Smith 2001) that progressing eutrophication can induce bacterial expansion of the methanotrophic and chemoautotrophic bacterial community in surface sediments which are the additional source of the isotopically light carbon to DIC reservoir. But such conditions are characteristic particularly at the small eutrophic lakes and with long residence time of water. In case of Lake Wigry, it is more likely that the formation of isotopic composition of sediments in the northern subbasin is influenced by the factors that alter the isotopic signal of biological productivity. As the oxygen isotope composition of river waters and its stable carbon isotope composition of DIC are depleted in heavier isotopes, the significant contribution of surface inflows in supplying the Northern Ploso can affect the lower $\delta^{13}C$, as well as $\delta^{18}O$ values of sediments from the profile ZH. The preliminary analyses of stable carbon isotope composition of dissolved inorganic carbon in Lake Wigry water and in its groundwater and surface inlets (performed in October 2006) show that the $\delta^{13}C$-DIC values of inflow waters are about 5‰ (groundwater) and 2‰ (river water) lower than $\delta^{13}C$-DIC of epilimnetic water in the southern subbasin which confirms the above assumption. Furthermore, higher trophic state of the Northern Ploso (Table 17.2) can contribute to the higher replenishment rate of lake water in ^{12}C-DIC derived from river and groundwater and can result in the ^{13}C-depletion of dissolved inorganic carbon compared to the basins with lower TSI values.

In addition, there are also differences between the upper fragments of five analyzed profiles with regard to the $\delta^{18}O$ values in carbonates (Table 17.4). Thus, one should presume that the isotopic signal of the local climate have not been clearly recorded in sediments of Lake Wigry. It denotes that the divergence of the oxygen isotope records is attributed to the different character of the main subbasins and also correlates with the distinct amounts of inflow waters and the various intensity of water exchange. Although the $\delta^{18}O$ values of sediments in the northern subbasin are lower than in the southern one, the oxygen isotope composition of carbonates from profile PS do not contribute to the rising $\delta^{18}O$ values of sediments toward the southwestern parts of Lake Wigry (Table 17.4). The higher $^{18}O/^{16}O$ ratio of carbonates from PS are likely related to some extent to the depth of sampling site—the lowest depth (1.5 m) relative to the other stations (10–20 m). Atmospheric temperature and evaporation are considered to predominantly influence the hydrological condition in shallow parts of the lake and in consequence to stronger affect the oxygen isotope composition of water, compared to the deepest sites. This is confirmed by the highest $\delta^{18}O$ values determined for the profile ZW which was also sampled at the littoral zone of the lake at water depth of 0.85 m.

The $\delta^{13}C$ and $\delta^{18}O$ records in sediments of Lake Wigry likewise show the vertical variability of isotopic data between studied profiles (Figure 17.2). An interesting result is the positive correlation between $^{18}O/^{16}O$ and $^{13}C/^{12}C$ ratios of carbonates in profile PS (Figure 17.3) and which is not observed for the rest of the sediment cores. Since the high R^2 coefficient is typical for closed-basin lakes or can be an indicator of change in regional climate, in case of Lake Wigry the positive $\delta^{13}C$-$\delta^{18}O$ covariance is more likely to be related also to the relatively small depth of the sampling site as an indirect factor. Such a small thickness of the water layer could have determined a substantial effect of climatic patterns—precipitation and evaporation, and lake level fluctuations on variability of $\delta^{13}C$ and $\delta^{18}O$ values over the studied profile. Regarding the stable carbon isotopic composition of carbonates, there is no clear evidence of the enhancement of lake eutrophication inferred from isotopic data of this sediment profile.

Considering temporal changes in Lake Wigry and its catchment environments, two periods of various conditions can be distinguished on the basis of isotopic composition of carbonates from the southern subbasin (ZS and K). The first section ranges from 12 to 34 cm and corresponds to the period before 1950. The second one is marked in the upper part of profiles from 1 to 11 cm, and spans the last 50 years (Figure 17.2). The isotopic results were compared with available archival data related to ecological changes and climatic patterns measured in the studied region (Figure 17.6).

The period related to the lower part of sediments can be described as a stage of increasing biological productivity in the southern part of Lake Wigry. It is presumed that evidence of increasing primary

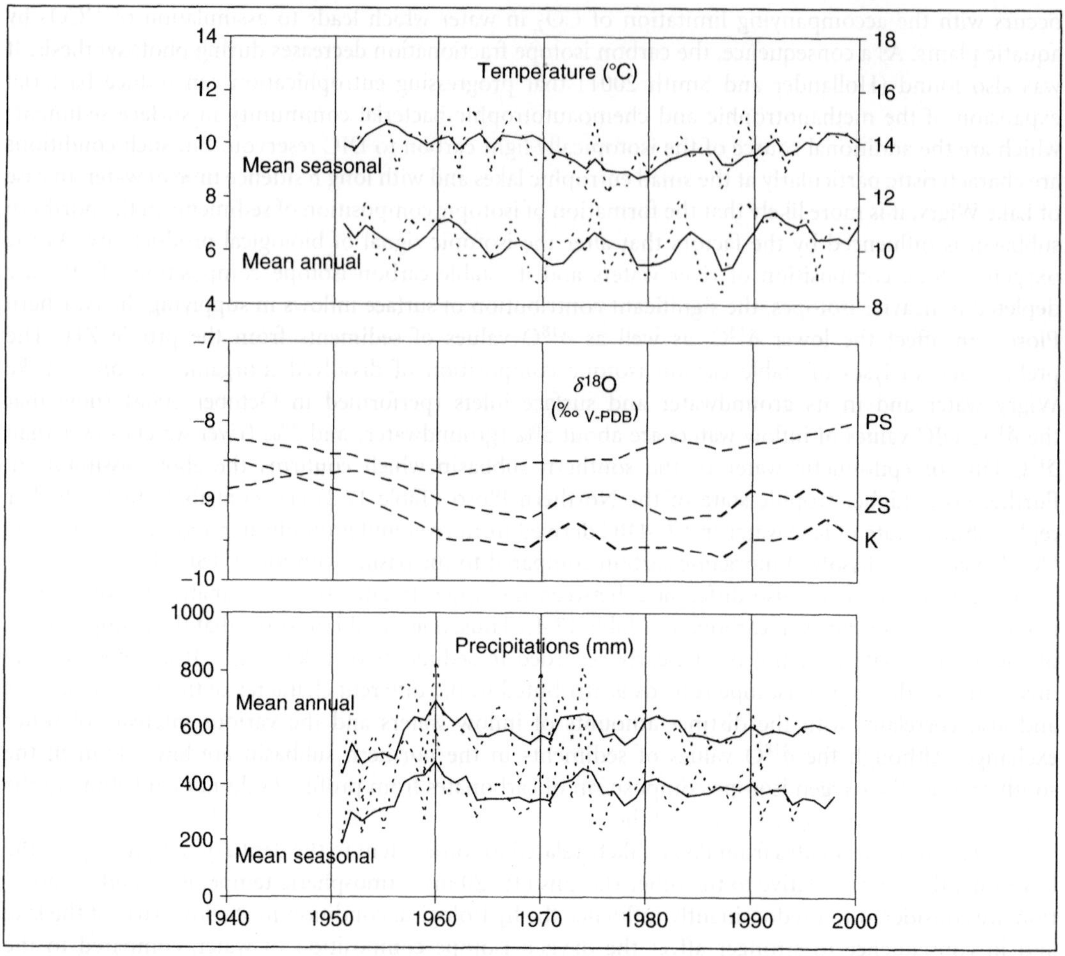

FIGURE 17.6 Comparison of $\delta^{18}O$ values of sediments from profile PS, K, and ZS with mean air temperature, precipitation and with their trends in the studied region. The data relate to spring–summer seasons and annual periods in the years 1950–2000. (Based on IMGW data).

productivity of the lake is displayed by the 1.5% and 2‰ increase of the $\delta^{13}C$ values in two carbonate sediments profiles from the southern subbasin. The substantial increase of the $\delta^{13}C$ values in these profiles is therefore in good agreement with the negative environmental changes in Lake Wigry catchment during recent centuries (Section II.B). Although the effects of slow eutrophication are mostly observed at the northern subbasin, the signs of changes in trophy have been recognized at the Southwestern Ploso (Zdanowski 1992, Górniak 2006). However, such a relatively distinct trend toward more heavy carbon isotope values in sediments, especially at Słupiańska Bay (profile ZS) (Figure 17.2), must have been also caused by the additional factors related to the hydrodynamic conditions in this part of Lake Wigry. The strong summer stratification of water in wind sheltered Słupiańska Bay could hamper the replenishment of isotopically light carbon from the decomposition of organic matter in sediments to the surface water layer. As a result, the DIC pool in the epilimnion, being the source of precipitated carbonates, became more enriched in ^{13}C. In addition, the longer residence time of water compared to the other subbasins of Lake Wigry might enhance the signals of trophic changes recorded in sediments. The characteristic for this stage is also a little variation of $\delta^{18}O$ values and a positive $\delta^{13}C$-$\delta^{18}O$ and $\delta^{13}C$-%OM covariance (Figures 17.3 and 17.4) The higher $\delta^{13}C$ values are consistent with decreasing calcium carbonate and increasing organic matter concentration in sediments which confirms slow increasing in bioproductivity of Lake Wigry in the described stage.

An important point is that there are distinct changes in the $\delta^{13}C$, $\delta^{18}O$ values and in the $CaCO_3$ and OM content observed in the upper part of sediments from the southern subbasin. These differences indicate that some changes in environmental conditions of the studied site occurred ca. 50 years ago. The different record of the $\delta^{13}C$ values in the upper part of both profiles from the southwestern subbasin (Figure 17.2) and the lack of the $d^{13}C$-$\delta^{18}O$ covariance (Figure 17.3) are coincidental with the major threat to Lake Wigry environment during the last few decades. This period is also marked by a decrease of calcium carbonates and a considerable increase of organic matter content in sediments from profile ZS (Figure 17.2) which indicates an elevated biological productivity at Słupiańska Bay water. The threat is related to the impact of tourism and more intensive nutrients load to the lake which could have resulted in changes in trophic state of the southern part of Lake Wigry. However, an increase in algal and macrophyte photosynthesis would normally produce higher $\delta^{13}C$ values in authigenic carbonates precipitated in the epilimnion. In turn, a slightly negative $\delta^{13}C$ shift in these two sediment profiles was determined. This effect can be interpreted as a result of oxidation of organic matter in the water column and disequilibrium fractionation of carbon isotopes during $CaCO_3$ precipitation. According to Fronval *et al.* (1995) and Hollander and Smith (2001), disequilibrium fractionation of oxygen and carbon isotopes as a consequence of an increased rate of carbonate precipitation may be influenced by rapid organic matter synthesis in lake water.

The results of the oxygen isotope analyses determined for the upper parts of three sediment profiles do not correlate with the air temperature and precipitations, as presented in Figure 17.6, which indicates that another processes and factors must have prevailed in forming the oxygen isotope composition of sediments. Thus, it is assumed that the lower $^{18}O/^{16}O$ isotope ratios determined for the upper parts of sediment profiles from the southwestern subbasin can also indicate that carbonates did not precipitate at equilibrium with ambient water in the last few decades. But the significant shift toward more negative $\delta^{18}O$ values must have been influenced by additional factors during that time. It is because the isotopic data of the upper sections of these profiles display a divergent trend of changes in $\delta^{13}C$ and $\delta^{18}O$ values that show that ^{13}C-depletion of carbonates occurred over 10 years later than the decrease of the $^{18}O/^{16}O$ ratio. The reasonable explanation is that the authigenic carbonate precipitation took place in earlier period of spring–summer season compared to previous decades. Thus, the surface lake water and subsequently $CaCO_3$ was then depleted in ^{18}O due to the lower $\delta^{18}O$ values of meteoric water and a lower evaporation effect in the early (colder) period of carbonate precipitation. As described by Neumann *et al.* (2002) the shift of the carbonate formation season can be determined by the increase of nutrient levels, which favor algal blooms even in colder waters.

Subtly different $\delta^{13}C$ values and organic matter concentration are denoted in the surface layer of sediments (1–4 cm) from the southern subbasin. Decreasing $\delta^{13}C$ values in this fragment of the two cores represent the last 12 years and correspond to the period of the creation of Wigry National Park in 1989 and increasing protection of Lake Wigry environment.

VI. SUMMARY AND CONCLUSIONS

The stable carbon and oxygen isotope records of the recent sediments from Lake Wigry were related to the diverse character of the main subbasins of the lake with respect to bathymetry, hydrological conditions, and trophic state. The temporal variations of carbon and oxygen isotope data were also compared with the environmental changes in Lake Wigry watershed. In general, the differences in oxygen isotope compositions between studied carbonates indicate that $\delta^{18}O$ values do not clearly reflect the local climate condition during the last few decades in studied region. This effect is assumed to result particularly from the various contribution of the inflow waters supplying the different part of

Lake Wigry which could alter the real isotopic signal of climatic changes preserved in sediments. In addition, the complex bathymetry and spatial variations in the intensity of biological productivity in the lake could also determine the discrepancy in the $\delta^{18}O$ values. Further, the stable carbon isotope compositions of studied carbonates are considered to have been modified by hydrodynamic condition within the lake system which is also a factor related to lake morphometry. Although the presented sediment profiles display an individual stable carbon isotope record, the sediments preserve the isotopic signal of primary productivity increase and changes in trophic state of Lake Wigry during the previous century.

The obtained isotopic data confirm the assumption that the in-lake processes and local factors, like the time of water exchange, circulation within the water column, different contribution of various sources of lake water supply, can alter or reduce the isotopic signals of ecological changes preserved in lake systems. This should be considered in analyses of stable carbon and oxygen isotope data of lake sediments. In order to establish the dependence of these local processes and indirect factors on the measured isotope signal, it is necessary to compare the stable isotope composition of carbonates from different sites of an individual lake, especially those that are large and morphologically complex one. The suitable interpretation of the stable isotope data also requires making some additional analyses. Of great importance is the knowledge about the surrounding hydrology as well as the isotopic composition of lake and inflow waters and biogenic materials in modern lake system. And finally, it is no doubt that further studies of the spatial and temporal stable isotope records in sediments and $\delta^{13}C$ and $\delta^{18}O$ of various lacustrine components should be undertaken to better understand the role of the stable carbon and oxygen isotopes as tracers of ecological changes in freshwater environments.

VII. REFERENCES

Bajkiewicz-Grabowska, E. 1999. Bilans wodny Jeziora Wigry w latach 1981–1995. Pages 113–128, B. Zdanowski, M. Kamiński, and A. Martyniak (Eds.) *Funkcjonowanie i Ochrona Ekosystemów Wodnych na Obszarach Chronionych.* Wyd. IRS Olsztyn.

Bajkiewicz-Grabowska, E. 2002. Obieg materii w systemach rzeczno-jeziornych. *Wyd. Wydz. Geogr. i Stud. Reg. Uniwersytetu Warszawskiego.*

Dansgaard, W. 1964. Stable isotopes in precipitation. *Tellus* **16:**1–120.

Dean, W. E. 2002. A 1500-year record of climatic and environmental change in Elk Lake, Clearwater County, Minnesota II: Geochemistry, mineralogy, and stable isotopes. *Journal of Paleolimnology* **27:**301–319.

Drummond, C. N., W. P. Patterson, and J. C. G. Walker. 1995. Climatic forcing of carbon-oxygen isotopic covariance in temperate-region marl lakes. *Geology* **23:**1031–1034.

Edwards, T. W. D., and B. B. Wolfe. 1996. Influence of changing atmospheric circulation on precipitation $\delta^{18}O$—temperature relations in Canada during the Holocene. *Quaternary Research* **46:**211–218.

Ekdahl, E. J., J. L. Teranes, T. P. Guilderson, C. L. Turton, J. H. McAndrews, C. A. Wittkop, and E. F. Stoermer. 2004. Prehistorical record of cultural eutrophication from Crawford Lake, Canada. *Geology* **32:**745–748.

Friedman, I., and J. R. O'Neil. 1977. Compilation of stable isotope fractionation factors of geochemical interest. *In* M. Fleischer (Ed.) *Data of Geochemistry,* 6th edn. *U. S. Geological Survey, Professional Paper* 440-KK, U.S. Gov. Printing office, Washington D.C.

Fronval, T., N. B. Jensen, and B. Buchardt. 1995. Oxygen isotope disequilibrium precipitation of calcite in Lake Arresø, Denmark. *Geology* **23:**463–466.

Górniak, A. (Ed.). 2006. Jeziora Wigierskiego Parku Narodowego. Aktualna jakość i trofia wód. *Zakład Hydrobiologii, Uniwersytet w Białymstoku.*

Hammarlund, D., S. Björck, B. Buchardt, C. Israelson, and C. T. Thomsen. 2003. Rapid hydrological changes during the Holocene revealed by stable isotope records of lacustrine carbonates from Lake Igelsjön, southern Sweden. *Quaternary Science Reviews* **22:**353–370.

Hollander, D. J., and J. A. McKenzie. 1991. CO_2 control on carbon-isotope fractionation during aqueous photosynthesis: A paleo-pCO_2 barometer. *Geology* **19:**929–932.

Hollander, D. J., and M. A. Smith. 2001. Microbially mediated carbon cycling as a control on the $\delta^{13}C$ of sedimentary carbon in eutrophic Lake Mendota (USA): New models for interpreting isotopic excursions in the sedimentary record. *Geochimica et Cosmochimica Acta* **65:**4321–4337.

Jinglu, W., M. K. Gagan, J. Xuezhong, X. Weilan, and W. Sumin. 2004. Sedimentary geochemical evidence for recent eutrophication of Lake Chenghai, Yunnan, China. *Journal of Paleolimnology* **32:**85–94.

Kamiński, M. 1999. Lake Wigry—the lake "adopted" by International Association of Theoretical and Applied Limnology (SIL "Lake Adoption" Project). *Polish Journal of Ecology* **47:**215–224.

Leng, M. J., and J. D. Marshall. 2004. Paleoclimate interpretation of stable isotope data from lake sediment archives. *Quaternary Science Reviews* **23:**811–831.

Leng, M. J., N. Roberts, J. M. Reed, and H. J. Sloane. 1999. Late Quaternary palaeohydrology of the Konya Basin, Turkey, based on isotope studies of modern hydrology and lacustrine carbonates. *Journal of Paleolimnology* **22:**187–204.

Li, H. C., and T. L. Ku. 1997. $\delta^{13}C$–$\delta^{18}O$ covariance as a paleohydrological indicator for closed-basin lakes. *Palaeogeography, Palaeoclimatology, Palaeoecology* **133:**69–80.

Makhnach, N., V. Zernitskaya, I. Kolosov, O. Demeneva, and G. Simakova. 2000. $\delta^{18}O$ and $\delta^{13}C$ in calcite of freshwater carbonate deposits as indicators of climatic and hydrological changes in the Late-Glacial and Holocene in Belarus. *Journal of Geochemical Exploration* **69–70:**435–439.

Mayer, B., and L. Schwark. 1999. A 15,000-year stable isotope record from sediments of Lake Steisslingen, Southwest Germany. *Chemical Geology* **161:**315–337.

Myrbo, A., and M. D. Shapley. 2006. Seasonal water-column dynamics of dissolved inorganic carbon stable isotopic compositions ($\delta^{13}C_{DIC}$) in small hardwater lakes in Minnesota and Montana. *Geochimica et Cosmochimica Acta* **70:**2699–2714.

Neumann, T., A. Stögbauer, E. Walpersdorf, D. Stüben, and H. Kunzendorf. 2002. Stable isotopes in recent sediments of Lake Arendsee, NE Germany: Response to eutrophication and remediation measures. *Palaeogeography, Palaeoclimatology, Palaeoecology* **178:**75–90.

Niewolak, S. 1999. Evaluation of pollution and the sanitary-bacteriological state of Lake Wigry, Poland. *Polish Journal of Environmental Studies* **8:**89–100.

Ogrinc, N., S. Lojen, and J. Faganeli. 2002. A mass balance of carbon stable isotopes in an organic-rich methane-producing lacustrine sediment (Lake Bled, Slovenia). *Global and Planetary Change* **33:**57–72.

Różański, K., L. Araguas-Araguas, and R. Gonfiantini. 1992. Relation between long-term trends of oxygen-18 isotope composition of precipitation and climate. *Science* **258:**981–985.

Rutkowski, J. 2004. Osady jeziora Wigry. *Rocznik Augustowsko-Suwalski* **4:**19–36.

Schelske, C. L., and D. A. Hodell. 1991. Recent changes in productivity and climate of Lake Ontario detected by isotopic analysis of sediments. *Limnology and Oceanography* **36:**961–975.

Schwalb, A., and W. E. Dean. 1998. Stable isotopes and sediments from Pickerel Lake, South Dakota, USA: A 12 ky record of environmental changes. *Journal of Paleolimnology* **20:**15–30.

Schwalb, A., and W. E. Dean. 2002. Reconstruction of hydrological changes and response to effective moisture variations from North-Central USA lake sediments. *Quaternary Science Reviews* **21:**1541–1554.

Siegenthaler, U., and H. Oeschger. 1980. Correlation of ^{18}O in precipitation with temperature and altitude. *Nature* **285:**189–223.

Talbot, M. R. 1990. A review of the palaeohydrological interpretation of carbon and oxygen isotopic ratios in primary lacustrine carbonates. *Chemical Geology* **80:**261–279.

von Grafenstein, U., H. Erlenkeuser, J. Muller, P. Trimborn, and J. Alefs. 1996. A 200 year mid-European air temperature record preserved in lake sediments: An extension of the $\delta^{18}O_p$-air temperature relation into the past. *Geochimica et Cosmochimica Acta* **60:**4025–4036.

Vreča, P. 2003. Carbon cycling at the sediment-water interface in a eutrophic mountain lake (Jezero na Planini pri Jezeru, Slovenia). *Organic Geochemistry* **34:**671–680.

Wachniew, P., and K. Różański. 1997. Carbon budget of a mid-latitude, groundwater-controlled lake: Isotopic evidence for the importance of dissolved inorganic carbon recycling. *Geochimica et Cosmochimica Acta* **61:**2453–2465.

Yu, Z., and U. Eicher. 1998. Abrupt climate oscillations during the last deglaciation in central North America. *Science* **282:**2235–2238.

Zdanowski, B. (Ed.). 1992. Jeziora Wigierskigo Parku Narodowego. Stan eutrofizacji i kierunki ochrony. *PAN. Kom. Naukowy "Człowiek i Środowisko." Zeszyty Naukowe* **3:**249.

Section 5

Humans, Isotopes and Ecological Change

CHAPTER 18

Stable Isotopes and Human Water Resources: Signals of Change

Gabriel J. Bowen,[*,‡] Thure E. Cerling,[†,‡] and James R. Ehleringer[‡]

*Earth and Atmospheric Sciences and Purdue Climate Change Research Center
†Geology and Geophysics Department
‡Department of Biology, University of Utah

Contents

I. INTRODUCTION

Freshwater is among the most critical natural resources that humans rely on, impact, and manage. Although stocks of freshwater are continuously renewed by the climate system, the rate of this renewal is finite and largely beyond human control. Humans do, however, have the capacity to exert powerful influence on a range of other hydrologic fluxes. Urban development creates impermeable surfaces, reducing infiltration and enhancing runoff. Surface impoundments capture runoff, raise the water table, and may increase infiltration and recharge. Impounded water can be diverted for uses such as agricultural irrigation that may increase the evapotranspiration flux to the atmosphere. Extraction of groundwater from deep aquifers circumvents established pathways of circulation within the water cycle, and may significantly reduce the residence time of groundwater while temporarily increasing runoff and/or evapotranspiration fluxes where the extracted water is used.

As stewards of, and stakeholders in, continental ecosystems, humans must learn to manage a finite supply of freshwater with the common goals of meeting our immediate needs and sustaining the many other natural systems that depend on this supply. In order to support the development and assessment of such management practices, a comprehensive understanding of the water cycle is needed, including the capacity to both monitor and model fluxes of water along the many pathways that define the continental water cycle. Because the water cycle interacts dynamically with a host of other Earth systems, including the atmosphere and climate system, the biosphere, and the lithosphere, the inventory and complexity of mechanisms and feedbacks that can affect hydrologic fluxes are enormous. Thus, the toolbox of the water cycle scientist must include tools for measuring and partitioning water fluxes in and among a range of systems. Natural variation in the stable isotope ratios of water represents one such tool.

II. ISOTOPE HYDROLOGY OF NATURAL SYSTEMS

A. Principles

The stable isotopes of hydrogen (^{1}H, ^{2}H) and oxygen (^{16}O, ^{17}O, ^{18}O) have been widely employed in hydrologic research for several decades, and the principles underlying that research can be applied to the study of human–hydrologic systems (Figure 18.1). Isotope ratios of both elements ($^{2}H/^{1}H$, reported as $\delta^{2}H$ values, and $^{18}O/^{16}O$, reported as $\delta^{18}O$ values, will be considered here) exhibit a wide range of variation in natural waters. In both cases, the range of values in natural hydrologic systems (−495 to +129‰ for $\delta^{2}H$, −62.8 to +31.3‰ for $\delta^{18}O$; Coplen *et al.* 2002) exceeds several hundred times the precision of mass spectrometric isotope ratio determinations (on the order of ± 1 and 0.1‰ for $\delta^{2}H$ and $\delta^{18}O$, respectively), meaning that for many applications measurement precision is not a limiting factor. A variety of analytical options for the determination of $\delta^{2}H$ and $\delta^{18}O$ values of water are now available to researchers, including traditional CO_2 equilibration methods ($^{18}O/^{16}O$) and catalyzed reduction methods ($^{2}H/^{1}H$) for the generation of pure gases to be measured via dual inlet isotope ratio mass spectrometry (Epstein and Mayeda 1953, Coleman *et al.* 1982), modifications of these methods designed for use with automated devices and/or continuous-flow mass spectrometers (Fessenden *et al.* 2002), and high-temperature pyrolysis of autoinjected samples followed by chromatographic separation of H_2 and CO gases and analysis via continuous-flow isotope ratio mass spectrometry (Sharp *et al.* 2001, Gehre *et al.* 2004).

At the root of the natural variation of H- and O-isotope ratios in water are the isotope effects accompanying the phase-change process between vapor and liquid water or ice. As shown by Craig and Gordon (1965) and reviewed by Gat (1996), these effects cannot be considered the result of a single reaction, but reflect a coupled set of equilibrium and kinetic processes that occur across an interface separating the gaseous and liquid (or solid) phases. These processes fractionate both hydrogen and oxygen isotopes simultaneously, with a relative fractionation of $\alpha^{2}H/\alpha^{18}O \leq 8$. The relative magnitude of these fractionations is variable and is determined by the balance of kinetic and equilibrium processes. This, in turn, is set by the physical conditions under which the process occurs, including the relative humidity in the gaseous phase and the boundary layer dynamics. The relative magnitude of the fractionation factors for ^{2}H and ^{18}O determines the slope of a line in $\delta^{2}H/\delta^{18}O$ space along which the phase-change reaction will proceed (Figure 18.1). For equilibrium processes, the slope of this line is near 8, which leads to the definition of a parameter, the deuterium excess ($d = \delta^{2}H - 8 \times \delta^{18}O$), which can be used to identify the influence of nonequilibrium phase-change reactions on the isotopic

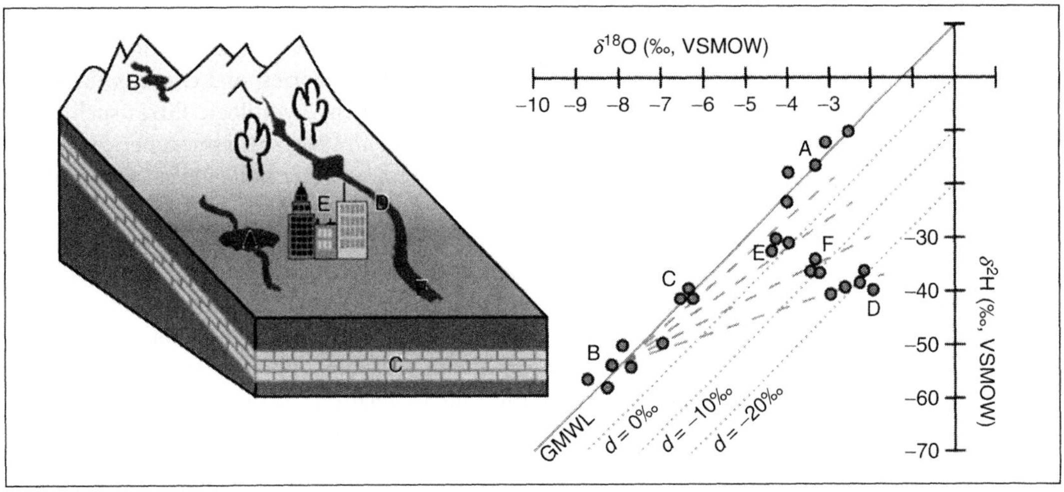

FIGURE 18.1 Patterns of isotope ratio variation in a hypothetical human-exploited hydrologic systems. Small surface water systems in the vicinity of a major city (A) are characterized by higher δ^2H and $\delta^{18}O$ values than those in the nearby mountains (B). Isotope ratios of each of these systems lie along the global meteoric water line (GMWL) and their deuterium excess (*d*) values of ca. +10‰ are similar to those of most precipitation suggesting that they carry precipitation-derived water that has not been subject to significant kinetic isotope fractionation (*i.e.*, during evaporation). The high level of isotopic variability might reflect variability in the composition of local precipitation over time, and suggests that these systems are characterized by relatively short residence times.Water in a subsurface reservoir (C), a mixture of mountain and basin recharge, has an isotope composition intermediate to A and B, the same *d* value, and a much lower level of isotopic variability due to its long residence time. Water in a mountain-fed river having traversed an arid landscape (D) has been subjected to intensive evaporation, and its isotopic composition has evolved from that of mountain water along an evaporation line (dashed lines). The slope of the evaporation line, which is determined by the physical conditions under which evaporation takes place, is less than that of the GMWL (8), so as the δ^2H and $\delta^{18}O$ values of the river water increase with evaporation the *d* value decreases (dotted lines). The isotopic offset between (B) and (D) along the evaporation line is related to the fractional amount of water lost to evaporation, which can be calculated if the fractionation factors for this reaction can be approximated.Tap water at the basinal city (E) is a mixture of aquifer water and water from local and mountain-fed runoff (A, C, and D), and has an isotopic composition that is a linear mixture of those of the three sources. The contribution of an evaporated source (D) to the tap water can be seen in that tap water values plot below the GMWL, giving *d* values less than 10. Effluent from the city's water treatment plant is reintroduced to the mountain-fed river, and downstream of the city the isotopic composition of this system is intermediate to (E) and (D). Dilution of the river water (D) with unevaporated aquifer and local surface water is indicated by higher *d* values at (F).

composition of water. Because condensation in clouds is effectively an equilibrium process, most precipitation has similar *d* values close to 10. Water evaporating from the land surface, however, undergoes kinetic fractionation. In such situations, the evaporated vapor has a *d* value greater than 10, and as the evaporation process proceeds, the *d* of the remaining continental water progressively evolves to values less than 10.

B. Processes and Patterns

The H- and O-isotope ratios of water in the ocean, as representative of the largest reservoir of free water at the Earth's surface, anchor the isotopic composition of other natural waters. An artificial, ideal, modern ocean water sample, Vienna standard mean ocean water (V-SMOW) defines the zero point of the scale against which the δ^2H and $\delta^{18}O$ values of other waters are reported. As a result of spatial and temporal variation in rates of evaporation and return of freshwater to or from the ocean, however, the isotopic composition of natural marine water is not actually uniform in space or time (Bigg and Rohling 2000). These variations in ocean water δ^2H and δ^{18}O values correspond closely with spatial and temporal changes in salinity, and have proven useful in studying hydrologic fluxes such as the buildup and breakdown of polar glaciers through time (Imbrie *et al.* 1984), net atmospheric transport of water among ocean basins (Haug *et al.* 2001), and the geometry of thermohaline circulation (Bigg and Rohling 2000).

The natural range of δ^2H and δ^{18}O values for precipitation is several times that observed in natural ocean waters (Figure 18.2). A small part of this variability is inherited from variability in the isotope ratios of marine water sources, but the majority results from the progressive depletion of the heavy isotopes ^{2}H and ^{18}O from atmospheric vapor as air masses traverse the continents, cool, and desiccate (Fig. 1) (Dansgaard 1964). This process leads to a continuous evolution of precipitation δ^2H and δ^{18}O values along trajectories of atmospheric circulation, and to relatively smooth patterns of spatial variation in water isotope ratios across the continents (Yurtsever and Gat 1981, Bowen and Wilkinson 2002). Regions lying near the low-latitude oceans, which constitute the main source of water to the atmosphere, are characterized by precipitation with relatively high H- and O-isotope ratios. High-latitude and -altitude regions and the continental interiors, which lie near the endpoints of moisture transport pathways, receive ^{2}H- and ^{18}O-depleted precipitation. Because the distillation process is strongly controlled by air mass temperature, spatial, seasonal, and temporal correlations exist between site temperature and average isotope ratios of precipitation at and across mid- to high-latitude sites where significant temperature variations exist (Siegenthaler and Oeschger 1978, Rozanski *et al.* 1992, 1993). These relationships have been applied to geologic materials to reconstruct paleoclimate across a range of timescales (Dansgaard *et al.* 1969, Hays and Grossman 1991, Feng and Epstein 1994).

Variation in precipitation isotope ratios is not a simple function of temperature alone, however, as changes in vapor sources, atmospheric trajectories, and meteorologic dynamics can each affect the isotopic composition of rainwater in ways that are not directly related to simple measures of site temperature. These stable isotope signals provide an additional range of opportunities for applying H- and O-isotopes to the study of the hydrologic cycle. For example, values of δ^2H, δ^{18}O, and deuterium excess are useful for characterizing atmospheric fluxes of water vapor, and have been used to demonstrate the importance of recycled water (*i.e.*, reevaporation and transpiration from the land surface) on the hydrologic budgets of several regions (Gat *et al.* 1994, 2003, Pierrehumbert 1999).

On the continents, the dominant processes modifying the isotope ratios of water stocks are mixing of waters from different sources and reevaporation from surface water bodies (Figure 18.2). A study by Dutton *et al.* (2005) has synthesized a large collection of surface water isotope ratio data and identified these processes reflected in regional differences between surface water and precipitation isotope ratios. A range of applications of H- and O-isotope ratios in surface and groundwater hydrology at catchment and smaller scales have been reviewed elsewhere (papers in Kendall and McDonnell 1996). These include a large number of studies that capitalize on the temporal and spatial variation in climatological water sources to distinguish sources of groundwater recharge (Dutton 1995, Scholl *et al.* 1996, Lee *et al.* 1999), sources and pathways of streamflow generation (Harris *et al.* 1995, Rose 1996), and pathways of soil water infiltration (Gazis and Feng 2004). Wherever the potential for natural or artificial hydrologic mixing of water from multiple sources of differing isotopic composition exists, H- and O-isotopes can be employed to help constrain water sources and their proportional contribution to the mixture. In addition, measured changes in δ^2H and δ^{18}O values, and particularly in *d*, of water may be useful for identifying the amount, conditions, and pathways of evapotranspiration in natural and managed landscapes.

In the remainder of this chapter, we will review existing and potential applications of water isotope ratios to study change in human-dominated systems. We adopt a large-scale, spatial analysis-based approach to documenting patterns and inferring processes that dominate the isotopic signature of

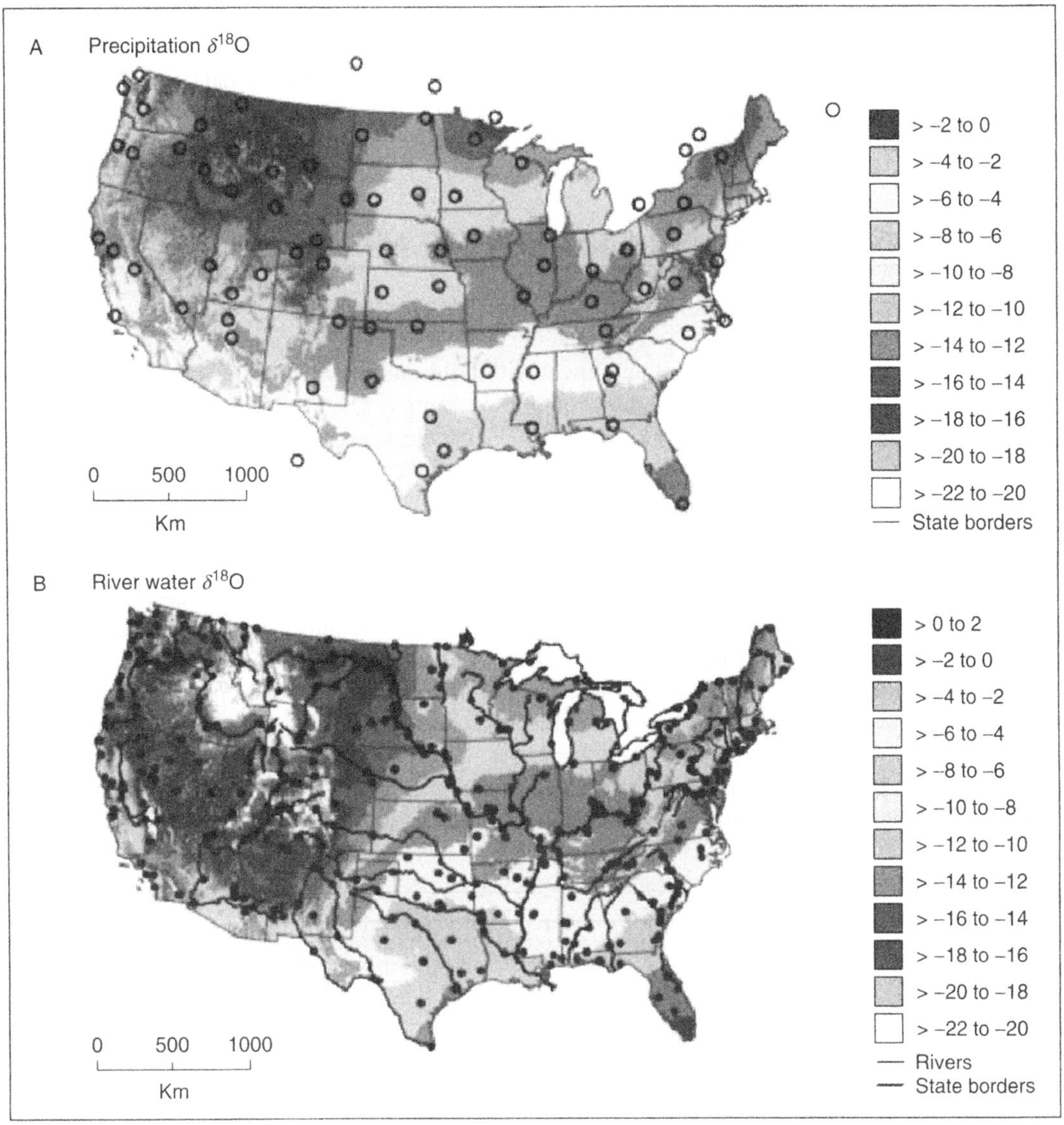

FIGURE 18.2 Reprinted from "Spatial distribution and seasonal variation in $^{18}O/^{16}O$ of modern precipitation and river water across the conterminous USA"; A. Dutton, *et al.*, *Hydrological Processes*; volume 19; © 2005 John Wiley & Sons Ltd.). Spatial patterns of stable O-isotope ratios in US precipitation (A) and river water (B). River water isotope data from Kendall and Coplen (2001). (**See Color Plate.**)

water supplies in the contiguous United States. This approach builds on those used by Bowen and Revenaugh (2003) and Dutton *et al.* (2005) in their work on precipitation and surface waters, and is one that has the power to reveal the regional prevalence and impact of processes that impact the isotopic composition of waters. Our assessment is based on a published dataset and analysis of US tap water (Bowen *et al.* 2007). Although this data provides a basis for a preliminary, large-scale analysis of patterns in US tap water data, several limitations of the data, most critically the limited temporal scope (only one sample at most sites), precludes a more comprehensive analysis of the data at this time. As a result, we focus here on only the strongest, clearest patterns observed in the dataset, in most cases patterns with a magnitude of several 10s of ‰ for δ^2H (many ‰ for $\delta^{18}O$) and those supported by data from many stations distributed across a region.

III. STABLE ISOTOPE RATIOS OF MODERN TAP WATERS

A. Spatial Patterns

Tap water isotope ratios from 496 cities and towns within the contiguous United States were reported by Bowen *et al.* (2007). The isotopic composition of these samples (Figure 18.3) ranged from −152‰ to +11‰ (δ^2H) and from −19.4‰ to +4.2‰ ($\delta^{18}O$). Values clustered near the Global Meteoric Water Line (GMWL) ($\delta^2H = \delta^{18}O \times 8 + 10$; Craig 1961), although 74% of tap water samples had *d* values less than 10 and the average value of *d* for the tap water samples was 5. The range of isotope ratio and *d* values for the tap water samples were similar to but slightly larger than the range of values for US mean annual precipitation (Bowen and Revenaugh 2003) or for annually averaged river waters (Kendall and Coplen 2001) in the coterminous United States. Taken together, these observations suggest that isotope ratios of tap water are those of "normal" environmental waters, and may preserve information related to source and/or history of these waters. The prevalence of samples with *d* lower than 10 may suggest some tendency for US tap water to be more evaporated on average than precipitation or river waters, a pattern which could easily reflect evaporation during residence in artificial or natural surface reservoirs.

Analysis of isotope ratio data for other anthropogenically influenced waters, such as bottled water (Bowen *et al.* 2005), has shown that reference to local values and spatial patterns of precipitation

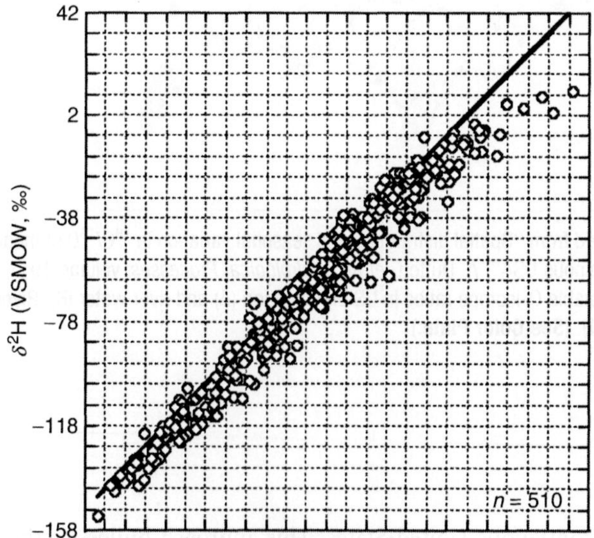

FIGURE 18.3 Tap water stable isotope ratios for 510 samples from 496 US cities and towns. The bold black line represents the GMWL ($\delta^2H = 8 \times \delta^{18}O + 10$). Modified from Bowen *et al.* (2007).

isotope ratios can be used to identify patterns of water source and history. For the much more extensive and spatially denser US tap water dataset, Bowen *et al.* (2007) were able to delineate regional patterns in water resource characteristics through comparative analysis of the tap water data and interpolated grids of precipitation isotope ratios. Spatial patterns in tap water isotope ratios were strong, with regional patterns accounting for about 85% of the total variation in H- and O-isotope ratios. Across many parts of the United States, tap water and local precipitation isotope ratios were similar within ~8‰ for δ^2H or ~1‰ for $\delta^{18}O$. Several regions stood out in their deviation from this relationship, however (Figure 18.4).

For both precipitation and tap water, H- and O-isotope ratios were lowest in the northern Rocky Mountains and Great Plains, locations that are distal from the coasts and characterized by seasonally cold climate and high elevations. Moreover, throughout the Rocky Mountains tap water isotope values were consistently ~30‰ lower than local precipitation for δ^2H and ~3‰ lower for $\delta^{18}O$. This pattern likely reflects regional patterns in water sources feeding local tap water stocks. Two potential factors could account for the observed pattern, and it is likely that each plays a role. First, recharge of water sources in this region is predominantly derived from high-elevation, mountain precipitation. Annually averaged precipitation amounts across the northern Rocky Mountains are commonly 400–800% greater over mountain ranges than in interspersed valleys (Daly *et al.* 1994). Recharge of valley aquifers and surface reservoirs in this region in known to derive largely from mountain runoff (Wilson and Guan 2004), and these water sources likely impact the δ^2H and $\delta^{18}O$ values of local tap water. Given estimates of the change in $\delta^{18}O$ values with elevation of 0.20–0.28‰ per kilometer (Poage and Chamberlain 2001, Bowen and Wilkinson 2002, Dutton *et al.* 2005), the isotope ratio data suggest that the average elevation of recharge water in this region may be between 1 and 1.5 km higher than the elevation of cities and towns that use this water. The tap water isotope ratios are likely also affected by a seasonal bias in reservoir recharge. ^2H- and ^{18}O-depleted winter precipitation, stored in mountain snowpack and released through winter and spring melting, may more strongly impact the volume and isotope ratios of recharge than summer storm precipitation, which is more likely to be lost through evapotraspiration and/or surface runoff. The primary indication of the comparison between tap water and precipitation isotope ratios for the northern Rocky Mountain and Great Plains regions, then, is that δ^2H and $\delta^{18}O$ values of tap water for this region provide a strong signal of the reliance of local communities on water sourced from high-elevation, wintertime precipitation and snowpack.

Tap water isotope ratios that were much lower than those of local precipitation also characterized areas near major rivers with their headwaters in mountainous regions. This pattern was strongest for cities along the Missouri River, parts of the Ohio River Valley, and a large area of southern California and Arizona including the lower reaches of the Colorado River, and it reflects the use of water from these riverways as a primary source of municipal water. In these areas, the δ^2H and $\delta^{18}O$ values of local tap water were often as much as ~35% and 4‰ lower than those of locally sourced waters (respectively). The isotopic compositions of the tap waters were similar to or greater than those of the major, mountain-sourced rivers that represented potential water sources for these communities (Kendall and Coplen 2001). The tap water also had low d values, as did the large rivers, suggestive of evaporative water loss. The distinctive isotopic signatures of the mountain-derived, river-transported, water represent a method for tracing the use and distribution of water from these river sources. The geographic distribution of tap water isotope ratios suggestive of river water usage indicate two distinct modes of distribution employed in these regions. River water use in Southern California and Arizona is characterized by active distribution of Colorado River water through aqueduct systems that cross distances of hundreds of kilometers, whereas in the Missouri and Ohio River valleys river sourced water use appears to be restricted to communities within ~25 km or less of the river source.

The highest δ^2H and $\delta^{18}O$ values for US tap water, in contrast to those of precipitation, occurred in a bull's eye region of North Central Texas and Oklahoma. H- and O-isotope ratios of tap water from communities in this region were as high as 11% and 4.2‰, respectively, and on average were about 44% and 7.0‰ greater than values for local precipitation. Interpretation of these data is improved by

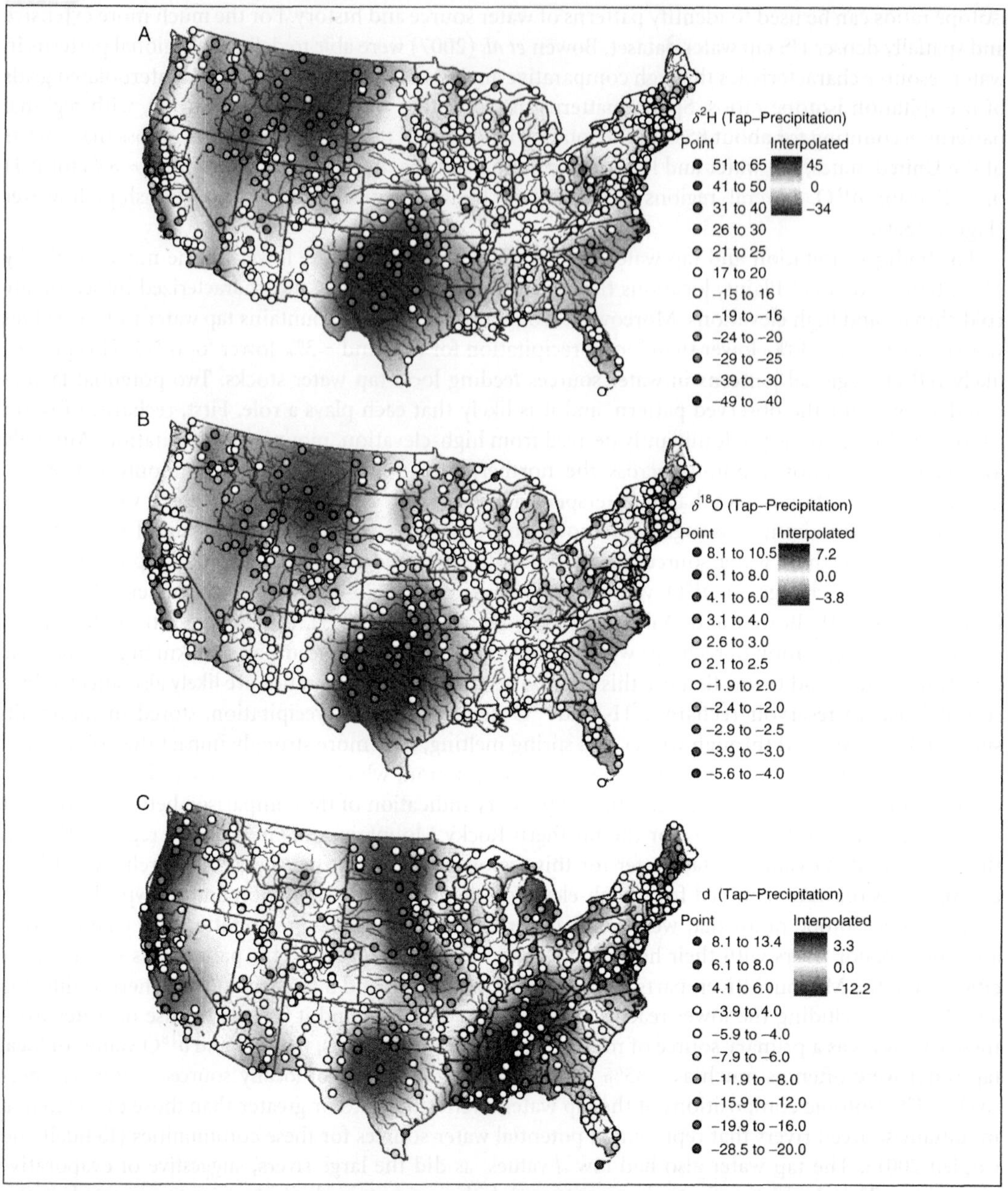

FIGURE 18.4 Isotopic differences between tap water and precipitation across the United States. Point data are shown for communities where the magnitude of the difference between tap water and local precipitation values was greater than 16‰ for δ^2H (A), 2‰ for $\delta^{18}O$ (B), and 4 for *d* (C). Other stations are shown with white dots. Background shows a smoothed field of tap–precipitation values obtained from all data by ordinary Kriging with a spherical semivariogram. Modified from Bowen *et al.* (2007). (**See Color Plate.**)

consideration of the associated *d* values, which in this region, as well as in the lower Missouri River Valley and parts of the lower Great Lakes region, were significantly lower (*e.g.*, by 10‰ or more) than those of local precipitation. These observations suggest that tap water isotope ratios in each of these regions have been affected by substantial evaporation of the source waters. Surface impoundments are widely used for water storage in North Central Texas, and it is likely that evaporative water losses of tap water sources in these areas include significant losses from impounded surface reservoirs. Along the margins of the Great Lakes and parts of the northern Great Plains characterized by low values of *d*, lake water represents a likely source for such water, whereas evaporation of water from the Missouri River may impart low *d* values to tap waters along its reach.

In some such cases, stable isotope data may be useful to monitor the extent of evaporative loss from water supply systems. For small open bodies of water such as a reservoir, the net isotope effect of evaporation (ε_t) can be approximated following Gonfiantini (1986):

$$\varepsilon_t = \varepsilon_e + \Delta\varepsilon_k \tag{18.1}$$

where the equilibrium isotope effect of the liquid–vapor phase change (ε_e) is a function of temperature (*t*; Majoube 1971) and the contribution of kinetic fractionation ($\Delta\varepsilon_k$) is a product of the isotope effect of diffusion (Merlivat 1978) and $(1 - h)$, where h is the relative humidity of the ambient atmosphere. Using the approximation $\varepsilon_t \approx 1000 \ln(\alpha)$ and the Rayleigh equation:

$$R = R_0 \times f^{(\alpha-1)} \tag{18.2}$$

where R_0 and R are the isotope ratios of the water before and after evaporation, the fraction of liquid water remaining following evaporative losses (f) can be estimated. North Central Texas, for example, is an arid region (mean annual $h \approx 62\%$, $t = 16$ °C; data for Lubbock, TX from http://www1.ncdc.noaa.gov/pub/data/ccd-data/) where tap water and estimated local precipitation isotope compositions at several sites differed by ~50‰ (δ^2H) and 8‰ (δ^{18}O). Calculations for this region based on the H and O isotope systems are consistent and suggest evaporative losses of up to 40–45% prior to tap water distribution and sampling.

B. Temporal Patterns

Bowen *et al.* (2007) also reported the preliminary results of a survey of temporal variation in tap water isotope ratios. Seasonal variability in the δ^2H, δ^{18}O, and *d* values of tap waters from single communities differed widely throughout the United States, but did not exceed 10‰ (δ^2H; 1 σ) or 1.7‰ (δ^{18}O) for any of the communities sampled. Data from the seasonal samples suggest at least three modes of intra-annual variation in tap water isotope ratios (Figure 18.5). At many sites, tap water δ^2H, δ^{18}O, and *d* values each are relatively invariant (within ~5‰ for δ^2H and 1‰ for δ^{18}O) throughout the seasonal cycle (Figure 18.5A). These temporally invariant isotope ratios suggest that neither the climatological source nor the proximal source of water changes with the seasons. Such a pattern could be caused either by use of water from a large, relatively static reservoir (*e.g.*, groundwater) or by use of water derived from a climatological source lacking significant isotopic seasonality.

At a second subset of sites, substantial seasonal variation in δ^2H and δ^{18}O values of tap water occurs without accompanying large changes in *d* (Figure 18.5B). This pattern of change is likely reflective of changes in the climatological source of water used either through natural, seasonal variation in sources of precipitation or through active switching of extraction efforts among multiple reservoirs. In the case shown in Figure 18.5B (Ithaca, New York), seasonal isotopic variability reflects natural climatological source changes between southern moisture sources during the summer and moisture largely derived from the Great Lakes during the winter (Burnett *et al.* 2004). Although the Ithaca data do not show large seasonal changes in *d*, they do reveal a subtle shifts between high *d* values (~15‰) during winter and spring and lower *d* (~10‰) during the summer and fall months. This pattern is consistent with that expected to result from seasonal shifts between winter precipitation derived in part from reevaporation of continental water (*e.g.*, from the Great Lakes) and summer moisture derived from the Atlantic and Gulf of Mexico (Gat *et al.* 1994, Burnett *et al.* 2004).

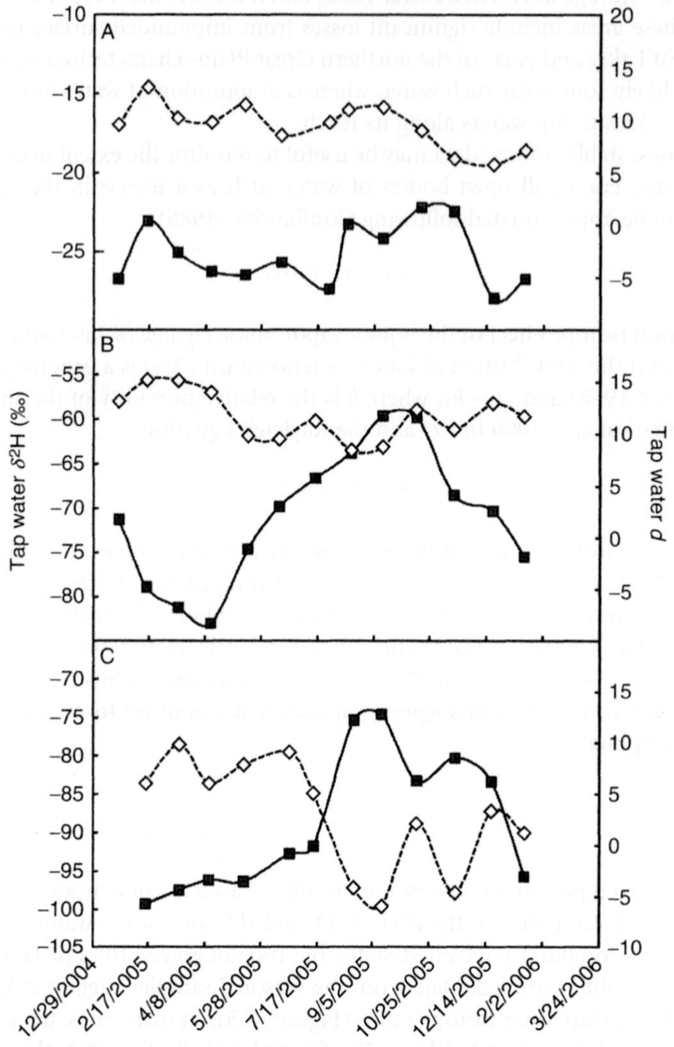

FIGURE 18.5 Time series of tap water stable isotope data for three US cities: Athens, GA (A); Ithaca, NY (B); and Reno, NV (C). Hydrogen isotope ratios are shown as squares with gray fill and solid black line, deuterium excess as diamonds with white fill and dotted black line.

Data from Reno, Nevada demonstrate a third common pattern of seasonal variation in tap water isotope ratios characterized by strong seasonal variation in both δ^2H (δ^{18}O) and d values (Figure 18.5C). The hydrogen isotopic composition of Reno tap water varied by more than 25‰ during 2005, with the highest values occurring during the late summer and fall months and the lowest during winter. The isotope ratio variation is accompanied by seasonal d variation of more than 15‰, which closely mirrors the pattern of δ^2H change. This pattern of variation reflects seasonal changes, either due to natural hydrologic processes or due to human selection, in the degree of evaporation of water in reservoirs tapped by the city. Low d values during the late summer and fall likely signal the use of water from surface reservoirs that have been subject to evaporation throughout the summer months. Values of d nearer to 10‰ during the winter months, and abrupt and irregular increases in d at several times during the year are indicative of inputs of less-evaporated water to the Reno water system.

These could reflect either rapid recharge of surface reservoirs with new precipitation or supplementation of surface water resources by extraction of groundwater.

IV. WATER ISOTOPES AND CHANGE

In many cases tap water isotope ratios vary spatially and temporally in ways that are distinct from the patterns of isotope ratios values characteristic of natural water resources. The distinctive isotopic signatures of human-modified and -exploited water provide a tool for probing our interaction with the water cycle. Stable isotope ratios of tap water carry information on water sources provisioning human use and provide markers for identifying human water use impacts on natural hydrologic and atmospheric systems. Although exploration of these signatures as a tool for probing human-induced ecological and hydrologic change is in its infancy, several developing and potential applications can be highlighted where the H- and O-isotope ratios of tap water represent promising tools for reconstructing, identifying, and forecasting past, present, and future interactions between humans and the water cycle.

A. Reconstructing

Modern monitoring efforts focused on understanding the spatial and temporal patterns of tap water isotope ratios provide a template for investigations into past human behavior and water use strategies. Biological materials that are commonly preserved at archaeological sites over timescales of hundreds to thousands of years, including discarded food items, bones, and hair, have been used as substrates for isotope analysis to derive information on the diet and behavior of archeological humans (White 1993, Araus *et al.* 1997, Schurr 1997, Iacumin *et al.* 1998, O'Connell and Hedges 1999, Sharp *et al.* 2003, Williams *et al.* 2005). Understanding modern tap water isotope distributions is relevant to interpreting such data if the archeological substrates record isotopic information related to the interaction of humans with environmental water, such as in cases where reconstruction of paleo-irrigation practices or human migration patterns is at question. Published research results indicate the promise and potential for stable isotope data from archaeological material to document characteristics of past human/water cycle interaction.

Williams *et al.* (2005) measured and modeled the O-isotope ratios of archeological maize cobs from southern and eastern Utah in order to gain information on the sources of water used by native societies to sustain maize agriculture. The relatively low values they reconstruct for water used by maize plants at a number of sites (as low as $\sim -16.5 \pm 2.5$‰) were interpreted to indicate use of irrigation water derived from high-altitude snowmelt and carried by local streams. The reconstructed isotope ratios of irrigation water are significantly lower than those of (modern) mean annual basinal precipitation at the study sites (~ -12.5‰), and more similar to those of modern tap waters from cities in southern and eastern Utah (~ -14.5‰). At other, upland sites, however, the authors found that maize was grown with isotopically heavy water, similar in composition to basinal precipitation. These archaeological data suggest that water use strategies employed by the Freemont and Puebloan cultures had begun to demonstrate the same pattern of reliance on water sourced from high elevations that is characteristic of modern cities and towns in this region. Modern water use throughout Utah, however, is characterized by pervasive use and distribution of high-elevation snowpack water across the landscape, as evidenced by uniformly low isotope ratios of modern tap water (Figure 18.4). This pattern was apparently not as prevalent within the Freemont and Puebloan cultures, which in some cases appear to have sustained maize agriculture using locally and/or summer season sourced water. Temporally resolved collections of data from archaeological samples, considered in comparison to modern tap water isotope data, could thus be used to examine to determine the level of similarity between modern and ancient water use practices, particularly where ancillary data are available to constrain climatically induced changes in environmental isotope ratios through time.

A second archaeological application for which understating spatiotemporal patterns of stable isotope ratio variation in human water resources is important is the reconstruction of past human migration patterns from archeological materials (*e.g.*, human bone, teeth, or hair). Isotope ratios of H and O in water represent one of the most spatially dynamic isotopic systems that influences the stable isotope ratios of body tissues, and isotope ratios of H and O in archaeological samples could potentially be used to reconstruct the location of origin and/or migration patterns of individuals along latitudinal, continental, or altitudinal gradients. Müller *et al.* (2003), for example, used $\delta^{18}O$ values of bone and tooth enamel to constrain the potential life location and migratory history of the mummified Alpine Iceman. In doing so, the authors considered a wide range of potential water sources available to prehistoric inhabitants in the region where the Iceman was found in order to identify robust geospatial patterns capable of constraining the origin of this individual. Modern humans are opportunistic and aggressive in their search for water resources, and it is likely that in most cases surveys of modern tap water isotope ratios such as that conducted here would provide liberal estimates of the range of potential water source stable isotope ratios for locations and regions. Tap water surveys, then, represent an efficient method of generating templates of geospatial patterns and local variability in water resource isotope ratios against which archeological data can be interpreted.

B. Identifying

The potential for isotopic studies to identify and quantify processes operating in human–hydrologic systems, as illustrated by the US tap water dataset, could be valuable to both water management and studies of hydrologic and ecological impacts of human water use. From the standpoint of management, isotope data offer the power to assess and monitor several processes that may impact the stability of water resources. When coupled with data on natural hydrologic cycle components such as precipitation, isotope ratios provide a powerful tool for identifying sources and quantifying mixing among sources used as water supplies. In systems where source contributions are otherwise difficult to discern, for example, in many groundwater systems, isotopic monitoring could provide baseline data establishing the relative importance of water derived from different climatological sources or hydrologic pathways to local water supplies. Such information might provide a basis for risk assessment or water resource planning decisions. Another management-relevant factor that is reflected in stable isotope ratios of tap water is evaporative water waste. Surface reservoirs provide ~63% of all US public supply water (Hutson *et al.* 2004), and quantification of evaporative loss is one important factor in planning and managing these systems. In complex storage and supply systems where a water losses occur simultaneously due to multiple factors (*e.g.*, evaporation, infiltration, leakage during distribution), more robust estimates of based on isotope monitoring and appropriate climate data may help to quantify the impact of evaporative loss from reservoirs and inform decisions that ensure water supply stability.

Isotope data from tap water can also play a role in identifying the downstream impacts of human water use on the water cycle and ecosystems. Humans now use an estimated 408 billion gallons of water daily in the United States (Hutson *et al.* 2004), but the net impact of this use on water cycle fluxes across the land surface, between the ocean and atmosphere, and between surface and groundwater reservoirs is poorly understood. Potential impacts could be both local or regional (*e.g.*, changing stream flow, groundwater availability, or humidity) and global (*e.g.*, changes in global partitioning between continental runoff and evapotranspiration). Data on isotope ratios of tap water and water used in other human activities is important in this context because it may reveal isotopic characteristics of human-exploited water resources that will be useful for identifying and tracing the impacts of water use throughout hydrologic systems. Although human consumption and household use represent only ~10% of the total volume of water used in the United States (Hutson *et al.* 2004), it is reasonable to expect that sources of tap water are in many cases similar to those available for more volumetrically significant uses such as irrigation or hydroelectric power generation. This should particularly be so in arid and semiarid regions, where the total range of water resources available is limited. Tap water survey data therefore provide a first indication of how stable isotope tracers may be useful for identifying human-induced change in the water cycle.

Two prevalent patterns stand out in the tap water dataset of Bowen *et al.* (2007) that may be useful for identifying human impacts on water cycle fluxes. First, throughout much of the country the stable isotope ratios of tap water reflect the preferential extraction, distribution, and use of water derived from nonlocal precipitation. This pattern is strongest across the arid and semiarid western United States, where the use of high-altitude and seasonally sourced water imparts a strong negative anomaly to tap water isotope ratios. Water with low H- and O-isotope ratios, extracted from mountain reservoirs, rivers, or groundwater, is distributed as tap water throughout much of the western United States, and the isotope ratios of this water are likely to distinguish it from natural precipitated water present in soils, plants, and locally sourced surface reservoirs throughout this region. Second, evaporation from impounded surface water reservoirs and distribution of evaporated water leads to negative anomalies in *d* values and, in some cases, to positive anomalies in δ^2H and $\delta^{18}O$ values across much of the United States. Strong negative anomalies in tap water across the Great Plains and in the vicinity of the Great Lakes, and more modest anomalies throughout the Rocky Mountain States, represent a distinctive isotopic signature of human-used waters throughout these regions.

The distinctive isotope ratio anomalies associated with these water use patterns provide a tool for tracing human-used water through hydrologic systems. With the development of intensive water isotope monitoring programs and improvements in models for isotopic fractionation of water during evaporation and transpiration, it is likely that these isotope tracers will be useful for identifying and quantifying a range of human-impacted water fluxes between the earth surface and atmosphere, including evaporation from surface reservoirs, crop and lawn irrigation, and thermoelectric power generation. The stage is already set for use of these tracers to identify human impacts on surface water systems, and basic monitoring programs operating in large watersheds should be able to capitalize on stable isotope tracers to identify the contribution of human-generated runoff and seepage to river systems. An additional capability related to the distinctive isotope ratios of human-accessed water in surface water systems may be the use of natural isotope labels incorporated into organic particulates to distinguish autochthonous and allochthonous productivity in streams and rivers impacted by human-generated runoff. An applied example of the use of natural water isotope tracers to identify human-induced changes in groundwater systems was provided by Muir and Coplen (1981), who mapped the extent of artificial recharge within a Santa Clara Valley (CA) aquifer using stable hydrogen and oxygen isotope ratios.

C. Forecasting

Programs designed to document and develop understanding of the isotopic composition of human–hydrologic systems are in their infancy, but through continued monitoring efforts and the development of a modeling framework for interpretation of this data stable isotope ratios of tap water may aid efforts to improve predictions of the future status of water resources. Two areas of contribution can be envisioned. First, isotopic data can improve our understanding of fundamental processes that affect, and are critical to forecasting, the status of water resources. A wide range of contributions of this type are possible, including improved understanding of water waste associated with irrigation, impoundment, and diversion; rates and patterns of artificial recharge; and the impact of water diversion and irrigation on regional precipitation climatology. Second, improved understanding of natural isotope ratio variation in human water systems and sources of this variation may allow the use of isotope ratios as early indicators of water system vulnerability to environmental changes (Chapter 24). Seasonal tap water data from Ithaca, New York (Figure 18.5B), for example, preserve an isotopic (*d*) measure of water inputs derived from reevaporated Great Lakes water. Monitoring of such signals through time could be used to identify ongoing changes in water inputs from such sources as they respond to climate and land-use shifts, providing preemptive indication of potentially detrimental changes in water resources.

V. CONCLUSIONS

Researchers have long recognized the power of H and O isotope ratios in water to contribute to the study of the natural water cycle, and preliminary data on stable isotope ratios of tap water suggest the potential contributions of isotope geochemistry to the study of human interaction with the water cycle. Exciting possibilities exist for the use of focused isotope data collection to generate new quantitative knowledge of the processes and mechanisms by which our water use choices impact natural hydrologic and atmospheric systems. Equally exciting potential exists for the use of isotopes as monitors of water use patterns and impacts, both in the past and in the present. These approaches to studying the interface of human society and the water cycle will benefit both from the continued collection of spatially and temporally resolved data on hydrologic fluxes associated with human water use and their isotopic composition and from improvements in and application of models for the physical, biological, and socioeconomic processes governing how we interface with and drive changes in water cycling.

Isotopic monitoring data, such as those discussed here, can be interpreted on many levels. Here we have entered at the lowest level of interpretation: identifying the broadest patterns in the data that beg immediately for mechanistic explanation, and, in some cases, producing potential explanations from an existing body of knowledge on principals and process. These patterns in and of themselves reveal targets for more focused study, however, with the potential to improve our models for the underlying processes in each of these case studies and in others that were not considered here. The data on tap water isotope ratios discussed here reflect the complex interaction of resource availability, socioeconomic decisions, and natural and human-altered physical and biological processes, and likely preserve information on each component of this system. Extracting this information from such a complex system is and will remain a challenge, but our ability to interpret and benefit from such data will only increase with continued study. The system, however, like most aspects of our modern world, is in the process of directional change. If we can envision today how data such as those discussed here will help us understand and forecast the trajectories of this change, we must continue to gather and archive such data both to inform the direction of future study and to preserve basic data on the interaction of humans and the water cycle that cannot be obtained retrospectively.

VI. REFERENCES

Araus, J. L., A. Febrero, R. Buxo, M. O. Rodriguez-Ariza, F. Molina, M. D. Camalich, D. Martin, and J. Voltas. 1997. Identification of ancient Irrigation practices based on the carbon isotope discrimination of plant seeds: A case study from the South-East Iberian Peninsula. *Journal of Archaeological Science* **24**(8):729–740.

Bigg, G. R., and E. J. Rohling. 2000. An oxygen isotope data set for marine water. *Journal of Geophysical Research* **105:**8527–8535.

Bowen, G. J., and B. Wilkinson. 2002. Spatial distribution of $\delta^{18}O$ in meteoric precipitation. *Geology* **30**(4):315–318.

Bowen, G. J., and J. Revenaugh. 2003. Interpolating the isotopic composition of modern meteoric precipitation. *Water Resources Research* **39:**1299, doi:10.1029/2003WR002086.

Bowen, G. J., D. A. Winter, H. J. Spero, R. A. Zierenberg, T. E. Cerling, and J. R. Ehleringer. 2005. Stable hydrogen and oxygen isotope ratios of bottled waters of the world. *Rapid Communications in Mass Spectrometry* **19:**3442–3450.

Bowen, G. J., J. R. Ehleringer, L. A. Chesson, E. Stange, and T. E. Cerling. 2007. Stable isotope ratios of tap water in the contiguous USA. *Water Resources Research* **43:**W03419, doi:10.1029/2006wr005186.

Burnett, A. W., H. T. Mullins, and W. P. Patterson. 2004. Relationship between atmospheric circulation and winter precipitation $\delta^{18}O$ in central New York State. *Geophysical Research Letters* **31:**L22209, doi:10.1029/2004GL021089: 1–4.

Coleman, M. L., T. J. Shepherd, J. J. Durham, J. E. Rouse, and G. R. Moore. 1982. Reduction of water with zinc for hydrogen isotope analysis. *Analytical Chemistry* **54:**993–995.

Coplen, T. B., J. A. Hopple, J. K. Böhlke, H. S. Peiser, S. E. Rieder, H. R. Krouse, K. J. R. Rosman, T. Ding, R. D. J. Vocke, K. M. Révész, A. Lamberty, P. Taylor, *et al.* 2002. Compilation of Minimum and Maximum Isotope Ratios of Selected Elements in Naturally Occurring Terrestrial Materials and Reagents, US Geological Survey, Reston, Virginia.

Craig, H. 1961. Isotopic variations in meteoric waters. *Science* **133:**1702–1703.

Craig, H., and L. I. Gordon. 1965. Deuterium and oxygen-18 variations in the ocean and the marine atmosphere. *In* E. Tongiorgi (Ed.) *Proceedings of a Conference on Stable Isotopes in Oceanographic Studies and Paleotemperatures.* Spoleto, Italy.

Daly, C., R. P. Neilson, and D. L. Phillips. 1994. A statistical-topographic model for mapping climatological precipitation over mountainous terrain. *Journal of Applied Meteorology* **33**(2):140–158.

Dansgaard, W. 1964. Stable isotopes in precipitation. *Tellus* **16:**436–468.

Dansgaard, W., S. J. Johnson, J. Moller, and C. C. Langway, Jr. 1969. One thousand centuries of climatic record from Camp Century on the Greenland Ice sheet. *Science* **166**(3903):377–381.

Dutton, A., B. H. Wilkinson, J. M. Welker, G. J. Bowen, and K. C. Lohmann. 2005. Spatial distribution and seasonal variation in $^{18}O/^{16}O$ of modern precipitation and river water across the conterminous United States. *Hydrological Processes* **19:**4121–4146.

Dutton, A. R. 1995. Groundwater isotopic evidence for paleorecharge in US High Plains aquifers. *Quaternary Research* **43**(2):221–231.

Epstein, S., and T. Mayeda. 1953. Variation of O^{18} content of waters from natural sources. *Geochimica et Cosmochimica Acta* **4:**213–224.

Feng, X., and S. Epstein. 1994. Climate implications of an 8000-year hydrogen isotope time series from bristlecone pine trees. *Science* **265**(5175):1079–1081.

Fessenden, J. E., C. S. Cook, M. J. Lott, and J. R. Ehleringer. 2002. Rapid ^{18}O analysis of small water and CO_2 samples using a continuous-flow isotope ratio mass spectrometer. *Rapid Communications in Mass Spectrometry* **16:**1257–1260.

Gat, J. R. 1996. Oxygen and hydrogen isotopes in the hydrologic cycle. *Annual Review of Earth and Planetary Sciences* **24:**225–262.

Gat, J. R., C. J. Bowser, and C. Kendall. 1994. The contribution of evaporation from the Great Lakes to the continental atmosphere; estimate based on stable isotope data. *Geophysical Research Letters* **21**(7):557–560.

Gat, J. R., B. Klein, Y. Kushnir, W. Roether, H. Wernli, R. Yam, and A. Shemesh. 2003. Isotope composition of air moisture over the Mediterranean Sea: An index of the air-sea interaction pattern. *Tellus* **55B:**953–965.

Gazis, C., and X. Feng. 2004. A stable isotope study of soil water: Evidence for mixing and preferential flow paths. *Geoderma* **119:**97–111.

Gehre, M., H. Geilmann, J. Richter, R. A. Werner, and W. A. Brand. 2004. Continuous flow $^{2}H/^{1}H$ and $^{18}O/^{16}O$ analysis of water samples with dual inlet precision. *Rapid Communications in Mass Spectrometry* **18**(22):2650–2660.

Gonfiantini, R. 1986. Environmental isotopes in lake studies. Pages 113–168 *in* P. Fritz and J. C. Fontes (Eds.) *Handbook of Environmental Isotope Geochemistry, Vol. 2, The Terrestrial Environment, B.* Elsevier, Amsterdam, The Netherlands.

Harris, D. M., J. J. McDonnell, and A. Rodhe. 1995. Hydrograph separation using continuous open-system isotope mixing. *Water Resources Research* **31:**157–171.

Haug, G. H., R. Tiedemann, R. Zahn, and A. C. Ravelo. 2001. Role of Panama uplift on oceanic freshwater balance. *Geology* **29** (3):207–210.

Hays, P. D., and E. L. Grossman. 1991. Oxygen isotopes in meteoric calcite cements as indicators of continental paleoclimate. *Geology* **19**(5):441–444.

Hutson, S. S., N. L. Barber, J. F. Kenny, K. S. Linsey, D. S. Lumia, and M. A. Maupin. 2004. *Estimated Use of Water in the United States in 2000.* United States Geological Survey Circular 1268, Reston, Virginia.

Iacumin, P., H. Bocherens, L. Chaix, and A. Marioth. 1998. Stable carbon and nitrogen isotopes as dietary indicators of ancient Nubian populations (Northern Sudan). *Journal of Archaeological Science* **25:**293–301.

Imbrie, J., J. D. Hays, D. G. Martinson, A. McIntyre, A. C. Mix, J. J. Morley, N. G. Pisias, W. L. Prell, and N. J. Shackleton. 1984. The orbital theory of Pleistocene climate: Support from a revised chronology of the marine $\delta^{18}O$ record. Pages 269–306 *in* A. Berger, J. Imbrie, J. Hays, G. Kukla, and B. Saltzman (Eds.) *Milankovitch and Climate.* Reidel, Dordrecht.

Kendall, C., and J. J. McDonnell (Eds.). 1996. *Isotope Tracers in Catchment Hydrology.* Elsevier, Amsterdam, 839 p.

Kendall, C., and T. B. Coplen. 2001. Distribution of oxygen-18 and deuterium in river waters across the United States. *Hydrological Processes* **15**(7):1363–1393.

Lee, K. S., D. B. Wenner, and I. Lee. 1999. Using H- and O-isotopic data for estimating the relative contributions of rainy and dry season precipitation to groundwater: Example from Cheju Island, Korea. *Journal of Hydrology* **222**(1–4):65–74.

Majoube, M. 1971. Fractionnement en oxygène-18 et en deutérium entre l'eau et sa vapeur. *Journal of Chemical Physics* **197:**1423–1436.

Merlivat, L. 1978. Molecular diffusivities of $H_2{}^{16}O$, $HD^{16}O$ and $H_2{}^{18}O$ in gases. *Journal of Chemical Physics* **69:**2864–2871.

Muir, K. S., and T. B. Coplen. 1981. Tracing Ground-Water Movement by Using the Stable Isotopes of Oxygen and Hydrogen, Upper Penitencia Creek Alluvial Fan, Santa Clara Valley, California, United States Geological Survey Water-Supply Paper 2075, Washington, DC.

Müller, W., H. Fricke, A. N. Halliday, M. T. McCulloch, and J.-A. Wartho. 2003. Origin and migration of the Alpine Iceman. *Science* **302:**862–866.

O'Connell, T. C., and R. E. M. Hedges. 1999. Isotopic comparison of hair and bone: Archaeological analyses. *Journal of Archaeological Science* **26:**661–665.

Pierrehumbert, R. T. 1999. Huascaran $\delta^{18}O$ as an indicator of tropical climate during the Last Glacial Maximum. *Geophysical Research Letters* **26**(9):1345–1348.

Poage, M. A., and C. P. Chamberlain. 2001. Empirical relationships between elevation and the stable isotope composition of precipitation and surface waters: Considerations for studies of paleoelevation change. *American Journal of Science* **301**(1):1–15.

Rose, S. 1996. Temporal environmental isotopic variation within the Falling Creek (Georgia) watershed: Implications for contributions to streamflow. *Journal of Hydrology* **174:**243–261.

Rozanski, K., L. Araguas-Araguas, and R. Gonfiantini. 1992. Relation between long-term trends of oxygen-18 isotope composition of precipitation and climate. *Science* **258**(5084):981–985.

Rozanski, K., L. Araguas-Araguas, and R. Gonfiantini. 1993. Isotopic patterns in modern global precipitation. Pages 1–36 *in* P. K. Swart, K. C. Lohmann, J. McKenzie, and S. Savin (Eds.) *Climate Change in Continental Isotopic Records. Geophysical Monograph 78.* American Geophysical Union, Washington, DC.

Scholl, M. A., S. E. Ingebritsen, C. J. Janik, and J. P. Kauahikaua. 1996. Use of precipitation and groundwater isotopes to interpret regional hydrology on a tropical volcanic island: Kilauea volcano area, Hawaii. *Water Resources Research* **32**(12):3525–3537.

Schurr, M. R. 1997. Stable nitrogen isotopes as evidence for the age of weaning at the Angel Site: A comparison of isotopic and demographic measures of weaning age. *Journal of Archaeological Science* **24:**919–927.

Sharp, Z. D., V. Atudorei, and T. Durakiewicz. 2001. A rapid method for determination of hydrogen and oxygen isotope ratios from water and hydrous minerals. *Chemical Geology* **178**(1–4):197–210.

Sharp, Z. D., V. Atudorei, H. O. Panarello, J. Fernández, and C. Douthitt. 2003. Hydrogen isotope systematics of hair: Archeological and forensic applications. *Journal of Archaeological Science* **30:**1709–1716.

Siegenthaler, U., and H. Oeschger. 1978. Correlation of ^{18}O in precipitation with temperature and altitude. *Nature* **285** (5763):314–317.

White, C. D. 1993. Isotopic determination of seasonality in diet and death from Nubian mummy hair. *Journal of Archaeological Science* **20:**657–666.

Williams, D. G., J. B. Coltrain, M. J. Lott, N. B. English, and J. R. Ehleringer. 2005. Oxygen isotopes in cellulose identify source water for archaeological maize in the American Southwest. *Journal of Archaeological Science* **32**(6):931–939.

Wilson, J. L., and H. Guan. 2004. Mountain-block hydrology and mountain-front recharge. Pages 113–137 *in* F. M. Phillips, J. F. Hogan, and B. R. Scanlon (Eds.) *Groundwater Recharge in a Desert Environment: The Southwestern United States.* American Geophysical Union, Washington, DC.

Yurtsever, Y., and J. R. Gat. 1981. Atmospheric waters. Pages 103–142 *in* J. R. Gat and R. Gonfiantini (Eds.) *Stable Isotope Hydrology: Deuterium and Oxygen-18 in the Water Cycle.* International Atomic Energy Agency, Vienna.

CHAPTER 19

The Use of Carbon and Nitrogen Stable Isotopes to Track Effects of Land-Use Changes in the Brazilian Amazon Region

Luiz Antonio Martinelli,* Jean Pierre Henry Balbauld Ometto,* Françoise Yoko Ishida,* Tomas Ferreira Domingues,*,[1] Gabriela Bielefeld Nardoto,* Rafael Silva Oliveira,* and James R. Ehleringer†

*CENA, University of São Paulo

†Department of Biology, University of Utah

Contents

I. INTRODUCTION

On average 18,000 km^2 of Brazilian Amazon terra firme forest vegetation is burned every year. Most of this area is replaced by pastures cultivated to support cattle ranching. Up to now, ~600,000 km^2 have already been burned. Such area is equivalent to 12% of the total area of the Brazilian Amazon region (5 million km^2) or equivalent to two times the area of Italy or almost seven times the area of Portugal. This is probably

[1] Current Address: Institute of Geography, School of GeoSciences, The University of Edinburgh, Edinburgh-EH8 9XP Scotland-UK.

one of the major land-use changes currently taking place on Earth. A change of this magnitude has the potential to alter significantly not only the landscape but also the vital biogeochemical processes that maintain the basic ecosystem functions (Ometto *et al.* 2005).

The replacement of tropical terra firme forests by cultivated pastures has introduced different species of African C_4 tropical grasses, mainly of the genus *Brachiaria*. This change opens up a unique opportunity for the use of carbon stable isotope composition to track the fate of introduced C_4 grass species since the main matrix of plant species in the Amazon terra firme forests is composed by C_3 plants. The $\delta^{13}C$ values of C_4 species occurring at the Brazilian Amazon vary from −14‰ to −11‰, while C_3 species vary from −38‰ to −26‰ at that biome (Ometto *et al.* 2006).

It has also been documented that tropical terra firme forests generally have an open nitrogen cycle, where losses of N are significantly higher in relation to the inputs of N (Martinelli *et al.* 1999). As a consequence of these high losses, soils and plants of tropical terra firme forests are characterized by distinctly high values of $\delta^{15}N$ (Ometto *et al.* 2006). As N availability increases, losses from the ecosystem, such as gaseous N losses and leaching of NO_3^- as a result of incomplete nitrification, should increase and lead to ^{15}N-enriched plants (Högberg 1997). Such fact opens up a second window of opportunity for using the N stable isotopic composition as a tool to investigate changes in the N cycle due to land-use changes that are occurring in the Amazon basin.

In this chapter, we explore how we have been using the stable C and N isotopic composition to evaluate changes in ecosystem functioning due to land-use changes that have been occurring in the Amazon region. We propose to document along this chapter isotopic shifts from individual components of the ecosystems, like plants, and show how the signals of the C_4 vegetation have already been incorporated in other reservoirs of C such as soils, rivers, and atmosphere.

Most of our studies were conducted in three main sites of the LBA project (large scale biosphere atmosphere experiment—http://lba.cptec.inpe.br/lba/site/) near the cities of Manaus (2°85′ S; 54°95′ W), Santarém (2°50′ S; 60°0′ W), and Ji-Paraná (10°08′S; 61°92′ W) (Figure 19.1). Manaus is located in the central part of the Amazon in the State of Amazonas and most of the work was conducted at Reserva do Cuieiras that belongs to the Instituto Nacional de Pesquisas da Amazônia. In Santarém, which is located in the central east region of the Amazon in the State of Pará, samples were collected at the Floresta Nacional do Tapajós, located ~70-km south of the city of Santarém. In addition, we used data in this chapter from several experiments conducted at Fazenda Nova Vida, a cattle ranch located in the southwest region of the Amazon in the State of Rondônia, ~200-km south of the city of Ji-Paraná. Finally, we compared terra firme forests with data collected in several savanna sites in Brazil. For detailed information about these studied sites refer to the following studies: Ometto *et al.* (2002) and Bernardes *et al.* (2004).

At the continental scale, the Amazon basin includes considerable variation in climatic regime, topography, and geography (Elsenbeer and Lack 1997). The mean precipitation in this region varies from less than 2000 to more than 3000 mm $year^{-1}$ (Liebmann and Marengo 2001) and the dry season varies from nonexistent to periods when there are seven consecutive months with less than 100 mm $month^{-1}$ of rain (Sombroek 2001). Terra firme forests (upland, nonseasonally flooded) represent the most common vegetation type distributed along the Amazonas and Orinoco basin (Cuevas 2001). These are dense evergreen forests with mean canopy height of 25 m and maximum canopy height of 55 m (Clark and Clark 1996).

Savanna ecosystems are characterized by the codominance of two contrasting plant life-forms—trees of low stature and grasses. Although the relative representation of these life-forms varies considerably across savanna types, they typically comprise a gradient of woody plant density ranging from open grasslands, through open shrubland to dense woodlands (Eiten 1972, Ratter *et al.* 2003). Savanna structure is undoubtedly the consequence of several interacting factors, including climate, resource competition, fire, and grazing, that operate at various spatial and temporal scales (Scholes and Archer 1997). In Brazil, the savannas occupy a core area in the central territory, locally named as Cerrado, but disjuncted areas are also found in the Amazon basin (Ratter *et al.* 2003).

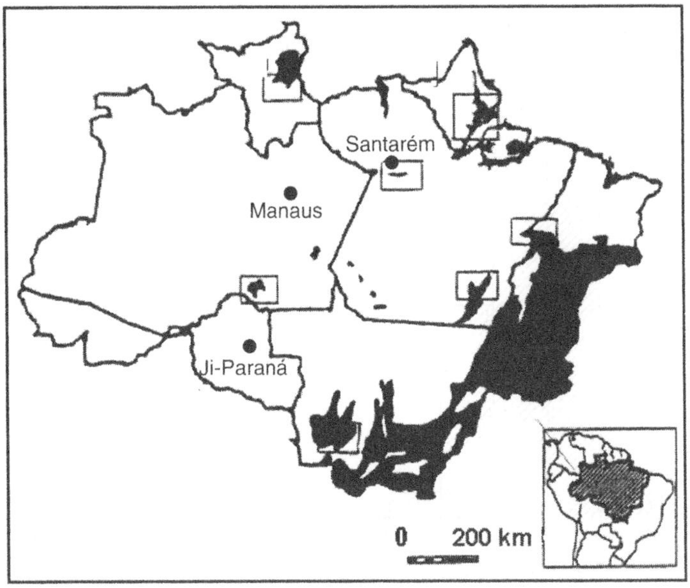

FIGURE 19.1 The Brazilian Amazon region showing the location of the cities where sampling was concentrated and the Brazilian savannas represented by the black areas.

II. STABLE ISOTOPE COMPOSITION OF PLANTS AND SOILS IN THE AMAZON REGION

The vegetation represents the first C reservoir in terrestrial ecosystems that suffer an abrupt change in plant types with the introduction of C_4 plants. This change pass along the soil organic matter through litter decomposition, and with time the soil will reflect a mixture of old C_3 C from the original terra firme forests and the rate of incorporation of the new C_4 plant added to the system. We also compared terra firme forests with savannas, which is the second most important vegetation type in the Amazon region. The Amazon forest and the savannas are the two dominant ecosystem types in Brazil comprising more than 60% of the territory. They are both characterized by a mosaic of vegetation types, a highly diversified and complex flora (Eiten 1972, Ratter *et al.* 1997, Pitman *et al.* 2001), growing on very weathered soils with low nutrient availability (Furley and Ratter 1988, Richter and Babbar 1991).

A. Plants

We grouped 1953 samples of plant species from terra firme forests collected in different areas of the Brazilian Amazon basin. The overall $\delta^{13}C$ average of these samples was $-32 \pm 2.3‰$, ranging between $-38.5‰$ and $-24.6‰$ (Figure 19.2).

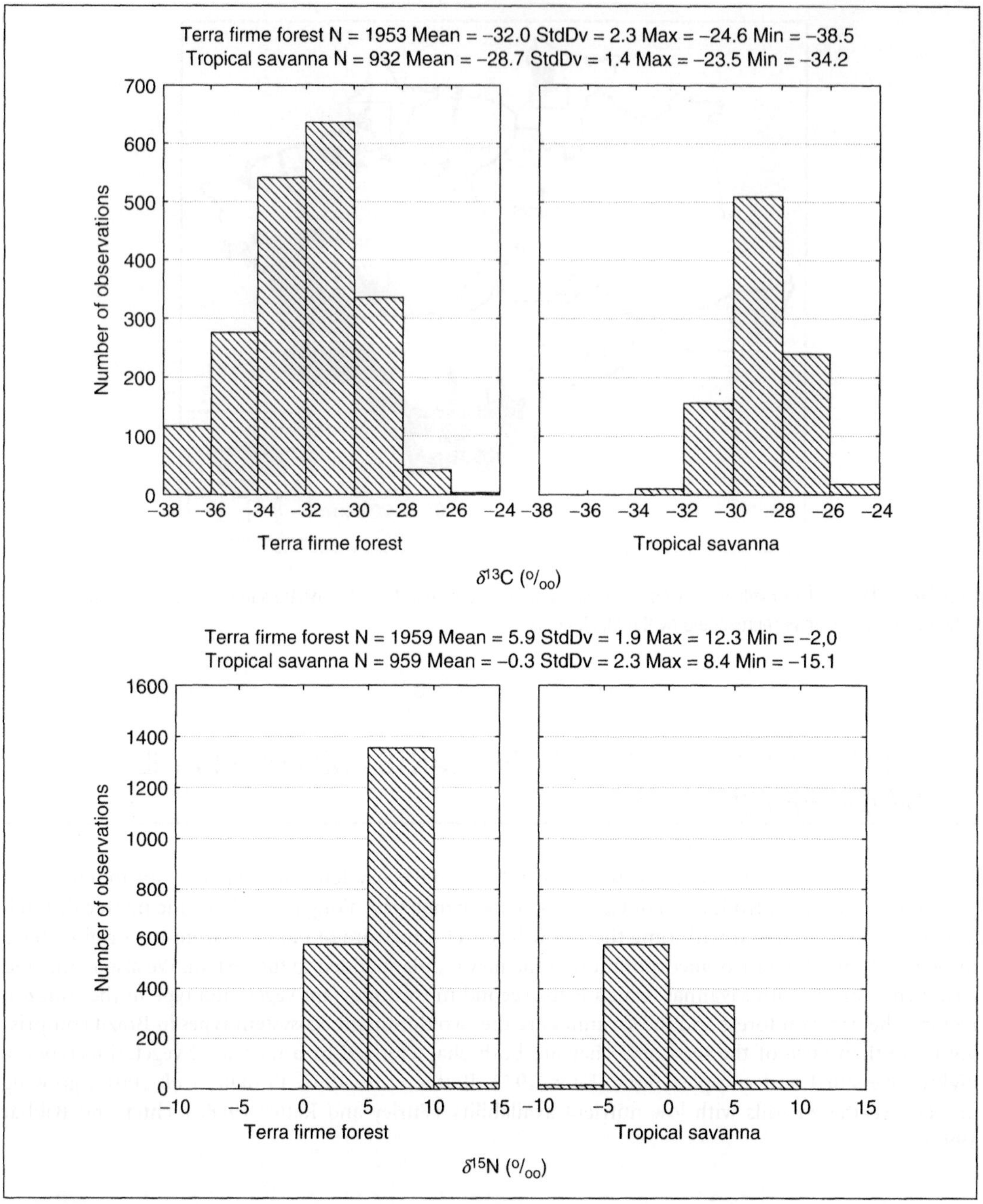

FIGURE 19.2 Histogram of δ^{13}C and δ^{15}N values of plants species from terra firme forests and savannas in the Amazon region. Above the graphics basic statistics are shown: number of samples (N), and average (Mean), standard deviation (StdDv), maximum (Max), and minimum (Min) values expressed in per mil (‰).

The average value of terra firme forests was significantly lower ($p < 0.01$) than the average found for 932 samples of plant species collected in savannas of the Brazilian Amazon region (Figure 19.2). The minimum value for the savannas was equal to -34.2‰ and the maximum equal to -23.5‰, with an average equal to -28.7 ± 1.4‰. The higher δ^{13}C values of savannas in relation to terra firme forests are probably due to the relatively higher water stress in the former than in the later. Generally, savannas are more open areas than forests given that the tree canopies are much denser in the later biome. As a response to water stress, savanna plants have evolved a suite of mechanisms to control water loss. Among these mechanisms, stomata regulation of whole-canopy water loss is one of the most critical expected responses of savanna plants to water deficits. Lower leaf conductance and lower c_i/c_a ratios will lead to a lower isotopic discrimination in relation to the atmospheric CO_2 by savanna species, leading to higher $\delta^{13}C$ values.

The average $\delta^{15}N$ value of leaves of terra firme forest species was equal to 5.9 ± 1.9‰, with a minimum and maximum values of −2.0‰ and 12.3‰, respectively (Figure 19.2). The average $\delta^{15}N$ value of savanna leaves (−0.3 ± 2.3‰) was significantly lower ($p < 0.01$) than the average value of terra firme forests. The range of $\delta^{15}N$ values was higher in the savanna vegetation when compared with the terra firme forest. The former had a range of 23.5‰ and the later 14.3‰ (Figure 19.2).

We attributed the lower $\delta^{15}N$ values of savanna plants in relation to the terra firme forests to differences on the relative openness of the N cycle (Martinelli *et al.* 1999). There is much evidence showing that N losses are lower in savannas than in the terra firme forests (Bustamante *et al.* 2004a, 2006). As a consequence there is less isotopic fractionation in the savannas than in the terra firme forests, leading to lower $\delta^{15}N$ values. As an indirect evidence of the higher N abundance in the terra firme forest, it was observed a significantly higher ($p < 0.01$) foliar N concentration in the terra firme forests (2.3 ± 0.7%) in relation to the foliar N concentration in the savannas (1.6 ± 0.7%) (Figure 19.3). Consequently, the average foliar C:N ratio of the terra firme forests (22.6 ± 7.4) was significantly lower ($p < 0.01$) than the foliar C:N ratio of the savannas (34.4 ± 12.1).

We compared the isotopic and elemental composition of the terra firme forests and savannas with the new plant communities that were established after the conversion of forest to pastures in the Amazon region (Figure 19.4). We considered three groups of new established plants: the C_4 African

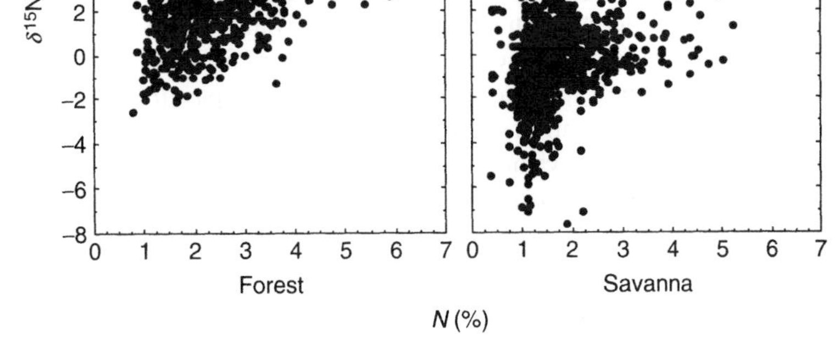

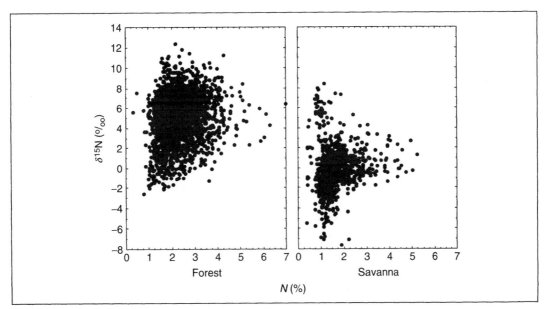

FIGURE 19.3 Relationship between foliar N and $\delta^{15}N$ values of plant species collected in terra firme forests and savannas of the Amazon region.

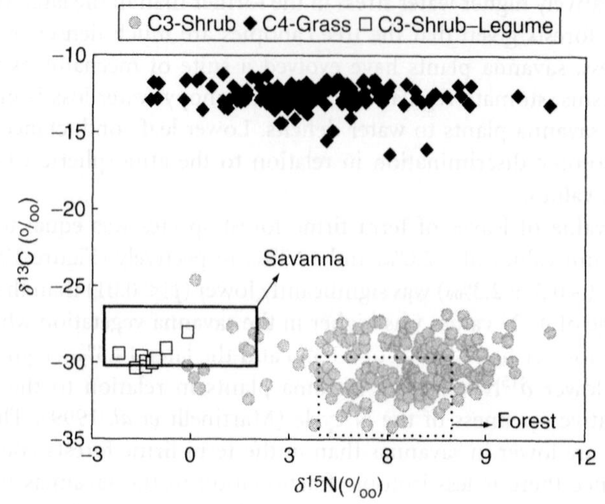

FIGURE 19.4 Relationship between foliar $\delta^{15}N$ and foliar $\delta^{13}C$ of plant species collected in pasture sites near the cities of Santarém and Manaus. For comparison, the solid box represents the isotopic values of savanna leaves and the dashed box represents the isotopic values of forest leaves.

TABLE 19.1 Average values of compositional parameters of new plants introduced after the conversion of forest to pasture in the Brazilian amazon region

	$\delta^{13}C$ (‰)	$\delta^{15}N$ (‰)	C (%)	N (%)	C:N
C_4 grass	−12.6	3.8	41.6	1.30	34.5
C_3 shrubs	−30.2	6.5	47.4	1.95	28.4
C_3 legumes	−29.7	−1.2	42.9	4.36	10.2

grasses, generally from the genus *Brachiaria*; the C_3 type shrubs encroached in the pastures; and leguminous C_3 shrubs, which potentially have the capability to fix N_2 from the atmosphere.

Most of *Brachiaria* species followed the C_4 photosynthetic pathway. Therefore, as expected, the average $\delta^{13}C$ values was −12.6 ± 0.6‰ (n = 146), which is significantly different ($P < 0.01$) than C_3 plants of terra firme forests and savannas (Figure 19.4).

The $\delta^{13}C$ values of C_3 shrubs from pastures were significantly higher ($P < 0.01$) than the $\delta^{13}C$ values of terra firme forests, but not different than the values of the savannas. This is probably due to the fact that pastures and savannas have more open canopies than the terra firme forests. Since they, in general, are more exposed to sunlight, the c_i/c_a ratio will be smaller in the pastures and savannas leading to higher $\delta^{13}C$ values. This explanation can be supported by the fact that leaves from the top of the canopy of terra firme forests had similar $\delta^{13}C$ values to C_3 shrubs of pastures (Ometto *et al.* 2002).

The $\delta^{15}N$ values of C_4 shrubs were extremely variable, varying from −2‰ to 12‰. The average was significantly smaller than the average of terra firme forests, but significantly higher than the values found in savannas (Table 19.1). It is difficult to explain such variability without knowing specific conditions of the site in which the grasses were sampled. However, in a broader perspective, it has been observed an impoverishment of N in pastures established in the Amazon basin, leading to lower mineralization and nitrification rates and smaller N gas losses to the atmosphere (Neill *et al.* 1997). There is a tendency of N-rich systems to have higher $\delta^{15}N$ values than N-poor systems. This is probably due to the fact that N-rich systems have more losses of N than N-poor systems. As N losses lead to an isotopic enrichment of the substrate (Högberg 1997), the N left behind will be isotopically enriched (Martinelli *et al.* 1999). Corroborating this hypothesis, there is the fact that the foliar N content of C_4 grasses was significantly lower than the foliar N content of the terra firme forests. As a consequence, the C:N ratio of C_4 grasses was significantly higher than the foliar C:N ratio of terra firme forests, indicating a N impoverishment in pastures when compared to these forests (Table 19.1).

Another factor that may contribute to the decrease of $\delta^{15}N$ values in tropical C_4 grasses is the fact that these plants are potentially able to fix N from the atmosphere (Reis *et al.* 2001). N fixation decreases the $\delta^{15}N$ value of the plant because the $\delta^{15}N$ value of atmospheric N is close to zero. However, in order to fix atmospheric N, plants need an adequate phosphorus supply, which is uncommon in the Amazon region due to the lack of available P in these weathered tropical soils (Richter and Babbar 1991).

Interestingly enough, the C_3 shrubs of Amazon pastures did not follow the same trend, and had $\delta^{15}N$ similar to values found in the terra firme forests (Figure 19.4). We expected that the C_3 shrubs would also have smaller $\delta^{15}N$ values in relation to terra firme forests due to the N impoverishment observed in pastures of the Amazon. One possibility is that C_3 shrubs are preferentially taking up NH_4^+, which is more abundant in pasture soils due to the decrease of nitrification (Neill *et al.* 1997). Ammonium tends to have higher $\delta^{15}N$ values than NO_3^- and as a consequence, plants taking up NH_4^+ will have higher $\delta^{15}N$ values (Högberg 1997). As an example, in the savanna area of the Central Brazil region, the species *Roupala montana* has distinctly higher $\delta^{15}N$ values than other species due to its preference for NH_4^+ (Bustamante *et al.* 2004b). Another possibility that cannot be ruled out is the fact that C_3 shrubs would have deeper roots than C_4 grasses, exploring deeper soil layers, where the $\delta^{15}N$ of the soil organic matter is higher than that in the surface (Bustamante *et al.* 2004b).

Finally, the C_3 legumes had significantly lower $\delta^{15}N$ values in relation to plants of tropical forests, C_3 nonlegumes shrubs and C_4 grasses (Table 19.1). This indicates that this group of plants is actively fixing N from the atmosphere. This fact corroborates with earlier studies that have hypothesized that N biological fixation would occur in tropical systems only when a disruption in the N cycle is large enough to trigger the energetically expensive N biological fixation process (Vitousek *et al.* 2002).

B. Soils

The $\delta^{13}C$ values of the soil organic matter under terra firme forests varied from −30‰ to −26‰ with a slight decrease with soil depth (Figure 19.5). This trend contrasts with soils under savannas. In general these soils had higher $\delta^{13}C$ values at soil surface with values varying from −26‰ to −14‰ (Figure 19.5). The variation with depth of $\delta^{13}C$ values of the soil organic matter from savannas was also more variable than from soils of terra firme forests. In a savanna area located in Central Brazil, near the city of Brasília, the variability with depth was not so high, with most of the values varying from −25‰ to −21‰. The largest variability with depth was observed in the savannas encroached in the Amazon region and surrounded by terra firme forests. In this case, most of the time it was observed a large decrease of the $\delta^{13}C$ values with depth (Figure 19.5). It is believed that the variability observed with soil depth in these savannas is caused by past vegetation changes that lead to changes in the vegetation type ($C_3 \times C_4$) (Martinelli *et al.* 1996). In the savanna of Central Brazil, these changes did not seem to be as pronounced as in the encroached savannas from the Amazon region, where it is generally observed that at deeper soil layers there was relatively more C_3 vegetation type (Sanaiotti *et al.* 2002).

In general, it was observed a slight increase of $\delta^{15}N$ with soil depth, and a high variability of $\delta^{15}N$ values of soil organic matter in soils under terra firme forests, from 4‰ to 12‰ (Figure 19.6). There was also some variability in soils under savanna, approximately from 2‰ to 8‰ (Figure 19.6). Five soil profiles of terra firme forests had higher $\delta^{15}N$ values than soil profiles under savanna vegetation, while six other soil profiles had similar values with those of the savanna soils (Figure 19.6). Differences between terra firme forest and savanna soils became evident when $\delta^{13}C$ values were plotted against

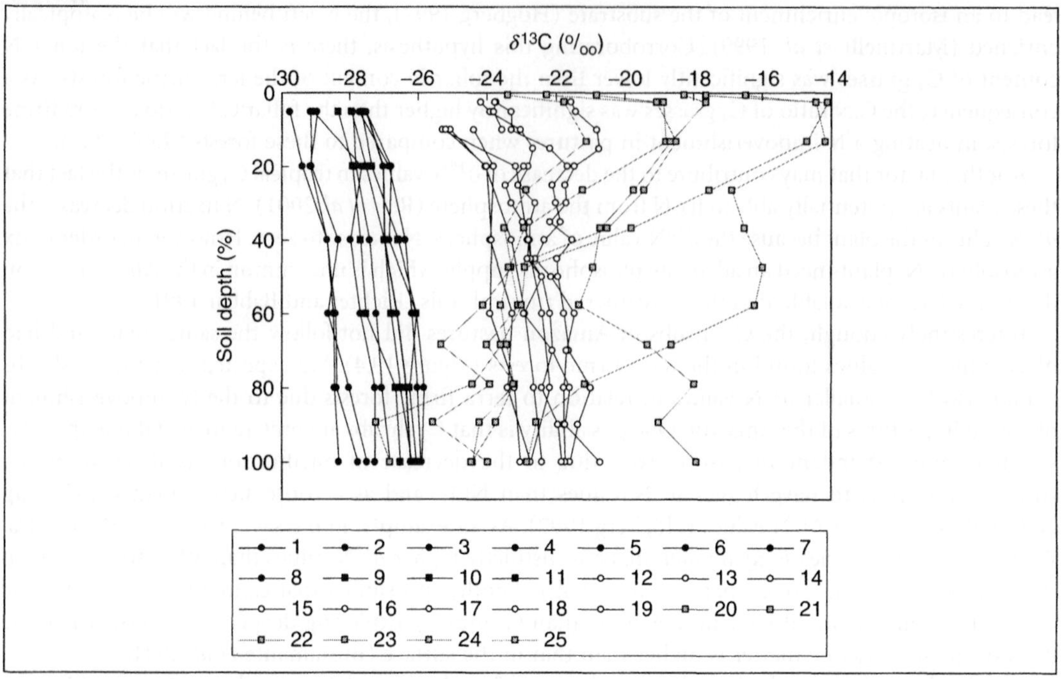

FIGURE 19.5 Variation of $\delta^{13}C$ of soil organic matter in relation to soil depth (expressed as percentage of the deepest sampling). Black symbols represent terra firme forest soil profiles from different sites: Manaus (1–3), Santarém (4–8), São Gabriel da Cachoeira (8–11). Open circles represent savanna soil profiles from different sites located in the Brazil central region; gray symbols represent savanna encroached in terra firme forests of the Amazon region.

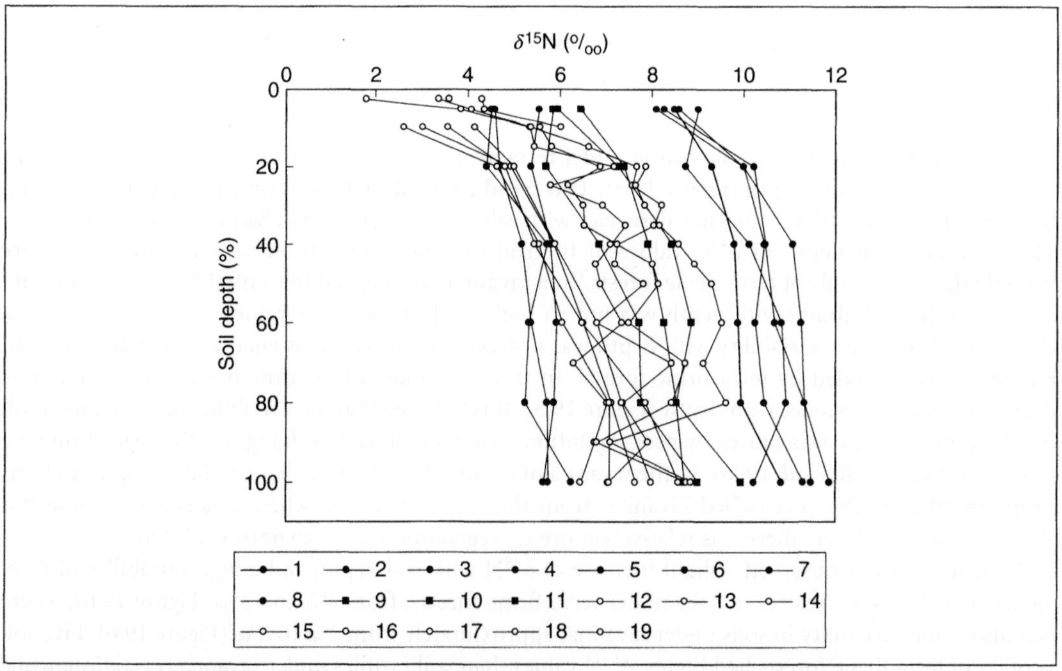

FIGURE 19.6 Variation of $\delta^{15}N$ of soil organic matter in relation to soil depth (expressed as percentage of the deepest sampling). Black symbols represent terra firme forests soil profiles from different sites: Manaus (1–3), Santarém (4–8), São Gabriel da Cachoeira (8–11). Open circles represents savanna soil profiles from different sites located in the Brazil central region.

$\delta^{15}N$ (Figure 19.7). While there was a direct correlation between $\delta^{13}C$ and $\delta^{15}N$ for soils under terra firme forests, there was an inverse correlation between $\delta^{13}C$ and $\delta^{15}N$ for soils under savanna (Figure 19.7). The direct correlation observed for soils of terra firme forests can be explained by a progressive isotopic enrichment in direction to deeper soil layer due to the decomposition of organic matter. $\delta^{15}N$ values in savanna soils also become higher with soil depth; however, due to the vegetation change, deeper soil layers are richer in C_3 organic material than the surface soil layer that is richer in C_4 organic material (Figure 19.7).

The C_4 material is also present when terra firme forests are replaced by pastures. This is the most common land cover change in the Amazon region during the last four decades. As a consequence the surface soil organic matter becomes progressively more enriched with C_4 material (Figure 19.8). In a chronosequence of pastures located in the State of Rondônia, the initial $\delta^{13}C$ of the surface soil organic matter was around −27‰. After 80 years of pasture implementation, the $\delta^{13}C$ of the surface soil

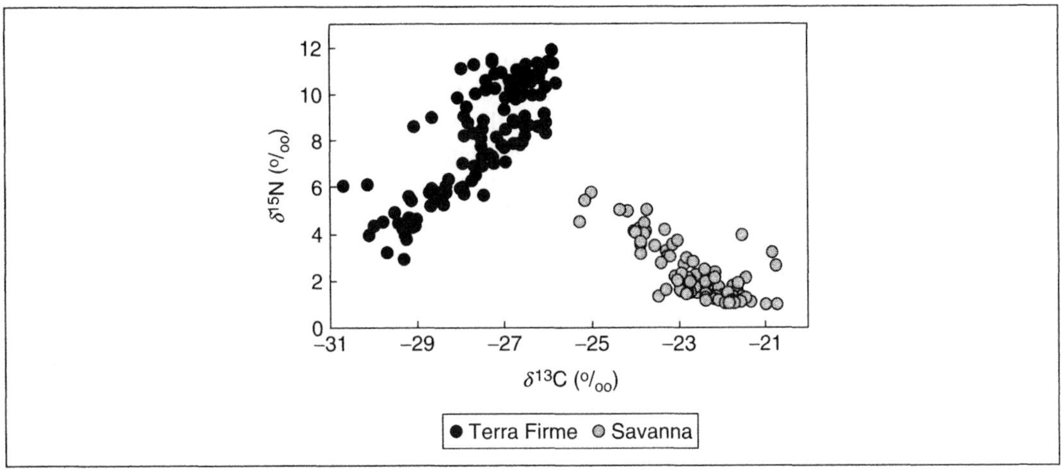

FIGURE 19.7 Relationship between foliar $\delta^{13}C$ and foliar $\delta^{15}N$ of plant species collected in several terra firme forests and in several savannas in the Amazon region.

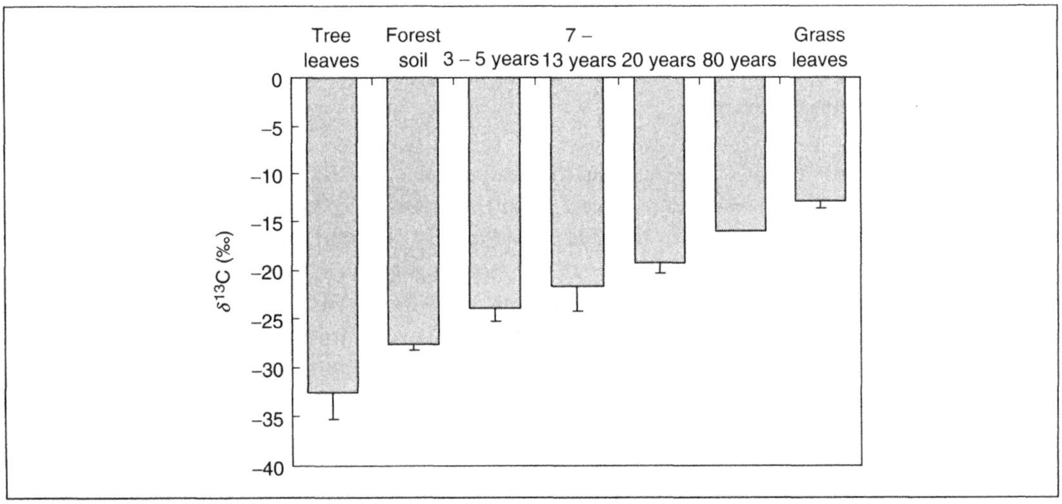

FIGURE 19.8 The $\delta^{13}C$ values of tree leaves, forest soil organic matter, pasture soil organic matter of different ages and grass leaves. Data of soil organic matter of pastures are from Cerri *et al.* (2004).

organic matter stabilized around −16‰, which means that almost 90% of the original C was replaced by C_4 organic matter from the pasture (Cerri *et al.* 2004).

III. STABLE ISOTOPE COMPOSITION OF RIVERINE CARBON IN THE AMAZON REGION

Due to soil erosion, the C_4 material produced by pasture grasses and attached to the soil particles may enter in the aquatic systems. We used the $\delta^{13}C$ values of particulate organic carbon (POC) and dissolved inorganic carbon (DIC) to show the influence of pasture cultivation on aquatic systems of the Amazon region.

A. Particulate Organic Carbon

We first compared the elemental and C stable isotope composition of the Amazon River (river of sixth order) and Ji-Paraná River (fourth order), which is the main tributary of the Madeira River (fifth order), one of the most important tributaries of the Amazon River, with a small stream located at the Nova Vida farm, near the city of Ariquemes, in the State of Rondônia, which have first- to second-order streams that belong to the watershed of the Ji-Paraná River (Table 19.2; Neill *et al.* 2001, Bernardes *et al.* 2004, Thomas *et al.* 2004). The Ji-Paraná River drains the State of Rondônia, where one of the most aggressive conversions of forest to pasture has been seen in the last decades.

The $\delta^{13}C$ values of the coarse (>63 μm) and the fine (<63 μm) POC of the forested reach of the Nova Vida stream were similar to the $\delta^{13}C$ values of large rivers POC from the Amazon and rivers of the Ji-Paraná basin (Table 19.2). These values are similar to surface soils of the terra firme forest, which acquired its organic matter from C_3 plants (Bernardes *et al.* 2004). On the other hand, the $\delta^{13}C$ values of both coarse and fine fractions of the pasture reach of the Nova Vida stream were almost 7‰ higher than the $\delta^{13}C$ values of the forested reach of the Nova Vida stream (Table 19.2). Such higher values suggest that the organic matter produced by C_4 grasses from the pasture is already being incorporated into the riverine particulate organic matter. Comparing the $\delta^{13}C$ values of the POC of the Nova Vida stream with the Amazon and Ji-Paraná rivers, but also with rivers located in the State of São Paulo, where there is a dominance of pasture and sugarcane, both C_4-type crops (Figure 19.9), it is clear that the pasture reach of this stream has a $\delta^{13}C$ of its POC comparable to rivers of the State of São Paulo, especially to those streams of similar size such as the Cabras and Pisca (Figure 19.9).

B. Dissolved Inorganic Carbon

The $\delta^{13}C$ of the DIC ($\delta^{13}C$-DIC) of forested reach streams of the Fazenda Nova Vida had values varying from near −20‰ to −8‰. It was also observed an inverse relationship between the $\delta^{13}C$-DIC and the DIC concentration (Figure 19.10). On the other hand, it was observed a direct relationship between $\delta^{13}C$-DIC and the DIC concentration (Figure 19.10). Besides, the $\delta^{13}C$-DIC values of the pasture were significantly higher than in the forested reach streams of Nova Vida. The values varied from a minimum of −8‰ to near 0‰ (Figure 19.10). Therefore, between the maximum values of the $\delta^{13}C$-DIC of the pasture reach and the minimum value observed in the forested reach of Nova Vida streams, there was a difference of ∼20‰, suggesting that the sources of DIC are different between the reaches of the Nova Vida streams (Figure 19.10).

Higher $\delta^{13}C$-DIC in river waters are generally related with the photosynthesis, air–water exchange, and carbonate sources, while river and soil respiration and groundwater input in silicate lithology drive the $\delta^{13}C$-DIC to more negative values (Atekwana and Krishnamurthy 1998, Wang and Veizer 2000,

TABLE 19.2 Averages of compositional parameters of the riverine coarse and fine particulate organic matter from the basins investigated in this study

	Amazon	Ji-Paraná	Nova Vida forest	Nova Vida pasture
Coarse, POC (%)	1.1	9.0	12.4	16.7
Coarse, PON (%)	0.04	0.50	0.90	1.20
Coarse, C:N	25	18	14	13
Coarse, $\delta^{13}C$ (‰)	−28.0	−29.1	−27.4	−20.9
Fine, POC (%)	1.2	8.4	6.6	9.6
Fine, PON (%)	0.10	0.72	0.60	1.10
Fine, C:N	11	11	12	13
Fine, $\delta^{13}C$ (‰)	−26.9	−28.1	−27.2	−20.5

Different letters indicate statistically significant differences.

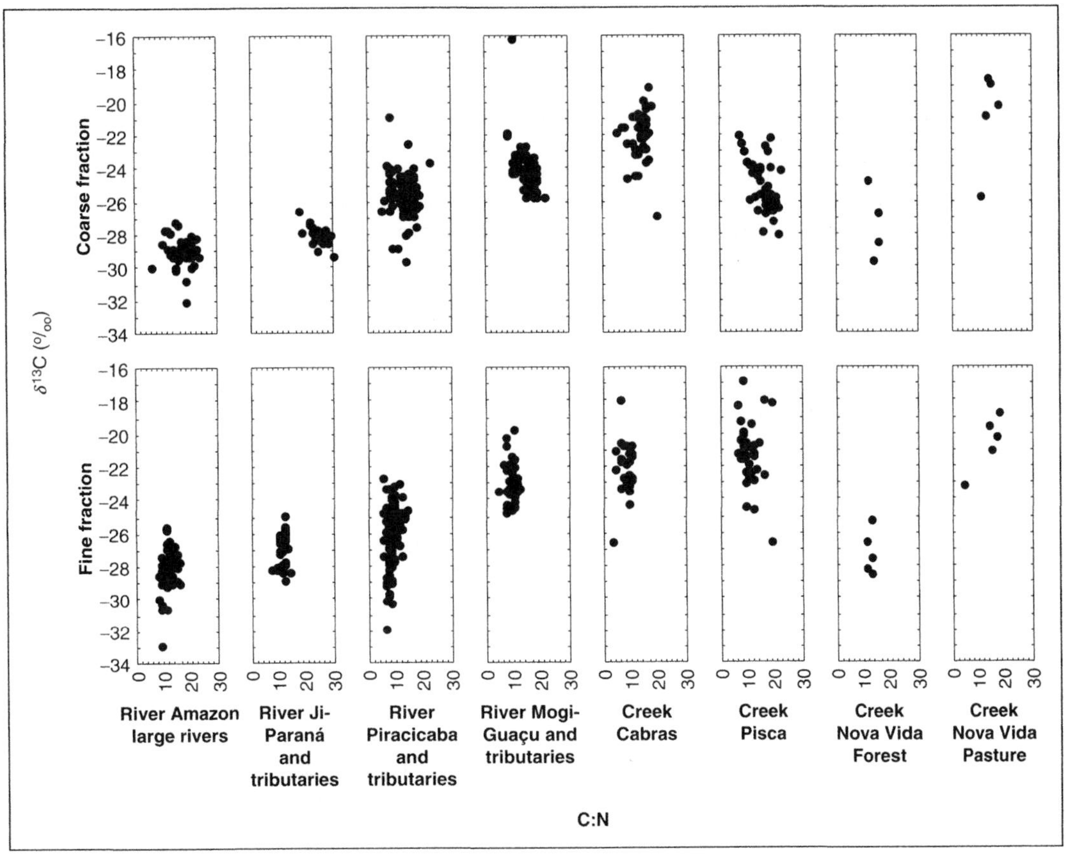

FIGURE 19.9 Relationship between C:N ratio and $\delta^{13}C$ of the fine and coarse POC of several rivers.

Helie *et al.* 2002). Because the soils of the Amazon region are generally very old, weathered, and deep, it can be assumed that there are no more carbonates left in the soil profile. On the other hand, it is well known that Amazon soils are rich in silicates that produce $\delta^{13}C$-DIC similar to the soil organic matter. The weathering of silicates usually produce DIC with an isotopic composition similar to that of the soil organic matter (Amiotte-Suchet *et al.* 1999, Telmer and Veizer 1999, Helie *et al.* 2002, Finlay 2003).

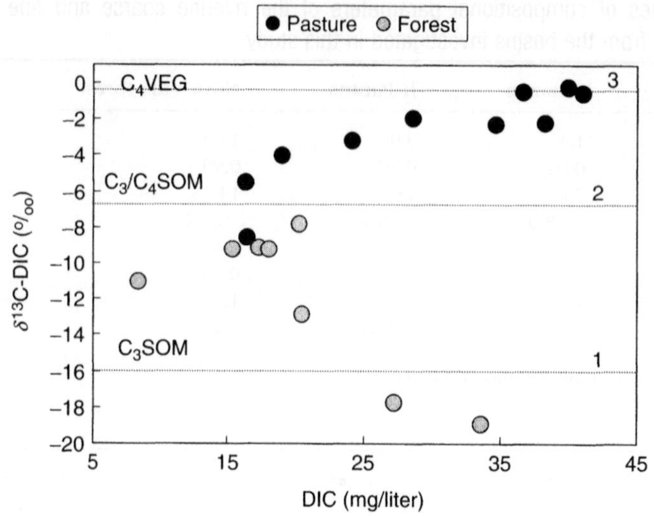

FIGURE 19.10 Relationship between DIC concentration and $\delta^{13}C$ of the DIC in waters collected in the forest reach stream and in the pasture reach stream at Fazenda Nova Vida Ranch. Line 1 indicates the $\delta^{13}C$-DIC produced by decomposition of a soil organic matter under dominium of C_3 plants. Line 2 indicates the $\delta^{13}C$-DIC produced by decomposition of a soil organic matter under a cover composed by a mixture of C_4 and C_3 plants. Line 3 indicates the $\delta^{13}C$-DIC produced by decomposition of a soil organic matter under dominium of C_4 plants.

On the other hand, the weathering of carbonates, depending of the type of weathering (closed × open systems), usually produces DIC with an isotopic composition between the isotopic composition of the carbonates and the soil organic matter (Yang *et al.* 1996, Amiotte-Suchet *et al.* 1999, Telmer and Viezer 1999, Helie *et al.* 2002, Barth *et al.* 2003, Finlay 2003).

The soil CO_2 derived from respiration should reflect the signal of the soil organic matter (Wang and Veizer 2000) because there is a small fractionation between the organic matter respired and the CO_2 produced. The diffusion of this gas in the soil may cause a fractionation up to 4.4‰ as suggested by Amiotte-Suchet *et al.* (1999). The isotopic signal of soil organic matter in the rivers basins varied according to the proportion of C_3 and C_4 plants present in the basin. Soil organic matter in basins under pure C_3 cover has a surface $\delta^{13}C$ value of ∼−30‰ to −28‰ (Figure 19.8). Assuming a $\delta^{13}C$ average value of −29‰ for soils under C_3 vegetation and considering a 4‰ fractionation for gas diffusion, it would produce a CO_2 gas in the soil with a $\delta^{13}C$ of −25‰. Assuming the fractionation factors between DIC and C species presented by Zhang *et al.* (1995) under a temperature of 25 °C, the $\delta^{13}C$-DIC produced by a soil organic matter composed only by C_3 plants would be equal to −17‰.

The $\delta^{13}C$ of soil organic matter samples under cultivated pastures in the Fazenda Nova Vida produced values from −16‰ to −23.5‰ (Cerri *et al.* 2004). Assuming an average of −20‰ and the gas diffusion fractionation of 4‰, the $\delta^{13}C$ of the gas in the soil would be equal to −16‰. Therefore, assuming the same value of fractionation from Zhang *et al.* (1995) used above, the $\delta^{13}C$-DIC produced by a soil organic matter composed by a mixture of C_3 and C_4 plants would be equal to −8‰.

In addition, we need to consider the decomposition of terrestrial C_4 grasses that float under the water surface and provide a complete cover of the stream in these reaches. In these cases, the source of organic matter to be decomposed and generate DIC has an average value of −12‰. By adding the gas diffusion fractionation of 4‰ and assuming the same fractionation factors between DIC and C species presented by Zhang *et al.* (1995) under a temperature of 25 °C assumed above, the $\delta^{13}C$-DIC produced by the terrestrial C_4 floating vegetation would be equal to ∼0‰.

It seems, therefore, that the main source of DIC from water samples collected in the pasture reach of the stream is a mixture of C_3–C_4 material. As the DIC increases, the influence of the C_4 material also increases. At the highest DIC concentration, C_4 material appears to be the only source of C to the dissolved inorganic fraction (Figure 19.10). It is interesting to note that in the terra firme forests stream reach, at the highest DIC concentration, C_3 material appears to be the main source of C. At lower DIC concentration, the δ^{13}C-DIC became less negative. Two processes that may lead to this change are equilibrium of the riverine DIC with the atmosphere, assuming a δ^{13}C-CO_2 of the atmosphere equal to −7‰ to −8‰ (Helie *et al.* 2002, Finlay 2003), or the effect of photosynthesis that would tend to enrich the DIC with ^{13}C atoms (Atekwana and Krishnamurthy 1998, Wang and Veizer 2000). Up to now we do not have enough information to further explore these two possibilities.

IV. STABLE ISOTOPE COMPOSITION OF RESPIRED CO_2

The first set of data obtained on a regular basis attempting to show a temporal variation of the isotopic values for the respired CO_2 (δ^{13}Cr) in the Amazon basin was presented by Ometto *et al.* (2002). This study was done in two terra firme forest sites, one near Manaus and the other near the city of Santarém. The authors used the Keeling Plot's intercept technique to evaluate the variation of ecosystem respiration, obtaining a good indication of the dominant C in the system being oxidized into CO_2, both in primary forests and cultivated pastures (Figure 19.11). This technique is particularly useful to analyze the relative contribution of C_3 and C_4 plants for the respired C in an ecosystem.

The δ^{13}C-CO_2 of a primary terra firme forest in Santarém site varied from −29.3‰ to −26‰. There was a sharp decrease of the δ^{13}C-CO_2 with the increase of monthly precipitation reaching a minimum value of −29.3‰ when the precipitation reached 450 mm per month. Subsequently, the δ^{13}C-CO_2 increased to near −27‰ when the monthly precipitation decreased to 300 mm. In contrast, there was no significant seasonal variation in δ^{13}C-CO_2 at the Manaus terra firme forests site (average δ^{13}Cr ~−28‰), consistent with a narrower range of variation in monthly precipitation compared to a more seasonal pattern that is observed at the Santarém site (Figure 19.12).

In spite of such seasonal variability in Santarém, the average δ^{13}C-CO_2 between Santarém and Manaus was similar to the δ^{13}C values of the foliage located in the upper canopy of these two terra firme forests. This pattern suggested that the major portion of recently respired carbon dioxide in these terra firme forests was the metabolized carbohydrate fixed by the sun leaves at the top of the forest canopy (Ometto *et al.* 2002).

As expected, the δ^{13}C-CO_2 of the pastures was enriched in ^{13}C compared to the forest ecosystems (Figure 19.13). The average δ^{13}C-CO_2 in the pastures was near −16‰ both in Manaus and Santarém, indicating that most of the organic matter being respired in these systems came from C_4 grasses (Ometto *et al.* 2002). After the pasture site in Manaus was burned, there was an increase in the dominance of C_4 grasses, indicated by an increase in the δ^{13}C-CO_2 (Figure 19.13). The fire might have removed from the pasture much of the encroached C_3 shrub vegetation (Ometto *et al.* 2002).

A more extensive dataset was recently obtained in the Santarém region (Figure 19.14). From March, 1999 to September, 2004, the δ^{13}C-CO_2 of the primary terra firme forests varied from almost −32‰ up to −26‰ with an average of −28.8 ± 1.3‰ ($n = 33$) (Figure 19.14). As expected, the average δ^{13}C-CO_2 of the pasture was higher, averaging −20.6 ± 2.8‰ ($n = 25$) for the entire period. However, from March, 1999 to March, 2002, the δ^{13}C-CO_2 values of the pasture (δ^{13}C-CO_2 = −17.9‰) were higher than the following period, from June, 2002 to September, 2004 (δ^{13}C-CO_2 = −22.7‰) (Figure 19.14). It is difficult to determine what happened from March to June in terms of pasture management, but the results indicated that the proportion of C_3-encroached shrubs increased, and the respiration of this organic matter led to lower δ^{13}C values (Figure 19.14).

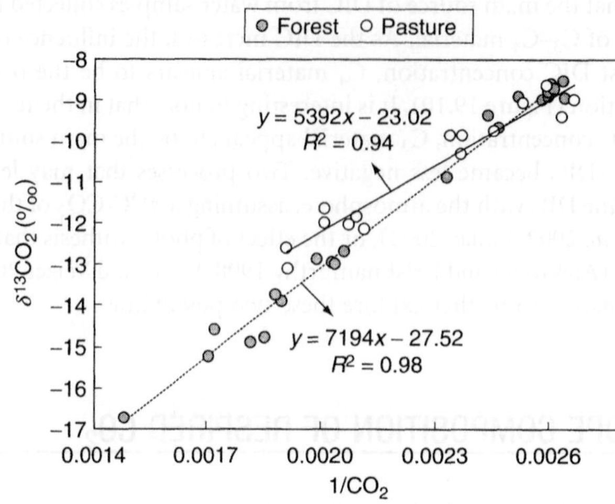

FIGURE 19.11 Examples of Keeling plots obtained in the terra firme forest and a pasture area near Santarém. Adapted from Ometto *et al.* (2002).

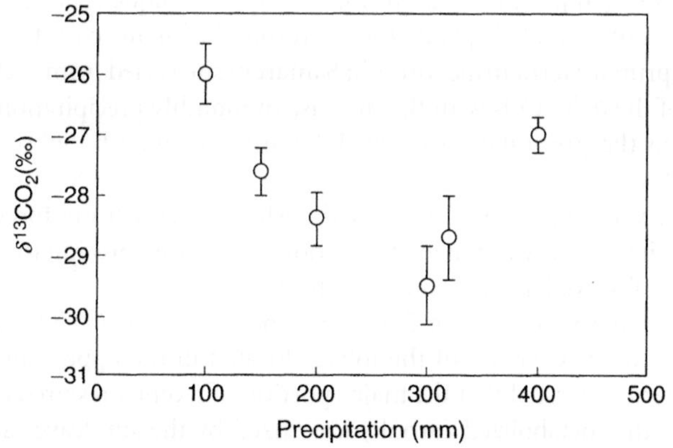

FIGURE 19.12 Monthly average $\delta^{13}C$-CO_2 in a primary terra firme forest of Santarém versus the monthly average precipitation. Adapted from Ometto *et al.* (2002).

V. CONCLUSIONS

We have tracked changes in the ecosystems structure and functioning associated with the conversion of forests to pastures in the Amazon region using stable isotopes. Figure 19.15 summarizes integrated values of $\delta^{13}C$ of different C compartments and how the conversion of terra firme forest to pasture changes their isotopic signal.

We observed that the recently introduced C_4 material is already present in all compartments and in different C fractions, like in the POC and DIC of small streams from regions where conversion of terra firme forest to pasture were observed. In the aquatic systems, several changes in the functioning of small stream have already been detected. The labile C_4 organic material increased the respiration rates

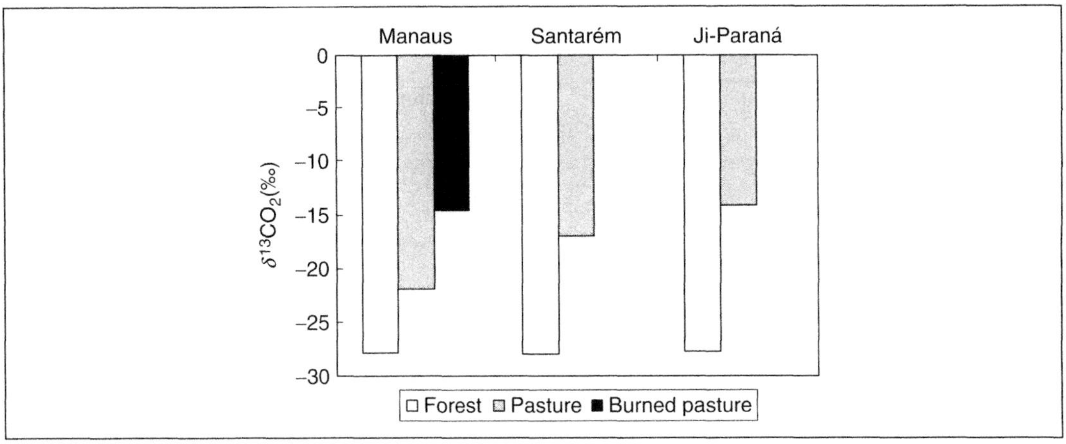

FIGURE 19.13 The δ^{13}C-CO_2 of terra firme forest and pastures from Manaus, Santarém, and Ji-Paraná.

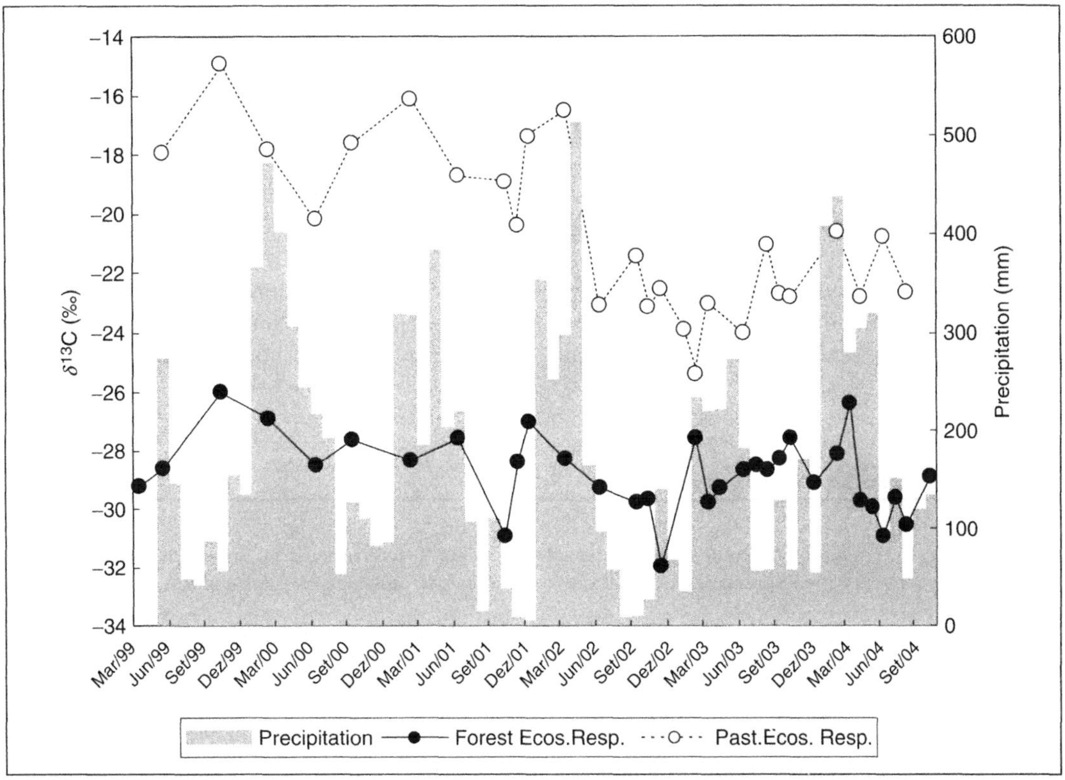

FIGURE 19.14 Temporal variation of the δ^{13}C-CO_2 (δ^{13}Cr) from the terra firme forest and pasture sites near the city of Santarém. The bars in the secondary axis represent monthly precipitation measured in Santarém during the period studied.

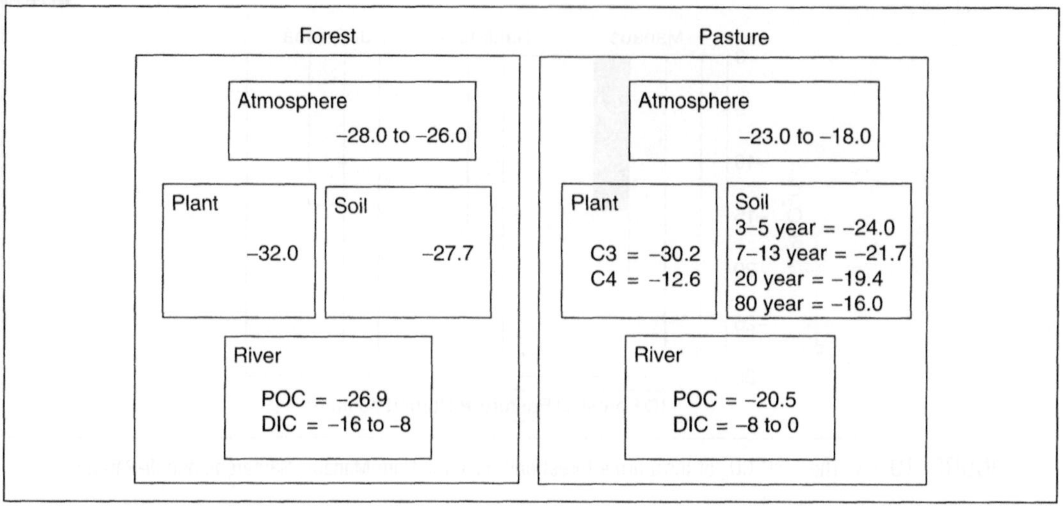

FIGURE 19.15 Conceptual framework of the $\delta^{13}C$ values of the main C reservoirs in a terra firme forest and in a pasture in the Amazon region.

in pasture streams, increasing the DIC concentration and decreasing the dissolved oxygen concentration. It is crucial to survey if these changes are propagating to streams of higher orders. In addition, changes in the aquatic biodiversity should be evaluated because changes of this magnitude in the functioning of the system most certainly will affect the distribution and composition of species.

It is also interesting to note that the soil chronosequence clearly showed that more C_4 material is incorporated to the soil organic matter as the pasture becomes older. The C_4 signal imprinted in the soil organic matter will allow us to build future scenarios of the fate of the soil organic matter with the rapid land cover changes taking place in that region. Such modeling has already started and can be seen in Cerri *et al.* (2004).

Finally, an average change of 10‰ in the respired CO_2 of the vegetation denotes an increase of the C_4 grasses component in the system. Such a change is probably altering the signal that the "Amazon entity" is sending to the atmosphere and this trend has to be incorporated in future modeling exercises.

VI. REFERENCES

Amiotte-Suchet, P., D. Aubert, J. L. Probst, F. Gauthier-Lafaye, A. Probst, F. Andreux, and D. Viville. 1999. $\delta^{13}C$ pattern of dissolved inorganic carbon in a small granitic catchment: The Strenbach case study (Vosges mountains, France). *Chemical Geology* **159:**129–145.

Atekwana, E. A., and R. V. Krishnamurthy. 1998. Seasonal variations of dissolved inorganic carbon and $\delta^{13}C$ of surface waters: Application of a modified gas evolution technique. *Journal of Hydrology* **205:**265–278.

Barth, J. A. C., A. A. Cronin, J. Dunlop, and R. M. Kalin. 2003. Influence of carbonates on the riverine carbon cycle in an anthropogenically dominated catchment basin: Evidence from major elements and stable carbon isotopes in the Lagan River (N. Ireland). *Chemical Geology* **200:**203–216.

Bernardes, M. C., L. A. Martinelli, A. V. Krusche, J. Gudeman, M. Z. Moreira, R. L. Victoria, J. P. H. B. Ometto, M. V. R. Ballester, A. K. Aufdenkampe, J. E. Richey, and J. I. Hedges. 2004. Riverine organic matter composition as a function of land use changes, southwest Amazon. *Ecological Applications* **14**(4):263–279.

Bustamante, M. M. C., G. B. Nardoto, and L. A. Martinelli. 2004a. Aspectos Comparativos Del Ciclaje De Nutrientes Entre Bosques Amazónicos De Terra-Firme Y Sabanas Tropicales (Cerrado Brasileiro). Pages 189–205 *in* Hernán Marino Cabrera (Ed.) *Fisiología Ecológica En Plantas. Mecanismos y Respuestas a Estrés en los Ecosistemas.* Ediciones Universitárias de Valparaiso, Pontificia Universidad Católica de Valparaíso, Valparaiso, Chile.

Bustamante, M. M., L. A. Martinelli, D. A. Silva, P. B. Camargo, C. A. Klink, T. F. Domingues, and R. V. Santos. 2004b. ^{15}N natural abundance in woody plants and soils of the savanna in Central Brazil (Cerrado). *Ecological Applications* **14**(4):200–213.

Bustamante, M. M. C., E. Medina, G. P. Asner, G. B. Nardoto, and D. C. Garcia-Montiel. 2006. Nitrogen cycling in tropical and temperate savannas. *Biogeochemistry* **79:**209–237.

Cerri, C. E. P., K. Paustian, M. Bernoux, R. L. Victoria, J. M. Melillo, and C. C. Cerri. 2004. Modeling changes in soil organic matter in Amazon terra firme forests to pasture conversion with the century model. *Global Change Biology* **10:**815–832.

Clark, D. B., and D. A. Clark. 1996. Abundance, growth and mortality of very large trees in neotropical lowland rain forest. *Forest Ecology and Management* **80:**235–244.

Cuevas, E. 2001. Soil versus biological controls on nutrient cycling in terra firme forests. Pages 53–67 *in* M. E. McClain, R. L. Victoria, and J. E. Richey (Eds.) *The Biogeochemistry of the Amazon Basin.* Oxford University Press, New York.

Eiten, G. 1972. The cerrado vegetation of Brazil. *Botanical Reviews* **38:**201–341.

Elsenbeer, H., and A. Lack. 1997. Hydrological pathways and water chemistry in Amazonian rain forests. Pages 939–959 *in* M. G. Anderson and S. M. Brooks (Eds.) *Advances in Hillslope Processes,* Vol. 2. Wiley, New York.

Finlay, J. C. 2003. Controls of streamwater dissolved inorganic carbon dynamics in a forested watershed. *Biogeochemistry* **62:**231–252.

Furley, P. A., and J. A. Ratter. 1988. Soil resources and plant communities of the Central Brazilian cerrado and their development. *Journal of Biogeography* **15:**97–108.

Helie, J. F., C. Hillaire-Marcel, and B. Rondeau. 2002. Seasonal changes in the sources and fluxes of dissolved inorganic carbon through the St. Lawrence River—isotopic and chemical constraint. *Chemical Geology* **186**(1–2):117–138.

Högberg, P. 1997. Tansley review no. 95: ^{15}N natural abundance in soil-plant systems. *New Phytologist* **137:**179–203.

Liebmann, B., and J. A. Marengo. 2001. Interannnual variability of the rainy season and rainfall in the Brazilian Amazon Basin. *Journal of Climate* **14:**4308–4318.

Martinelli, L. A., L. C. R. Pessenda, E. Espinoza, P. B. Camargo, E. C. Telles, C. C. Cerri, R. L. Victoria, R. Aravena, J. Richey, and S. Trumbore. 1996. Carbon-13 depth variation in soil of Brazil and relations with climate changes during the Quaternary. *Oecologia* **106:**376–381.

Martinelli, L. A., M. C. Piccolo, A. R. Townsend, P. M. Vitousek, E. Cuevas, W. Mcdowell, G. P. Robertson, O. C. Santos, and K. Treseder. 1999. Nitrogen stable isotopic composition of leaves and soil: Tropical versus temperate forests. *Biogeochemistry* **46:**45–65.

Neill, C., J. M. Melillo, P. A. Steudler, C. C. Cerri, J. F. L. de Moraes, M. C. Piccolo, and M. Brito. 1997. Soil carbon and nitrogen stocks following forest clearing for pasture in the southwestern Brazilian Amazon. *Ecological Applications* **7** (4):1216–1225.

Neill, C., L. A. Deegan, S. M. Thomas, and C. C. Cerri. 2001. Stream characteristics and water chemistry following deforestation for pasture in small Amazonian watersheds. *Ecological Applications* **11**(6):1817–1828.

Ometto, J. P. H. B., L. Flanagan, L. A. Martinelli, M. Z. Moreira, N. Higuchi, and J. Ehleringer. 2002. Carbon isotope discrimination in forest and pasture ecosystems of the Amazon Basin, Brazil. *Global Biogeochemical Cycles* **16** (4):1109–1123.

Ometto, J. P. H. B., A. D. Nobre, H. R. Rocha, P. Artaxo, and L. A. Martinelli. 2005. Amazonia and the modern carbon cycle: Lessons learned. *Oecologia* **143**(4):483–500.

Ometto, J. P. H. B., J. R. Ehleringer, T. F. Domingues, J. A. Berry, F. O. Y. Ishida, E. Mazzi, N. Higuchi, L. B. Flanagan, G. B. Nardoto, and L. A. Martinelli. 2006. The stable carbon and nitrogen isotopic composition of vegetation in tropical forests of the Amazon Basin, Brazil. *Biogeochemistry* **79:**251–274.

Pitman, N. C. A., J. W. Terborgh, M. R. Silman, P. Nunez, D. A. Neill, C. E. Ceron, W. A. Palacios, and M. Aulestia. 2001. Dominance and distribution of tree species in upper Amazonian terra firme forests. *Ecology* **82**(8):2101–2117.

Ratter, J. A., J. F. Ribeiro, and S. Bridgewater. 1997. The Brazilian Cerrado vegetation and threats to its biodiversity. *Annals of Botany* **80:**223–230.

Ratter, J. A., S. Bridgewater, and J. F. Ribeiro. 2003. Analysis of the floristic composition of the Brazilian Cerrado vegetation III: Comparison of the woody vegetation of 376 areas. *Edinburgh Journal of Botany* **60**(1):57–109.

Reis, V. M., F. B. dos Reis, D. M. Quesada, O. C. A. de Oliveira, B. J. R. Alves, S. Urquiaga, and R. M. Boddey. 2001. Biological nitrogen fixation associated with tropical pasture grasses. *Australian Journal of Plant Physiology* **28**(9):837–844.

Richter, D. D., and L. I. Babbar. 1991. Soil diversity in the tropics. *Advances in Ecological Research* **21:**315–389.

Sanaiotti, T., L. A. Martinelli, R. L. Victoria, S. E. Trumbore, and P. B. Camargo. 2002. Past vegetation changes in Amazon savannas by using carbon isotopes of soil organic matter. *Biotropica* **34:**2–16.

Scholes, R. J., and S. R. Archer. 1997. Tree–grass interactions in savannas. *Annual Reviews in Ecological Systems* **28:**517–544.

Sombroek, W. 2001. Spatial and temporal patterns of Amazon rainfall—consequences for the planning of agricultural occupation and the protection of primary forests. *Ambio* **30**(7):388–396.

Telmer, K., and J. Veizer. 1999. Carbon fluxes, pCO and substrate weathering in a large northern river basin, Canada: Carbon isotope perspectives. *Chemical Geology* **159:**61–86.

Thomas, S. M., C. Neil, L. A. Deegan, A. V. Krusche, M. V. Ballester, and R. L. Victoria. 2004. Influences of land use and stream size on particulate and dissolved materials in a small Amazonian stream network. *Biogeochemistry* **68**(2):135–151.

Vitousek, P. M., K. Cassman, C. Cleveland, T. Crews, C. B. Field, N. B. Grimm, R. W. Howarth, R. Marino, L. Martinell, E. B. Rastetter, and J. Sprent. 2002. Towards an ecological understanding of biological nitrogen fixation. *Biogeochemistry* **48:**1–45.

Yang, C., T. Kevin, and J. Veizer. 1996. Chemical dynamics of the "St. Lawrence" riverine system: δDH_2O $\delta^{18}OH_2O$ $\delta^{13}CDIC$, δ^{34}Ssulfate and dissolved $^{87}Sr/^{86}Sr$. *Geochimica et Cosmochimica Acta* **60**(5):851–866.

Wang, X. F., and J. Veizer. 2000. Respiration-photosynthesis balance of terrestrial aquatic ecosystems, Ottawa area, Canada. *Geochimica et Cosmochimica Acta* **64**(22):3775–3786.

Zhang, J., P. D. Quay, and D. O. Wilbur. 1995. Carbon isotope fractionation during gas-water exchange and dissolution of CO_2. *Geochimica et Cosmochimica Acta* **59**(1):107–114.

CHAPTER 20

Reconstruction of Climate and Crop Conditions in the Past Based on the Carbon Isotope Signature of Archaeobotanical Remains

Juan Pedro Ferrio,* Jordi Voltas,* Natàlia Alonso,† and José Luis Araus‡

*Department of Crop and Forest Sciences, University of Lleida
†Department of History, University of Lleida
‡Department of Plant Biology, University of Barcelona

Contents

I. OVERVIEW

During the last 10,000 years, agriculture has been adopted in many parts of the world as a result of independent processes: so far, archaeological evidence has been found for plant domestication in the Near East, Ethiopia, sub-Saharan Africa, China, Mesoamerica, and South America (Harlan 1998). Agricultural societies appeared a few millennia after the retreat of Pleistocene ice, spreading during what had been the most stable warm period in almost half a million years: the Holocene. Therefore, the origins and subsequent expansion of agriculture cannot be isolated from the particular environmental conditions in which it evolved. Conversely, agriculture has modified radically the social and demographic structure of human groups, as well as their interaction with the environment. Its adoption is probably the first example of reciprocal interaction between the environment and the humanity. Thus, reconstructing the environmental conditions (either climatic or anthropogenic) that characterized this process is of great interest in order to assess the potential causes behind its embracing, as well as to understand the long-term effects of agriculture economy on the environment.

In this chapter, we will describe a methodology used to infer climate and growing conditions in ancient agriculture from the analysis of carbon isotope composition ($\delta^{13}C$) in carbonized plant remains (wood charcoal and charred seeds) routinely recovered in the course of archaeological excavations. After a general introduction about the characteristics of archaeobotanical remains, the chapter illustrates the approach followed by means of different case studies showing the application of carbon isotopes in such material as a paleoenvironmental tool. We will focus on Mediterranean agriculture (Figure 20.1) because it is, up to now, one of the most extensively studied areas in the field of archaeobotany and main crops are C_3 plants, which show strong climate responses as reflected in leaf $\delta^{13}C$ (Chapters 1–3). Moreover, crop performance in this region is strongly limited by water availability and thus the suitability of using $\delta^{13}C$ to infer plant water status is of particular interest in such geographic context.

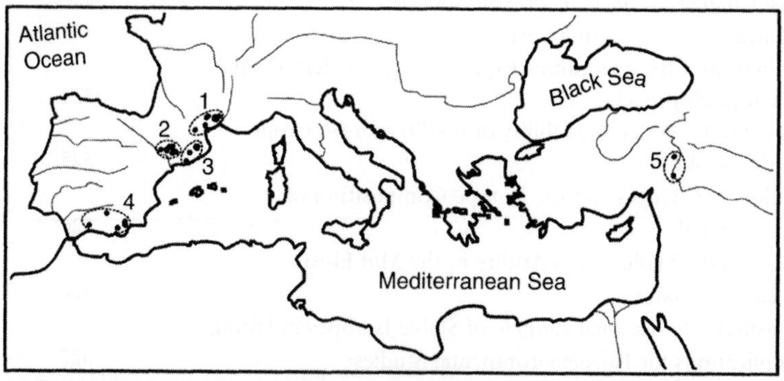

FIGURE 20.1 Map of the Mediterranean region showing the origin of the archaeological data discussed in this work (encircled black dots). 1. Languedoc (SE France); 2 and 3. Mid Ebro Basin and Eastern Catalonia, respectively (NE Spain); 4. Eastern Andalusia (SE Spain); and 5. Middle Euphrates (SE Turkey/NW Syria).

II. BRIEF PHYSIOLOGICAL BACKGROUND

There are two stable carbon isotopes in the air (^{12}C and ^{13}C). During photosynthesis, plants take-up preferentially ^{12}C instead of ^{13}C. The magnitude of this event is affected by several physiological processes, and is expressed in terms of carbon isotope discrimination ($\Delta^{13}C$), as defined by Farquhar *et al.* (1982):

$$\Delta^{13}C = \frac{\delta^{13}C_{air} - \delta^{13}C_{plant}}{(1 + \delta^{13}C_{plant}/1000)} \tag{20.1}$$

where $\delta^{13}C_{air}$ and $\delta^{13}C_{plant}$ denote air and plant $\delta^{13}C$, respectively. In C_3 plants (*e.g.*, most tree species and winter cereals), $\Delta^{13}C$ in plant tissues constitutes an integrated record of the ratio of intercellular to atmospheric concentration of CO_2 (c_i/c_a) during the period in which carbon atoms were fixed, which in turn reflects the balance between assimilation rate (A) and stomatal conductance (g_s). As thoroughly discussed in other chapters from this volume, plants typically react against water stress through stomata closure and, although A may also decline as a result of stomatal limitation to photosynthesis, g_s is usually affected to a larger extent, originating a reduction in $\Delta^{13}C$. This is the basis for the extensively reported relationships between $\Delta^{13}C$ and environmental variables related to plant water status, such as precipitation, relative humidity, or potential evapotranspiration (Chapter 3).

III. THE STUDY OF ARCHAEOLOGICAL PLANT REMAINS: ARCHAEOBOTANY

In the process of excavating archaeological sites one comes across plant matter of different kinds. Exceptionally, plant remains can be preserved in desiccated form inside dry caves or in dessert environments, as well as in very cold sites with permafrost and in relatively fresh state within waterlogged (anaerobic) sediments. However, in temperate and semiarid climates, the usual way of preservation of plant material is through carbonization (Renfrew 1973, Van Zeist and Casparie 1984). When plant materials are combusted under poor oxygen availability they become carbonized, which prevents fungal or bacterial decay, while keeping identifiable morphological traits. During this process, however, soft plant parts (*e.g.*, leaves) are usually lost, and thus the most common archaeobotanical remains are wood charcoal and charred seeds.

Although some early studies date from the end of the nineteenth century, the interest on plant remains did not began to generalise until the works of Helbaek (Helbaek and Schultze 1981) in the Near East. Since then, and due to its particular relevance for the origin of Mediterranean agriculture, the Near East is still the region most extensively studied from archaeobotany (Renfrew 1973, Van Zeist and Casparie 1984).

A. Recovery of Plant Remains

Small charred seeds and charcoal fragments are not easily identified by the naked eye and, as a consequence, specialized means are necessary to obtain them. The most common method by which these smaller macrobotanical plant fragments are concentrated and recovered is through the technique known as flotation. Many different flotation systems have been devised, but all of them are based on the same principle: if archaeological sediment is released into a container filled with water, then the sediment sinks and the charred plant remains float (Renfrew 1973, Van Zeist and Casparie 1984). Subsequently, floating seeds and charcoal fragments are retrieved with a submerged sieve and, once dried, stored and identified.

B. Dating of Samples

The dating of archaeobotanical remains is not a single issue. The radiocarbon dating method, based on the rate of decay of the unstable isotope ^{14}C, is today the most widely applied dating technique for the Late Pleistocene and Holocene periods (Taylor and Aitken 1997). Radiocarbon dating, however, must be calibrated externally to account for past changes in atmospheric ^{14}C composition, and the final accuracy varies from several hundred years to a few decades, depending on the period considered (Stuiver *et al.* 1998). Besides, the elevated cost of these analyses strongly limits the amount of samples to be dated. Consequently, most plant macroremains are assigned an age according to the dating obtained from other charcoals, seeds, or bones from the same (or the nearest) stratigraphic layer. In most cases, a combination of stratigraphic and archaeological methods (*e.g.*, based on ceramic or tool styles) can provide more accurate dating than radiocarbon analyses alone.

IV. ANALYSIS OF $\delta^{13}C$ IN ARCHAEOBOTANICAL REMAINS: METHODOLOGICAL ASPECTS

A. Removal of Soil Contaminants

Plant material recovered from archaeological sediments often carries soil substances which might alter $\delta^{13}C$. The most common contaminants are carbonate precipitates and humic/fulvic acids, although their amounts strongly depend on soil conditions. De Niro and Hastorf (1985), after evaluating different methods, provided a standard protocol for contaminant removal without altering the charred plant material: carbonate crusts and humic/fulvic acids are usually removed by soaking samples with 6-M HCl for 24 h at room temperature, rinsing to neutrality, soaking in 1-M NaOH for 24 h, rinsing and then soaking again in 6-M HCl for 10 min.

B. Past Changes in the $\delta^{13}C$ of Atmospheric CO_2

The source of carbon for most terrestrial plants is atmospheric CO_2. Consequently, changes in the $\delta^{13}C$ of atmospheric CO_2 ($\delta^{13}C_{air}$) are somehow reflected in plant tissues [Eq. (20.1)]. As $\delta^{13}C_{air}$ is not constant over time (*e.g.*, currently due to fossil fuel emissions), we need to estimate past $\delta^{13}C_{air}$ before aiming to compare modern and archaeological data, as well as fossil samples from different ages. At the local scale, the analysis of $\delta^{13}C$ plant remains from either C_4 plants (Marino *et al.* 1992, Toolin and Eastoe 1993) or peat mosses (White *et al.* 1994) might be used as a proxy for ancient $\delta^{13}C_{air}$. This method assumes $\Delta^{13}C$ constancy for these species, regardless of environmental conditions, which has been denied by several studies (Buchmann *et al.* 1996, Rice and Giles 1996). Alternatively, Antarctic ice cores provide a direct and extensive record of global atmospheric changes, but they cannot account for local variations. Nevertheless, we consider ice core data as the most suitable reference value, up to now, for $\delta^{13}C_{air}$ in archaeological material. In Ferrio *et al.* (2006a), we fitted a $\delta^{13}C_{air}$ curve covering the whole Holocene by interpolating data from Antarctic ice cores (see references in Chapter 14), together with modern data from the two Antarctic stations of the CU-INSTAAR/NOAA-CMDL network for atmospheric CO_2 (ftp://ftp.cmdl.noaa.gov/ccg/co2c13/flask/readme.html).

C. Changes in $\delta^{13}C$ During Carbonization

As pointed out in Section III, most archaeobotanical remains are found in carbonized state. Carbonization prevents microbial degradation, yet the process might involve significant shifts in the original plant $\delta^{13}C$. DeNiro and coworkers (De Niro and Hastorf 1985, Marino and De Niro 1987) performed detailed studies on the effect of temperature and carbonization in several types of seeds and plant parts, finding little change in $\delta^{13}C$ due to carbonization and chemical treatments. Similarly, Araus *et al.* (1997b) found that the $\delta^{13}C$ of wheat grains did not vary significantly when carbonized at 400 °C over different times (15–120 min) and under two atmospheric conditions (aerobic and anaerobic). However, temperatures over 300 °C are unrealistic as grains become so brittle and distorted that they would not be identified in the course of archaeological excavations (Ferrio *et al.* 2004). In another experiment (Ferrio, Voltas, Araus, unpublished results), we compared the ^{13}C signal of intact grains cultivated under a wide range of environmental conditions across Spain with their corresponding values once carbonized at 250 °C under both aerobic and anaerobic conditions (Figure 20.2A). The slope and intercept of the relationship between intact and carbonized grains did not differ significantly from unity and zero, respectively, further suggesting that $\delta^{13}C$ in cereal grains remains unaffected after charring.

Unlike for cereal grains, the effect of carbonization on wood tissues might be considerable and shows several uncertainties. This is probably due to the extraordinary chemical complexity of wood if compared to that of cereal grains (mostly composed of carbohydrates). Although numerous studies on $\delta^{13}C$ of coals and subfossil wood have been carried out, especially for deep geological times (see references in Gröcke 1998),

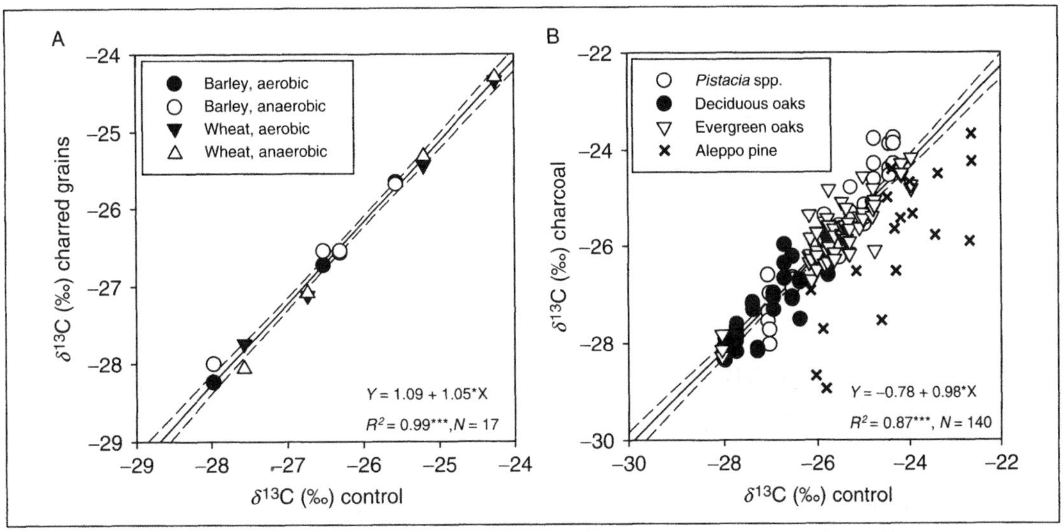

FIGURE 20.2 (A) Relationship between $\delta^{13}C$ values in control (intact) and experimentally carbonized (250 °C) grains of wheat (*Triticum aestivum* L.) and barley (*Hordeum vulgare* L.). From Ferrio, Voltas, and Araus (unpublished data). (B) Relationship between $\delta^{13}C$ values in control and experimentally carbonized (300–500 °C, following Ferrio *et al.* 2006) wood samples of several *Quercus* (evergreen: *Q. ilex, Q. calliprinos* and deciduous: *Q. brantii, Q. cerris*) and *Pistacia* species (*P. atlantica, P. vera, P. palestina, P. lentiscus*). Original data for Aleppo pine from Ferrio *et al.* (2006) and from Alonso, Ferrio, and Voltas (unpublished data) for the rest of species.

those focused on wood charcoal are still scarce. According to a variety of morphological and physical variables (Jones and Chaloner 1991, Guo and Bustin 1998, Edwards and Axe 2004, Ferrio *et al.* 2006a), the usual range of temperatures for charcoal formation appears to be relatively restricted in the fossil record (around 350–450 °C) as a compromise between increasing resistance to chemical/biological degradation and decreasing toughness. Some works performing experimental carbonization in wood fragments (Jones and Chaloner 1991, Czimczik *et al.* 2002, Ferrio *et al.* 2006a) have shown that wood charred at low temperatures (150–300 °C) becomes slightly enriched in $\delta^{13}C$, but exhibits decreasing $\delta^{13}C$ values when carbonized at higher temperatures (300–600 °C). Consequently, $\delta^{13}C$ shifts during charcoal formation may vary with charring temperature, which in turn determines the degree of carbonization. Fortunately, the $\delta^{13}C$ of charcoal ($\delta^{13}C_C$) is still strongly related with the original wood $\delta^{13}C$ ($\delta^{13}C_W$) across a rainfall gradient in conifers such as *Pinus halepensis*, and changes due to carbonization can be successfully corrected by taking carbon concentration in charcoal ($\%C_C$) as an indicator of the degree of carbonization (Ferrio *et al.* 2006a):

$$\delta^{13}C_w = 0.706 \times \delta^{13}C_c + 0.031 \times \%C_c - 8.074 \quad N = 18, R^2 = 0.72, P < 0.001 \tag{20.2}$$

This remains to be tested for angiosperms, which have different chemical and physical wood properties. Experimental carbonizations performed on homogenized *Quercus* wood showed a progressive depletion in $\delta^{13}C$ with temperature, as occurred in conifers (Poole *et al.* 2004, Turney *et al.* 2006). However, we have found contrasting results after testing for the effect of carbonization at different temperatures on the $\delta^{13}C$ of wood fragments from several species of Mediterranean oaks and the genus *Pistacia* that had been sampled across a rainfall gradient (Figure 20.2B). Unlike in conifers, these fragments showed little changes in $\delta^{13}C$ after carbonization, despite reaching comparable levels of carbonization (up to 90% of carbon concentration). Moreover, the relationship between control and carbonized samples followed a 1:1 relationship (Figure 20.1B), suggesting that, for these species, $\delta^{13}C$ in charcoal might be directly comparable with the values of intact wood.

Overall, we can conclude that the environmental signal from wood is at least partly preserved in charcoal, but further work is still needed to understand the fate of isotope fractionation during carbonization. Combining stable isotopes with analytical techniques, such as pyrolysis gas chromatography–mass spectrometry (Py-GC–MS), nuclear magnetic resonance (NMR) or thermal analysis (TG-DMS) might provide the necessary clues regarding the chemical changes involved in this process (Czimczik *et al.* 2002, Poole *et al.* 2004, Lopez-Capel *et al.* 2005). However, up to now analytical studies were based on homogenized wood, often derived from a unique source, and thus did not account for between-tree or between-site natural variability. In this context, it would be desirable to perform surveys across environmental gradients with known $\delta^{13}C$ variability aimed to characterize changes due to carbonization relative to the degree of preservation of the original environmental signal.

V. AN INSIGHT INTO THE AGRONOMIC CONDITIONS IN ANCIENT CROPS: THE ISOTOPE APPROACH

Winter cereals, such as wheat and barley, are of particular relevance in the context of the origins of Mediterranean agriculture, as they constituted the "staple food" among the first cultivated crops in the Near East. Moreover, wheat and barley are C_3 species that show strong changes in $\delta^{13}C$ in response to environmental variables and, particularly, water status. This, together with a spreading interest in plant remains among archaeologists, provided the necessary background for the first attempts relating $\delta^{13}C$ of archaeological cereal grains to the water status of ancient crops (Araus and Buxó 1993, Araus *et al.* 1997a, 1999).

A. Case Study: Water Availability of Mediterranean Crops in the Neolithic

An example of the application of $\delta^{13}C$ to estimate past water availability in cereal crops is shown in Figure 20.3. Charred wheat and barley grains were recovered from several archaeological sites in NE and SE Spain (Araus and Buxó 1993, Araus *et al.* 1997a) and in two early neolithic sites from the Near East (Araus *et al.* 1999, 2007). To account for past changes in $\delta^{13}C_{air}$, $\Delta^{13}C$ in archaeological and modern samples was calculated following Eq. (20.1) after inferring $\delta^{13}C_{air}$ from ice core data. A model relating total water inputs (WI, rainfall + irrigation) and $\Delta^{13}C$ was fitted using data from crops covering a wide range of environmental conditions (Araus *et al.* 1997a, 1999, modified by Ferrio *et al.* 2005a):

$$WI_{barley}(mm) = 0.175 \times e^{(0.376 \times \Delta^{13}C)} \quad N = 22, R^2 = 0.73, P < 0.001 \tag{20.3}$$

$$WI_{wheat}(mm) = 0.225 \times e^{(0.364 \times \Delta^{13}C)} \quad N = 25, R^2 = 0.73, P < 0.001 \tag{20.4}$$

Finally, the models were applied to archaeobotanical $\Delta^{13}C$ data to estimate past WI. By comparing estimated values for archaeological material with current average values for precipitation during the grain filling period (approximately from April to the first half of May, Figure 20.3), we found a consistent decrease in water availability from neolithic to present, particularly evident for the driest sites such as those from SE Spain and the Near East. These results indicate that, during early agriculture, wheat and barley were cultivated under better water status than that expected from present-day (rainfed) conditions. Archaeobotanical evidence (based on species distribution) also supports the possibility that environmental conditions during early agriculture were cooler and moister than today both in the Near East (Willcox 1996, Harlan 1998) and Iberian Peninsula (Vernet and Thiebault 1987, Vernet 1990).

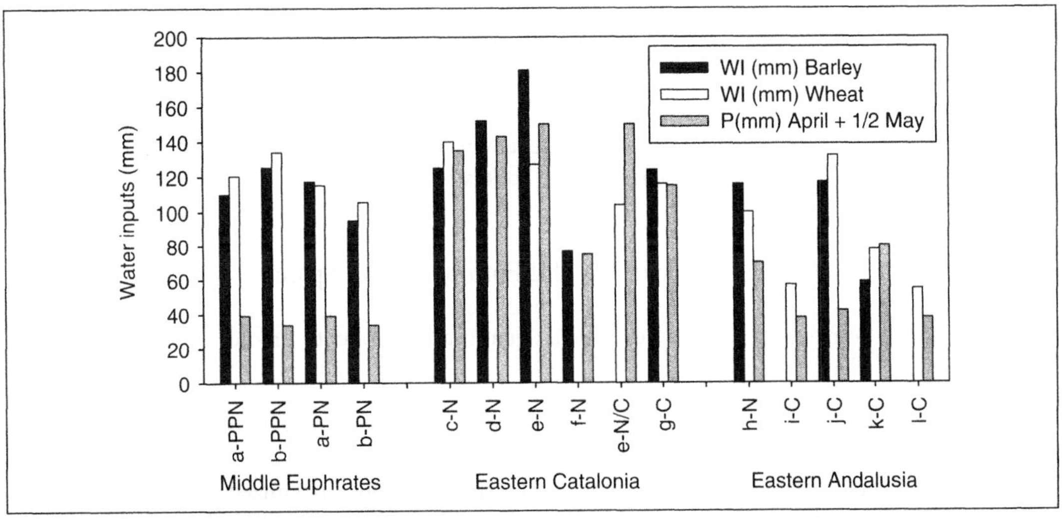

FIGURE 20.3 Water inputs (WI) during grain filling as estimated from the $\Delta^{13}C$ of wheat (*Triticum aestivum/durum*) and barley (*Hordeum* spp.) grains recovered from several archaeological sites in the two extremes of the Mediterranean Basin, compared with current average values for precipitation during the period of grain filling (approximately from April to the first half of May). Drawn from original data in Araus and Buxó (1993) and Araus *et al.* (1997a, 1999, 2007). Archaeological sites: a—Akarçay Tepe, b—Tell Halula, c—La Draga, d—Plansallosa, e—Cova 120, f—Can Tintorer, g—Institut Manlleu, h—Cueva del Toro, i—Campos, j—Millares, k—Malagón, l—Pilas. Cultural periods: PPN-Pre-Pottery Neolithic (ca. 8000–7000 BCE), PN-Pottery Neolithic (ca. 7000–6000 BCE), N-Late Neolithic (ca. 6000–3000 BCE), and C-Calcolithic (ca. 3000–2000 BCE).

Alternatively, some cultural practices, like irrigation or planting in naturally wet alluvial soils, may have enhanced the differences between neolithic and present samples (Bar-Yosef *et al.* 1989, Araus *et al.* 1997b, Ferrio *et al.* 2005a). Although the latter explanation might suffice for such a great change in water availability, it also points out to the need of an independent, local source of climatic information (not affected by changes in crop management) in order to rule out or otherwise substantiate a nonanthropogenic origin of environmental changes. For this purpose, wood charcoal appears as a good counterpart, as will be discussed later.

At this point, we should bear in mind that modern cultivars have been selected to attain better performance under optimal conditions, being less conservative in terms of water expense, and displaying shorter growing cycles (Araus *et al.* 2002). This translates into a lower stomatal limitation of photosynthesis, and thus a higher $\Delta^{13}C$ (with differences of up to 1‰) than traditional landraces (Fischer *et al.* 1998, Muñoz *et al.* 1998, Araus *et al.* 2007). Consequently, past WI displayed in Figure 20.3 might have been underestimated and, hence, the difference between past and present conditions would be even greater. Thus, it is recommendable now to expand reference $\delta^{13}C$ data with landraces and wild relatives grown under various environmental conditions (including the core areas for domestication) to aid solving this particular uncertainty.

VI. PAST CLIMATE CHANGES AND THE ISOTOPE COMPOSITION OF WOOD CHARCOAL

Tree-ring variability in wood or cellulose $\delta^{13}C$ has been successfully related to different environmental variables (Chapter 3). In seasonally dry climates, the most consistent associations can be found with variables determining plant water status such as precipitation or evaporative demand (see references in Ferrio *et al.* 2005b). Although some exceptionally long tree-ring chronologies cover almost the whole Holocene, vast regions of the world have been deforested for a long time, and thus long tree-ring records are no longer available. Some authors suggested in the 1990s the use of $\delta^{13}C$ from wood charcoal as alternative (February and Van der Merwe 1992, Vernet *et al.* 1996). Despite its lower temporal resolution (as compared with tree-ring chronologies) and the uncertainties associated with the carbonization event, the use of charcoal as paleoenvironmental indicator might be the best way to get local information on climate in archaeological contexts.

A. Case Study: Evolution of Aridity in the Mid Ebro Basin (NE Spain)

In a recent study, Ferrio *et al.* (2006a) developed a quantitative model to predict annual precipitation from the $\Delta^{13}C$ of Aleppo pine wood:

$$P_{an}(mm) = 119.9 \times \Delta^{13}C - 1510 \quad N = 29, R^2 = 0.79, P < 0.001 \qquad (20.5)$$

This model combined data from both intact and experimentally carbonized wood, as regression parameters did not differ significantly using either intact or carbonized wood. The resulting model was applied to a series of archaeological samples from the Mid Ebro Basin (NE Spain) in order to reconstruct the evolution of aridity in this area during the last 4000 years (Figure 20.4). This region is among the most arid zones in Europe (annual rainfall between 300 and 400 mm); however, it remains unclear whether present conditions are due to recent environmental changes or to a progressive aridification beginning in prehistoric times. According to Ferrio *et al.* (2006a), estimated water availability in the past was generally higher than present values throughout the period studied (from ca. 2000 BCE to the eighteenth century), indicating that latter-day (semiarid) conditions are mostly

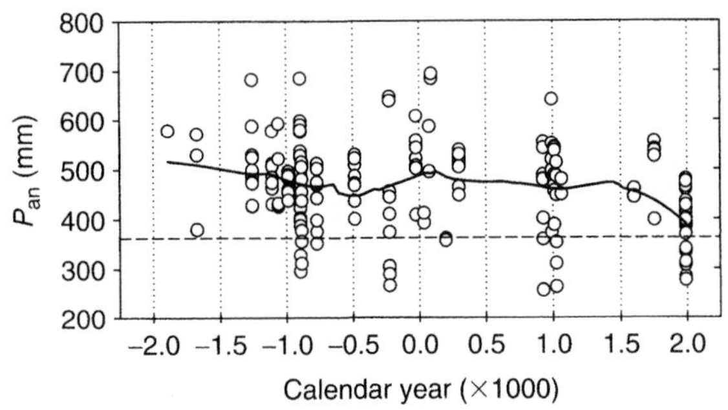

FIGURE 20.4 Evolution of estimated mean annual precipitation (P_{an}) in the Mid Ebro Basin (NE Spain). P_{an} was estimated from $\Delta^{13}C$ of charcoal remains of Aleppo pine recovered from archaeological sites. Drawn from original data in Ferrio *et al.* (2006a). The trend line depict a LOESS regression curve (Cleveland 1979) fitted to the data with *span* = 0.4. The dashed line indicates current average values for the sites.

due to recent climate changes. In addition to the overall differences between past and present-day values, two main phases of greater water availability (1500–900 BCE; 300 BCE–300 CE) alternating with drier periods (900–300 BCE; 900–1100 CE) were detected. Overall, these findings were supported by other paleoenvironmental evidences, mostly derived from pollen and geomorphologic studies (Davis 1994, Gutiérrez-Elorza and Peña-Monné 1998, Jalut *et al.* 2000, Riera *et al.* 2004). Moreover, the evolution of $\Delta^{13}C$ in charcoal remains from other co-occuring species, such as the Mastic tree (*Pistacia lentiscus* L.) and both evergreen and deciduous oaks (*Quercus* spp.), is consistent with the results obtained for Aleppo pine (Figure 20.5). Interestingly, even the $\delta^{13}C$ values reported by Vernet *et al.* (1996) for archaeological sites from SE France showed similar trends (Figure 20.5D).

B. Sensitivity to Seasonal Climate of Stable Isotopes in Wood: Implications for Paleoenvironmental Studies

Despite the overall agreement among species, the climate reconstruction derived from Aleppo pine differ for some periods from that achieved using oaks. This might be caused by differences in species habitat, but also by a differential response to changes in climate seasonality. Indeed, Ferrio *et al.* (2003) showed that, throughout the year, the periods in which precipitation events had a significant effect on the wood, ^{13}C signal differed substantially between Holm oak and Aleppo pine (Figure 20.6A). For Aleppo pine, the greatest sensitivity of $\Delta^{13}C$ to climate variables occurs during the expected periods of active growth, in agreement with its nearly continuous, indeterminate growth (Liphschitz and Lev-Yadun 1986, Klein *et al.* 2005). For Holm oak, however, the seasonal variation in the sensitivity of $\Delta^{13}C$ to precipitation contrasts with the phenology of this species. Unlike the pine, Holm oak has a true dormancy period during winter (Liphschitz and Lev-Yadun 1986), the season in which precipitation is the greatest. However, winter rains, although occurring during cambium dormancy, can feed deep-soil water reservoirs, which in turn increase effective water availability during the growing period for a deep-rooting species, such as Holm oak, therefore increasing $\Delta^{13}C$ of the new wood. Nevertheless, differences in the response to seasonal rainfall between species should not be considered as a handicap for the use of stable isotopes in wood/charcoal as paleoenvironmental indicators. Quite the opposite, this is an interesting feature that can provide complementary information by combining data from different species. Thus, if responses to seasonal rainfall are compared in mixed stands of Aleppo pine and Holm oak, then the potential divergences between species may be used as an indicator of changes in seasonality. In particular, since Aleppo pine is a thermophilous species and has a poor ability to

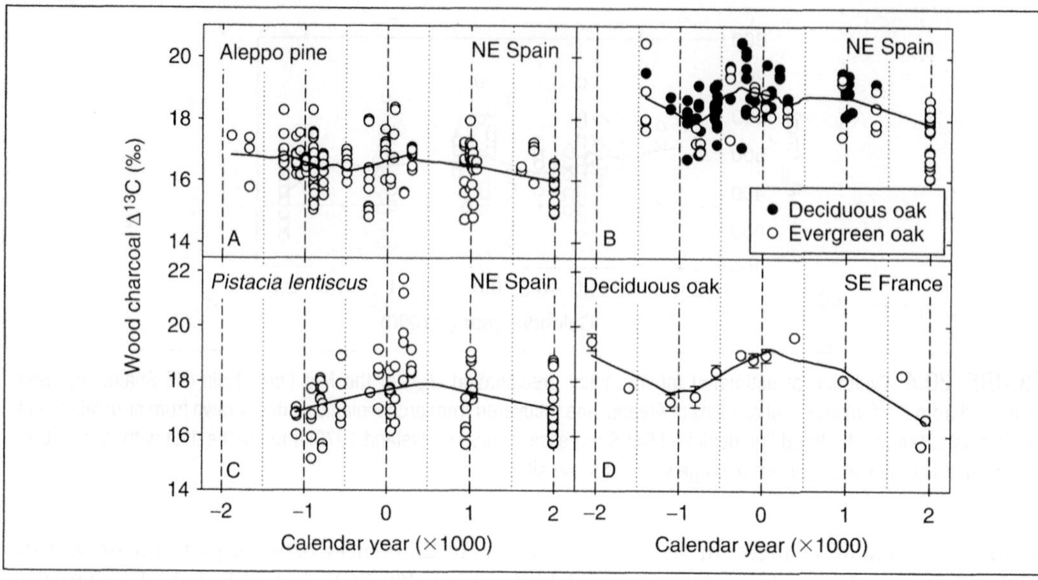

FIGURE 20.5 (A), (B), and (C) Evolution of $\Delta^{13}C$ in wood charcoal recovered in archaeological sites from the Mid Ebro Basin (NE Spain) and Languedoc (SE France) over the last four millennia. Modern wood (experimentally carbonized) is also included for reference. (A) Drawn from original data in Ferrio *et al.* (2006a). (B) and (C) Alonso, Ferrio, and Voltas (unpublished results). (D) Calculated $\delta^{13}C$ values from original $\delta^{13}C$ data from Vernet *et al.* (1996). Trend lines depict locally weighted least squares regression curves (LOESS, Cleveland 1979) fitted to the data with *span* = 0.4 for A and B, *span* = 0.8 for C, and *span* = 0.6 for D.

A

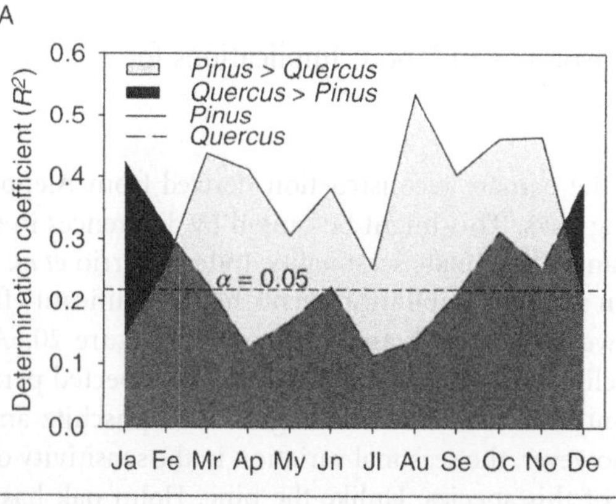

B

Stable isotopes		P_{spr}	P_{sum}	P_{aut}	P_{win}
$\uparrow \Delta^{13}C_P$	$\uparrow \Delta^{13}C_Q$	↑	↑	↑	↑
$\approx \Delta^{13}C_P$	$\approx \Delta^{13}C_Q$	≈	≈	≈	≈
$\downarrow \Delta^{13}C_P$	$\downarrow \Delta^{13}C_Q$	↓	↓	↓	↓
$\uparrow \Delta^{13}C_P$	$\approx \Delta^{13}C_Q$	≈	↑	≈	≈
$\approx \Delta^{13}C_P$	$\uparrow \Delta^{13}C_Q$	≈	≈	≈	↑
$\uparrow \Delta^{13}C_P$	$\downarrow \Delta^{13}C_Q$	↓	↑↑	↓	↓
$\downarrow \Delta^{13}C_P$	$\uparrow \Delta^{13}C_Q$	↓	↓	↓	↑↑

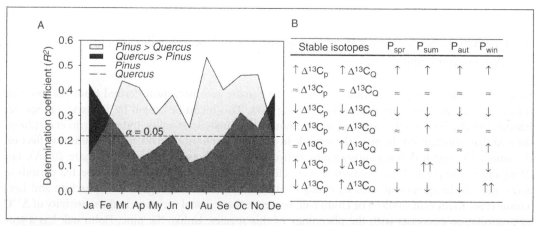

FIGURE 20.6 (A) Determination coefficients (R^2) between $\Delta^{13}C$ in wood and monthly precipitation for Holm oak (*Quercus ilex* L.) and Aleppo pine (*Pinus halepensis* Mill.). (B) Proposed conceptual model for the interpretation of $\Delta^{13}C$ values in co-occuring oaks and pines, according to their differential response to seasonal precipitation. $\Delta^{13}C_P$, $\Delta^{13}C_Q$, $\Delta^{13}C$ in wood of pines and oaks, respectively; P_{spr}, P_{sum}, P_{aut}, and P_{win} total seasonal precipitation in spring, summer, autumn, and winter, respectively; ↑, increasing variable; ≈, steady variable; ↓, decreasing variable. Original data from Ferrio *et al.* (2003).

exploit winter-fed water reserves, an increase in winter precipitations, if accompanied with a descent in temperature, would bring about a decrease in bulk $\Delta^{13}C$, as a greater portion of wood is synthesized during spring/summer. In contrast, Holm oak would be favored by this situation: the augment in deep-soil water would increase effective water availability, leading to higher wood $\Delta^{13}C$. Under such scenario, we might find relatively high $\Delta^{13}C$ in the oak, but not in the pine (Figure 20.6B). Similarly, an increase in summer storms would cause a positive response in the opportunistic Aleppo pine, while having little effect on the Holm oak. Such a conceptual approach can be useful to clarify the real nature of certain climatic anomalies found in the paleoenvironmental record.

C. Reinterpreting the "Cold Iron Age Epoch" in the Ebro Basin in the Light of $\Delta^{13}C$ Data

We can take advantage of this theoretical model to further discuss the environmental conditions in the Mid Ebro Basin during the so-called *Cold Iron Age Epoch* (ca. 900–300 BCE). Gribbin and Lamb (1978) estimated a decrease in temperature of about 2 °C for this period, evidenced by an expansion of the glaciers among other signals of an abrupt climate cooling in Europe (Van Geel *et al.* 1996). According to $\Delta^{13}C$ analysis of Aleppo pine charcoal (Figure 20.4), this period appears as a relatively arid phase in the Mid Ebro Basin. Such increase in aridity is confirmed by pollen data from archaeological sites in the area (Alonso *et al.* 2002) and from other sites in NE Spain (Riera 1994), as well as by the outcome of hydrological studies from other areas in NW Mediterranean (Magny and Richard 1992). However, other paleoenvironmental proxies provide (apparently) contradictory evidences for this period. Geomorphological studies performed throughout the Ebro Basin (Gutiérrez-Elorza and Peña-Monné 1998) indicate a generalized increase in sediment accumulation on slope deposits around 1000–400 BCE, which has been interpreted as an increase in vegetation cover, and a colder/wetter climate. Moreover, pollen analyses from a salt lake in a nearby area of the Ebro Basin (Salada Pequeña) suggest an increase in water levels around 2600–2000 BP (650–50 BCE) (Davis 1994). Are these lines of evidence compatible with an actual water shortage in Aleppo pine stands? The answer to this question might come from the role of seasonality. Anthracological evidences (Ros 1997, Alonso 1999) show the existence of mixed forests of evergreen oaks, pines, and deciduous oaks in some sites, but a decrease in vegetation cover and increased continentality in others, involving the loss of cold-sensitive species (*e.g.*, *P. lentiscus*). Thus, the spread of evergreen and deciduous oaks in the area could be interpreted as the result of an increase in continentality, rather than a consequence of a wetter climate. Although $\Delta^{13}C$ values in oaks from the same sites as well as from SE France (see Figure 20.5B and D) also indicate an arid episode ca. 3000–2700 BP (1050–750 BCE), it is followed by a recovery in $\Delta^{13}C$ from ca. 2600–1600 BP (650 BCE–350 CE), more than three centuries before the recovery of Aleppo pine (ca. 300 BCE). Conversely, *P. lentiscus*, which is also a thermophilous species, shows nearly the same time course found for Aleppo pine (Figure 20.5C). According to Figure 20.5B, the divergence between pines and oaks can be interpreted as an increase in winter precipitation, along with a decrease in spring–summer rainfall. In regard of this, $\Delta^{13}C$ analyses of cereal grains indicate that WI during grain filling (late-spring) were significantly reduced during this period in the area (Ferrio *et al.* 2006b), as well as in other NE Spain sites (Araus and Buxó 1993, Araus *et al.* 1997a, Ferrio *et al.* 2005a). On the other hand, as deciduous oaks are less adapted than evergreen oaks to summer drought, a decrease in summer precipitation may also explain the relative prevalence of the latter reported by Jalut *et al.* (2000). The observed increase in lake levels (as well as glaciers) could be derived exclusively from increased winter recharges. Similarly, a reduction in soil erosion might arise from the absence of summer convective storms (Gutiérrez-Elorza and Peña-Monné 1998), together with an increase in deep-rooted, cold-resistant species.

VII. CONCLUDING REMARKS AND FUTURE PROSPECTS

For decades, the use of stable isotopes in environmental archaeology has been mostly focused on the study of animal and human diets. Nowadays, archaeobotanists exhibit an increasing interest in these methods, and the number of studies dealing with the archaeoecological significance of plant remains is increasing steadily. In particular, the analysis of carbon isotopes in cereal grains has shown to provide direct evidences on the water availability of ancient crops, while charcoals can give reliable estimations of past precipitation regimes. The latter may contribute to the characterization of past climate dynamics, moreover to being an additional *in situ* source of environmental data for archaeologists.

This chapter focused on carbon isotopes, as they have a well-established methodology and are relatively easy to interpret. However, the most desirable scenario for the future would be the combination of multiple approaches to face specific archaeological queries. Oxygen isotopes, for example, can be used to determine the origin of source water in cultivated crops (Williams *et al.* 2005). However, they are strongly affected by carbonization (Marino and De Niro 1987, Williams *et al.* 2005) and, consequently, its use is restricted to desiccated or waterlogged material. On the other hand, nitrogen isotopes show little changes due to carbonization, and thus are potentially applicable to most archaeobotanical remains (De Niro and Hastorf 1985, Bogaard *et al.* 2007). Despite the complexity of the nitrogen cycle, nitrogen isotopes might be valuable tools to assess nutrient management practices (*e.g.*, manuring, rotation with legumes) in ancient crops (Bol *et al.* 2005, Bogaard *et al.* 2007). But a multi-isotopic approach alone is probably not enough to bridge the gaps left by single stable isotope measurements. In the paleoclimate context, we have shown that a careful analysis of the divergences between stable isotope data from different species, as well as with other paleoclimate proxies, provides much more insight than any individual method alone. For the study of crops, the comparison with charcoal data offers the possibility to identify events in which early farmers started to develop a conscious water management to improve crop performance. This can be complemented, for example, with the study of weed floras, given that they vary with water and nutrient availability (Jones *et al.* 2000). Finally, we should bear in mind that, in the archaeological context, paleoenvironmental reconstructions, although interesting per se, become particularly useful when related to the structure and dynamics of the settlement. This allows a depiction of the complex relationships between ancient societies and their environment. Population changes associated to paleoenvironmental data (*e.g.*, derived from stable isotopes) might help to understand how early farming societies developed in a changing environment. It is unlikely, however, that at present a complete answer for the most intriguing question on the Origins of Agriculture, that is the reason that pushed hunter-gatherers to initiate such a self-defeating lifestyle, can be provided. In words of a Bushman (Lee and DeVore 1968):

> "Why should I farm when there are so many mongongo nuts?"

VIII. ACKNOWLEDGMENTS

This study was supported by the projects DGI (CGL2005-08175-C02/BOS) and MENMED (INCO-MED-ICA3-CT-2002-10022). We thank M. Ros and R. Piqué for performing charcoal determinations, and A. Florit, A. Vega, J. Coello, and A. Ameztegui for assistance in sampling preparation and data processing. J. P. Ferrio is a recipient of a postdoctoral fellowship *Beatriu de Pinós* from the Generalitat de Catalunya.

IX. REFERENCES

Alonso, N. 1999. De la llavor a la farina: Els processos agrícoles protohistòrics a la Catalunya Occidental. UMR 154-CNRS, Lattes.

Alonso, N., M. Gené, E. Junyent, A. Lafuente, J. B. López, A. Moya, and E. Tartera. 2002. Recuperant el passat a la línia del Tren d'Alta Velocitat. *L'assentament protohistòric, medieval i d'època moderna de El Vilot de Montagut (Alcarràs, Lleida).* GIF, Generalitat de Catalunya, Lleida.

Araus, J. L., and R. Buxó. 1993. Changes in carbon isotope discrimination in grain cereals from the north-western Mediterranean basin during the past seven millennia. *Australian Journal of Plant Physiology* **20:**117–128.

Araus, J. L., A. Febrero, R. Buxó, M. D. Camalich, D. Martin, F. Molina, M. O. Rodriguez-Ariza, and I. Romagosa. 1997a. Changes in carbon isotope discrimination in grain cereals from different regions of the western Mediterranean basin during the past seven millennia. Palaeoenvironmental evidence of a differential change in aridity during the late Holocene. *Global Change Biology* **3:**107–118.

Araus, J. L., A. Febrero, R. Buxó, M. O. Rodriguez-Ariza, F. Molina, M. D. Camalich, D. Martin, and J. Voltas. 1997b. Identification of ancient irrigation practices based on the carbon isotope discrimination of plant seeds: A case study from the South-East Iberian Peninsula. *Journal of Archaeological Science* **24:**729–740.

Araus, J. L., A. Febrero, M. Catala, M. Molist, J. Voltas, and I. Romagosa. 1999. Crop water availability in early agriculture: Evidence from carbon isotope discrimination of seeds from a tenth millennium BP site on the Euphrates. *Global Change Biology* **5:**201–212.

Araus, J. L., G. A. Slafer, M. P. Reynolds, and C. Royo. 2002. Plant breeding and drought in C_3 cereals: What should we breed for? *Annals of Botany* **89:**925–940.

Araus, J. L., J. P. Ferrio, R. Buxó, and J. Voltas. 2007. The historical perspective of dryland agriculture: Lessons learned from 10000 years of wheat cultivation. *Journal of Experimental Botany* **58:**131–145.

Bar-Yosef, O., M. E. Kislev, D. R. Harris, and G. C. Hillman. 1989. Early farming communities in the Jordan Valley. Pages 632–642 *in* D. R. Harris and G. C. Hillman (Eds.) *Foraging and Farming: The Evolution of Plant Exploitation.* Unwin Hyman, London.

Bogaard, A., T. H. E. Heaton, P. Poulton, and I. Merbach. 2007. The impact of manuring on nitrogen isotope ratios in cereals: Archaeological implications for reconstruction. *Journal of Archaeological Science* **34:**335–343.

Bol, R., J. Eriksen, P. Smith, M. H. Garnett, K. Coleman, and B. T. Christensen. 2005. The natural abundance of C-13, N-15, S-34 and C-14 in archived (1923–2000) plant and soil samples from the Askov long-term experiments on animal manure and mineral fertilizer. *Rapid Communications in Mass Spectrometry* **19:**3216–3226.

Buchmann, N., J. R. Brooks, K. D. Rapp, and J. R. Ehleringer. 1996. Carbon isotope composition of C4 grasses is influenced by light and water supply. *Plant, Cell and Environment* **19:**392–402.

Cleveland, W. S. 1979. Robust locally weighted regression and smoothing scatterplots. *Journal of the American Statistical Association* **74:**829–836.

Davis, B. A. S. 1994. Palaeolimnology and Holocene environmental change from endorreic lakes in the Ebro Basin, N.E. Spain. Ph.D. Thesis, University of Newcastle upon Tyne.

Czimczik, C. I., C. M. Preston, M. W. I. Schmidt, R. A. Werner, and E. D. Schulze. 2002. Effects of charring on mass, organic carbon, and stable carbon isotope composition of wood. *Organic Geochemistry* **33:**1207–1223.

De Niro, M. J., and C. A. Hastorf. 1985. Alteration of $^{15}N/^{14}N$ and $^{13}C/^{12}C$ ratios of plant matter during the initial stages of diagenesis: Studies utilizing archaeological specimens from Peru. *Geochimica et Cosmochimica Acta* **49:**97–115.

Edwards, D., and L. Axe. 2004. Anatomical evidence in the detection of the earliest wildfires. *PALAIOS* **19:**113–128.

Farquhar, G. D., M. H. O'Leary, and J. A. Berry. 1982. On the relationship between carbon isotope discrimination and the intercellular carbon dioxide concentration in leaves. *Australian Journal of Plant Physiology* **9:**121–137.

February, E. C., and N. J. Van der Merwe. 1992. Stable carbon isotope ratios of wood charcoal during the past 4000 years: Anthropogenic and climatic influences. *South African Journal of Science* **88:**291–292.

Ferrio, J. P., A. Florit, A. Vega, L. Serrano, and J. Voltas. 2003. $\Delta^{13}C$ and tree-ring width reflect different drought responses in *Quercus ilex* and *Pinus halepensis. Oecologia* **137:**512–518.

Ferrio, J. P., N. Alonso, J. Voltas, and J. L. Araus. 2004. Estimating grain weight in archaeological cereal crops: A quantitative approach for comparison with current conditions. *Journal of Archaeological Science* **31:**1635–1642.

Ferrio, J. P., J. L. Araus, R. Buxó, J. Voltas, and J. Bort. 2005a. Water management practices and climate in ancient agriculture: Inference from the stable isotope composition of archaeobotanical remains. *Vegetation History and Archaeobotany* **14:**510–517.

Ferrio, J. P., V. Resco, D. G. Williams, L. Serrano, and J. Voltas. 2005b. Stable isotopes in arid and semi-arid forest systems. *Investigación Agraria, Sistemas y Recursos Forestales* **14:**371–382.

Ferrio, J. P., N. Alonso, J. B. López, J. L. Araus, and J. Voltas. 2006a. Carbon isotope composition of fossil charcoal reveals aridity changes in the NW Mediterranean basin. *Global Change Biology* **12:**1253–1266.

Ferrio, J. P., N. Alonso, J. Voltas, and J. L. Araus. 2006b. Grain weight changes over time in ancient cereal crops: Potential roles of climate and genetic improvement. *Journal of Cereal Science* **44:**323–332.

Fischer, R. A., D. Rees, K. D. Sayre, Z. M. Lu, A. G. Condon, and A. L. Saavedra. 1998. Wheat yield progress associated with higher stomatal conductance and photosynthetic rate, and cooler canopies. *Crop Science* **38:**1467–1475.

Gribbin, J., and H. H. Lamb. 1978. Climatic change in historical times. Pages 68–82 *in* J. Gribbin (Ed.) *Climatic Change.* Cambridge University Press, Cambridge.

Gröcke, D. R. 1998. Carbon-isotope analyses of fossil plants as a chemostratigraphic and palaeoenvironmental tool. *Lethaia* **31:**1–13.

Guo, Y., and R. M. Bustin. 1998. FTIR spectroscopy and reflectance of modern charcoals and fungal decayed woods: Implications for studies of inertinite in coals. *International Journal of Coal Geology* **37:**29–53.

Gutiérrez-Elorza, M., and J. L. Peña-Monné. 1998. Geomorphology and late Holocene climatic change in northeastern Spain. *Geomorphology* **23:**205–217.

Harlan, J. R. 1998. *The living fields: Our agricultural heritage.* Cambridge University Press, Cambridge.

Helbaek, H., and M. J. Schultze. 1981. The scientific work of Hans Helbaek. *Kulturpflanze* **29:**443–446.

Jalut, G., A. Esteban-Amat, L. Bonnet, T. Gauquelin, and M. Fontugne. 2000. Holocene climatic changes in the Western Mediterranean, from south-east France to south-east Spain. *Palaeogeography, Palaeoclimatology, Palaeoecology* **160:**255–290.

Jones, G., A. Bogaard, M. Charles, and J. G. Hodgson. 2000. Distinguishing the effects of agricultural practices relating to fertility and disturbance: A functional ecological approach in archaeobotany. *Journal of Archaeological Science* **27:**1073–1084.

Jones, T. P., and W. G. Chaloner. 1991. Fossil charcoal, its recognition and palaeoatmospheric significance. *Palaeogeography, Palaeoclimatology, Palaeoecology* **97:**39–50.

Klein, T., D. Hemming, T. Lin, J. M. Grunzweig, K. Maseyk, E. Rotenberg, and D. Yakir. 2005. Association between tree-ring and needle delta13C and leaf gas exchange in Pinus halepensis under semi-arid conditions. *Oecologia* **144:**45–54.

Lee, R. B., and I. DeVore. 1968. *Man the hunter.* Aldine, Chicago.

Liphschitz, N., and S. Lev-Yadun. 1986. Cambial activity of evergreen and seasonal dimorphics around the Mediterranean. *IAWA Bulletin* **7:**145–153.

Lopez-Capel, E., R. Bol, and D. A. C. Manning. 2005. Application of simultaneous thermal analysis mass spectrometry and stable carbon isotope analysis in a carbon sequestration study. *Rapid Communications in Mass Spectrometry* **19:**3192–3198.

Magny, M., and H. Richard. 1992. Essai de synthèse vers une courbe de l'évolution du climat entre 500BC et 500AD. *Les nouvelles de l'archéologie* **50:**58–60.

Marino, B. D., and M. J. De Niro. 1987. Isotope analysis of archaeobotanicals to reconstruct past climates: Effects of activities associated with food preparation on carbon, hydrogen and oxygen isotope ratios of plant cellulose. *Journal of Archaeological Science* **14:**537–548.

Marino, B. D., M. B. MacElroy, R. J. Salawitch, and W. G. Spaulding. 1992. Glacial-to-interglacial variations in the carbon isotopic composition of atmospheric CO_2. *Nature* **357:**461–466.

Muñoz, P., J. Voltas, J. L. Araus, E. Igartua, and I. Romagosa. 1998. Changes over time in the adaptation of barley releases in north-eastern Spain. *Plant Breeding* **117:**531–535.

Poole, I., P. F. Van Bergen, J. Kool, S. Schouten, and D. J. Cantrill. 2004. Molecular isotopic heterogeneity of fossil organic matter: Implications for delta(13) C-biomass and delta(13) C-palaeoatmosphere proxies. *Organic Geochemistry* **35:**1261–1274.

Renfrew, J. 1973. *Palaeoethnobotany.* Methuen, London.

Rice, S. K., and L. Giles. 1996. The influence of water content and leaf anatomy on carbon isotope discrimination and photosynthesis in Sphagnum. *Plant, Cell and Environment* **19:**118–124.

Riera, S. 1994. Evolució del paisatge vegetal holocè al Pla de Barcelona, a partir de dades pol.liniques. Ph.D. thesis, University of Barcelona.

Riera, S., G. Wansard, and R. Julià. 2004. 2000-year environmental history of a karstic lake in the Mediterranean pre-Pyrenees: The Estanya lakes (Spain). *Catena* **55:**293–324.

Ros, M. T. 1997. La vegetació de la Catalunya Meridional i territoris propers de la Depressió de l'Ebre, en la Prehistòria recent i Protohistòria, a partir dels estudis antracològics. *Gala* **3–5:**19–32.

Stuiver, M., P. J. Reimer, E. Bard, J. W. Beck, G. S. Burr, K. A. Hughen, B. Kromer, G. McCormac, J. van der Plicht, and M. Spurk. 1998. INTCAL98 radiocarbon age calibration, 24,000-0 cal BP. *Radiocarbon* **40:**1041–1083.

Taylor, R. E., and M. J. Aitken. 1997. Chronometric dating in archaeology. *Advances in Archaeological and Museum Science,* Vol. 2. Oxford University Press, Oxford.

Toolin, L. J., and C. J. Eastoe. 1993. Late pleistocene recent atmospheric delta-C-13 record in C4 grasses. *Radiocarbon* **35:**263–269.

Turney, C. S. M., D. Wheeler, and A. R. Chivas. 2006. Carbon isotope fractionation in wood during carbonization. *Geochimica et Cosmochimica Acta* **70:**960–964.

Van Geel, B., J. Buurman, and H. T. Waterbolk. 1996. Archaeological and palaeoecological indications of an abrupt climate change in The Netherlands, and evidence for climatological teleconnections around 2650 BP. *Journal of Quaternary Science* **11:**451–460.

Van Zeist, W., and W. A. Casparie. 1984. *Plants and Ancient Man: Studies in Palaeoethnobotany.* Balkema, Rotterdam.

Vernet, J. L. 1990. The bearing of phyto-archaeological evidence on discussions of climatic change over recent millennia. *Philosophical Transactions of the Royal Society of London Series A* **330:**671–677.

Vernet, J. L., and S. Thiebault. 1987. An approach to northwestern Mediterranean recent prehistoric vegetation and ecologic implications. *Journal of Biogeography* **14:**117–127.

Vernet, J. L., C. Pachiaudi, F. Bazile, A. Durand, L. Fabre, C. Heinz, M. E. Solari, and S. Thiebault. 1996. Le δ^{13} C de charbons de bois préhistoriques et historiques méditerranéens, de 35000 BP a l'àctuel. Premiers resultats. *Comptes Rendus de l'Academie des Sciences, série II* **323:**319–324.

White, J. W. C., P. Ciais, R. A. Figge, R. Kenny, and V. Markgraf. 1994. A high resolution record of atmospheric CO_2 content from carbon isotopes in peat. *Nature* **367:**153–156.

Willcox, G. 1996. Evidence for plant exploitation and vegetation history from three Early Neolithic pre-pottery sites on the Euphrates (Syria). *Vegetation History and Archaeobotany* **5:**143–152.

Williams, D. G., J. B. Coltrain, M. Lott, N. B. English, and J. R. Ehleringer. 2005. Oxygen isotopes in cellulose identify source water for archaeological maize in the American Southwest. *Journal of Archaeological Science* **32:**931–939.

CHAPTER 21

Change of the Origin of Calcium in Forest Ecosystems in the Twentieth Century Highlighted by Natural Sr Isotopes

Thomas Drouet,* Jacques Herbauts,* and Daniel Demaiffe†

**Laboratoire de Génétique et d'Ecologie Végétales, Université Libre de Bruxelles (ULB)*
†Laboratoire de Géochimie Isotopique et Géodynamique Chimique, Université Libre de Bruxelles (ULB)

Contents

I. INTRODUCTION

Acidification of forest soils by atmospheric deposition involving depletion of alkaline earth cations is still an active subject of research. Factors implicated in this process are numerous and strongly interconnected so that a lot of questions still remains unresolved today. However, the understanding of the soil acidification process is essential to maintain the environment quality and to guarantee forest sustainability for commercial forestry.

Works undertaken in Western Europe and North America reported increasing depletion of exchangeable base cations (mainly Ca and Mg) in sensitive soils (*i.e.*, poor base status), caused by inorganic acid inputs (Falkengren-Grerup *et al.* 1987, Thimonier *et al.* 2000) that can cause tree nutritional deficiencies. Excess of anion deposition affects the ecosystem balance by depleting the pool of cation buffers more rapidly than the buffers that can be replaced, leading to an aluminization of the exchange complex, which interferes with the uptake of calcium by trees. In soils with a low weatherable mineral reserve, weathering inputs may not be sufficient to replace the accelerated losses of basic cations associated with disturbances such as acidic deposition or intensive forest harvest.

In order to counteract this effect, international environmental regulations were carried out and have led to widespread decline in the rate of acidic deposition across large areas of Europe and North America (Hedin *et al.* 1994). This change was mainly due to reduction in SO_2 emissions through the installation of filter systems in power plants and industry. However, the decline in ecosystem inputs of the acid anion SO_4 is accompanied by the reduced deposition of base cations so that changes in net acidity may be small (Hedin *et al.* 1994). As a consequence, the base cation status of the soils would not be expected to improve and soil acidification could actually continue.

Only few methods are able to reconstruct accurately the steps of the acidification mechanism in soils and its impact on the tree mineral nutrition. Decrease of soil base saturation (BS) during different periods of the last century has been highlighted by several methods, for example soil resampling (Falkengren-Grerup *et al.* 1987) and long-term site observation (Blake *et al.* 1999).

The depletion of Ca is a reliable indicator of overall cation depletion in an ecosystem because of its relative prevalence in the vegetation and its status as a major base cation on the soil exchange complex. Indeed, calcium plays a key role in these ecosystems, both as plant nutrient and in soils as buffer of acidic inputs. Therefore, the determination of the sources of Ca into the soil–vegetation–atmosphere system allows to estimate the real danger of soil impoverishment process on the mineral nutrition of forest stands. The two main sources of calcium in forest ecosystems are soil mineral weathering and the atmospheric inputs through dissolved cations in rainwater and dust fall. Our approach was to use strontium (Sr) as tracer of calcium. Strontium acts as an analogue for Ca because both are alkaline earth elements with similar ionic radius and the same valence. Because of their similar chemical structure, Ca^{2+} and Sr^{2+} behave similarly in the soil–plant system (Drouet 2005), although some fractionation of Sr/Ca ratio may occur once elements have passes through some plant compartments (Poszwa *et al.* 2000).

Stable Sr isotopes offer a powerful tool to discriminate atmospherically derived Ca from Ca originating from minerals. Isotope techniques do not require to determine element fluxes (extensive variables), instead using intensive-type variables (isotopic ratios) which do not impose the knowledge of structure or boundaries of the system studied. Strontium 86 is nonradiogenic whereas ^{87}Sr is produced by the radioactive β^- decay of ^{87}Rb (half-life $\sim 48.8 \times 10^9$ year; decay constant $\lambda = 1.42 \times 10^{-11}$ year^{-1}) so that their proportions vary between different natural materials. The principle of this isotopic method is to measure the natural variation of the Sr isotopic composition, expressed by the $^{87}Sr/^{86}Sr$ ratio, as tracer of the sources of Ca in forest ecosystems. Because of the slight mass ratio between Sr isotopes, biological and chemical processes cause neglected isotopic fractionation compared to low-mass isotopic systems such as O, C, N, and S. Variations in the Sr isotopic composition in the different components of the forest ecosystem are therefore caused entirely by the mixing of Sr derived from sources with specific $^{87}Sr/^{86}Sr$ ratio. For this reason, the $^{87}Sr/^{86}Sr$ ratio can be used to identify and quantify the contribution of different Sr (and Ca) sources to the forest vegetation.

The determination of the origin of Ca in the vegetation is based on the measurement of the $^{87}Sr/^{86}Sr$ ratio in the vegetation and the supplying sources, providing sufficient contrast in isotopic composition between the two sources. Strontium released by weathering of ancient geological substrates is characterized by a high isotopic ratio (rich in radiogenic ^{87}Sr). By contrast, atmospherically derived Sr has a low isotopic ratio, generally closed to that of seawater, which is constant through the timescale studied. Vegetation has an intermediate $^{87}Sr/^{86}Sr$ ratio between the end-members so that the relative contributions to the binary mixing can be calculated. Atmospheric signature can be estimated by direct measurement on bulk precipitation samples. Sea-salt aerosols, which have constant marine $^{87}Sr/^{86}Sr$

(~0.709), dominate the sites studied, but continental dust sources may impart spatial and temporal variations on this value in other locations (Nakano and Tanaka 1997). Whereas the isotopic signature of the atmospheric inputs and the plant materials are relatively easy to determine, the estimation of the weathering end-member remains more problematic. Weathering involves combination of different minerals, each with a particular weathering rate and isotopic signature. The minerals $^{87}Sr/^{86}Sr$ ratios depend on the Rb/Sr ratio and the age of the parent material because ^{87}Sr is produced by the decay of ^{87}Rb. Therefore, bulk soil $^{87}Sr/^{86}Sr$ ratio cannot be used as weathering signature in old substrates because individual minerals of the parent material have different weathering rates, and their $^{87}Sr/^{86}Sr$ ratios have diverged over geologic timescale. Some studies estimated this weathering end-member on the basis of the seasonal variation of chemistry and $^{87}Sr/^{86}Sr$ ratio of stream waters (Clow *et al.* 1997, Kennedy *et al.* 2002). Natural weathering can also be simulated by extraction of soil samples with acidic solutions (Blum *et al.* 2002, Drouet *et al.* 2005a). The soil exchangeable fraction, which corresponds to the plant-available pool, can be leached from the soil by salt solution as ammonium acetate.

II. THE USE OF STRONTIUM ISOTOPE TECHNIQUE

Sr isotopic ratios were measured in several ecosystem compartments: plant material, precipitation, exchangeable, and acid extractable fractions (Figure 21.1). Chemical separation of Sr from these samples was carried out by cation exchange chromatography. Sr isotopic compositions were measured on a VG Sector 54 thermal ionization mass spectrometer with solid source, housed at the Laboratoire de Géochimie isotopique, Université Libre de Bruxelles. Samples were analyzed in static multicollector mode, with 100 repeated measurements per sample. The measured $^{87}Sr/^{86}Sr$ ratios were normalized to $^{86}Sr/^{88}Sr = 0.1194$ in order to correct for the effects of any fractionation of Sr isotopes that occurs during thermal ionization over the course of analysis in the mass spectrometer. The NBS-987 Sr standard, measured to determine instrument bias, yielded on average an $^{87}Sr/^{86}Sr$ value of 0.710270 ± 0.000009 (2σ, $n = 25$). The proportions of Sr in a mixture (in this case, vegetation) derived from two sources (atmosphere and soil weathering) are calculated using a mixing equation (Capo *et al.* 1998):

$$X(\mathrm{Sr})_1 = \frac{(^{87}\mathrm{Sr}/^{86}\mathrm{Sr})_{\mathrm{Mix}} - (^{87}\mathrm{Sr}/^{86}\mathrm{Sr})_2}{(^{87}\mathrm{Sr}/^{86}\mathrm{Sr})_1 - (^{87}\mathrm{Sr}/^{86}\mathrm{Sr})_2} \tag{21.1}$$

where $X(\mathrm{Sr})_1$ represents the mass fraction of Sr derived from source no. 1. Subscripts 1 and 2 refer to the two sources. The "Mix" subscript indicates the mixture component. The calculated percentage of atmospheric Sr in vegetation cannot be directly transposed to that of Ca because of the different proportion of Ca and Sr (Sr/Ca ratio) between the two sources in the ecosystems. The relative contribution of Ca from two sources to a mixing component is given by a two-component mixing equation (Capo *et al.* 1998):

$$X(\mathrm{Ca})_1 = \frac{\left[(^{87}\mathrm{Sr}/^{86}\mathrm{Sr})_{\mathrm{Mix}} - (^{87}\mathrm{Sr}/^{86}\mathrm{Sr})_2\right](\mathrm{Sr}/\mathrm{Ca})_2}{\left[(^{87}\mathrm{Sr}/^{86}\mathrm{Sr})_{\mathrm{Mix}} - (^{87}\mathrm{Sr}/^{86}\mathrm{Sr})_2\right](\mathrm{Sr}/\mathrm{Ca})_2 + \left[(^{87}\mathrm{Sr}/^{86}\mathrm{Sr})_1 - (^{87}\mathrm{Sr}/^{86}\mathrm{Sr})_{\mathrm{Mix}}\right](\mathrm{Sr}/\mathrm{Ca})_1}, \tag{21.2}$$

where $X(\mathrm{Ca})_1$ represents the mass fraction of Ca derived from the atmospheric source. Sr/Ca ratios are correcting factors used to take into account the different proportions of Ca and Sr between the two sources. The distinct isotopic signatures of Sr contained in atmospheric precipitations and of Sr derived from mineral weathering allow to quantify the respective contributions of these two sources to the vegetation.

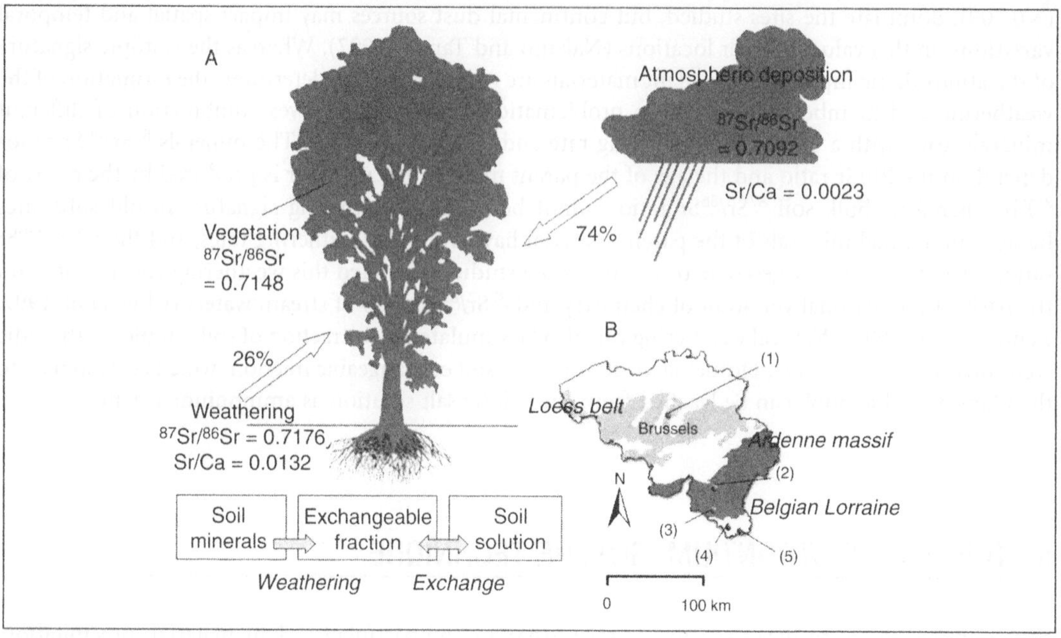

FIGURE 21.1 (A) Schematic showing the different compartments and sources of Ca (mineral weathering and atmospheric deposition) of a forest ecosystem. Calculation of the relative contributions of these sources to vegetation is illustrated for a mature beech stand (*F. sylvatica*) in high Belgium. (B) Localization of the sites studied in Belgium. (1) Soignes Regional Forest (nine sites), (2) Herbeumont State Forest, (3) Smuid Wood, (4) Meix-devant-Virton Wood, and (5) Côte Wood (two sites).

III. SEPARATION OF ATMOSPHERIC AND WEATHERING SOURCES OF CALCIUM IN FOREST ECOSYSTEMS

The isotopic approach was first applied on two temperate forest ecosystems (Drouet *et al.* 2005a). In this preliminary study we shown that, unexpectedly, some beech forests in Belgium were supplied by more than 70% of Ca from atmospheric origin (Figure 21.1A). This illustrated the low amount of Ca released by weathering. Moreover, low annual inputs of Ca by atmospheric precipitation (<10 kg Ca ha^{-1} year^{-1}) relative to tree growth requirements may cause nutritional deficiencies enhanced by intensive logging and acid deposition. These first results emphasize the importance of the atmospherically derived Ca for tree mineral nutrition in the stands studied, suggesting that these ecosystems on acid soils are sensitive to changes in the atmospheric chemistry, for example, acid deposition associated with decreasing input of atmospheric cations. This fundamental finding has important implications regarding forest stand management.

Second, the calculation of the contributions of the two above-mentioned sources was extended to 14 forest ecosystems in central and high Belgium (Figure 21.1B) growing in soils displaying a wide range of Ca reserves (Drouet 2005). All study sites were located on broad upland topography (mainly plateaus) and supported mature beech trees as dominant species.

Criteria for site selection were: (1) Sr isotopic composition of the parent material sufficiently different from that of the atmospheric source, (2) beech (*Fagus sylvatica* L.) as a dominant species of the forest stand, and (3) for all the sites, a large range of soil exchangeable Ca concentration. The majority of the sites were situated in central Belgium and have developed a Dystric Podzoluvisol

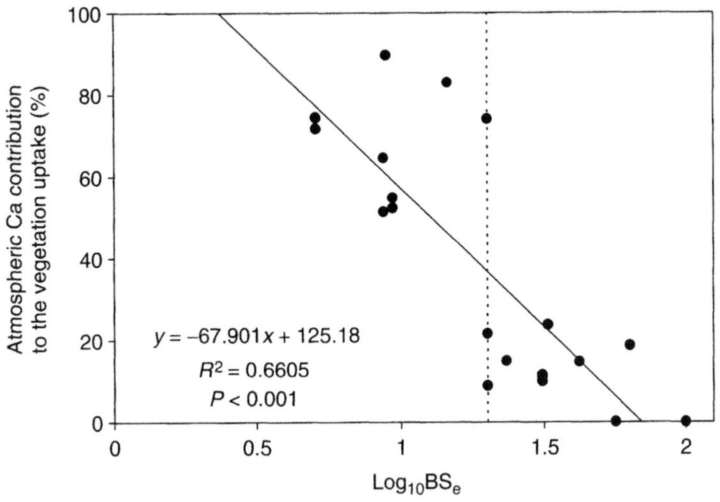

FIGURE 21.2 Relationship between the percentage of atmospherically derived Ca in the vegetation and the log value of the mean effective BS_e of the soil in the first meter depth. Decreasing BS_e is associated with concomitant decrease of Ca concentrations in foliage. Vertical dotted line indicates the 20% BS beneath which exchange complex and soil solution are dominated by cationic Al (Reuss 1983).

(FAO classification) on Pleistocene loess. A carbonated horizon appears at different depth (from 180 to 400-cm depth) related to the sites, delimiting a soil sequence of contrasting Ca reserve. Other sites were chosen in central and high Belgium on soils with large range of Ca content, from Ca poor Orthic Podzol (FAO classification) on Tertiary sands (400-kg exchangeable Ca ha^{-1} 0.5 m^{-1}) to Ca saturated Mollic Leptosol on outcropping of Jurassic limestone ($>4 \times 10^6$ kg exchangeable Ca ha^{-1} 0.5 m^{-1}). This large-scale investigation showed that Ca supply of the trees was very variable, from stands getting their nutrients mainly from mineral weathering to stands principally supplied by the atmospheric source. Data from this study pointed out a general inverse relationship between the percentage of atmospheric Ca in the vegetation and the soil reserves in exchangeable Ca or BS rate (Figure 21.2). The close connection between the proportions of the two sources on the soil exchange complex and in the vegetation was also confirmed.

IV. EXCHANGE PROCESSES IN THE SOIL

Measurement of the Sr isotopic composition in precipitation, the vegetation and the exchangeable fraction and soil solution in different soil horizons allow to gain insight into the exchanges of Ca between the compartments of the soil. Reserves of bioavailable cations in the soil are mainly in exchangeable form, adsorbed on the mineral and organic matter surface (exchange complex). The advantage of the use of the Sr isotope technique is that, although concentrations in major elements released by weathering can be modified by the formation of secondary minerals or exchange processes on the adsorbing complex, soil $^{87}Sr/^{86}Sr$ remains a robust signature.

The study of a large number of forest stands suggests that the soil exchange complex does not always reflect an equilibrium between the present mineral weathering and the atmospheric inputs. In some cases, the exchange complex can conserve the isotopic signature of ancient weathering processes or a

TABLE 21.1 Sr isotopic signature of the exchangeable fraction (CH_3COONH_4 extract) and 0.1 N HCl extracts (after the exchangeable cation removed) of soils

Site	Soil horizon	Depth (cm)	Soil type	$^{87}Sr/^{86}Sr$ exchangeable fraction	$^{87}Sr/^{86}Sr$ HCl extract	Clay (%)	Free Fe[d] (%)
MES	B2t	55–75	Dystric Podzoluvisol	0.715291[a]	0.712067	19.1	0.81
	B3t	175–200		0.712234[b]	0.713571	16.5	–
	C	230–240		0.710707[b]		15.0	–
MLN	B2t	55–75	Dystric Podzoluvisol	0.721618[a]	0.71639	16.2	–
DIEP	B2tg	55–75	Stagnic Podzoluvisol	0.71816[a]	0.713607	37.7	2.19
COT II	B2t	55–65	Eutric Luvisol	0.710056[c]	0.712396	65.9	11.8
LND	E	5–25	Orthic Podzol	0.714254	0.720862	0.2	0.05
	C	80–100			0.721445	11.8	1.11

[a] Soil horizons influenced by highly radiogenic Sr from ancient sources.
[b] Soil horizons influenced by a removed carbonated fraction still present at 250-cm depth ($^{87}Sr/^{86}Sr = 0.708255$).
[c] Isotopic value of the exchangeable fraction is assumed to be that of the vegetation; influence of the ancient carbonates still present at 200-cm depth ($^{87}Sr/^{86}Sr = 0.708020$).
[d] Extracted with acid ammonium oxalate and Na dithionite at 60 °C.

signature inherited from transfers of cations through the soil profile (Table 21.1). For example, in soils developed on loess, Ca and Sr derived from the dissolution of carbonate phase, present in the original loessic material but totally moved out now, were retained on the exchange complex of B_{3t} and C horizons (MES site, Table 21.1 and Drouet *et al.* 2007). In another soil on Tertiary sandy-clayed substrate (DIEP site, Table 21.1), the exchange complex is constituted of Sr issued from the early weathering of glauconite, an ^{87}Sr-rich mineral, which is now strongly weathered and does not influence the weathering product (Drouet 2005). This phenomenon of exchange complex "memory" seems particularly active in soil horizons with large amounts of clay and/or Fe oxyhydroxide, the major part of the "free" Fe (Table 21.1). Such observations are in agreement with those reported in other studies (Blum and Erel 1997, Bullen *et al.* 1997). Soils without enriched clay horizon have lower exchangeable $^{87}Sr/^{86}Sr$ ratio than that of the weathering source, and reflects the equilibrium between atmospheric and weathering sources of Sr (*e.g.*, LND soil in Table 21.1).

V. EVOLUTION OF THE Ca SOURCES THROUGH THE TIME HIGHLIGHTED BY TREE-RINGS RECORDS

Finally, Sr isotopes were associated to dendrochemistry, which uses archive properties of the tree growth rings. This technique allows a retrospective analysis of the origin of Ca in the tree nutrition (Åberg 1995, Poszwa *et al.* 2003, Drouet *et al.* 2005b). This last method is based on the hypothesis that the radial distribution of an element in the tree rings reflects the chemistry of the environment at the year of ring formation. Isotopic dendrochemistry can potentially give information on the evolution of Ca nutrition of trees on long periods of time (>100 years).

In this study (Drouet *et al.* 2005b), the validity of the tree-ring recorder was first evaluated on beech wood samples from a stand that experienced a large nutritional perturbation, namely a liming. The stand was limed in 1972 by means of 12 t of crushed limestone for soil status improvement. This liming material ($CaCO_3$) has a Sr isotopic signature (very low $^{87}Sr/^{86}Sr$ ratio: 0.707857) clearly distinct from

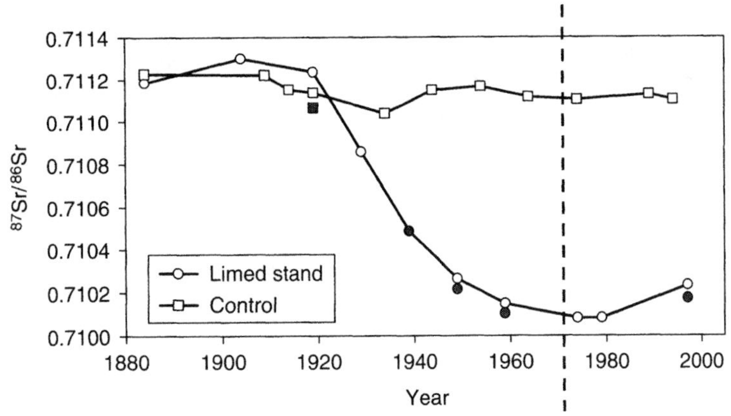

FIGURE 21.3 Sr isotopic trend of a limed beech stand of central Belgium. Open circles indicate the limed stand chronology, solid squares show the $^{87}Sr/^{86}Sr$ values of a control stand. The stand was limed by a blower with ~12 t ha^{-1} of crushed Frasnian limestone ($^{87}Sr/^{86}Sr$ ratio = 0.707857). The dotted line indicates the liming year (1972). Filled symbols show isotopic measurements on salt extractible Sr. [Redrawn from Drouet *et al.* (2005b) with permission and from data on wood exchangeable forms (CH_3COONH_4 extracts): Drouet, Herbauts, and Demaiffe, unpublished data].

that of the two other sources and produces a dramatic shift of the wood $^{87}Sr/^{86}Sr$ ratio. Surprisingly, the Sr isotopic mark of the $CaCO_3$ appears about 50 years before the liming application date (Figure 21.3). This can be explained by the fact that tree rings remain active during several years after their formation. Therefore, the sap circulation through the living wood maintains the isotopic equilibrium between wood and the soil solution. This process has previously been found by other authors (Houle *et al.* 2002) and is called "lateral reequilibration." The evaluation of this phenomenon is crucial for the interpretation of other dendrochemical profiles. Another explanation could be that Sr can migrate from a ring to another, for example, through parenchyma rays or by diffusion. Different wood fractions of cations with distinct mobility may be considered: water-soluble, salt-extractible, and residual (Herbauts *et al.* 2002). Most of the residual (not extractible with salt solution) calcium and strontium in wood tissues is probably occulted in pectic compounds of galacturonic acids cross-linked by Ca and Sr bridges (Buvat 1989). Following the hypothesis of element migration, residual fraction of Sr should be a more robust signature of the Sr fixed during the year of ring formation. By contrast, more labile fractions might display another isotopic signal. However, some exploratory Sr isotopic measurements on the mobile fraction of wood (CH_3COONH_4 extracted) do not differed from that on the whole ring sample (Figure 21.3).

Second, Sr isotopic measurements were performed on tree rings of European beech and pedunculate oak from other forest stands of central and high Belgium (Figure 21.4). We measured a steep decrease of the $^{87}Sr/^{86}Sr$ ratio in tree-rings dated from ~1870 to ~1920, except for sites of central Belgium with higher soil Ca reserve. The modification of the isotopic signal occurs in a relatively short period (1870–1920). This clearly shows a decrease of the weathering contribution (high $^{87}Sr/^{86}Sr$ ratio) to the vegetation supply that benefits to the atmospheric source, which is characterized by a low $^{87}Sr/^{86}Sr$ ratio. Other tree-ring chronologies of sensitive stands (*i.e.*, growing on base-poor soils) in Scandinavia (Åberg 1995) and United States (Bullen and Bailey 2005) display the same isotopic pattern and show the geographic extent of this perturbation. But foremost, the comparison between these dendrochemical studies enlightens a synchronic decrease of the Sr isotopic ratio in the growth rings of different tree species and arguably point out to a global environmental effect.

We can consider several mechanisms to explain this major change of Ca nutrition of forest sites with critical Ca-status. A modification of the weathering isotope signal by variation of the relative

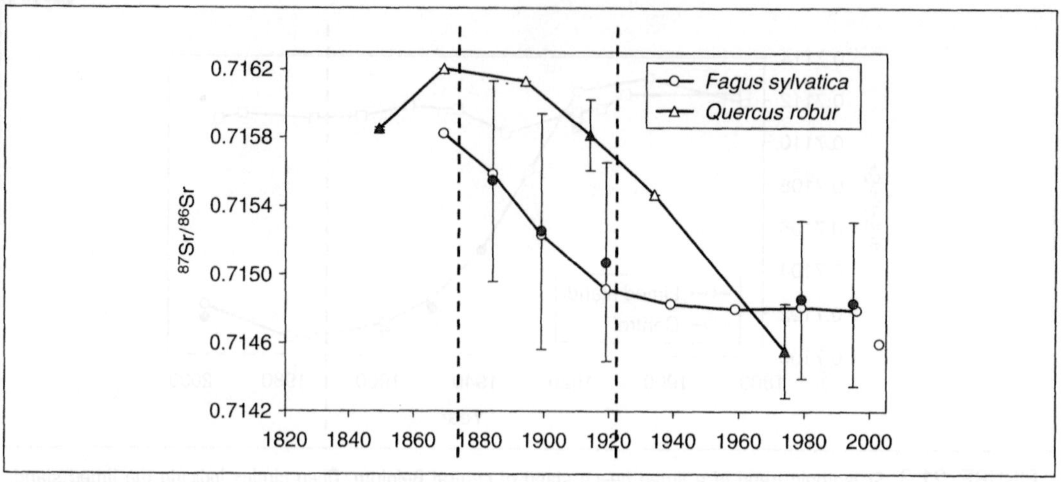

FIGURE 21.4 Sr isotopic chronology for beech (*F. sylvatica* L.) (circles) and oak trees (*Quercus robur* L.) (triangles) in a High Belgium site. Gray symbols indicate the mean ± SE of four individual tree measurements. Open symbols indicate composite samples ($n = 4$). Vertical dotted lines enclose the timescale of rapid evolution of the Sr isotopic ratio.

weathering contributions of individual minerals within a period of ~100 years is very unlikely. An hypothetical effect of tree aging implying change in root prospecting layers (with different isotope compositions) or differences in organic acid production enhancing may also be rejected in view of the very similar isotope signature measured in plantlets, in young and mature trees of two different stands. Our study highlighted the striking parallelism between the beginning of the $^{87}Sr/^{86}Sr$ decrease and the beginning of the increase of sulfur emission corresponding to the Industrial Revolution at the end of the nineteenth century. Time trends of these anthropic activities could be reconstructed by means of estimations of SO_4^{2-} deposition (Ulrich 1987) or by records of nonmarine SO_4^{2-} deposition in ice core from Greenland or Alpine glaciers (Mayewski *et al.* 1990, Schwikowski *et al.* 1999). Deposition of sulfates is known to increase the loss of basic cations from the rooting zone by leaching (Reuss 1983). Cations are displaced from the soil exchange complex at a rate faster than the replenishment of the cation pool by mineral weathering. As a result, the Sr isotope composition of the soil solution taken up by trees moves toward that of the atmospheric source. This explanation is also in agreement with the finding of a soil desaturation in base cation through the time in the same regions (Falkengren-Grerup *et al.* 1987, Thimonier *et al.* 2000). This suggests that forest ecosystems were affected by atmospheric inputs of strong acids earlier than previously thought.

As shown for trees of the limed stand, the Sr isotopic trend is influenced by a lateral reequilibration process in beech rings, which implies that ring activity can last about 50 years after the ring formation. So, this isotope recorder foredates the environmental signal. It may therefore be expected that the effect of acid deposition has not occurred gradually from 1870 to 1920, but more abruptly around the 1920s.

An alternative hypothesis implicates the increase of alkaline earth content (low isotopic ratio) in the atmosphere to explain the drop of the $^{87}Sr/^{86}Sr$ ratio growth rings. This change of the isotopic signature in wood would not be due to a rapid desaturation of the exchange complex—mainly supplied by radiogenic Sr from weathering—but to a replacement of the exchangeable cations by increasing atmospheric inputs with low Sr isotopic ratio. Indeed, besides its role in the acid deposition, Industrial Revolution was also at the origin of an increasing emission of Ca-bearing particles (coal burning, cement manufacturing and so on). However, this hypothesis cannot be confirmed because no precise time trends on total deposition of basic cations and the emission of fine air particulate (PM-10) are available for such early dates. Nevertheless, it must be kept in mind that the annual input of Ca from mineral weathering and the atmosphere are far lower compared to the soil bioavailable reserves. These calcium reserves induce therefore inertia of the system against change in the amount of atmospheric

Ca inputs. So, such a change would probably come into sight through a progressive modification of the $^{87}Sr/^{86}Sr$ ratio in the tree rings and is not compatible with an abrupt change (if we take into account the lateral reequilibration process) reconstituted from the sensitive sites.

The lack of a decrease of the Sr isotopic ratio in tree rings of beech in the central Belgium sites was a very instructive case to complete the explanation of the above-discussed processes. For this stands, we could calculate that an increase of 3% and 9% of the Ca atmospheric contribution should be detectable at a significant to highly significant level, respectively (paired *t*-test). So, the change in the Ca source proportion of 20–30% that could be necessary for the isotopic shift in the High Belgium trees, following the second hypothesis, would have been also detectable in central Belgium. The stability of the Sr isotopic signature in the beech wood of central Belgium could rather be explained by a soil-effective base saturation (BS_e), on average higher than the critical value of 20% in the upper meter. Above these values, the soil solution is considered to be buffered mainly by release of base cations from the exchange complex (Reuss 1983) with higher $^{87}Sr/^{86}Sr$ ratio. Beneath this value, neutralization of acidity is provided by Al and the soil solution could be dominated by atmospheric Sr (Figure 21.2).

VI. PERSPECTIVES

These different studies show that the $^{87}Sr/^{86}Sr$ ratio is a useful tool as tracers of the origin of Ca in several levels of the forest ecosystem. Nevertheless, the measurement of the $^{87}Sr/^{86}Sr$ ratio does not give information on the internal cycle of the Ca (OM mineralization, cycling through the vegetation, and so on). However, these internal fluxes are generally dominant in forest compared to weathering and atmospheric inputs. The use of other isotopes (Ca, Mg) could constitute a complementary tool to determine the effect of environmental changes on the nutrient cycle and the possibility of the ecosystem to recover from acidification. Despite some technical difficulties, Ca isotopes can be measured with precision today (DePaolo 2004). Several studies showed that biotic processes fractionate Ca isotopes (Schmitt and Stille 2005). In forest ecosystems, light Ca isotopes are privileged in the absorption by the vegetation. As a consequence, soil is enriched in heavy isotopes at each element cycling through the vegetation (Wiegand *et al.* 2005). Given the isotopic fractionation induced by each cycling of Ca through the biological material, $^{44}Ca/^{40}Ca$ ratio could be used in the forest stands as a direct measurement of the recycling intensity of nutrients by the vegetation and offers new perspectives of research.

VII. SUMMARY

Calcium is essential in forest ecosystems, both as nutrient for trees and for its role in the neutralization of acid inputs. The decline in bioavailable Ca reserve in the soil, caused by acid deposition produced by industrial activity, was reported for a lot of forest sites in Europe and North America. The understanding of the causal mechanisms requires a precise knowledge of the supplying sources of Ca. Measurements of natural Sr isotopes can be used to determine accurately the origin of Ca in the vegetation. By means of this method, we highlighted the extreme dependence of some forest ecosystems to the Ca input by atmospheric precipitation. The use of this isotopic tracer permits also to characterize the Ca exchanges between the ecosystem components. Finally, the coupling of this method with dendrochemistry adds an important piece to the knowledge of the soil acidification story. Change of the Sr isotopic ratio in the tree rings of several Belgian sites, and the comparison with similar trends in other regions suggest that the soil acidification in industrial regions started in the beginning of the twentieth century, that is, earlier than it is generally admitted.

VIII. ACKNOWLEDGMENTS

This research was supported by the Fonds National de la Recherche Scientifique (FNRS, Belgium, convention no. 2.457002 F) and by student research grants from the FRIA for the first author.

IX. REFERENCES

Åberg, G. 1995. The use of natural strontium isotopes as tracers in environmental studies. *Water Air and Soil Pollution* **79**:309–322.

Blake, L., K. W. T. Goulding, C. J. B. Mott, and A. E. Johnston. 1999. Changes in soil chemistry accompanying acidification over more than 100 years under woodland and grass at Rothamsted Experimental Station, UK. *European Journal of Soil Science* **50**:401–412.

Blum, J. D., and Y. Erel. 1997. Rb-Sr isotope systematics of a granitic soil chronosequence: The importance of biotite weathering. *Geochimica et Cosmochimica Acta* **61**:3193–3204.

Blum, J. D., A. Klaue, C. A. Nezat, C. T. Driscoll, C. E. Johnson, T. G. Siccama, C. Eagar, T. J. Fahey, and G. E. Likens. 2002. Mycorrhizal weathering of apatite as an important calcium source in base-poor forest ecosystems. *Nature* **417**:729–731.

Bullen, T. D., and S. W. Bailey. 2005. Identifying calcium sources at an acid deposition-impacted spruce forest: A strontium isotope, alkaline earth element multi-tracer approach. *Biogeochemistry* **74**:63–99.

Bullen, T. D., A. White, A. Blum, J. Harden, and M. Schulz. 1997. Chemical weathering of a soil chronosequence on granitoid alluvium: II. Mineralogic and isotopic constraints on the behavior of strontium. *Geochimica et Cosmochimica Acta* **61**:291–306.

Buvat, R. 1989. *Ontogeny, Cell Differentiation, and Structure of Vascular Plants.* 581 pp. Springer-Verlag, Berlin.

Capo, R. C., B. W. Stewart, and O. A. Chadwick. 1998. Strontium isotopes as tracers of ecosystem processes: Theory and methods. *Geoderma* **82**:197–225.

Clow, D. W., M. A. Mast, T. D. Bullen, and J. T. Turk. 1997. Strontium 87/strontium 86 as a tracer of mineral weathering reactions and calcium sources in an alpine/subalpine watershed, Loch Vale, Colorado. *Water Resources Research* **33**:1335–1351.

DePaolo, D. J. 2004. Calcium isotopic variations produced by biological, kinetic, radiogenic and nucleosynthetic processes. *Reviews in Mineralogy and Geochemistry* **55**:255–288.

Drouet, T. 2005. Etude de l'origine du calcium dans les écosystèmes forestiers par les méthodes de géochimie et de dendrochimie isotopiques du strontium. Ph.D. Thesis (Université Libre de Bruxelles), 151 pp.

Drouet, T., J. Herbauts, W. Gruber, and D. Demaiffe. 2005a. Strontium isotope composition as a tracer of calcium sources in two forest ecosystems in Belgium. *Geoderma* **126**:203–223.

Drouet, T., J. Herbauts, and D. Demaiffe. 2005b. Long-term records of strontium isotopic composition in tree-rings suggest changes in forest calcium sources in the early 20th century. *Global Change Biology* **11**:1926–1940.

Drouet, T., J. Herbauts, W. Gruber, and D. Demaiffe. 2007. Natural strontium isotope composition as a tracer of weathering patterns and of exchangeable calcium sources in acid leached soils developed on loess of central Belgium. *European Journal of Soil Science* **58**:302–319.

Falkengren-Grerup, U., N. Linnermark, and G. Tyler. 1987. Changes in acidity and cation pools of south Swedish soils between 1949 and 1985. *Chemosphere* **16**:2239–2248.

Hedin, L. O., L. Granat, G. E. Likens, T. A. Buishand, J. N. Galloway, T. J. Butler, and H. Rodhe. 1994. Steep declines in atmospheric base cations in regions of Europe and North America. *Nature* **367**:351–354.

Herbauts, J., V. Penninckx, W. Gruber, and P. Meerts. 2002. Radial variations in cation exchange capacity and base saturation rate in the wood of pedunculate oak and European beech. *Canadian Journal of Forest Research* **32**:1829–1837.

Houle, D., L. Duchesne, J.-D. Moore, M. R. Laflèche, and R. Ouimet. 2002. Soil and tree-ring chemistry response to liming in a sugar maple stand. *Journal of Environmental Quality* **31**:1993–2000.

Kennedy, M. J., L. O. Hedin, and L. A. Derry. 2002. Decoupling of unpolluted temperate forests from rock nutrient sources revealed by natural $^{87}Sr/^{86}Sr$ and ^{84}Sr tracer addition. *Proceedings of the National Academy of Sciences of the United States of America* **99**:9639–9644.

Mayewski, P. A., W. B. Lyons, M. J. Spencer, M. S. Twickler, C. F. Buck, and S. Whitlow. 1990. An ice-core record of atmospheric response to anthropogenic sulphate and nitrate. *Nature* **346**:554–556.

Nakano, T., and T. Tanaka. 1997. Strontium isotope constraints on the seasonal variation of the provenance of base cations in rain water at Kawakami, central Japan. *Atmospheric Environment* **31**:4237–4245.

Poszwa, A., E. Dambrine, B. Pollier, and O. Atteia. 2000. A comparison between Ca and Sr cycling in forest ecosystems. *Plant and Soil* **225**:299–310.

Poszwa, A., T. Wickman, E. Dambrine, B. Ferry, J.-L. Dupouey, G. Helle, G. Schleser, and N. Breda. 2003. A retrospective isotopic study of spruce decline in the Vosges mountains (France). *Water, Air and Soil Pollution: Focus* **3**:201–222.

Reuss, J. O. 1983. Implications of the Ca-Al exchange system for the effect of acid precipitation on soils. *Journal of Environmental Quality* **12**:591–595.

Schmitt, A.-D., and P. Stille. 2005. The source of calcium in wet atmospheric deposits: Ca-Sr isotope evidence. *Geochimica et Cosmochimica Acta* **69**:3463–3468.

Schwikowski, M., A. Döscher, H. W. Gäggeler, and U. Schotterer. 1999. Anthropogenic versus natural sources of atmospheric sulphate from an Alpine ice core. *Tellus* **51B**:938–951.

Thimonier, A., J. L. Dupouey, and F. Le Tacon. 2000. Recent losses of base cations from soils of *Fagus sylvatica* L. stands in northeastern France. *Ambio* **29**:314–321.

Ulrich, B. 1987. Impact on soils related to industrial activities: Part IV. Effect of air pollutants on the soil. Pages 299–310 *in* H. Barth and P. L'Hermite (Eds.) *Scientific Basis for Soil Protection in the European Community.* Elsevier Applied Science, London.

Wiegand, B. A., O. A. Chadwick, P. M. Vitousek, and J. L. Wooden. 2005. Ca cycling and isotopic fluxes in forested ecosystems in Hawaii. *Geophysical Research Letters* **32**:L11404.

Section 6

New Challenges and Frontiers: Biodiversity, Ecological Change and Stable Isotope Networks

CHAPTER 22

Addressing the Functional Value of Biodiversity for Ecosystem Functioning Using Stable Isotopes

Ansgar Kahmen* and Nina Buchmann†

*Department of Integrative Biology, University of California
†Institute of Plant Sciences, ETH Zurich, Universitaetsstr. 2

Contents

I. INTRODUCTION

The earth's biosphere is altered by human beings at an unprecedented rate. Conversion of natural ecosystems to agricultural land and urban developments have already changed one third to one half of the earth's ice-free terrestrial surface, impacting ecosystem services on a continental scale (Turner *et al.* 1990). Also, the breakdown of biogeographic barriers by international travel and trade has caused species to extend their range far beyond their natural range, invading and changing the properties of entire ecosystems (Brooks *et al.* 2004). Probably the most dramatic of all man caused direct changes of the biosphere is, however, the rapid decline of biological diversity (Chapin *et al.* 2000). In contrast to agricultural land that can be reconverted to forests or invading species that can—with large efforts—be eradicated from an invaded area, the genetic code of extinct species is irreversibly lost.

Extinction rates of plants, fungi, and animals that are observed today already exceed natural background extinction levels by several orders of magnitude and are predicted to further increase in the future (MA 2005). In fact, studies predict that of the estimated ~10 million species presently living on earth (May 1990), a large proportion might go extinct in the next 100 years (Pimm *et al.* 1995, Thomas *et al.* 2004). Other than for the five mass extinction events in the geologic past, the causes of current extinctions are well understood and have been described extensively in the scientific and nonscientific literature (Ehrlich and Ehrlich 1981). What remains largely unclear, however, are the consequences this decline in biological diversity will have for ecosystem functioning and the Earth system and consequently for the well-being of humans and societies.

The unknown consequences of declining biodiversity for the Earth system have created an outstanding research interest in the ecological community during the last 15 years. In fact, research addressing the relationship between biodiversity and what is termed "ecosystem functioning" is currently one of the most widely discussed topics in the scientific ecological community. The interest in the effect of species diversity on ecosystem functioning is, however, not new but actually one of the oldest questions in ecology. In 1859, Charles Darwin postulated in "The origin of Species" that "*It has been experimentally proven that if a plot of ground be sown with one species of grasses, a greater number of plants and a greater weight of dry herbage can thus be raised.*" (Darwin 1859, Hector and Hooper 2002). In the last decade, modern "Biodiversity and Ecosystem Functioning Research" was largely triggered by a state of the art conference in Germany in 1991, with the proceedings published in 1993 (Schulze and Mooney 1993). Since then a large number of experimental studies have been conducted mainly in microcosms (Naeem *et al.* 1994, McGrady-Steed *et al.* 1997) and experimental grasslands (Hector *et al.* 1999, Tilman *et al.* 2001, Roscher *et al.* 2004). [For a complete review of experimental biodiversity and ecosystem functioning studies, see Loreau *et al.* (2002) and Hooper *et al.* (2005)]. In these experiments, ecosystem functions are typically tested along artificial diversity gradients. Despite a vivid debate in the scientific community about the experimental design of many of the above studies and thus the general applicability of the observed results (Grime 1997, Huston 1997, Aarssen 2001), a general positive but asymptotic effect of biodiversity on ecosystem functioning has now largely been agreed on (Loreau *et al.* 2001, Hooper *et al.* 2005).

Three functional mechanisms have been proposed to explain the observed positive effect of biodiversity on ecosystem functioning: selection effect (Huston 1997), resource facilitation (Cardinale *et al.* 2002, Spehn *et al.* 2002), and niche complementarity (Tilman *et al.* 1996, Hector 1998, Loreau 1998, Loreau and Hector 2001). The niche complementarity hypothesis states that plant species in an ecosystem occupy distinct ecological niches and use resources in a complementary manner, so that increasing numbers of species result in a more effective resource exploitation, leading in turn to enhanced ecosystem functions. Resource facilitation suggests that species alter the environment such that it benefits another co-occurring species, leading to a greater resource use and thus positive and nonadditive effects of biodiversity on ecosystem functioning. Alternatively, the sampling or selection effect suggests that the chance of a community to hold a species with a particular functional trait that drives the overall function of an ecosystem increases with increasing diversity. Consequently, positive biodiversity effects on ecosystem functioning should largely result of statistical likelihood rather than "true" diversity effects.

Despite the growing body of scientific literature addressing the effect of biodiversity on ecosystem functioning, only a few studies have directly tested the functional mechanisms described above. In these studies, the application of stable isotopes as tools—both as tracers and at natural abundance levels—has significantly advanced the understanding of the functional mechanisms underlying the biodiversity and ecosystem functioning relationship. In the chapter presented here, we highlight studies that have directly addressed the functional mechanisms of biodiversity and ecosystem functioning using stable isotopes. Our chapter reveals that stable isotopes have been established as a valuable and indispensable tool that allows to link community and ecosystem ecology, but that research on biodiversity and ecosystem functioning could benefit much more by combining classical methodology with stable isotope analyses.

II. NICHE COMPLEMENTARITY

The niche complementarity hypothesis states that plant species occupy functionally distinct fundamental niches and thus use resources in a complementary way. As a result, a decline in species diversity would lead to a decreased efficiency in resource exploitation and thus impacts on ecosystem functioning. In most temperate ecosystems, nitrogen is a critical nutrient that limits plant growth and thus ecosystem productivity (Vitousek and Howarth 1991). Consequently, complementary exploitation of different nitrogen pools by different co-occurring plant species has been suggested to increase overall nitrogen uptake and thus effects ecosystem productivity (Spehn *et al.* 2002, Scherer-Lorenzen *et al.* 2003). First evidence that plant species might in fact partition nitrogen in nitrogen-limited ecosystems comes from several early isotope tracer studies. In these studies, nitrogen partitioning was tested for co-occurring plant species as a mechanism avoiding competitive exclusion and thus explaining species coexistence. McKane *et al.* (1990), for example, have shown relative differences in spatial and temporal ^{15}N uptake among six herbaceous species that co-occur in an old-field community. Likewise, spatial partitioning of nitrogen uptake was shown for four European plant species that co-occur in temperate grasslands (Jumpponen *et al.* 2002). In addition, data from a ^{15}N tracer experiment in a tundra ecosystem revealed that different plant species in a community not only differ in their relative spatiotemporal nitrogen uptake patterns but also show contrasting preferences for different chemical forms of nitrogen such as nitrate, ammonium, and amino acids (McKane *et al.* 2002).

An alternative set of studies has used naturally occurring differences in N isotope ratios among co-occurring plant species to infer nitrogen partitioning in nitrogen-limited ecosystems (Schulze *et al.* 1994, Nadelhoffer *et al.* 1996, Bustamante *et al.* 2004). Such investigations are possible since nitrogen sources, such as nitrate and ammonium, often differ naturally in their $\delta^{15}N$ values as a result of fractionation during mineralization and nitrification (Högberg 1997). Also, the isotopic signature of soil nitrogen changes with soil depth where nitrogen in deeper soil layers is typically enriched in ^{15}N compared to shallow soil layers (Nadelhoffer and Fry 1994). Finally, different nitrogen uptake strategies such as uptake of nitrogen via mycorrhizal symbionts also effect the $\delta^{15}N$ values of plants (Read 1994). As a result, differences in the natural abundance of ^{15}N isotopes among different plants (species) indicate different chemical and spatial sources or strategies of nitrogen acquisition among plants.

Studies utilizing the natural abundance of ^{15}N isotopes to compare nitrogen uptake patterns among co-occurring plant species have now been performed in several different ecosystems. Schulze *et al.* (1994), for example, investigated the $\delta^{15}N$ values of spruce trees (*Picea glauca* and *P. mariana*) and other accompanying species in a boreal forest in Alaska. From their data the authors conclude that in the same habitat spruce trees use a different nitrogen source (inorganic N from shallow soil layers) than *Vaccinium vitis-idaea* (organic nitrogen) and the grass *Calamagrostis canadensis* (inorganic nitrogen from deeper soil layers). In a similar study, the variability in ^{15}N isotopes among co-occurring plant species in an Alaskan tundra ecosystem indicates that Tundra plants partition the limiting soil nitrogen pool to avoid competitive exclusion (Nadelhoffer *et al.* 1996). Bustamante *et al.* (2004) investigated differences in $\delta^{15}N$ isotope ratios among 45 different plant species that grow in the cerrado of central Brazil. The range of $\delta^{15}N$ values described spanned from −5‰ to +7.9‰. The authors relate these differences to several factors that influence the nitrogen nutrition of the investigated species, including the presence of nitrogen-fixing legumes, variability in different organic and inorganic nitrogen sources, associations with mycorrhizal fungi, or variability with depth.

In summary, the studies described above give strong evidence that co-occurring plant species partition nitrogen sources and thus differ in their realized niches. While resource partitioning is an important mechanism avoiding competitive exclusion and thus explaining species coexistence, it cannot be automatically concluded that resource partitioning as described in the studies above also leads to enhanced resource use on the ecosystem level as suggested by the niche complementarity hypothesis. For complementary use of a limiting resource to increase overall resource use, species have to differ not only in their realized niches but must differ in their fundamental niches. Moreover, characterizing the species' fundamental niches has to be based on comparisons of quantitative (absolute resource uptake from a specific resource pool) rather than relative resource uptake (relative contribution of a resource pool to the species total resource uptake).

In a ^{15}N tracer study, Kahmen *et al.* (2006) addressed the differences in relative and absolute resource uptake for co-occurring plant species from five functional groups across different temperate European grasslands with respect to spatial, chemical and temporal form of nitrogen acquisition and tested the plasticity of these patterns. While this study confirmed the relative differences with regard to nitrogen uptake among co-occurring plant species observed in earlier studies, absolute nitrogen uptake of the investigated plants differed consistently across functional groups but irrespective of spatial, chemical, or temporal form of nitrogen uptake (Figure 22.1). As a result, Kahmen and co-workers concluded that although different co-occurring plants partition the soil nitrogen pool to avoid competitive exclusion, total nitrogen uptake on the ecosystem level – irrespective of the nitrogen pool considered – depends in these grassland communities on the identity of the species or the functional group and not on plant diversity.

While the nitrogen uptake patterns described by Kahmen *et al.* (2006) give no functional evidence for a positive biodiversity effect on ecosystem functioning, an interesting mechanism with respect to different and potentially complementary nitrogen strategies among different functional groups was detected in the same study (Figure 22.2). Nitrogen concentrations of the investigated species across the different functional groups were negatively correlated with nitrogen uptake from the soil (Figure 22.2), suggesting that plants rely on additional nitrogen sources other than soil nitrogen with different intensities: Legumes use nitrogen derived from nitrogen-fixing bacteria and therefore seem independent of the soil nitrogen pool, while spring herbs, small herbs, grasses, and tall herbs are likely to depend to a

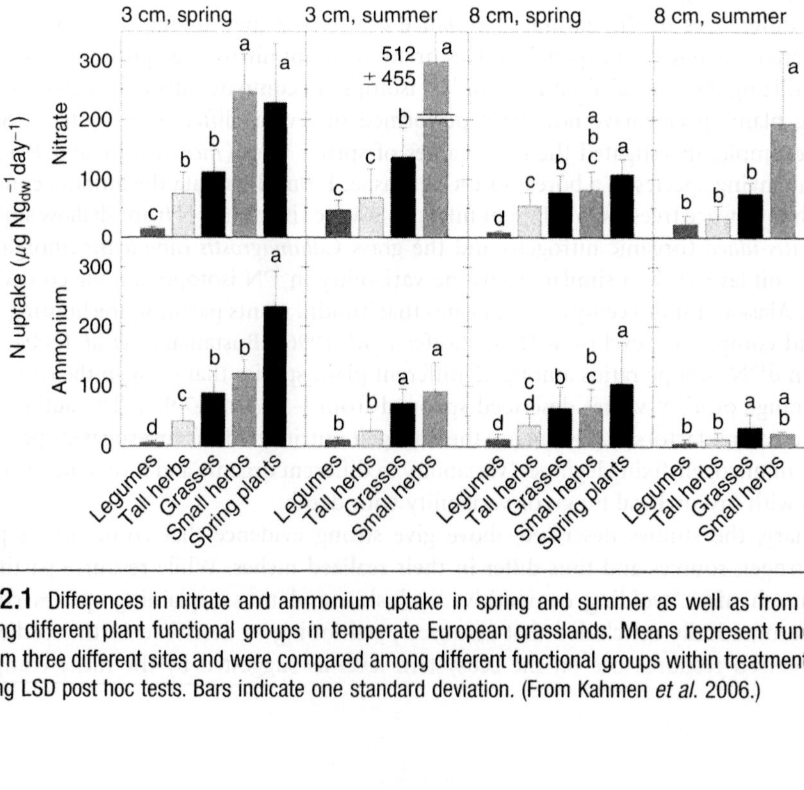

FIGURE 22.1 Differences in nitrate and ammonium uptake in spring and summer as well as from different soil depths among different plant functional groups in temperate European grasslands. Means represent functional group averages from three different sites and were compared among different functional groups within treatments in one way ANOVAs using LSD post hoc tests. Bars indicate one standard deviation. (From Kahmen *et al.* 2006.)

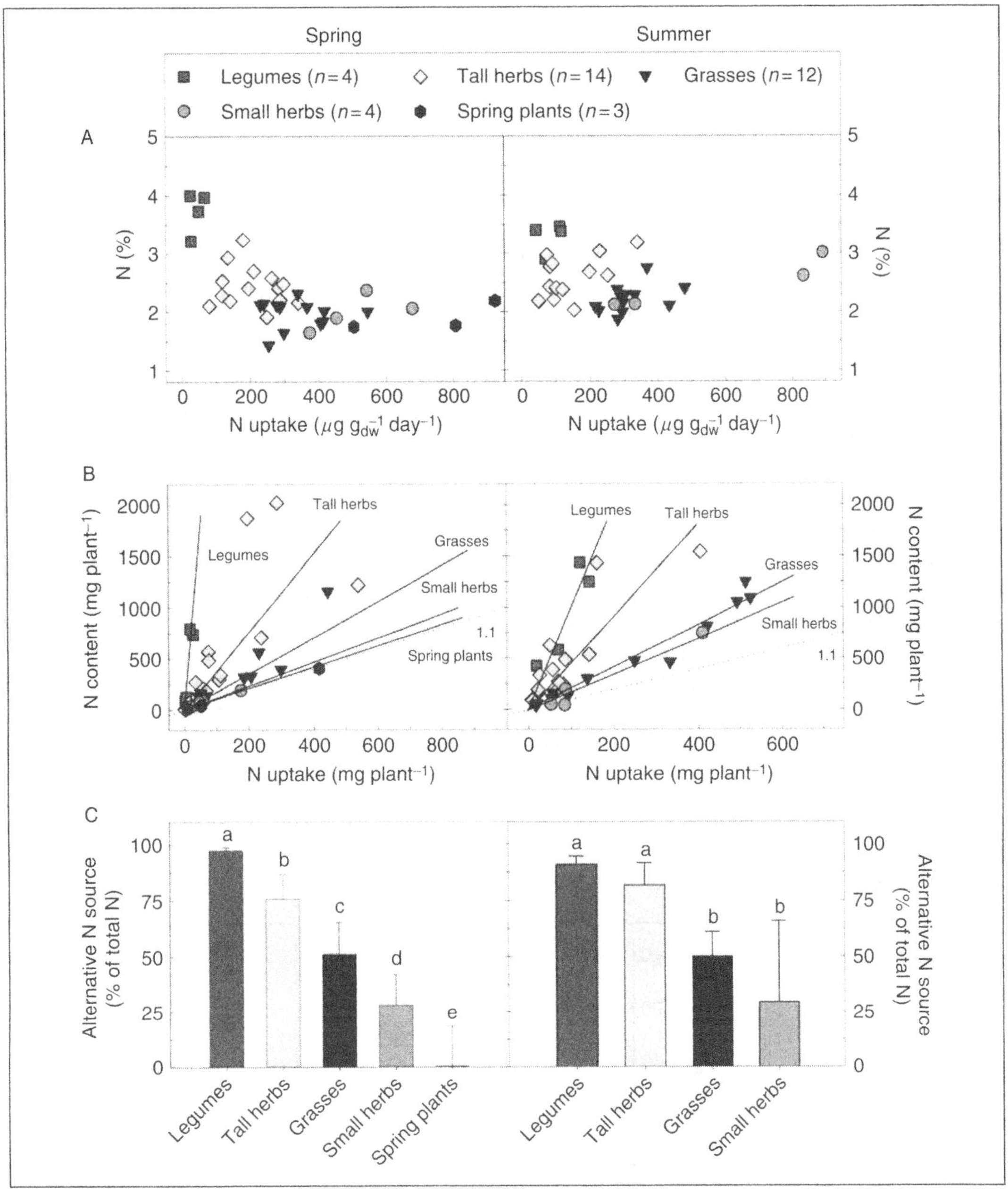

FIGURE 22.2 The relationship between nitrogen uptake and nitrogen concentration in aboveground biomass for 19 different plant species in five functional groups from temperate European grasslands: Average daily nitrogen uptake per gram aboveground biomass of plants in relation to aboveground nitrogen concentration (A). Total seasonal nitrogen uptake per plant correlates positively ($p < 0.05$) with total aboveground nitrogen contents, but slopes are different for functional groups (B). Estimated additional nitrogen sources other than soil nitrogen taken up during the growing season for the five functional groups (C). Means were compared among different functional groups in one way ANOVAs using LSD post hoc tests. Bars indicate one standard deviation. (From Kahmen *et al.* 2006.)

varying degree on the remobilization of accumulated nitrogen to meet their N requirements (Figure 22.2). Since these patterns were consistent across different sites, the observed differences in total nitrogen uptake and nitrogen strategy among functional groups seemed to have little plasticity and are thus likely to characterize different fundamental niches. According to the theory of niche complementarity, co-occurring species or functional groups with different and specific ecological strategies, that is different fundamental niches, will lead to increased resource exploitation and thus to a positive relationship between biodiversity and ecosystem functioning. Kahmen and coworkers therefore suggested that the different functional groups are complementary in their specific nitrogen strategies and that a positive effect of functional group diversity on ecosystem functioning is thus to be expected.

Most biodiversity and ecosystem functioning research in the last decade has focused on temperate grasslands where nitrogen is believed to be the critical and limiting resource (Hooper *et al.* 2005). Consequently, research addressing niche complementartity and resource partitioning has largely focused on nitrogen. Partitioning and complementary use of limiting resources among different co-occurring plant species was, however, also detected for water in a desert ecosystem (Ehleringer *et al.* 1991). Using the natural abundance of hydrogen isotopes in xylem water, Ehleringer and coworkers were able to show that utilization of different water pools by plants was strongly life form dependent (Figure 22.3). While annuals and succulent perennial species used exclusively summer precipitation, herbaceous and woody perennials used both, summer and winter precipitation. Finally, deep-rooted perennials exploited only winter precipitation and groundwater sources. The clear distinction of the different water pools exploited by plants of different life forms suggests significantly different niches with respect to water use. Thus, the observed life form diversity in the investigated desert ecosystem is likely to affect overall functioning of the ecosystem, although this was not explicitly tested in this study.

Similar studies on water partitioning have been conducted in a seasonally dry tropical forest (Meinzer *et al.* 1999), a dry neotropical savanna (Jackson *et al.* 1999), and a Mediterranean Macchia community (Valentini *et al.* 1992). In all these ecosystems, a comparison of natural abundance of hydrogen isotopes in soil water and xylem water revealed that co-occurring shrubs and trees utilize water from different soil depths. As for most studies addressing nitrogen partitioning these studies were

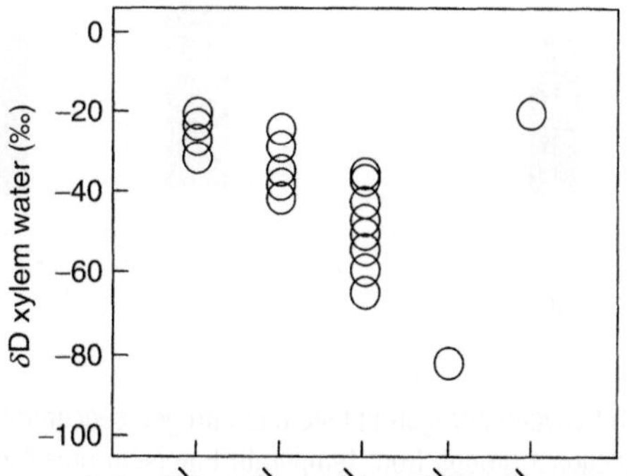

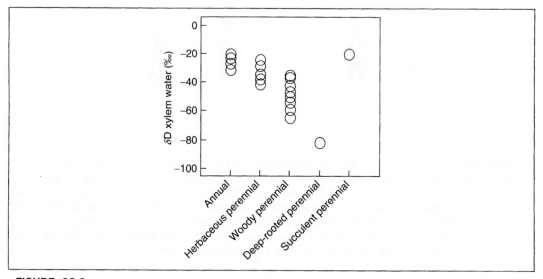

FIGURE 22.3 Natural abundance of hydrogen isotopes in xylem water of plants in a desert shrub ecosystem in southern Utah reveals that—depending on the life form—plants exploit different water pools. (From Ehleringer *et al.* 1991.)

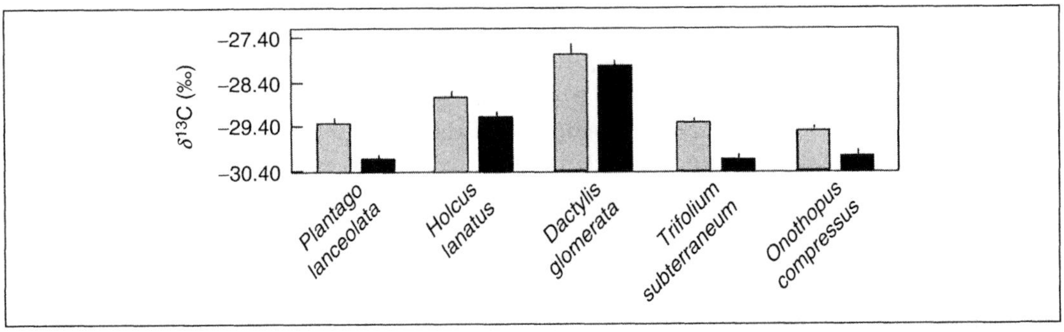

FIGURE 22.4 Effects of plant diversity on bulk leaf $\delta^{13}C$ values of five different plant species in a Mediterranean grassland. Gray bars indicate plants growing in monocultures, black bars indicate plants from mixtures. (Modified after Caldeira *et al.* 2001.)

intended to investigate resources partitioning as a mechanism to avoid competitive exclusion and explaining species coexistence. It therefore remains unclear if the differential water uptake patterns observed for different plant species are the result of different realized or fundamental niches. If the contrasting water uptake patterns were the result of the plants' different fundamental niches, increased water uptake with functionally distinct species should lead to a positive biodiversity effect on ecosystem functioning.

Finally, Caldeira *et al.* (2001) investigated effects of plant diversity on ecosystem functioning in a dry Mediterranean grassland. In their study, the effects of plant diversity not only on biomass production but also on the leaf carbon isotope ratio ($\delta^{13}C$) of plant species growing in either monocultures or multispecies mixtures was investigated. The leaf carbon isotope ratio serves as an index of the ratio of intercellular to ambient CO_2 concentrations (c_i/c_a) when carbon in the leaf is assimilated, and can be related to stomatal behavior (Farquhar and Richards 1984). As a result, $\delta^{13}C$ values of plant material can be used to reflect plant water use. Caldeira and coworkers found that plants growing in species mixtures had significantly lower $\delta^{13}C$ values than plants growing in monocultures. This suggests that plants in mixtures suffered less water stress than plants growing in monocultures (Figure 22.4). Higher soil moisture levels in plots with species mixtures supported this finding. Caldeira and coworkers gave two possible interpretations of the data. First, reduced water stress in mixed communities, resulting from complementary water use of different water pools, reduced interspecific competition and thus water stress. Alternatively, the increased diversity, which also caused increased community biomass, could have let to facilitative effects such as reduced evaporation through increased ground cover or dew interception. Although the mechanisms were not directly addressed, $\delta^{13}C$ values in plant tissue gave strong evidence of a positive water-related effect of biodiversity on ecosystem functioning.

III. RESOURCE FACILITATION

Interactions among different plant species in a community are largely viewed to be driven by competition. In face of the biodiversity and ecosystem functioning debate, however, positive interactions among plant species, where one species facilitates resources to another, have received increasing attention among plant ecologists. A classical example of resource facilitation among plant species comes from agricultural and intercropping experiments, addressing the interactions of legumes and grass species (Vandermeer 1989). Legumes with their ability to directly use atmospheric nitrogen via symbiotic nitrogen-fixing bacteria have been shown in a vast number of experiments to affect ecosystem nitrogen dynamics such as soil fertility, litter quality, or nitrogen mineralization. In addition, legumes increase nitrogen supply rates into the soil and thus increase the nitrogen availability for nonnitrogen-fixing neighboring plants. Using a pool dilution method, where soil nitrogen is experimentally enriched in ^{15}N isotopes, several studies have shown a direct transfer of legume-fixed nitrogen to neighboring grass species (Broadbent *et al.* 1982, Goodman and Collison 1986, Boller and Nösberger 1987). Nitrogen transfer from legumes to nonnitrogen-fixing plant species in a community is thus an important facilitation mechanism, where the trait of one species or functional group alters the environment to the benefit of another species, leading to nonadditive effects of biodiversity on ecosystem functioning.

Despite ample evidence that legumes facilitate nitrogen to neighboring plant species, most of these data come from agricultural and species poor ecosystems; the facilitative effects of legumes in naturally diverse ecosystems have received less attention (Vitousek *et al.* 2002). Therefore, the role of legumes as keystone species for ecosystem functioning was investigated with much detail in the context of the pan-European BIODiversity and Ecological Processes in Terrestrial Herbaceous ecosystems (BIODEPTH) project (Mulder *et al.* 2002, Scherer-Lorenzen *et al.* 2003, Spehn *et al.* 2005). Similar to previous studies in agricultural systems, the BIODEPTH project revealed that legumes had major effects on nitrogen accumulation in the total community biomass, leading to increased productivity (Spehn *et al.* 2002). Further $\delta^{15}N$ signatures revealed that nitrogen that was fixed symbiotically by legumes was transferred to other plant species (Figure 22.5). Although the degree of N_2 fixation varied among different legume

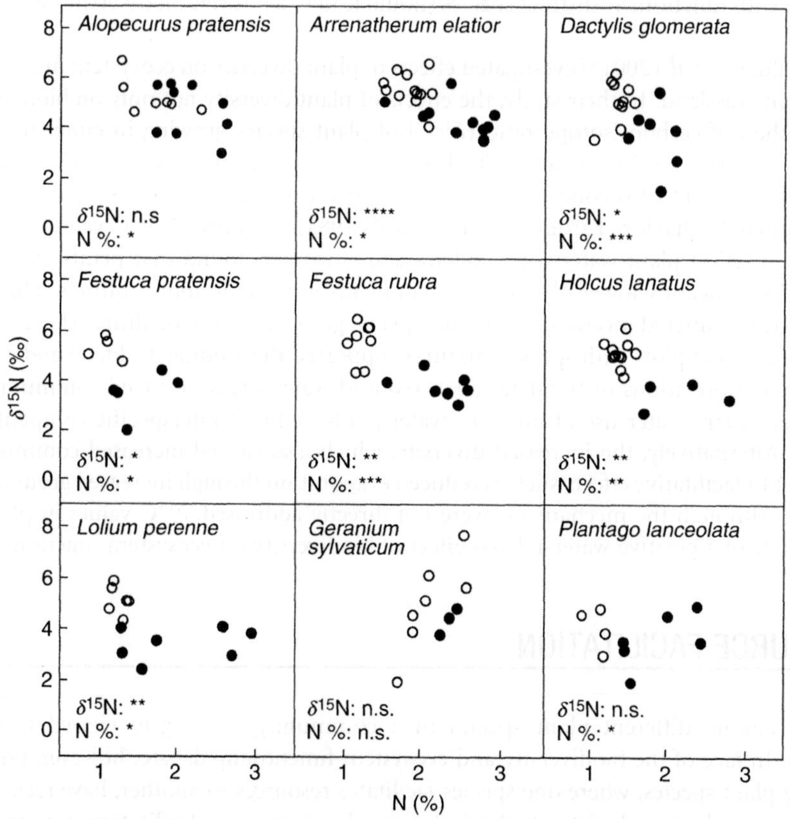

FIGURE 22.5 Presence of legumes affects $\delta^{15}N$ and %N values of co-occurring plant species in temperate grasslands. Closed symbols represent plants co-occurring with legumes, open symbols indicate plants growing without legumes. (From Spehn *et al.* 2002.)

species, the authors concluded that nitrogen fixation by legumes was one of the major functional traits that affected ecosystem functioning in their study. Plant neighbors to a legume plant seem, however, benefit to very different degrees from this additional nitrogen source (Temperton *et al.* 2007). In the Jena Experiment, Temperton *et al.* found the largest increase in biomass due to this additional N input for the grass *Festuca pratense* and a lesser biomass gain for *Plantago lanceolata* and *Knautia arvensis* while another legume species (*Trifolium pratense*) was negatively affected. These results agree with the findings of Kahmen *et al.* (2006), where the N use strategies of these species also differed greatly. Mulder *et al.* (2002) followed the effect of legumes on neighboring plants for several years in the BIODEPTH experiment, showing that the legume effect decreased with the duration of the experiment. This pattern was explained with the competitive advantage of legumes in newly established communities as well as the relatively short life span of legumes and the difficulty of legume reestablishment in mature communities.

Another classic example of resource facilitation among plant species that was detected and investigated with the help of stable isotopes is hydraulic lift (Mooney *et al.* 1980, Richards and Caldwell 1987, Caldwell and Richards 1989). Hydraulic lift describes the nocturnal movement of water from deep moist soil layers or groundwater through roots to upper and dryer soil layers. The water released from the roots to dryer soil layers is typically resorbed the following day, when transpirational demand exceeds water supply. Hydraulic lift can therefore significantly improve the water supply of a plant. This process was first described for *Prosopis tamarugo*, a small tree in the Atacama Desert in northern Peru by Mooney *et al.* (1980) and later identified for the desert shrub *Artemisia tridentata* (Richards and Caldwell 1987) and other species from arid and semiarid ecosystems (Caldwell *et al.* 1998). Interestingly, hydraulic lift is not a unique property of species from arid and semiarid ecosystems, but has also been described for trees from mesic temperate forests (Dawson 1993), suggesting that it is a widespread phenomenon.

While the initial interpretation of hydraulic lift emphasized beneficial effects for the water-lifting plant itself, facilitative effects for neighboring plants were also investigated. In a study using stable hydrogen isotopes as tracers, Caldwell and Richards (1989) showed that water hydraulically lifted by *A. tridentata* was recovered in *Agropyron desertorum.* While this finding suggested an important mechanism regarding plant–plant interactions, the quantitative importance for ecosystem structure and function remained unclear. The magnitude of hydraulically lifted water utilization by neighboring plants was first demonstrated by Dawson (1993), using the natural abundance of hydrogen in different ecosystem water pools in a temperate *Acer saccharum* forest (Figure 22.6). This isotope study revealed that understory plant species in the vicinity of the tree received 3–60% of their xylem water from water facilitated by *A. saccharum*, having significant effects on the water status and growth of these understory plants. As outlined for legume species, water facilitation via hydraulic lift illustrates that the presence of one species with a particular functional trait can have beneficial and nonadditive effects on other species in a community and can influence the overall function of the ecosystem.

IV. TRUE DIVERSITY EFFECTS VERSUS SAMPLING EFFECTS

The effects of biodiversity on ecosystem functioning were largely tested along experimental diversity gradients, where species diversity was manipulated artificially while other environmental parameters were controlled (Hooper *et al.* 2005). The quality of this experimental design and the applicability of the results generated from these experiments triggered a vivid and occasionally emotional debate in the scientific literature (Kaiser 2000). In summary, several authors have questioned the applicability of the experimental studies for natural ecosystems, where biodiversity will be insignificant compared to the overwhelming influences of environmental and anthropogenic factors on ecosystem functioning (Grime 1997, Wardle *et al.* 1997, Huston and McBride 2002). More importantly, critics have argued that the positive

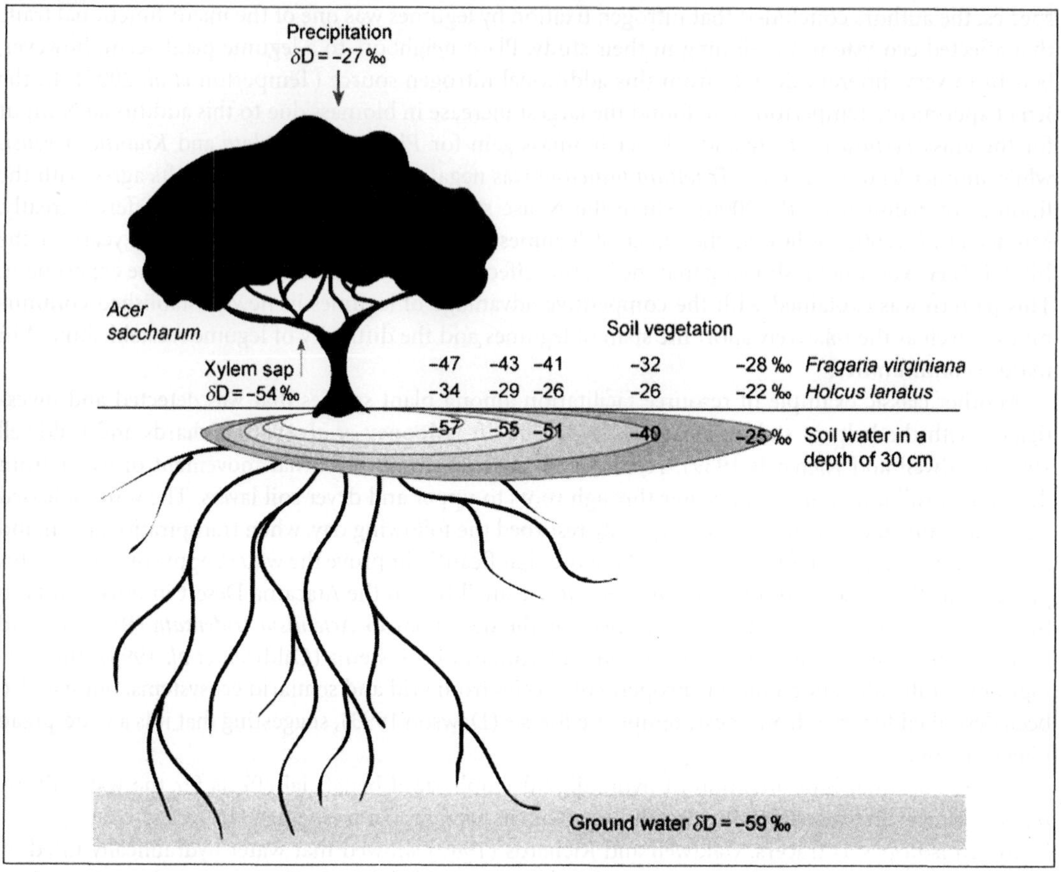

FIGURE 22.6 Natural abundance of hydrogen isotopes in precipitation, groundwater, soilwater, and xylem of trees and understory vegetation reveals the facilitative effects of hydraulically lifted groundwater for understory herbs in *A. saccharum* forests. (From Dawson 1993, modified by Schulze *et al.* 2002.)

relationship between biodiversity and ecosystem functioning detected in experimental studies was based on statistical artifacts such as sampling effects: the higher the diversity in a plot, the greater the chance of highly productive species to be present in the plot, leading to a greater overall productivity in the system (Aarssen 1997, Huston 1997).

In the light of this discussion, several statistical tools have now been developed and deployed that separate "true" or nonadditive diversity effects from statistical artifacts (Loreau and Hector 2001, Fox 2006). While these tests quantify diversity effects based on statistical computations, they provide no information on the functional or physiological mechanisms that drive the observed biodiversity effects. Throughout this chapter we have illustrated a number of examples, where stable isotopes have been used to illustrate that the presence of one species has beneficial and nonadditive effects on the performance of other species in the community. Most of these investigations have, however, not been performed in the context of biodiversity and ecosystem functioning and can therefore only give indirect evidence of potential mechanistic biodiversity effects. Such mechanistic investigations as illustrated here are, however, essential in the context of biodiversity and ecosystem functioning studies to uncover the functional processes behind statistically detected biodiversity effects and to settle the debate about "true" and "sampling" effects in biodiversity and ecosystem functioning studies.

V. CONCLUSIONS AND OUTLOOK

A meta-analysis of almost 450 studies of terrestrial and aquatic systems as well as microcosms (over the last 50 years until mid 2004) about biodiversity effects on ecosystem functioning and services provided further evidence that biodiversity has a positive effect on most ecosystem services such as productivity, nutrient cycling, or pest control (Balvanera *et al.* 2006). However, studies using stable isotopes were still very rare exceptions.

This is even more surprising since many of the still open research questions, for example, about the physiological mechanisms underlying facilitation or resource partitioning or higher level trophic interactions can be tackled perfectly well using isotopes at natural abundance or tracer levels. For example, why do some plants benefit more than others from additional legume N available in diverse communities, where does this N go once it is taken up, how do allocation or storage of this N differ when plants grow in communities of different diversities, how do higher trophic level organisms benefit? Such questions on the fate, use, and pathways of certain resources such as nitrogen, but also water or carbon, can be addressed much better using stable isotope techniques than relying solely on classical ecophysiological or plant population approaches. A much closer link between ecophysiological research where stable isotopes are quite established as methodological tools and plant diversity research seems urgently needed.

VI. REFERENCES

Aarssen, L. W. 1997. High productivity in grassland ecosystems: Effected by species diversity or productive species? *Oikos* **80:**183–184.

Aarssen, L. W. 2001. On correlations and causations between productivity and species richness in vegetation: Predictions from habitat attributes. *Basic and Applied Ecology* **2:**105–114.

Balvanera, P., A. B. Pfisterer, N. Buchmann, J. S. He, T. Nakashizuka, D. Raffaelli, and B. Schmid. 2006. Quantifying the evidence for biodiversity effects on ecosystem functioning and services. *Ecology Letters* **9:**1146–1156.

Boller, B. C., and J. Nösberger. 1987. Symbiotically fixed nitrogen from field-grown white and red clover mixed with ryegrasses at low levels of ^{15}N fertilization. *Plant and Soil* **104:**219–226.

Broadbent, F. E., T. Nakashima, and G. Y. Chang. 1982. Estimation of nitrogen fixation by isotope dilution in field and greenhouse experiments. *Agronomy Journal* **74:**625–628.

Brooks, M. L., C. M. D'Antonio, D. M. Richardson, J. B. Grace, J. E. Keeley, J. M. DiTomaso, R. J. Hobbs, M. Pellant, and D. Pyke. 2004. Effects of invasive alien plants on fire regimes. *BioScience* **54:**677–688.

Bustamante, M. M. C., L. A. Martinelli, D. A. Silva, P. B. Camargo, C. A. Klink, T. F. Domingues, and R. V. Santos. 2004. ^{15}N natural abundance in woody plants and soils of central Brazilian savannas (cerrado). *Ecological Applications* **14:**S200–S213.

Caldeira, M. C., R. J. Ryel, J. H. Lawton, and J. S. Pereira. 2001. Mechanisms of positive biodiversity-production relationships: Insights provided by delta ^{13}C analysis in experimental Mediterranean grassland plots. *Ecology Letters* **4:**439–443.

Caldwell, M. M., and J. H. Richards. 1989. Hydraulic lift: Water efflux from upper roots improves effectiveness of water uptake by deep roots. *Oecologia* **79:**1–5.

Caldwell, M. M., T. E. Dawson, and J. H. Richards. 1998. Hydraulic lift: Consequences of water efflux from the roots of plants. *Oecologia* **113:**151–161.

Cardinale, B. J., M. A. Palmer, and S. L. Collins. 2002. Species diversity enhances ecosystem functioning through interspecific facilitation. *Nature* **415:**426–429.

Chapin, F. S., E. S. Zavaleta, V. T. Eviner, R. L. Naylor, P. M. Vitousek, H. L. Reynolds, D. U. Hooper, S. Lavorel, O. E. Sala, S. E. Hobbie, M. C. Mack, and S. Diaz. 2000. Consequences of changing biodiversity. *Nature* **405:**234–242.

Darwin, C. 1859. *The Origin of Species by means of Natural Selection.* John Murray, London, UK.

Dawson, T. E. 1993. Hydraulic lift and water-use by plants - Implications for water-balance, performance and plant-plant interactions. *Oecologia* **95:**565–574.

Ehleringer, J. R., S. L. Phillips, W. S. F. Schuster, and D. R. Sandquist. 1991. Differential utilization of summer rains by desert plants. *Oecologia* **88:**430–434.

Ehrlich, P. R., and A. H. Ehrlich. 1981. Extinction. *The Causes and Consequences for the Disappearance of Species.* Random House, New York, USA.

Farquhar, G. D., and R. A. Richards. 1984. Isotopic composition of plant carbon correlates with water use efficiency of wheat genotypes. *Australian Journal of Plant Physiology* **11:**539–552.

Fox, J. W. 2006. Using the Price Equation to partition the effects of biodiversity loss on ecosystem function. *Ecology* **87:**2687–2696.

Goodman, P. J., and M. Collison. 1986. Effect of three clover varieties on growth, ^{15}N uptake and fixation by ryegrass white clover mixtures at three sites in Wales. *Grass and Forage Science* **41:**191–198.

Grime, J. P. 1997. Biodiversity and ecosystem function: The debate deepens. *Science* **277:**1260–1261.

Hector, A. 1998. The effect of diversity on productivity: Detecting the role of species complementarity. *Oikos* **82:**597–599.

Hector, A., and R. Hooper. 2002. Darwin and the first ecological experiment. *Science* **295:**639–640.

Hector, A., B. Schmid, C. Beierkuhnlein, M. C. Caldeira, M. Diemer, P. G. Dimitrakopoulos, J. A. Finn, H. Freitas, P. S. Giller, J. Good, R. Harris, P. Högberg, *et al.* 1999. Plant diversity and productivity experiments in European grasslands. *Science* **286:**1123–1127.

Hooper, D. U., F. S. Chapin, J. J. Ewel, A. Hector, P. Inchausti, S. Lavorel, J. H. Lawton, D. M. Lodge, M. Loreau, S. Naeem, B. Schmid, H. Setala, *et al.* 2005. Effects of biodiversity on ecosystem functioning: A consensus of current knowledge. *Ecological Monographs* **75:**3–35.

Huston, M. A. 1997. Hidden treatments in ecological experiments: Re-evaluating the ecosystem function of biodiversity. *Oecologia* **110:**449–460.

Huston, M. A., and A. C. McBride. 2002. Evaluating the relative strengths of biotic versus abiotic controls on ecosystem processes. Pages 47–60 *in* M. Loreau, S. Naeem, and P. Inchausti (Eds.) *Biodiversity and Ecosystem Functioning: Synthesis and Perspectives.* Oxford University Press, Inc., New York.

Högberg, P. 1997. Tansley review No 95 ^{15}N natural abundance in soil-plant systems. *New Phytologist* **137:**179–203.

Jackson, P. C., F. C. Meinzer, M. Bustamante, G. Goldstein, A. Franco, P. W. Rundel, L. Caldas, E. Igler, and F. Causin. 1999. Partitioning of soil water among tree species in a Brazilian Cerrado ecosystem. *Tree Physiology* **19:**717–724.

Jumpponen, A., P. Högberg, K. Huss-Danell, and C. P. H. Mulder. 2002. Interspecific and spatial differences in nitrogen uptake in monocultures and two-species mixtures in north European grasslands. *Functional Ecology* **16:**454–461.

Kahmen, A., C. Renker, S. Unsicker, and N. Buchmann. 2006. Niche complementarity for nitrogen: An explanation for the biodiversity and ecosystem functioning relationship? *Ecology* **87:**1244–1255.

Kaiser, J. 2000. Rift over biodiversity divides ecologists. *Science* **289:**1282–1283.

Loreau, M. 1998. Separating sampling and other effects in biodiversity experiments. *Oikos* **82:**600–602.

Loreau, M., and A. Hector. 2001. Partitioning selection and complementarity in biodiversity experiments. *Nature* **412:**72–76.

Loreau, M., S. Naeem, P. Inchausti, J. Bengtsson, J. P. Grime, A. Hector, D. U. Hooper, M. A. Huston, D. Raffaelli, B. Schmid, D. Tilman, and D. A. Wardle. 2001. Biodiversity and ecosystem functioning: Current knowledge and future challenges. *Science* **294:**804–808.

Loreau, M., S. Naeem, and P. Inchausti. 2002. *Biodiversity and Ecosystem Functioning: Synthesis and Perspectives.* Oxford University Press, Inc., New York.

May, R. M. 1990. How many species? *Philosophical Transactions of the Royal Society of London Series B-Biological Sciences* **330:**293–304.

McGrady-Steed, J., P. M. Harris, and P. J. Morin. 1997. Biodiversity regulates ecosystem predictability. *Nature* **390:**162–165.

McKane, R. B., D. F. Grigal, and M. P. Russelle. 1990. Spatiotemporal differences in ^{15}N uptake and the organization of an old-field plant community. *Ecology* **71:**1126–1132.

McKane, R. B., L. C. Johnson, G. R. Shaver, K. J. Nadelhoffer, E. B. Rastetter, B. Fry, A. E. Giblin, K. Kielland, B. L. Kwiatkowski, J. A. Laundre, and G. Murray. 2002. Resource-based niches provide a basis for plant species diversity and dominance in arctic tundra. *Nature* **415:**68–71.

MA. 2005. *Ecosystems & Human Well-Being—Synthesis Report.* Island Press, Washington DC, USA.

Meinzer, F. C., J. L. Andrade, G. Goldstein, N. M. Holbrook, J. Cavelier, and S. J. Wright. 1999. Partitioning of soil water among canopy trees in a seasonally dry tropical forest. *Oecologia* **121:**293–301.

Mooney, H. A., S. L. Gulmon, P. W. Rundel, and J. Ehleringer. 1980. Further observations on the water relations of *Prosopis tamarugo* of the northern Atacama desert. *Oecologia* **44:**177–180.

Mulder, C. P. H., A. Jumpponen, P. Högberg, and K. Huss-Danell. 2002. How plant diversity and legumes affect nitrogen dynamics in experimental grassland communities. *Oecologia* **133:**412–421.

Nadelhoffer, K. J., and B. Fry. 1994. Nitrogen isotope studies in forest ecosystems. Pages 22–44 *in* K. Lajtha and R. H. Michener (Eds.) Stable isotopes in ecology and environmental sciences. Blackwell Scientific Publications, Oxford, UK.

Nadelhoffer, K., G. Shaver, B. Fry, A. Giblin, L. Johnson, and R. McKane. 1996. ^{15} N natural abundances and N use by tundra plants. *Oecologia* **107:**386–394.

Naeem, S., L. J. Thompson, S. P. Lawler, J. H. Lawton, and R. M. Woodfin. 1994. Declining biodiversity can alter the performance of ecosystems. *Nature* **368:**734–737.

Pimm, S. L., G. J. Russell, J. L. Gittleman, and T. M. Brooks. 1995. The future of biodiversity. *Science* **269:**347–350.

Read, D. 1994. Plant-microbe mutualism and community structure. Pages 181–203 *in* E. D. Schulze and H. A. Mooney (Eds.) *Biodiversity and Ecosystem Functioning.* Springer, Berlin, Heidelberg, New York.

Richards, J. H., and M. M. Caldwell. 1987. Hydraulic lift—substantial nocturnal water transport between soil layers by *Artemisia tridentata* roots. *Oecologia* **73:**486–489.

Roscher, C., J. Schumacher, J. Baade, W. Wilcke, G. Gleixner, W. W. Weisser, B. Schmid, and E. D. Schulze. 2004. The role of biodiversity for element cycling and trophic interactions: An experimental approach in a grassland community. *Basic and Applied Ecology* **5:**107–121.

Scherer-Lorenzen, M., C. Palmborg, A. Prinz, and E. D. Schulze. 2003. The role of plant diversity and composition for nitrate leaching in grasslands. *Ecology* **84:**1539–1552.

Schulze, E.-D., and H. A. Mooney. 1993. *Biodiversity and Ecosystem Function.* Springer Verlag, Berlin Heidelberg.

Schulze, E.-D., F. S. Chapin, and G. Gebauer. 1994. Nitrogen nutrition and isotope differences among life forms at the northern treeline of Alaska. *Oecologia* **100:**406–412.

Schulze, E.-D., E. Beck, and K. Müller-Hohenstein. 2002. *Pflanzenökologie.* Spectrum Verlag Akademischer Verlag, Heidelberg.

Spehn, E. M., M. Scherer-Lorenzen, B. Schmid, A. Hector, M. C. Caldeira, P. G. Dimitrakopoulos, J. A. Finn, A. Jumpponen, G. O'Donnovan, J. S. Pereira, E. D. Schulze, A. Y. Troumbis, *et al.* 2002. The role of legumes as a component of biodiversity in a cross-European study of grassland biomass nitrogen. *Oikos* **98:**205–218.

Spehn, E. M., A. Hector, J. Joshi, M. Scherer-Lorenzen, B. Schmid, E. Bazeley-White, C. Beierkuhnlein, M. C. Caldeira, M. Diemer, P. G. Dimitrakopoulos, J. A. Finn, H. Freitas, *et al.* 2005. Ecosystem effects of biodiversity manipulations in European grasslands. *Ecological Monographs* **75:**37–63.

Temperton, V. M., P. Mwangi, M. Scherer-Lorenzen, B. Schmid, and N. Buchmann. 2007. Positive interactions between nitrogen-fixing legumes and four different neighboring species in a biodiversity experiment. *Oecologia* **151:**190–205.

Thomas, C. D., A. Cameron, P. E. Green, M. Bakkenes, L. Z. Beaumont, Y. C. Collingham, B. F. N. Erasmus, M. F. De Silqueira, A. Grainger, L. Hama, L. Hughes, B. Huntley, *et al.* 2004. Extinction risk from climate change. *Nature* **427:**145–148.

Tilman, D., D. Wedin, and J. Knops. 1996. Productivity and sustainability influenced by biodiversity in grassland ecosystems. *Nature* **379:**718–720.

Tilman, D., P. B. Reich, J. Knops, D. Wedin, T. Mielke, and C. Lehman. 2001. Diversity and productivity in a long-term grassland experiment. *Science* **294:**843–845.

Turner, B. L. I., W. C. Clark, R. W. Kates, J. F. Richards, J. T. Meathews, and W. B. Meyer. 1990. *The Earth as Transformed by Human Action.* Cambridge University Press, Cambridge, England.

Valentini, R., G. E. S. Mugnozza, and J. R. Ehleringer. 1992. Hydrogen and carbon isotope ratios of selected species of a Mediterranean Macchia ecosystem. *Functional Ecology* **6:**627–631.

Vandermeer, J. 1989. *The Ecology of Intercropping.* Cambridge University Press, Cambridge, UK.

Vitousek, P. M., and R. W. Howarth. 1991. Nitrogen limitation on land and in the sea - How can it occur. *Biogeochemistry* **13:**87–115.

Vitousek, P. M., K. Cassman, C. Cleveland, T. Crews, C. B. Field, N. B. Grimm, R. W. Howarth, R. Marino, L. Martinelli, E. B. Rastetter, and J. I. Sprent. 2002. Towards an ecological understanding of biological nitrogen fixation. *Biogeochemistry* **57:**1–45.

Wardle, D. A., O. Zackrisson, G. Hornberg, and C. Gallet. 1997. The influence of island area on ecosystem properties. *Science* **277:**1296–1299.

CHAPTER 23

The Future of Large-Scale Stable Isotope Networks

D. Hemming,* H. Griffiths,† N. J. Loader,‡ A. Marca,§ I. Robertson,‡ D. Williams,¶ L. Wingate,|| and D. Yakir**

*Hadley Centre, Met Office, Exeter. Devon. EX1 3PB
†Department of Plant Sciences, University of Cambridge
‡Department of Geography, University of Wales Swansea
§Stable Isotope Laboratory, School of Environmental Sciences, University of East Anglia
¶Departments of Renewable Resources and Botany, University of Wyoming
||Institute of Atmospheric and Environmental Science, School of GeoSciences, University of Edinburgh
**Department of Environmental Sciences and Energy Research, The Weizmann Institute of Science

Contents

I. INTRODUCTION

Understanding large-scale ecological changes will become increasingly important in the future as increases in population, energy consumption, and changes in climate exacerbate the pressures on ecosystem services, including fresh water, forests, and agricultural products. Skilful management of global ecological resources will be required to advance economic development and reduce poverty worldwide. Addressing and quantifying the nature of these global challenges will require large-scale policy solutions that are informed by multinational and multidisciplinary research projects.

Funding for large-scale ecosystem research has been notoriously difficult to secure and sustain over more than a few years. Yet it is this information that is, and will become even more vital in order to understand and tackle global environmental changes and their impacts on societies. Atmospheric carbon dioxide (CO_2) monitoring provides a good example of this problem; without the insight and tireless efforts of Charles Keeling and colleagues to maintain CO_2 measurements from background atmospheric sites [see Section II.A, and Weart (2003) "Money for Keeling: Monitoring CO_2 levels" essay online: http://www.aip.org/history/climate/Kfunds.htm], we would be ignorant of probably the most convincing record of the impact of human activities on the global climate system.

Measurements of stable isotope ratios present in the important elements of life (hydrogen, carbon, oxygen, nitrogen, and sulphur) provide unique information on the structure and processes associated with ecosystem and biome functioning. They have played a central role in understanding many ecological processes from molecular to global scales. At large scales, encompassing the continents up to the globe, networks of stable isotope measurements offer quantitative information on the status of the Earth's natural, agricultural and urban ecosystems, and the services they supply to humans. They have provided key ecological information which has significantly increased our understanding of the complex integrated processes linking natural exchanges and human activities with the global climate system (Section III). Without these networks, relatively little would be known about many of the recent global changes that threaten the habitability of Earth, such as the sources of rising atmospheric greenhouse gas (GHG) concentrations. This information has played a vital role in guiding and informing international policy issues, for example the United Nations Framework Convention on Climate Change (UNFCCC), Montreal Protocol, and Kyoto Protocol, and is well suited to informing on future global environmental challenges (Section IV).

Despite the clear utility of maintaining large-scale networks of stable isotope measurements for furthering scientific understanding of global environmental issues and informing international policy, there is considerable concern among the isotope research community as to how these networks can be developed and sustained into the future. Conventional sources of funding for scientific research, such a research councils, tend to be focused on relatively short-term projects (three to five years) that do not depend on large spatial scales or international collaborations. It is significant to note that nearly all of the longest running existing large-scale stable isotope networks are maintained only by international nonacademic organizations like the International Atomic Energy Agency (Section II.A.2), or by governments (Section II.A.1), and they are reliant on voluntary sampling at sites around the globe.

To sustain large-scale isotopic monitoring into the future, it is important to identify the key advantages of these networks, particularly when combined with socioeconomic indicators, and to explore how these advantages may be maximized in order to secure long-term funding. In this chapter, we address these points by giving a brief overview of existing large-scale stable isotope networks (Section II), emphasizing some of the key information available from these (Section III) and discussing the possible application of these data to key international policy issues (Section IV). While the importance of targeting research toward the current international policy agenda is emphasized, it is also considered essential to balance this with fundamental research that provides vision toward the future potential applications of large-scale isotope data, and a focus for the development of new technologies that introduce novel and more accurate information. By successfully integrating these aspects, it is hoped that large-scale stable isotope networks will be an important part of a well-developed and vibrant ecological research field that will be sustained into the future.

II. LARGE-SCALE ISOTOPE NETWORKS

In this section, we provide background information on recent large-scale isotope networks. This is not intended as an exhaustive list, rather a synopsis of the range of environments and timescales sampled and the nature of data available. For ease of reference, we describe these according to the general source of samples analyzed, although it should be noted that most networks and laboratories analyze samples from multiple sources.

A. Atmospheric Gases

1. SIO/CIO[1]

In addition to the well-known measurements of background atmospheric CO_2 initiated by Dr Charles Keeling in 1958, Keeling and coworkers at the Scripps Institute of Oceanography (SIO), United States and Dr Willem Mook of Centrum voor Isotopen Onderzeok (CIO), University of Gröningen, The Netherlands, began the first regular isotopic analyses of the carbon ($\delta^{13}C$) and oxygen ($\delta^{18}O$) of atmospheric CO_2 in 1977. Monthly flask samples of background air were collected from their network, which consisted of 10 stations spanning a north–south transect of the Pacific Ocean (Figure 23.1); (Keeling *et al.* 1979, 2005). Since 1992, the samples have been analyzed by Dr Martin Wahlen, also at SIO.

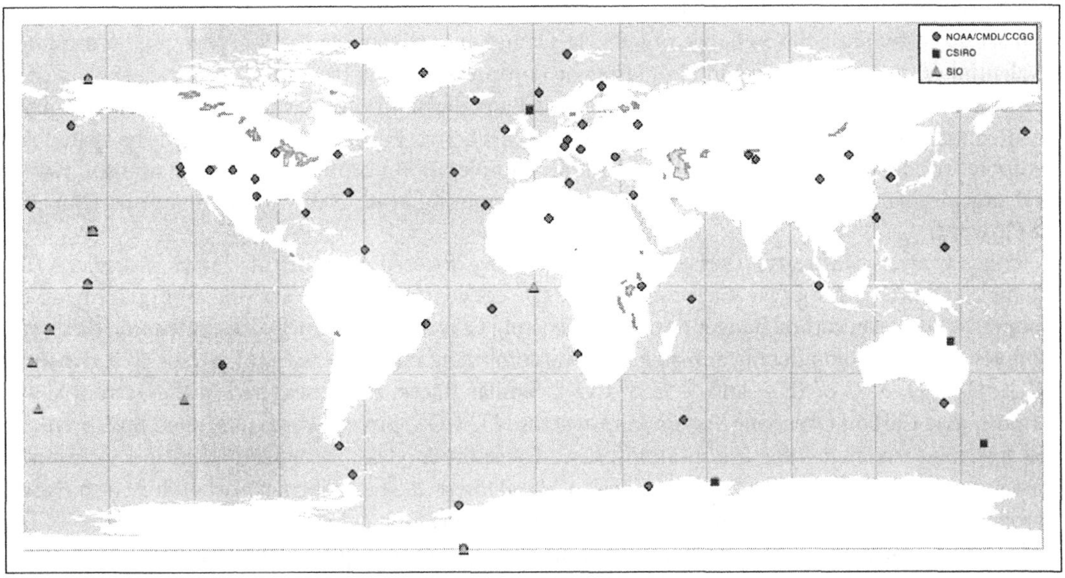

FIGURE 23.1 Location of atmospheric sampling sites for the National Oceanic and Atmospheric Administration Carbon Cycle Greenhouse Gases group (NOAA/CMDL/CCGG), Commonwealth Scientific and Industrial Research Organisation (CSIRO), and Scripps Institute of Oceanography (SIO) networks.

[1] http://cdiac.ornl.gov/trends/co2/iso-sio/iso-sio.html

2. NOAA-CMDL-CCGG/INSTAAR[2]

Stable isotope analyses of atmospheric gases were introduced in 1989 as part of the US Government's National Oceanic and Atmospheric Administration (NOAA), Earth System Research Laboratory (ESRL), formerly Climate Monitoring and Diagnostics Laboratory (CMDL), cooperative air sampling network coordinated by the Carbon Cycle Greenhouse Gases (CCGG) group (Conway *et al.* 1988, Trolier *et al.* 1996). Currently, atmospheric concentration measurements are made by CCGG for a suite of radiatively important trace gases: CO_2, CH_4, CO, H_2, N_2O, and SF_6, and the $\delta^{13}C$, $\delta^{18}O$, and hydrogen isotope (δD) analyses of atmospheric CO_2 and CH_4 are carried out at the University of Colorado's Institute of Arctic and Alpine Research (INSTAAR), Stable Isotope Laboratory (SIL). From an initial selection of six sites and two ships in 1990, these isotope measurements have grown, and now include samples from all the NOAA-CMDL network sites, including a network of very tall towers (Bakwin *et al.* 1998) and aircraft sampling campaigns. This has effectively extended the global coverage of CO_2 stable isotope and concentration measurements by over 55 stations (Figure 23.1).

3. CSIRO-GASLAB[3]

The Commonwealth Scientific and Industrial Research Organisation (CSIRO) Global Atmospheric Sampling LABoratory (GASLAB) was initiated in 1990. It operates a weekly to monthly flask sampling network which monitors key atmospheric trace gases (CO_2, CH_4, CO, and H_2) and the $\delta^{13}C$ of CO_2 from nine sites around the globe, including five in the Southern Hemisphere, and one roving aeroplane sampling unit operating over Cape Grim, Tasmania, and Bass Strait, between the Australian continent and Tasmania (Francey *et al.* 1995, Allison *et al.* 2003) (Figure 23.1).

Other large-scale isotope networks have existed over the years for relatively short periods as a result of fixed term sources of funding. CarboEurope, an EU Framework V Integrated Project (1998–2002), was borne initially in the mid-1990s from a collaborative network of 17 flux sites dispersed across Europe and set up to monitor the exchange of CO_2, H_2O, and energy between the biosphere and atmosphere (Valentini *et al.* 2000). It grew into a cluster of European projects (see http://www.carboeurope.org/ for details), one of which, Carboeuroflux, included a dedicated isotope work package which complimented the network of forest ecosystem measurements across Europe with analyses of the spatial and temporal variability of $\delta^{13}C$ and $\delta^{18}O$ in monthly samples of atmospheric CO_2, plant and soil waters, and organic matter from 17 of the existing sites (Hemming *et al.* 2005, http://www.weizmann.ac.il/ESER/wp5/).

Other projects within the CarboEurope cluster have focused on particular tasks or regions. For example, Aerocarb (http://www.aerocarb.cnrs-gif.fr/) concentrated on monitoring and quantifying the temperate European carbon budget by unifying existing CO_2 monitoring networks, extending these with samples collected from aircraft campaigns, and by combining multiple tracers of carbon cycle dynamics (CO_2, O_2, CO, $\delta^{13}C$ of CO_2, and $\delta^{14}C$ of CO_2). Similar tracers were measured in the Terrestrial and Atmospheric Carbon Observing System infrastructure (TACOS) project which integrated flux tower and tall tower measurements for use in atmospheric transport models. To better understand interannual changes in the net carbon balance of the Siberian boreal forest, the EuroSiberian Carbonflux (http://www.carboeurope.org/) and TCOS-Siberia (http://www.bgc-jena.mpg.de/bgc-systems/projects/web_TCOS/index.html) projects examined seasonal and annual changes in trace gases (CO_2, CO, CH_4, O_2/N_2) and the $\delta^{13}C$ of CO_2 within the convective boundary layer across Siberia.

Efforts to understand regional-scale isotope signals were extended further during the EU-Framework VI funding round (2002–2006) with the initiation of the new Continuous HI-precisiOn Tall Tower Observations (CHIOTTO) of GHG network of tall towers strategically situated across

[2] http://cmdl.noaa.gov/ccgg/index.html, and http://instaar.colorado.edu/sil/research/index.php

[3] http://www.dar.csiro.au/gaslab/index.html

Europe (http://www.chiotto.org/). This network is based on and extends earlier research projects (AEROCARB, TCOS, and TACOS) by coupling isotopic and trace gas measurements from existing surface flux towers with measurements made in the atmospheric boundary layer from tall towers. Further to this, the CERES (CarboEurope Regional Experiment Strategy—http://carboregional.mediasfrance.org/projet/index) campaign strategically linked these local- and regional-scale measurements with discrete multiscale experimental initiatives coordinating measurements at the forest scale with those taken at tall towers and by aircraft.

The Global Change and Terrestrial Ecosystem (GCTE) project (http://www.gcte.org/about.htm) is a core part of the International Geosphere-Biosphere Program (IGBP), an international scientific research program established in the United States in 1986 by the International Council of Scientific Unions (ICSU). Stable isotope analyses of atmospheric gases are one of the tools used to achieve its key objectives which are: (1) to predict the effects of changes in climate, atmospheric composition, biodiversity, and land use on terrestrial ecosystems (including agriculture, forestry, soils) and (2) to determine how these effects lead to feedbacks to the atmosphere and the physical climate system. This has included a network of measurements of the $\delta^{13}C$ of atmospheric CO_2 collected from a range of ecosystems across North America, from which "Keeling" plots have been used to derive the $\delta^{13}C$ of ecosystem respired CO_2 (Pataki *et al.* 2003, Lai *et al.* 2004, 2005), which was conducted under the auspices of the GCTE Focus 1 Biosphere-Atmosphere Stable Isotope Network (BASIN).

B. Water

1. GNIP[4]

The longest continuous dedicated global isotope network is the Global Network of Isotopes in Precipitation (GNIP) which commenced in 1958. It maintains analyses of the $\delta^{18}O$ and δD of monthly precipitation samples, and is currently composed of over 500 sampling stations in 93 countries and territories (Figure 23.2). GNIP is a collaboration project between the International Atomic Energy Agency and the World Meteorological Organisation and is partly composed of individual national networks (http://isohis.iaea.org/userupdate/national networks.html for details) in France (BDISO), Switzerland (NISOT), Canada (CNIP), and United States (USNIP).

Additional to the core monthly monitoring, supplementary campaigns are conducted by GNIP members toward specific research projects. One large-scale isotope measurement project associated with the USNIP program is currently analyzing the growing store of over 70,000 archived samples of weekly precipitation collected for the National Atmospheric Deposition Program (NADP) from over 200 sites spanning the continental United States, Puerto Rico, Alaska, and the Virgin Islands for the NADP (http://nadp/sws/uiuc/edu/). While the principle aim of the NADP network is to monitor precipitation chemistry, isotopic analyses of these samples (Welker 2000, Harvey 2001) provide one of the few valuable sources of recent information on precipitation isotopes across the United States.

2. GNIR[5]

The Global Network of Isotopes in Rivers (GNIR) is an IAEA supported network that became operational in 2002 as a result of a coordinated research project entitled "Isotope tracing of hydrologic processes in large river basins." The purpose of this network is to monitor changes in the global water cycle, specifically the impacts of global climate conditions, land-use changes, dams, and diversion schemes on large-scale river basins. The sampling and isotopic analyses required for the GNIR is

[4] http://isohis.iaea.org/
[5] http://www-naweb.iaea.org/napc/ih/IHS_resources_GNIR2.html

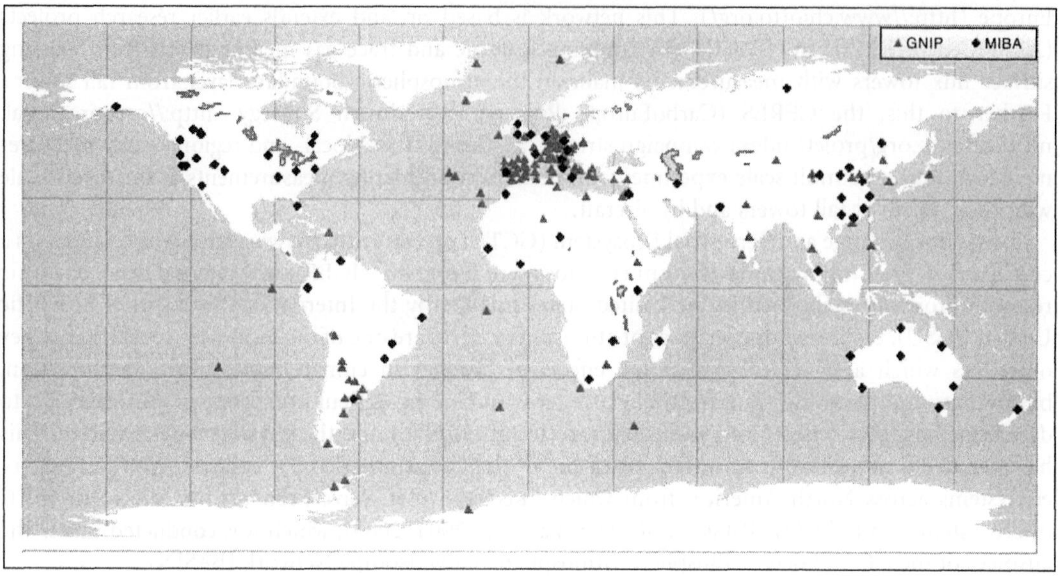

FIGURE 23.2 Location of sampling sites for the Global Network of Isotopes in Precipitation (GNIP) and the Moisture Isotopes in Biosphere and Atmosphere (MIBA) network.

supported by national research institutes and universities, and in most locations it is complimentary to GNIP monitoring.

3. MIBA[6]

The Moisture Isotopes in Biosphere and Atmosphere (MIBA) network is a relatively new IAEA project initiated in 2004. It is composed of a global network of sites, many of which are currently making preparations for sampling, and some have already commenced (Figure 23.2). In this network, monthly or bimonthly samples of atmospheric water vapour and leaf, stem, and soil water from the dominant plant type, mainly forests, are collected, and the $\delta^{18}O$ and δD analyzed.

4. GSO-18[7]

The Global Seawater Oxygen-18 (GSO-18) database is a collection of over 22,000 seawater $\delta^{18}O$ values assembled by Gavin Schmidt, Grant Bigg, and Eelco Rohling from numerous individual sampling networks and campaigns spread across the globe since 1950 (Figure 23.3). Additional information is available with the samples, including salinity, temperature, and depth, and a preliminary $1° \times 1°$ resolution gridded dataset has been constructed to facilitate comparison with global modeled fields (LeGrande and Schmidt 2006).

[6] http://www-naweb.iaea.org/napc/ih/MIBA/IHS_MIBA.html
[7] http://data.giss.nasa.gov/o18data/

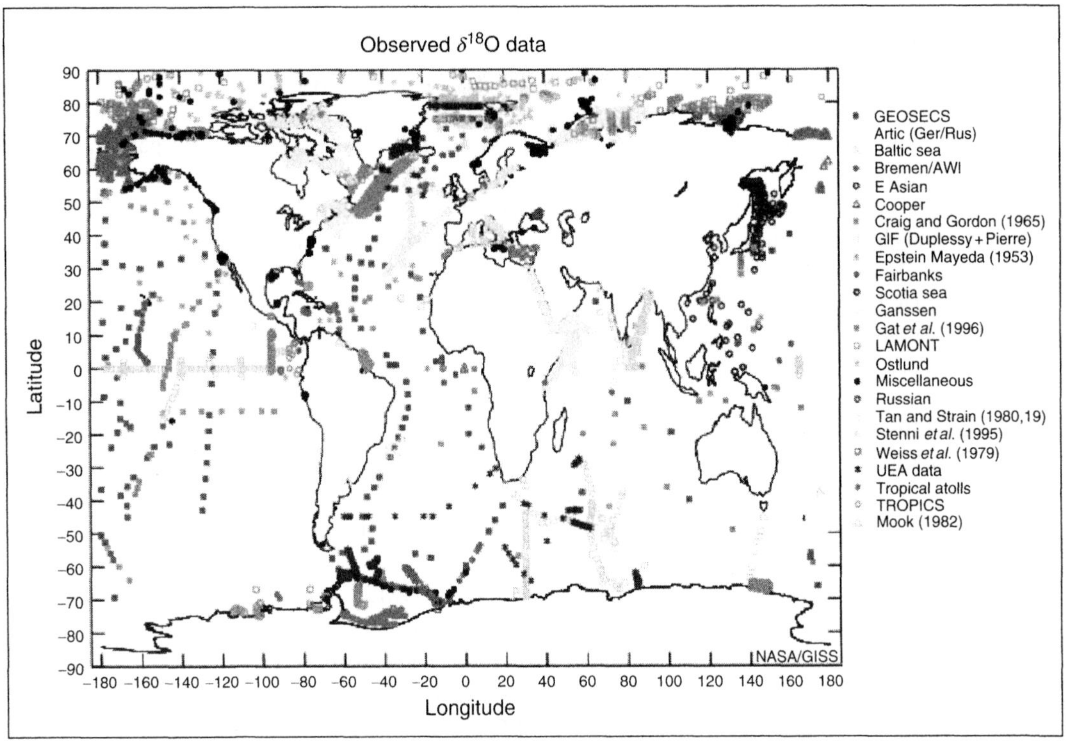

FIGURE 23.3 The Global Seawater 0-18 database (Schmidt *et al.* 1999). (**See Color Plate.**)

C. Other Environmental Archives

Considerable isotope data exists from studies that have utilized stable isotope measurements on a wide range of natural archives (Peterson and Fry 1987, Darling 2004, Nave and Gough 2006). These include ocean, lake, and peat bog cores (Shackleton and Opdyke 1977, Brännvall *et al.* 1997, Grafenstein *et al.* 1998), ice cores (Dansgaard *et al.* 1993, Augustin *et al.* 2004), corals (Correge 2006), speleothem (Sinha *et al.* 2005, Cheng *et al.* 2006), animal bones (Patterson *et al.* 1993), pollen (Loader and Hemming 2001), and tree rings (McCarroll and Loader 2004).

Some of the longest records of environmental change have been obtained from the analyses of the δ^{18}O of foraminifera in ocean sediment cores (Shackleton *et al.* 1977, Spivack *et al.* 1993, Waelbroeck *et al.* 2002). These records revolutionized the understanding of very long-term (multimillennial) climate changes, providing information on such aspects as ocean circulation, isotopic composition of sea water and land ice volume. Also from the marine environment, δ^{18}O records from corals provide a wide range of environmental information, including sea surface temperature and precipitation-induced salinity changes (Quinn *et al.* 1996), and changes in large-scale weather patterns such as the Monsoon (Charles *et al.* 2003).

Other multimillennial isotope records of environmental changes have been obtained from various terrestrial-based archives. Ice core isotope records have provided significant long-term information on large-scale environmental changes and their relationship with high-latitude and high-altitude locations (Petit *et al.* 1999), particularly regarding the interactions between land ice volume changes and variations in the concentration and isotopic composition of atmospheric gases such as CO_2 and NH_4. Furthermore, isotopic records from lake and bog sediments have extended the spatial coverage of multimillennial environmental change records, and enabled the impacts of these changes on bog and lake environments to be assessed (Brännvall *et al.* 1997, Mayer and Schwark 1999). Also, the $\delta^{18}O$ of speleothem calcite from numerous cave locations around the globe have provided a range of paleoenvironmental information. For example, Bar-Matthews *et al.* (1997) reconstructed Late Quaternary paleoclimate in the eastern Mediterranean using stable isotope analysis of speleothem, and Sinha *et al.* (2005) combined speleothem $\delta^{18}O$ records across southeast Asia to reconstruct past changes in monsoon precipitation.

On shorter timescales, absolutely dated tree rings offer the potential to reconstruct climate and environmental changes at an annual resolution without the dating ambiguities that are often present in other proxies. Networks of tree-ring isotope measurements have been used to examine past environmental changes in many locations across the globe (Switsur and Waterhouse 1998, McCarroll and Loader 2004), including, for example, in the southwest United States to infer the $\delta^{13}C$ of past atmospheric CO_2, drought variability, and plant water-use efficiency changes over the last 400–500 years (http://www.ncdc.noaa.gov/paleo/treering/isotope/iso-drought.html, Leavitt and Long 1988, 1989, Feng 1999), across Europe to reconstruct spatiotemporal climate changes throughout the last 400 years (ISONET project, http://www.isonet-online.de/), in northern Eurasia to examine recent changes in the tree line (Saurer *et al.* 2002), over northern Pakistan to reconstruct environmental changes during the last millennium (Treydte *et al.* 2006), across south east United States to provide a paleorecord of the frequency of tropical cyclones (Miller *et al.* 2006), and in various forest ecosystems to examine the impacts of pollutants on forest development (IRISALP network—http://www.wsl.ch/forest/wus/irisalp/iris-en.htm). Tree trunk isotope records have also been used to examine tropical environmental changes, where annual tree rings are not clearly defined (Evans and Schrag 2004).

A number of large-scale networks that include stable isotope measurements have been established to focus research on key ecosystems or processes. For example, Boreal Ecosystem-Atmosphere Study (BOREAS) is an interdisciplinary project that focuses experiments on the northern Boreal forests of Canada. The stable isotope work within BOREAS combines isotopic analyses of air, water, and tree organic matter with other ecosystem measurements in order to improve the understanding of vegetation–atmosphere CO_2 and H_2O exchange processes within this environment (http://www-eosdis.ornl.gov/BOREAS/bhs/Groups/TE/TE-05.html). Another network with regional focus is the Africa Carbon Exchange (ACE) project (http://www.nrel.colostate.edu/projects/ace/about.html), which uses a wide range of techniques, including isotopic, to improve the understanding of spatial and temporal changes in carbon exchange across Africa.

Other isotope information is available from the masses of studies that have focused on small spatial and/or temporal scales or controlled environment experiments. While these are too numerous and diverse to attempt to summarize here, they include, for example, experiments to understand plant physiological and leaf-gas exchange responses to environmental changes (Farquhar and Richards 1984, Robinson *et al.* 2000), isotope fractionation processes in marine systems (Johnson and Kennedy 1998), isotopic variability among wood constituents (Barbour *et al.* 2001, Loader *et al.* 2003), the origins of materials such as bacteria, honey, or illicit drugs for forensic investigations (Martin *et al.* 1998, Ehleringer *et al.* 1999, Kreuzer-Martin *et al.* 2004), the influence of a wide range of pollutants on environments (Weiss *et al.* 1999), the origins and migration of wildlife (Cerling and Harris 1999, Hobson 1999), the spatial and temporal variations in the stable isotope ratios of tap water (Bowen *et al.* 2007), and geochemical studies of ocean regions [*e.g.*, GEOSECS project (Östlund *et al.* 1987), http://iridl.ldeo.columbia.edu/SOURCES/.GEOSECS/].

While many of the individual records or small networks highlighted in this section are not large scale in their focus, when collated they can provide significant information on environmental changes at large scales that can be used along side existing networks. Such potential is clearly demonstrated by the GSO-18 database (Section II.A.4) which integrated some of the wealth of seawater oxygen isotope data available from a wide range of studies. In some areas of research, focused research programs have already enabled large databases of individual records to be collated, for example paleoenvironmental data (for details see Webb *et al.* 1993, Overpeck *et al.* 1996). Other projects have concentrated on mapping spatially explicit observations using Global Information Systems (GIS), which allows these data to be compared more easily with other observational and modeled data (GNIP maps and animations—http://isohis.iaea.org/userupdate/Waterloo/index.html; IsoScapes—http://ecophys.biology.utah.edu/Research/Isoscapes/about.html). Further, the scale of information provided has also been enhanced by integrating different archives, for example, with tree ring and coral records (D'Arrigo *et al.* 2006), tree ring and bog records (Jedrysek *et al.* 2003), and with ice core and ocean sediment records (Oppo *et al.* 2003). However, there is still considerable scope for integration of small-scale records into regional and global databases that can be applied to large-scale issues in environmental change.

D. Collaboration Networks

1. BASIN[8]

The Biosphere-Atmosphere Stable Isotope Network (BASIN) was initially a 5-year program (2001–2006), now been extended for a further 5 years (BASIN II), funded by the US National Science Foundation (2001–2006) under the auspices of the GCTE Focus 1 program (Section II.A.1). Its main objectives are to provide a foundation for drawing together scientific groups, establishing isotope links for ongoing and future isotope-related networks, developing a framework and harmonizing sampling methods, and facilitating data comparison between different networks and current/future experimental and modeling programs. It operates through workshops, model intercomparisons, student training and mobility support, development of appropriate working standards for isotope analyses, dataset construction, and promotion of collaborative research.

Within the BASIN framework there are multiple projects analyzing the isotopic composition of various atmospheric and ecosystem components. In particular, studies undertaken through BASIN aim to provide a bridge between ecosystem and regional flux studies, in order to generate more realistic ecological inputs for regional and global climate models. Many of the ecosystem measurements are conducted at FLUXNET (http://www.fluxnet.ornl.gov/fluxnet/index.cfm) sites where micrometeorologic and eddy covariance methods are used to measure the exchanges of carbon dioxide, water vapor, and energy between the ecosystem and atmosphere.

2. SIBAE

The Stable Isotopes in Biospheric-Atmospheric Exchange (SIBAE) program is a 5-year European Science Foundation Scientific Program project which began in 2002. Like BASIN, SIBAE is a collaboration network which operates through exchange visits, workshops, conferences, and summer schools. The aims of SIBAE are to bring together European researchers in order to promote multidisciplinary applications of stable isotope methods, focusing on the role of terrestrial ecosystem exchanges of CO_2 and H_2O and their influence on the global carbon budget.

Importantly, BASIN and SIBAE in many respects are sister networks with mutually shared scientific objectives. From the outset of these two networks, communication has been facilitated through a series of coordinated conferences drawing the two networks together in such meetings as held in Banff, Orvieto, Interlaken (http://sibaebasin.web.psi.ch/), Marconi and Tomar. The future of BASIN has been secured for a further 5 years, but sadly the future of SIBAE remains uncertain.

[8] http://basinisotopes.org

III. INFORMATION AVAILABLE FROM LARGE-SCALE ISOTOPE NETWORKS

Isotopes have proved one of the most versatile tools for understanding complex ecological processes across a very wide range of temporal and spatial scales (from seconds to millennia and atomic to global) (Griffiths 1998). While small-scale isotopic measurements have been vital for understanding the nature of specific ecosystem processes (Section II.C), large-scale networks of measurements are required to understand and evaluate the relative significance and contributions of these small-scale processes to integrated regional and global signals.

One of the key strengths of stable isotopes over other kinds of environmental data lies in the selective nature of many natural processes against heavier isotopes. Not only does this provide unique information on the specific processes involved, but it also means that such information is available from a multitude of different environmental samples and across a range of spatial scales.

The biosphere exerts a significant influence on the concentration and isotopic composition of many gases present in the atmosphere. This is largely as a result of photosynthesis and respiration processes which regulate the magnitude and timing of fluxes of biospherically active gases, that is, CO_2, H_2O, CH_4, N_2O, and C_5H_8. Analyses of air samples from stations that are dominated by long-range transport of air masses, so-called "background" stations, provide an integrated picture of the dominant sources and sinks of atmospheric gases across large spatial scales (Ciais *et al.* 1995). Such samples also provide unique constraints for understanding the scaling of specific ecological processes from local to regional and global levels. Furthermore, seasonal cycles in the isotopic composition and concentration of many of these gases also provide additional information on the timing and spatial distribution of terrestrial ecosystem processes (Fung *et al.* 1997, Randerson *et al.* 2002, Randerson 2005).

Measurements of the concentration and $\delta^{13}C$ of atmospheric CO_2 from large-scale air sampling networks have proved a breakthrough for understanding and quantifying the dominant global sources and sinks of atmospheric CO_2 and their distribution throughout the global atmosphere. Increases in the annual atmospheric CO_2 concentration and corresponding decreases in its $\delta^{13}C$, observed from these networks over the last decades, have provided clear evidence that human activities have a significant influence on atmospheric CO_2 concentration (Pataki *et al.* 2006). This information, combined with the evidence of seasonal cycles, that are opposite in the northern and Southern Hemispheres and smaller in magnitude in the Southern Hemisphere, indicate that the vast majority of fossil fuel emissions originate from Northern Hemisphere sources, and there is a time lag of 6–12 months before these emissions are transported through the atmosphere to the various monitoring stations in the Southern Hemisphere.

Utilizing the records of background atmospheric CO_2 concentration and $\delta^{13}C$ with "inverse models," which trace back the transport processes of atmospheric gases, has enabled the relative net contributions of ocean and terrestrial plant CO_2 fluxes to atmospheric CO_2 and their regional biases to be quantified (Keeling *et al.* 1989, Tans *et al.* 1990, 1993, Ciais *et al.* 1995, Enting *et al.* 1995, Francey *et al.* 1995, Trolier *et al.* 1996). A key result from this work is that the terrestrial biosphere accounts for approximately half of the total sink of atmospheric CO_2 (comparable with the net oceanic CO_2 sink), which is considerably larger than previously believed. This information has been instrumental in informing many global policy decisions, particularly those concerned with controlling rising atmospheric CO_2 concentrations and land use.

Enhancements in the spatial and temporal resolution of the CO_2 and $\delta^{13}C$ networks over time has enabled more detailed information to be obtained on regional CO_2 sources and sinks (Helliker *et al.* 2005). For example, Ciais *et al.* (1995) provided data which indicates that the northern temperate zone terrestrial sink in 1992/1993 was 3.5 Pg C, which was considerably larger than previously believed (Houghton *et al.* 1987). They also indicated a large CO_2 source in the same year (2–3 Pg C) across the northern tropics (from the equator to 30°N). Fan *et al.* (1998) used the differences in CO_2

measurements between Atlantic and Pacific stations to infer a time-specific carbon sink of 1.7 Pg year^{-1} over North America. Furthermore, Francey *et al.* (1995) suggested that oceanic changes in CO_2 fluxes were responsible for the inverse relationship between El Niño-Southern Oscillation events and atmospheric CO_2 growth rate changes in the Cape Grim, Tasmania, time series.

The $\delta^{13}C$ and δD of methane is now also measured routinely for the NOAA-CMDL-CCDD/INSTAAR network (Section II.A.1). These data provide information on the partitioning of atmospheric methane, an important GHG, into its source components, particularly biogenic (*e.g.*, bacteria) and pyrogenic (from the burning of trees, grasses, and so on), and they also allow identification of the general location of the methane source, that is high-latitude tundra or low-latitude swamps and rice paddies (Gupta *et al.* 1996, Ferretti *et al.* 2005). Diurnal variations in the methane concentration and $\delta^{13}C$ have also been used to identify the urban sources of emissions and verify inventories (Lowry *et al.* 2001).

Variations in the $\delta^{18}O$ of biospherically active gases in the atmosphere carry sensitive information on ecological changes relating to both carbon and water cycling by the terrestrial biosphere. The $\delta^{18}O$ of atmospheric CO_2 from remote background stations can provide information on the sources of large-scale CO_2 fluxes, the $\delta^{18}O$ of specific pools of water that have been in contact with the CO_2, and large-scale atmospheric mixing processes, although at present more work is needed to simulate accurately the spatial and temporal variations in the $\delta^{18}O$ of atmospheric CO_2 observed by the flask network.

Since the earliest use of water stable isotopes to interpret long-term temperature trends (Dansgaard 1964), large-scale networks of the δD and $\delta^{18}O$ values of water have considerably advanced our understanding of past and present hydrologic cycles (Clark and Fritz 1997). They are particularly powerful for identifying phase changes and diffusion processes which are usually characterized by clear isotopic fractionations (Fritz and Fontes 1989). Utilizing this feature, stable isotope measurements of water and water vapor in the atmosphere have been used to partition evapotranspiration into transpiration from vegetation and evaporation from the surface (Salati and Vose 1984, Yepez *et al.* 2003), improve the quantification of regional- and local-scale water recycling in basins (McGuffie and Henderson-Sellers 2004), differentiate ecosystem water sources and responses to moisture stress (Yakir and Wang 1996), and improve the understanding of hydrologic processes in global climate models (Sturm *et al.* 2005). For example, recent results of $\delta^{18}O$ and δD from the NADP US network of water samples (see GNIP—Section II.B.1) have been used to understand the contributions of precipitation from different sources: the Gulf of Mexico, North Pacific Ocean, and Atlantic Ocean, and to estimate the significance of recycling of water across the continent. Spatial differences between the isotopic ratios of tap water and precipitation have also been used to examine modifications along the chain between water sources and the consumer (Bowen *et al.* 2007).

Isotopes in river water are excellent environmental tracers of the runoff portion of the hydrologic cycle, especially including information on the surface water budget, water flow paths in the basin, and water sources to rivers (Karim and Veizer 2002, Theakstone 2003, Su *et al.* 2004). This information cannot typically be obtained by gross water budget calculations or global climate models alone, yet it is crucial for the assessment of changes in basin hydrology related to climate or land-use changes or basin management practices, and may be used to understand the source of pollutants.

Isotopes in sea water are used to indicate the sources of water masses and freshwater inputs to the oceans, and have been combined with other tracers to reveal, for example, the balance between inorganic phosphate (an essential nutrient for life) transport and biological turnover rates in marine ecosystems (Colman *et al.* 2005), which is closely linked with oceanic carbon fluxes. Sea water isotopes can also be used to better understand the processes influencing the isotopic composition of carbonates found in ocean sediments, which have provided significant information on the nature of long-term environmental changes such as glacial–interglacial cycles.

Paleoenvironmental isotope records in general provide often unique information on many aspects of long-term environmental change research, including, the patterns and causes of decade- to century-scale environmental variability, the response of the climate system to large changes in forcing,

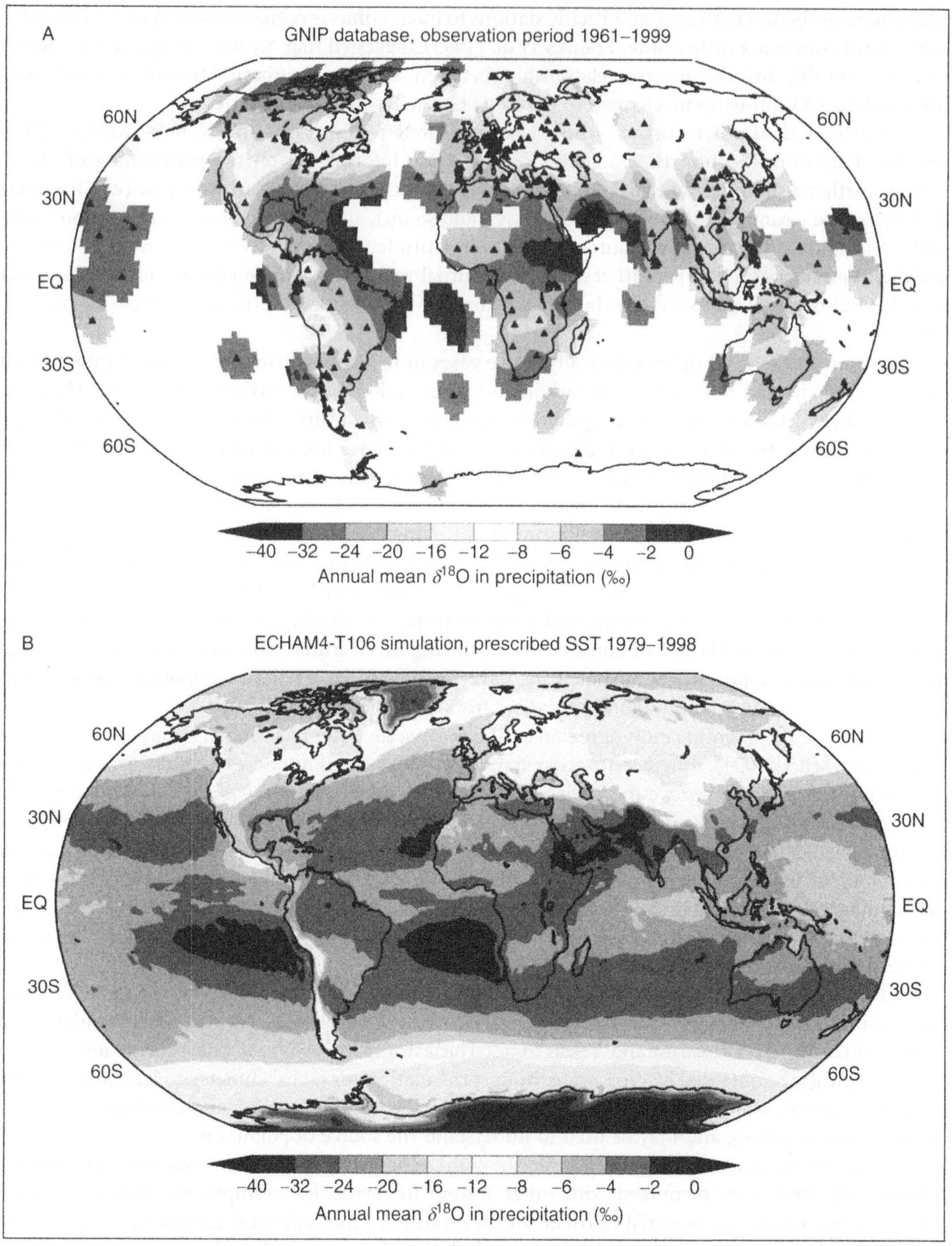

FIGURE 23.4 $\delta_{18}O$ in precipitation from: (A) The GNIP observation database 1961–1999 and (B) Modeled using the ECHAM4-T106 simulation for the period 1979–1998 (Werner, personal communication, see SWING web site: http://www.bgc-jena.mpg.de/bgc-synthesis/projects/SWING/). (**See Color Plate.**)

information to parameterize and verify the performance of the atmospheric, ocean, biosphere, and trace gas models, and understanding on the baseline of past natural environmental variability from which human-induced global changes can be recognized.

In general, the processes involved with isotopic fractionation in the global water cycle are now reasonably well understood, largely as a result of large-scale water monitoring networks (Section II.B). As a result, Global Climate Models (GCMs) are now able to describe the general geographical distribution of $\delta^{18}O$ and δD in precipitation (Figure 23.4). Furthermore, work is underway via the SWING (Stable Water Isotopes INtercomparison Group, http://www.bgc-jena.mpg.de/bgc-synthesis/projects/SWING/index.shtml) and iPILPS (isotopes in Project for Intercomparison of Land-surface Parameterization Schemes; Henderson-Sellers 2006) projects to improve the simulation of global hydrologic processes by combining the information from large-scale isotopic networks, including GNIP and GNIR with international model intercomparison projects, such as the World Climate Research Program (WCRP) and the Integrated Land Ecosystem-Atmosphere Processes Studies (iLEAPS).

The SWING project highlights the important role of large-scale isotope networks for constraining and validating the parameters and processes defined in large-scale climate models. Synthesizing model and observation data from a wide range of environmental sources and time periods is vital for quantifying and reducing model uncertainties. Understanding the major uncertainties in such models will help guide future research and also inform decision makers on strategies to avoid the most serious and likely adverse impacts of environmental changes. Raupach *et al.* (2005) provide a detailed examination of model data synthesis and its implications for quantifying and reducing data and model uncertainties.

One important aspect that has led to significant advances in the quality and availability of stable isotope data has been technological developments in the methods employed to collect and analyze samples. These include new techniques that have improved the analysis accuracy or reduced the sample size (Ferretti *et al.* 2000, Ribas-Carbo *et al.* 2002, Huber and Leuenberger 2005, Lee *et al.* 2005), allowed isotopic analyses to be measured on different gases or types of samples (Ghosh and Brand 2003, Loader and Hemming 2000), and enabled direct stable isotopic analyses to be made at high temporal resolution in fieldwork locations (Bowling *et al.* 2003, Lee *et al.* 2005, Weidmann *et al.* 2005). Particularly exciting for large-scale stable isotope measurements has been the advances in laser techniques for measuring the stable isotopes of CO_2 and H_2O (Bowling *et al.* 2005). These systems enable rapid, *in situ* isotopic analyses to be performed in field locations, and are now being installed at many of the long-term ecosystem flux monitoring sites within FLUXNET (Section II.D.1).

IV. RELEVANCE TO GLOBAL CHANGE POLICY

Appropriate policy decisions are necessarily dependent on the quality of information available to decision makers. As we have demonstrated in Sections II and III, large-scale isotope networks are well placed to provide some of the key information required by policymakers, particularly in the fields of carbon and water cycle management. Here, we aim to emphasize the specific relevance of some of this information to contemporary global change policy objectives.

Growing international concern over the adverse impacts of climate change and associated environmental stresses on sustainable development and poverty has ensured that global change policy issues remain a central focus of most large funding agencies. These key challenges cross international and academic boundaries and require truly multinational and multidisciplinary approaches to understand and address them.

One major global policy agreement that has clearly highlighted the utility of large-scale isotope networks for informing global policy is the United Nations Framework Convention on Climate Change (UNFCCC), which was adopted in May 1992 and came into force on March 21, 1994. The UNFCCC was the result of an international political response to global change concerns. It provides a structure for participating countries to stabilize the concentrations of GHG in the atmosphere in order to avoid "dangerous anthropogenic interference with the climate system." In December 1997, a protocol to the UNFCCC, the Kyoto Protocol, was agreed which committed developed and transition countries (known as Annex 1 Parties) to achieving reductions in their emissions of 6 GHGs by an average of 5.2% below 1990 levels between 2008 and 2012.

Central to the UNFCCC and Kyoto Protocol is a requirement to quantify the regional sources and sinks of GHGs over time, in order to monitor progress and highlight new methods for manipulation of these systems. This has been determined mainly using regional and global-scale networks which have monitored characteristics of various GHGs in the atmosphere.

While atmospheric concentration measurements of GHGs are a key measure to quantify the net fluxes of these gases from and to the atmosphere, they provide little indication of the nature of their sources and sinks. In this context, stable isotopes have proved a vital tool for policymakers. As outlined in Sections II and III, measurements of $\delta^{13}C$ in atmospheric CO_2 are regularly used to partition net CO_2 fluxes into their gross oceanic or biospheric components, and are also able to determine variations in regional terrestrial fluxes. Furthermore, stable isotopes of oxygen provide an important synergy for understanding the interactions between hydrologic and carbon cycles. The $\delta^{18}O$ of atmospheric CO_2 can provide an additional constraint on the sources of CO_2 fluxes. This is particularly suited to understanding the relative fluxes of CO_2 originating from assimilation and respiration by terrestrial ecosystems, which is an important consideration for managing the impacts of land-use change on global CO_2 and CH_4 emissions.

For global change policymaking, the last decade has seen a considerable increase in the coordination of national governments, nongovernmental organizations (NGOs), and international donors toward common policies, strategies, and practices. Nearly all large-funding agencies now focus on similar development agendas which are based on the objectives of major international agreements, in particular the Millennium Development Goals (MDGs).

Eight MDGs were agreed on by nearly 190 countries at the United Nations Millennium Summit in September 2000. They were designed to encourage the international community to work together to make tangible improvements in the developing world, and consist of:

1. Eradicate extreme poverty and hunger
2. Achieve universal primary education
3. Promote gender equality and empower women
4. Reduce child mortality
5. Improve maternal health
6. Combat HIV/AIDS, malaria, and other diseases
7. Ensure environmental sustainability
8. Develop a global partnership for development

Within this framework, 18 key targets and 48 indicators were set with the intention that these are achieved by 2015. MDG 7, ensure environmental sustainability, has clear applications for large-scale stable isotope networks. However, it is also possible to identify applications of stable isotope data and networks that would address other MDGs, for example, identifying polluted water sources or encouraging global cooperation and exchange of expertise.

The central focus of the MDGs is human well-being and how to manage natural and anthropogenic resources to sustain and improve societies. To assess the role of ecosystems in this fundamental focus, the United Nations Secretary General, Kofi Annan, commissioned the Millennium Ecosystem Assessment (MA) in 2000. This involved over 1360 experts worldwide and has been peer reviewed by both policymakers and scientists. Its main objective was:

> "...to assess the consequences of ecosystem change for human well-being and the scientific basis for actions needed to enhance the conservation and sustainable use of those systems and their contribution to human well-being."
>
> Millennium Ecosystem Assessment (MA) 2005, Preface

Figure 23.5 summarizes the main linkages that the MA identified between ecosystem services and constituents of human well-being.

Many of the MA findings can be directly studied using large-scale isotope networks. For example, one of the key messages of the MA was:

> "...the intrinsic vulnerability of the two billion people living in dry regions to the loss of ecosystem services, including water supply."
>
> Millennium Ecosystem Assessment (MA) 2005, p.3

Large-scale isotope networks which provide data on the primary components of the hydrologic cycle (precipitation, river water, atmospheric vapor, water in ecosystems) inform on a broad range of water resource management issues. The GNIP record of isotopes in precipitation (Section II.A.1) has provided information on the processes affecting geographical and temporal variation in precipitation. This is a valuable resource for modelers to validate large-scale climate and hydrologic models, which

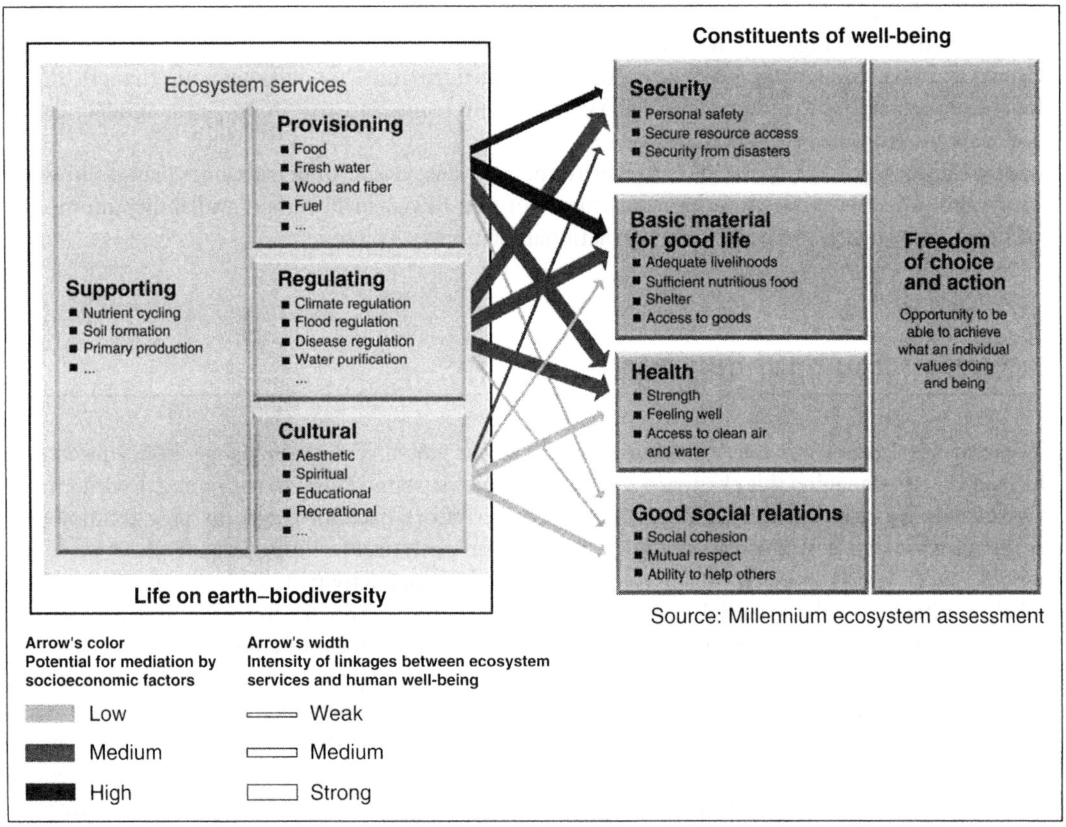

FIGURE 23.5 Linkages between ecosystem services and human well-being (Millennium Ecosystem Assessment 2005).

can now model these distributions reasonably accurately (Figure 23.4). Such isotope models can also be used for future projections of water cycle changes, and are therefore very useful as early warning tools for policymakers.

Further policy-relevant issues focusing on the world's freshwater resources are well highlighted in the World Water Development Reports (World Water Development Report 2003, 2006). These were the result of a collaboration of 24 UN agencies and secretariats, and they aim to provide decision makers with the tools to implement sustainable use of our water. They highlight many of the key areas and causes for concern, for example:

> "Because of the increase in water withdrawals, the pressure on water resources will grow significantly in more than 60% of the world. . .including large parts of Africa, Asia, and Latin America."
>
> World Water Development Report 2003, p.27

As $\sim$35% of precipitation over continents is lost *via* rivers, there is considerable scope for further improvements in surface water management. Large-scale monitoring of isotopes in rivers, such as in the GNIR (Section II.A.2), can be used to examine river basin water balance issues and how these could change under scenarios of global climate and environment change. Furthermore, large-scale monitoring of precipitation and tap water stable isotopes can inform on regional water supply stability and provide insight into impending water resource changes (Bowen *et al.* 2007).

Evaporation and transpiration are also key components of the hydrologic cycle as they account for $\sim$65% of precipitation losses from continents. The recently formed MIBA network (Section II.A.3), which surveys isotopes of atmospheric moisture and ecosystem waters, can provide considerable information for improving estimates of large-scale water balance and therefore assessments of the impacts of climate change on water availability. This isotopic information is also useful for indicating regional to global-scale changes in evaporation, perhaps in response to global climatic changes, that is global warming and global dimming. The MIBA measurements can also improve the modeling, and therefore future prediction of these processes.

Considering the future possible issues surrounding water crises, understanding changes in water supply, and how these relate to other constituents of well-being, such as food availability and health, will be vital for peaceful national and international relations.

V. CONCLUSIONS AND RECOMMENDATIONS

The twenty-first century is likely to be a period of major global ecological change. Policymakers are challenged with managing this change in a way that ensures sustained human well-being. In order to do this, they require multiscale, multidisciplinary scientific efforts facilitated by emerging technologies, data integration, whole-system modeling, and networked analytical facilities. Large-scale isotope networks are well placed to play a major role in this international activity, both in terms of long-term global monitoring of the essential elements of life, and as tools for assessing ecological processes in relation to other environmental and socioeconomic changes from local to global scales.

Developing and sustaining a large-scale monitoring network requires considerable coordination of multinational resources, personnel, and institutions. To ensure these networks are well resourced in the future, it is important to focus network research outputs toward the current international policy agenda, while ensuring the development of fundamental isotopic measurements and core scientific research that has vision to future potential applications. Central to these activities are improvements in collection, measurement and analytical techniques, and frameworks to quantify and reduce uncertainties.

In this chapter, we have described some of the existing large-scale isotope networks and highlighted key environmental and policy-relevant information that is available from these. To encourage a future sustained development of large-scale isotope networks, the following recommendations draw together the main positive features of past and existing large-scale isotope networks and suggest potential approaches to build on these. They are not listed in any order of priority.

VI. RECOMMENDATIONS

- Focus research on current policy-relevant issues.
- Promote integration of isotope results with socioeconomic indicators of human development and environmental stress.
- Collate existing isotope records into new and larger networks.
- Integrate isotope records and networks from different archives and disciplines.
- Coordinate sampling for isotopic measurements with other large-scale monitoring.
- Promote new technologies for the collection and measurement of isotopes.
- Focus on experiments to better understand how environmental processes scale up from small to large scales (Osmond *et al.* 2004).
- Formalize procedures for quantifying errors in all aspects of isotope measurements (Raupach *et al.* 2005).
- Promote the incorporation of stable isotope theory in global-climate models.
- Provide large-scale isotopic datasets that are suitable for direct comparison with climate model results.

VII. REFERENCES

Allison, C. E., R. J. Francey, and P. B. Krummel. 2003. $\delta^{13}C$ in CO_2 from sites in the CSIRO Atmospheric Research GASLAB air sampling network. *Trends: A Compendium of Data on Global Change.* Carbon Dioxide Information Analysis Center, Oak Ridge National Laboratory, US Department of Energy, Oak Ridge, TN, USA.

Augustin, L., C. Barbante, P. R. F. Barnes, J. M. Barnola, M. Bigler, E. Castellano, O. Cattani, J. Chappellaz, D. Dahl-Jensen, B. Delmonte, G. Dreyfus, and G. Durand. 2004. Eight glacial cycles from an Antarctic ice core. *Nature* **429:**623–628.

Bakwin, P. S., P. P. Tans, D. F. Hurst, and C. Zhao. 1998. Measurements of carbon dioxide on very tall towers: Results of the NOAA/CMDL program. *Tellus B* **50:**401–415.

Barbour, M. M., T. J. Andrews, and G. D. Farquhar. 2001. Correlations between oxygen isotope ratios of wood constituents of *Quercus* and *Pinus* samples from around the world. *Australian Journal of Plant Physiology* **28:**335–348.

Bar-Matthews, M., A. Ayalon, and A. Kaufman. 1997. Late Quaternary paleoclimate in the Eastern Mediterranean Region from stable isotope analysis of Speleothems at Soreq Cave, Israel. *Quaternary Research* **47:**155–168.

Bowen, G. J., J. R. Ehleringer, L. A. Chesson, E. Stange, and T. E. Cerling. 2007. Stable isotope ratios of tap water in the contiguous USA. *Water Resources Research,* **43,**W03419 doi:10.129/2006WR005186.

Bowling, D. R., S. D. Sargent, B. D. Tanner, and J. R. Ehleringer. 2003. Tunable diode laser absorption spectroscopy for stable isotope studies of ecosystem-atmosphere CO_2 exchange. *Agricultural and Forest Meteorology* **118:**1–19.

Bowling, D. R., S. P. Burns, T. Conway, R. Monson, and J. W. C. White. 2005. Extensive observations of CO_2 carbon isotope content in and above a high-elevation subalpine forest. *Global Biogeochemical Cycles* **19,** doi:10.1029/2004GB002394.

Brännvall, M.-L., R. Bindler, O. Emteryd, M. Nilsson, and I. Renberg. 1997. Stable isotope and concentration records of atmospheric lead pollution in peat and lake sediments in Sweden. *Water, Air and Soil Pollution* **100**(3–4):243–252.

Cerling, T. E., and J. M. Harris. 1999. Carbon isotope fractionation between diet and bioapatite in ungulate mammals and implications for ecological and paleoecological studies. *Oecologia* **120**(3):347–363.

Charles, C. D., K. Cobb, M. D. Moore, and R. G. Fairbanks. 2003. Monsoon-tropical ocean interaction in a network of coral records spanning the 20th century. *Marine Geology* **201**(1):207–222.

Cheng, H., R. L. Edwards, Y. Wang, X. Kong, Y. Ming, M. J. Kelly, X. Wang, C. D. Gallup, and W. Liu. 2006. A penultimate glacial monsoon record from Hulu Cave and two-phase glacial terminations. *Geology* **34:**217–220.

Ciais, P., P. Tans, M. Trolier, J. White, and R. Francey. 1995. A large northern hemisphere terrestrial sink indicated by the $^{13}C/^{12}C$ ratio of atmospheric CO_2. *Science* **269:**1098–1102.

Clark, I., and P. Fritz. 1997. *Environmental Isotopes in Hydrogeology.* Lewis Publishers, New York. 328 pp.

Colman, A. S., R. E. Blake, D. M. Karl, M. L. Fogel, and K. K. Turekian. 2005. Marine phosphate oxygen isotopes and organic matter remineralisation in the oceans. *Proceedings of the National Academy of Sciences of the United States of America* **102**(37):13023–13028.

Conway, T. J., L. S. Waterman, P. Tans, K. W. Thoning, and K. A. Masarie. 1988. Atmospheric carbon dioxide measurements in the remote global troposphere, 1981–1984. *Tellus B* **40:**81–115.

Correge, T. 2006. Sea surface temperature and salinity reconstruction from coral geochemical tracers. *Palaeogeography, Palaeoclimatology, Palaeoecology* **232:**408–428.

Dansgaard, W. 1964. Stable isotopes in precipitation. *Tellus* **16:**436–468.

Dansgaard, W., S. J. Johnsen, H. B. Clausen, D. Dahl-Jensen, N. S. Gundestrup, C. U. Hammer, C. S. Hvidberg, J. P. Steffensen, S. V. Sveinbjornsdottir, J. Jouzel, and G. Bond. 1993. Evidence for general instability of past climate from a 250-kyr ice-core record. *Nature* **364:**218–220.

Darling, W. G. 2004. Hydrological factors in the interpretation of stable isotopic proxy data present and past: A European perspective. *Quaternary Science Reviews* **23:**743–770.

D'Arrigo, R., R. Wilson, J. Palmer, P. Krusic, A. Curtis, J. Sakulich, S. Bijaksana, S. Zulaikah, L. O. Ngkoimani, and A. Tudhope. 2006. The reconstructed Indonesian warm pool sea surface temperatures from tree rings and corals: Linkages to Asian monsoon drought and El Niño-Southern Oscillation. *Paleoceanography* **21**, doi:10.1029/2005PA001256.

Ehleringer, J. R., D. A. Cooper, M. J. Lott, and C. S. Cook. 1999. Geo-location of heroin and cocaine by stable isotope ratios. *Forensic Science International* **106**(1):27–35.

Enting, I. G., C. M. Trudinger, and R. J. Francey. 1995. A synthesis inversion of the concentration and $\delta^{13}C$ of atmospheric CO_2. *Tellus* **47B:**35–52.

Evans, M. N., and D. P. Schrag. 2004. A stable isotope-based approach to tropical dendroclimatology. *Geochimica et Cosmochimica Acta* **68**(16):3295–3305.

Fan, S.-M., M. Gloor, J. Mahlman, S. Pacala, J. Sarmiento, T. Takahashi, and P. Tans. 1998. A large terrestrial carbon sink in North America implied by atmospheric and oceanic carbon dioxide data and models. *Science* **282:**442–446.

Farquhar, G. D., and R. A. Richards. 1984. Isotopic composition of plant carbon correlates with water use efficiency of wheat genotypes. *Australian Journal of Plant Physiology* **11:**539–552.

Feng, X. 1999. Trends in intrinsic water-use efficiency of natural trees for the past 100–200 years: A response to atmospheric CO_2 concentration. *Geochimica et Cosmochimica Acta* **63:**1891–1903.

Ferretti, D. F., D. C. Lowe, R. J. Martin, and G. W. Brailsford. 2000. A new gas chromatograph-isotope ratio mass spectrometry technique for high precision, N_2O-free analysis of $\delta^{13}C$ and $\delta^{18}O$ in atmospheric CO_2 from small air samples. *Journal of Geophysical Research* **105:**6709–6718.

Ferretti, D. F., J. B. Miller, J. W. C. White, D. M. Etheridge, K. R. Lassey, D. C. Lowe, M. C. MacFarling Meure, M. F. Dreier, C. M. Trudinger, T. M. van Ommen, and R. L. Langenfelds. 2005. Unexpected changes to the global methane budget over the Past 2000 Years. *Science* **309:**1714–1717.

Francey, R. J., P. P. Tans, C. E. Allison, I. G. Enting, J. W. C. White, and M. Trolier. 1995. Changes in oceanic and terrestrial carbon uptake since 1982. *Nature* **373:**326–330.

Fritz, M., and J. Fontes (Eds.). 1989. *Handbook of Environmental Isotope Geochemistry,* Vol. 3: The Marine Environment, A. Elsevier, New York, 428 pp.

Fung, I. Y., J. A. Berry, C. B. Field, M. V. Thompson, J. T. Randerson, P. M. Vitousek, G. J. Collatz, P. Sellers, C. M. Malmstrom, and J. John. 1997. ^{13}C exchange between the atmosphere and the biosphere. *Global Biogeochemical Cycles* **11:**507–535.

Ghosh, P., and W. A. Brand. 2003. Stable isotope ratio mass spectrometry in global climate change research. *International Journal of Mass Spectrometry* **228:**1–33.

Grafenstein, U. von, H. Erlenkeuser, J. Müller, J. Jouzel, and S. Johnsen. 1998. The cold event 8200 years ago documented in oxygen isotope records of precipitation in Europe and Greenland. *Climate Dynamics* **14**(2):73–81.

Griffiths, H. 1998. *Stable Isotopes: Integration of Biological, Ecological and Geochemical Processes?* Bios Scientific Publishers, Oxford, 438 pp.

Gupta, M., S. Tyler, and R. Cicerone. 1996. Modeling atmospheric $\delta^{13}CH_4$ and the causes of recent changes in atmospheric CH_4 amounts. *Journal of Geophysical Research* **101:**22923–22932.

Harvey, F. E. 2001. Use of NADP archive samples to determine the isotope composition of precipitation: Characterizing the meteoric input function for use in ground water studies. *Ground Water* **39**(3):380–390.

Helliker, B. R., J. A. Berry, A. K. Betts, P. S. Bakwin, K. J. Davis, J. R. Ehleringer, M. P. Butler, and D. M. Ricciuto. 2005. Regional-scale estimates of forest CO_2 and isotope flux based on monthly CO_2 budgets of the atmospheric boundary layer. Pages 77–92 *in* H. Griffiths and P. G. Jarvis (Eds.) *The Carbon Balance of Forest Biomes.* Taylor and Francis Group, Oxford.

Hemming, D., D. Yakir, P. Ambus, M Aurela, C. Besson, K. Black, N. Buchmann, R. Burlett, A. Cescatti, R. Clement, P. Gross, A. Granier, *et al.* 2005. Pan-European delta C-13 values of air and organic matter from forest ecosystems. *Global Change Biology* **11:**1065–1093.

Henderson-Sellers, A. M. 2006. Improving land-surface parameterization schemes using stable water isotopes: Introducing the 'iPILPS' initiative. *Global and Planetary Change* **51**:3–24.

Hobson, K. A. 1999. Tracing origins and migration of wildlife using stable isotopes: A review. *Oecologia* **120**(3):314–326.

Houghton, R. A., R. D. Boone, J. R. Fruci, J. E. Hobbie, J. M. Melillo, C. A. Palm, B. J. Peterson, G. R. Shaver, G. M. Woodwell, B. Moore, D. L. Skole, and N. Myers. 1987. The flux of carbon from terrestrial ecosystems to the atmosphere in 1980 due to changes in land use: Geographic distribution of the global flux. *Tellus* **39B**:122–139.

Huber, C., and M. Leuenberger. 2005. On-line systems for continuous water and gas isotope ratio measurements. *Isotopes in Environmental and Health Studies* **41**(3):189–205.

Jedrysek, M. O., M. Krapiec, G. Skrzypek, and A. Kaluzny. 2003. Air-pollution effect and paleotemperature scale versus $\delta^{13}C$ records in tree rings and in a peat core (Southern Poland). *Water Air and Soil Pollution* **145**(1–4):359–375.

Johnson, A. M., and H. Kennedy. 1998. Carbon stable isotope fractionation in marine systems: Open ocean studies and laboratory experiments. Pages 239–253 *in* H. Griffiths (Ed.) *Stable Isotopes: Integration of Biological, Ecological and Geochemical Processes,* Vol. 15. Bios Scientific Publishers, Oxford.

Karim, A., and J. Veizer. 2002. Water balance of the Indus River Basin and moisture source in the Karakoram and western Himalayas: Implications from hydrogen and oxygen isotopes in river water. *Journal of Geophysical Research* **107**(D18), doi:10.1029/2000JD000253.

Keeling, C. D., W. G. Mook, and P. P. Tans. 1979. Recent trends in the 13C/12C ratio of atmospheric carbon dioxide. *Nature* **277**:121–123.

Keeling, C. D., R. B. Bacastow, A. F. Carter, S. C. Piper, T. P. Whorf, M. Heimann, W. G. Mook, and H. Roeloffzen. 1989. Three dimensional model of atmospheric CO_2 transport based on observed winds: 1. Analysis of observational data. Pages 165–236 *in* D. H. Peterson (Ed.) *Aspects of climate variability in the Pacific and the Western Americas.* American Geophysical Union, Washington, D.C.

Keeling, C. D., A. F. Bollenbacher, and T. P. Whorf. 2005. Monthly atmospheric $^{13}C/^{12}C$ isotopic ratios for 10 SIO stations. *Trends: A Compendium of Data on Global Change,* Carbon Dioxide Information Analysis Center, Oak Ridge National Laboratory, US Department of Energy, Oak Ridge, TN, USA.

Kreuzer-Martin, H. W., L. A. Chesson, M. J. Lott, J. V. Dorigan, and J. R. Ehleringer. 2004. Stable isotope ratios as a tool in microbial forensics II. Isotopic variation among different growth media as a tool for sourcing origins of bacterial cells or spores. *Journal of Forensic Sciences* **49**(5):961–967.

Lai, C.-T., J. R. Ehleringer, P. Tans, S. C. Wolfsy, S. P. Urbanski, and D. Y. Hollinger. 2004. Estimating photosynthetic ^{13}C discrimination in terrestrial CO_2 exchange from canopy to regional scales. *Global Biogeochemical Cycles* **18**:1041, doi:10.1029/2003GB002148.

Lai, C.-T., J. R. Ehleringer, A. J. Schauer, P. P. Tans, D. Y. Hollinger, K. T. Paw, J. W. Munger, and S. C. Wolfsy. 2005. Canopy-scale $\delta^{13}C$ of photosynthesis and respiratory CO_2 fluxes: Observations in forest biomes across the United States. *Global Change Biology* **11**:633–643.

Leavitt, S. W., and A. Long. 1988. Stable carbon isotope chronologies from trees in the southwestern United States. *Global Biogeochemical Cycles* **2**:189–198.

Leavitt, S. W., and A. Long. 1989. Drought indicated in carbon-13/carbon-12 ratios of southwestern tree rings. *Water Resources Bulletin* **25**:341–347.

Lee, X., S. Sargent, R. Smith, and B. Tanner. 2005. *In-situ* measurement of water vapour $^{18}O/^{16}O$ isotope ratio for atmospheric and ecological applications. *Journal of Atmospheric and Oceanic Technology* **22**:555–565.

LeGrande, A. N., and G. A. Schmidt. 2006. Global gridded data set of the oxygen isotopic composition in seawater. *Geophysical Research Letters* **33**:2604, doi:10.1029/2006GL026011.

Loader, N. J., and D. L. Hemming. 2000. Preparation of pollen for stable carbon isotope analyses. *Chemical Geology* **165**:339–344.

Loader, N. J., I. Robertson, and D. McCarroll. 2003. Comparison of stable carbon isotope ratios in the whole wood, cellulose and lignin of oak tree-rings. *Palaeogeography, Palaeoclimatology, Palaeoecology* **196**:395–407.

Lowry, D., C. W. Holmes, N. D. Rata, P. O'Brien, and E. G. Nisbet. 2001. London methane emissions: Use of diurnal changes in concentration and delta C-13 to identify urban sources and verify inventories. *Journal of Geophysical Research-Atmosphere* **106**(D7):7427–7448.

Martin, G. I., E. M. Marcias, J. S. Sanchez, and B. G. Rivera. 1998. Detection of honey adulteration with beet sugar using stable isotope methodology. *Food Chemistry* **61**(3):281–286.

Mayer, B., and L. Schwark. 1999. A 15,000-year stable isotope record from sediments of Lake Steisslingen, Southwest Germany. *Chemical Geology* **161**(1):315–337.

McCarroll, D., and N. J. Loader. 2004. Stable isotope in tree rings. *Quaternary Science Reviews* **23**:771–801.

McGuffie, K., and A. Henderson-Sellers. 2004. Stable water isotope characterization of human and natural impacts on land-atmosphere exchanges in the Amazon Basin. *Journal of Geophysical Research* **109**, doi:10.1029/2003JD004388.

Miller, D. L., C. I. Mora, H. D. Grissino-Mayer, C. J. Mock, M. E. Uhle, and Z. Sharp. 2006. Tree-ring isotope records of tropical cyclone activity. *Proceedings of the National Academy of Sciences of the United States of America* **103**(39):14294–14297.

Millennium Ecosystem Assessment (MA). 2005. Ecosystems and Human Well-Being: General Synthesis. Island Press, Washington, DC, USA.

Nave, L. E., and C. M. Gough. 2006. Quantifying ecological change using stable isotopes: Digging deep into the past to predict the future. *New Phytologist* **171**(1):3–5.

Oppo, D. W., J. F. McManus, and J. L. Cullen. 2003. Palaeo-oceanography: Deepwater variability in the Holocene epoch. *Nature* **422**(6929):277.

Osmond, B., G. Ananyev, J. Berry, C. Langdon, Z. Kolber, G. Lin, R. Monson, C. Nichol, U. Rascher, U. Schurr, S. Smith, and D. Yakir. 2004. Changing the way we think about global change research: Scaling up in experimental ecosystem science. *Global Change Biology* **10**(4):393–407.

Östlund, H. G., H. Craig, W. S. Broecker, and D. Spencer. 1987. *GEOSECS Atlantic, Pacific, and Indian Ocean expeditions: Shorebased Data and Graphics 7.* National Science Foundation, Washington, DC.

Overpeck, J., R. Webb, and D. Anderson. 1996. Teaming up to meet IGBP paleoenvironmental data needs. *IGBP Global Changes Newsletter* **27**:28–29.

Pataki, D. E., J. R. Ehleringer, L. B. Flanagan, D. Yakir, D. R. Bowling, C. Still, N. Buchmann, J. O. Kaplan, and J. A. Berry. 2003. The application and interpretation of Keeling plots in terrestrial carbon cycle research. *Global Biogeochemical Cycles* **17**(1):1022, doi:10.1029/2001GB001850.

Pataki, D. E., R. J. Alig, A. S. Fung, N. E. Golubiewski, C. A. Kennedy, E. G. McPherson, D. J. Nowak, R. V. Pouyat, and P. Romero Lankao. 2006. Urban ecosystems and the North American carbon cycle. *Global Change Biology* **12**:2092–2101.

Patterson, W. P., G. R. Smith, and K. C. Lohmann. 1993. Continental paleothermometry and seasonality using the isotopic composition of aragonitic otoliths of freshwater fishes. Pages 191–215 *in* P. K. Swart, K. C. Lohmann, J. McKenzie, and S. Savin (Eds.) *Climate Change in Continental Isotopic Records.* American Geophysical Union, Washington, DC.

Peterson, B. J., and B. Fry. 1987. Stable Isotopes in Ecosystem Studies. *Annual Review of Ecology and Systematics* **18**:293–320.

Petit, J. R., J. Jouzel, D. Raynaud, N. I. Barkov, J.-M. Barnola, I. Basile, M. Bender, J. Chappellaz, M. Davis, G. Delayque, M. Delmotte, V. M. Kotlyakov, *et al.* 1999. Climate and atmospheric history of the past 420,000 years from the Vostok ice core, Antarctica. *Nature* **399**:429–436.

Quinn, T. M., T. J. Crowley, and F. W. Taylor. 1996. New stable isotope results from a 173-year coral from Espiritu Santo, Vanuatu. *Geophysical Research Letters* **23**(23):3413–3416.

Randerson, J. T. 2005. Terrestrial ecosystems and interannual variability in the global atmospheric budgets of $^{13}CO_2$ and $^{12}CO_2$. *In* L. B. Flanagan, D. E. Pataki, and J. Ehleringer (Eds.) *Stable Isotopes and Biosphere-Atmosphere Interactions: Processes and Biological Controls.* Academic Press, San Diego.

Randerson, J. T., G. J. Collatz, J. E. Fessenden, A. D. Munoz, C. J. Still, J. A. Berry, I. Y. Fung, N. Suits, and A. S. Denning. 2002. A possible global covariance between terrestrial gross primary production and ^{13}C discrimination: Consequences for the atmospheric ^{13}C budget and its response to ENSO. *Global Biogeochemical Cycles* **16**(4):1136, doi: 10.1029/2001GB001845.

Raupach, M. R., P. J. Raynet, D. J. Barrett, R. S. DeFries, M. Heimann, D. S. O. Jima, S. Quegan, and C. C. Schmullius. 2005. Model-data synthesis in terrestrial carbon observation: Methods, data requirements and data uncertainty specifications. *Global Change Biology* **11**(3):278–297.

Robinson, D., L. L. Handley, C. M. Scrimgeour, D. C. Gordon, B. P. Forster, and R. P Ellis. 2000. Using stable isotope natural abundances ($\delta^{15}N$ and $\delta^{13}C$) to integrate the stress responses of wild barley (Hordeum spontaneum C. Koch.) genotypes. *Journal of Experimental Botany* **51**(342):41–50.

Ribas-Carbo, M., C. Still, and J. A. Berry. 2002. Automated system for simultaneous analysis of $\delta^{13}C$, $\delta^{18}O$ and CO_2 concentrations in small air samples. *Rapid Communications in Mass Spectrometry* **16**:339–345.

Salati, E., and P. B. Vose. 1984. Amazon Basin: A system in equilibrium. *Science* **225**:129–138.

Saurer, M., F. Schweingruber, E. A. Vaganov, S. G. Shiyatov, and R. T. W. Siegwolf. 2002. Spatial and temporal oxygen isotope trends at the northern tree-line in Eurasia. *Geophysical Research Letters* **29**(9), doi:10.1029/2001GL013739.

Schmidt, G. A., G. R. Bigg, and E. J. Rohling. 1999. "Global Seawater Oxygen-18 Database." http://data.giss.nasa.gov/o18data/

Shackleton, N. J., and N. D. Opdyke. 1977. Oxygen isotope and palaeomagnetic evidence for early Northern Hemisphere glaciation. *Nature* **270**:216–219.

Shackleton, N. J., H. H. Lamb, B. C. Worssam, J. M. Hodgson, A. R. Lord, F. W. Shotton, D. J. Schove, and L. H. N. Cooper. 1977. The oxygen isotope stratigraphic record of the late Pleistocene [and discussion]. *Philosophical Transactions of the Royal Society of London. B* [A discussion on the changing environmental conditions in Great Britain and Ireland during the Devensian (last) Cold Stage (August 17, 1977)] **280**(972):169–182.

Sinha, A., K. G. Cannariato, L. D. Stott, H.-C. Li, C.-F. You, H. Cheng, R. L. Edwards, and I. B. Singh. 2005. Variability of southwest Indian summer monsoon precipitation during the Bølling-Allerød. *Geology* **33**(10):813–816.

Spivack, A. J., C.-F. You, and H. J. Smith. 1993. Foraminiferal boron isotope ratios as a proxy for surface ocean pH over the past 21 Myr. *Nature* **363**:149–151.

Sturm, K., G. Hoffmann, B. Langmann, and W. Stichler. 2005. Simulation of $\delta^{18}O$ in precipitation by the regional circulation model $REMO_{iso}$. *Hydrological Process* **19**:3425–3444.

Su, X., X. Lin, Z. Liao, and J. Wang. 2004. The main factors affecting isotopes of Yellow River water in China. *Water International* **29**(4):475–482.

Switsur, V. R., and J. S. Waterhouse. 1998. Stable isotopes in tree ring cellulose. Pages 303–321 *in* H. Griffiths (Ed.) *Stable Isotopes: Integration of Biological, Ecological and Geochemical Processes.* BIOS Scientific Publishing Ltd., Oxford.

Tans, P. P., I. Y. Fung, and T. Takahashi. 1990. Observational constraints on the global atmospheric CO_2 budget. *Science* **247**:1431–1438.

Tans, P. P., R. F. Keeling, and J. A. Berry. 1993. Oceanic ^{13}C data. A new window on CO_2 uptake by the oceans. *Global Biogeochemical Cycles* **7**:353–368.

Theakstone, W. H. 2003. Oxygen isotopes in glacier-river water, Austre Okstindbreen, Okstindan, Norway. *Journal of Glaciology* **49**(165):282–298.

Treydte, K. S., G. H. Schleser, G. Helle, D. C. Frank, M. Winiger, G. H. Haug, and J. Esper. 2006. The twentieth century was the wettest period in northern Pakistan over the past millennium. *Nature* **440**:1179–1182.

Trolier, M., J. W. C. White, P. P. Tans, K. A. Masarie, and P. A. Genery. 1996. Monitoring the isotopic composition of atmospheric CO_2: Measurements from NOAA global air sampling network. *Journal of Geophysical Research* **101**:25897–25916.

Valentini, R., G. Matteucci, A. J. Dolman, E.-D. Schulze, C. Rebmann, E. J. Moors, A. Granier, P. Gross, N. O. Jensen, K. Pilegaard, A. Lindroth, A. Grelle, *et al.* 2000. Respiration as the main determinant of carbon balance in European forests. *Nature* **404**:861–865.

Waelbroeck, C., L. Labeyrie, E. Michel, J. C. Duplessy, J. F. McManus, K. Lambeck, E. Balbon, and M. Labracherie. 2002. Sea-level and deep water temperature changes derived from benthic foraminifera isotopic records. *Quaternary Science Reviews* **21**(1):295–305.

Weart, S. R. 2003. *The Discovery of Global Warming.* Harvard University Press, Cambridge Massachusetts, London England.

Webb, R. S., J. T. Overpeck, D. M. Anderson, B. A. Baurer, M. K. England, W. S. Gross, E. A. Meyers, and M. M. Worobec. 1993. World Data Center-A for paleoclimatology at the NOAA Paleoclimatology Program. *Journal of Paleolimnology* **9**:69–75.

Weidmann, D., G. Wysocki, C. Oppenheimer, and F. K. Tittel. 2005. Development of a compact quantum cascade laser spectrometer for field measurement of carbon isotopes. *Applied Physics B* **80**:255–260.

Weiss, D., W. Shotyk, and O. Kempf. 1999. Archives of atmospheric lead pollution. *Naturwissenschaften* **86**(6):262–273.

Welker, J. M. 2000. Isotopic ($\delta^{18}O$) Characteristics of weekly precipitation collected across the United States: An initial analysis with application to water source studies. *Hydrological Processes* **14**:1449–1464.

World Water Development Report. 2003. The United Nations World Water Development Report 1: *Water for People, Water for Life. UN.*

World Water Development Report. 2006. The United Nations World Water Development Report 2: *Water, a shared responsibility. UN.*

Yakir, D., and X. F. Wang. 1996. Fluxes of CO_2 and water between terrestrial vegetation and the atmosphere estimated from isotope measurements. *Nature* **380**:515–517.

Yepez, E. A., D. G. Williams, R. L. Scott, and G. Lin. 2003. Parititioning overstory and understory evapotranspiration in a semiarid savanna woodland from the isotopic composition of water vapour. *Agricultural and Forest Meteorology* **119**:53–68.

CHAPTER 24

Applications of Stable Isotope Measurements for Early-Warning Detection of Ecological Change

David G. Williams,* R. David Evans,† Jason B. West,‡ and James R. Ehleringer‡

**Departments of Renewable Resources and Botany, University of Wyoming*
†School of Biological Sciences, Washington State University
‡Department of Biology, University of Utah

Contents

I. INTRODUCTION

Changes in land use and climate, ground water depletion, release of organic and inorganic pollutants, loss of biodiversity, and spread of invasive species challenge society to quantify and mitigate potentially damaging alterations to natural, agricultural, and urban environments. The challenges to scientists are to understand the interactions, feedbacks, and consequences of these changes and predict their impact on ecological functions at large scales. Ultimately, human well-being and health and the economic systems that sustain our way of life depend on our abilities to address these critical environmental issues (Daily 1997). The field of ecology has progressed such that scientists now can quantitatively address large temporal and spatial scale questions such as:

- To what extent are pollutants entering human food supplies and water sources, and what are their origins?
- What sources of pollutants in the environment are critical to reduce?
- How do aboveground ecosystems and human uses affect ground water levels?
- How do invasive species alter productivity and economic value of the landscape?
- What roles do individual species play in ecosystems that make them critical for preservation?
- What is the geographic origin and migration pattern of disease vectors?
- To what extent is climate change altering the productivity of natural and manmade ecosystems?

One important approach now widely available for helping to address such large-scale ecological questions and detect impacts of human activities at early stages involves measurement of stable isotope ratios present at natural abundances in the environment. Isotope measurements on key elements (H, C, N, O, and S) are now an important component of ecological monitoring because they integrate source and process information and are often more sensitive to ecological perturbations than are elemental or compound concentrations or fluxes in nature. Indeed, stable isotope measurements capture a fundamentally different aspect and dimension of ecosystem change, not realized with other conventional types of environmental measurements. Therefore, combining stable isotope measurements with other approaches could serve as an early-warning system to quantify how human activities are affecting ecosystem functions.

Natural variation in the stable isotope ratios of light elements in both biotic and abiotic components of ecological systems occur as a result of biological and physical fractionation events within ecosystems. One important consequence of this is that sources of element and material fluxes can be traced at large scales since different sources often have different isotope ratios based on natural fractionations in the environment. The isotope ratios of organic and inorganic substances also provide a temporal integration of significant metabolic and geochemical processes on the landscape. Further, the isotope ratios of well-mixed environmental reservoirs, as is reflected, for example, in the δ^2H and δ^{18}O of water, the δ^{13}C and δ^{15}N of dissolved compounds in stream discharge from watersheds and in the tissues of organisms, represent an integration of source inputs that extend over large spatial scales and the processing (mixing, losses, biogeochemical transformations) of elements within ecosystems. Whereas measurements of resource pool sizes and fluxes in the biogeosphere–atmosphere system are critical for understanding the functioning of ecosystems and their responses to perturbations, these measurements alone often do not identify the origins of particular species (*e.g.*, NO_3^-), or of the processes effecting any observed changes. Because challenges confronting society in the areas of ecology and the environment are addressed uniquely with stable isotope ratio measurements, isotope analyses have become essential components of continental and global environmental monitoring networks (Chapter 23).

In this chapter, we describe the rationale and framework for stable isotope monitoring to assess ecological condition and change at ecosystem to global scales. Isotope ratios of compounds in aquatic, terrestrial, and atmospheric environments are very sensitive to changes in ecological processes. As such, isotope measurements may serve as an early-warning signal of ecological changes related to ecosystem functions. This unique application and role for stable isotope monitoring can greatly assist management efforts and inform environmental policy. We go on to propose a specific framework for developing an isotope-monitoring network and the spatial modeling necessary to detect and understand ecological change at a continental scale. The framework is based on isotopic measurements of atmospheric inputs, ecosystem outputs, the changes between inputs and outputs as elements are cycled within ecosystems, and sentinel organisms as integrators and indicators of ecological change. Such a framework has the capacity to provide unique insight into how climate and land-use changes and associated biotic and abiotic disturbances impact ecological functioning and connectivity across large regions to continents.

II. ADDRESSING CRITICAL ISSUES IN ECOLOGY AND ENVIRONMENT WITH ISOTOPE MEASUREMENTS

Isotope analyses provide cross-cutting information on the structure, function, and changes in ecosystems. When used in combination with other environmental measurements, isotope ratios provide important knowledge needed to address ecological challenges facing society. The National Research Council (2001) identified seven high-priority areas in environmental sciences requiring research investment. Many of these high-priority areas are congruent with those highlighted by the UN Millennium Ecosystem Assessment (http://www.maweb.org), and stress the critical role science must play in addressing environmental problems affecting human health and well being. Within the context of ecology, the NRC report highlighted the need to better understand: (1) ecological aspects of biogeochemical cycles, (2) biological diversity and ecosystem functioning, (3) ecological impacts of climate variability, (4) ecological–hydrologic interactions, (5) ecology and evolution of infectious diseases, (6) ecological consequences of invasive species, and (7) land use and habitat alteration. Isotope ratio measurements uniquely address scientific challenges in all of these areas of ecological concern, and thus should be an essential component of ecological monitoring to detect changes at early stages.

A. Imprint of Invasive Species and of Changes in Biodiversity

Ecologists now recognize the important linkages between biodiversity and ecosystem function, as well as the potentially destabilizing impact of invasive species on these relationships (D'Antonio and Vitousek 1992, Tilman 1999). Stable isotope ratio analyses provide important information about the importance of biological diversity within ecosystems and the role single species (native and nonnative) play in ecosystem function and change (Vitousek *et al.* 1987, Evans and Ehleringer 1993, Brooks *et al.* 1997, Evans and Belnap 1999, Wolf and Martinez del Rio 2003, Sperry *et al.* 2006, Chapter 20).

The establishment and spread of the Eurasian annual grass *Bromus tectorum* (cheatgrass) following disturbance in arid and semiarid regions of western North America serves as an important example of how nonnative species invasion and disturbance can fundamentally alter ecosystem processes and how stable isotopes are sensitive indicators of those alterations. Invasions of nonnative plant species into native plant communities often decrease ecosystem stability by altering the availability of nitrogen for plant growth through changes in litter quality, rates of N_2 fixation, or rates of nitrogen cycling and loss. A Rayleigh distillation relationship between soil $\delta^{15}N$ and soil nitrogen content indicates the impact of cheatgrass invasion on soil nitrogen (Figure 24.1). Invaded sites have consistently greater $\delta^{15}N$ values, indicating that loss of nitrogen was greater than new inputs (Evans and Ehleringer 1993, Evans and Belnap 1999). Furthermore, the Rayleigh relationship was linear for each invasion regime, allowing it to

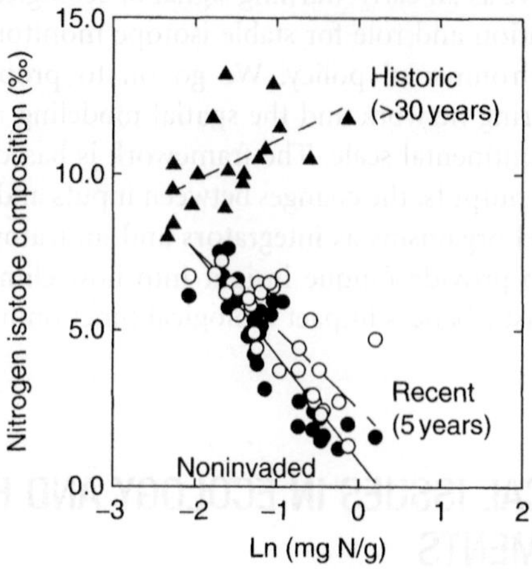

FIGURE 24.1 Soil nitrogen isotope composition ($\delta^{15}N$) plotted against the natural log of soil N content for three sites with different histories of invasion by cheatgrass. This example shows how isotope ratio measurements of soils reveal fundamental changes to ecosystem N cycling associated with organism invasion and disturbance. Redrawn from Sperry *et al.* (2006).

be used as a "clock" to integrate the effects of invasion and disturbance on soil nitrogen dynamics, and as a tool to indicate the health of ecosystems following invasion (Sperry *et al.* 2006). Such isotopic analyses quantify and detect, in a unique way, the impact of invaders on ecosystem processes. This example is particularly relevant because it reveals how long-term sampling at a single site would detect shifts in N cycling.

B. Quantifying Source Changes in Hydroecology

The most critical resource needed to sustain human health and welfare is fresh water. Understanding ecological impacts of drought and human use and alteration of freshwater resources is a key challenge to scientists. Isotopic measurements have the potential to detect important changes in hydroecological interactions and processes that would not be detectable in amount or flux measurements. Stable isotopes of hydrogen and oxygen, for example, are excellent tracers for detecting the origin, recharge and cycling of water within hydroecological systems (Dawson and Ehleringer 1998, Snyder and Williams 2000, Dawson *et al.* 2002), and are thus useful for constraining regional land-surface models (Henderson-Seller *et al.* 2006). Isotopic measurements integrate evaporation losses and mixing across scales (Gat and Airey 2006) and thus provide critical information for land managers and policymakers charged with allocating water for sustaining natural ecosystems and human needs (Goodrich *et al.* 2000, Williams and Scott, in press).

C. Partitioning Sources and Revealing Processes in Biogeochemical Cycles

Probably the greatest contribution of isotope analyses to ecology is in studies of terrestrial, aquatic, and atmospheric biogeochemical cycling. Isotope ratio analyses are essential in studies of ecosystem N and C cycling (Robinson 2001, Pendall 2002) and play a dominant role in our understanding of trace element cycling. Society is challenged to mitigate increasing levels of pollutants in the environment. Isotope analyses are particularly useful for identifying sources of these pollutants and understanding their movement and transformations (Gebauer *et al.* 1994, McClelland *et al.* 1997, Stewart *et al.* 2002).

D. Identifying Geographical Origins of Infectious Diseases and Ecology and Movement of Their Vectors

At least two important aspects of the ecology and spread of infectious diseases can be addressed using isotope ratio analyses. Systematic geographic variation in isotopic signals present in the environment (Bowen and Wilkinson 2002) is the basis for tracing the movement of animal disease vectors and identifying the origin of disease propagules. Isotope variation present in hair, bones, feathers, and so on provides a record of dietary and source water information revealing movement across the landscape (Hobson *et al.* 2004, Cerling *et al.* 2006). Such information is useful for modeling the population ecology and migratory behavior of disease vectors and thus for understanding the epidemiology of infectious disease. Migratory birds encounter diseases and parasites throughout their migratory range. Identifying geographic patterns of migration by individual birds is needed to understand how avian diseases are acquired and spread (Smith *et al.* 2004) and to identify point of origin of bacterial spores (Kreuzer-Martin *et al.* 2003).

E. Assessing the Impact of Regional Climate Change

Many animals and plants record physiological responses to climate changes within their tissues. Tree rings are the best example for plants, but such information is recorded also in animal hair, tooth enamel, claws, nails, and horns. In addition to the excellent information recorded in ring-width variation from annual growth increments of wood, the $\delta^{13}C$ of cellulose of each ring records the tree's physiological condition in relation to changes in climate (Francey and Farquhar 1982, Leavitt and Long 1989). The $\delta^{13}C$ variations are due to variation in the ratio of leaf internal to ambient CO_2 concentration (c_i/c_a) during photosynthesis, which is a measure of stomatal "openness" (Farquhar *et al.* 1989). Annual fluctuations of c_i/c_a calculated from $\delta^{13}C$ of cellulose extracted from wood rings show remarkable correspondence with annual fluctuations in precipitation. Such information allows researchers to quantify variation in the physiological response of vegetation to climate changes.

F. Quantification of Land-Use Change Impacts on Ecological Processes

Land-use and land-cover changes interact with other global change factors and have impacts on terrestrial, aquatic, and atmospheric processes and the services they supply to humans (Houghton 1994, Vitousek *et al.* 1997). Urbanization and the intensification of agriculture, mining, water development, and other land uses will require careful management if natural ecosystems and their benefits are to be sustained. Isotope ratio measurements are useful for quantifying impacts of land-use changes at regional scales. In a comparative study of 16 watersheds in the northeastern United States, Mayer *et al.* (2002) showed how $\delta^{15}N$ and $\delta^{18}O$ values of stream nitrate reflected impacts of land-use changes. Nitrate concentrations and $\delta^{15}N$ values were lowest in streams draining forested watersheds, whereas watersheds that had large areas of urban and agricultural land had high $\delta^{15}N$-nitrate values and high nitrate concentrations (Figure 24.2). Nitrate from urban wastewater and agricultural manure contributed significantly to stream nitrate in these latter watersheds. However, $\delta^{15}N$ values alone are not sufficient to distinguish terrestrial pollutants and soil denitrification from atmospheric nitrate deposition sources to watershed outflow. However, the combination of $\delta^{15}N$ and $\delta^{18}O$ values allows these sources to be distinguished (Mayer *et al.* 2002). The isotopic composition of stream nitrate, therefore, is a sensitive indicator of land use and its impact on the nitrogen cycle, and therefore would be useful as an early-warning signal of ecological changes.

Urban ecology, a relatively new area in ecology focused explicitly on ecological impacts of human land use, is attracting attention from a variety of perspectives related to biodiversity, invasive species

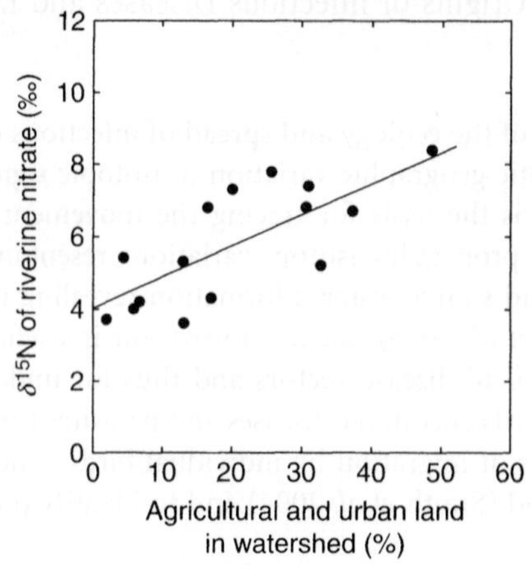

FIGURE 24.2 Mean nitrogen isotope ratio of nitrate in rivers draining 16 watersheds in the mid-Atlantic and New England states, United States plotted against percent agricultural plus urban land within each watershed. This example reveals how isotope monitoring of stream discharge provides a sensitive indicator of land-use changes altering nitrogen sources and processing. Redrawn from Mayer *et al.* (2002).

ecology, hydroecology, and local and regional climate processes. The expansion of urban areas in developing and developed countries is creating unique challenges for the sustainable management of clean water and air. Stable isotope ratio analyses are proving to be useful for understanding sources of atmospheric pollutants in urban settings. Pataki *et al.* (2003) demonstrate the application of $\delta^{13}C$ and $\delta^{18}O$ analyses for identifying seasonal changes and sources of atmospheric CO_2 in the Salt Lake City airshed. Because the CO_2 originating from gasoline combustion, natural gas combustion, and respiration from plants and soils each carry unique isotopic signals, their individual contributions to airshed CO_2 mixing ratios can be distinguished. A surprisingly large contribution of CO_2 from this urban area was from biogenic respiration.

The above examples highlight how isotope ratio analyses provide unique and critical information on high-priority issues in the environmental and ecological sciences. Indeed, isotope ratios of most compounds and substances in the environment are fairly easy and inexpensive to measure with current technologies. Unfortunately, like most ecological parameters, isotope ratios are typically measured at one location, or a small set of locations for a short period; typically 2–3 years, the time necessary to generate sufficient data for a graduate research thesis or dissertation or the typical grant cycle. There are notable exceptions to this, but generally funding from government or private sources is short-term and cannot support long-term monitoring of isotopic ratios in the environment (Chapter 23). This is in spite of the unique insights isotope monitoring can provide into ecological processes and its potential use as an early warning of important changes to ecological systems and their sustainable supply of ecosystem goods and services.

III. ISOTOPE RATIO MEASUREMENTS FOR LARGE-SCALE ECOLOGICAL MONITORING

Although historically not the case, coordinated networks involving long-term ecological monitoring are now increasingly being relied on to provide information to policymakers and resource managers on changes to Earth's ecological support systems (Vaughan *et al.* 2001, Hopkin 2006). The capacity for isotope analyses to integrate source and process information in ecological systems presents an unparalleled opportunity to develop monitoring networks capable of detecting ecological change at a fundamental level and at relatively early stages. The potential benefits and applications of isotope monitoring are tremendous and rival those obtained from satellite remote sensing of land-cover changes and real-time monitoring of surface microclimate. There is tremendous benefit for such continental-scale ecological monitoring networks to include isotope monitoring of terrestrial, aquatic, and atmospheric environmental parameters. Isotope monitoring has great potential as an early-warning system of important ecological changes affecting the sustainability of ecosystem services. Isotope measurements made within this context have the capacity to transcend the many diverse challenge areas in ecology, described above.

Frequent sampling and isotopic analysis of key terrestrial, aquatic, and atmospheric environmental parameters can identify anomalies in system behavior that relate to changes in significant ecological processes, similar to how human heart rate, blood pressure, and body temperature are screened routinely to detect anomalies of human health. Continental-scale ecological states can be monitored and screened using isotope measurements. Such monitoring would serve as an early warning of ecological changes that could be investigated as necessary in detail using more intensive sampling and a broader spectrum of methods. Isotopes are useful in this way because they are extremely sensitive to changes ecological processes and source inputs, as the examples above attest.

In spite of the tremendous potential in such measurements, there are few examples of how isotopic monitoring signaled important changes to Earth's ecological condition and produced substantial insight into large-scale ecological processes. However, important examples are evident in isotope data initially collected from a single or a few sites and now globally that have driven global carbon and hydrologic research and policy for the last several decades. In 1957, Dave Keeling and colleagues began monitoring carbon dioxide concentrations on Mauna Loa, and in 1977, they also began measuring carbon isotope composition of the CO_2 (Keeling 1960, Keeling *et al.* 2005). Because it signaled a global increase in atmospheric CO_2 concentration, and simultaneously indicated the source of this increase, this isotope-monitoring work *at a single site* has been central to understanding the global carbon budget and detecting anthropogenic perturbations to the biosphere. It also alerted the scientific community and policymakers to an important ecological change and resulted in a global wave of research and policy actions that continue today. This dramatic insight was possible because the site chosen represented a location that alone allowed an integration of the atmosphere signal. Ecologists are currently charged with understanding very complex changes occurring at discrete points and across landscapes. Such a challenge requires a spatially distributed network and methods of observation that allow an integrated understanding of these changes.

In another important example from the hydrologic sciences, the International Atomic Energy Agency and World Meteorological Organization (IAEA-WMO) sponsors and manages the Global Network of Isotopes in Precipitation (GNIP). This very successful effort has supported monitoring of δ^2H and $\delta^{18}O$ values in precipitation around the world since 1961. The data are used in the fields of hydrology, oceanography, hydrometeorology, and ecology and in investigations related to the earth's water cycles and climate changes (Dansgaard 1964). The improved understanding of Earth's hydrologic cycle resulting from GNIP has been dramatic. One key outcome of long-term observations from GNIP is the detection of climate warming-related impacts on precipitation processes (Rozanski *et al.* 1992). Long-term changes in isotopic composition of precipitation over mid- and high-latitudes track changes of surface air temperature providing unique insight into climate dynamics.

Other more regionally focused isotope-monitoring efforts have provided substantial insight into ecological change as well. The Biosphere-Atmosphere Stable Isotope Network (BASIN) in North America and a parallel program in Europe, Stable Isotopes in Biospheric-Atmospheric Exchange (SIBAE), are important examples from the ecological sciences. These measurement and training networks have been active only since the mid 1990s, but have already generated tremendous new insight into carbon and water cycle processes from ecosystem to continental spatial scales and seasonal and interannual timescales. At the same time, BASIN and SIBAE have focused heavily on training young isotope scientists. Training is critical to the success of ecological monitoring networks, and establishes a legacy of knowledge critical for continued conceptual and technical innovation. Such isotope-monitoring networks offer the capacity to detect fundamental changes in the ecological systems that sustain human societies and the ecological resources on which they depend. As such, the continued expansion of isotope-monitoring networks will further enhance our understanding of Earth's ecosystems and ability to adapt to or affect ongoing changes.

IV. A CONCEPTUAL DESIGN FOR USING ISOTOPES TO DETECT ECOLOGICAL CHANGE AT THE CONTINENTAL SCALE

Here we describe the conceptual framework for a monitoring plan [Isotope Network of Ecological Warning Signals (INEWS)] that if implemented would serve to track ecosystem changes at a continental scale. The INEWS design addresses fundamental interactions and connections between intensively managed (urban, agricultural) and wildland ecosystems and the likelihood that interactions between these systems will be intensified or altered with an accelerated pace of land-use modification and continued global climate change. INEWS is being considered as a component of the National Ecological Observatory Network (NEON) for the United States by the National Science Foundation (http://www.neoninc.org; Hopkin 2006), which will establish a dense network of observation sites across the United States. INEWS is only one potential approach for employing isotope measurements within an early-warning system for detection of ecological change within such a network and we describe it only as an example. However, the requirements for an effective system highlighted within the INEWS scheme are broadly applicable.

An early-warning monitoring system to detect ecological change using isotope measurements as one central component is likely to be successful if: (1) the substances and isotope parameters to be monitored are easy and inexpensive to collect and analyze; (2) the parameters integrate ecological change over large and definable areas; (3) the measured parameters provide useful insight into processes related to ecological functions affecting ecosystem goods and services; (4) other key non-isotope variables, such as temperature, element concentrations, are simultaneously measured; and (5) the isotope data can be incorporated into process-based ecological or biogeochemical models useful for interpreting and forecasting ecological change within a spatial context. Isotope monitoring would be best accomplished through continuous, real-time measurements of compounds *in situ*. Fortunately, the technology associated with isotope ratio measurements continues to evolve rapidly. Demands for innovation from the fields of ecology, medical science, atmospheric science, biogeoscience, and forensics have been important in the development of techniques for automated and rapid isotopic analysis of organic compounds and trace gases. Continuous-flow inlet systems coupled with gas chromatographs and multisample preparation devices are now widely available. Despite these important advancements, there is a need for continued innovation to meet demands in these very active fields of research, and especially in ecology.

Many important questions in ecology and near real-time assessment of ecological change cannot be addressed with available techniques. Innovation is needed to meet limitations relating to: time-scale and/or time resolution of measurements, reactivity of compounds, storage of compounds, and sampling disturbance. Some exciting advancements have been made over the last several years involving use of laser-based spectroscopic techniques for continuous field measurement of isotopic ratios in CO_2 and water vapor (Bowling *et al.* 2003, Lee *et al.* 2005). Similar advancements are needed for continuous, noninvasive monitoring of other organic and inorganic trace gases (*e.g.*, CH_4, N_2O, NO_x, O_2, N_2, isoprenes) and compounds in aquatic environments (*e.g.*, water, DOC,

DIC, nitrate, sulfate). Such advancements are already developing and will fundamentally alter the way we investigate ecological phenomena and use isotope information. More importantly, detection of ecological change using isotope signals will greatly expand as a routine environmental-monitoring approach and provide a sensitive indicator or warning of perturbations to ecological functions.

The conceptual framework for INEWS is based on three types of warning systems involving isotopic measurements on ecosystem inputs, ecosystem outputs, and sentinel organisms. The stable isotope composition of precipitation and wet and dry deposition inputs is an effective way to detect changes in moisture and nitrogen sources driving ecosystem dynamics. The isotopic signatures of these inputs are subsequently modified within an ecosystem due to physical and biological processes, and these processes are integrated in the isotope ratio of the ecosystem outputs of stream water, and dissolved organic and inorganic compounds. These changes are also detectable in the stable isotope ratios of sentinel organisms (soil microbes, field mice, and birds) that acquire nutrients and water from ecosystems and landscapes. Stable isotope ratio measurements on sentinel organisms provide additional, critical information about ecosystem function, and more importantly, detect how ecosystems are changing across large regions because they are the result of changes in metabolism, diet, nutritional stress, and movement patterns.

A. Ecological Change Detected in Atmospheric Inputs

The isotope ratios of inputs to wildland and urban ecosystems provide critical information about the sources of environmental pollutants and air mass origins (Figure 24.3). Monitored over time, shifts in the isotope ratios of inputs reveal how large-scale environmental changes are affecting patterns of production, transport, and deposition of key ecosystem drivers. The isotope ratios of precipitation (δ^2H and δ^{18}O) vary depending on moisture source, the degree of rainout and the amount of evapotranspiration across the continent (Dansgaard 1964, Araguas-Araguas *et al.* 2000, Welker 2000). These are important signals of global change related to climate warming and shifts in air mass circulation (Rozanski *et al.* 1992, Amundson *et al.* 1996). These measurements also represent an

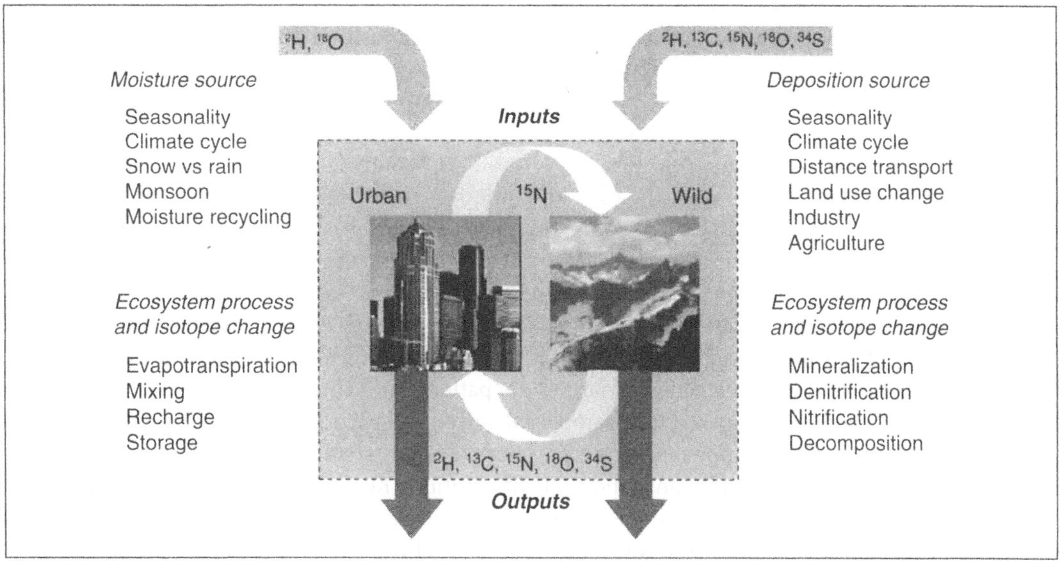

FIGURE 24.3 Conceptual framework for isotope monitoring to detect ecological change at large scales. The isotope ratio of material inputs and outputs reveal changes in sources and ecosystem processing on a continental scale, warning of important ecological changes over time.

important starting point for tracing changes in ecosystem and watershed hydroecology (Dawson and Ehleringer 1998, Gat and Airey 2006). The isotope ratio of inputs to ecosystems, if collected across a network such as that proposed by INEWS, will be useful for modeling, visualizing, and understanding the sources of precipitation and nitrogen pollution and associated climatic and land-use drivers. These measurements are critical to ecological monitoring also because they can account for shifts in the type and origin of N deposition to wildland and urban ecosystems (Kendall 1998). Although somewhat more complicated in practice, the isotope ratio of these potential nitrogen inputs is determined primarily by the amount of reduced (agricultural source) versus oxidized (industrial source) nitrogen (Bragazza *et al.* 2005). Unambiguous identification of sources of pollutants in wet and dry deposition would require measurement of additional isotopic species. Specifically, measurements of total sample $\delta^{34}S$ and $\delta^{13}C$ in deposition inputs would have the potential to more clearly distinguish industrial from urban sources of pollutants and identify inputs derived from beyond continental boundaries.

B. Ecological Change Detected in Stream Discharge

The isotope composition of dissolved organic and inorganic constituents in streams integrates and indicates catchment-scale ecological processes and conditions, providing, for example, information on the sources and terrestrial ecological processes associated with stream chemistry changes (Mitchell *et al.* 1998, Mayer *et al.* 2002). These measurements therefore permit an integrated understanding of watershed-scale ecological condition from a sample taken from one or a few points on the landscape. Geographically distributed, regional to continental scale monitoring networks involving isotope measurements on stream discharge would warn of potentially important large-scale changes taking place in the ecological condition of critical watersheds, and the pattern of these changes at a continental scale. Key measurements that would be useful in this early-warning system are the δ^2H and $\delta^{18}O$ of stream water (Benson 1994, Turner and Barnes 1998), the $\delta^{13}C$ of dissolved inorganic carbon, the $\delta^{13}C$ and $\delta^{15}N$ of stream particulate and dissolved organic matter (Ziegler and Fogel 2003), and the $\delta^{15}N$ and $\delta^{18}O$ of stream nitrate (Kendall 1998, Mayer *et al.* 2002). Other measurements are possible, and could be very useful, but those listed here are easy to collect and analyze and provide an integrated picture of important ecological changes within a watershed.

Long-term changes in the hydrogen and oxygen isotope composition of stream discharge relative to that in precipitation reveals changes in hydroecological processes that are likely also to affect nitrogen cycling and export at a watershed scale. An appropriate unit of integration for an isotope monitoring effort such as INEWS is a low-order (primary or secondary) watershed, which captures material inputs, and through internal biogeochemical processing modifies isotopic signatures that are detected in stream discharge (Figure 24.3). Focusing on a low-order watershed for isotopic measurements would reveal primarily how terrestrial processes are responding to external drivers, rather than how the aquatic system has affected these signals. The isotope ratios of stream water and dissolved organic and inorganic compounds represent the integration of ecological processes associated with mixing, biogeochemical transformations, and loss at the watershed scale. Ecosystem processes recorded in the isotope signals of stream discharge are site specific, but sampling multiple sites at the regional to continental scale would allow detection of large-scale patterns and changes.

The $\delta^{15}N$, $\delta^{34}S$, and $\delta^{13}C$ values of total sample (dissolved and particulate) and the $\delta^{15}N$ and $\delta^{18}O$ of dissolved inorganic nitrate fractions from stream water are highly sensitive to fractionations associated with uptake by soil microbes and plants and to denitrification (Kendall 1998). Measurements of $\delta^{15}N$ and $\delta^{18}O$ in stream discharge integrate these fundamental element-cycling processes and can therefore detect biogeochemical changes in the watershed. They can also distinguish changes in natural versus anthropogenic nitrate inputs over time. Because cycling, retention, and export of nitrogen are critically tied to the hydrologic cycle, measurements of the isotopic composition of stream water (δ^2H and $\delta^{18}O$) should be included in monitoring schemes to detect integrated changes in storage, mixing, evaporation, and flow paths affecting discharge. These processes are highly sensitive to changes in hydroecology, which is likely to be altered substantially in a rapidly urbanizing landscape and in wildland ecosystems altered by changes in species composition (plant invasions, vegetation shifts) or disturbances related to climate change (drought- or insect-induced mortality, fire).

Urban ecosystems process carbon, water, nitrogen, and sulfur cycles differently than wildland ecosystems and serve as point sources of pollutants with a predictable stable isotope ratio composition. Climate change is likely to modify continental-scale patterns of transport and deposition of these pollutants. Associated with these changes is an increase in the relative abundance of anthropogenic versus natural sources of nitrogen output in stream systems from low-order watersheds (Mayer *et al.* 2002).

C. Ecological Change Detected in Isotope Ratios of Organisms

Shifts in metabolism, resource use, and movement over the short and long term are likely to be reflected in progressive changes in the isotopic ratios of organisms. The isotopic composition of organism biomass provides the link between the previously described biogeochemical measurements and biological response (West *et al.* 2006). The isotope ratios of organism biomass integrate food and water sources and the modifications that occur from metabolic processes and stress. Three groups of sentinel organisms can be incorporated into an isotope-monitoring network to detect and integrate biotic responses to changes in environmental drivers (Figure 24.4). The isotopic composition of soil microbial biomass reflects rapid changes in source inputs, functional group diversity, and soil biogeochemical cycling. These measures necessarily integrate processes at a small spatial scale but provide critical information about the functioning of microbial communities that control element processing within ecosystems. Field mice integrate larger spatial footprints than do soil microbes and because they have generation times of months record information about source inputs and metabolism reflective of seasonal change. Resident and migratory birds interact with the landscape over much larger footprints and greater temporal scales than do field mice, and thus capture in their tissue isotope composition changes in dietary sources and stresses that are the result of ecological conditions at the landscape, region, and continental scales. The integration of these observations could provide previously unavailable information about critical ecosystem responses to continental scale environmental changes.

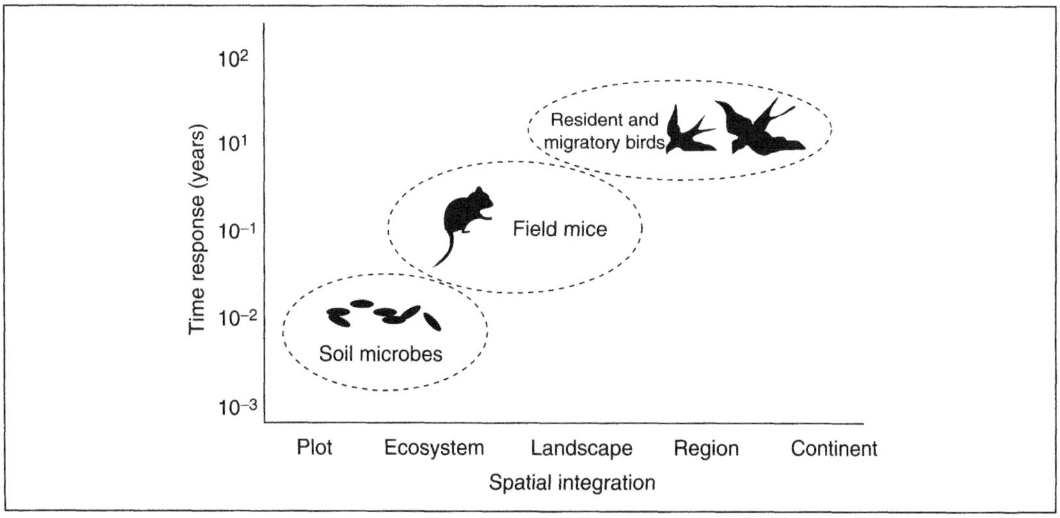

FIGURE 24.4 Conceptual framework for monitoring isotope ratios of organisms as integrators of ecological changes. The isotope ratio of organism biomass integrates changes in ecosystem metabolism, organism function, and organism movement over time.

1. Soil Microbe Sentinels

The relative proportion of different phospholipid fatty acids (PLFA) in soils is characteristic of different functional groups of soil microbes, and over time can detect changes in the relative productivity of these groups (White and Ringelberg 1998, Abraham *et al.* 1998). Specific PLFA biomarkers have been developed for diverse groups of soil organisms such as $Gram^+$ and $Gram^-$ bacteria, fungi, and actinomycetes allowing continental scale tracking of changes in the relative abundance of these groups in response to global change. Further, measuring the $\delta^{13}C$ composition of these PLFA compounds gives insight into patterns and potential changes in substrate use within the soil. This complements nucleic acid methods to achieve even greater differentiation of treatment effects on microbial dynamics and functions.

2. Field Mice Sentinels

Tissue sampled from mice and other rodents carries isotopic information about environmental changes and metabolic responses integrated across the landscape (Stapp and Polis 2003). Models explaining variation in $\delta^{13}C$, $\delta^{15}N$, $\delta^{34}S$, $\delta^{18}O$, and $\delta^{2}H$ values of the organic component of hair in animals are now sufficiently developed such that isotopic shifts can be accurately interpreted within an ecological context (Drever *et al.* 2000). Mice, like birds (below), incorporate into their tissues the isotopic signals from their food and water, but with modifications dependent on metabolic stress and food quality. For example, predictable shifts in $\delta^{15}N$ occur when the animal is fasting, or undergoing food limitation (Hobson *et al.* 1993). Changes in aridity sensed by the animal can also be detected through tissue $\delta^{18}O$, identifying changes in habitat quality (Levin *et al.* 2006). Because mice are present almost in all biomes worldwide, they can serve as isotopic sentinels of ecological changes integrating spatial footprints representative of the collective home ranges of individuals sampled within populations (ecosystem to landscape). Mice obtain much if not all of their dietary water (and associated isotopic signals of H and O) from the food they eat. The isotopic signature of precipitation inputs and subsequent fractionations associated with evaporation establishes a climate signal that can be interpreted from the $\delta^{2}H$ and $\delta^{18}O$ of mouse hair; long-term shifts are attributable to climate changes impacting local to continental-scale hydroecology, and to a lesser degree, what food sources are being consumed to obtain water. Hair sampled from mice twice annually (once in late spring following the winter moisture signal and once in fall following the summer moisture signal) for analysis of $\delta^{13}C$, $\delta^{15}N$, $\delta^{34}S$, $\delta^{18}O$, and $\delta^{2}H$ would capture integrated changes in biotic function and hydroecology at the ecosystem to landscape scales. This information will be invaluable for studying the spread and outbreak of infectious diseases because the isotope ratios of mice are a direct result of organism metabolism linked to global change, and reflects shifts in the way mice interact with their resource base and habitat.

3. Bird Sentinels

Isotopic analyses on tissues collected from migratory and nonmigratory birds, in addition to providing insight into metabolic and nutritional stress and diet, reveal patterns of movement across regions and continents. As climate and land-use changes alter habitat quality and availability, migratory birds are likely to reveal at very early stages impacts of these changes; nonmigratory birds will reflect more locally relevant dietary sources. As in field mice, the $\delta^{2}H$ and $\delta^{18}O$ values in organic constituents of keratin-based tissues (feathers) incorporate signals of body water, which are reflective largely of environmental water. The H and O isotopic composition of feathers is fixed at the time of molting establishing a fingerprint of location at this critical stage of the bird's life history. The $\delta^{13}C$, $\delta^{15}N$, and $\delta^{34}S$ values in feathers integrate information about diet and metabolic condition, also at this period of the birds migratory cycle (Mizutani *et al.* 1990, Hobson and Wassenaar 1997, Hobson 1999, Hobson *et al.* 2003).

D. Predictive Models, Visualization, and Forecasting

An effective early-warning system for ecological change is one that would alert scientists, land managers, and policymakers of imminent damage or threats to essential ecosystem functions. The system must take advantage of a spatially distributed and temporally extensive network of isotopic information and build a continental-scale understanding of ecological states and functions from that network. An isotope-monitoring design, such as described for INEWS, would be uniquely poised to alert scientists and others about the timing, location, and extent of changes to the hydrologic, biogeochemical, and ecological systems on which we all depend. Not only could such a system provide stand-alone information on ecosystem states and functions (*e.g.*, changing sources of nitrogen inputs or altered food webs in response to increasing aridity), the production of derived products such as spatial maps of shifting hydrologic regimes or sources of pollutants could facilitate direct interaction with other products such as those derived from satellites or simulation model outputs. The strength of isotope-monitoring networks can then be leveraged with significant international investments in other efforts such as NASA's Earth Observing System (http://eospso.gsfc.nasa.gov/). Hydrologic, biogeochemical, and ecological models that incorporate an isotope component provide the basis for interpreting isotope signals within the context of ecosystem attributes that are valuable to society. A number of models that simulate isotope exchange between the land surface and atmosphere (Riley *et al.* 2002, Baldocchi and Bowling 2003, Braud *et al.* 2005) and describe hydrogeochemical responses of stream discharge (Turner and Barnes 1998) have been developed, which could be useful within this context.

Such mechanistic models should be coupled with spatial interpolation methods for interpretation and visualization of isotopic variation across large regions to continents (Figure 24.5) (J. B. West,

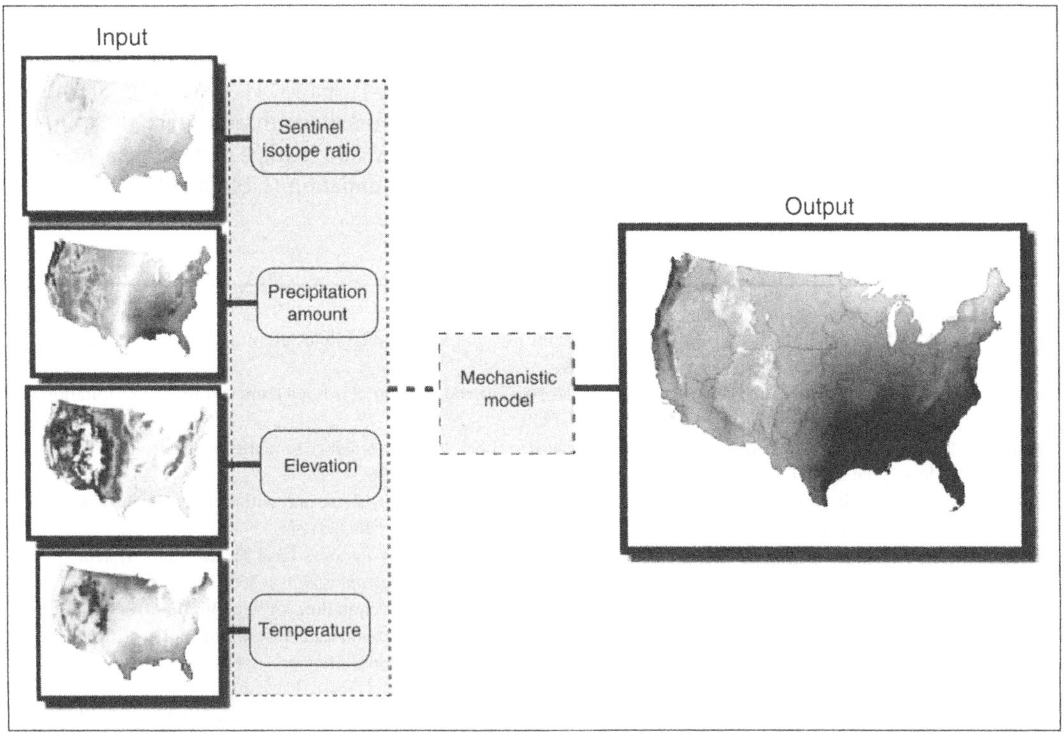

FIGURE 24.5 Spatial integration of isotopic data and forecasting through mechanistic modeling and spatial interpolation. (**See Color Plate.**)

unpublished observations). Ecologists have traditionally investigated ecological responses to external drivers at a single or a few sites and this work remains crucial to the continued development of mechanistic understanding of ecological processes. However, the demands imposed by the global-scale alteration of the biosphere by humans and the scientific questions that follow from that require a new, large-scale approach to ecology. Isotope data collected at multiple sites will allow ecologists to understand continental-scale responses, and importantly, to detect and interpret changes. A paper from Helliker and Griffiths (2007) illustrates the utility of understanding isotope fractionation in plants and then using that understanding to reconstruct atmospheric water vapor that is itself related to changes in the climate system across a wide latitudinal gradient. This type of understanding, applied to a widely distributed network monitoring the isotopic composition of critical components of the biosphere, can then be used to quantify ecosystem services both spatially (between sampling locations) and temporally (in response to future changes in the drivers), making it possible to link processes such as the input and export of nitrogen to variability in climate, topography, wind direction, and other drivers such as proximities to various human and natural disturbances. All of these components must be integrated into models of ecosystem dynamics that will be developed, utilized, and modified by a broad range of scientists with potentially very different backgrounds and interests. The demands necessitated by these large-scale changes are significant and require new ways of doing science. Networks of data collection, spatial integration methods, new observation technologies and multidisciplinary groups will be required as we continue to grapple with the wide-reaching changes to the biosphere that have already occurred and will continue to occur into the foreseeable future.

V. ACKNOWLEDGMENTS

The following individuals participated in valuable discussions that formed the basis for this chapter: S. Beaupre, S. Billings, G. Bowen, D. Bowling, D. Breshears, R. Brooks, J. Chanton, D. Dettman, R. Doucett, B. Fry, C. Kendall, J. King, G. Lin, S. Macko, J. Marshall, N. McDowell, D. Murnick, N. Ostrom, D. Pataki, B. Popp, D. Sandquist, L. Saito, A. Sayer, J. Sparks, H. Steltzer, L. Sternberg, P. Sullyvan, V. Terwilliger, B. Vaughn, J. Welker, and B. Wolf. Workshops that brought these individuals together were funded by BASIN (Biosphere-Atmosphere Stable Isotope Network), a Research Coordination Network supported by the US National Science Foundation (DEB 0090135).

VI. REFERENCES

Abraham, W.-R., C. Hesse, and O. Pelz. 1998. Ratios of carbon isotopes in microbial lipids as an indicator of substrate usage. *Applied and Environmental Microbiology* **64:**4202–4209.

Araguas-Araguas, L., K. Froehlich, and K. Rozanski. 2000. Deuterium and oxygen-18 isotope composition of precipitation and atmospheric moisture. *Hydrological Processes* **14:**1341–1455.

Amundson, R., O. Chadwick, C. Kendall, Y. Wang, and M. DeNiro. 1996. Isotopic evidence for shifts in atmospheric circulation patterns during the late Quaternary in mid-North America. *Geology* **24:**23–26.

Baldocchi, D. D., and D. R. Bowling. 2003. Modelling the discrimination of $^{13}CO_2$ above and within a temperate broad-leaved forest canopy on hourly to seasonal time scales. *Plant, Cell and Environment* **26:**231–244.

Benson, L. V. 1994. Stable isotopes of oxygen and hydrogen in the Truckee River-Pyramid Lake surface-water system. 1. Data analysis and extraction of paleoclimatic information. *Limnology and Oceanography* **39:**344–355.

Bowen, G. J., and B. Wilkinson. 2002. Spatial distribution of $\delta^{18}O$ in meteoric precipitation. *Geology* **30:**315–318.

Bowling, D. R., S. D. Sargent, B. D. Tanner, and J. R. Ehleringer. 2003. Tunable diode laser absorption spectroscopy for stable isotope studies of ecosystem–atmosphere CO_2 exchange. *Agricultural and Forest Meteorology* **118:**1–19.

Bragazza, L., J. Limpens, R. Gerdol, P. Grosvernier, M. Hájek, T. Hájek, P. Hajkova, I. Hansen, P. Iacumin, L. Kutnar, H. Rydin, and T. Tahvanainen. 2005. Nitrogen concentration and $\delta^{15}N$ signature of ombrotrophic *Sphagnum* mosses at different N deposition levels in Europe. *Global Change Biology* **11:**106–114.

Braud, I., T. Bariac, J. P. Gaudet, and M. Vauclin. 2005. SiSPAT-Isotope, a coupled heat, water and stable isotope (HDO and $H_2{}^{18}O$) transport model for bare soil. Part I. Model description and first verifications. *Journal of Hydrology* **309:**277–300.

Brooks, J. R., L. Flanagan, N. Buchmann, and J. R. Ehleringer. 1997. Carbon isotope composition of boreal plants: Functional grouping of life forms. *Oecologia* **110:**301–311.

Cerling, T. E., G. Wittemyer, H. B. Rasmussen, F. Vollrath, C. E. Cerling, T. J. Robinson, and I. Douglas-Hamilton. 2006. Stable isotopes in elephant hair document migration patterns and diet changes. *Proceedings of the National Academy of Science* **103:**371–373.

Daily, G. C. 1997. *Nature's services.* Island Press, Washington, D.C.

D'Antonio, C. M., and P. M. Vitousek. 1992. Biological invasions by exotic grasses, the grass/fire cycle, and global change. *Annual Review of Ecology and Systematics* **23:**63–87.

Dansgaard, W. 1964. Stable isotopes in precipitation. *Tellus* **16:**436–468.

Dawson, T. E., and J. R. Ehleringer. 1998. Plants, isotopes and water use: A catchment-scale perspective. Pages 165–202 *in* C. Kendall and J. J. McDonnell (Eds.) *Isotope Tracers in Catchment Hydrology.* Elsevier Science B.V., Amsterdam.

Dawson, T. E., S. Mambelli, A. H. Plamboeck, P. H. Templer, and K. P. Tu. 2002. Stable isotopes in plant ecology. *Annual Review of Ecology and Systematics* **33:**507–559.

Drever, M. C., L. K. Blight, K. A. Hobson, and D. F. Bertram. 2000. Predation on seabird eggs by Keen's mice (*Peromyscus keeni*): Using stable isotopes to decipher the diet of a terrestrial omnivore on a remote offshore island. *Canadian Journal of Zoology* **78:**2010–2018.

Evans, R. D., and J. Belnap. 1999. Longterm consequences of disturbance on nitrogen dynamics in an arid grassland ecosystem. *Ecology* **80:**150–160.

Evans, R. D., and J. R. Ehleringer. 1993. A break in the nitrogen cycle of aridlands: Evidence from $\delta^{15}N$ of soils. *Oecologia* **94:**314–317.

Farquhar, G. D., J. R. Ehleringer, and K. T. Hubick. 1989. Carbon isotope discrimination and photosynthesis. *Annual Review of Plant Physiology and Plant Molecular Biology* **40:**503–537.

Francey, R. J., and G. D. Farquhar. 1982. An explanation of $^{13}C/^{12}C$ variations in tree rings. *Nature* **297:**28–31.

Gat, J. R., and P. L. Airey. 2006. Stable water isotopes in the atmosphere/biosphere/lithosphere interface: Scaling-up from the local to continental scale, under humid and dry conditions. *Global and Planetary Change* **51:**25–33.

Gebauer, G., A. Giesemann, E.-D. Schulze, and H.-J. Jäger. 1994. Isotope ratios and concentrations of sulfur and nitrogen in needles and soils of *Picea abies* stands as influenced by atmospheric deposition of sulfur and nitrogen compounds. *Plant and Soil* **164:**267–281.

Goodrich, D. C., R. Scott, J. Qi, B. Goff, C. L. Unkrich, M. S. Moran, D. G. Williams, S. Schaeffer, K. Snyder, R. MacNish, T. Maddock, D. Pool, *et al.* 2000. Seasonal estimates of riparian evapotranspiration using remote and *in-situ* measurements. *Agricultural and Forest Meteorology* **105:**281–309.

Helliker, B. R., and H. Griffiths. 2007. Towards a plant-based proxy for the isotope ratio of atmospheric water vapor. *Global Change Biology* **13:**723–733.

Henderson-Seller, A., M. Fischer, I. Aleinov, K. McGuffie, W. J. Riley, G. A. Schmidt, K. Sturm, K. Yoshimura, and P. Irannejad. 2006. Stable water isotope simulation by current land-surface schemes: Results of iPILPS Phase 1. *Global and Planetary Change* **51:**34–58.

Hobson, K. A. 1999. Stable-carbon and nitrogen isotope ratios of songbird feathers grown in two terrestrial biomes: Implications for evaluating trophic relationships and breeding origins. *The Condor* **101:**799–805.

Hobson, K. A., and L. I. Wassenaar. 1997. Linking breeding and wintering grounds of neotropical migrant songbirds using stable hydrogen isotopic analysis of feathers. *Oecologia* **109:**142–148.

Hobson, K. A., R. T. Alisauskas, and R. G. Clark. 1993. Stable-nitrogen isotope enrichment in avian tissues due to fasting and nutritional stress: Implications for isotopic analyses of diet. *The Condor* **95:**388–394.

Hobson, K. A., L. I. Wassenaar, B. Milá, I. Lovette, C. Dingle, and T. B. Smith. 2003. Stable isotopes as indicators of altitudinal distributions and movements in an Ecuadorean hummingbird community. *Oecologia* **136:**302–308.

Hobson, K. A., G. J. Bowen, L. I. Wassenaar, Y. Ferrand, and H. Lormee. 2004. Using stable hydrogen and oxygen isotope measurements of feathers to infer geographical origins of migrating European birds. *Oecologia* **141:**477–488.

Hopkin, M. 2006. Spying on nature. *Nature* **444:**420–421.

Houghton, R. A. 1994. The world wide effect of land-use change. *BioScience* **44:**305–313.

Keeling, C. D. 1960. The concentration and isotopic abundances of carbon dioxide in the atmosphere. *Tellus* **12:**200–203.

Keeling, C. D., S. C. Piper, R. B. Bacastow, M. Whalen, T. P. Whorf, M. Heimann, and H. A. Meijer. 2005. Atmospheric CO_2 and $^{13}CO_2$ exchange with the terrestrial biosphere and oceans from 1978 to 2000: Observations and carbon cycle implications. *In* J. R. Ehleringer, T. E. Cerling, and M. D. Dearing (Eds.) *A History of Atmospheric CO_2 and its Effects on Plants, Animals, and Ecosystems, Ecological Studies*, Vol. 177. Springer Science, Business Media, Inc., New York.

Kendall, C. 1998. Tracing nitrogen sources and cycles in catchments. Pages 519–576 *in* C. Kendall and J. J. McDonnell (Eds.) *Isotope Tracers in Catchment Hydrology.* Elsevier Science B.V., Amsterdam.

Kreuzer-Martin, H. W., M. J. Lott, J. Dorigan, and J. R. Ehleringer. 2003. Microbe forensics: Oxygen and hydrogen stable isotope ratios in *Bacillus subtilis* cells and spores. *Proceedings of the National Academy of Science* **100:**815–819.

Leavitt, S., and A. Long. 1989. Drought indicated in carbon-13/carbon-12 ratios of SW tree rings. *Water Resources Bulletin* **25:**341–347.

Lee, X., S. Sargent, R. Smith, and B. Tanner. 2005. *In situ* measurement of the water vapor $^{18}O/^{16}O$ isotope ratio for atmospheric and ecological applications. *Journal of Atmospheric and Oceanic Technology* **22:**555–565.

Levin, N. E., T. E. Cerling, B. H. Passey, J. M. Harris, and J. R. Ehleringer. 2006. A stable isotope aridity index for terrestrial environments. *Proceedings of the Natonal Academy of Science* **103:**11201–11205.

Mayer, B., E. W. Boyer, C. Goodale, N. A. Jawarski, N. van Breeman, R. W. Howarth, S. Seitzinger, G. Billen, K. Lajtha, K. Nadelhoffer, D. van Dam, L. J. Hetling, *et al.* 2002. Sources of nitrate in rivers draining sixteen watersheds in the northeastern U.S.: Isotopic constraints. *Biogeochemistry* **57/58:**171–197.

McClelland, J. W., I. Valiela, and R. H. Michener. 1997. Nitrogen-stable isotope signatures in estuarine food webs: A record of increasing urbanization in coastal watersheds. *Limnology and Oceanography* **42:**930–937.

Mitchell, M. J., H. R. Krouse, B. Mayer, A. C. Stam, and Y. Zhang. 1998. Use of stable isotopes in evaluating sulfur biogeochemistry of forest ecosystems. *In* C. Kendall and J. J. McDonnell (Eds.) *Isotope Tracers in Catchment Hydrology.* Elsevier Science B.V., Amsterdam.

Mizutani, H., M. Fukuda, Y. Kabaya, and E. Wada. 1990. Carbon isotope ratio of feathers reveals feeding behavior of cormorants. *Auk* **107:**400.

National Research Council. I2001. *Grand Challenges in Environmental Sciences.* National Academy Press, Washington, DC.

Pataki, D. E., D. R. Bowling, and J. R. Ehleringer. 2003. Seasonal cycle of carbon dioxide and its isotopic composition in an urban atmosphere: Anthropogenic and biogenic effects. *Journal of Geophysical Research* **108**(D23):4735, doi:10.1029/2003JD003865.

Pendall, E. 2002. Where does all the carbon go? The missing sink. *New Phytologist* **153:**199–211.

Riley, W. J., C. J. Still, M. S. Torn, and J. A. Berry. 2002. A mechanistic model of $H_2{}^{18}O$ and $C^{18}OO$ fluxes between ecosystems and the atmosphere: Model description and sensitivity analysis. *Global Biogeochemical Cycles* **16**, doi:10.1029/2002GB001878.

Robinson, D. 2001. Delta N-15 as an integrator of the nitrogen cycle. *Trends in Ecology and Evolution* **16:**153–162.

Rozanski, K., L. Araguás-Araguás, and R. Gonfiantini. 1992. Relation between long-term trends of oxygen-18 isotope composition of precipitation and climate. *Science* **258:**981–985.

Smith, R. B., E. C. Greiner, and B. O. Wolf. 2004. Migratory movements of sharp-shinned hawks (*Accipitor striatus*) captured in New Mexico in relation to prevalence, intensity, and biogeography of avian hematozoa. *The Auk* **121:**837–846.

Snyder, K., and D. G. Williams. 2000. Water sources used by riparian trees varies among stream types on the San Pedro River, Arizona. *Agricultural and Forest Meteorology* **105:**227–240.

Sperry, L., J. Belnap, and R. D. Evans. 2006. *Bromus tectorum* invasion alters nitrogen dynamics in an undisturbed grassland ecosystem. *Ecology* **87:**603–615.

Stapp, P., and G. A. Polis. 2003. Marine resources subsidize insular rodent populations in the Gulf of California, Mexico. *Oecologia* **134:**496–504.

Stewart, G. R., M. P. Aidar, C. A. Joly, and S. Schmidt. 2002. Impact of point source pollution on nitrogen isotope signatures ($\delta^{15}N$) of vegetation in SE Brazil. *Oecologia* **131:**468–472.

Tilman, D. 1999. Ecological consequences of biodiversity: A search for general principals. *Ecology* **80:**1455–1474.

Turner, J. V., and C. J. Barnes. 1998. Modeling of isotopes and hydrogeochemical responses in catchment hydrology. Pages 723–760 *in* C. Kendall and J. J. McDonnell (Eds.) *Isotope Tracers in Catchment Hydrology.* Elsevier Science B.V., Amsterdam.

Vaughan, H., T. Brydges, A. Fenech, and A. Lumb. 2001. Monitoring long-term ecological changes through the ecological monitoring and assessment network: Science-based and policy relevant. *Environmental Monitoring and Assessment* **67:**3–28.

Vitousek, P. M., L. R. Walker, L. D. Whiteaker, D. Mueller-Dombois, and P. A. Matson. 1987. Biological invasion by *Myrica faya* alters ecosystem development in Hawaii. *Science* **238:**802–804.

Vitousek, P. M., H. A. Mooney, J. Lubchenco, and J. M. Melillo. 1997. Human domination of Earth's ecosystems. *Science* **277:**494–499.

Welker, J. M. 2000. Isotopic ($\delta^{18}O$) characteristics of weekly precipitation collected across the USA: An initial analysis with application to water source studies. *Hydrological Processes* **14:**1449–1464.

West, J. B., G. J. Bowen, T. E. Cerling, and J. R. Ehleringer. 2006. Stable isotopes as one of nature's ecological recorders. *Trends in Ecology and Evolution* **21:**408–414.

White, D. C., and D. B. Ringelberg. 1998. Signature lipid biomarker analysis. Pages 255–272 *in* R. D. Burlage, R. Atlas, D. Stahl, G. Geesey, and G. Sayler (Eds.) *Techniques in Microbial Ecology.* Oxford University Press, New York.

Williams, D. G., and R. Scott. in press. Vegetation-hydrology interactions: Dynamics of riparian plant water use along the San Pedro River, Arizona. *in* J. Stromberg and B. Tellman (Eds.) *Riparian Area Conservation and Ecology in a Semi-arid Region: The San Pedro River example.* University of Arizona Press, Tucson.

Wolf, B. O., and C. Martinez del Rio. 2003. How important are columnar cacti as sources of water and nutrients for desert consumers? A review. *Isotopes in Environmental and Health Studies* **39:**53–67.

Ziegler, S., and M. L. Fogel. 2003. Seasonal and diel relationships between the isotopic compositions of dissolved and particulate organic matter in a freshwater ecosystem. *Biogeochemistry* **64:**25–52.

CHAPTER 25

Stable Isotopes as Indicators, Tracers, and Recorders of Ecological Change: Synthesis and Outlook

Kevin P. Tu,* Gabriel J. Bowen,† Debbie Hemming,‡ Ansgar Kahmen,§ Alexander Knohl,¶ Chun-Ta Lai,|| and Christiane Werner**

*Center for Stable Isotope Biogeochemistry, Department of Integrative Biology University of California
†Earth and Atmospheric Sciences and Purdue Climate Change Research Center Purdue University
‡Hadley Center for Climate Prediction and Research, Met Office
§Department of Integrative Biology, University of California
¶ETH Zürich, Institute of Plant Sciences, Universitaetsstr. 2
||Department of Biology, San Diego State University
**Experimental and Systems Ecology, University of Bielefeld

Contents

I. INTRODUCTION

The chapters presented in this volume explore a diversity of ways in which stable isotopes inform present and future changes in a wide variety of ecological systems. Recent advances in contemporary research are highlighted, focusing on the knowledge and methodologies that have been developed during the past two decades. These approaches cover a wide range of time and space scales and levels of biological organization (Figure 25.1). As summarized by Ehleringer and Dawson (Chapter 2), stable isotopes are powerful tools in ecological studies for tracing, recording, sourcing, and integrating different ecological parameters. Further, as noted by Dawson and Siegwolf (Chapter 1), with environmental changes

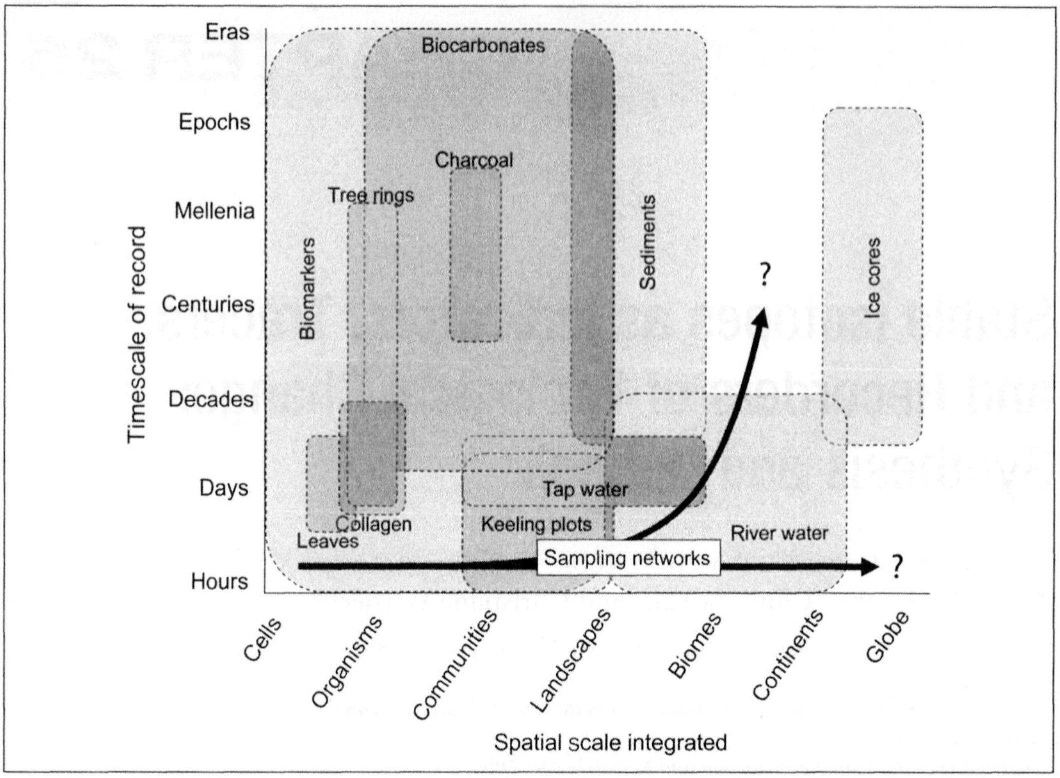

FIGURE 25.1 Timescale and space scale covered by stable isotope analyses of different samples. Timescales refer to the timescale informed from the existing record, for example tree rings integrate seasonal tree growth, but the record extends back thousands of years. Further, spatial scales refer to the spatial integration inherent to the sample, for example river water sampled at the catchment outlet provides spatial integration of the whole catchment. Notably, the temporal integration of samples may be relatively short (*e.g.*, H-isotopes of lipid biomarkers reflect source water during biosynthesis), but the existing record can be quite long (*e.g.*, 600 million years). Further, while samples can differ greatly in the spatial scale they integrate (*e.g.*, leaves reflect their canopy microenvironments, whereas river water can reflect entire catchments), the spatial coverage of most sample records is limited only by the spatial extent of the collections (*e.g.*, leaves can be collected in global sampling network). These issues highlight the need for sampling networks to extend collections and measurements of short-term integrators (*e.g.*, leaves, collagen, and Keeling plots) through time to provide long-term records of short-term processes (along the direction of the upturned arrow in the figure). In addition, geographically distributed sampling networks are needed to extend the spatial scale of samples that inherently integrate over small areas to continental and global scales (along the direction of the horizontal arrow in the figure).

occurring at unprecedented rates in Earth's history, stable isotope methods and applications will undoubtedly serve a critical role in documenting the nature and magnitude of change, as well as highlighting solutions for mitigating ecological impacts that threaten the future of all organisms.

In this final chapter, we provide a synthesis of the major findings, knowledge gaps, and outlooks for future research on the use of stable isotopes as indicators, tracers, and recorders of ecological change.

II. ISOTOPES IN PLANTS

Tree rings are well known to be sensitive recorders of past climate variations, offering the opportunity for climate reconstruction over the last few centuries. Recent advances now allow isotopic analysis of carbon, oxygen, hydrogen, and nitrogen in tree-ring cellulose (Chapter 1). These isotopes can be used to reveal information on past growing season temperatures, precipitation, humidity (Chapter 3), photosynthesis and stomatal conductance (Chapter 4), and atmospheric circulation patterns, including cyclone events (Chapter 5). Similar information can also be retrieved from stable isotopes in lichens (Chapter 6).

Multiparameter approaches will likely be key to future efforts to detect ecological change (Chapters 3 and 4). These approaches simultaneously consider information from multiple isotopes (*e.g.*, C, O, N) as well as other available parameters (*e.g.*, wood density and ring width) to provide significant additional constraint for identifying and separating physiological and climatological signals. For example, Saurer and Siegwolf (Chapter 4) used tree-ring C, O, and N isotopes to assess how human-induced changes such as increasing CO_2 concentrations and air pollution have altered conditions of tree growth over the past centuries. Carbon isotopes were used to indicate how the increased level of CO_2 has influenced the balance of stomatal versus biochemical limitation of photosynthesis and hence has led to an increase in intrinsic water-use efficiency by 30–50% at continental scales. The additional information from the oxygen isotope signal can bring further insights about the cause of this widespread increase in water-use efficiency by disentangling the effect of a reduced stomatal conductance to water loss versus an increased photosynthetic capacity. The stable nitrogen isotope identifies the influence of nitrogen deposition originating from traffic or agriculture on tree growth. In a case study, the combination all three stable isotopes revealed that high nitrogen deposition from traffic led to an increase in water-use efficiency caused by stimulated photosynthesis, while stomatal conductance remained almost unchanged.

While stable isotopes in tree rings provide a unique terrestrial archive for reconstructing ecological change from the perspective of individual trees with an absolute timescale at seasonal to annual resolution, calibration of the methods, including separating climatic and physiological isotopic effects, as well as accounting for between-tree variability remains problematic. A solution will most likely involve the development of replicated, seasonally to annually resolved tree-ring isotope measurements as part of a sampling network like ISONET (http://www.isonet-online.de/). A sampling network would additionally allow stable isotopes in tree rings collected across the globe to help identify regions where recent changes are most pronounced and most accelerated compared to natural variation over the last centuries. The advent of laser-based spectroscopy for continuous stable isotope monitoring of photosynthesis of an entire forest will also be critical for developing a mechanistic understanding of the biochemical processes of wood formation and the role of stored carbohydrates in the interpretation of stable isotopes in tree rings.

In the face of globally declining biodiversity, the relationship between biodiversity and ecosystem functioning has emerged as a central question in ecological research during the last 15 years. Most experiments that have addressed this question have detected a positive, asymptotic relationship between biodiversity and ecosystem functioning. Several mechanisms have been proposed that explain this positive relationship, mainly complementary resource use among different co-occurring species or resource facilitation. With the help of stable isotopes, both in natural abundance and as tracers, several studies now give convincing evidence for the physiological mechanisms behind the proposed mechanisms driving the biodiversity and ecosystem function relationship. Kahmen and Buchmann (Chapter 22) present an overview of studies from a range of different ecosystems that have used C, N, O, and H isotopes to address complementary resource use and resource facilitation among co-occurring plant species in a range of different ecosystems. These studies provide the foundation for future hypothesis building and experimental planning to directly address biodiversity effects in ecosystem studies and to implement and validate these effects in process-based ecosystem models.

III. ISOTOPES IN ANIMALS

Animal ecologists have used naturally occurring stable isotope ratio measurements in studies of dietary reconstruction, trophic level and body conditions of consumers, bioaccumulation of contaminant, and nutrient allocation to reproduction (Chapters 9, 10, and 12). Emerging observations suggest that stable isotope analyses of consumer tissues can be used to determine isotope patterns of trophic level and at the base of the food web. The latter covaries with environmental conditions and as a result, there are growing interests of using birds or other organisms as isotopic indicators of ecological change in relation to future climate.

Applications of stable isotopes in animal ecology rely heavily on assumptions of diet-tissue fractionation and temporal information of isotope distribution across landscape (given the name, isoscapes). The need for more laboratory experiments to test the assumptions that the majority of field studies are based on continues to be the most critical to isotope animal ecology. Progress has been made with respect to the development of terrestrial isoscapes for carbon isotopes on the basis of better predictions of global photosynthetic pathway distribution (Still *et al.* 2003), and for water isotopes (^{18}O and ^{2}H) on the basis of GNIP precipitation database and tap water collection (Bowen *et al.* 2003, 2007; Chapter 18). Nevertheless, refinement in our understanding of terrestrial isoscapes will involve increased spatial coverage of isotope measurements and the continual development of algorithms for predicting their spatial and temporal patterns. Currently, a predictive model for marine isoscapes does not exist, but future trophic level studies in marine systems will certainly benefit from the establishment of baseline phytoplankton isotope values.

Animal tissues from commercially linked sampling programs and material archived in museums or archeological collections provide a resource for isotopic investigations (Chapter 10). However, sample sizes from such resources are often very small. Development of protocols to use absolute minimal amounts of material for analysis is almost certainly required.

Interpretation of stable isotopes in ecological materials is often limited by the mixing of process-level signals and the resulting integration of physiological, ecological, and/or environmental information. However, in cases where samples record changes in the isotopic composition change over time, Cerling *et al.* (Chapter 11) describe a generalized reaction process model based on first-order kinetics that can be used to quantitatively reconstruct discrete resource change events (such as a change in dietary food source of an animal) from the effects of other pools that contribute to the isotope ratio of the "recorder." The model's strength lies in its simplicity and wide applicability and can be applied to a range of systems that involve mixing among more than two pools with distinct turnover times. A key requirement of the model is that the archive of samples documents both "equilibrium" end member isotope compositions (*i.e.*, before change and after change) or that these values be known independently. In controlled experiments with well-studied systems, this should usually be the case, but in complex natural systems such information may not always be available. Thus, sampling networks and museum archives are needed to ensure the availability of this essential information.

IV. ISOTOPES IN AIR, ICE CORES, AND SEDIMENTS

At the ecosystem scale, short-term nocturnal variations of atmospheric CO_2, both vertically and temporally, reflect the contributions of ecosystem components to respiratory fluxes (Chapter 2). The ecosystem-scale measures of CO_2 and isotopes, known as Keeling plots (Dawson *et al.* 2002), can be used to partition the opposing fluxes of respiration and photosynthesis and assess their response to ecological changes. Keeling plot information can also be integrated across continental scales to provide a broad picture of how ecosystem C and O isotope fractionation varies across ecosystems and in response to climate drivers. While it is generally assumed that there is minimal short-term variation in $\delta^{13}C_R$, Werner *et al.* (Chapter 13) report high temporal variability in $\delta^{13}C_R$ at annual, seasonal, and nocturnal time-scales. Such variation underscores the need for future research to fill the current gap in our understanding of processes controlling the respiratory signals and dynamics in $\delta^{13}C_R$. Important processes to consider include respiratory fractionation in different organs (roots, shoots), plant species (functional groups), and ecosystem compartments (heterotrophic, autotrophic respiration), and how they feedback at the ecosystem level. Further, considering the high short-term variation in $\delta^{13}C_R$, improved sampling protocols for nocturnal Keeling plots are needed in global sampling networks. Technological advances are also needed that enable high temporal resolution of $\delta^{13}C$ respired from different ecosystem components, such as that provided by field mass spectrometers or tuneable laser diodes (Bowling *et al.* 2003).

Ice cores provide one of the best archives for recording global isotopic composition of atmospheric CO_2 (Chapter 14) and CH_4 (Chapter 15). Like tree rings, ice cores constitute sensitive recorders of past climate variation while providing the opportunity for climate reconstruction over much greater timescale and space scale (Figure 25.1). Long-term records of $\delta^{13}C$ in atmospheric CO_2 are crucial for interpreting ecological change from carbon isotope signals recorded in plant tissues (see above), as the latter reflects variations in both the atmospheric source (*i.e.*, the CO_2) and the biochemical fractionation during the process of photosynthesis. Long-term records of carbon isotopes of CH_4 provide information that is not apparent in observations of CH_4 concentrations. White *et al.* (Chapter 15) argue that this unique CH_4 isotope signal provides evidence that humans have long been important players in the atmospheric methane budget. It is noted that there is a clear need to better understanding variability in methane sinks and their associated isotope fractionations. There is also a need to extend analyses farther back in time, include measurements at both poles, and to combine information from multiple isotopes (both C and H).

Lake sediments provide an excellent record of paleoclimatic information. Stable isotope ratios of carbonate minerals have been used to indicate both natural and anthropogenic changes and the biological responses to these changes within lake systems (Chapter 17). Sediments also contain biomarkers, or fossil molecules derived from biotic sources. These biomarkers are stable over geologic timescales (Figure 25.1) and their C and H isotope composition can be used as records of environmental conditions during their formation (Chapter 16). As with tree rings, there is a clear need for consistent data on biochemical fractionations in plants to better understand of the mechanisms regulating the geographic and temporal variability in stable isotope signals with environmental conditions and plant functional type.

V. HUMAN IMPACTS

Stable carbon isotopes provide a sensitive tool to monitor human impacts on vegetation and ecosystems over time. For example, stable isotopes of particulate organic carbon (POC) and dissolved inorganic carbon (DIC) in rivers can provide a spatially integrated signal of land-use change over large areas between relatively ^{13}C-depleted native C_3 forests and relatively ^{13}C-enriched cultivated C_4 pastures (Chapter 19). Carbon isotopes in carbonized archeobotanical remains (*e.g.*, wood and seeds) can provide a record of past climate and crop conditions and in turn, irrigation and agricultural practices of early human societies (Chapter 20). Drouet *et al.* (Chapter 21) used strontium isotopes in tree rings to separate sources of calcium into forest ecosystems between mineral weathering and atmospheric deposition. Strontium isotopes were also used to indicate the timing of soil acidification associated with acid deposition. Vallano and Sparks (Chapter 7) highlight the use of nitrogen isotopes in leaves as indicators of the proportion of foliar nitrogen incorporated from atmospheric nitrogen pollution. As previously mentioned, nitrogen isotopes in tree rings can be used to identify the influence of nitrogen deposition originating from traffic or agriculture on tree growth (Chapter 4). Further, isotopic analyses of short-term atmospheric collections can be used to partition anthropogenic from natural CO_2 sources (Pataki *et al.* 2003). Finally, Bowen *et al.* (Chapter 18) explore the potential for using hydrogen and oxygen isotope data gathered through spatiotemporal monitoring networks to serve as a tracer of water in the study and management of human/water cycle interactions.

Stable isotopes could be used to quantify connectivity between sources, uses, and fates of water and provide a means of looking backward in time at the prehistorical development of modern water use practices. In addition, water isotopes could be used to forecast water system status using the concept of isotopes as early-warning signals (Chapter 24). Major challenges remain, however. In particular, the logistical challenges associated with the routine, continued collection and analysis of large numbers of samples from spatially distributed sampling networks and the need for accurate, spatiotemporally explicit climatological and hydrologic data to constrain the interpretation of isotope data represent factors that currently limit such work.

VI. EMERGING THEMES AND NEW FRONTIERS

Several common themes emerged from the chapters in the use of stable isotopes as indicators, tracers, and recorders of ecological change. These include the (1) use of multiparameter approaches, (2) need for sampling networks, and (3) need for museum archiving.

Multiparameter approaches involve the simultaneous use of multiple isotopes (*e.g.*, C and O), as well as other available parameters (*e.g.*, temperature, precipitation, CO_2 concentration) to constrain a problem and identify and separate process-level effects (*e.g.*, photosynthesis versus stomatal conductance) on the isotopic signal that often integrates multiple processes (*e.g.*, $\delta^{13}C$ reflects the ratio of photosynthesis to conductance). This approach has proven useful for extracting climate (Chapter 3) and physiological (Chapter 4) information from isotopes in tree rings, and the approach appears promising for interpreting methane sinks from carbon isotopes signals in CH_4 (Chapter15).

Sampling networks are needed to provide coordinated and routine collection and measurements using consistent methods with comprehensive geographic coverage (Figure 25.1). As noted by Ehleringer and Dawson (Chapter 2), short-term isotopic signals are often important (*e.g.*, in atmospheric trace gases), but are not preserved in time. Further, many ecological patterns are only revealed through long-term monitoring, such as changes in nutrient loading, disturbance/recovery dynamics, and effects of land-use history.

Sampling networks can provide the needed replication of tree ring observations (*e.g.*, ISONET) and allow tree rings isotopes to help identify regions where recent changes are most pronounced compared to recent natural variation. They could also provide the needed temporal information on isotope distribution across landscapes (isoscapes) essential for studies in both terrestrial and aquatic environments. While several terrestrial isoscapes have been developed (Bowen, Chapter 18), a predictive model for marine isoscapes does not currently exist. Future trophic level studies in marine systems will certainly benefit from the establishment of baseline phytoplankton isotope values. Coordinated networks of isotopes in precipitation (*e.g.*, GNIP), rivers (*e.g.*, GNIR), leaf water and atmospheric water vapor (*e.g.*, MIBA), ecosystem respiration (*e.g.*, http://co2.utah.edu; Chapter 2), and CO_2 (*e.g.*, CMDL's Cooperative Air Sampling Network) can provide the data needed for mechanistic studies of variation in isotope fractionations. For example, a sampling network could provide data from a wide range of climatic and vegetation conditions needed to develop and test process-based models that link isotopes, physiology and climate (*e.g.*, in tree rings, biomarkers, sediments) and generalized isotope-based approaches such as that based on the use of the triple isotopic composition of oxygen in leaf water for quantifying past biosphere productivity (Chapter 8).

Isotopes in aquatic, terrestrial, and atmospheric compounds are highly sensitive to changes in ecological processes. Williams *et al.* (Chapter 24) suggest that isotopes may serve as an early-warning signal of ecological changes that are related to ecosystem functioning such as the spread of invasive species, changing biodiversity, source changes in hydroecology, ecological impacts of climate change, or alterations of biogeochemical cycles. They present a framework for stable isotope monitoring in a network context that will allow for the detection of ecological changes and addressing their effects at ecosystem to the global scales. This framework is based on isotopic measurements of C, N, O, H, and S of atmospheric inputs, ecosystem outputs, and sentinel organisms. In combination with spatial modeling an isotopic-monitoring network has the capacity to provide insight to ecological functioning across large regions and to provide essential information to policy makers and resource managers on the earth's ecological support systems. Recent advances in technology that are needed for the implementation of a stable isotope-monitoring network such as isotope lasers and the development of new tools that allow the rapid and continuous analyses of isotopes in various compounds are now in place. In addition, the emergence of the large-scale National Ecological Observatory Network (NEON) in the United States now provides the ideal platform to implement a stable isotope monitoring network for an early detection of ecological change at the continental scale.

While sampling networks are a critical component to future stable isotope research, securing the success of such large-scale ecological networks and sustaining funding are not trivial matters. Hemming *et al.* (Chapter 23) highlight potential approaches for sustaining these networks into the future. Particular emphasis is placed on focusing research outcomes from large-scale stable isotope networks toward the current international policy agenda that reflects a recent increase in the coordination of national governments, Nongovernmental organizations (NGOs) and international donors toward common policies, strategies, and practices, many of which are now directed toward achieving the Millennium Development Goals (Chapter 1).

Finally, in addition to the need for sampling networks, there is also a critical need for preservation of biological materials through time. For example, current museum archives have been an extremely valuable source of materials to reconstruct temporal patterns (Chapter 10). As noted by Ehleringer and Dawson (Chapter 2), future efforts to use isotopes for process-level changes (*e.g.*, trophic-level dynamics, impacts of invasive species, and human activities) will most likely benefit from a taxonomically diverse and geographically distributed network of collections.

VII. REFERENCES

Bowen, G. J., and J. Revenaugh. 2003. Interpolating the isotopic composition of modern meteoric precipitation. *Water Resources Research* **39:**1299, doi: 10.1029/2003WR002086.

Bowen, G. J., J. R. Ehleringer, L. A. Chesson, E. Stange, and T. E. Cerling. 2007. Stable isotope ratios of tap water in the contiguous USA. *Water Resources Research* **43:**W03419, doi: 10.1029/2006WR005186.

Bowling, D. R., S. D. Sargent, B. D. Tanner, and J. R. Ehleringer. 2003. Tunable diode laser absorption spectroscopy for stable isotope studies of ecosystem-atmosphere, CO_2 exchange. *Agricultural and Forest Meteorology* **118**(1–2):1–19.

Dawson, T. E., S. Mambelli, A. H. Plamboeck, P. H. Templer, and K. P. Tu. 2002. Stable isotopes in plant ecology. *Annual Review of Ecology and Systematics* **33:**507–559.

Pataki, D. E., D. R. Bowling, and J. R. Ehleringer. 2003. Seasonal cycle of carbon dioxide and its isotopic composition in an urban atmosphere: Anthropogenic and biogenic effects. *Journal of Geophysical Research* **108**(D23):4735, doi:4710.1029/2003JD003865.

Still, C. J., J. A. Berry, G. J. Collatz, and R. S. DeFries. 2003. The global distribution of C3 and C4 vegetation: Carbon cycle implications. *Global Biogeochemical Cycles* **17**(1):1006.

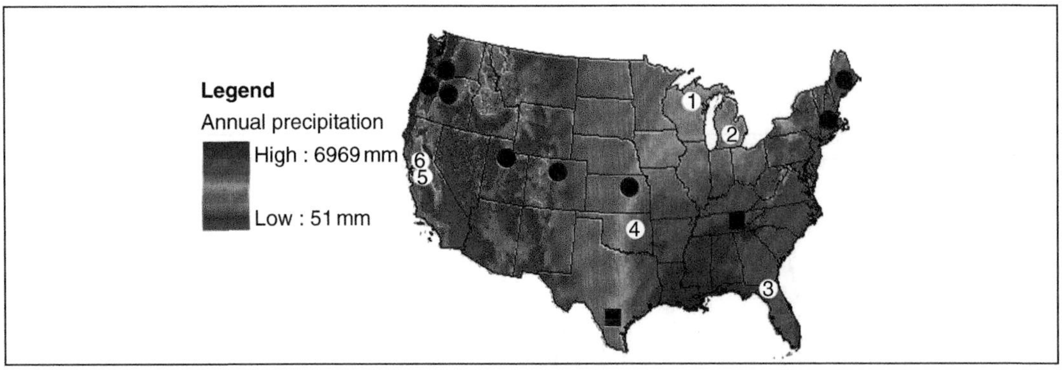

Figure 2.1, page 21

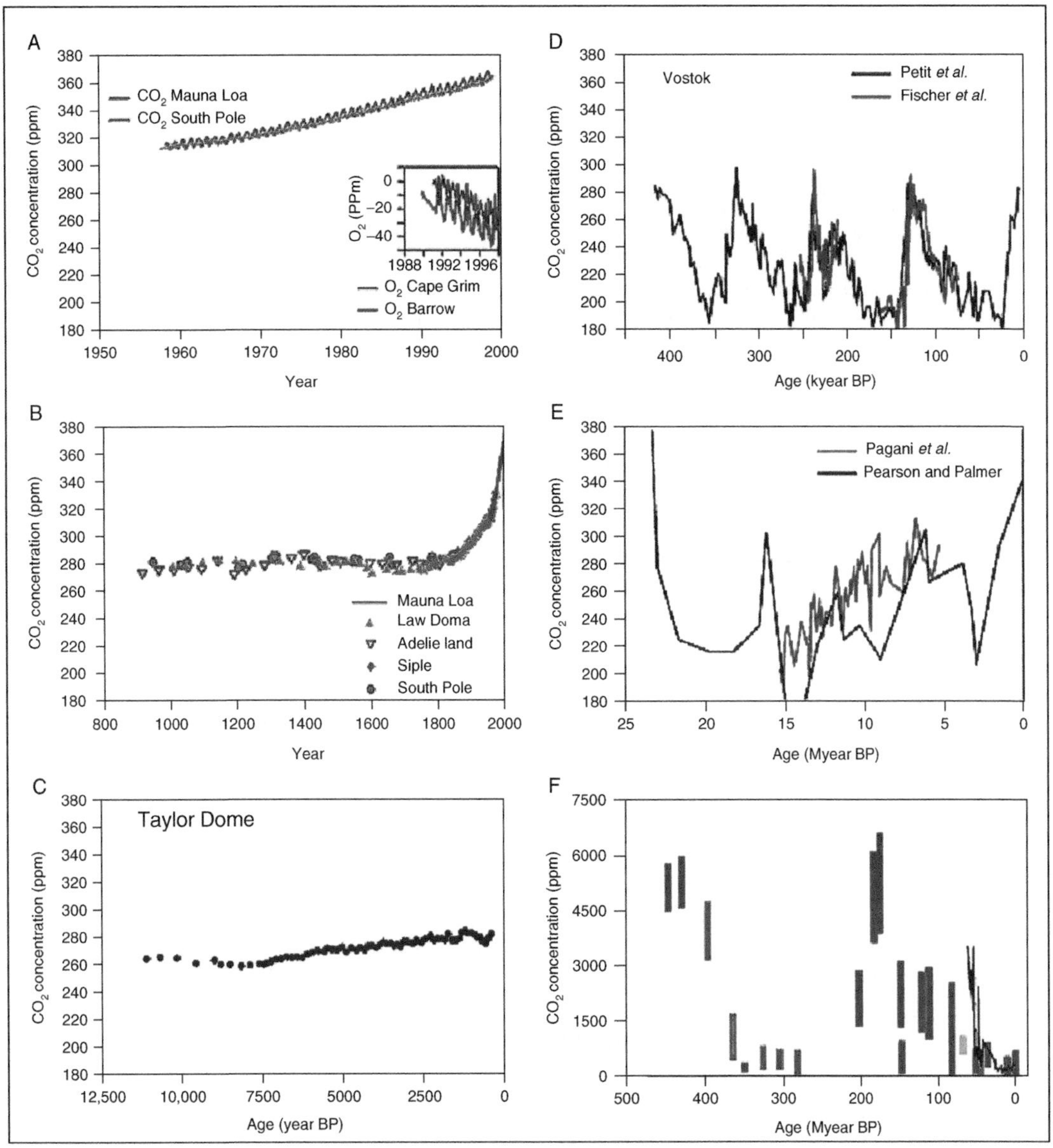

Figure 14.2, page 214

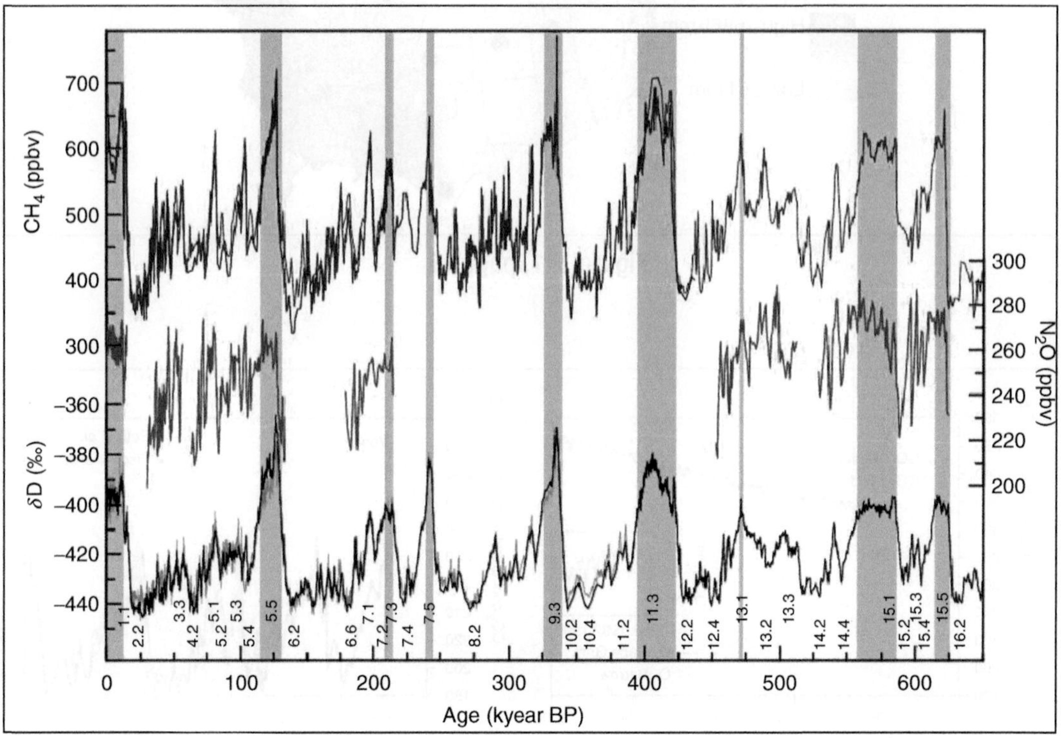

Figure 14.4, page 216

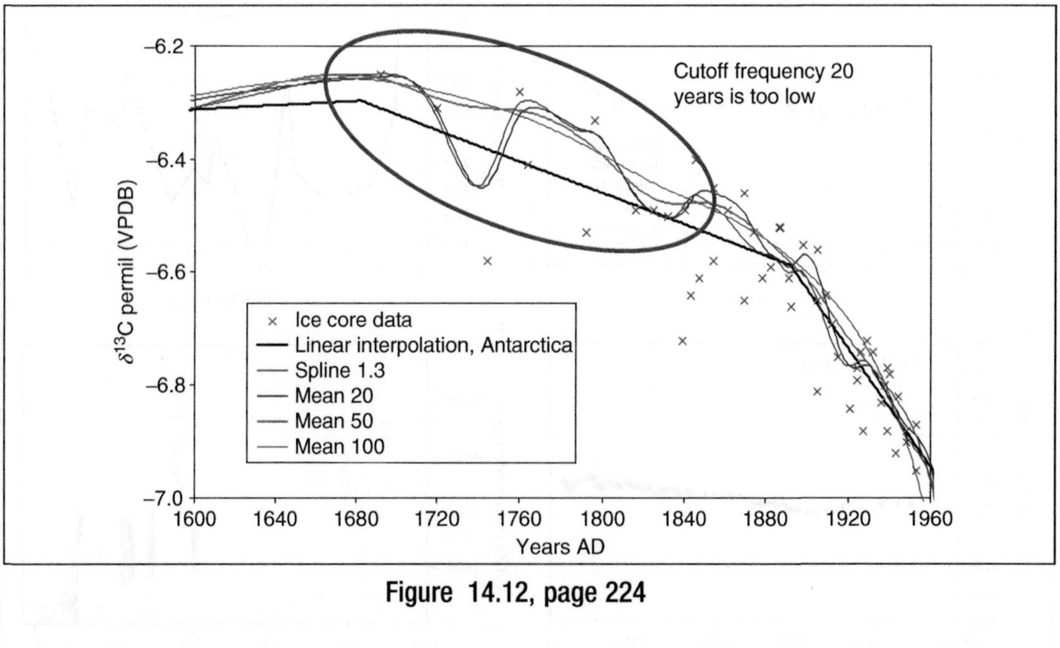

Figure 14.12, page 224

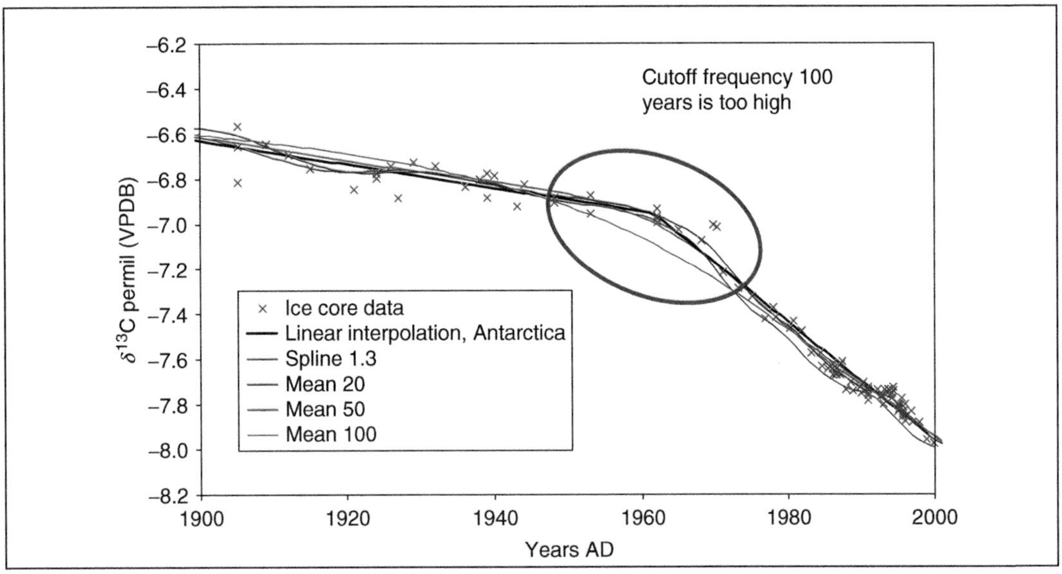

Figure 14.13, page 225

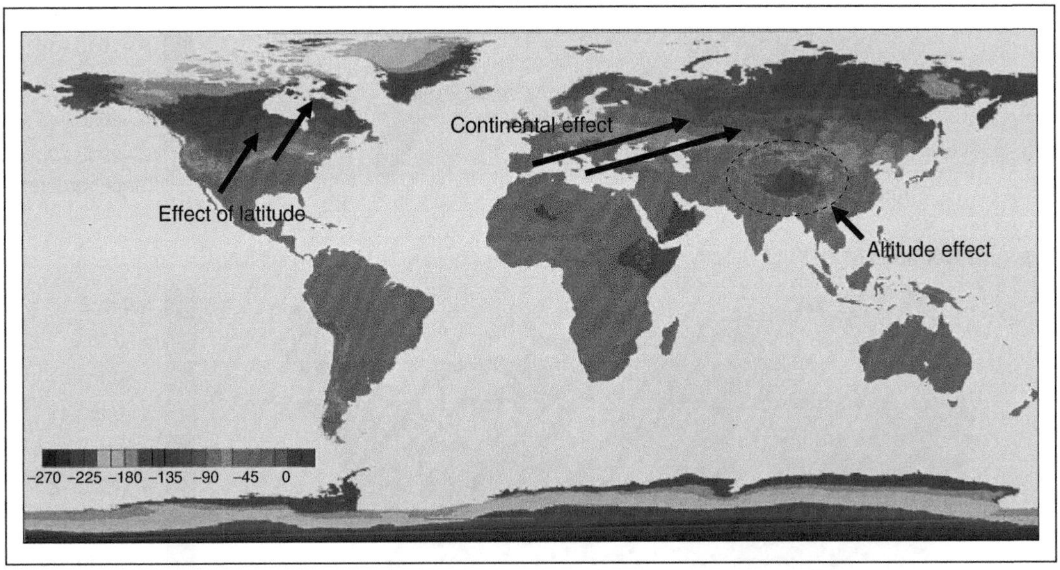

Figure 16.5, page 254

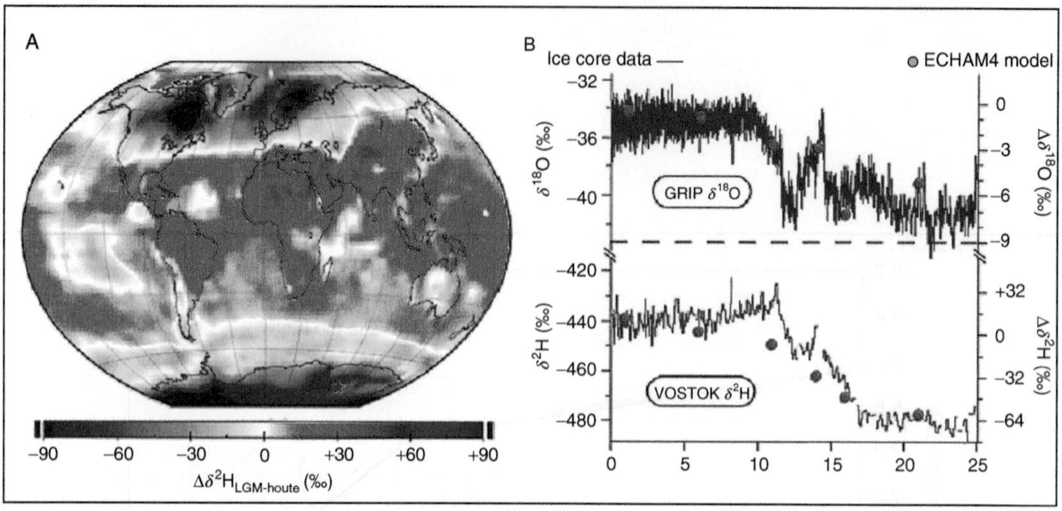

Figure 16.10, page 261

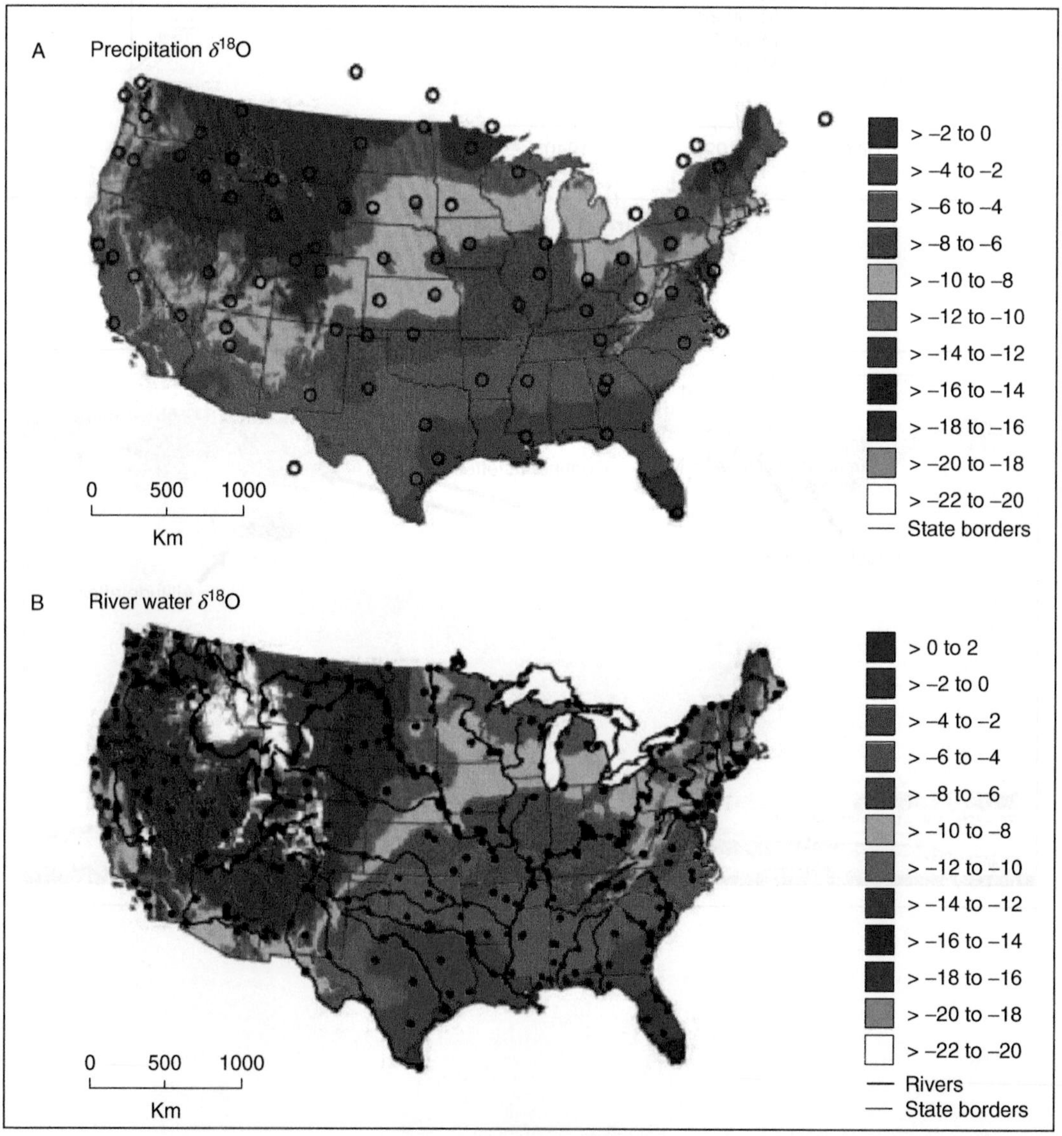

Figure 18.2, page 289

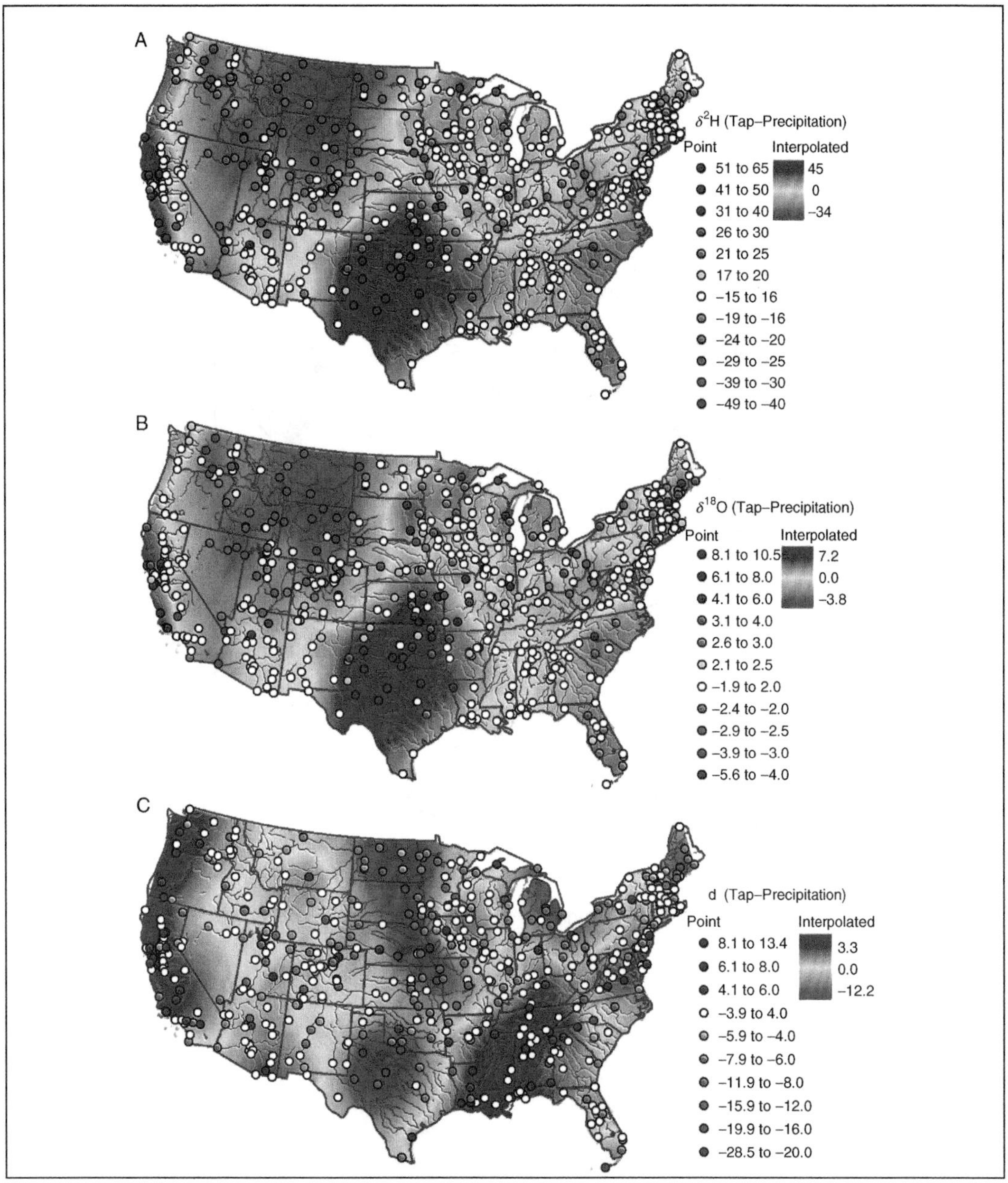

Figure 18.4, page 292

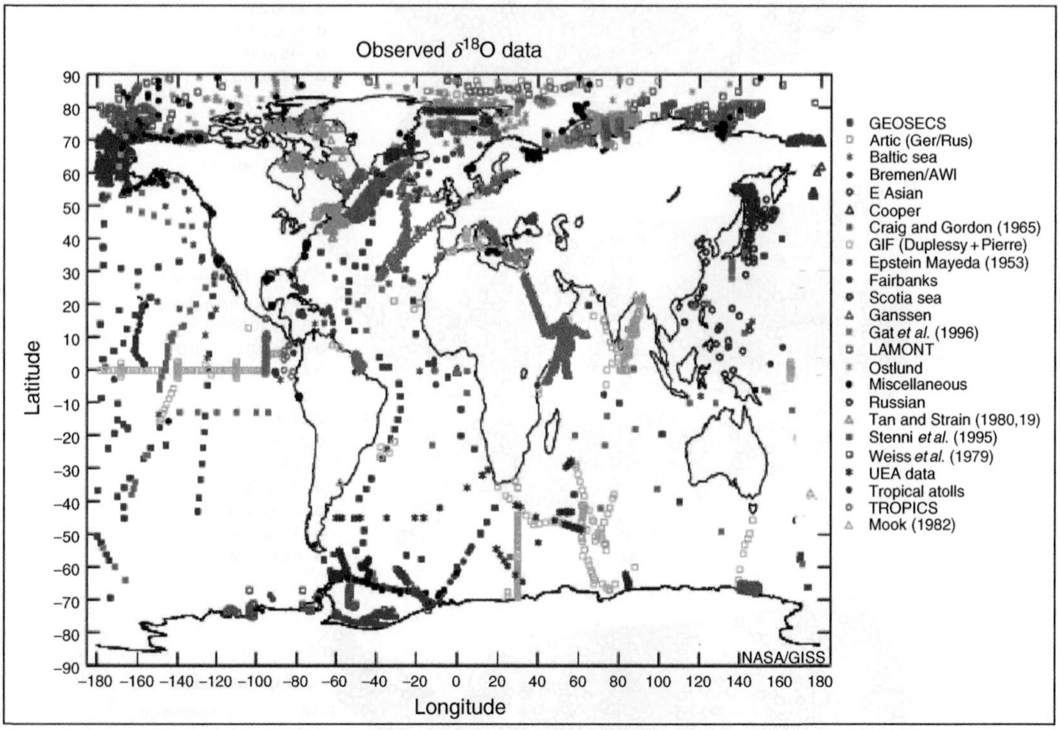

Figure 23.3, page 367

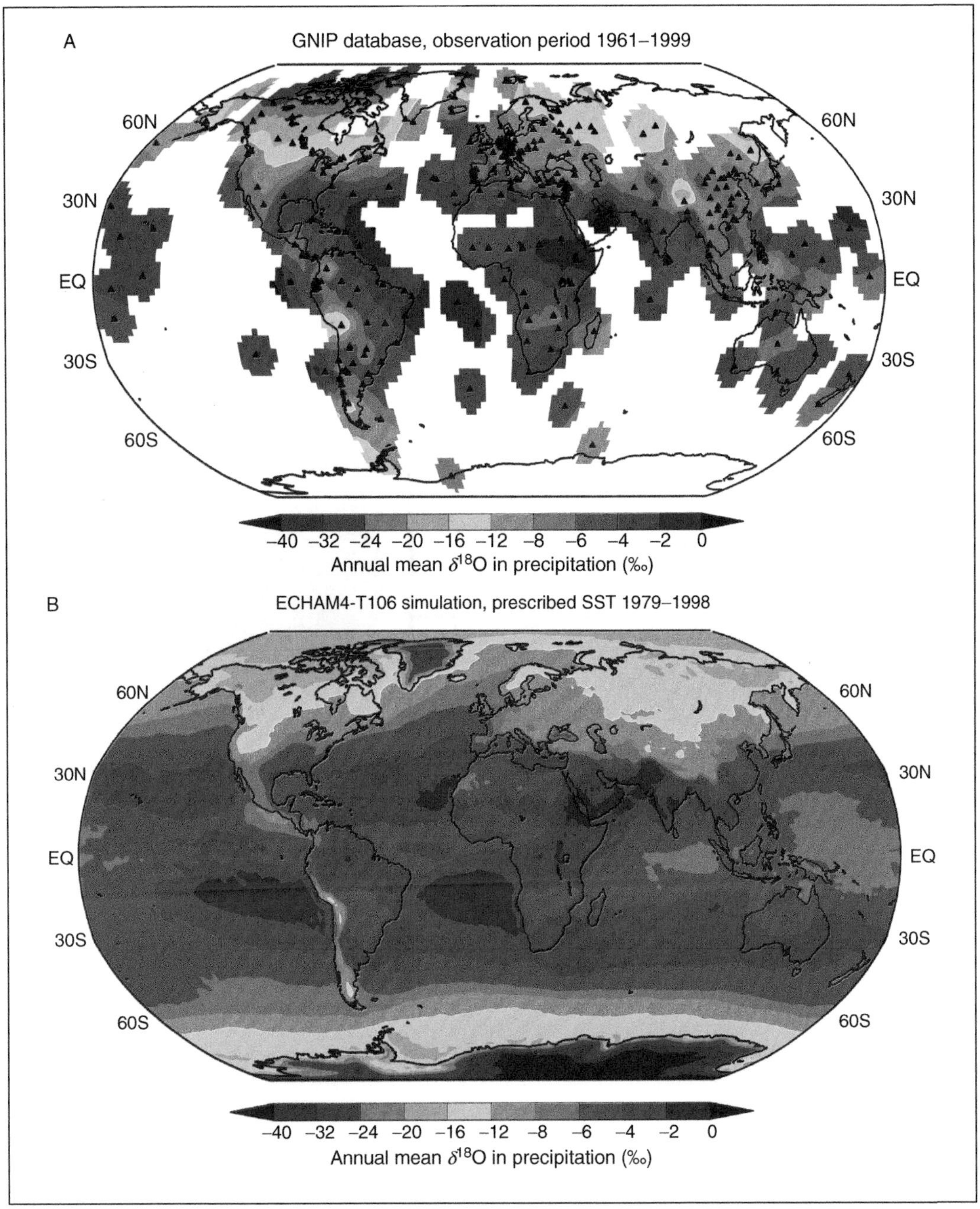

Figure 23.4, page 372

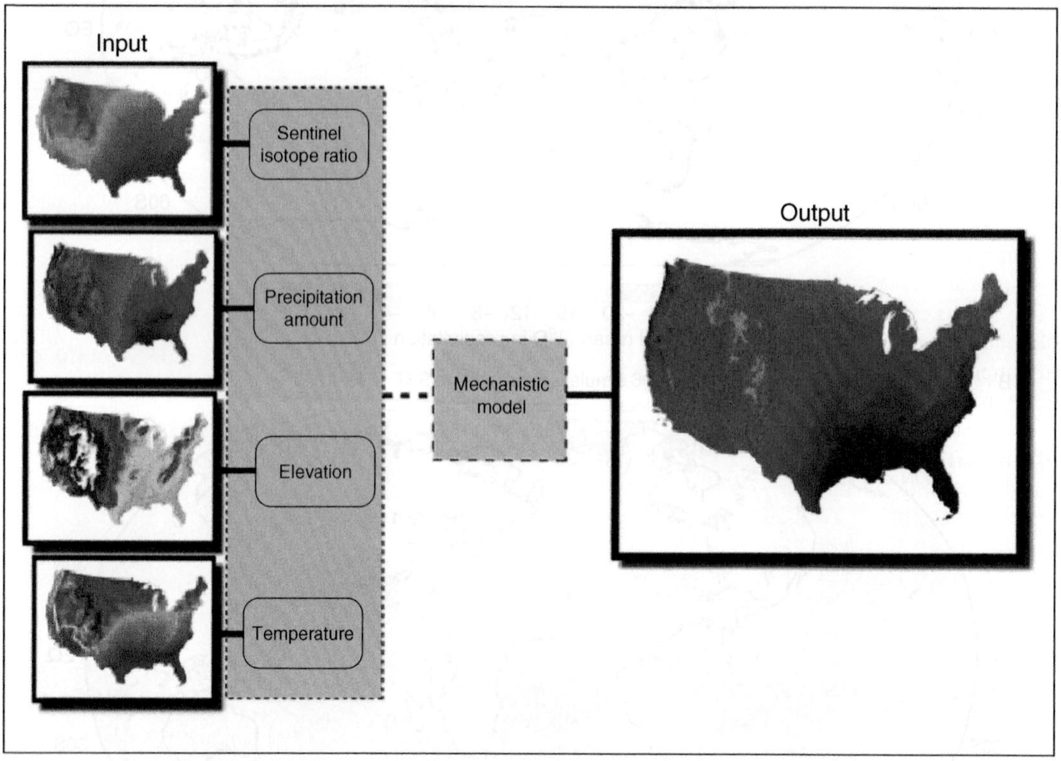

Figure 24.5, page 395

Index

L

M

N

O

T

Printed and bound by CPI Group (UK) Ltd, Croydon, CR0 4YY
03/10/2024
01040312-0009